This book presents five sets of pedagogical lectures by internationally respected researchers on nonlinear instabilities and the transition to turbulence in hydrodynamics.

The book begins with a general introduction to hydrodynamics covering fluid properties, flow measurement, dimensional analysis and turbulence. Chapter 2 reviews the special characteristics of instabilities in open flows. Chapter 3 presents mathematical tools for multiscale analysis and asymptotic matching applied to the dynamics of fronts and localized nonlinear states. Chapter 4 gives a detailed review of pattern forming instabilities. The final chapter provides a detailed and comprehensive introduction to the instability of flames, shocks and detonations. Together, these lectures provide a thought provoking overview of current research in this important area.

For graduate students and researchers in statistical physics, condensed matter physics, fluid dynamics, and applied mathematics.

COLLECTION ALÉA-SACLAY:
MONOGRAPHS AND TEXTS IN STATISTICAL PHYSICS

General editor: Claude Godrèche

HYDRODYNAMICS AND NONLINEAR INSTABILITIES

Hydrodynamics and Nonlinear Instabilities

EDITED BY

CLAUDE GODRÈCHE
Centre d'Etudes de Saclay

PAUL MANNEVILLE
Ecole Polytechnique

CAMBRIDGE
UNIVERSITY PRESS

CAMBRIDGE UNIVERSITY PRESS
Cambridge, New York, Melbourne, Madrid, Cape Town, Singapore, São Paulo

Cambridge University Press
The Edinburgh Building, Cambridge CB2 2RU, UK

Published in the United States of America by Cambridge University Press, New York

www.cambridge.org
Information on this title: www.cambridge.org/9780521455039

© Cambridge University Press 1998

First published 1998
This digitally printed first paperback version 2005

A catalogue record for this publication is available from the British Library

Library of Congress Cataloguing in Publication data

Hydrodynamics and nonlinear instabilities / Claude Godrèche, Paul Manneville.
p.   cm. – (Collection Alèa-Saclay ; 3)
Includes bibliographical references and index.
ISBN 0 521 45503 0 (hardcover)
1. Hydrodynamics. 2. Stability. 3. Nonlinear theories.
I. Godrèche, C. II. Manneville, P. (Paul), 1946– . III. Series.
QC151.H93 1996
532'.5 – dc20 95–4916 CIP

ISBN-13 978-0-521-45503-9 hardback
ISBN-10 0-521-45503-0 hardback

ISBN-13 978-0-521-01763-3 paperback
ISBN-10 0-521-01763-7 paperback

Contents

Preface

Fluid mechanics, one of the oldest branches of continuum mechanics, is still full of life. However, it is too often taught in a very formal way, with a heavy insistence on the analytical formulation and on peculiar solutions of the equations. But it has been recognized for a long time that many of the "classical results" are of little or no help for understanding real flows. This was to the great disappointment of G.I. Taylor who realized that the niceties of Rayleigh's stability theory of parallel flows were off the mark most of the time, but that made him happier at last to see his theory of the Taylor–Couette instability explain his experiments. Perhaps the best example of the irrelevance of classical stability theory is that it predicts the most simple shear flow —the plane Couette flow— to remain always linearly stable, although it becomes experimentally highly turbulent as soon as the Reynolds number goes beyond a few hundreds.

There is a curious example of what might be called conservatism in the exposition of fluid mechanics. Around the time of World War I, Henri Bénard in Paris did very careful experiments on the wake behind a cylinder. He basically showed that this wake changes around Reynolds 40–50 from stationary to time periodic, something now called a Poincaré–Andronov bifurcation. Surprisingly, although Bénard had Poincaré as a colleague at the University of Paris at the time, he never referred to the idea of a limit cycle that his famous fellow-professor had developed in order to explain this type of transition. Even more, when looking at the modern treaties of fluid mechanics, I have found only one, *Fluid Mechanics* by Landau and Lifschitz, with a "limit cycle" entry, as well as an explanation of what it is, Landau style, straight to the point. This is all the more surprising, because most text books dwell at length on the description of the Bénard–von Karman wake. Usually, they "explain" it by the linear stability analysis of von Karman for a row of alternating vortices in an inviscid fluid. A little reflection shows that the relationship between von Karman's analysis and Bénard's observations is at best quite loose, since a non inviscid system (inviscid = without dissipation) can have a limit cycle.

These remarks illustrate the need for books in fluid mechanics incorporating, in a wide sense, modern (and sometimes not so modern...) developments on the subject. The present book does an excellent job in this direction. It stems from a 1991 summer school taught in the beautiful and active setting of the "École Nationale de Voile de Beg-Rohu" (sailing school in South Brittany, on the French Atlantic shore). These courses represented a very successful attempt to teach fluid mechanics by aiming mostly at understanding real physical phenomena, whence its comprehensive character, although it is a collective work. The balance between "advanced" topics and "less advanced" ones is very well kept. I enjoyed reading the introduction to fluid dynamics, full of clever remarks and yielding orders of magnitude estimates of various effects, something that we, teachers, all know so well to be a difficult job!

One area of fluid mechanics witnessed explosive growth during the last ten or fifteen years, this is the study of instabilities and of nonlinear phenomena. It has brought lasting changes in our understanding of many basic situations. Nonlinearities are now seen as a source of new phenomena, and not as a mere source of troubles for the elementary analytic approach. We have almost seen an old dream coming true, the building of the "thermodynamics of nonequilibrium systems." This does not mean of course that the ideas of classical thermodynamics have to be applied to fluid flows after a little effort of translation. It means that broad categories of phenomena may be understood qualitatively within a common framework, like, for instance, the formation of patches of turbulence when instabilities are subcritical, or the effect of the mean flow on the growth of instabilities in parallel flows. Perhaps the domain where this resemblance with thermodynamics is the most spectacular is the theory of pattern, which has borrowed many concepts and ideas from (equilibrium) condensed matter theory. Nowadays it is fair to say that the tide has even reversed, and we are witnessing a flow of ideas from nonequilibrium systems theory towards more classical areas of condensed matter physics.

Perhaps the most fascinating field of investigation in fluid mechanics is combustion theory. Even the structure of a simple flame in a kitchen gas burner requires a sophisticated analysis to be understood. When one adds complications from fluid motion, from compressibility, from chemistry,... an almost unlimited domain of problems reveals itself. Because this domain has very important and obvious applications, the pressure from "downstream" has the effect that combustion is too often an exclusive playground for programmers and their kin, although physical intuition should remain at the forefront instead. Needless to say that combustion is covered in this "physical" spirit in this book. It is a pity that such an important and interesting field is not more developed in most countries.

Hopefully the chapter here will inspire future teaching and research in combustion and related phenomena.

At the turn of the century, L. Prandtl created the notion of boundary layer, probably the deepest contribution of fluid dynamics to mathematics. This subject is still alive and well, and it has recently produced outstanding results in areas unlinked to fluid dynamics. It is amazing to see how really important ideas, like that of boundary layer, have such a broad range of applications, and this is rightly seen here as a subject in itself.

The book is written by outstanding scientists, who have all made important contributions to the covered subjects. They have all been able to explain, in the most efficient and direct way, the heart of the matter, skipping useless formalism. This book will be most helpful to those who are interested in the modern developments in fluid mechanics, and there are plenty of those new developments. Its scope is very broad, as it goes from problems in nonequilibrium patterns (like the dynamics of rolls in Rayleigh–Bénard convection) to cellular flows, to bifurcation theory, linear and nonlinear, to combustion, detonation, and so on. Its style is, on average, at the level of a graduate course, although many parts do not require such an advanced knowledge, and can be read profitably by engineers or by scientists of other areas with a fair understanding of college physics, mechanics, and calculus.

Yves Pomeau

Contributors

Dr. Paul Manneville

Laboratoire d'Hydrodynamique de l'École Polytechnique,
Route de Saclay, 91128 Palaiseau cedex, France.

Prof. Bernard Castaing

Centre de Recherches sur les Très Basses Températures,
25, avenue des Martyrs, B.P. 166, 38042 Grenoble cedex, France.

Prof. Patrick Huerre[1] and Dr. Maurice Rossi[2]

[1]Laboratoire d'Hydrodynamique de l'École Polytechnique,
Route de Saclay, 91128 Palaiseau cedex, France.

[2]Laboratoire de Modélisation en Mécanique de l'Université Pierre
et Marie Curie, 4, place Jussieu, 75252 Paris cedex 05, France.

Dr. Vincent Hakim

Laboratoire de Physique Statistique, École Normale Supérieure
24, rue Lhomond, 75231 Paris cedex 05, France.

Prof. Stephan Fauve

Laboratoire de Physique, École Normale Supérieure
46, allée d'Italie, 69364 Lyon cedex 07, France.

Dr. Guy Joulin and Dr. Pierre Vidal

Laboratoire de Combustion et de Détonique, ENSMA
B.P. 109, Poitiers, 86960 Futuroscope cedex, France.

Overview

Paul Manneville

Hydrodynamics plays a ubiquitous role in our current environment. In the preamble of his lectures notes on fluid dynamics, K. Moffatt (1973) gives a wheel-shaped picture of the world as conceived by an egocentric hydrodynamicist. His picture strikingly illustrates the interrelations between this scientific domain and others, from the most basic fields such as pure mathematics to applied engineering or medicine, all organized in concentric circles. In our Fig. 1, we display an adaptation of his viewpoint, focused more on the material contained in this book, namely topics involving the innermost levels relating physics and mathematics to fluid mechanics. But we hope that this will not hide the fact that enrichments are often motivated by problems raised in fields possibly remote from the "center." This Overview chapter is therefore a long digression about this figure, intended to provide the reader with a documented guide on how the material in the rest of the book is organized.*

As implied by the first part of its title, the book begins with a general introduction to hydrodynamics under the leadership of B. Castaing. From a physical point of view, hydrodynamics accounts for the low-frequency, long-wavelength properties of matter, i.e., its macroscopic behavior as related to conserved or quasi-conserved quantities (cf. Martin *et al.*, 1972). The usual understanding is more restrictively centered on continuous media that "flow," the description of which will be introduced in C1-1. To be more specific, let us consider the simplest case of a one-component Newtonian fluid. Its motion is governed by a set of partial differential

---

* In the following, we shall refer to a given section n of some Chapter m as Cm-n.

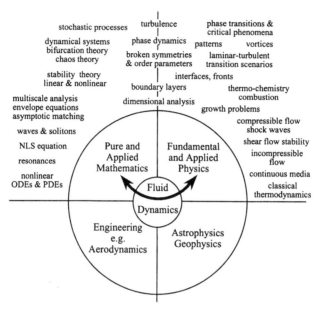

Fig. 1. Meandering to and fro at the border between mathematics and physics, through fluid dynamics.

equations, called the Navier–Stokes (NS) equations that reads

$$(\partial_t + \mathbf{v} \cdot \nabla)\,\mathbf{v} = -\rho^{-1}\,\nabla\,p + \nu\,\nabla^2\mathbf{v}\,, \tag{1}$$

where $\mathbf{v}$ is the velocity field, ρ the density, and ν the kinematic viscosity. At low flow speeds, the pressure p looses its thermodynamic meaning and is just a Lagrange multiplier associated with the incompressibility condition

$$\nabla \cdot \mathbf{v} = 0\,. \tag{2}$$

At this stage some additional complexity enters through non-trivial thermodynamic properties (energy balance equation and equation of state), the simplest consequences of which are associated with compressibility effects (C1-2). Further intricacies arising from the possibility of chemical reactions occurring in flowing mixtures are postponed to Chapter 5 by G. Joulin and P. Vidal. At this stage, external conditions applied to the medium remain to be specified i.e., relevant boundary conditions at the frontiers of the physical domain of interest, for example the "no-slip" condition that a solid wall imposes its velocity on the fluid directly in contact with it. Once the problem has been so formulated, the next step is to search for solutions corresponding to flow configurations of interest.

Let us adopt the framework of *dynamical systems theory* and write down the evolution problem as

$$\dot{V} = F(V) \,, \tag{3}$$

where V denotes the *degrees of freedom* describing the system in its *phase space* and F defines a vector field on phase space accounting for the dynamics, while the dot designates the time derivative. For incompressible hydrodynamics, the phase space is the functional space of three-dimensional solenoidal vector fields — condition (2). At this formal level, we can state the problem as a search for solutions of (3), given some initial condition $V_0 = V (t = t_0)$. In fact, we are more concretely interested in solutions which correspond to regimes observed in the long term and not to specific orbits in phase space since, usually, we do not control initial conditions precisely. This is a quite delicate matter owing to the nonlinear character of (3) which generically allows for multiple solutions. In (1) nonlinearities enter the problem through the so-called advection term $\mathbf{v} \cdot \boldsymbol{\nabla} \mathbf{v}$ whose role decreases as the imposed gradients become small everywhere, in practice, close to thermodynamic equilibrium. Any initial condition is then expected to relax towards the unique solution of the resulting linear problem (C1-1.7). The strength of the applied constraints can be appreciated from the ratio of the time constant governing the diffusive relaxation of velocity gradients over a typical distance ℓ, ℓ^2 / ν, to the time constant for advection by the velocity field with typical velocity U over the same distance, ℓ / U, hence: $U\ell / \nu$. This ratio, called the *Reynolds number* and abbreviated as Re, is thus the natural *control parameter* in flow problems.

As a first step in the search for solutions, it is always rewarding to notice that differently defined problems are in fact mathematically similar. This notion of similarity goes beyond simple geometrical considerations and incorporates the full physics of the problem (C1-4) as exemplified by the Reynolds number just introduced, which refers to the geometry (ℓ), the conditions of the experiment (U), and the properties of the fluid (ν). Introducing appropriate scales, we can then write down the primitive equations in dimensionless form, e.g., the rescaled NS equation:

$$(\partial_t + \mathbf{v} \cdot \boldsymbol{\nabla}) \mathbf{v} = -\nabla p + \mathrm{Re}^{-1} \nabla^2 \mathbf{v} \,, \tag{1'}$$

which depends only on the Reynolds number. In practice, different regimes can be expected from the simple consideration of its order of magnitude and, in more complicated problems, of the order of magnitude of similarly introduced dimensionless numbers, which can help us considering simplified model equations valid in specific limits. Similarity theory and dimensional analysis are presented in C1-4 using simple examples and are further applied to a scaling analysis of turbulent boundary layers. Their importance cannot be underestimated and their influence can be found in all other chapters (see Table 1).

Flow at low Reynolds numbers (*Stokes limit* mostly considered in C1-1.7) is dominated by the presence of the diffusive term on the r.h.s. of (1'). By contrast, the opposite case Re $\to \infty$, often called the *inviscid limit*, is governed by the truncated system:

$$(\partial_t + \mathbf{v} \cdot \nabla) \mathbf{v} = -\nabla p, \qquad (1'')$$

called the Euler equations. This simple truncation hides severe mathematical difficulties because the regularizing effect of viscous friction is suppressed and, accordingly, the largest differentiation order of the equations is decreased. Inviscid flow is characterized by the absence of transfer of momentum between adjacent flow lines, so that discontinuous velocity fields become acceptable. Furthermore, allowed discontinuities of the tangential velocity at solid walls correspond to a reduction of the number of boundary conditions. But this is only a limit; in practice, one must distinguish different regions in the flow according to whether viscosity effects may or may not be neglected. For example, consider the flow of a real fluid close to a solid boundary at rest in the laboratory frame. Since fluid particles are expected to stick to it while velocity is different from zero at some distance, some amount of momentum has to be transferred between the wall and the fluid. This transfer takes place through a *boundary layer* where dissipation cannot be neglected. Mathematically, one has to face the problem of deriving the profile of the velocity field inside the layer (Blasius profile, C1-2.3) which is part of a larger program of asymptotically matching an *outer* solution with an *inner* solution to get a *uniformly valid* approximation to a given problem of this class. In Chapter 3 by V. Hakim the problem of the coating of a wall by a fluid film will be treated in a similar way which nicely takes advantage of similarity analysis (C3-1).

Table 1. Some dimensionless numbers appearing in other chapters

Name	Definition	Physical meaning
Reynolds number	$Re = \ell U/\nu$	see text
Mach number	$Ma = (U/c_s)^2$	(velocity/sound velocity) squared
Lewis number	$Le = \kappa/D_{mol}$	thermal diffusivity/molecular diffusivity
Prandtl number	$Pr = \nu/\kappa$	vorticity diffusivity/thermal diffusivity
Capillary number	$Ca = \rho \nu U/\gamma$	viscous forces/capillary forces
Aspect ratio	$\Gamma = \ell/\lambda_c$	size/instability lengthscale

When searching for solutions of specific problems we find it advisable to consider separately "bulk-flow" and "free-boundary" problems. In the first case, the walls bounding the system, though playing a role in the physics (cf. boundary layer) have no non-trivial dynamics so that the actual degrees of freedom are directly associated with bulk fields (velocity, pressure,...); boundaries, e.g., "rigid walls," then enter the problem only through the definition of dimensionless parameters (dynamical and geometrical). The situation is different for flows bounded by "soft," "moving" interfaces. A preliminary step is then the elimination of bulk variables yielding an equation with reduced dimensional dependence accounting for the geometry and the dynamics of the interface (e.g., C3-2). For the moment, let us concentrate our attention on bulk flows.

In mathematics, problems are often posed in terms of the existence and uniqueness of solutions. In physics, one is usually tempted to take the existence of a solution for granted, without always being fully aware of the implicit assumptions (a counter-example where a solution with intuitively natural characteristics does not exist will be introduced later and discussed at depth in C3-4). But, at any rate, one cannot evade the question of the uniqueness of solutions under given external conditions. In fact, non-uniqueness is the rule as soon as nonlinearities are important (i.e., Re finite), the more as a theoretically possible solution may happen to be unobservable in practice. This is the well-known *stability* problem: if a given solution does not withstand the introduction of perturbations, then it is "unstable." A formal introduction to basic stability concepts is given in Chapter 2 by Huerre and Rossi (C2-3.1). For fluid flows this yields the distinction between *laminar* and *turbulent* regimes: the term "laminar" is associated with the idea of a smooth, regular and predictable behavior, in contrast with that of "turbulent" which implies wild stochastic fluctuations over a wide range of space-time scales (C1-5.3). Usually, turbulence sets in, not in a single-step process, but more likely at the end of a bifurcation cascade. The understanding of the transition to turbulence, a subject in itself, has made much progress recently at the pace of the theory of dynamical systems. So, let us return to this framework for a while and make explicit the fact that the governing equation (3) depends on control parameters by writing it as

$$\dot{V} = F_r(V) \,, \tag{3'}$$

where the subscript r denotes the set of control parameters.

The *basic state* V_0 is a special solution of (3') that we suppose time-independent for simplicity. The stability problem is then the initial value problem for the perturbation V' defined by $V = V_0 + V'$. By substitution one gets:

$$\dot{V}' = F_r(V_0 + V') - F_r(V_0) = G_{r|V_0}(V'),\tag{4}$$

which, by construction, has $V' \equiv 0$ as a solution. *Linear stability theory* is then concerned with the fate of infinitesimal solutions that fulfill

$$\dot{V}' = L_r V',\tag{5}$$

where the linear operator L_r is obtained from a Taylor expansion of (4) around the null solution. Solutions to (5) can be searched for in the form:

$$V' = A\overline{V}\exp(st),\tag{6}$$

where A is the (infinitesimal) amplitude of the eigenvector $\overline{V}$ associated with the eigenvalue s solution of

$$L_r\overline{V} = s\overline{V}.$$

Generically, the growth rate s of mode $\overline{V}$ is a complex number: $s = \sigma - i\omega$. Instability occurs when σ is positive, which depends on the considered mode and the value of the control parameter r. These so-called *normal modes* encode the space-time coherence implied by the existence of a definite instability mechanism which couples the different fluctuations, amplifying specific scales and damping others. A classical example is given by thermal convection studied in Chapter 4 (C4-4) by S. Fauve. This instability manifests itself as a spatially organized overturning of a fluid layer initially at rest and heated from below. RB convection develops as a regular pattern of *stationary* rolls. The eigenvalue corresponding to this normal mode has no imaginary part ($\omega = 0$) but this is not the case for, e.g., convection in binary mixtures which may be *oscillatory* ($\omega \neq 0$).

Up until now, nothing has been said about the spatial structure of the modes. For systems of hydrodynamic type, L_r is a partial differential operator in the space variables and, in translationally invariant domains, the space dependence of perturbations implicitly contained in (6) can be analyzed by a Fourier transform:

$$\overline{V} = \overline{V}(x) = \overline{\overline{V}}\exp(ikx).\tag{7}$$

The (generically complex) growth rate s is then related to the wavevector k of the perturbation through a *dispersion relation*:

$$\mathcal{D}(s, k; r) = 0\tag{8}$$

that can be solved for s:

$$s = s(k; r) = \sigma(k; r) + i\omega(k; r).\tag{8'}$$

The typical aspect of the dispersion relation is sketched in Fig. 2. The real part σ of a given branch of the dispersion relation usually displays a maximum σ_{max} for some $k = k_{max}$. The basic state is stable as long as $\sigma_{max} < 0$ for all branches. An instability develops when the most dangerous branch, i.e., that with the largest σ_{max}, becomes marginal, which happens for some $r = r_c$ called the *threshold*. This defines a set of critical conditions: r_c, k_c, and $\omega_c = \omega(k_c; r_c)$. Different cases are possible according to whether the most unstable mode, i.e., the mode with largest σ has non-vanishing k_c and/or ω_c. A *cellular* instability occurs when $k_c \neq 0$ and $\omega_c = 0$ (e.g., plain Rayleigh–Bénard convection). A homogeneous *oscillatory* instability corresponds to $k_c = 0$ and $\omega_c \neq 0$ (e.g., Belousov–Zhabotinsky reaction in chemistry). *Waves* are associated with k_c and ω_c both non-vanishing; the phase velocity of a given mode k is then defined as $c = \omega(k)/k$ and the group velocity of a wavepacket centered at k as $c_g = d\omega/dk$.

From a physical point of view, the nature of the couplings that sustain the normal modes has much interest. The mechanism leading to RB convection is appealingly simple: the layer with its colder and denser fluid initially situated on top of a hotter and lighter fluid is a reservoir of gravitational energy that may be released by overturning the layer. This gain of potential energy is further consumed by dissipation (viscous friction

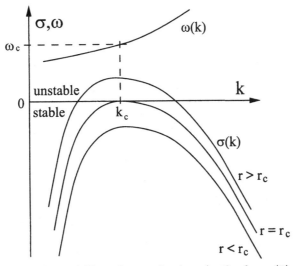

Fig. 2. At fixed r, the stability of perturbations in the form (7) can be decided from the value of the real part σ of their complex growth-rate $s = \sigma - i\omega$. As a function of k, σ usually displays a maximum. Unstable modes correspond to $\sigma > 0$. The threshold in $r = r_c$ is reached for some $k = k_c$ where the curve $\sigma = \sigma(k; r)$ is tangent to the "$\sigma = 0$" axis. By convention, we assume henceforth that destabilization is obtained upon increasing a single control parameter.

and heat diffusion). The fluid layer becomes unstable when the former overcomes the latter and the pattern which sets in achieves a kind of optimal balance between them for some specific wavelength. RB cells are the prototype of *dissipative structures* (Nicolis & Prigogine, 1977; see also Manneville, 1990, Chapters 1 and 3). This concept is mostly appropriate for *closed* systems. By contrast most flows of hydrodynamical interest are *open*, i.e., fluid particles are advected by a mean flow and ultimately leave the domain of interest. Their phenomenology is summarized in C2-2.

In open flows, the situation is in some senses more subtle. Dissipation may or may not play a trivial role: some instability modes, of mechanical nature, may already exist at the inviscid limit, and viscosity then plays its intuitive damping role, or the flow may remain stable at the inviscid limit so that instability can only result from a more complex mechanism. As shown in C2-4.3, for two-dimensional parallel flows, the purely mechanical instability is connected with the presence of an inflection point in the unperturbed velocity profile (Rayleigh theorem). The corresponding prototype flow is the *mixing layer* between two fluid veins flowing side-by-side and experiencing a Kelvin–Helmholtz instability (C2-5). By contrast, the parabolic profile of the classical *Poiseuille flow* driven by a pressure gradient between two parallel plates, lacking an inflection point, is mechanically stable but unstable against viscous instability modes called *Tollmien–Schlichting* waves, the heuristic analysis of which is developed in C2-7. As a rule, mechanically unstable flows of a real (i.e., $\nu \neq 0$) fluid are unstable at low Reynolds numbers, whereas inviscidly stable flows become unstable at large* Reynolds numbers.

Before leaving the realm of linear stability, i.e., stability with respect to infinitesimal perturbations for which the general problem (4) reduces itself strictly to the linear problem (5), let us point out that the approach presented so far is deliberately situated in a *temporal* context, as is apparent from the definition (6–7) of the perturbation modes and the solution (8′) of the dispersion relation (8). For open flows, owing to the presence of a mean flow that tends to advect perturbations, it may be advisable to develop stability theory also from a *spatial* point of view. In simple terms, instead of considering the stability problem as an initial value problem for a perturbation with a specific spatial dependence, we must worry about the spatial growth of a perturbation introduced at a given point in the laboratory frame, with a given temporal signature. The dispersion relation (8) has thus to be re-analyzed assuming, e.g., $s = i\omega$ and solving for k complex (instead of k real and s complex). With $k = k' + ik''$, spatial

---

* Possibly arbitrary large Re, as in the case of the Poiseuille flow in a tube with circular section, or the plane Couette flow, a simple shear flow produced by two parallel plates in relative motion at velocity $\pm U/2$.

amplification occurs whenever $k'' \neq 0$. As explained in C2-3.2, a rigorous formulation is in terms of Green's functions expressing the response to a perturbation considered as a superposition of delta functions localized in space and time. In the presence of a mean flow, this leads to the intuitively clear distinction between so-called *convective* instabilities whose normal modes grow while traveling downstream but ultimately disappear when observed at a given fixed location in the laboratory frame, and *absolute* instabilities that are sufficiently strong to invade the whole system in spite of the mean flow.

Let us now turn to nonlinear aspects. Normal modes can be used as a basis to expand arbitrary fluctuations of finite amplitude and a meaningful separation between damped and excited modes can be made on the basis of the value of the real part of their growth rate. Furthermore, damped modes are *enslaved* to excited modes and can thus be "adiabatically eliminated" (see Haken, 1983). At the end, one is left with a usually small number of *central modes* living their own life on some subset of the phase space called the *center manifold* (see Iooss, 1987). This is essentially how the theory of low-dimensional dissipative dynamical systems enters the scene. However, the picture is valid only in the case of *confined systems*, i.e., when lateral boundary effects are strong enough to maintain coherence everywhere throughout the system (see Manneville, 1990, Chapter 8). If this is not the case, one will have to deal with modulated patterns as discussed heuristically later.

Having restricted the problem to such a finite set of ordinary equations one can then study the bifurcations of the asymptotic regimes they account for (*attractors*) as the control parameters are varied. At this stage it is interesting to introduce a simple model valid for a single mode. Following Landau (1944) let us admit that the amplitude of such a mode is governed by an equation that reads:

$$\dot{A} = rA - \mathcal{N}(A), \tag{9}$$

where the nature of A and the structure of the nonlinear term $\mathcal{N}(A)$ depend on the considered instability, stationary or oscillatory.[*]

The study of the bifurcation diagram of (9) is elementary but helps one to categorize the possible cases.[†] Assuming first a stationary instability

---

[*] In (9), the control parameter r is redefined such that $r_c = 0$ and $r > 0$ corresponds to the unstable range. This also implies a rescaling of time and, in the case of oscillatory behavior, a change to a rotating A-frame that suppresses a term $-i\omega_c A$ on the right hand side.

[†] According to Joseph (1983) more than sixty per cent of the ideas of bifurcation theory can be drawn from such a study.

accounted for by a real amplitude A one has:

$$\dot{A} = rA - a_2 A^2 - a_3 A^3 - \dots, \tag{10}$$

which may be further constrained by symmetry considerations. The amplitude of the bifurcated state is the relevant *order parameter* in the sense of Landau for thermodynamic phase transitions. There is a strong parallel between Landau's theory of phase transitions (Landau, 1937) and bifurcation theory. If the problem is invariant with respect to the change $A \mapsto -A$, all even powers disappear from the expansion and two different situations occur depending on the sign of the first nonlinear coefficient a_3. Scaling the amplitude so as to make this coefficient disappear, at lowest order one gets:

$$\dot{A} = rA - \epsilon A^3 \qquad \text{with} \qquad \epsilon = \pm 1, \tag{11}$$

which has *fixed point* solutions $A = 0$ and $A = \pm\sqrt{\epsilon r}$ and account for what is known as the *pitchfork bifurcation*. The trivial solution is stable for $r < 0$ and unstable for $r > 0$. The nontrivial solution exists for $r > 0$ ($r < 0$) when $\epsilon = +1$ (-1) and can be shown to be stable (unstable); the bifurcation is said to be *supercritical* (*subcritical*), see Fig. 3a,b.

Higher order terms neglected in (11) are inessential in the supercritical case, as long as one is interested in low amplitude solutions close to the threshold. By contrast, they must imperatively be taken into account in the subcritical case in order to explain the saturation that may be observed at large amplitude (see Fig. 3b). In the absence of symmetry, a supplementary quadratic term is present in (11), which describes a *transcritical* bifurcation (see Fig. 3c for $\epsilon = +1$). Finally the amplitude A can sometimes be coupled with an external quantity, H say, which ends in an *imperfect bifurcation* as in Fig. 3d again with $\epsilon = +1$.

An important feature of (10) is that it derives from a potential called a *Lyapunov functional*:

$$\dot{A} = -\frac{\partial G}{\partial A}, \tag{10'}$$

with

$$G(A) = -\tfrac{1}{2}rA^2 + \tfrac{1}{3}a_2 A^3 + \tfrac{1}{4}a_3 A^4 + \dots, \tag{10''}$$

so that, for G considered as a function of time through A, one gets:

$$\frac{dG}{dt} = \frac{\partial G}{\partial A}\frac{dA}{dt} = -\left(\frac{dA}{dt}\right)^2 \le 0.$$

The evolution problem is therefore equivalent to the search for minima of G (if it is bounded from below). Generalizations of (10'–10'') introduced later on are considered in depth in C3 and C4.

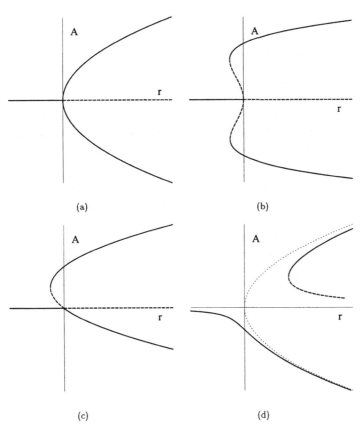

(a)

(b)

(c)

(d)

Fig. 3. Bifurcation diagram for model (10) in the real case. Solid (broken) lines correspond to stable (unstable) solutions. (a) Supercritical pitchfork bifurcation: $\dot{A} = rA - A^3$. (b) Subcritical bifurcation: $\dot{A} = rA + A^3 - A^5$. (c) Transcritical bifurcation: $\dot{A} = rA + A^2 - A^3$. (d) Imperfect bifurcation: $\dot{A} = rA - A^3 - H$, which corresponds to the unfolding of the cubic singularity (notice that $A = 0$ is no longer a fixed point; the perfect case (a) is indicated by the thin broken line).

For an oscillatory instability the potential property (10′) does not hold. The amplitude A is now a complex variable and the relevant form of (9) reads:

$$\dot{A} = rA - a_3|A|^2A - a_5|A|^4A - \dots , \qquad (12)$$

where coefficients a_n are generically complex. Equation (12) is the *normal form* for a *Hopf bifurcation*, i.e., the transition from a time-independent attractor to a periodic attractor represented by a *limit cycle* in phase space (Fig. 4, see C2-2.2 and C2-6 for a hydrodynamical example; see also C4-2). The bifurcated limit cycle is locally stable when the bifurcation is supercritical, which is the case when the real part of a_3 is positive (with the sign convention in (12)). If not, the limit cycle is unstable and the system bifurcates discontinuously towards a finite amplitude solution controlled by higher order terms.

Only so-called *resonant* nonlinear terms take part in (12), i.e., not all possible combinations of A and A^* but just those which "rotate" at the same speed as the linear term (see C4-2). All non-resonant terms can be eliminated by nonlinear changes of variables. This important technical issue is usually skipped when formulating phenomenological models of mode interactions (C4-5).

As conceived by Landau (1944), turbulence was the result of an indefinite accumulation of bifurcations, each introducing a new space-time scale in the system. In the present framework, one would then describe *laminar–laminar* transitions between time-independent attractors by (10′–10″), then to periodic behavior by (12) (Hopf bifurcation). The transition to two-periodic behavior would further be analyzed by means of Poincaré maps, i.e., iterations obtained from the original (continuous time) dynamical system by performing a *Poincaré section*, a technique that generalizes to self-oscillating systems the stroboscopic analysis of periodically forced systems (see Fig. 5). This leads onto a theory of Hopf bifurcation for first-return maps (Iooss, 1979). Further steps in the cascade would follow from a more and more involved description in terms of discrete-time dynamical systems. Fortunately, Ruelle and Takens (1971) showed that the indefinite accumulation of bifurcations conjectured by

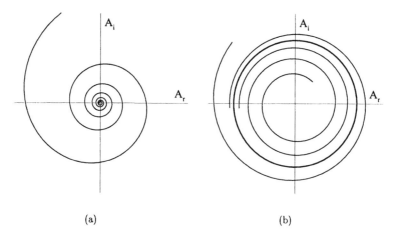

(a)   (b)

Fig. 4. Birth of a limit cycle *via* (supercritical) Hopf bifurcation governed by $\dot{A} = (r + i)A - |A|^2A$, i.e., (12) truncated above cubic terms, with $a_3 = 1$, but including the term iA to account for rotation in the laboratory frame (see a previous footnote): (a) $r < 0$, the trivial solution is stable; (b) $r > 0$, periodic behavior accounted for by a limit cycle with radius $|A| = \sqrt{r}$.

Landau was not necessary to get chaotic motion but that a finite number of them was generically sufficient, owing to the instability of trajectories on *strange attractors*, i.e., the divergence of neighboring trajectories due to the *sensitivity to initial conditions and small perturbations*. In this way, one can understand how *stochasticity* is made compatible with *determinism* at least for low-dimensional dissipative dynamical systems. The example of the well-known Lorenz model is illustrated in C1-5; for a more detailed presentation, consult Bergé *et al.* (1987), Schuster (1988), or Manneville (1990), Chapters 4–7.

In the preceding presentation, the modes were explicitly supposed to form a discrete set and it was mentioned that this was appropriate for confined systems. Confinement effects can be appreciated on the basis of *aspect ratios*, i.e., the ratios of the size of the system in different directions to the length scale generated by the instability mechanism already mentioned in Table 1. When the aspect-ratio is large, the system is called *extended* and the bifurcated state can be viewed as an assembly of individual cells in interaction with their neighbors. However, the strong coherence implied by the mechanism within a cell is not incompatible with the presence of *collective modes* accounting for the presence of fluctuations in local amplitude and position of the cells. Now if A represents the amplitude of the mode excited in some cell, the assembly of cells, called a *pattern* can be described by turning the scalar variable A function of time into an *envelope*, i.e., a field variable function of time and space (C2-3; C3-2; C4-4.3). This unfolding of the space dependence is meaningful only if modulations are slow when compared to the intrinsic

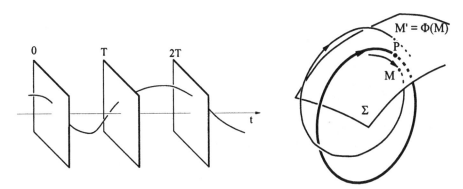

Fig. 5. Left: Stroboscopic analysis of a periodically forced two-dimensional system interpreted in extended phase space $\{t \in \mathbb{R}\} \times \{A \in \mathbb{R}^2\}$. Right: Poincaré section of phase space transverse to a limit cycle and *first return map* $\mathbf{M} \mapsto \mathbf{M}' = \Phi(\mathbf{M})$. The location where the limit cycle pierces the surface is a fixed point of the map.

time-space dependence of the considered mode. At lowest order, modulations are expected to behave in a diffusive way, which transforms (9) into a nonlinear diffusion equation. For one-dimensional modulations along the x-axis, this extension can be written as

$$\partial_t A = rA + \xi_0^2 \partial_{xx} A - \mathcal{N}(A)\,, \tag{13}$$

where ξ_0 is the *coherence length*, the value of which can be derived from the expression for the linear dispersion relation $s = s(k;r)$ at marginality ($\sigma(k;r) = 0$). ξ_0^2 is real (complex) for a stationary (oscillatory) instability. Equations like (13) are known under the generic name of *Ginzburg–Landau* (GL) equations. They play a ubiquitous role in this book as equations derived by a systematic multiscale expansion, presented in detail in C3-2, or more frequently, as models issued from a phenomenological approach constrained by physical arguments and symmetry considerations (see, e.g., C4-5).

The envelope formalism is the appropriate framework to account for pattern selection (C4-4.3) and the transition to turbulence understood as *spatio-temporal* chaos in extended systems. As a matter of fact, besides (extrinsic) modulations imposed by boundary conditions, large-aspect-ratio systems may experience modulations originating from the development of secondary instabilities or the presence of topological defects such as dislocations in a roll pattern. Here we concentrate our attention on long wavelength secondary instabilities (C4-6).

Returning to the definition (6) of the mode and its form (7) appropriate for the laterally unbounded limit, one easily understands that, even in the case of a stationary pattern, the amplitude A can be written as a complex function $|A| \exp(i\phi)$, whose modulus $|A|$ is associated with the intensity of the instability mode while the phase ϕ accounts for *translational invariance* (a change $x \mapsto x + x_0$ amounts to a "gauge change" $\phi \mapsto \phi + k\,x_0$). At finite distance from the threshold, in the absence of modulations, $|A|$ generically saturates[*] to a value $A_{\mathrm{sat}} \simeq \sqrt{r}$. While fluctuations around A_{sat} can be shown to relax at a finite rate $\sim 1/r$, the slow phase fluctuations ϕ are *nearly neutral*, as follows from the fact that A and $A\exp(i\phi)$ describe one and the same solution up to a translation, provided ϕ is constant. In the simplest cases the relaxation of phase fluctuations is merely diffusive:[†]

$$\partial_t \phi = D\partial_{xx}\phi\,. \tag{14}$$

The explicit expression of the effective phase diffusion coefficient can be obtained after fluctuations of the modulus (enslaved to the phase gradi-

---

[*] In the supercritical case, the only one we consider here.
[†] The phase is a hydrodynamic mode in the sense of Martin *et al.* (1972).

ent) have been adiabatically eliminated. D depends on the underlying wavevector of the pattern $k = k_c + q$ and the control parameter r. As long as D is positive, phase fluctuations relax and the pattern is stable but when D becomes negative, phase fluctuations diverge at the linear stage described by (14). For two-dimensional patterns one can distinguish phase diffusion parallel and perpendicular to the wavevector of the underlying pattern, hence two phase diffusion coefficients and, accordingly, two phase instabilities respectively called the Eckhaus instability (C2-3.3; C4-6.3) and the zigzag instability (C4-6.4).

In the case of a stationary primary instability, the nonlinear evolution of unstable phase fluctuations which develop for a pattern with an ill-adapted underlying wavevector, always result in a stationary state with a better adapted wavevector (C4-6). For a supercritical oscillatory instability, the space dependent extension of (12) is a GL equation with complex coefficients and a *cubic* nonlinearity (henceforth denoted as ^{3}CGL equation) here written in rescaled one-dimensional form

$$\partial_t A = A + (1 + i\alpha)\partial_{xx} A - (1 + i\beta)|A|^2 A. \tag{15}$$

The Benjamin–Feir (BF) phase instability occurs for $1 + \alpha\beta < 0$ (C4-3) and degenerates into *phase turbulence*, at least sufficiently close to the threshold where the modulus $|A|$ is expected to remain bounded away from 0.[*]

The equation governing phase turbulence can be obtained explicitly by a long wavelength expansion for phase fluctuations. At lowest significant order this yields the Kuramoto–Sivashinsky (KS) equation here written in rescaled one-dimensional form [see Kuramoto (1984) and references therein]:

$$\partial_t \phi + \partial_{xx}\phi + \partial_{xxxx}\phi + \tfrac{1}{2}\left(\partial_x\phi\right)^2 = 0, \tag{16}$$

where the second term on the left hand side, obviously anti-diffusive, is the rescaled version of that in (14) when $D \sim (1 + \alpha\beta)$ is small and negative. The third term regularizes solutions at small scale and the nonlinear term ensures a redistribution of energy among different scales. The trivial solution to (16) is easily seen to be unstable on any interval of sufficient length. Moreover it is known to possess "turbulent" solutions as exemplified in Fig. 6.

---

[*] This is no longer the case farther from the BF threshold and the system enters a regime of *amplitude turbulence* also called *defect-mediated turbulence*, Coullet *et al.* (1989). For a brief review of phase turbulence, see the corresponding entry by Chaté & Manneville in the "dictionary" edited by Cardoso & Tabeling (1995).

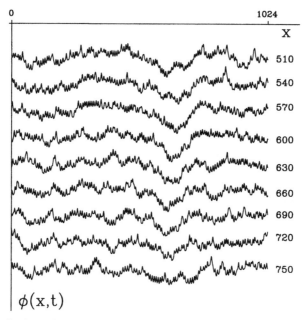

Fig. 6. Simulation of the KS equation showing the propagating-interface behavior of a turbulent solution.

The Russian-doll picture at which we arrive, with instability modes generating cells, then envelope equations describing modulated patterns, and phase dynamics governed by KS-like equations, leads us to understand the emergence of turbulence as an —essentially supercritical— growth of spatio-temporal chaos [see also Wesfreid *et al.* (1988) and, for a recent review, Cross & Hohenberg (1993)]. Though this scenario is appropriate for a large class of systems including convection and some open flows, its validity depends crucially on the fact that, at every step, the successive branchings of new regimes are "smooth," the bifurcated states gently saturating beyond threshold. This is no longer the case if one step is markedly subcritical. For example, as already mentioned, inviscidly stable flows become linearly unstable at large Reynolds numbers but this leaves the possibility of a coexistence of the laminar regime continuously connected to the Stokes flow (Re $\to$ 0) with different, more strongly nonlinear, regimes at Reynolds numbers well below the linear threshold. This is precisely what happens for the plane Poiseuille and Couette flows or the Poiseuille flow in a tube with circular section (C2-1.1) which experience the subcritical growth of *turbulent spots*, i.e., patches of turbulence amid laminar flow (C2-2.3).

Let us now turn to problems arising when the domain where the flow takes place has deformable boundaries, either material and determined by the external conditions of the experiment (e.g., the free surface of a liquid, a surface defined by the property that the fluid cannot go through it) or immaterial and characterized as interfaces between regions where the macroscopic properties of the medium change rapidly. Moving internal boundaries will be called *fronts* in the following: a classical example is the *flame front* separating the fresh from the burnt in combustion (C5).

To deal with interfaces in the wide sense, preliminary theoretical efforts are most often devoted to the elimination of the flow degrees of freedom and the determination of an effective equation with reduced space-dimensionality governing the interface itself. A good example of this procedure is given by the derivation of the equation that governs the envelope of a wavetrain of deep-water gravity waves by a multiscale method in C3-2, yielding a *nonlinear Schrödinger* (NLS) equation, the (conveniently rescaled) fully dispersive version of ^{3}CGL equation:

$$\partial_t A = i\partial_{xx} A + 2i|A|^2 A . \tag{17}$$

The Korteweg–de Vries equation governing long-wavelength waves at the surface of a shallow-water channel is another celebrated example (see, e.g., Whitham, 1974).

To approach the case of "internal boundaries" let us consider a one-dimensional *reaction-diffusion* system

$$\partial_t A = \partial_{xx} A - \frac{\partial V}{\partial A}, \tag{18}$$

where $V(A)$ is a potential with two minima V_1 and V_2 at A_1 and A_2 (Fig. 7, left).

As shown in detail in C3-3, this RGL equation ('R' for 'real') admits locally stable uniform solutions with $A \equiv A_1$ or A_2 but also propagating

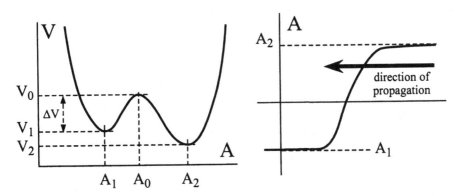

Fig. 7. Left: Potential V having two minima: the *metastable* state A_1 ($V_1 = V(A_1)$) is separated from the *stable* state A_2 with potential V_2 by the *unstable* state A_0 ($\Delta V = V_0 - V_1$ is the potential barrier). Right: Typical front solution.

inhomogeneous solutions called *fronts* with $A \to A_1$ (A_2) for $X \to -\infty$ $(+\infty)$ (Fig. 7, right).

Solutions moving at speed c without deformation can be expressed in terms of a single variable $\tilde{x} = x - ct$ so that the partial differential equation (18) is transformed into an ordinary differential equation in $\tilde{x}$:

$$\frac{d^2 A}{d\tilde{x}^2} + c\frac{dA}{d\tilde{x}} - \frac{\partial V}{\partial A} = 0, \tag{19}$$

which describes the motion of a particle at position A in a potential $U(A) = -V(A)$ as a function of time $\tilde{x}$. This mechanical analogy is used at numerous places throughout C3 and C4, e.g., in the search for solitary wave solutions of the NLS equation (17) in C4-3.3. It helps us to build a bridge between partial and ordinary differential dynamical systems. Here, if $V_2 < V_1$ the region with $A = A_2$ (the stable state) grows at the expense of the region where $A = A_1$ (the metastable state). The velocity of the front is then just the value of the parameter c which allows for the existence of a *heteroclinic trajectory* between the two fixed points A_1 and A_2 of system (19). The evolution of small domains of one state immersed in larger domains of the other can further be studied as the interaction of distant fronts (C3-1, C4-8.2). In such a simple case, bounded states of two fronts at some fixed distance, localized solutions of (18), are not possible and the least stable state always recedes so as to minimize the global potential. This is no longer the case for the ^{5}CGL equation, i.e., the spatial extension of (12) with a fifth-order nonlinearity appropriate for a subcritical oscillatory instability, an equation that does not derive from a potential and allows for localized solutions called *Fauve–Thual pulses* (C3-3.1, C4-8.3).

To continue with fronts between stable and metastable states, let us point out a suggestion by Pomeau (1986) according to whom the transition to turbulence *via* turbulent spots in subcritically unstable flows would result from a kind of *contamination* process by stochastic front propagation, the chaotic regime inside the spot being in competition with the laminar regime outside. The corresponding route to turbulence is called *spatio-temporal intermittency*. This suggestion implies a new connection between statistical physics and the theory of turbulence which offers an interesting framework that works well for, e.g., convection in elongated containers.* It seems appealing also for open-flow cases already mentioned and evoked in C2-2.3. No definite results have been obtained in this context yet but relevant investigations, still at their very beginning (Dauchot & Daviaud, 1994), are encouraging. At this stage,

---

* Consult the brief reviews by Chaté & Manneville (theory), and Daviaud (experiments) in Cardoso & Tabeling (1995).

it may be appropriate to mention that the statistical approach to spatio-temporal intermittency is in line with the theory of critical phenomena in statistical physics (see Kinzel, 1983), and somewhat different in spirit from the statistical theory of *developed turbulence* introduced in C1-5. In fact, not many of the recent advances from dynamical systems theory and the transition to temporal and spatio-temporal chaos have been incorporated in the theory of developed turbulence while some progress could be expected from a better understanding of the dynamics of, e.g., coherent structures (Cantwell, 1981; Aubry *et al.*, 1988) and the interaction with small scale fluctuations in these terms.

Paradoxically, the problem of the invasion of an unstable state by a stable state, e.g., with previous notations, the space-time growth of a domain of A_2 in a system initially quenched in state A_0, is much more subtle. Within the same framework, it is readily shown that for such problems a continuum of speeds is possible, which immediately raises the problem of velocity selection. This problem is often solved by admitting a "marginal stability principle" which states that the selected front has an analytic structure neutral with respect to infinitesimal perturbations in a frame moving at the selected speed. However, this linear criterion may fail in some nonlinear cases (C3-3.2).

In the sense of (deterministic) dynamical systems, the minima of the potential $V(A)$ appearing in (18) correspond to locally stable states, i.e., states that are stable against infinitesimal perturbations. However, in a thermodynamical context, the state with the lowest potential is absolutely stable, whereas the other state is only metastable. This means that, owing to spontaneous thermal fluctuations, it can decay with an exponentially small rate given by an Arrhenius law $\sim \exp(-\Delta V/T)$, where ΔV is the height of the potential barrier between the two states (see Fig. 7, left). As suggested by the Arrhenius dependence of the spontaneous decay rate, actual chemical reactions are usually strongly influenced by the temperature and, since they usually release or absorb heat, one may expect complicated feedback couplings between the temperature and the concentrations of chemicals. Models more complicated than (18) with its simple polynomial "reaction" term directly drawn from the bifurcation approach (9–10) are therefore expected. Flames studied in C5 serve as a good example. The fresh reacting mixture at ordinary temperature is in the metastable state. The "burnt" gases represent the stable state but the spontaneous decay towards it takes an "astronomically" long time as long as the mixture remains uniformly cold. Ignition is achieved by heating the fresh mixture locally, which strongly increases the reaction rate. The neighboring fresh mixture is then set on fire by the heat previously released, and so on. As a result, a thin sheet-like reaction zone called a *flame front* is produced, propagating from the burned region

to the fresh mixture (Fig. 8). The sharp variation of the reaction rate
through the flame front makes it appropriate to look for the concentration and temperature profiles in boundary-layer form by an asymptotic
expansion (C5-4.3) of the same nature as those developed in C3-1, matching expressions valid in inner and outer regions. This naive description
is appropriate in a strictly diffusive context where density variations do
not play a dynamical role. Things can be notably more involved when
this is no longer the case (see C5).

Up until now we have considered a strictly one-dimensional medium
so that the front position could be characterized by a single variable.
This much-idealized situation must now be spatially unfolded in spaces
of larger dimensionality, yielding a front line (surface) in two (three)
dimensions. Such an unfolding immediately raises the problem of the
stability of flat fronts and, further, of the motion of curved fronts. Let
us consider a nearly flat front in an isotropic medium. First of all, whatever its orientation, the strictly flat front must propagate at the same
velocity, which means that only the velocity component parallel to the
local normal to the front can be controlled by the physics of the reaction.
But this velocity can be modified by the curvature of the front which, in
the flame context, modulates the gradients driving the diffusive transport
of heat and reagents (see below the discussion about stability). In two
dimensions, let us consider the propagation in the y-direction of a nearly
flat front parallel to the x-axis whose instantaneous position is denoted
as $\phi(x,t)$. Assuming a simple proportionality of the velocity correction

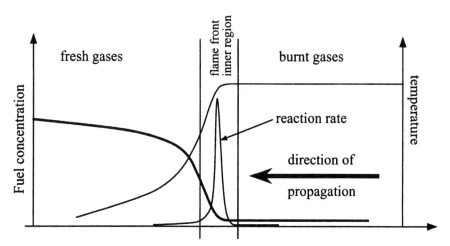

Fig. 8.   Typical temperature (thin line) and concentration (thick line) profiles
through a flame front. The reaction rate is high in a thin region separating the
fresh mixture from the burnt gases.

to the curvature, we get:

$$\partial_t \phi = \lambda \frac{\partial_{xx}\phi}{\sqrt{1+(\partial_x\phi)^2}} \tag{20}$$

where the term $1/\sqrt{1+(\partial_x\phi)^2}$ is the cosine of the angle made by the normal to the front with the y-axis (Fig. 9, left). The sign of the proportionality constant depends on the details of the process. The flat front is stable as long as $\lambda > 0$ but becomes unstable if $\lambda < 0$. In flames, the modulation of the temperature field by the front deformation is stabilizing since heat is concentrated in regions where the front is behind its mean position, which favors combustion and locally increases the speed so that the interface makes up lost time. Of course the reverse holds for portions ahead of the front's mean position (Fig. 9, right). However, the response of the reagent field depends on the Lewis number measuring the ratio of the thermal diffusivity to the molecular diffusivity (see Table 1): if the concentration gradient induced by the modulation implies a relative depletion of the fresh mixture in regions over-heated by the temperature gradient, then the expected stabilization does not happen and the coupled processes end up in an amplification of the distortion (see C5-4.5). Expanding equation (20) in powers of ϕ and completing it by a regularizing fourth-order derivative term one re-obtains the KS equation (16) as derived originally by Sivashinsky (1977)* and accounting for turbulent front propagation as already illustrated in Fig. 6. Here, we cannot go much beyond this naive formulation of the problem of flame stability developed at length in C5 where additional processes, including compressibility effects at the origin of detonations, are included in a step-by-step approach.

The motion of fronts that are no longer nearly flat but present large scale curvature poses delicate analytical problems (C3-4). Let us consider a (one-dimensional) curved front that propagates in a (two-dimensional) medium at velocity V in some direction. The equation generalizing (20) for this *geometrical model* then reads:

$$V\cos(\theta) = \lambda \times \text{curvature}, \tag{21}$$

where θ is the angle made by the normal to the front with the direction of motion. The shape of the front can be defined by $\theta(s)$ where s is the curvilinear coordinate along the front. θ and s are linked together through the local radius of curvature R simply by $ds = R d\theta$ (Fig. 9, left)

---

* See the review by Sivashinsky (1983).

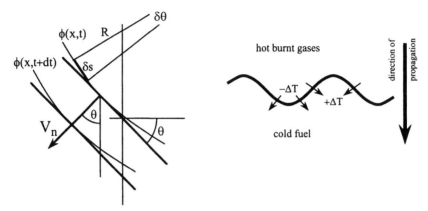

Fig. 9.   Left: Propagation of a front $\phi(x,t)$ at an angle θ with the y-axis. The normal velocity is insensitive to the local orientation but may depend on the curvature.   Right: focusing and defocusing of heat due to curvature of a modulated front is expected to modify the local normal velocity.

so that, after rescaling, (21) reads

$$d\theta/ds = \cos\theta. \tag{22}$$

Whereas a symmetrical solution (called a "needle front") can easily be obtained, yielding $\cos\theta = 1/\cosh s$ and $\theta \to \pm\pi/2$ as $s \to \pm\infty$, it comes as a surprise that the perturbation of (22) by the next relevant higher order term in the curvature, $\epsilon^2 d^3\theta/ds^3$, breaks this nice property so that the complete equation has no symmetrical solution with the same boundary conditions. The proof given in C3-4.2 is an interesting piece of asymptotic analysis in the complex plane that can be further adapted to physical problems such as viscous fingering or dendritic crystal growth.* For example, viscous fingering is the flow regime adopted by a thin viscous fluid layer pushed along a channel by a less viscous fluid (usually a gas). At steady state, a needle-like solution (the *finger*) sets in with the viscous fluid far from the tip occupying exactly half the width of the channel. While an analytic solution neglecting surface tension exists for all widths, this small perturbation of higher differential order is here responsible for the selection in the same way as the higher order curvature term in the previous geometric model led to the break-down of the symmetrical solution. Involving exponentially small terms of the form $\exp(-1/\epsilon)$, these effects are outside the scope of perturbation theory and require the development of an analysis "beyond all orders" (C3-4.3) involving boundary

---

* The first book in this series edited by Godrèche (1992), is entirely devoted to growth problems, see also the introductory chapter of and the reprints in Pelcé (1988).

layers and asymptotic matching in the vicinity of singularities in the complex plane.

To conclude, throughout this introduction we have favored the physical approach inscribed in the point of view developed by the different contributors. That is, we have read connections between physics and applied mathematics through fluid dynamics in Fig. 1 mostly from right to left. Our review of the topics dealt with was arranged according to three key-stages: 1) formulation of the problems, from primitive equations to heuristic models; 2) search for basic solutions, either direct or by perturbation; 3) stability study, linear and nonlinear, including little of genuinely "turbulent" aspects. We hope that our views will motivate the most mathematically-oriented readers while encouraging others to go deeper into the mathematical subtleties that flourish in the field of nonlinear dynamics.

References

Aubry N., Holmes P., Lumley J.L., Stone E. (1988). The dynamics of coherent structures in the wall region of a turbulent boundary layer, *J. Fluid Mech.* **192**, 115.

Bergé P., Pomeau Y., Vidal Ch. (1987). *Order within Chaos* (Wiley, New York).

Cantwell B.J. (1981). Organized motion in turbulent flow, *Ann. Rev. Fluid Mech.* **13**, 457.

Cardoso O., Tabeling P. (1995). *Turbulence: a Tentative Dictionary* (Plenum Press, New York).

Coullet P., Gil L., Lega J. (1989). A form of turbulence associated with defects, *Physica D* **37**, 91.

Cross M.C., Hohenberg P. (1993). Pattern formation outside of equilibrium, *Rev. Mod. Phys.* **65**, 851.

Dauchot O., Daviaud F. (1994). Finite amplitude perturbations in plain Couette flow, *Europhysics Letters* **28**, 225.

Godrèche C. (1992). *Solids far from Equilibrium* (Cambridge University Press, Cambridge).

Gukenheimer J., Holmes Ph. (1983). *Nonlinear Oscillations, Dynamical Systems, and Bifurcation of Vector Fields* (Springer Verlag, New York).

Haken H. (1983). *Advanced Synergetics* (Springer Verlag, Berlin).

Iooss G. (1979). *Bifurcation of Maps and Applications* (North-Holland, Amsterdam).

Iooss G. (1987). Reduction of the dynamics of a bifurcation problem using normal forms and symmetries, in *Instabilities and Nonequilibrium Structures*, ed. E. Tirapegi and D. Villaroel (Reidel, Dordrecht).

Joseph D.D. (1983). Stability and bifurcation theory, in *Chaotic Behaviour of Deterministic Systems*, ed. G. Iooss, R.H.G. Helleman, R. Stora (North-Holland, Amsterdam).

Kinzel W. (1983). Directed percolation, in *Percolation Structures and Processes*, ed. G. Deutcher *et al.*, *Annals of the Israel Physical Society* **5**, 425.

Kuramoto Y. (1984). *Chemical Oscillations, Waves, and Turbulence* (Springer Verlag, Berlin).

Landau L.D. (1937). On the theory of phase transitions, reprinted in Landau (1965).

Landau L.D. (1944). On the nature of turbulence, reprinted in Landau (1965).

Landau L.D. (1965). *Collected Papers*, ed. D. ter Haar (Pergamon Press, Oxford).

Manneville P. (1990). *Dissipative Structures and Weak Turbulence* (Academic Press, Boston).

Martin P.C., Parodi O., Pershan P.S. (1972). Unified hydrodynamics theory for crystals, liquid crystals, and normal fluids, *Phys. Rev. A* **6**, 2401.

Moffatt H.K. (1973). Six lectures on general fluid dynamics and two on hydromagnetic dynamo theory, in *Fluid Dynamics*, ed. R. Balian and J.L. Peube (Gordon and Breach, London).

Nicolis G., Prigogine I. (1977). *Self-organization in nonequilibrium systems, from dissipative structures to order through fluctuations* (Wiley, New York).

Pelcé P. (1988). *The Dynamics of Curved Fronts* (Academic Press, New York).

Pomeau Y. (1986). Front motion, metastability and subcritical bifurcations in hydrodynamics, *Physica D* **23**, 3.

Ruelle D., Takens F. (1971). On the nature of turbulence, *Comm. Math. Phys.* **20**, 167.

Schuster H.G. (1988). *Deterministic Chaos, an Introduction* (VCH, Weinheim).

Sivashinsky G.I. (1983). Instabilities, pattern formation, and turbulence in flames, *Ann. Rev. Fluid Mech.* **15**, 179.

Wesfreid J.E., *et al.* (1988). *Propagation in Systems Far from Equilibrium* (Springer, New York).

Whitham G.B. (1974). *Linear and Nonlinear Waves* (Wiley, New York).

1

An introduction to hydrodynamics

Bernard Castaing

1 What is a fluid?

1.1 Introduction

This chapter is an introduction to hydrodynamics. There exist superb books on fluid mechanics, with prestigious authors and the goal here is not to challenge them. Even hydrodynamics may not be the exact term to use, as it is clear that air flows will also interest us. But it could significantly reveal our spirit. What is fascinating for physicists is that the basic equations in this field have been known for more than a century. However the fundamentals of what is turbulence, for example, are not elucidated.

This chapter will thus be a promenade into fluid mechanics, setting aside the exact derivation of simple flows, keeping only from them some intuition of the behavior of a fluid. Going through a great variety of examples, some of them will lack the long and rigorous introduction they need. We realize that only the reader with knowledge of the subject will find some interest in the corresponding remarks. But we hope that every reader will find at least one interesting remark.

Let us first examine what we shall consider as a fluid. For us it will be a continuous medium with a state equation connecting the pressure p, the temperature T and the density ρ:

$$p = f(\rho, T). \tag{1.1}$$

It implies that volume elements we shall have to consider will have sufficient dimensions for this equation to be meaningful. Second, our fluid will "take the shape of the container." It thus allows permanent deformations without stress. What is the corresponding limitation?

Let us imagine that the fluid is submitted to a continuous shear starting suddenly at $t = 0$. The instantaneous behavior, before any atomic rearrangement, is a solid one and the stress will grow linearly with time. Then a saturation will occur and the stress will remain a constant, viscous one (Fig. 1.1).

The time of cross-over, τ, is given by:

$$\rho c_t{}^2 \tau \left(\frac{dv}{dx}\right) = \mu \left(\frac{dv}{dx}\right),$$

where $\rho c_t{}^2$ is the shear modulus and μ the viscosity. It gives the limitation. Elementary time intervals must be large compared to τ. To speak as in condensed matter physics, hydrodynamics is valid at low ω and k.

An example can be useful. On a scale of years, a glacier flows, while ice is clearly a solid. Let L be the transverse dimension. For a steep slope, the weight of a volume L^3 is balanced by the viscous forces:

$$\rho g L^3 \simeq \mu \frac{U}{L} L^2 \qquad \text{and} \qquad \frac{\mu}{\rho} \simeq \frac{g L^2}{U}.$$

The velocity U is such that a kilometer is traveled in a few years. Viscous flow is not the only way a glacier progresses. Melting and freezing processes are important too, but we look for an order of magnitude:

$$\frac{\mu}{\rho} \simeq 10^{10} \text{ m}^2/\text{s} \qquad \text{and} \qquad \tau \simeq \frac{\mu}{\rho c_t{}^2} \simeq 10^4 \text{ s}.$$

For times shorter than a day, ice behaves like a solid. On larger time scales it is a fluid.

1.2 Viscosity and Reynolds number

Everybody has intuition of what a fluid is and what viscosity is. For instance we say that honey is a viscous fluid, water and mercury are not.

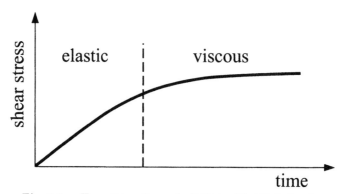

Fig. 1.1.   Transition from "solid" to "fluid" behavior.

Table 1. Typical kinematic viscosities ν in m^2/s.

Material	ν	Material	ν
Earth mantle	10^{22}	Mercury	10^{-7}
Honey	10^{-2}	Liquid nitrogen	10^{-7}
Air	10^{-5}	Gaseous ^{4}He, 1 bar, 4 K	10^{-7}
Water	10^{-6}	Liquid ^{4}He, 4 K	10^{-8}

However, we shall see that this impression of viscosity is not characteristic of the fluid, but of the flow.

Indeed, let a fluid be flowing along a plate, and consider a volume of this fluid of dimension L, within a distance L of the plate. It can be roughly materialized by a boat in a channel whose size is of the order of the depth of the channel.

If this volume initially moves at velocity U, it will stop in a distance ℓ, due to internal dissipation in the fluid. Intuitively we could say that the fluid is viscous if $\ell \ll L$ and not if $\ell \gg L$.

The kinetic energy of the volume is of order $\rho U^2 L^3$. In a laminar approximation, the damping force is $F \simeq \mu U L$. Then:

$$\ell \simeq \frac{E_c}{F} = \frac{\rho U L^2}{\mu}$$

and

$$\frac{\ell}{L} = \frac{\rho U L}{\mu} = \frac{U L}{\nu} = \text{Re} .$$

Re is the Reynolds number and $\nu = \mu/\rho$ is the kinematic viscosity of the fluid. Let us give some characteristic values for ν (Table 1) and some examples of Reynolds numbers (Table 2).

Thus, the same fluid appears as viscous or not depending on the situation. Looked at in a microscope, microorganisms like Paramecia seem to move in a viscous fluid which is water! Here is our first aside remark. We

Table 2. Typical Reynolds numbers.

	U (m/s)	L (m)	ν (m^2/s)	Re
Steamer	20	50	10^{-6}	10^9
Plane	300	10	10^{-5}	$3\,10^8$
Swimmer	1	1	10^{-6}	10^6
Paramecia	10^{-5}	10^{-4}	10^{-6}	10^{-3}

have introduced the Reynolds number as the ratio of the stopping length to the size of the object. For a steamer it would give 50 million kilometers as stopping length. A dream for navigation companies (at least if we forgot waves)! What is wrong here is the laminar assumption. The flow will be turbulent. Turbulence thus must be very efficient in dissipating the large scale energy.

1.3 The basic equations

There are six unknown quantities: ρ, p, T and the three velocity components v_i. We need six equations. The first one is the equation of state characterizing the fluid. The second one accounts for mass conservation:

$$\partial_t \rho + \partial_i (\rho v_i) = 0 . \tag{1.2}$$

We have used several useful conventions. ∂_t is the time derivative, ∂_i is the derivative versus the i coordinate. Summation on repeated indices is implicit. This equation has the general aspect of budget equations:

$$\partial_t (\text{density}) + \text{div} (\text{flux}) = (\text{sources}) .$$

Here there are no sources: no spontaneous creation of matter.

The following three equations correspond to Newton's law of mechanics applied to a fluid element. For calculating the acceleration γ, recall that the fluid element moves by $\mathbf{v}dt$ in a time dt.

$$\gamma dt = \mathbf{v}(\mathbf{r} + \mathbf{v}dt, t + dt) - \mathbf{v}(\mathbf{r}, t)$$
$$= (\partial_t \mathbf{v} + v_j \partial_j \mathbf{v})dt = D_t \mathbf{v}\, dt .$$

$D_t = \partial_t + v_j \partial_j$ is the time derivative "following the flow," or "Lagrangian" derivative.

Forces will be separated into volume ones (ρF per unit volume) and surface ones given by the stress tensor σ_{ij}.

σ_{ij} is the i component of the force per unit surface on a plane normal to the j axis, applied by the largest x_j side on the lowest one. Taking as the elementary volume $d\tau$, a small cube whose faces are normal to coordinate axes, it is easy to find the i component of the surface forces:

$$\partial_j \sigma_{ij} dt .$$

The Newton equation is then:

$$\rho \left(\partial_t + v_j \partial_j \right) v_i = \rho F_i + \partial_j \sigma_{ij} . \tag{1.3}$$

These basic considerations invite several remarks.

- The first one concerns the nonlinear term $v_j \partial_j v_i$. It is responsible for most of the difficulties in fluid mechanics. At first sight, however, it seems not to respect Galilean invariance. Changing frame adds $\mathbf{V}_0$ to all velocities. Time is unchanged but the time derivative is changed becoming: $\partial_t - V_{0j}\partial_j$ and the equation is invariant.
- The second remark concerns the separation between volume forces and surface ones. It is important to note that this is a largely artificial one. Consider, for instance, the electrostatic stress tensor which contains the term $E_i D_j$ where $\mathbf{E}$ is the field and $\mathbf{D}$ the electric induction. Taking its divergence gives $q\mathbf{E}$, where q is the free charge density, which looks like a volume force.
- The third remark concerns the traditional way of writing the stress tensor:

$$\sigma_{ij} = -p\delta_{ij} + \sigma'_{ij} . \tag{1.4}$$

Here it is important to note that p does not represent the normal force on a surface. Such a confusion would deny the existence of a second viscosity. Indeed, here p is the function of density and temperature which is equal to the pressure at equilibrium, and must be used as such.

1.4 Momentum budget: two examples

Equation (1.3) can be presented as a budget equation for the momentum. As the density of momentum is ρv_i we look for an equation like:

$$\partial_t (\rho v_i) + \partial_j \Pi_{ij} = \text{``sources''} ,$$

where Π_{ij} will be the (tensor) flux of momentum. Using the mass budget:

$$\partial_t (\rho v_i) = \rho \partial_t v_i + v_i \partial_t \rho = \rho \partial_t v_i - v_i \partial_j (\rho v_j) ,$$

we can transform the left hand side of (1.3):

$$\partial_t (\rho v_i) + v_i \partial_j (\rho v_j) + \rho v_j \partial_j v_i = \partial_j \sigma_{ij} + \rho F_i ,$$

which gives the desired equation:

$$\partial_t (\rho v_i) + \partial_j (\rho v_i v_j - \sigma_{ij}) = \rho F_i .$$

There are again several remarks to make. It is reasonable that volume forces appear as momentum sources. On the other hand, a stress tensor appears as a flux of momentum, which is obvious in one dimension. Pulling an object with a rope, you can consider your arm as a source of momentum which propagates along the rope thanks to its tension, and accumulates in the object.

However we saw just before that the distinction between volume and surface forces is artificial. Indeed the choice is equivalent to deciding what is part of the fluid and what is not. In the example we mentioned above (section 1.3) when electric field and charges are present, we can either add a term $E_i D_j$ to σ_{ij}, or add a force $q\mathbf{E}$ to the volume forces. The total force on the fluid element is the same. In the first case the field enters σ_{ij} as it is considered as part of our system (fluid + field). In the second case we consider an "external" field $\mathbf{E}$ acting on our system (fluid alone).

The momentum budget is particularly useful when you can neglect the viscosity, that is the σ'_{ij} part of the stress tensor. We shall treat two examples.

The first one is the hydraulic jump (see Fig. 1.2). It is a sea wave at the end of its life! Far from the jump, the velocity is parallel to the bottom. The height is h_1 (h_2), and the velocity zero (v_2) in front of (behind) the jump. In these regions the pressure p verifies:

$$\partial_z p = -\rho g \, .$$

Let us work in a frame moving at the speed c of the jump and calculate the horizontal momentum flux across the dashed surface:

$$\int_0^{h_2} \left[p + \rho(v_2 - c)^2 \right] dz - p_0\,(h_2 - h_1) - \int_0^{h_1} \left[p + \rho c^2 \right] dz = 0$$

(no contribution at the bottom since viscosity is neglected). This yields:

$$\tfrac{1}{2}\rho g h_2^2 + \rho\,(v_2 - c)^2\,h_2 = \tfrac{1}{2}\rho g h_1^2 + \rho c^2 h_1 \, .$$

On the other hand, mass conservation implies:

$$\rho c h_1 = \rho\,(c - v_2)\,h_2 = \rho Q \, .$$

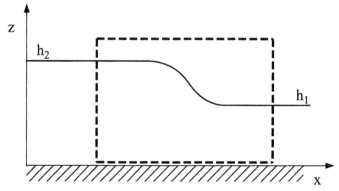

Fig. 1.2.   Hydraulic jump moving toward the right.

Then

$$Q^2 \left(\frac{1}{h_1} - \frac{1}{h_2} \right) = g \frac{h_2^2 - h_1^2}{2}$$

and

$$c^2 = \frac{Q^2}{h_1^2} = g \frac{h_2}{h_1} \frac{h_1 + h_2}{2}.$$

Using budget equations is the most practical process for "shock waves." Again we have some aside remarks:

- For $h_1 \simeq h_2$ we obtain the velocity of small amplitude waves in shallow water: $c = \sqrt{gh}$.
- $c^2 = g(h_2/h_1)(h_1 + h_2)/2 > gh_1$ means that the shock goes faster than small perturbations in front of it and "eats" them. In the same way $gh_2 > (c - v_2)^2$ and small perturbations behind the shock can merge with it. This contributes to the stability of the shock.
- For periodic perturbations of wave number k the induced movement of the fluid cannot exceed a depth of order $1/k$. When kh becomes larger than 1 the velocity is $c = \sqrt{g/k}$.

Our second example is the turbulent round jet at large Reynolds number (see Fig. 1.3). Far from the nozzle the jet will be conical. Any other shape would introduce a characteristic length which does not exist. The axis parallel momentum budget can be written:

$$d_1^2 \rho \left\langle v_1^2 \right\rangle = d_2^2 \rho \left\langle v_2^2 \right\rangle .$$

Thus, the Reynolds number $d \left\langle v^2 \right\rangle^{1/2} / \nu$ is conserved along the jet and the flux of matter involved, proportional to vd^2, increases, which reveals the pumping action of the jet on the surrounding fluid.

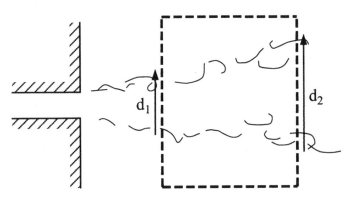

Fig. 1.3. Turbulent jet.

1.5 Energy and entropy budget

The last equation is the energy budget. It can be established for a fluid element of mass $\rho d\tau$. Calling ϵ the internal energy per unit mass, the change of energy per unit time of this element is:

$$\rho D_t \left(\epsilon + \tfrac{1}{2}v^2\right) d\tau .$$

On the one hand this energy comes from the heat flux (defined in the fluid frame):

$$-J_j dS_j = (-\partial_j J_j) d\tau$$

and, on the other hand, from the work of applied forces, volume forces

$$\rho F_i v_i d\tau$$

and surface forces

$$v_i \sigma_{ij} dS_j = \partial_j \left(v_i \sigma_{ij}\right) d\tau .$$

Before collecting all contributions, let us note that for any "density" per unit mass α we can write

$$\partial_t (\rho\alpha) = \rho\partial_t\alpha + \alpha\partial_t\rho = \rho D_t\alpha - \rho v_j\partial_j\alpha - \alpha\partial_j(\rho v_j)$$

thus:

$$\rho D_t\alpha = \partial_t(\rho\alpha) + \partial_j(\rho\alpha v_j) . \tag{1.5}$$

The energy budget is then:

$$\partial_t \left(\epsilon + \tfrac{1}{2}v^2\right) + \partial_j \left(\rho v_j \left(\epsilon + \tfrac{1}{2}v^2\right) + J_j - \sigma_{ij}v_i\right) = \rho F_i v_i . \tag{1.6}$$

It has the general form of budget equations. The energy flux appears as the sum of a "convective" term, of the heat flux (in the fluid frame), and of a term corresponding to the stress tensor. Volume forces give the "source." We see again here that the distinction between volume and surface forces depends upon choosing what is part of the fluid and what is not.

We have thus our six equations. It is however interesting to transform the last one in an entropy budget. The need for a precise definition of entropy is even clearer than for the pressure p (section 1.3). The fluid elements are not at equilibrium. What we call "entropy per unit mass" s, and what we use as it, is simply a function of density ρ and temperature T, which would be equal to the entropy at equilibrium with the same ρ and T. The well known relation:

$$d\epsilon = Tds + \left(\frac{p}{\rho^2}\right) d\rho$$

gives

$$\rho D_t s = \left(\frac{\rho}{T}\right) D_t \epsilon - \left(\frac{p}{\rho T}\right) D_t \rho .$$

The mass budget reads:

$$\partial_t \rho + v_j \, \partial_j \rho + \rho \partial_j v_j = D_t \rho + \rho \partial_j v_j = 0 .$$

From (1.3) one gets:

$$\rho D_t \left(\tfrac{1}{2} v^2\right) = \rho v_i D_t v_i = \rho v_i F_i + v_i \partial_j \sigma_{ij}$$

so that the internal energy budget reads:

$$\rho D_t \epsilon = -\partial_j J_j + \sigma_{ij} \partial_j v_i$$

yielding:

$$\rho D_t s = -\left(\frac{1}{T}\right) \partial_j J_j + \frac{1}{T} \left(\sigma_{ij} + p\delta_{ij}\right) \partial_j v_i$$

$$= -\partial_j \left(\frac{J_j}{T}\right) - \frac{J_j}{T^2} \partial_j T + \frac{\sigma'_{ij}}{T} \partial_j v_i ,$$

which can be written:

$$\partial_t \left(\rho s\right) + \partial_j \left(\rho s v_j + \frac{J_j}{T}\right) = -\left(\frac{J_j}{T^2}\right) \partial_j T + \frac{\sigma'_{ij}}{T} \partial_j v_i . \qquad (1.7)$$

The entropy flux thus contains a convection term, plus $\mathbf{J}/T$ which corresponds to the reversible entropy exchange across a surface. The entropy sources are first the irreversible heat flow, second the viscous friction. Both must be positive. When $\mathbf{J}$ is taken as linear in $\mathrm{grad}\,T$, the corresponding source is a quadratic form of $\mathrm{grad}\,T$. In the same spirit, for Newtonian fluids, σ'_{ij} is a linear function of $\partial_j v_i$.

In isotropic fluids, the only scalars quadratic in $\partial_j v_i$ are: $\partial_j v_i \partial_j v_i$ (sum of the squared components), $\partial_j v_i \partial_i v_j$, and $(\partial_i v_i)^2$ (i.e. $(\mathrm{div}\,\mathbf{v})^2$). The first two can be grouped in $e_{ij} e_{ij}$ or $\omega_{ij} \omega_{ij}$ where $e_{ij} = \tfrac{1}{2} \left(\partial_i v_j + \partial_j v_i\right)$ and $\omega_{ij} = \tfrac{1}{2} \left(\partial_i v_j - \partial_j v_i\right)$.

In a solid-like rotation there is no stress, $e_{ij} = 0$ but not ω_{ij}. Thus σ'_{ij} does not depend on ω_{ij}. In order to have independent quantities one introduces:

$$e'_{ij} = e_{ij} - \tfrac{1}{3} \delta_{ij} \left(\partial_l v_l\right) .$$

The (positive) dissipation will be a combination of $e'_{ij} e'_{ij}$ and $(\partial_i v_i)^2$:

$$\sigma'_{ij} \partial_j v_i = \mu e'_{ij} e'_{ij} + \zeta \left(\partial_i v_i\right)^2 ;$$

μ is the shear viscosity, and ζ the bulk viscosity. Both must be positive. Without going into details let us simply point out that we then obtain:

$$\sigma'_{ij} = \mu \left(\partial_i v_j + \partial_j v_i \right) + \left(\zeta - \tfrac{2}{3}\mu \right) \delta_{ij} \left(\partial_l v_l \right) . \tag{1.8}$$

1.6 The Navier–Stokes equation

The Navier–Stokes equation is simply the dynamic equation for incompressible Newtonian fluids. We must first define what is incompressible for us. It will always be written:

$$\operatorname{div} \mathbf{v} = 0 \, ,$$

which in turn means that div $\mathbf{v}$ is much smaller than any other combination of velocity derivatives. From the mass conservation equation

$$\operatorname{div} \mathbf{v} = -\frac{1}{\rho} D_t \rho \, . \tag{1.9}$$

Several factors can cause large relative changes in density. Applied forces, like gravity can do it, and it is clear that large atmospheric vertical motions can hardly be considered as incompressible. Very large temperature change can also be a cause, and, of course, pressure fluctuations. The order of magnitude of pressure fluctuations in a flow is the stopping pressure ρv^2. The corresponding relative change in density is:

$$\frac{1}{\rho} \rho v^2 \left(\frac{\partial \rho}{\partial p} \right) = \frac{v^2}{c^2} = \mathrm{Ma}^2 \, ,$$

where Ma is the Mach number. Thus incompressible *flows* (better than *fluids*) are first and foremost subsonic flows (Ma $\ll$ 1).

Using div $\mathbf{v} = 0$ the dynamic equation can be written as:

$$\partial_t v_i + v_j \partial_j v_i = -\partial_i (p/\rho) + \nu \nabla^2 v_i \tag{1.10}$$

This is the Navier–Stokes equation. $\nu = \mu/\rho$ is the kinematic viscosity.

Again a remark. Taking the divergence of the Newtonian stress tensor, we have considered the viscosity as constant. This is not simply considering a homogeneous fluid. Indeed, taking the derivative of the viscosity would be inconsistent in some sense with the assumptions.

To be specific, imagine that the viscosity varies from place to place due to its temperature dependence. Writing the stress tensor as

$$\sigma_{ij} = \mu(T) \left(\partial_i v_j + \partial_j v_i \right) = 2\mu(T) e_{ij}$$

implicitly assumes that $\sigma_{ij} = 0$ when $e_{ij} = 0$ even when the temperature is inhomogeneous. This is not true. A term can exist like

$$\lambda \frac{\partial^2 T}{\partial x_i \partial x_j} \, .$$

We shall now try to estimate λ and compare the above term with those resulting from differentiating μ. Let us call c the sound velocity. We have $\mu \sim \rho c \ell$ where ℓ is a kind of mean free path. Put in other words, to a shear $e_{ij} \sim c/\ell$ corresponds a stress of order of the compression modulus ρc^2.

In the same spirit let us call θ the temperature range over which the fluid properties noticeably change. Then $\partial \mu / \partial T \sim \mu / \theta$. Assuming that $\lambda\, \partial^2 T / \partial x_i \partial x_j$ is of order ρc^2 when $\partial^2 T / \partial x_i \partial x_j \sim \theta / \ell^2$ gives:

$$\lambda \sim \frac{\rho c^2 \ell^2}{\theta}.$$

Now, if T is a typical temperature range, L a typical size, and V a typical velocity, the correction due to the viscosity temperature dependence is of order

$$T \frac{\partial \mu}{\partial T} e_{ij} \simeq \frac{\rho c \ell}{\theta} \frac{T V}{L}$$

and the new term:

$$\lambda \frac{\partial^2 T}{\partial x_i \partial x_j} \simeq \frac{\rho c^2 \ell^2}{\theta} \frac{T}{L^2}.$$

Their ratio is $VL/c\ell \simeq$ Re which means that taking account of $\mu(T)$ dependence while dropping the pure temperature term is not consistent at moderate Reynolds numbers.

1.7 Life at low Reynolds numbers

Though the book is rather devoted to high Reynolds number situations, low Reynolds numbers have intriguing and amazing aspects which afford to be mentioned. Indeed, our everyday experience with water or with air at human scales, corresponds to high Reynolds numbers, and our intuition often fails when faced with low Reynolds cases.

On the contrary, low Reynolds numbers are a fact of everyday life for microscopic animals like bacteria. "Life at low Reynolds numbers" is also the title of a celebrated paper by E.M. Purcell (1977) by which this paragraph is largely inspired.

The first important remark is the simplification which occurs in the Navier–Stokes equation. It becomes:

$$\mu \nabla^2 \mathbf{v} = \operatorname{grad} p. \tag{1.11}$$

Indeed the missing term is $\rho D_t \mathbf{v}$ whose order of magnitude is $\rho (V/L) V$ where V and L are typical velocity and length in the flow. The order of magnitude of $\mu \nabla^2 \mathbf{v}$ is $\mu \left(V/L^2 \right)$ and the ratio of these terms is the Reynolds number, which is assumed small. Note that this is true for "confined" low Reynolds flow but not for large distances to a body in slow motion. We shall come back to it at the end of the paragraph.

The nice thing with this equation is its invariance with respect to dilatation and time reversal. Putting $t = \lambda t'$, $v = v'/\lambda$ and $p = p'/\lambda$ leaves the equation and boundary conditions unchanged, whatever the sign of λ. This should not be surprising. Low Reynolds means "close to equilibrium" where evolutions are reversible.

However true reversibility needs more than this symmetry. The time reversal transformed solution could be unstable! The stability is ensured if there exists a unique solution for given boundary conditions. This is the case here.

Let us begin by a lemma to show this uniqueness. Consider two velocity fields, $\mathbf{v}$ and $\mathbf{v}'$, coherent with boundary conditions, one of them $(\mathbf{v})$ verifying the equation. We shall show that

$$I = \int \partial_j \left(v'_i - v_i \right) \partial_j v_i \, d^3\mathbf{r} = 0 \, .$$

If $\mathbf{A}$ is the vector whose components are:

$$A_j = \left(v'_i - v_i \right) \partial_j v_i \, ,$$

then

$$\partial_j A_j = \partial_j \left(v'_i - v_i \right) \partial_j v_i + \left(v'_i - v_i \right) \nabla^2 v_i$$

$$= \partial_j \left(v'_i - v_i \right) \partial_j v_i + \frac{1}{\mu} \left(v'_i - v_i \right) \partial_i p \, .$$

Owing to the incompressibility $\partial_i p \left(v'_i - v_i \right) = \left(v'_i - v_i \right) \partial_i p$. Thus:

$$I = \int \mathrm{div} \left(\mathbf{A} - \frac{p}{\mu} \left(\mathbf{v}' - \mathbf{v} \right) \right) d^3\mathbf{r}$$

$$= \oint_S \left[\mathbf{A} - \frac{p}{\mu} \left(\mathbf{v}' - \mathbf{v} \right) \right] d\mathbf{S} = 0$$

as $\mathbf{v} = \mathbf{v}'$ on boundaries. Moreover

$$\partial_j v'_i \partial_j v'_i = \partial_j v_i \partial_j v_i + \partial_j \left(v'_i - v_i \right) \partial_j \left(v'_i - v_i \right) + 2 \partial_j v_i \partial_j \left(v'_i - v_i \right)$$

and we obtain two results:

a) If $\mathbf{v}'$ is also a solution of Eq. 1.11, then:

$$\int \partial_j \left(v'_i - v_i \right) \partial_j \left(v'_i - v_i \right) d^3\mathbf{r} = 0 \, ,$$

which implies $\mathbf{v}' = \mathbf{v}$.

b) The solution of Eq. 1.11 minimizes the entropy production:

$$\frac{\mu}{T} \int \partial_j v_i \partial_j v_i d^3\mathbf{r} \, .$$

Unfortunately this result is not valid for larger Reynolds numbers and the hopes it raised have been dashed.

As discussed by Purcell, this reversibility has impressive consequences for the way of swimming at low Reynolds numbers. A simple "scallop-like" back and forth movement cannot result in any propulsion. Indeed the displacement is the same whatever the velocity, and at the end of the cycle, the scallop would get back its original position.

According to Purcell the simplest swimmer would be a three "branched" animal whose swimming cycle could be that described in Fig. 1.4 (left). What would be the direction of the displacement? x or y? The best method is to schematize the cycle on a diagram whose axes represent the two arms, the up position corresponding to +1 (Fig. 1.4, right).

A $y \mapsto -y$ symmetry results in exchanging up and down for each arm, but the rotation on the diagram is always clockwise. This only changes the time origin, position (3) becoming position (1). On the contrary $x \mapsto -x$ changes the rotation to counter-clockwise. The animal would be moving in the x direction.

Is the velocity positive or negative? It may be more prudent to avoid the question as Purcell did, but we can try anyway. The first movement $(1)\rightarrow(2)$ should push the animal up. Due to the position of the right arm, this movement should be deviated to the right. The second movement should also make the animal slip to the right due to the new position of the left arm. The velocity thus should be positive. But these conclusions could depend on the length of the middle arm!

This is only a short biased view of the interesting physical questions Purcell raised about life at low Reynolds numbers, and I strongly encourage everybody to read this paper.

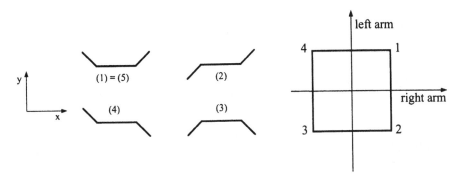

Fig. 1.4. Left: Purcell's low Reynolds swimming lesson. Right: diagrammatic sketch of the swimming steps.

Let us turn now to a discussion of what is occurring far from a body in slow motion. For an exact calculation the body should be an ellipsoid but for our discussion it can have any shape. Call simply a its characteristic size. In the body's frame, let us consider a shell of fluid between r and $r + \delta r$ from the body. As the flow is stationary ($\partial_t = 0$), the shell is in equilibrium under the opposing viscous action on both its sides. The viscous stress per unit area is of order $\mu \, \partial v / \partial r$ and the area grows like r^2. Thus $r^2 \partial v / \partial r$ is constant.

Writing the velocity $-\mathbf{U} + \mathbf{v}$ where $\mathbf{U}$ is the body velocity, is equivalent to coming back into the laboratory frame. Then $r^2 \partial v / \partial r = \text{cst}$ means that v is decreasing like $1/r$:

$$v \simeq \frac{Ua}{r}. \tag{1.12}$$

The Navier–Stokes equation now reads:

$$-U_j \partial_j v_i \quad + \quad v_j \partial_j v_i \quad = \quad -\rho^{-1} \partial_i p \quad + \quad \nu \nabla^2 v_i .$$

$$(\alpha) \qquad\qquad (\beta) \qquad\qquad\qquad\qquad (\gamma)$$

The reasoning we just made was equivalent to neglecting (α) and (β) with respect to (γ). From (1.12) one gets:

$$(\alpha) \simeq \frac{U^2 a}{r^2} \qquad (\beta) \simeq \frac{U^2 a^2}{r^3} \qquad (\gamma) \simeq \nu \frac{Ua}{r^3} ;$$

$(\beta)/(\gamma) = Ua/\nu$ which is always small, but $(\alpha)/(\gamma) = Ur/\nu$. Far from the body, at distances $r > \nu/U$, the (α) term becomes dominant. That is the outer inertial range (see Fig. 1.5).

However, the estimate we have made of (γ) is not true everywhere. On the body trajectory, the fluid has been put in motion. The transverse length over which this motion has diffused, a time $\tau = r/U$ later, is $(\nu\tau)^{1/2} = (\nu r/U)^{1/2} = \lambda$. In this "wake" the fluid momentum is due to

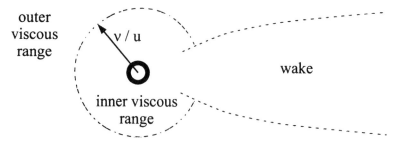

outer
viscous
range

v / u

inner viscous
range

wake

Fig. 1.5. The three regions around a moving body.

the drag on the body. On a length ℓ, the drag force F acted during a time ℓ/U. The momentum is:

$$\rho v \left(\lambda^2 \ell \right) = F \frac{\ell}{U}.$$

The order of magnitude of (γ) is then:

$$\nu \frac{v}{\lambda^2} = \frac{\nu F}{\ell \lambda^4 U} = \frac{FU}{\rho \nu r^2},$$

which can balance with (α) at any distance r as $F \simeq \rho \nu U a$.

In the laboratory frame, the fluid is dragged through the wake in the inner viscous range. It must go out in the outer inertial range. From above, the flux is $F/\rho U \simeq \nu a$. The velocity is thus decreasing in the inertial range as

$$v \simeq \frac{\nu a}{r^2},$$

which fits nicely at the boundary with the viscous range (Fig. 1.6).

Several lessons can be drawn from what precedes. First, the method we have used, i.e. cutting the space into regions where only two terms balance in the equation, is often used in fluid mechanics. Matching the solution at the boundary of these regions is then the main problem.

Next, the separation between a viscous range, an inertial range and a wake is not limited to low Reynolds numbers. At large Reynolds numbers the viscous range will become the boundary layer.

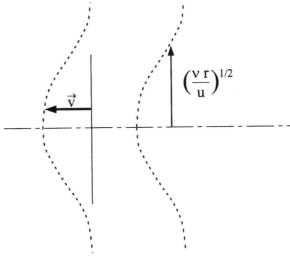

Fig. 1.6.   Velocity profile in the wake showing the impulse given to a layer.

Finally one could try to use the same method for two-dimensional flows (past a cylinder). Then one gets into problems. Two-dimensionality is critical for diffusion-like equations. The problem is that the (β) term is not negligible. It results in a logarithmic correction for the drag per unit length of a cylinder:

$$F = \frac{\rho \nu U}{\ln(\nu/Ua)} \, .$$

2 The mechanisms

2.1 Vorticity: diffusion, freezing, sources

Vorticity is the curl of the velocity:

$$\omega = \operatorname{curl} \mathbf{v} \, . \tag{2.1}$$

Its behavior is a central problem in turbulence. We shall study this behavior from simple examples. What are the basic equations? First the divergence of ω is zero:

$$\partial_j \omega_j = 0 \, .$$

We shall start from the following form of the velocity equation:

$$\partial_t v_i + v_j \partial_j v_i = -\partial_i h + \nu \nabla^2 v_i + \frac{1}{\rho}\left(\zeta - \tfrac{2}{3}\mu\right)\partial_i\left(\partial_l v_l\right) , \tag{2.2}$$

valid for barotropic fluids, where the pressure is a function of the density only. Then $h = \int dp/\rho$ is the enthalpy per unit mass. Taking the curl of equation (2.2):

$$\partial_t \omega_i + \epsilon_{ijk}\partial_j\left(v_l\partial_l v_k\right) = \nu \nabla^2 \omega_i + \text{(neglected terms)} \, .$$

The neglected terms correspond to the variation of ρ in the viscous terms. Let us estimate them. Call η the interesting scale and v_η the velocity variations at this scale. Density variations are of order of v_η^2/c^2 where c is the sound velocity. As remarked before $\nu = c\ell$ defines a "mean free path" ℓ. The neglected terms are v_η^2/c^2 times the viscous one. Their ratio to the inertial terms at the scale η we consider is thus:

$$\frac{\nu}{v_\eta\eta}\frac{v_\eta^2}{c^2} = \frac{v_\eta\ell}{c\eta} \, .$$

Such a ratio can be significant only in shock waves. Beyond this case we have to neglect the derivatives of ν.

Let us come back to the equation. We used the antisymmetric tensor:

$$\left\{ \begin{array}{lll} \epsilon_{ijk} = & 1 & \text{if } (ijk) \text{ is an even permutation of } (xyz) \, , \\ \epsilon_{ijk} = & -1 & \text{if the permutation is odd} \, , \\ \epsilon_{ijk} = & 0 & \text{if two indices are equal} \, . \end{array} \right.$$

For example:

$$(\text{curl }\mathbf{A})_i = \epsilon_{ijk}\partial_j A_k, \qquad (\mathbf{A}\times\mathbf{B})_i = \epsilon_{ijk}A_j B_k.$$

Remark that:

$$\epsilon_{ijk}\partial_j\left(v_l\partial_k v_l\right) = \left[\text{curl}\left(\text{grad }\tfrac{1}{2}\mathbf{v}^2\right)\right]_i = 0.$$

On the other hand

$$\partial_l v_k - \partial_k v_l = \epsilon_{mlk}\omega_m$$

thus

$$\epsilon_{ijk}\partial_j\left(v_l\partial_l v_k\right) = \epsilon_{ijk}\epsilon_{mlk}\partial_j\left(v_l\omega_m\right).$$

The product $\epsilon_{ijk}\epsilon_{mlk}$ is:

$$\begin{cases} 1 & \text{if } (m,l) = (i,j), \\ -1 & \text{if } (m,l) = (j,i), \\ 0 & \text{otherwise}. \end{cases}$$

Then:

$$\begin{aligned}
\epsilon_{ijk}\partial_j\left(v_l\partial_l v_k\right) &= \partial_j\left(v_j\omega_i - v_i\omega_j\right) \\
&= v_j\partial_j\omega_i + \omega_i\partial_j v_j - \omega_j\partial_j v_i.
\end{aligned}$$

From mass conservation:

$$\rho\partial_j v_j = -\partial_t\rho - v_j\partial_j\rho = -D_t\rho.$$

Finally:

$$D_t\omega_i - \frac{\omega_i}{\rho}D_t\rho = \omega_j\partial_j v_i + \nu\nabla^2\omega_i. \tag{2.3}$$

This is the basic equation. From now on we shall only comment on it. Several remarks can be made.

First the diffusion of vorticity. Without the nonlinear terms this equation is simply a diffusion equation (2.3). The simplest example is Rayleigh flow: a plate is suddenly put in motion parallel to itself (see Fig. 2.1, top). Then ω is parallel to y axis $\omega_y = \partial_z v_x$. All the nonlinear terms are zero:

$$v_x\partial_x\omega = 0; \qquad \partial_j v_j = \partial_x v_x = 0; \qquad \omega_y\partial_y\mathbf{v} = 0.$$

Only the diffusion remains:

$$\partial_t\omega_y = \nu\nabla^2\omega_y.$$

It can be considered as the principal property of fluids when compared to solids, that transverse motions are diffusing and not propagating. A consequence is that the drag on the plate, which is infinite at $t = 0$, is continuously decreasing as the motion propagates more and more slowly in the fluid. In the elastic case (solid), the matter already in motion is continuing at the plate velocity, but a new shell of $c_t \delta t$ depth is put in motion during δt (c_t is the transverse sound velocity, see Fig. 2.1, bottom).

The Rayleigh flow gives an intermediate case when the fluid conducts electricity and a magnetic field is present. We shall qualitatively discuss this case without even writing the equations. The velocity profile is as shown in Fig. 2.2. By dimensional arguments we can obtain the various parameters. The depth a on which viscosity acts (the Hartmann layer) could depend on the magnetic permeability μ_0, the conductivity σ, B and μ. ρ does not enter as no acceleration of the fluid occurs in this range. Then $a \simeq (\mu/\sigma B^2)^{1/2}$. Out of this viscous range, the movement is propagating at the Alfven wave velocity $c_A = B/(\mu_0 \rho)^{1/2}$ as in the elastic case. The velocity u of the fluid in this range is given by:

$$\rho c_A u = \mu \frac{U - u}{a},$$

which expresses that the viscous drag finally results in putting in motion new shells of fluid at distance $c_A t$. D is the total diffusion coefficient, sum of the magnetic and vorticity coefficients: $D = \nu + 1/\mu_0 \sigma$. A paradox is that, for $B \to 0$, u is finite! But the intermediate range where it is defined would appear after an infinite time ($a \to \infty$).

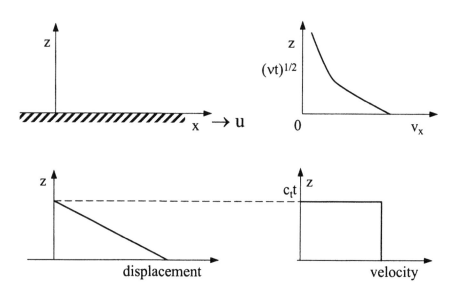

Fig. 2.1.   Top: the Rayleigh flow geometry and its velocity profile. Bottom: displacement and velocity profiles in the "solid Rayleigh flow" case.

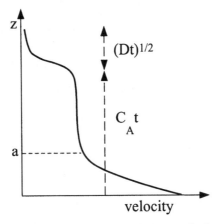

Fig. 2.2. The magnetic Rayleigh flow velocity profile.

The second remark bears on the sources of vorticity or of enstrophy, which is the vorticity squared: ω^2. In the preceding example, vorticity was clearly injected into the fluid from the wall. Indeed the total amount was:

$$\int \omega_y\, dz = -U$$

per unit area of the plate, injected at $t = 0$. It is a general result that vorticity is generated at the boundaries. The basic equation reads:

$$\partial_t \omega_i + \partial_j \left(v_j \omega_i - \omega_j v_i - \nu \partial_j \omega_i \right) = 0 \,.$$

Remember that we neglect derivatives of ν, which are important only in shock waves, where creation of vorticity can occur. Apart from this extreme case, the equation gives the budget of a conserved quantity whose flux is:

$$\Omega_{ji} = v_j \omega_i - \omega_j v_i - \nu \partial_j \omega_i \,.$$

Now consider the enstrophy. We obtain its evolution, multiplying the basic equation by ω_i:

$$\partial_t \omega^2 + \partial_j \left(v_j \omega^2 - \nu \partial_j \omega^2 \right) = 2\omega_i \omega_j \partial_j v_i - 2\nu \partial_j \omega_i \partial_j \omega_i \,.$$

The last term is the viscous damping. Let us analyze the preceding source term $2\omega_i \omega_j \partial_j v_i$. The $\omega_j \partial_j v_i$ part indicates that velocity must vary along ω for this term not to be zero. The ω_i factor means that the $\delta \mathbf{v}$ variation itself must be parallel to ω. Thus, creation of enstrophy means stretching of vorticity tubes. Out of this there is only convection or diffusion of enstrophy, and damping through viscosity.

In two dimensions stretching cannot occur. Here, considering a fluid element, vorticity is directly related to its kinetic momentum, which is exchanged between neighboring fluid elements by viscosity, and ω^2 is related to its kinetic energy, which disappears through viscous damping.

The final remark is about the "freezing" of potential vorticity: ω/ρ. In the inviscid limit $\nu \to 0$, dividing Eq. 2.3 by ρ gives:

$$\frac{1}{\rho} D_t \omega_i - \frac{\omega_i}{\rho^2} D_t \rho = D_t \frac{\omega_i}{\rho} = \frac{\omega_j}{\rho} \partial_j v_i \tag{2.4}$$

We shall show that the elementary vector $\delta\boldsymbol{\ell}$ between two neighboring points obeys the same equation as ω/ρ. Indeed:

$$D_t \delta\boldsymbol{\ell} = \mathbf{v}(\mathbf{r} + \delta\boldsymbol{\ell}) - \mathbf{v}(\mathbf{r}) = \delta\ell_j \partial_j \mathbf{v} \,.$$

$\delta\boldsymbol{\ell}$ is clearly frozen in the fluid. So is ω/ρ. The physical significance is simply angular momentum conservation. Consider a cylinder of fluid whose axis is parallel to ω, with radius r and length d. The proportionality of ω/ρ and $\delta\boldsymbol{\ell}$ means that $\omega/\rho d$ remains constant. On the other hand $m = \rho r^2 d$ is constant (constant mass). This is thus equivalent to $m\omega r^2$ being constant, i.e., conservation of the angular momentum.

It is then obvious that we have to neglect the viscosity. But why is the barotropic assumption important? In the barotropic case, constant density surfaces are identical to constant pressure surfaces. The resulting pressure force passes through the center of mass. In the opposite case, a torque would appear.

2.2 Energy transfer in turbulence

Turbulent flows are very complex. Distinguishing mean flow and fluctuations could be a good way to simplify them. This idea can be traced back to Reynolds. It is far from a panacea, but, as shown by Tennekes and Lumley (1974), it helps in clarifying some problems.

First, it is not so simple to define a good averaging process. Assuming that the quantity $A(\mathbf{x}, t)$ can be Fourier transformed by

$$A(\mathbf{k}, t) = \frac{1}{2\pi} \int e^{-i\mathbf{k}\mathbf{x}} A(\mathbf{x}, t)\, d^3\mathbf{x} \,,$$

we shall take:

$$\overline{A} = \overline{A}_{k_0}(\mathbf{x}, t) = \int_{|\mathbf{k}| > k_0} A(\mathbf{k}, t)\, e^{i\mathbf{k}\mathbf{x}}\, d^3\mathbf{k} \,.$$

This is not a very good averaging, as the averaged value in $\mathbf{x}$ is sensitive to the whole flow. Moreover the averaged quantity does not fulfill the boundary conditions. We can thus only consider "homogeneous" flows, "far" from boundaries. However this averaging process ensures:

$$\overline{\overline{A}} = \overline{A}.$$

For the velocity we can write:

$$\mathbf{v} = \mathbf{U} + \mathbf{u} \qquad \text{with} \qquad \mathbf{U} = \overline{\mathbf{v}} \quad \text{and} \quad \overline{\mathbf{u}} = 0.$$

Let us restrict ourselves to incompressible flows. Then div $\mathbf{v} = 0$. Averaging we have:

$$\text{div } \mathbf{U} = 0 \qquad \text{and thus} \qquad \text{div } \mathbf{u} = 0.$$

The averaged NS equation reads:

$$\rho \partial_t U_i + \rho U_j \partial_j U_i + \rho \overline{u_j \partial_j u_i} = -\partial_i \overline{p} + \mu \nabla^2 U_i, \tag{2.5}$$

which can take the form of an average momentum budget:

$$(\partial_t + U_j \partial_j) \rho U_i = \partial_j \left\{ -\overline{p}\,\delta_{ij} + \rho \nu \left(\partial_j U_i + \partial_i U_j \right) - \rho \overline{u_i u_j} \right\}; \tag{2.6}$$

$\tau_{ij} = -\rho \overline{u_i u_j}$ is the Reynolds stress tensor. Small scale fluctuations give a contribution to the momentum flux. It could be phenomenologically seen as a "turbulent viscosity" in the same way that individual atomic velocities result in the ordinary viscosity tensor $\mu \left(\partial_j v_i + \partial_i v_j \right)$ when averaged. But μ does not depend on the scale of averaging as soon as this scale is larger than the mean free path. This is not true in turbulence where the "turbulent viscosity" depends on the scale of averaging k_0^{-1}. Indeed, the kinematic turbulent viscosity must be of order $\nu_T \simeq v\ell$ where velocity v and scale ℓ are characteristic of small scale fluctuations. It turns out that ℓ is always of order k_0^{-1}.

From an empirical point of view, the "mixing length" ℓ and the turbulent viscosity have been useful notions. However they do not shed light on mechanisms in turbulence. Moreover, as we shall see later, it is indeed in full contradiction with the very spirit of turbulence.

Let us look now at energy exchange between large and small scales. The large scale kinetic energy is $E_{\mathrm{m}} = \rho \frac{1}{2} U^2$ and the small scale one is $E_{\mathrm{f}} = \frac{1}{2} \rho \overline{u^2}$. The evolution of E_{m} is obtained, multiplying the averaged NS equation (2.5) by U_i:

$$\frac{\partial E_{\mathrm{m}}}{\partial t} = \rho U_i \frac{\partial U_i}{\partial t}$$
$$= -\rho U_i U_j \partial_j U_i - U_i \partial_i \overline{p} + \mu U_i \nabla^2 U_i - \rho U_i \overline{u_j \partial_j u_i}.$$

Owing to div $\mathbf{U} = \text{div } \mathbf{u} = 0$, this can be written:

$$\partial_t E_{\mathrm{m}} + \partial_j \Phi_j = \rho \overline{u_i u_j} \partial_j U_i - \mu \partial_j U_i \partial_j U_i$$
$$= -\epsilon^* - \mu \partial_j U_i \partial_j U_i, \tag{2.7}$$

where

$$\Phi_j = \rho U_j E_{\rm m} + \langle p \rangle U_i \delta_{ij} - \mu \partial_j E_{\rm m} + \rho U_i \overline{u_i u_j}$$

is the flux of large scale energy. The same procedure, applied to the non averaged NS equation gives:

$$\rho \partial_t \tfrac{1}{2} v^2 + \partial_j \phi_j = -\mu \partial_j v_i \partial_j v_i \,, \tag{2.8}$$

where ϕ_j will not be detailed and $\mathbf{v} = \mathbf{U} + \mathbf{u}$. The new term is thus $-\epsilon^* = \rho \, \overline{u_i u_j} \, \partial_j U_i$ and represents the transfer towards small scales. We can see this by averaging the last equation (2.8):

$$\partial_t \left(E_{\rm m} + E_{\rm f} \right) + \partial_j \overline{\phi}_j = -\mu \partial_j U_i \partial_j U_i - \mu \, \overline{\partial_j u_i \partial_j u_i} \,,$$

where $E_{\rm f} = \tfrac{1}{2} \rho \overline{u^2}$, and subtracting the preceding one (2.7):

$$\partial_t E_{\rm f} + \partial_j \left(\overline{\phi}_j - \Phi_j \right) = -\rho \, \overline{u_i u_j} \, \partial_j U_i - \mu \, \overline{\partial_j u_i \partial_j u_i} \,. \tag{2.9}$$

The second term in the right hand side is the viscous damping corresponding to the small scales. The first one is the exchange with the large scales.

Two remarks can be made. The first one lies again in the difference between the Reynolds stresses and molecular viscosity. The problem is that the main contribution to $\overline{u_i u_j}$ comes from the largest of the small scales, that is of order k_0^{-1}. If the main contribution came from the smallest scales, we could speak again of viscosity. But, on the other hand, $\partial_j U_i$ is dominated by the smallest of the large scales, of order k_0^{-1} again. Thus the energy transfer is local in scales. We insist on the point that it is the exact counterpart of the impossibility of reliably defining a turbulent viscosity. On the contrary the molecular viscosity results in a direct transfer of large scale energy towards molecular agitation, without feeding the intermediate scales.

The second remark bears on the relation between the energy transfer and vortex stretching. It has been pointed out by Richardson (1922) and can be illustrated as follows (see Tennekes and Lumley, 1974). Consider a uniform shear in the xy plane:

$$\partial_x U_x = s \,, \qquad \partial_y U_y = -s \,, \qquad \partial_z U_z = 0 \,.$$

Stretching in the x direction enhances the energy of vortices whose axis is parallel to x, and thus tends to enhance the u_y and u_z fluctuations. The compression in the y direction has the opposite effect and tends to reduce u_x and u_z fluctuations. On a whole u_y is enhanced, u_x is reduced and u_z is neutral. Starting from an isotropic situation, the energy transfer rate:

$$-\rho \, \langle u_i u_j \rangle \frac{\partial U_i}{\partial x_j} = \rho \left(\overline{u_y^2} - \overline{u_x^2} \right) s$$

is positive.

Stretching of vortices is certainly an effective process for transporting energy toward small scales. Indeed, in two dimensions, where stretching cannot occur, this transfer is much less efficient, but it does not vanish completely, which shows that stretching is not all.

It is time now to say a few words about Kolmogorov's K41 theory (see Monin and Yaglom, 1975). Notice first that in the large scale energy budget, the ratio between the transfer term and the viscous damping one is of order:

$$\rho \langle u_i u_j \rangle \frac{\partial U_i}{\partial x_j} \bigg/ \mu \left(\frac{\partial U_i}{\partial x_j} \frac{\partial U_i}{\partial x_j} \right) \simeq \frac{U_{k_0}^3 k_0}{\nu k_0^2 U_{k_0}^2} = \frac{U_{k_0}}{k_0 \nu} \, .$$

This is the Reynolds number at scale k_0^{-1}. When this number is large, viscous damping is negligible. It is important only at small scales. At relatively large scales, energy is simply transferred from scale to scale. The Kolmogorov hypothesis is that an intermediate range of scales exists, the inertial range, sufficiently large for viscosity to be negligible, but small enough for the boundary conditions to be forgotten. Thus, the average transfer rate $\langle \epsilon^* \rangle$ (averaged in time here) is independent of the scale and is the only relevant information about the very large scale imposed flow. In this range, turbulence would be homogeneous and isotropic.

Consider, for instance, the velocity difference between two points whose distance is $\mathbf{r}$:

$$\mathbf{v}(\mathbf{x} + \mathbf{r}) - \mathbf{v}(\mathbf{x}) = \mathbf{u}_r \, . \tag{2.10}$$

If r is in the inertial range, the statistics of $\mathbf{u}_r$ should be insensitive to the large scale anisotropy. It will depend only on $\langle \epsilon^* \rangle$ and r. By dimensional analysis the only possibility for the mean square is for instance:

$$\left\langle u_r^2 \right\rangle = C_2 \left\langle \epsilon^* \right\rangle^{2/3} r^{2/3} \, , \tag{2.11}$$

where C_2 should be a universal number.

About this statistics, an important result has been derived by Kolmogorov. Assuming isotropy, and neglecting the viscous term, for u_{lr} being the component of $\mathbf{u}_r$ parallel to $\mathbf{r}$:

$$\left\langle u_{lr}^3 \right\rangle = -\tfrac{4}{5} \left\langle \epsilon^* \right\rangle r \tag{2.12}$$

(see Landau and Lifshitz, 1959). The consequence is that the distribution of u_{lr} is not symmetric. It is skewed, large negative (inward) fluctuations being more probable than positive ones (outward).

We have already spoken of the lowest interesting scale η in turbulent flows, without a precise definition. Following the Kolmogorov idea, this scale must only depend on $\langle \epsilon^* \rangle$ and ν. The following expressions:

$$\eta = \left(\nu^3 \big/ \langle \epsilon^* \rangle \right)^{1/4}, \qquad v_\eta = \left(\langle \epsilon^* \rangle \, \eta \right)^{1/3} \qquad (2.13)$$

are the only possible ones, out of a universal factor, for this scale and the corresponding velocity fluctuations. Following Kolmogorov, velocity statistics are universal in the inertial range if velocities are expressed in units of v_η and length in units of η.

Kolmogorov's K41 theory fits nicely with experiments. For instance the $r^{2/3}$ law (Eq. 2.1) corresponds to the famous $k^{-5/3}$ law for spectra. There are strong objections however. Landau soon remarked that, if ϵ^* has different average values on different parts of the flow, the C_2 coefficient cannot be universal, the same for the parts and for the whole flow: $\langle \epsilon^{*2/3} \rangle$ is not equal to $\langle \epsilon^* \rangle^{2/3}$. Fluctuations of ϵ^* are related to the intermittency problem in turbulence. It is still an open question.

2.3 Boundary layers and their separation

The boundary layer concept is a very important one for several reasons.

The method first. It consists in separating small regions where an a priori small term becomes important, solving the new equation in this region, and then connecting the solution with the outside one. This method will be developed in Chapter 3. The Blasius boundary layer profile is a classical example.

Separation of boundary layers from the wall is also an important phenomenon. It introduces vorticity in the main flow, where an inviscid approximation would forbid it to appear.

Last but not least, instabilities of boundary layers, and their own turbulence are often triggering or controlling the main flow turbulence.

Let us start from the simplest example: a semi-infinite plate parallel to the main flow. U is the velocity far away from the plate. It is zero on the plate (Fig. 2.3).

Close to the plate the velocity gradient implies a vorticity $\omega_z = -\partial_y v_x$ which was not present in the flow before the plate. As seen previously it has been created at the plate. After a time x/U following the flow, this vorticity has diffused on a width $\delta \simeq (\nu x/U)^{1/2}$. Out of this layer, the flow remains inviscid.

At very large Reynolds number $\mathrm{Re} = Ux/\nu$, δ is much smaller than x. Prandtl's idea (see Landau and Lifshitz, 1959) is to expand this region to obtain an equation valid inside the layer, the outside inviscid flow acting as a boundary condition at infinity. We shall use for this purpose

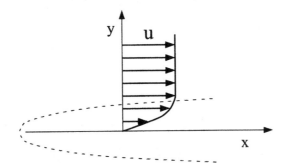

Fig. 2.3. The boundary layer profile for a semi-infinite plate.

nondimensional coordinates x', y', v'_x, v'_y, defined as follow:

$$x = Lx', \quad y = \left(\frac{\nu L}{U}\right)^{1/2} y', \quad v_x = U v'_x, \quad v_y = \left(\frac{U\nu}{L}\right)^{1/2} v'_y.$$

Thus

$$\partial_x \to \frac{1}{L}\partial_{x'} \quad \text{and} \quad \partial_y \to \left(\frac{U}{\nu L}\right)^{1/2} \partial_{y'}.$$

The incompressibility condition now reads:

$$\partial_{x'} v'_x + \partial_{y'} v'_y = 0. \tag{2.14}$$

For a stationary motion, the x-component of the Navier–Stokes equation reads:

$$v_x \partial_x v_x + v_y \partial_y v_x = \nu \partial_{y^2} v_x,$$

where we have neglected in the Laplacian the x derivatives compared to the y ones. Using the new coordinates:

$$v'_x \partial_{x'} v'_x + v'_y \partial_{y'} v'_x = \partial_{y'^2} v'_x. \tag{2.15}$$

In these new coordinates, within the high Reynolds number assumption, no parameter enters the equation. The solution only depends on the boundary conditions in these coordinates:

$$v'_x = f(x', y') \quad \text{and} \quad v'_y = g(x', y')$$

Or:

$$v_x = U f\left(\frac{x}{L}, y\left(\frac{U}{\nu L}\right)^{1/2}\right), \quad v_y = \left(\frac{U\nu}{L}\right)^{1/2} g\left(\frac{x}{L}, y\left(\frac{U}{\nu L}\right)^{1/2}\right).$$

If the plate were finite, its length would introduce a natural scale. Changing L would change the boundary conditions in the new coordinates. With an infinite plate however, at given x and y the solution should be independent of L. Thus x' and y' must combine in such a way that L disappears, hence

$$v_x = U f\left(\left(\frac{U}{\nu}\right)^{1/2} \frac{y}{x^{1/2}}\right), \quad v_y = \left(\frac{\nu U}{x}\right)^{1/2} g_1\left[\left(\frac{U}{\nu}\right)^{1/2} \frac{y}{x^{1/2}}\right],$$

or

$$v'_x = f(\zeta), \quad v'_y = \frac{1}{x'^{1/2}} g_1(\zeta)$$

with $\zeta = y'/x'^{1/2}$. g_1 can be eliminated thanks to the incompressibility relation. This yields the Blasius equation for f:

$$f'' + \frac{1}{2} f' \int_0^\zeta f(\zeta') \, d\zeta' = 0, \tag{2.16}$$

with the boundary conditions:

$$f(\infty) = 1; \qquad f(0) = 0.$$

The solution is the Blasius profile for the boundary layer flow (Fig. 2.4). Close to the plate:

$$\partial_y v_x = U \left(\frac{U}{\nu x}\right)^{1/2} f'(0).$$

The viscous drag on a plate is thus proportional to $U^{3/2}$. But this only concerns laminar flows.

Let us now turn to a "real" body (Fig. 2.5). If the radius of curvature of its surface is always much larger than the boundary layer thickness δ the plate approximation is valid. But the inviscid external flow U varies along the body and the pressure is no longer uniform along x.

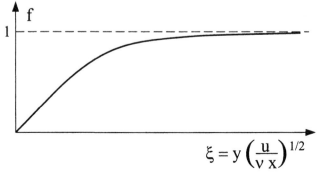

Fig. 2.4.   Blasius velocity profile.

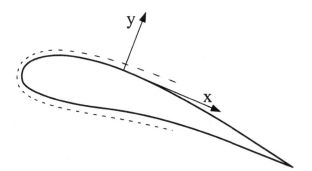

Fig. 2.5. Boundary layer profile around a real body.

The first important remark is that pressure variations are negligible in the y direction perpendicular to the body. The reason is the following. Neglect first the curvature of the surface. The (stationary) Navier–Stokes equation gives:

$$\partial_i p = \mu \nabla^2 v_i - \rho \left(v_j \partial_j \right) v_i \, .$$

As the v_y component is much smaller than the v_x one, $\partial_y p$ will be smaller than $\partial_x p$. Indeed:

$$\partial_x v_x = -\partial_y v_y$$

gives in order of magnitude $v_x/x \simeq v_y/\delta$ and

$$\partial_y p \simeq \frac{\delta}{x} \partial_x p \, .$$

Then the variation of pressure across the layer is:

$$p' - p \simeq \frac{\delta^2}{x} \partial_x p$$

and

$$\partial_x \left(p' - p \right) \simeq \frac{\delta^2}{x^2} \partial_x p \, .$$

This is not true with a curvature due to the rotation of x and y axes. The pressure difference across the layer then balances with centrifugal force. If r is the radius of curvature ($x \simeq r$):

$$p' - p \simeq \rho \frac{U^2}{r} \delta$$

and

$$\partial_x (p' - p) \simeq \frac{\delta}{r} \rho U \, \partial_x U \, .$$

The difference is of first order and not second order in δ/x, but remains negligible. We thus can calculate the pressure gradient in the inviscid flow:

$$\rho U \, \partial_x U = -\partial_x p \,,$$

giving the pressure term in our v_x equation:

$$v_x \partial_x v_x + v_y \partial_y v_x = -U\partial_x U + \nu \partial_{y^2} v_x \,. \tag{2.17}$$

We already know that vorticity is injected at the boundary and then diffuses in the fluid. For diffusion, considering positive quantities helps our intuition. Consider thus:

$$w = \partial_y v_x = -\omega_z$$

and its flux on the wall:

$$J_w = -\nu \partial_y w = U \partial_x U \,.$$

Thus, for a growing external velocity, the surface is a source of w. For a decreasing U it is a sink. This is the basic idea which shows the necessity for the boundary layer to separate from the wall. A stagnation point lies at the tip of the body. From this point, along the body, U first increases: the wall is creating w which then diffuses in the flow.

At some point, with an inviscid flow, the velocity U begins to decrease down to zero at the downstream stagnation point. Correlatively the pressure increases and this is the origin of the zero drag (d'Alembert paradox). The increases of pressure at the two stagnation points exactly compensate. The situation is different in our real fluid. As U decreases, the wall becomes a sink of w and the vorticity injected in the flow is pumped back. One clearly obtains $w < 0$ at the wall with $\int_0^\infty w \, dy > 0$, thus before U being zero (Fig. 2.6). The boundary layer is thus separated from the wall.

Separation in fact occurs earlier. An inflection point in the v_x profile makes the boundary layer unstable (see Chapter 2). Also the separation itself influences the outside inviscid flow which makes the separation occur sooner. But the main idea is here and we can draw some qualitative conclusions from it.

i) Sharp edge effect. Close to a sharp edge, an inviscid flow has extreme velocity variations. This is a natural point for separation.

ii) Convergent and divergent channels. In a convergent channel, the velocity increases and the boundary layers are stable. Their width is even reduced as most of the vorticity has been created soon before. This is the most popular way of obtaining a uniform flow in a tube.

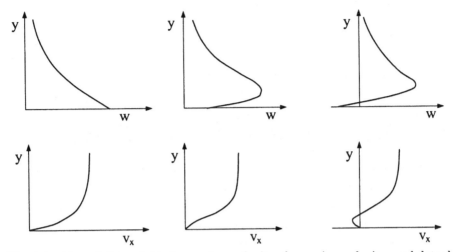

Fig. 2.6. From left to right: increasing velocity, decreasing velocity, and detachment. Top: vorticity. Bottom: velocity profiles.

On the contrary a divergent flow is unstable, the boundary layer detaching from the wall on one side or the other (see Fig. 2.7).

iii) Drag. The separation suppresses the downstream stagnation point and thus kills d'Alembert's paradox. A drag force appears equal to the stagnation pressure $\frac{1}{2}\rho U^2$ times the cross section defined by the separation line on the body. This drag dominates the viscous one, proportional to $U^{3/2}$. Controlling and delaying the separation is thus an important matter.

iv) A first way to avoid the separation is to give time for the wall to pump back the vorticity. This is why the cross section of streamlined bodies decreases slowly at the rear. A second way is to enhance the vorticity diffusion at the back. This can be made by provoking the transition to a turbulent boundary layer and this is the reason for the roughness of golf balls. Vortex generators on plane wings resort to the same idea.

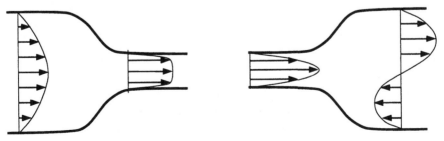

Fig. 2.7. Convergent and divergent channels.

2.4 Compressibility

Compressibility can appear in many ways in a flow. We shall enumerate the most important ones, borrowing this classification from Tritton's book. We shall then discuss some points where it can alter the mechanisms discussed earlier.

We have first to remark that the way we shall take the compressibility does not exactly fit with our intuition. Indeed, up to now, we have drawn two different consequences from incompressibility. For instance we quietly replaced $(1/\rho)\,\mathrm{grad}\,p$ by $\mathrm{grad}(p/\rho)$, which is correct if $\partial_i p/\rho \gg p\,(\partial_i\rho/\rho^2)$. This condition never holds for a gas and this corresponds to the intuition that gases are compressible.

However, in most of the derivations we only need that $\rho^{-1}\,\mathrm{grad}\,p$ be a gradient. This is the case for barotropic flows (section 2.1). Indeed, even in some convection problems, the density variation with temperature is only kept in the buoyancy force (Boussinesq approximation) and the flow can be considered as "incompressible."

A better definition of incompressibility will thus be:

$$\frac{1}{\rho}D_t\rho \simeq 0$$

or, more precisely:

$$\left|\frac{1}{\rho}D_t\rho\right| = |\,\mathrm{div}\,\mathbf{v}\,| < \frac{U}{L}, \tag{2.18}$$

which does not exclude gases.

Let us examine this condition. As usual U, L, t_0 represent orders of magnitude for velocity, size and typical evolution time. In general $t_0 \simeq L/U$ but it can be much shorter if an oscillatory motion is present. In a "barotropic" frame we estimate the density variations by:

$$\frac{1}{\rho}D_t\rho = \frac{1}{\rho c^2}D_t p \simeq \frac{1}{\rho c^2}\frac{\delta p}{t_0}$$

where c is the sound velocity, and the pressure variations amplitude is estimated from the Navier–Stokes equation

$$\rho^{-1}\partial_i p \;=\; (\delta p/\rho L) \;=\; -\partial_t v_i \;-\; v_j\partial_j v_i \;+\; F_i \;+\; \nu\nabla^2 v_i$$
$$\qquad\qquad\qquad\qquad\quad (a)\qquad\qquad (b)\qquad\quad (c)\qquad\quad (d)$$

Consider successively each term as dominant:

(a) $(\delta p/\rho) \simeq (UL/t_0)$. Condition (2.18) then gives: $(U/L) \gg (UL/c^2 t_0^2)$ or $L^2 \ll c^2 t_0^2$. The size L must be much shorter than the sound wavelength at frequency t_0^{-1}.

(b) $(\delta p/\rho) \simeq U^2$. Then: $(U/L) \gg (U^2/c^2 t_0)$ as $t_0 < L/U$ it gives: $U^2 \ll c^2$. The typical velocity must be much smaller than the sound velocity.

(c) $(\delta p/\rho) \simeq FL$. Applied forces must not make the density vary.

(d) $(\delta p/\rho) \simeq (\nu U/L)$. Then $(U/L) \gg (\nu U/L t_0 c^2)$ or $t_0 \gg (\nu/c^2)$. The typical evolution time must be larger than the viscoelastic relaxation time (section 1.1). This seems obvious but, in the Earth mantle for instance, this is not verified.

The most popular condition is obviously that given by (b). Conversely supersonic flows are often present in compressible flow studies. We shall not enter deeply in this subject but only present an example which shows how counter-intuitive they can be.

Consider a tube whose section $A(x)$ has a minimum, in such a way that the flow first converges, then diverges (Fig. 2.8).

In a one-dimensional approximation, mass conservation reads:

$$\rho v A = \text{cst} ,$$

which gives:

$$\frac{1}{\rho}\frac{d\rho}{dx} + \frac{1}{v}\frac{dv}{dx} + \frac{1}{A}\frac{dA}{dx} = 0 .$$

Neglecting the viscosity:

$$v\frac{dv}{dx} + \frac{1}{\rho}\frac{dp}{dx} = 0 .$$

As $dp/dx = c^2 \, d\rho/dx$ we have finally the Hugoniot equation:

$$\left(\frac{v^2}{c^2} - 1\right)\frac{1}{v}\frac{dv}{dx} = \frac{1}{A}\frac{dA}{dx} .$$

The surprising implication is that the fluid cannot reach the sound velocity, except at the minimum of A. Then dv/dx does not change sign at

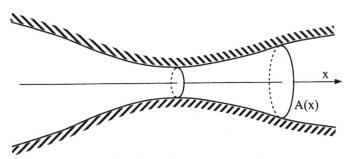

Fig. 2.8.   Laval nozzle for supersonic flows.

the minimum. The velocity increases further downstream, and pressure and density decrease adiabatically. For a gas, one has to strongly preheat it to avoid liquefaction.

The attentive reader will have remarked that we have assumed the divergent flow to be stable, which is not obvious as seen before. Again this is due to compressibility and to the considerations below.

To base our intuition on what we already know, let us consider a poorly compressible flow and use again the NS equation to estimate the pressure (and thus the density) variations.

We start from an equation for the pressure evolution that I know from M.E. Brachet. Take the divergence of the NS equation. We have:

$$\operatorname{div} \partial_t v = \partial_t \operatorname{div} v \simeq 0 , \qquad \operatorname{div}\left(\nabla^2 v\right) \simeq 0 .$$

Finally:

$$\nabla^2 p = -\rho\, \partial_i \left(v_j \partial_j v_i\right) = -\rho\left(\partial_i v_j\right)\left(\partial_j v_i\right) ,$$

but:

$$- \left(\partial_i v_j\right)\left(\partial_j v_i\right) = \frac{1}{4}\sum_{ij}\left\{\left(\partial_i v_j - \partial_j v_i\right)^2 - \left(\partial_i v_j + \partial_j v_i\right)^2\right\}$$

and

$$\nabla^2 p = \tfrac{1}{2}\rho\omega^2 - \rho e_{ij}e_{ij} .$$

The second term on the right hand side is the dissipation divided by ν. This equation looks like the Laplace equation in electrostatics:

$$\nabla^2 V = -\frac{q}{\epsilon_0}$$

where V is the potential and q the charge density.

Following this analogy the positive charge density is proportional to the dissipation and the negative one to the enstrophy. Pressure maxima are thus in dissipation zones and pressure minima on vorticity concentrations. In a laminar flow both are mixed. As recently shown by numerical simulations, this is not the case in turbulent flows. In a way, the transition to turbulence can be interpreted as a demixing transition, where vorticity and dissipation are separating.

Let us now come back to vortex stretching. Indeed we have seen that it results in amplifying ω/ρ. For incompressible fluids this process mainly ensures the energy transfer towards small scales, by amplifying ω. In compressible flows, as pressure minima coincide with vortex cores, ρ falls down, which limits this amplification of ω. The energy transfer is thus reduced. The same is true for flow instabilities, mainly based on vorticity amplification. This explains why divergent flows are stabilized when compressibility appears.

3 Flow measurement methods

3.1 Introduction

From flow visualization to point measurements, the panel of methods is very large. Velocity is not the only quantity to measure. Vorticity, pressure, temperature, contaminant concentration, currents and potentials in electro- or magnetohydrodynamics... are also obviously important. For each quantity many methods have been developed. However, we shall focus here on velocity measurement methods and mainly on those giving the highest signal-to-noise ratio (see Comte-Bellot, 1976).

Let us begin with a discussion on the expected characteristics of the measure (band width, signal-to-noise,...) depending on the type of study. Statistics seem the most demanding. Using Eqs. 2.11 and 2.13 the smallest interesting scale in turbulence η is related to the largest one L as:

$$\eta = L \operatorname{Re}^{-3/4} .$$

At this scale, typical velocity differences are of order:

$$v_\eta \simeq (\epsilon^* \eta)^{1/3} \simeq u \operatorname{Re}^{-1/4} ,$$

where u gives the size of velocity fluctuation, typically:

$$u = 0.1 \, U ,$$

where U is the mean flow velocity.

Asking only for 10% precision for v_η gives for the signal-to-noise ratio (SNR):

$$\frac{U}{\delta u} = 10^2 \operatorname{Re}^{1/4} \simeq 10^4$$

for $\operatorname{Re} \simeq 10^8$. It is clear that only a few methods will meet this performance. As for the band width it is given by:

$$\Delta f = \frac{U}{\eta} .$$

Some studies focus on averaged quantities (scaling laws, velocity profile). They are clearly less demanding. The measurement noise need not be much smaller than natural fluctuations and a SNR of 10^2 is sufficient. However absolute precision is necessary here. On the other hand the band width can be small only if the signal depends linearly on the velocity.

Coherent structures studies are difficult to discuss as a good definition of what they are is still to be found. Let us take them as vorticity concentrations. Simultaneous measurements must be made at many points, the accuracy being sufficient to distinguish between vorticity profiles given by concurrent theories. That is difficult and numerical simulations seem best placed for such studies.

Among traditional methods, the Pitot tube is mainly used for calibration or for average velocity measurements. It measures the stopping pressure. It is a slow detector, for relatively large velocities.

The most popular tools are the hot wire and the Doppler laser anemometer. Before discussing them in the next paragraphs let us mention some new or old useful methods concerning velocity or other quantities.

There is a lack of sufficiently local pressure measurements. Temperature measurements on the contrary are easy and fast. Temperature can act as a simple marker or as an active scalar.

Let us also mention sound diffraction which gives direct access to the Fourier transform of the vorticity field. But this is a very recent method.

3.2 Hot wire anemometer

As suggested by the name this is a Joule heated wire cooled by the flow. For a sufficient wire-diameter based Reynolds number (Ud/ν), a boundary layer would form, with thickness d inversely proportional to $U_n^{1/2}$ where U_n is the component of U normal to the wire. The heat path being reduced, the cooling is enhanced. The following relation for the dissipated power:

$$RI^2 = \left(A + BU_n^{1/2}\right)(T - T_0)$$

that is King's relation, captures the spirit of what precedes. Here T is the wire temperature, and T_0 the main flow one. R is the resistance per unit length of the wire, depending on T. Two methods are thus possible:

i) At constant I, the resistance is measured; the signal is the resistance, thus the temperature.
ii) The intensity I is adjusted to maintain a constant resistance (constant temperature). The signal is I.

To be specific, for inert gases, the hot wire is often made from a Wollaston wire: it is a platinum wire forming the core of a silver wire. Typical dimensions are 1 micron for the platinum core diameter d and 50 microns for the silver coating. This wire is placed between two arms and the platinum wire is revealed by chemical dissolution of silver on a length ℓ of 0.3 to 1 mm. This length must not be too small: the heat would mainly flow at extremities through the silver and the cooling would depend on the parallel component of the velocity if the ratio ℓ/d of the length to the diameter of the platinum wire were too small. A good value is:

$$\frac{\ell}{d} \simeq 200$$

For liquids, hot films are used, made by vacuum deposition of Pt on a quartz fiber, then coated by quartz deposition.

With method (i) the thermal inertia of the wire limits the band width. The time constant is the product of the wire heat capacity and the thermal resistance of the surrounding fluid. Let us estimate it.

If the fluid is a monoatomic gas, its thermal conductivity is:

$$K \simeq \tfrac{1}{3} c \lambda C_p,$$

c is the thermal velocity, C_p the heat capacity of the gas: $C_p = \tfrac{5}{2} n k_B$. n is the atomic density and the mean free path $\lambda \simeq 1/n\sigma$ where σ is the atomic cross section. Thus:

$$K \simeq \frac{c k_B}{\sigma}.$$

The heat capacity per unit length of the wire can be estimated through the Dulong and Petit law:

$$C_s = 3 n_m k_B \frac{\pi d^2}{4},$$

where n_m is the platinum atomic density. The time constant is then:

$$\tau = \frac{C_s \delta}{K \pi d} = \frac{\sigma n_m}{c} d\delta.$$

This gives $\tau \simeq 10^{-4}$ s at low velocity $(\delta \simeq d)$.

We see that there are several reasons for the wire to be thin:

- reducing the time constant,
- avoiding spurious cooling, and
- ensuring a good directivity, by a large ℓ/d ratio while keeping ℓ smaller than the smallest interesting scales.

Notice that, in very low temperature (helium) experiments, solid heat capacities are vanishing (small time constant) and the side cooling is reduced by using superconductors whose heat conductivity is low.

The second method, i.e., constant temperature measurement, has the advantage of reducing the time constant. To understand it let us linearize the relation between the current (i), velocity (u) and temperature (θ) fluctuations:

$$i = \alpha u + \beta \left(\theta + \tau \partial_t \theta\right).$$

Let us impose the feedback:

$$i = -G\beta\theta,$$

where G is the gain of the loop. Neglecting the second order term in τ we obtain:

$$\theta \simeq -\frac{\alpha}{\beta(G+1)}u + \frac{\tau}{\beta(G+1)^2}\alpha\partial_t u,$$

which shows how the temperature fluctuations are reduced. On the other hand:

$$i = -G\beta\theta = \alpha\frac{G}{G+1}\left(u + \frac{\tau}{G+1}\partial_t u\right).$$

The new time constant is $\tau' = \tau/(G+1)$. There is a drawback however, which appears when taking account of a noise term b:

$$i = -G\beta(\theta + b).$$

The preceding relation becomes:

$$i = \alpha\frac{G}{G+1}\left(u + \frac{\tau}{G+1}\partial_t u\right) + \frac{G\beta}{G+1}\left(b + \frac{\tau G}{G+1}\partial_t b\right).$$

The reduction of the time constant does not occur for the noise which increases as soon as its frequency is larger than $(G+1)/G\tau$.

Among others, two noise origins are temperature fluctuations and Nyquist noise. Indeed a hot wire is intrinsically sensitive to the temperature. Using King's relation one can estimate both the signal:

$$S = U\left(\frac{dI}{du}\right)$$

and the temperature noise:

$$B_t = \Delta T\left(\frac{dI}{dT}\right)$$

if ΔT is the temperature fluctuation amplitude. Then:

$$\frac{S}{B_t} \simeq \frac{T - T_0}{2\Delta T},$$

which shows the importance of overheating the wire. It is clear that this noise is both the most important and the most dangerous one as its spectrum is roughly similar to the velocity one.

As for Nyquist noise, its amplitude ΔV is given by:

$$(\Delta V)^2 = 4Rk_BT\Delta f \simeq 4Rk_BT\frac{U}{\ell},$$

ℓ being the length of the wire. From King's relation the corresponding "velocity" noise Δu is:

$$\frac{\Delta u}{U} = 4 \frac{\Delta V}{RI} .$$

On the other hand, with our estimation of the gas heat conductivity, the dissipated power is:

$$RI^2 \simeq \frac{\pi \ell d}{\delta} \frac{ck_{\mathrm{B}}}{\sigma} (T - T_0) ,$$

where d/δ is the Nusselt number of the wire. Finally, we get:

$$\left(\frac{\Delta u}{U} \right)^2 \simeq \left(\frac{64}{\pi} \right) \left(\frac{T}{T - T_0} \right) \left(\frac{U}{c} \right) \left(\frac{\delta \sigma}{\ell^2 d} \right) .$$

Again, overheating is important. Under usual conditions, the corresponding signal-to-noise ratio is about 10^6. Serious limitations could only come from much smaller dimension ℓ (1 micron).

There are other warnings. King's relation is only an indicative one and a calibration must be made of each new wire. Drifts are observed, which could be due to small dust in the flow. Close to a wall, an additional cooling perturbs the measurements.

Finally, the arms which support the wire disturb the flow. Differences of 10% can be observed for the measured velocity, depending on their orientation. Absolute measurements are poorly reliable with a hot wire.

On the other hand, hot wire probes have been associated in a single device for measuring derivatives of the velocity field (vorticity...). Four wire probes are commonly used and a 20 wire probe has been constructed.

3.3 Doppler laser anemometer

Light scattered by a particle experiences a Doppler frequency shift proportional to its velocity. Two different set-ups are generally used:

i) The diffracted light is collected in a well defined direction, coming from a small length of the laser beam. Mixing with a reference beam gives a signal whose frequency gives the velocity. This set-up requires a large number of particles in the measurement volume. The coherence time, which bounds the precision, is the time a given particle stays in this volume.

ii) Two beams cross in the fluid and a fringe system is formed at their intersection. A particle crossing these fringes gives a modulated diffracted light. Here, only one particle at a time must be in the measurement volume.

In the first case, if $\mathbf{k}_i$ $(\mathbf{k}_d)$ is the incident (scattered) wave-vector, the observed frequency is $\omega = \mathbf{u}\,(\mathbf{k}_i - \mathbf{k}_d)$ where $\mathbf{u}$ is the particle velocity. The same relation holds in the second case if $\mathbf{k}_i$ and $\mathbf{k}_d$ are the wave-vectors of both beams.

This method has many advantages: no flow disturbance, linear response, which is a precious advantage for large fluctuations, absolute measurement, good definition of the measured component, positive and negative fluctuations can be distinguished by imposing an artificial drift to the fringes, suitable from small (less than 1 cm/s) to large velocities (70 m/s).

There are drawbacks however: In case ii) discontinuous measurements are made. The sampling is random, yielding difficulties in statistical treatments. Index fluctuation gives noise. Finally, spatial and temporal resolution are competing with the measurement precision. The signal-to-noise ratio is given by the number of fringes. It thus cannot go much beyond 10^2. Laser Doppler anemometry is better than hot wires in liquids but not in gases.

4 Dimensional analysis

4.1 Buckingham's theorem

Dimensional analysis can seemingly be traced back to Galileo. It was used long ago for solving fluid mechanics problems, and is now a common tool in physics. We have already used it in the previous paragraphs. Here we shall be more systematic and carefully analyze its interest and its limitations.

The basic idea is well known. Assume an experiment has been made with particular boundary conditions, and everything has been expressed in MKS units. To express it in CGS units we simply multiply numbers measuring lengths by 10^2, masses by 10^3, densities by 10^{-3}. But assume we forget that and simply change the names of units. Our result will be that of a new problem where lengths are 10^2 times smaller, masses 10^3 times smaller and densities 10^3 times larger. Solving our problem, we thus have solved a whole class of equivalent problems. In fact this may not be so useful since few liquids have 10^3 times the density of water for instance! Some quantities must be maintained constant (light velocity for instance) if they have some importance.

Let us formalize this following Buckingham (1914). Let $y_1, \ldots y_n$ be the parameters (boundary conditions, important quantities) and y the unknown quantity. We are searching for a mathematical relation:

$$y = f\,(y_1, \ldots, y_n)\,.$$

Let $A_1, \ldots, A_m$ be the independent dimensions (M for mass, T for time, ...). We shall see that their number is not so obviously determined but assume this is done. Then the dimensions of the y_i express themselves in the A_j:

$$[y_i] = A_1^{\alpha_{1i}} \ldots A_m^{\alpha_{mi}} .$$

The expression:

$$y_1^{x_1} \ldots y_n^{x_n}$$

will be of zero dimension if the m equations:

$$\alpha_{ji} x_i = 0$$

are fulfilled. We can thus form $n - m$ "independent" zero dimension quantities: $\pi_1, \ldots, \pi_{n-m}$. Let us take them as new parameters, and let $y_1', \ldots y_m'$ be the remaining ones. These y_i' have independent dimensions and there exist exponents β_i such that

$$[y] = [y_1']^{\beta_1} \ldots [y_m']^{\beta_m} .$$

Then

$$\pi = y y_1'^{-\beta_1} \ldots y_m'^{-\beta_m}$$

has zero dimension and is a function of all the parameters:

$$\pi = h \left(y_1', \ldots, y_m', \pi_1, \ldots, \pi_{n-m} \right) .$$

Neither π, nor the π_i values depend on the unit system. We can thus choose these units such that all $y_i' = 1$ and:

$$\pi = h \left(1, \ldots, 1, \pi_1, \ldots, \pi_{n-m} \right) = g \left(\pi_1, \ldots, \pi_{n-m} \right)$$

That is Buckingham's theorem: a zero-dimensional unknown quantity can only depend on the zero-dimensional numbers formed from the parameters.

The most interesting case is that where no zero-dimensional parameter can be formed. Then the function g is a constant g_0 and the problem is fully solved within this multiplication constant:

$$y = g_0 y_1'^{\beta_1} \ldots y_m'^{\beta_m} .$$

4.2 A simple example

Consider a V-shaped aperture in a dam (Fig. 4.1). In the nonviscous limit the bulk flow $\dot{V}$ a priori depends on the fluid density ρ, the angle α, the height h, and the free fall acceleration g. α is the only zero-dimensional parameter which can be formed. The dimension of $\dot{V}$ is $L^3 T^{-1}$, the same as $g^{1/2} h^{5/2}$. Thus:

$$\dot{V} = g^{1/2} h^{5/2} f(\alpha) .$$

Let us come back in another way. Without Buckingham's theorem we could write:

$$\frac{\dot{V}}{g^{1/2}h^{5/2}} = f(\rho, g, h, \alpha) \, .$$

Changing the mass unit changes the number ρ in this mathematical relation, without changing any of the other variables g, h, α, nor f. Thus f does not depend indeed on ρ:

$$\frac{\dot{V}}{g^{1/2}h^{5/2}} = f(g, h, \alpha) \, .$$

Changing the time unit changes g but neither h, α, nor f. Thus f does not depend on g:

$$\frac{\dot{V}}{g^{1/2}h^{5/2}} = f(h, \alpha) \, .$$

The same is true with the length unit and we come back to:

$$\dot{V} = g^{1/2}h^{5/2}f(\alpha) \, .$$

4.3 A less simple case

Consider now the cooling rate $\dot{Q}$ of a hot sphere (temperature T_1 diameter d) by a fluid of velocity U and temperature $T_0 < T_1$. We neglect the viscous heating and the fluid is incompressible. We shall see the importance of these assumptions later. The parameters are U, $(T_1 - T_0)$, d, the heat conductivity of the fluid K, its density ρ, viscosity μ and specific heat C_p (per unit volume).

Rayleigh (see Sedov, 1959) was the first to treat the problem by dimensional analysis. He considered as independent dimensions the three

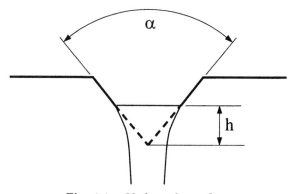

Fig. 4.1.   V-shaped overflow.

mechanical ones (mass, time, length), temperature and heat. Then only two zero-dimensional numbers can be formed:

$$\frac{\rho U d}{\mu} = \text{Re} \qquad \text{the Reynolds number},$$

$$\frac{\mu}{\rho} \frac{C_p}{K} = \text{Pr} \qquad \text{the Prandtl number}.$$

The Prandtl number is the ratio of two diffusion constants, the momentum one ($\nu = \mu/\rho$) and the heat one ($\chi = K/C_p$). It is of order one in a gas where all diffusivities are equal. It is lower in liquid metals (mercury $\text{Pr} \simeq 4\,10^{-2}$) where heat is transported by electrons or in the Sun ($\text{Pr} \simeq 10^{-8}$) where it is transported by light. It is larger in oils.

To come back to our problem, following Rayleigh

$$\dot{Q} = K d\,(T_1 - T_0)\,f(\text{Re}, \text{Pr}).$$

There is a difficulty however, soon pointed out by Riabouchinski. Heat is an energy, and temperature also through the Boltzmann constant k_B. Can we consider them as independent dimensions? If not, there exist two other zero-dimensional numbers:

$$\frac{C_p\,(T_1 - T_0)}{\rho U^2} \qquad \text{and} \qquad \frac{k_\text{B}\,(T_1 - T_0)}{\rho U^2}.$$

The good result is Rayleigh's, but the reason took more than twenty years to be understood. In this problem, as we neglect viscous heating, there is no transformation of mechanical work into heat. Then we can consider them as distinct quantities. In the same way, k_B appears in general in the state equation of the fluid, which is replaced here by $\rho = \text{cst}$.

The lesson from this example is that we can choose as many independent dimensions as we want. This simply introduces conversion factors which act as new parameters. There is no gain unless we know that these factors cannot enter the problem. For instance, we usually consider that time and length have distinct dimensions. According to the theory of relativity however, this is artificial and introduces a "conversion factor" which is the velocity of light. In classical mechanics we know that this parameter will not appear, which gives all its interest to distinguishing time and length.

4.4 Turbulent boundary layer

As a final example of the possibilities and limits of dimensional analysis let us look at the average velocity profile of a turbulent flow close to the wall of a pipe. The distance to the wall is y, small compared to the pipe dimension L. The fluid layer between y and $y + dy$ has no average acceleration in steady state. Thus the flux of the momentum component parallel to the plate is conserved. Let us call it ρu_*^2. Note how the problem is posed: the friction is given and we shall infer the velocity profile $U_x(y)$. The gradient of U_x normal to the wall *a priori* depends on the parameters y, L, u_*, ν. The density ρ could not enter in any zero dimensional number:

$$\frac{dU_x}{dy} = f(y, L, u_*, \nu) .$$

Two zero-dimensional numbers can be built with the chosen parameters:

$$\text{Re} = \frac{u_* y}{\nu} \qquad \text{and} \qquad \frac{L}{y} .$$

In zero-dimensional terms the equation becomes:

$$\frac{y}{u_*} \frac{dU_x}{dy} = f(\text{Re}, L/y) .$$

In the classical treatment, due to von Kármán (see Landau and Lifshitz,1959), it is assumed that f has a finite limit $1/\kappa$ when Re and L/y tend to infinity. Then:

$$\frac{dU_x}{dy} = \frac{1}{\kappa} \frac{u_*}{y}$$

and

$$U_x = \frac{u_*}{\kappa} \ln\left(\frac{y}{y_0}\right) .$$

The experiments give $\kappa \simeq 2.5$. The above formula is the famous logarithmic profile in turbulent boundary layers. We shall come back to it later.

It cannot be valid for $y < y_0$, and the hypotheses which yield it fail for Re < 1, i.e., $y < \nu/u_*$. Identifying these two limits:

$$y_0 \simeq \frac{\nu}{u_*} ,$$

which gives the width of the "viscous sublayer."

The pressure drop can be estimated from what precedes. For $y \simeq L$, U is nearly independent of y and has the order of magnitude:

$$U \simeq \frac{u_*}{\kappa} \ln\left(\frac{L u_*}{\nu}\right) ,$$

which reciprocally allows to estimate u_*:

$$u_* \simeq \kappa \frac{U}{\ln(LU/\nu)} .$$

The volume flow is:

$$\dot{V} \simeq \frac{\pi L^2}{4} U .$$

The pressure difference on a length ℓ equilibrates the friction:

$$\frac{\pi L^2}{4} \ell \frac{dP}{dx} = \pi L \ell \rho u_*^2$$

or

$$L \frac{dP}{dx} = 4\rho \frac{\kappa^2 U^2}{\ln^2 (LU/\nu)} .$$

Qualitatively the pressure drop is ρU^2 on a length of order L.

4.5 Barenblatt's second kind self-similarity

The logarithmic law is reasonably well verified for the profile. However, fine measurements suggest that κ could depend on the global Reynolds number $\mathrm{Re}_* = u_* L/\nu$. On the other hand, power law profiles can give an even better agreement than the logarithmic one. Could this have a theoretical basis?

Indeed, this problem is one of the examples Barenblatt chooses in his book (Barenblatt, 1979) to illustrate the second kind self-similarity concept. However the basic example is the asymmetric diffusion problem we present now:

$$\begin{cases} \partial n/\partial t &=& (1+\epsilon)D\partial^2 n/\partial x^2 & \text{if} \quad \partial n/\partial t > 0, \\ \partial n/\partial t &=& D\partial^2 n/\partial x^2 & \text{if} \quad \partial n/\partial t < 0. \end{cases}$$

The initial distribution $n(x,0)$ has a width ℓ:

$$N = \int_{-\infty}^{+\infty} n(x,0) \, dx ,$$

$$\frac{1}{N} \int x^2 n(x,0) \, dx = \ell^2 .$$

Dimensional analysis gives three zero-dimensional parameters:

$$\xi = x/\sqrt{Dt} , \qquad \eta = \ell/\sqrt{Dt} , \qquad \epsilon ,$$

and thus:

$$n = \frac{N}{\sqrt{Dt}} F(\xi, \eta, \epsilon) .$$

In the trivial case, $\epsilon = 0$, there is a limit for F when $\eta \to 0$. The solution is completely self-similar versus η (independent of the width ℓ). In other words there is a solution of the diffusion equation for an initial Dirac distribution $\delta(x)$. It is not the case when $\epsilon \neq 0$. It can be shown however that there exists an exponent α such that:

$$\eta^{-\alpha} F(\xi, \eta, \epsilon)$$

has a finite limit when $\eta \to 0$. α depends on ϵ and $\alpha(\epsilon) \to 0$ when $\epsilon \to 0$. Barenblatt calls such a solution a second kind self-similar one, or incompletely self-similar versus η.

The same idea is applied by Barenblatt to the turbulent boundary layer profile. In place of Re and L/y, he takes as variables

$$\text{Re} = \frac{u_* y}{\nu} \quad \text{and} \quad \text{Re}_* = \frac{u_* L}{\nu} .$$

The equation for the profile is then written:

$$\frac{y}{u_*} \frac{\partial U_x}{\partial y} = g\,(\text{Re}, \text{Re}_*) .$$

Barenblatt then assumes incomplete self-similarity versus Re. So when Re goes to infinity:

$$g\,(\text{Re}, \text{Re}_*) = \text{Re}^\lambda \, \phi\,(\text{Re}_*) .$$

The solution is then:

$$U_x = \frac{u_*}{\lambda} \phi\,(\text{Re}_*) \left[\left(\frac{y u_*}{\nu}\right)^\lambda - \psi\,(\text{Re}_*) \right]$$

with $|\psi| \simeq 1$, i.e., a power law behavior with power λ (yet to be determined) in replacement of the logarithmic law. This is rather confusing! Which is the correct law? Can we be confident in simple arguments yielding the logarithmic law? Why and how must we adapt them? The best idea is to come back critically to these arguments.

We argued that: $(y/u_*)\,(\partial U_x/\partial y) = g\,(y u_*/\nu, L u_*/\nu)$ should be independent of y when $y u_*/\nu = \text{Re}$ is sufficiently large. Such arguments are often used. However there are obvious limits here: y must be large compared to ν/u_*, but also small compared to L (Re $=$ Re$_*$). Thus we should restrict our argument in the following way: $\ln g$ as a function of $\ln \text{Re}$ should not have noticeable variation on any small interval in the middle of $[0, \ln \text{Re}_*]$. Within such a small interval the extreme values 0 and $\ln \text{Re}_*$ are too distant to be relevant. Thus the slope $d \ln g / d \ln y$ must be of order $1/\ln \text{Re}_*$ for the total variation to be at most of order 1 on the whole $[0, \ln \text{Re}_*]$ interval (Fig. 4.2).

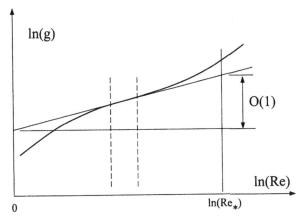

Fig. 4.2. General arguments ensure negligible variation of $\ln g$ only for $\ln \mathrm{Re}$ values between the dashed lines.

Barenblatt's hypothesis is equivalent to taking this slope constant and equal to λ. We can thus assess that, in the high Re_* limit:

$$\lambda = \frac{\lambda_0}{\ln\left(\mathrm{Re}_*/R_0\right)}$$

where λ_0 and R_0 are some constants.

As a final remark let us discuss the pressure drop in Barenblatt's approach. Indeed the functions $\psi\left(\mathrm{Re}_*\right)$ and $\phi\left(\mathrm{Re}_*\right)$ are unknown, but we can postulate that Barenblatt's law tends toward the logarithmic one when $\mathrm{Re}_* \to \infty$. This gives a finite limit for ψ and ϕ and the pressure drop keeps the same behavior.

5 Modern approaches in turbulence

In this last section I present an overview of some approaches which have been or will be important for our understanding of turbulence.

5.1 The Lorenz model

The first time somebody told me he was studying turbulence by looking at mappings of the $[0, 1]$ interval, I must admit I thought he was joking. I should have read Lorenz's paper (Lorenz, 1963), which I recommend to everybody. It is a very nice introduction to this approach and some other ideas.

The subject is two-dimensional Rayleigh–Bénard convection, that is buoyancy of a fluid layer heated from below (for positive thermal expansion). The first idea illustrated by this paper is the truncation of equations.

The idea is to find the most reduced orthonormal basis functions for the fields (here the stream function and the temperature) and derive coupled equations between amplitudes.

The three "modes" kept by Lorenz are all sinusoidal versus the two coordinates (vertical and one horizontal axes). X represents the velocity amplitude in the most simple convection mode. Y is the temperature difference between rising and falling fluid. Z is the departure from the linear vertical temperature profile. It has a double vertical spatial frequency but does not depend on the horizontal coordinate. With only these modes, coupling equations are:

$$\dot{X} = -\sigma X + \sigma Y,$$
$$\dot{Y} = -Y + rX - XZ, \qquad (5.1)$$
$$\dot{Z} = -bZ + XY.$$

$\dot{X} = -\sigma X$ corresponds to the viscous damping, while $\dot{Y} = -Y$ and $\dot{Z} = -bZ$ accounts for thermal conduction damping. It is more rapid for Z whose spatial period is shorter. σY is the convection forcing term. Nonlinear terms correspond to heat transport by the flow. σ is the Prandtl number and r the normalized Rayleigh number ($r = 1$ at the convection threshold).

Lorenz finds that a spontaneous time dependence appears when $r > r_c$ with $r_c \simeq 25$. For such a value, as he remarks, his equations no longer adequately represent the convection, due to truncation. He studies it however for:

$$r = 28, \qquad \sigma = 10, \qquad b = \frac{8}{3}.$$

Here comes the second idea, which is not new, as it can be traced back to Hadamard, but is simply and nicely illustrated in this paper: the extreme sensitivity to initial conditions. Looking at its profound origin, Lorenz studies the succession of the maxima of Z: M_i. He remarks that, to a good approximation M_{n+1} only depends on M_n. The corresponding curve has a lambda shape (Fig. 5.1).

Consider then two initial conditions giving two close values for the n-th maxima:

$$M'_n = M_n + \epsilon.$$

Then $M'_{n+1} = M_{n+1} + \Lambda\epsilon$ where Λ is the slope of the curve. Since the modulus of this slope is everywhere larger than 1, the distance between M and M' points increases exponentially. The trajectory's instability thus comes from this stretching which must be followed by a folding since the

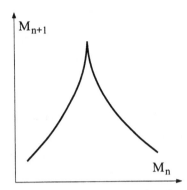

Fig. 5.1.   Sketch of the return map for the Lorenz model.

amplitude is bounded. This is the sketch of the famous baker map who stretches the dough then folds it to come back to the initial shape. Two initially close points in this dough will finish at arbitrary distance.

For real flows, this stretching-and-folding mechanism is obvious when looking at the dispersion of a dye in turbulent convection.

5.2 Statistical mechanics for turbulence

This paragraph and the following one concern my favorite subject and probably I cannot be totally unprejudiced here. A simple remark can be made however. Everybody agree that high Reynolds number flows have a very large number of degrees of freedom. But the fact that we can identify some structures in the flow suggests that we can use a much smaller number of parameters to classify them.

Large number of degrees of freedom, small number of parameters characterizing the state (the flow). This is a situation similar to that prevailing in thermodynamics.

We certainly will find that different sets of parameters (different states) do not have the same probability, as they do not correspond to the same number of microscopic states. The "observed" flow will be that which maximizes this probability.

In this paragraph we shall concentrate on inviscid two-dimensional flows. The use of statistical mechanics for these flows is not new. It can at least be traced back to Onsager who tried to represent a flow by a superposition of quantized vortex flows. He has shown that, for negative temperatures of this vortex gas, coherent structures appear, like big vortices. But the result was dependent on the circulation quantum chosen for the elementary vortex. Recent progress has been made in this subject. I shall only present here the approach of Robert (1990).

The main difficulty is to take into account every invariant quantity. For an inviscid two-dimensional flow, vorticity is frozen in the flow (see section 2.1), and only its distribution differs from the initial one. It means that the sum of any function of the vorticity:

$$\int f(\omega)\, d^2\mathbf{x}$$

is an invariant quantity. Another one is the total energy:

$$\int v^2\, d^2\mathbf{x}. \tag{5.2}$$

The method is easier to explain if the vorticity is initially uniform and equal to ω_i in domains D_i of area A_i. During the evolution following Euler equation, frontiers of domains are deforming and eventually become singular but their area remains constant. It is better then to define the probability $e_i(\mathbf{x})$ for a point $\mathbf{x}$ to belong to domain D_i as it is transformed by the flow. The observed value of ω will then be:

$$\overline{\omega}(\mathbf{x}) = \sum_i e_i(\mathbf{x})\omega_i \tag{5.3}$$

with:

$$\int e_i(\mathbf{x})\, d^2\mathbf{x} = A_i. \tag{5.4}$$

As usual, the corresponding entropy will be:

$$S = -\int \sum_i e_i(\mathbf{x}) \ln e_i(\mathbf{x})\, d^2\mathbf{x}. \tag{5.5}$$

A short discussion here could make things clearer. One could naively think that, when following a point in the flow, the e_i value on this point remains constant, equal to 1 or 0 (the point is or is not in domain D_i). Then S would be always zero. However, after a finite time, domains become so mixed that their frontiers themselves take a finite area. The same situation prevails in statistical mechanics: the growth of entropy seems contradictory with Liouville's theorem. We can physically think of a "coarse graining" to picture this effect and explain that e_i can differ from 0 or 1. But the result is independent of the size of the grains.

Introduce now the stream function $\psi(\mathbf{x})$ such that:

$$v_x = \partial_y \psi \qquad \text{and} \qquad v_y = -\partial_x \psi$$

and ψ is zero on the boundaries of the flow. Thus:

$$\mathbf{v}^2 = (\partial_x \psi)^2 + (\partial_y \psi)^2 \qquad \text{and} \qquad \omega = -\nabla^2 \psi.$$

Integrating by parts, if ψ is zero on the boundaries:

$$E = \int \mathbf{v}^2 \, d^2\mathbf{x} = \int \omega\psi \, d^2\mathbf{x} \,. \tag{5.6}$$

ψ is a continuous function. Only its second derivatives are discontinuous. We thus can write (by "coarse graining"):

$$E = \int \overline{\omega}\psi \, d^2\mathbf{x}$$

Integrating twice in parts:

$$E = \int \overline{\omega}\psi d^2\mathbf{x} = \int \omega\overline{\psi} \, d^2\mathbf{x}$$

and thus:

$$E = \int \overline{\omega}\,\overline{\psi} \, d^2\mathbf{x} \,. \tag{5.7}$$

Fluctuations do not contribute to the energy. This will be surprising for the specialists of superfluid helium hydrodynamics. However, here there are no quantized vortices. ψ is continuous.

For a variation $\{\delta e_i(\mathbf{x})\}$, all the required invariances are fulfilled if:

$$\int \delta e_i(\mathbf{x}) \, d^2\mathbf{x} = 0$$

and

$$dE = 2 \int \overline{\psi}\delta\overline{\omega} \, d^2\mathbf{x} = 2\sum_i \int \overline{\psi}\omega_i\delta e_i(\mathbf{x}) \, d^2\mathbf{x} = 0 \,.$$

Moreover, for each point $\mathbf{x}$: $\sum_i e_i(\mathbf{x}) = 1$ and thus $\sum_i \delta e_i(\mathbf{x}) = 0$. If the state is among the most probable ones:

$$dS = -\sum_i \int \left(\ln e_i(\mathbf{x}) + 1\right) \delta e_i(\mathbf{x}) \, d^2\mathbf{x} = 0$$

for any $\delta e_i(\mathbf{x})$ satisfying the invariances. The method of Lagrange multipliers then gives the solutions e_i:

$$e_i(\mathbf{x}) = \frac{e^{-\alpha_i - \beta\omega_i\overline{\psi}(\mathbf{x})}}{\sum_i e^{-\alpha_i - \beta\omega_i\overline{\psi}(\mathbf{x})}} \,, \tag{5.8}$$

where the α_i are chosen such that $\int e_i(\mathbf{x}) \, d^2\mathbf{x} = A_i$ and β such that the flow has the required energy. Then:

$$\overline{\omega}(\mathbf{x}) = -\nabla^2\overline{\psi}(\mathbf{x}) = \frac{\sum_i \omega_i \, e^{-\alpha_i - \beta\omega_i\overline{\psi}(\mathbf{x})}}{\sum_i e^{-\alpha_i - \beta\omega_i\overline{\psi}(\mathbf{x})}} = f\left(\overline{\psi}\right) \,. \tag{5.9}$$

It is known that such a relation ensures the stability of the flow. Some experimental support has been found in mercury layer flows and in numerical simulations, for a small number of domains.

What has been presented here is only the trivial part of the recent work I mentioned. The difficult part lies in the theorems showing that most of the flows are accumulated in the neighborhood of this most probable one, but that is another story!

5.3 Developed turbulence

In the previous example, the entropy was calculated thanks to the knowledge of the appropriate phase space which was the space of probability functions $e_i(\mathbf{x})$. In what follows for three-dimensional developed turbulence, the phase space is not known. However, from symmetry arguments, the behavior of the "entropy" can be deduced, and some nontrivial conclusions can be drawn.

The goal here is to understand the shape of the probability density functions (pdf) of velocity differences

$$\delta \mathbf{v} = \mathbf{v}(\mathbf{x} + \mathbf{r}) - \mathbf{v}(\mathbf{x}) \, .$$

In isotropic developed turbulence, this shape only depends on the separation distance r. For large values of r, close to the integral scale, the pdf is nearly Gaussian. Close to the dissipation scale it is strongly non Gaussian. There is as yet no full agreement on the interpretation of these shapes. There seems to be an agreement on the fact that the pdf at distance r depends in a well defined way on the large scale (L) pdf. Most of the proposed relations can be grouped in the following way:

$$\frac{1}{\sigma_r} P_r \left(\frac{\delta v}{\sigma_r} \right) = \int G_r \left(\frac{\sigma}{\sigma_r} \right) P_L \left(\frac{\delta v}{\sigma} \right) \frac{d\sigma}{\sigma^2} , \qquad (5.10)$$

that is, a superposition of pdf having the shape of the large scale pdf (P_L), with different variances σ. Here δv is some component of $\delta \mathbf{v}$, for example that parallel to $\mathbf{r}$. $G_r \left(\sigma / \sigma_r \right)$ can be seen as the probability distribution of variances σ. We shall not discuss Eq. 5.10, but only the way to determine $G_r \left(\sigma / \sigma_r \right)$.

Indeed this distribution can be seen as a functional of a function $\sigma(r)$ and gives probability of having $\sigma = \sigma(r)$ for the distance r. Instead of this random function $\sigma(r)$ we equivalently can use:

$$\epsilon(r) = \frac{\sigma^3(r)}{r} \, . \qquad (5.11)$$

We intentionally use the same letter for this variable and for the energy transfer term (section 2.2, Eq. 2.7). The reason for this is Kolmogorov's relation, Eq. 2.12:

$$\left\langle \delta v^3 \right\rangle = -\tfrac{4}{5} \left\langle \epsilon^* \right\rangle r$$

valid in the inertial range. Due to the above form for the distribution of δv, $\left\langle \delta v^3 \right\rangle$ is proportional to $\left\langle \sigma^3 \right\rangle$, the coefficient being independent of r (it only depends on P_L). Thus $\left\langle \sigma^3 \right\rangle / r = \left\langle \epsilon(r) \right\rangle$ is independent of r and proportional to $\left\langle \epsilon^* \right\rangle$. However it is important to note that ϵ and ϵ^* are distinct quantities.

As is usually done (in statistical thermodynamics for instance), we shall look for the opposite of the logarithm of the probability for the random function $\epsilon(r)$: $Q\{\epsilon(r)\}$. For reasons, to be exposed just below, we take it as:

$$Q = V \int q\left(\epsilon, d\epsilon/dr, r\right) dr . \tag{5.12}$$

The presence of the volume V corresponds to the idea that we discuss of homogeneous turbulence, where parts far away are independent. The probability per unit volume is the intrinsic quantity and not the probability itself. The sum over the scales r corresponds to the idea that different scales should be statistically independent.

However, for very close scales r and $r + dr$, values of σ and thus values of ϵ should be close. This is ensured by the presence of $d\epsilon/dr$ in the density of probability q. An infinite value for $d\epsilon/dr$ certainly will push Q out of its minimum value.

Another assumption is often made that we can use as an input: for the inertial range scales, the viscosity should not appear explicitly. This is a stringent assumption. Its consequence is that only one zero-dimensional quantity can be formed with the remaining parameters ϵ, $d\epsilon/dr$, r; namely:

$$\psi = \frac{r}{\epsilon} \frac{d\epsilon}{dr} = \frac{d\ln \epsilon}{d\ln r} . \tag{5.13}$$

As Q has zero dimension, q must behave like:

$$q(\epsilon, d\epsilon/dr, r) = r^{-\theta-1} f(\psi) \tag{5.14}$$

where θ is the space dimension ($\theta = 3$).

From all these general assumptions it can be shown that:

i) ϵ has log-normal fluctuation, that is $\ln \epsilon$ and thus $\ln \sigma$, are Gaussian;
ii) the variance of $\ln \epsilon$, $\Lambda^2 = \left\langle (\delta \ln \epsilon)^2 \right\rangle$ is proportional to the most probable $\psi(r) = \psi_m(r)$;
iii) ψ_m depends on r as a power law: $\psi_m(r) = K r^{-\beta}$;
iv) β asymptotically depends on the Reynolds number as: $\beta = \beta_0 / \ln \mathrm{Re}$.

To derive these properties let us write the random function $\psi(r)$ as:

$$\psi(r) = \psi_m(r) + \delta\psi(r)$$

and expand $f(\psi)$ around ψ_m for each r:

$$f(\psi) = f(\psi_m) + f'(\psi_m)\,\delta\psi + \tfrac{1}{2}f''(\psi_m)\,\delta\psi^2 + \dots$$

ψ_m is the function ψ which minimizes the integral

$$Q = V \int_{r_1}^{r_2} r^{-\theta-1} f(\psi)\,dr$$

at constant $\epsilon(r_1)$ and $\epsilon(r_2)$. Using the above expansion of $f(\psi)$ and calling Q_m the minimum value of Q:

$$Q - Q_m = V \left\{ \int_{r_1}^{r_2} r^{-\theta-1} f'(\psi_m)\,\delta\psi\,dr \right.$$
$$\left. + \frac{1}{2} \int_{r_1}^{r_2} r^{-\theta-1} f''(\psi_m)\,(\delta\psi)^2\,dr + \dots \right\}.$$

The first term in the brackets, linear in $\delta\psi$, is zero by definition of the extremum, for any $\delta\psi$ satisfying:

$$\int_{r_1}^{r_2} \frac{\delta\psi}{r}\,dr = \int_{r_1}^{r_2} \delta\left(\frac{d\ln\epsilon}{d\ln r}\right) d\ln r = 0$$

as $\epsilon(r_1)$ and $\epsilon(r_2)$ remain constant in the variation. Following the method of Lagrange multipliers:

$$f'(\psi_m) = -Ar^\theta. \tag{5.15}$$

We shall not show here that the constant A must be positive for Q_m being a minimum and not a maximum.

We dropped orders higher than 2 in the expansion due to the presence of the factor V. Q is the logarithm of the probability. Thus $\langle \delta\psi^2 \rangle$ is proportional to $1/V$ and in the following term $\delta\psi^3$ is of order $V^{-3/2}$. This makes the third order term vanishing for large V as $V^{-1/2}$. Thus, in the large V limit, $\delta\psi$ is Gaussian. But

$$\delta\ln\epsilon(r) = \int_r^{r_2} \frac{\delta\psi}{r'}\,dr' = \sum_{r'=r}^{r_2} \frac{\delta\psi\,\Delta}{r'}$$

writing the integral as a Riemann sum, r' taking discrete values spaced by Δ. Thus $\delta\ln\epsilon$ itself is Gaussian, as a sum of the Gaussian quantities $\Delta\,\delta\psi/r$. Property i) is shown.

We can also estimate the variance $\Lambda^2(r) = \langle (\delta\ln\epsilon)^2 \rangle$ which is the sum of variances of the $\Delta\,\delta\psi/r'$. Using Eq. 5.15:

$$f''(\psi_m)\frac{d\psi_m}{dr} = -\theta Ar^{\theta-1}$$

and we can write:

$$Q - Q_m = \frac{V}{2} \int_{r_1}^{r_2} r'^{-\theta-1} f''(\psi_m) \, \delta\psi^2 \, dr'$$

$$= \frac{-\theta V A}{2} \int_{r_1}^{r_2} \left(\frac{dr}{d\psi_m}\right) \left(\frac{\delta\psi}{r'}\right)^2 dr'$$

$$= -\sum_{r'=r_1}^{r_2} \frac{\theta V A}{2\Delta \, (d\psi_m/dr)} \left(\frac{\Delta \, \delta\psi}{r'}\right)^2 .$$

Thus the variance of $\Delta \, \delta\psi/r'$ is $(\Delta/\theta V A)(d\psi_m/dr)$ and:

$$\Lambda^2 = \left\langle (\delta \ln \epsilon)^2 \right\rangle = -\sum_{r'=r}^{r_2} \frac{\delta\psi_m}{dr} \frac{\Delta}{\theta V A}$$

$$= -\int_{r}^{r_2} \frac{1}{\theta V A} \frac{d\psi_m}{dr} \, dr' = \frac{2}{\beta} \psi_m(r) ,$$

(5.16)

where β is some constant assuming that ψ_m and Λ^2 are zero at the largest scale (r_2). This shows Property ii).

Now, we have to remember that $\langle \epsilon \rangle$ is independent of r. As ϵ is a log-normal random variable:

$$\langle \epsilon \rangle = \frac{\displaystyle\int_{-\infty}^{+\infty} \epsilon \exp\left(-\frac{\ln^2 (\epsilon/\epsilon_m)}{2\Lambda^2}\right) d\ln\epsilon}{\displaystyle\int_{-\infty}^{+\infty} \exp\left(-\frac{\ln^2 (\epsilon/\epsilon_m)}{2\Lambda^2}\right) d\ln\epsilon}$$

$$= \epsilon_m \frac{\displaystyle\int_{-\infty}^{+\infty} \exp\left(y - \frac{y^2}{2\Lambda^2}\right) dy}{\displaystyle\int_{-\infty}^{+\infty} \exp\left(-\frac{y^2}{2\Lambda^2}\right) dy} = \epsilon_m \exp(\Lambda^2/2) .$$

This gives another relation between ψ_m and Λ^2. Namely, since $\langle \epsilon \rangle$ is a constant:

$$\frac{d \ln \epsilon_m}{d \ln r} = -\frac{1}{2} \frac{d\Lambda^2}{d \ln r} = \psi_m = \frac{\beta}{2} \Lambda^2 .$$

This gives Property iii) which is far from trivial. The solution of the above equation is:

$$\Lambda^2 = \Lambda_0^2 \left(\frac{r}{\eta}\right)^{-\beta} , \tag{5.17}$$

where η is the dissipation scale.

Let us concentrate now on the moments of the distribution of δv: $\langle (\delta v)^n \rangle = F_n(r)$ which can be experimentally measured. In this approach, this moment is proportional to (5.10) and (5.11):

$$\langle \sigma^n \rangle = \langle \epsilon^{n/3} \rangle r^{n/3} = \epsilon_m^{n/3} r^{n/3} e^{\frac{n^2}{18} \Lambda^2}$$

$$= \langle \epsilon \rangle^{n/3} r^{n/3} e^{\frac{n(n-3)}{18} \Lambda^2} \ . \tag{5.18}$$

In the large Reynolds limit these moments are generally believed to vary as a power law of r:

$$F_n(r) = F_{0n} \, r^{\zeta_n} \ . \tag{5.19}$$

This is not the case here, as following Eq. 5.17 $d\Lambda^2/d\ln r$ is not constant. It asks for some discussion. The argument for (5.19) is as follows: if r is in the inertial range, as far from the large integral scale L as from the dissipation scale η, "there is no characteristic scale." A ratio such as $F_n(r_1)/F_n(r_2)$ can then only depend on (r_1/r_2):

$$\frac{F_n(r_1)}{F_n(r_2)} = \chi \left(\frac{r_1}{r_2} \right) \ .$$

Function χ has the obvious property

$$\chi(x)\chi(y) = \chi(xy)$$

and thus depends as a power law of its argument.

So far so good, but we can apply here the same argument as for the turbulent boundary layer (section 4.5). Consider the function

$$\zeta_n \left(\ln \frac{r}{\eta} \right) = \frac{\partial \ln F_n}{\partial \ln r} \ .$$

Strictly speaking, the above argument applies on a short interval in the middle of $[0, \ln(L/\eta)]$. On this short interval, ζ_n can be considered as constant. But this cannot exclude a slope $\partial \ln \zeta_n/\partial \ln r$ of order $1/\ln(L/\eta) \sim 1/\ln \text{Re}$. ζ_n contains the term $(n(n-3)/18)(d\Lambda^2/d\ln r)$ which is proportional to:

$$\left(\frac{r}{\eta} \right)^{-\beta} = \exp \left(-\beta \ln \frac{r}{\eta} \right) \ .$$

The derivative of the logarithm of this quantity versus $\ln(r/\eta)$ is simply β. It shows that such behavior is acceptable if, in the large Reynolds number limit:

$$\beta \ln \text{Re} \to \text{cst} \ . \qquad\qquad \text{Property (iv)}$$

We shall not discuss the experimental support to this theory which can be found in some recent papers (for example Castaing, Gagne and Hopfinger, 1990). We shall stop here and remark that, as in most of the topics discussed in this chapter, the interest is also in the method which could be used in other problems. Hydrodynamics has always been a wonderful workshop where a large number of the physicist's tools have been fabricated.

References

Barenblatt, G.I. (1979). *Similarity, Self-similarity and Intermediate Asymptotics* (Consultant Bureau, Washington).

Buckingham, D. (1914). *Phys. Rev.*, **14**, 345.

Castaing, B., Gagne, Y., Hopfinger, E. (1990). *Physica* **D46**, 177.

Comte-Bellot, G. (1976). *J. Physique (Paris) Col.* **37 C1**, C1.

Landau, L.D., Lifshitz, E.M. (1959). *Fluid Mechanics* (Pergamon Press, New York).

Lorenz, E.N. (1963). *J. Atmos. Sci.* **20**, 130.

Monin, A.S., Yaglom, A.M. (1975). *Statistical Fluid Mechanics, 1 and 2* (M.I.T. Press, Cambridge).

Purcell, E.M. (1977). *Am. J. Phys.* **45**, 3.

Richardson, L.F. (1922). *Proc. Royal Soc.* **A97**, 354.

Robert, R. (1990). *C.R. Acad. Sci. Paris* **311**, Série I, 575.

Sedov, L.I. (1959). *Similarity and Dimensional Methods in Mechanics* (Academic Press, Orlando).

Tennekes, H., Lumley, J.L. (1974). *A First Course in Turbulence* (M.I.T. Press, Cambridge).

Tritton, D.J. (1988). *Physical Fluid Dynamics* (2nd edition) (Clarendon Press, Oxford).

2
Hydrodynamic instabilities in open flows

P. Huerre and M. Rossi

1 Introduction

The objective of the following lecture notes is to provide a consistent account of the development of hydrodynamic instabilities in open flows such as mixing layers, wakes and boundary layers. These prototype shear flows are commonly encountered in a variety of technological applications in aerodynamics, mechanical and chemical engineering, and in many geophysical processes in the oceans and atmospheres. A physical understanding of the transition to turbulence in "simple" shear flows gives rise to fundamental issues that constitute the core of this course.

Flows are typically classified according to their open or closed nature and to their laminar or turbulent character. In *open flows*, fluid particles are not recycled within the physical domain of interest but leave it in a finite time (Huerre and Monkewitz, 1990), as in the mixing layer generated downstream of a splitter plate by the merging of two parallel streams in relative motion. By contrast, in *closed flows*, fluid particles always remain within the same physical region, as in Taylor–Couette flow between two counter-rotating cylinders (Andereck *et al.*, 1986). The distinction between laminar and turbulent flows is related to the degree of spatial and temporal coherence of a given flow (Tennekes and Lumley, 1972; Monin and Yaglom, 1975; Landau and Lifshitz, 1987a; Lesieur, 1991; Frisch, 1995). Typically, *turbulent flows* arise at high Reynolds numbers and are characterized by the presence of a wide range of spatial and temporal scales, three-dimensional vorticity fluctuations and a certain degree of unpredictability. However, this apparent state of disorder does not preclude the existence of large-scale *coherent structures* (Ho and Huerre, 1984; Robinson, 1991; Bonnet and Glauser, 1993) as strikingly exemplified in the experimental observations of Crow and Champagne (1971) in jets and Brown and Roshko (1974), Winant and Browand (1974) in mixing layers. From an engineering point of view, transition from laminar flow to turbulence is characterized by drastic changes in global properties such as drag (Schlichting, 1979), aeroacoustic noise generation (Crighton *et al.*, 1992) and mixing (Tennekes and Lumley, 1972).

The transition from laminar to turbulent states is preferentially analyzed within the context of *hydrodynamic stability theory* (Lin, 1955; Chandrasekhar, 1961; Betchov and Criminale, 1967; Drazin and Reid, 1981; Swinney and Gollub, 1981). In essence, one seeks to determine under which conditions a given configuration described as the *basic flow*, ceases to be observed when externally imposed constraints such as shear, temperature or pressure gradient cross specific critical values. In general, these external constraints are quantified in terms of dimensionless *control parameters* such as the Reynolds number (Tritton, 1988). The concept of stability to perturbations is then invoked to account for this qualitative change of behavior: at given control parameter settings, the temporal evolution of infinitesimal disturbances linearly superposed on the basic flow at time $t = 0$ is investigated. If all possible initial perturbations are temporally damped, the basic flow is said to be *stable*. In the reverse case, some perturbations grow in amplitude and bring the flow towards a new equilibrium state possessing a qualitatively distinct spatio-temporal behavior: the basic flow is said to be *unstable*. Two domains corresponding to either stability or instability and separated by a so-called *neutral surface* may then be identified in control parameter space. As the external constraints are varied, control parameters may cross the neutral surface, i.e. reach a critical value beyond which the basic flow becomes unstable. Under these conditions, perturbations evolve as a combination of growing instability waves with different frequencies and wavenumbers which are the theoretical counterparts of the vortical structures observed experimentally.

The sensitivity of many open flows to external perturbations, makes it essential to introduce the concepts of absolute or convective instabilities (Briggs, 1964; Bers, 1983; Landau and Lifshitz, 1987b; Huerre and Monkewitz, 1990). If initially localized disturbances spread upstream and downstream, thereby invading the entire physical domain of interest, the basic flow is said to be *absolutely unstable*. If, by contrast, these disturbances are swept away, the basic flow is said to be *convectively unstable*. Open shear flows may then be classified according to the nature of the instabilities they can support. Convectively unstable flows, such as spatially developing mixing layers, essentially display *extrinsic dynamics*: the flow response is determined by the *spatial* evolution of instability waves resulting from the amplification of an external input, as would be the case for a *noise amplifier*. *A contrario*, absolutely unstable flows, such as bluff-body wakes, may display *intrinsic dynamics* in the form of a *global mode* beating at a specific frequency. This *hydrodynamic oscillator* is then insensitive to low-level external noise.

The above notions are first typically developed within the context of *linear instability theory* whereby the governing equations of fluid motion are linearized around the basic flow of interest. In order to characterize the new state emerging in the unstable range of parameters, finite amplitude effects must be taken into account within a nonlinear framework.

Weakly nonlinear formulations carried out near the onset of instability reduce the dynamical description to an *amplitude equation* for the evolution of the perturbations, e.g. the celebrated Ginzburg–Landau equation. *Strongly nonlinear* approaches, which are usually implemented numerically, lead to a more complete knowledge of finite amplitude states far away from instability threshold. The finite amplitude solutions resulting from this *primary instability* may themselves be chosen as new basic states in order to undertake a *secondary instability* analysis. This process may be repeated to arrive at a *scenario* of successive bifurcations leading from the initial laminar state to turbulence. Instabilities may further be classified according to the nature of the underlying physical mechanism. In these lecture notes, the discussion is restricted to primary instabilities arising either from a purely *inviscid* redistribution of the basic vorticity field in the form of *Kelvin–Helmholtz instability waves*, or from a more subtle destabilizing mechanism paradoxically due to viscosity and giving rise to *Tollmien–Schlichting waves*. The resulting finite amplitude vortical states are shown to be subjected to a secondary inviscid *elliptical instability*. It should be emphasized that the above theoretical framework is not only relevant to understand the transition process but it may also describe the evolution of the coherent part of the motion in fully turbulent flows although this contention is not universally accepted in the turbulence community.

The course is structured in such a way as to provide the newcomer with the basic material on linear as well as nonlinear theory. Preliminary examples of open and closed flow instabilities are presented in sections 1.1 and 1.2 for pipe and Taylor–Couette flow respectively. A phenomenological description of the dynamics of various open flows such as mixing layers, bluff-body wakes and plane channel flow is given in section 2. Section 3 introduces the main linear and nonlinear instability concepts in a general setting without referring to a specific instability mechanism. The general formulation and properties of inviscid instabilities in parallel flows are presented in section 4. Section 5 applies the previous notions to an archetype of inviscidly and convectively unstable open flow giving rise to a noise amplifier, namely the spatial mixing layer. The wake behind a bluff body, which constitutes an example of inviscidly and globally unstable open flow, i.e. a hydrodynamic oscillator, is discussed in section 6. In section 7, the general formulation and properties of viscous-dominated instabilities in parallel flows are outlined with particular emphasis on the features that distinguish them from inviscid instabilities. Section 8 applies these notions to the primary instability in plane Poiseuille flow. By contrast, secondary instabilities arising in the same geometry are shown to be inviscid in character and to exhibit universal features that arise in a wide variety of bounded or free shear flows.

For general references to instability theory, the reader is advised to consult selected chapters in fluid dynamics texts (Craik, 1985; Landau and Lifshitz, 1987a; Tritton, 1988; Acheson, 1990; Kundu, 1990; Guyon *et al.*, 1991; Saffman, 1992; Lamb, 1995), general treatises on dynamical systems in the context of hydrodynamics (Bergé *et al.*, 1984; Lyapunov, 1988; Manneville, 1990; Chapter 4 by Fauve), monographs on hydrodynamic stability theory (Lin, 1955; Chandrasekhar, 1961; Betchov and Criminale, 1967; Drazin and Reid, 1981; Swinney and Gollub, 1981) or review articles (Maslowe, 1981; Orszag and Patera, 1981; Ho and Huerre, 1984; Maslowe, 1986; Bayly *et al.*, 1988; Herbert, 1988; Huerre and Monkewitz, 1990; Monkewitz, 1990; Cross and Hohenberg, 1993).

1.1 An open flow example: the pipe flow experiment of Reynolds

A seminal contribution to instability theory was made by Reynolds (1883). Though it was not —it is not yet— elucidated on theoretical grounds, this experiment (Fig. 1.1) has generated numerous questions and introduced new concepts such as the dimensionless Reynolds number, amplitude dependence, intermittency, criticality. Reynolds experimentally studied, in his Manchester laboratory, the flow in a tube of diameter d and length ℓ where the aspect ratio ℓ/d was chosen large enough to prevent any end effects (Fig. 1.2). A finite flow rate was generated by applying a constant pressure difference Δp between both extremities.

To perform this experiment, tubes of different diameters were used, various flow rates were selected and the flow was visualized by means of a streak of dye tracer entering the pipe at the trumpet shaped inlet. Whenever the pressure gradient $-dp/dx = \Delta p/\ell$ was set below a certain specific value, a steady flow was observed and found to be the axisymmetric *Hagen–Poiseuille velocity profile* known from classical hydrodynamics (Batchelor, 1967; Tritton, 1988). In this solution to the

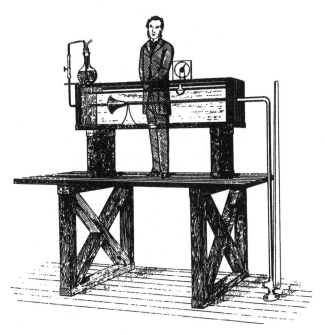

Fig. 1.1. Diagram of the experimental apparatus used by Reynolds. After Reynolds (1883).

Navier–Stokes equations and no-slip boundary conditions at the pipe wall, the convective terms vanish while the pressure gradient and diffusive term balance in the x-momentum equation to yield the parabolic streamwise velocity profile (Fig. 1.3a):

$$\mathbf{U}(r) = V\left[1 - \left(\frac{2r}{d}\right)^2\right]\mathbf{e}_x,\qquad(1.1)$$

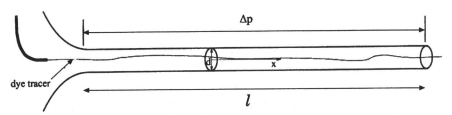

Fig. 1.2. Sketch of the pipe flow configuration. For clarity, the diameter d of the pipe has been enlarged with respect to its length ℓ.

with

$$V \equiv -\frac{1}{4\mu}\frac{dp}{dx}d^2 \, . \tag{1.2}$$

Here r is the radial coordinate, μ denotes the fluid shear viscosity while V stands for the velocity at the center of the pipe. According to (1.2), V varies linearly with pressure gradient. Moreover, by introducing a dye tracer (Fig. 1.4a), Reynolds demonstrated the high level of spatial and temporal coherence of the Hagen–Poiseuille flow: the dye is seen to move in straight lines as predicted by the unidirectional solution (1.1). This extremely regular motion is a typical example of *laminar flow*.

While fair agreement was found between experiments and theoretical predictions for low values of the pressure gradient, Reynolds observed that the flow was unexpectedly and profoundly altered when the pressure gradient exceeded a critical value: the dye streak suddenly appeared to waver and diffuse in the transverse direction (Fig. 1.4b). This change of behavior simply mirrors the distortion of particle paths and more

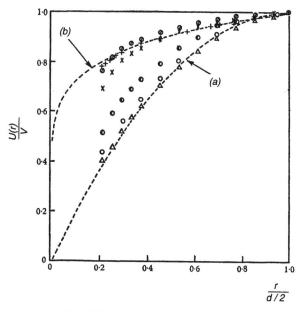

Fig. 1.3. Velocity profiles $U(r)/V$ in pipe flow: (a) laminar case: Hagen–Poiseuille velocity profile; (b) turbulent case. Open circles: Re $= 1,400$, triangles: Re $= 2,015$, half full circles: Re $= 2,440$, multiply signs: Re $= 2,680$, barred circles: Re $= 3,070$, plus signs: Re $= 4,060$. The Reynolds number is defined in equation (1.3). After Patel and Head (1969).

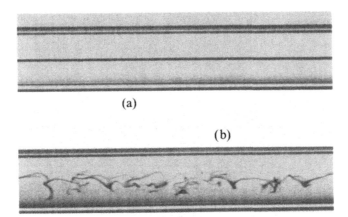

(a)

(b)

Fig. 1.4. Dye streaks in Reynolds' experiment: (a) laminar case; (b) turbulent case. After Van Dyke (1982).

generally the loss of temporal and spatial coherence in the underlying fluid motion: the flow is said to have reached a *turbulent regime* characterized by highly irregular spatio-temporal structures and enhanced mixing. As seen in hot-wire traces (Fig. 1.5), such changes are not necessarily taking place uniformly throughout the whole tube but they may occur in patches called *turbulent slugs* (Lindgren, 1969) that are delineated by fairly sharp boundaries while the rest of the fluid remains laminar. Turbulent slugs are generated intermittently within the tube and convected downstream. Visualizations indicate the presence of numerous eddies within each slug. As the pressure gradient is increased, the rate of production of slugs increases as well as their rate of expansion into the surrounding laminar regions. At sufficiently high $-dp/dx$ all slugs merge to produce a fully turbulent flow throughout the entire apparatus concurrently with the disappearance of intermittency.

Turbulent mixing affects the mean velocity distribution: in contrast with the laminar solution, the shape of the mean axial velocity profile within the slugs (Fig. 1.3b) becomes more uniform away from the pipe boundary and steep gradients are present near the wall to satisfy the no-slip boundary condition. The appearance of such high gradients leads to a dramatic increase in the magnitude of the wall shear stress $\tau = \mu(dU/dr)$, when compared to its laminar value at the same flow rate. For a fixed pressure gradient, i.e. a fixed wall shear stress, the volume flow rate in the turbulent state becomes almost proportional to $\sqrt{-dp/dx}$ instead of varying linearly with dp/dx as in equation (1.2). Turbulent flows are hence much more demanding in energy consumption. When the pipe diameter d or kinematic viscosity $\nu = \mu/\rho$, where ρ is the fluid

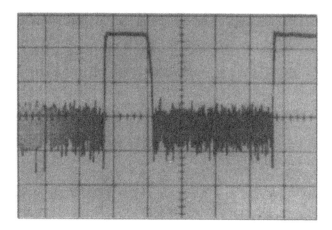

Fig. 1.5. Hot-wire traces left by turbulent slugs at Re $= 4,200$ (the signal is proportional to the instantaneous velocity). The Reynolds number Re is defined in equation (1.3). After Wygnanski and Champagne (1973).

density, are altered, distinct curves are obtained for the pressure gradient as a function of volume flow rate. Reynolds realized, however, that all these curves collapse on a single curve when the so-called nondimensional *Reynolds number*

$$\mathrm{Re} \equiv \frac{\bar{V}d}{\nu} \qquad (1.3)$$

and the nondimensional frictional resistance $\lambda \equiv -(2d/\rho\bar{V}^2)dp/dx$ are introduced (Fig. 1.6). Here $\bar{V}$ denotes the mean velocity averaged over the tube cross section and time. The Reynolds number (1.3) may be interpreted as a measure of the ratio between convective terms $\bar{V}^2/d$ and diffusive terms $\nu\bar{V}/d^2$. Alternatively, this parameter is the ratio between the viscous time scale $\tau_v = d^2/\nu$ necessary for diffusion to be felt over a distance d, and the inertial time $\tau_i = d/\bar{V}$ necessary for convective transport over the same distance.

Laminar–turbulent transition takes place at a Reynolds number Re_{tr}, but Reynolds himself showed this value to be highly sensitive to the level of external noise and to the nature of the experimental set-up. Transition values as high as $\mathrm{Re}_{tr} = 13,000$ were reached in the most careful set of experiments. This observation does not contradict the above dimensional argument since this reasoning assumes that the velocity $\bar{V}$ (or equivalently the pressure gradient), the diameter d, the fluid viscosity and density are sufficient to completely describe the flow. Additional hidden parameters such as the amplitude level of incoming disturbances, pipe wall rugosity, residual rotational motion contribute to significant variations in the

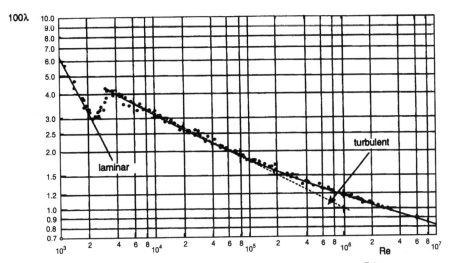

Fig. 1.6. Nondimensional frictional resistance $\lambda = -(2d/\rho \bar{V}^2)dp/dx$ versus Reynolds number Re in smooth pipe flow experiment. After Schlichting (1979).

transitional value Re_{tr} and in the frictional resistance curve (Schlichting, 1979). The transitional Reynolds number associated with the departure from the laminar frictional resistance law, may decrease to values as low as $Re_{tr} = 2,000$ in a noisy experimental environment, e.g. in the presence of high intensity disturbances, while "clean" experiments manage to sustain laminar flow for Reynolds numbers as large as 10^5.

Qualitative differences between laminar and turbulent states are rather profound and are not restricted to the pipe flow experiment. After Reynolds, various flow configurations have been found to exhibit features that are similar to this historical example: spatial and temporal disorder with an extended spectrum of spatial scales, the presence of localized regions of three-dimensional vorticity, intense mixing.

It should be emphasized that the parabolic velocity profile is a solution of the Navier–Stokes equations for arbitrary Reynolds numbers. Since we are pretty confident that the governing equations of fluid dynamics remain valid for all Reynolds numbers, one should explain why such a solution is not observed at large Reynolds numbers. The goal of stability theory is precisely to determine under which conditions, i.e. in which range of Reynolds numbers such a solution is observable and to predict the transitional value Re_{tr} where transition to another solution occurs. Additionally, stability theory should account for the growth of perturbations above Re_{tr}, and it should predict the main features of the observed state beyond transition. For instance, it should explain why steady constraints are capable of producing a highly irregular evolution in space and time.

These objectives are delicate to achieve even for simple configurations such as Hagen–Poiseuille flow. First it is sometimes difficult to compute a realistic basic flow solution even for simple geometries. In the pipe flow experiment, the parabolic velocity distribution is clearly not pertinent close to the pipe inlet where the wall vorticity has not diffused yet into the bulk of the flow (Schlichting, 1979; Tritton, 1988). A second difficulty of open flows, which may jeopardize any comparison between experimental and theoretical results, stems from their sensitivity to the choice of experimental (numerical) outflow and inflow boundary conditions (Buell and Huerre, 1988). It is amazing that Reynolds' pipe flow experiment from which originated many investigations in stability theory as well as key theoretical concepts, is still lacking a satisfactory explanation. For this reason, Hagen–Poiseuille flow is not considered further in this course!

1.2 A closed flow example: Taylor–Couette flow between rotating cylinders

Although the present lecture focuses on open shear flows, we briefly describe some of the results available in a typical closed system, namely Taylor–Couette flow between two rotating concentric cylinders (Taylor, 1923; Drazin and Reid, 1981). The reader is referred to Iooss and Adelmeyer (1992), Chossat and Iooss (1994) for the mathematical aspects of the problem and to Fenstermacher *et al.* (1979), Di Prima and Swinney (1981), Andereck *et al.* (1986) for an account of experimental studies. The first major success of hydrodynamic instability theory was obtained in this configuration by Taylor (1923) who predicted the critical value for transition as well as the structure of the bifurcated flow. More generally, many fundamental studies of turbulence chaos and transition have been performed in this context or on the problem of natural convection in a closed cell, i.e. the celebrated Rayleigh–Bénard experiment (Busse, 1981). These configurations are particularly attractive from an experimental and theoretical viewpoint since they are not plagued by the ambiguities affecting most open flows as discussed in section 1.1. In particular, boundary conditions are well-defined and experimental observations do not depend on "hidden parameters" such as incoming perturbations from outflow or inflow boundaries.

The Taylor–Couette experimental set-up (Fig. 1.7) is the following: an incompressible viscous fluid of kinematic viscosity ν is contained between two concentric cylinders of diameter a_1, a_2 which rotate respectively with angular velocity Ω_1, Ω_2. The governing equations as well as the boundary

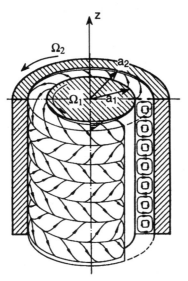

Fig. 1.7. Taylor–Couette experimental set-up. The flow regime shown here corresponds to toroidal Taylor vortices. After Schlichting (1979).

conditions are time-invariant under $t \to t + t_0$ and space invariant with respect to any rotation $\theta \to \theta + \theta_0$ about the cylinder axis, where θ denotes the azimuthal angle. If top and bottom boundary layer effects are neglected, the configuration is also invariant with respect to translations $z \to z + z_0$ along the cylinder axis, where z is the axial coordinate. In this infinite cylinder approximation,[*] the flow is completely defined by three nondimensional parameters:

$$\eta = \frac{a_1}{a_2}, \quad \mathrm{Re}_1 = \frac{\Omega_1 a_1 (a_2 - a_1)}{\nu}, \quad \mathrm{Re}_2 = \frac{\Omega_2 a_2 (a_2 - a_1)}{\nu}. \qquad (1.4)$$

For a fixed radius ratio η, the experimentally determined solid line in Fig. 1.8 delineates distinct flow regimes in the Re_2–Re_1 plane. Below this curve (hatched area in Fig. 1.8), the classical circular Couette flow solution of the Navier–Stokes equations is observed in the bulk. The associated velocity field reads

$$\mathbf{U}(r) = (Ar + B/r)\mathbf{e}_\theta, \qquad (1.5)$$

where r and $\mathbf{e}_\theta$ respectively denote the radial coordinate and the azimuthal unit vector. The constants A and B are prescribed by the no-slip boundary conditions on each cylinder and depend upon the three control parameters defined in (1.4). This regime satisfies all the above

---

[*] This is valid when the ratio between the height and the cylinder gap is sufficiently large.

symmetries: axisymmetry, homogeneity in the axial coordinate, and stationarity in time.

A bove the solid curve in Fig. 1.8, the observed behavior evolves towards various organized flows (Andereck *et al.*, 1986). These patterns do not display all the symmetries that are characteristic of the system itself. In different regions of control-parameter space, toroidal Taylor vortices, spiral structures or wavy vortex flows are observed which do not satisfy rotational symmetry, axial translational invariance or continuous time invariance. The flow has experienced symmetry breaking *bifurcations*. As stated in section 1.1, the goal of instability theory is precisely to determine the domains in control-parameter space where a given solution such as the Couette flow (1.5) becomes unstable and to predict the flow features beyond transition, e.g. symmetry breaking, amplitudes and frequency spectra.

The nature of the transitions very much depends on the path followed in control-parameter space (Fig. 1.8): a given sequence of bifurcations then defines a scenario, a concept that has proven to be useful in many other closed flows. For instance, let the inner cylinder rotation rate Re_1 be gradually increased while the outer cylinder is kept at rest ($Re_2 = 0$). As Re_1 crosses the solid curve at Re_{1c}, circular Couette flow (1.5) becomes unstable and is replaced by an array of regularly spaced steady toroidal

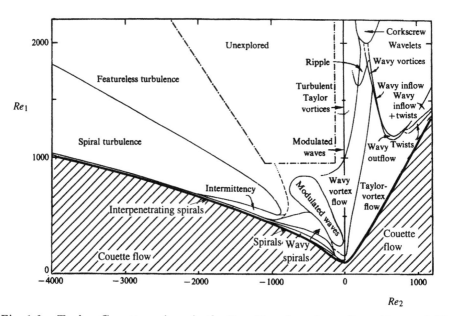

Fig. 1.8. Taylor–Couette regimes in the Re_2–Re_1 plane for radius ratio $\eta = 0.88$. After Andereck *et al.* (1986).

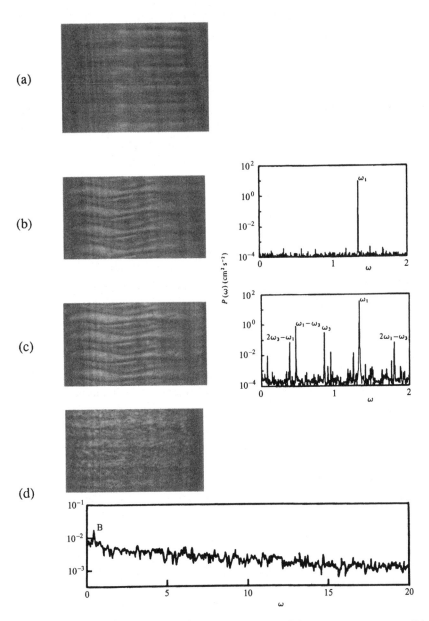

Fig. 1.9. Taylor–Couette regimes (flow structure and frequency spectrum $P(\omega)$) as Re_1 is increased for $Re_2 = 0$ (stationary outer cylinder). (a) Steady toroidal Taylor vortices; (b) wavy vortex flow; (c) modulated waves; (d) chaotic regime. After Fenstermacher *et al.* (1979).

Taylor vortices (Fig. 1.9a) which break the continuous translational axial invariance. The elementary mechanism responsible for this instability may be attributed to centrifugal effects (Drazin and Reid, 1981). Close to Re_{1c} the vortex strength is observed to vary as the square root of the threshold distance $\sqrt{Re_1 - Re_{1c}}$, a feature typical of many bifurcations. As Re_1 crosses a second critical value, this new structure itself becomes unstable and leads to an azimuthally traveling wave solution referred to as wavy vortex flow (Fig. 1.9b). Note that this second transition breaks the continuous time invariance $t \to t + t_0$ and the azimuthal invariance $\theta \to \theta + \theta_0$. If one keeps increasing Re_1, a quasi-periodic (Fig. 1.9c) and then chaotic regime (Fig. 1.9d) is reached as confirmed by high resolution power spectra.

In the wavy vortex flow state, the spectrum only contains a single fundamental frequency and its harmonics (Fig. 1.9b) while two incommensurate frequencies and their combination harmonics are present in the quasi-periodic flow regime pertaining to modulated waves (Fig. 1.9c). Finally, *the onset of chaos is characterized by the emergence of a broadband spectrum* (Fig. 1.9d). In contrast with many open flow situations, this broadband component is not *extrinsically* generated by ambient noise but it is *intrinsic* to the dynamics itself. Furthermore, in the Taylor–Couette experiment, measured frequency spectra display the same features at all spatial locations although the relative amplitudes of various Fourier components may differ from point to point. On the contrary in many open flows, frequency spectra will be shown to undergo qualitative changes with spatial location along the stream.

2 Phenomenology of open flows

Among open flows, *shear flows* (Fig. 2.1) are generally characterized by velocity-gradient regions where most of the vorticity is concentrated. Note that shear flows are present in numerous natural phenomena and technological devices: combustors, aircrafts, meteorological events.

In this course, we examine for the most part parallel or quasi-parallel shear flows. Such configurations are characterized by an almost unidirectional velocity field exhibiting large variations in the cross-stream direction. In the sequel, x, y, and z respectively denote the streamwise, cross-stream, and spanwise directions. Three major examples of almost parallel shear flows are reviewed: the *mixing layer* (Fig. 2.1b), the *wake* behind a bluff body (Fig. 2.1c), and *plane Poiseuille flow* (Fig. 2.1d). The attached boundary layer case (Fig. 2.1a) being analogous to plane Poiseuille flow is not specifically considered here. Finally jets (Fig. 2.1e) are not discussed since they share common properties with both mixing

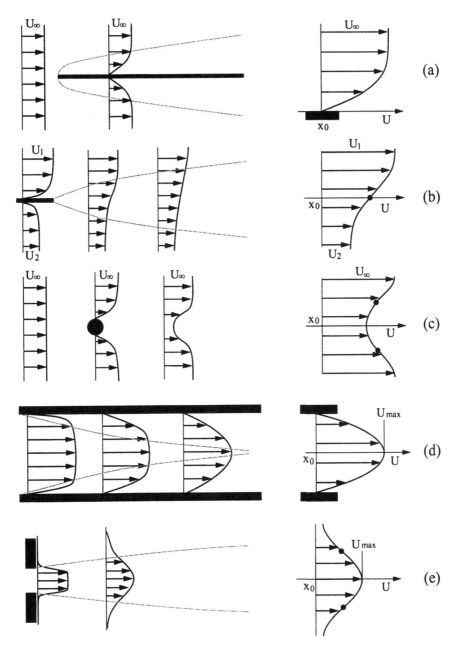

Fig. 2.1. Different shear flows. Spatially developing flows (left) and idealized parallel flows at downstream location x_0 (right). Full circles denote inflection points. (a) Boundary layer. (b) Mixing layer. (c) Wake behind a bluff body. (d) Plane Poiseuille flow. (e) Jet.

layers and wakes. Because of their extreme mathematical complexity, inherently three–dimensional (Reed *et al.*, 1996) or unsteady basic flows (Davis, 1978) are not addressed although they play a major part in engineering applications (boundary layers over swept aircraft wings or in a corner, pulsating Poiseuille flow, etc.).

In the present section, three specific shear flows are selected to illustrate the major experimental features of various classes of open flows. The mixing layer acts as a noise amplifier: it is the prototype of a flow undergoing an inviscid convective instability. The wake behind a bluff body exhibits intrinsic hydrodynamic oscillations that are insensitive to low levels of external noise: it is the prototype of a globally unstable flow somewhat akin to closed flow systems. Finally, plane channel flow is chosen to discuss a convective instability that is paradoxically induced by viscous diffusion.

2.1 The mixing layer as a prototype of noise amplifier

The paradigmatic mixing layer is formed when two uniform streams of respective velocity U_1 and $U_2 < U_1$, initially separated by a thin splitter plate, merge downstream of its trailing edge (Fig. 2.1b). Viscous diffusion between the two coflowing streams then generates, in the cross-stream direction, a region with steep gradients of streamwise velocity. This zone contains most of the spanwise vorticity. At a fixed streamwise coordinate x_0 downstream of the trailing edge, the streamwise component of the mean velocity profile is described by a smooth function of the cross-stream variable y that necessarily presents at least one *inflection point* (Fig. 2.1b) corresponding to a maximum of spanwise vorticity. Such a feature is the essential ingredient necessary to trigger an *inviscid instability* evolving on an inertial time scale τ_i as discussed in section 4. For a thorough review, consult Ho and Huerre (1984).

Three major features are observed experimentally (Browand and Weidman, 1976; Ho and Huang, 1982; Ho and Huerre, 1984). First, orderly large scale spanwise vortical structures are observed within the mixing layer (Fig. 2.2). More precisely the shear layer initially rolls up into spanwise co-rotating *Kelvin–Helmholtz vortices* that are convected in the streamwise direction at the average velocity $\bar{U} \equiv (U_1 + U_2)/2$ between the two streams. In that reference frame, they grow proportionally to the magnitude of the shear $\Delta U \equiv (U_1 - U_2)$ and they scale with the local shear layer thickness. Further downstream neighboring vortices start to merge to create larger structures (Fig. 2.3). This recurrent process called *pairing* also affects the newly-formed vortices. This brings us to the second point: the shear layer thickness continuously increases in the

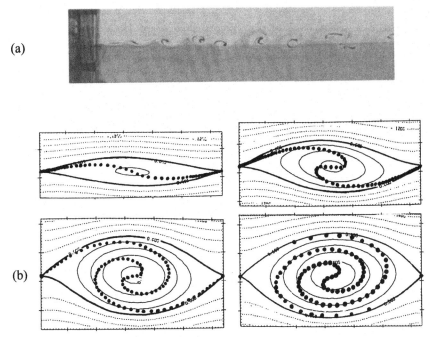

Fig. 2.2. Kelvin–Helmholtz spanwise vortices in mixing layers: (a) real spatially developing mixing layer experiment by Lasheras and Choi (1988); (b) numerical temporal mixing layer experiment by Corcos and Sherman (1984). Dots in (b) denote position of interfacial markers and continuous curves represent streamlines.

downstream direction primarily as a result of sequential mergings of the aforementioned large structures.

Finally these patterns are almost quasi two-dimensional in the x-y plane although a more careful study reveals the presence of *secondary streamwise vortices* (Fig. 2.4) developing on the primary spanwise vortices (Bernal and Roshko, 1986; Lasheras and Choi, 1988). It is enlightening to view this evolution as a scenario of successive symmetry breakings: first the appearance of primary structures breaks the translational invariance along the stream while the presence of secondary vortices breaks the translational invariance along the span.

The experimentally measured mean flow is not parallel. Nevertheless, the streamwise velocity U experiences fast variations in the cross-stream direction y while it slowly evolves over a typical separation distance between two vortices. As a result the streamwise component U is much larger than the cross-stream component V, which is the fundamental assumption underlying the WKBJ formulation to be presented in section 5.2. This approximation does not hold in the immediate vicinity of

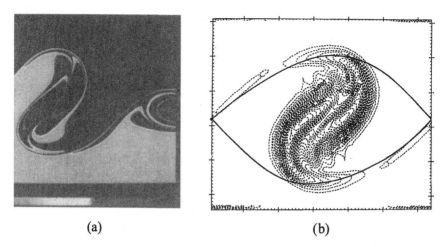

(a)   (b)

Fig. 2.3. Merging of Kelvin–Helmholtz vortices in mixing layers: (a) digitized picture from a real spatially developing mixing layer experiment by Koochesfahani and Dimotakis (1986); (b) numerical temporal mixing layer experiment by Corcos and Sherman (1984). Dashed curves represent vorticity contours.

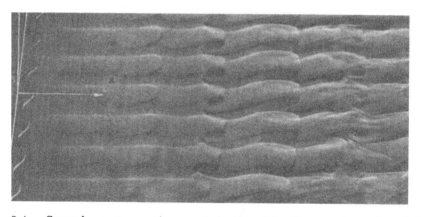

Fig. 2.4. Secondary streamwise vortices in mixing layers, here forced by sinusoidal indentation of splitter plate trailing edge. After Lasheras and Choi (1988).

the trailing edge where boundary layers on either side of the splitter plate have just merged. This small zone called the receptivity region, where streamwise and cross-stream gradients become comparable very close to the trailing edge (Goldstein, 1983; Kerschen, 1989) is not considered further in these lecture notes. It is convenient to quantify the thickness of the rotational region at the specific station x or else the magnitude of the velocity gradient. For instance the *vorticity thickness* (Fig. 2.5) is defined as

$$\delta_\omega(x) \equiv \frac{(U_1 - U_2)}{(dU/dy)_{\max}} . \tag{2.1}$$

Immediately downstream of the splitter plate, $\delta_\omega(x)$ increases as the square root of x (Figs. 2.8, 2.9), as one would expect through viscous diffusion. Further downstream, inviscid processes are observed to dominate the streamwise evolution of $\delta_\omega(x)$: *vortex roll-up* first leads to an increase in thickness and, once completed, to a local saturation, as evidenced by the first plateau in Fig. 2.8b. Beyond this phase recurring *merging events* lead either to sudden increases in $\delta_\omega(x)$ as in Fig. 2.9b, or to a more gradual spreading rate as in Fig. 2.8b (Ho and Huang, 1982). Far downstream, in the fully developed turbulent region, $\delta_\omega(x)$ ultimately becomes linear in x (Brown and Roshko, 1974).

From the four external dimensional quantities U_1, U_2, ν and the initial vorticity thickness $\delta_\omega(0)$, one may construct two dimensionless control parameters. In addition to the *Reynolds number*

$$\mathrm{Re} \equiv \frac{(U_1 - U_2)\delta_\omega(0)}{\nu} , \tag{2.2}$$

one defines the *velocity ratio*

$$R \equiv \frac{U_1 - U_2}{U_1 + U_2} = \frac{\Delta U}{2\bar{U}} , \tag{2.3}$$

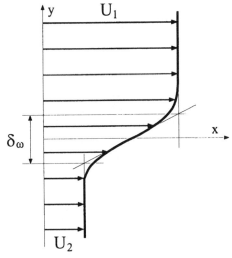

Fig. 2.5. Explanatory sketch of vorticity thickness definition.

which quantifies the relative magnitude of the shear ΔU with respect to the average velocity $\bar{U}$. When the two velocities U_1 and U_2 are assumed to be positive with $U_1 \geq U_2$, the parameter R varies in the range $0 \leq R \leq 1$ (Fig. 2.6a). When $R = 0$ there is no net shear: this corresponds to the case of the wake behind a flat plate (Fig. 2.6b). When $R = 1$, the lower stream is at rest (Fig. 2.6c). The spreading rate $d\delta_\omega(x)/dx$ of the vorticity layer thickness is observed to increase proportionally to the relative intensity of the shear R. When $R \ll 1$, the convection velocity $\bar{U}$ of the structures is much larger than the destabilizing shear ΔU. As a consequence the streamwise development of the flow is very slow. A contrario, when R increases towards unity, typical events such as roll-up and sequential pairings occur closer and closer to the trailing edge. In other words, the velocity ratio R acts as a dilatation parameter for the flow in the streamwise direction.

As in Hagen–Poiseuille flow, viscous effects are measured by the Reynolds number Re defined in (2.2). The overall large scale dynamical behavior is observed to be insensitive to this parameter at least for values greater than Re = 100. These observations illustrate the fundamentally inviscid nature of the large scale dynamics which are to be examined from the stability theory point of view. As seen in Figs. 2.7a and b, large coherent structures persist at very high Reynolds numbers in the fully turbulent regime. However, small scales superimposed on the large vortices are now present. Note, by comparing Figs. 2.7a and b, that small scales are enhanced as Re is doubled while large scales remain qualitatively unchanged (Brown and Roshko, 1974). Indeed, small scales

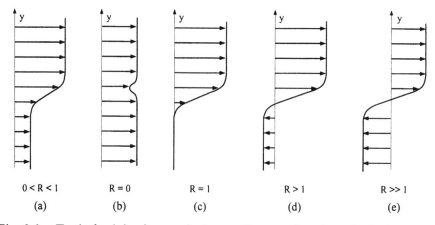

| $0 < R < 1$ | $R = 0$ | $R = 1$ | $R > 1$ | $R \gg 1$ |
| (a) | (b) | (c) | (d) | (e) |

Fig. 2.6. Typical mixing layer velocity profiles as a function of velocity ratio R when $U_1 \geq |U_2|$: (a) $0 < R < 1$; (b) $R = 0$; (c) $R = 1$; (d) $R > 1$; (e) $R \gg 1$.

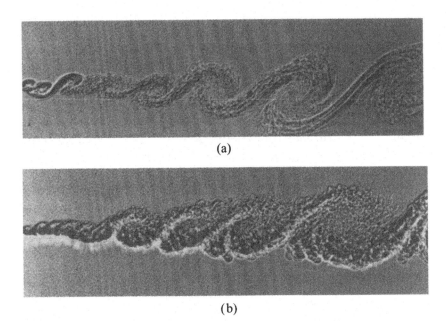

(a)

(b)

Fig. 2.7. Large scale structures in mixing layers at high Reynolds numbers. Re in (b) is doubled with respect to Re in (a). After Van Dyke (1982).

may be assumed to affect only the effective viscosity perceived by large scales.

The previous considerations fail to capture by far the most important control parameter, namely *external noise*. Spatially developing mixing layers have been observed to be extremely sensitive to the characteristics of perturbations whether they exist naturally as noise or they are applied intentionally as forcing. In the latter situation, external excitations can be induced by mechanical means (oscillating ribbons, trailing edge flaps, corrugated trailing edges), acoustical devices (loudspeakers) or flow injection. In all instances, it is of paramount importance to specify the intensity and frequency content of such perturbations just downstream of the splitter plate in order to determine the subsequent evolution of the flow. Indeed the mixing layer acts as a perturbation or *noise amplifier*.

In *unforced* mixing layers, the various scales of natural white noise are differentially amplified: the power spectra in the roll-up region then exhibit a large bandwidth of amplified frequencies with a maximum at a natural frequency f_n. This quantity is observed to scale with the vorticity thickness $\delta_\omega(0)$ and the average velocity $\bar{U}$ so that the *Strouhal number* $\mathrm{St}_n \equiv f_n \delta_\omega(0)/\bar{U}$ remains constant and approximately equal to 0.03. In the roll-up region, the response therefore takes the form of vortices characterized by a predominant shedding rate f_n and associated wavelength $\lambda = \bar{U}/f_n$. Further downstream, frequency spectra successively display peaks at the subharmonics $f_n/2$, $f_n/3 \ldots$ which are associated with sequential pairing, tripling... of vortices. Corresponding streamwise merging locations experience spatial jitter around some mean value.

In order to avoid such a lack of phase reference, most experiments are performed in the presence of *controlled forcing*, for instance at a given frequency f_f (Ho and Huang, 1982). A spatially organized pattern is then observed: primary vortices created just after the initial roll-up phase periodically evolve in time and an unambiguous phase reference can be defined. If $f_f \approx f_n$ (Fig. 2.8), the energy of the perturbations, immediately downstream of the splitter plate, is primarily contained

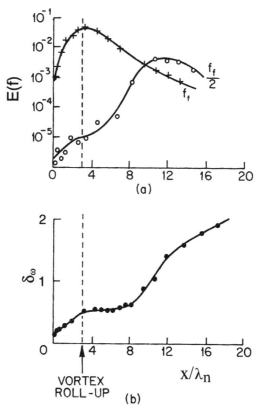

Fig. 2.8. Streamwise evolution of (a) energy $E(f)$ of spectral components f_f and $f_f/2$ on a semilog plot and (b) vorticity thickness δ_ω in forced mixing layers with $f_f \approx f_n$. Streamwise distance x from the trailing edge is normalized with respect to the natural Kelvin–Helmholtz wavelength λ_n. After Ho and Huerre (1984) based on measurements by Ho and Huang (1982).

in the fundamental frequency f_f and reaches a maximum at the roll-up location. Around the same station, the energy of the subharmonic $f_f/2$ (Fig. 2.8) starts to increase dramatically and becomes maximum at the pairing location. A different scenario is observed when the flow is forced at $f_f \approx f_n/2$ (Fig. 2.9): the energy contained in f_f increases rapidly right from the trailing edge and its maximum is reached further upstream. In physical space, the pairing process appears much earlier. This phenomenology clearly displays the extrinsic nature of mixing layer dynamics as opposed to the intrinsic nature of instability processes in closed flows. Note that variations in forcing frequency significantly affect the shear layer spreading rates as evidenced by comparing Fig. 2.8b and Fig. 2.9b. If one seeks to control mixing processes, input perturbations should therefore be carefully monitored.

It is worthwhile to briefly mention the case of counterflow mixing layers obtained for values of the velocity ratio $R > 1$ (Fig. 2.6d,e). Such

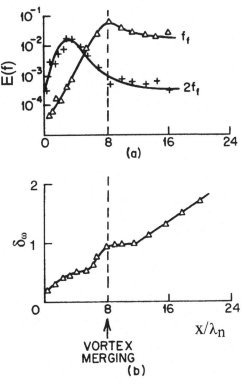

Fig. 2.9. Streamwise evolution of (a) energy $E(f)$ of spectral components f_f and $2f_f$ on a semilog plot and (b) vorticity thickness δ_ω in forced mixing layers with $f_f \approx f_n/2$. After Ho and Huerre (1984) based on measurements by Ho and Huang (1982).

a configuration is most easily generated by tilting a tube filled with two immiscible fluids of different density or with a stable continuously stratified fluid (Reynolds, 1883; Thorpe, 1971; Pouliquen *et al.*, 1994). The accelerating shear layer thus produced gives rise to Kelvin–Helmholtz instability waves rolling up into vortices (Fig. 2.10). Contrary to the case of spatially evolving mixing layers generated downstream of a splitter plate, these experiments give rise, when $R \gg 1$ (Fig. 2.6e), to *temporally* evolving Kelvin–Helmholtz vortices with a well-defined natural wavenumber k_n. Such a behavior is reminiscent of closed flows. This essential difference calls for radically distinct theoretical formulations as discussed in sections 3.2 and 5.1.

2.2 The wake behind a bluff body as a prototype of hydrodynamic oscillator

The wake formed behind an obstacle such as a circular cylinder, the axis of which is orthogonal to the upstream flow, constitutes yet another example of a free shear flow (Fig. 2.1c). In that case, a single characteristic velocity is present: the uniform upstream velocity U_∞. The obstacle cross-section

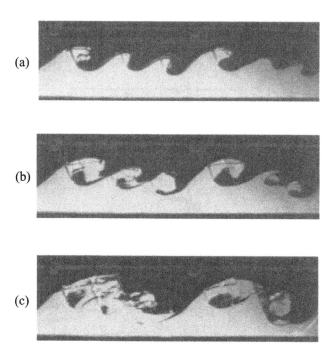

(a)

(b)

(c)

Fig. 2.10. Kelvin–Helmholtz instability waves produced at the interface between two immiscible fluids in a tilting tank experiment (courtesy of O. Pouliquen, LadHyX, École Polytechnique).

provides a first length scale, e.g. the cylinder diameter d, while a second scale is given by the cylinder length ℓ. Based on the above dimensional quantities, two dimensionless parameters may be defined: the *Reynolds number*

$$\mathrm{Re} \equiv \frac{U_\infty d}{\nu} \qquad (2.4)$$

and the *aspect ratio* ℓ/d. As in the case of mixing layers, one may invoke the locally-parallel approximation: at each downstream location, the mean velocity field in the wake is described by a streamwise velocity distribution which primarily depends on the cross-stream variable y. At this level of approximation, the wake velocity profile displays *two inflection points* or two vorticity maxima (Fig. 2.1c) as opposed to a single inflection point for mixing layers (Fig. 2.1b). The basic instability mechanism is therefore likely to be *inviscid* in character as demonstrated in section 4. An excellent and comprehensive review of wake instabilities is given in Williamson (1996).

Let us first consider the purely two-dimensional cylinder wake for which the aspect ratio ℓ/d goes to infinity. In such an instance, end effects are insignificant and there is a single dimensionless control parameter left: the Reynolds number. In the creeping steady flow regime $\mathrm{Re} \ll 1$, viscous diffusion dominates most of the flow and the Stokes solution (Batchelor, 1967; see also Chapter 1 by Castaing) applies. A symmetrical unseparated flow surrounds the body and no significant changes take place until Re exceeds 5 or 6 (Fig. 2.11). Above this value, the flow

Fig. 2.11. Steady wake flow behind a circular cylinder, $\mathrm{Re} = 1.54$. After Van Dyke (1982).

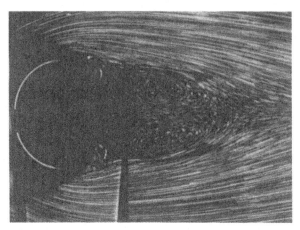

Fig. 2.12.   Steady attached counter-rotating vortices in circular cylinder wake, Re = 24.3. After Van Dyke (1982).

remains steady but displays a recirculation bubble immediately behind the obstacle composed of two counter-rotating vortices whose strength and spatial extent increase with Reynolds number (Fig. 2.12). As Re crosses a second critical value $Re_{G_c} = 48.5$, a new transition takes place leading to an unsteady periodic state commonly referred to as the *Bénard–Kármán vortex street* (Fig. 2.13) in which two traveling rows of

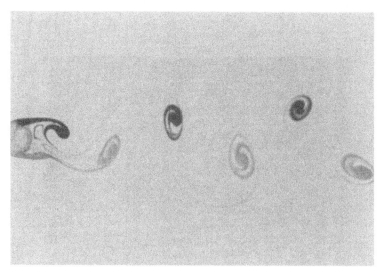

Fig. 2.13.   Bénard–Kármán vortex street in circular cylinder wake, Re = 80. After Perry *et al.* (1982).

counter-rotating vortices appear behind the cylinder. Instead of the wide frequency spectra typical of mixing layers, one observes that the energy is predominantly concentrated in a sharp peak at a fundamental frequency f (Fig. 2.14). In general, the associated Strouhal number $\mathrm{St} \equiv fd/U_\infty$ depends on the Reynolds number and aspect ratio. In the purely two-dimensional case, a least square curve fit (Williamson, 1989) gives the relationship

$$\mathrm{St} = 0.1816 - \frac{3.3265}{\mathrm{Re}} + 1.6 \times 10^{-4}\,\mathrm{Re} \; . \tag{2.5}$$

In the vicinity of the critical Reynolds number $\mathrm{Re}_{G_c} = 48.5$, the above relation yields as a first approximation the constant value $\mathrm{St} = 0.2$. These oscillations are observed to be insensitive to low levels of external noise. The wake in this range constitutes an example of *intrinsic* flow behavior in striking contrast to the *extrinsic* dynamics of mixing layers. It is shown in section 6 that *global-mode* concepts which are the hydrodynamical counterparts of bound states in quantum mechanics (Dirac, 1984), provide the theoretical framework for the understanding of such *self-sustained hydrodynamic oscillators*.

The onset of the Bénard–Kármán vortex street regime, where the continuous time invariance of the system has been broken, is associated with a *Hopf bifurcation*, as demonstrated in the thorough experimental investigation of Provansal *et al.* (1987). More specifically the results of forced and transients experiments may be summarized in the following way: any physical perturbation field $\psi(\mathbf{x}, t)$ is of the form

$$\psi(\mathbf{x}, t) = \mathcal{R}e\left\{\phi(\mathbf{x})a(t)exp(-i\omega_g t)\right\}, \tag{2.6}$$

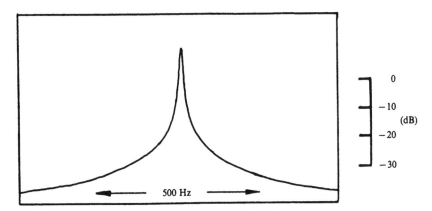

Fig. 2.14. Wake velocity spectrum at $\mathrm{Re} = 120.2$ in the Bénard–Kármán vortex street range. After Williamson (1989).

where ω_g is a real frequency, $\phi(\mathbf{x})$ a shape function characterizing the spatial distribution of the oscillator, $\mathcal{R}e$ denotes the real part, and $a(t)$ is a complex amplitude function effectively governed by the supercritical *Landau equation*

$$\frac{da}{dt} = \sigma a - b|a|^2 a, \tag{2.7}$$

as one would expect in a Hopf bifurcation (Guckenheimer and Holmes, 1983; Fauve in these lecture notes). The real part σ_r of the complex coefficient σ is found to be proportional to the distance $\mathrm{Re} - \mathrm{Re}_{G_c}$ from threshold and the real part of b is positive. According to equation (2.7), the amplitude $a(t)$ then asymptotically reaches a saturation amplitude $a(\infty)$ which is proportional to $\sqrt{\sigma_r} \propto \sqrt{\mathrm{Re} - \mathrm{Re}_{G_c}}$. This result is corroborated by wake measurements as exhibited in Fig. 2.15.

The above considerations nominally apply to the case of infinite aspect ratio ℓ/d. Recent experimental investigations of the cylinder wake below $\mathrm{Re} = 180$ (Williamson, 1988, 1989) have indicated that the finite length of the cylinder brings about non-trivial *three-dimensional effects* in the vortex street regime: *oblique shedding* takes place whereby vortices

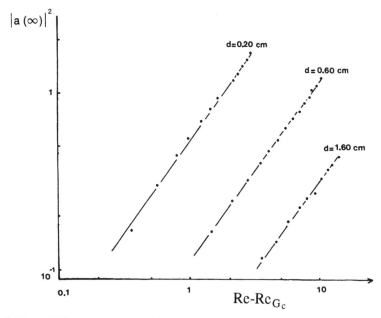

Fig. 2.15. Wake transverse kinetic energy, i.e. square of the saturation amplitude $|a(\infty)|$, versus threshold distance $\mathrm{Re} - \mathrm{Re}_{G_c}$ near onset of Bénard–Kármán vortex street on a log-log plot: the slope is seen to be close to unity which implies that the amplitude $a(\infty)$ is proportional to $\sqrt{\mathrm{Re} - \mathrm{Re}_{G_c}}$. After Mathis *et al.* (1984).

are produced in a slanted direction to the cylinder axis as displayed in Fig. 2.16. As a result, Strouhal number variations with Reynolds number present a discontinuity below Re = 180 (Fig. 2.17) and measured power spectra exhibit quasi-periodic behavior with two incommensurate frequencies and their respective beat frequencies.

These features have been attributed in the past to various factors such as nonuniformities of the flow field or coupling between the flow and the cylinder vibrational modes. They have now been clearly linked to *end effects* (Williamson, 1989): near the cylinder extremities, there exists a region where shedding frequencies are observed to be lower than in the bulk flow. Such domains are of finite extent and their size is independent of ℓ/d. When $\ell/d \leq 20$, they are large enough to directly influence the entire flow field and cannot be distinguished from the bulk flow. When $\ell/d \geq 20$, the effect of finite aspect ratio is more subtle. In the transient regime, *parallel shedding*, i.e. in the same direction as the cylinder axis, takes place throughout the flow but frequency mismatch between the end regions and the bulk must be accommodated through dislocations and bending of vortex lines. As time evolves these bends propagate into the bulk from the tips (Fig. 2.18) so as to give rise to slanted shedding throughout. If end boundary conditions are symmetric, slanted shedding takes the form of a symmetric *chevron pattern* as shown in Fig. 2.16. If they are asymmetric, the chevron is also asymmetric or may be replaced by uniformly oblique shedding throughout. This phenomenology accounts for the presence of a discontinuity in the Strouhal number versus Reynolds

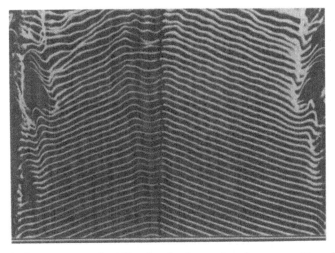

Fig. 2.16. Oblique vortex shedding in the form of a chevron pattern induced by end boundary conditions, Re = 85. After Williamson (1989).

number curve below Re = 180 (Fig. 2.17). Let us assume that the
wavelength λ and convection velocity of slanted vortices along the stream
are identical to the parallel shedding case at the same Reynolds number.
The measured Strouhal number St_0 in the oblique shedding case is then
related to its parallel counterpart St through $St_0 = St \cos \theta$ where θ is
the shedding angle (Fig. 2.18). Using this purely empirical relationship,
the experimentally observed discontinuity in Strouhal number St_0 may
be collapsed onto one universal Strouhal number $St = St_0 / \cos \theta$ versus
Reynolds number curve (Fig. 2.17). Note that it is possible to manipulate
the flow and induce parallel shedding by introducing suitably positioned
end plates (Hammache and Gharib, 1989; Williamson, 1989). Finally,
the various incommensurate frequencies appearing in the power spectra
are related to the oblique shedding modes prevailing in different parts of
the wake.

In the range Re = 190–260, three-dimensional effects do not solely
arise as a result of end effects: two distinct intrinsic modes of instability
take place in the bulk corresponding to markedly different spanwise

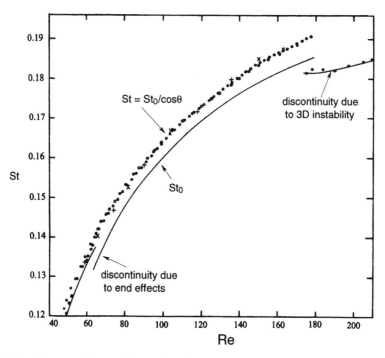

Fig. 2.17. Measured Strouhal number St_0 versus Reynolds number Re in circular
cylinder wake (continuous line) and renormalized universal curve $St = St_0 / \cos \theta$
(dots). After Williamson (1989).

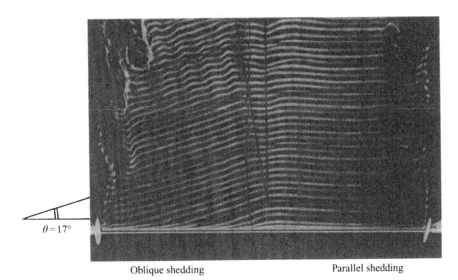

$\theta = 17°$

Oblique shedding   Parallel shedding

Fig. 2.18. Propagation of the front separating oblique vortex shedding from parallel vortex shedding, Re = 90. After Williamson (1989)

wavelengths (Williamson, 1996). One such mode is seen in Fig. 2.17 as a discontinuity appearing in the continuous Strouhal number curve. In closing, it is essential to bear in mind that a complicated sequence of transitions takes place as the Reynolds number is further increased. They successively involve transition to turbulence in the separated shear layer issuing from the cylinder walls and in the boundary layer on the cylinder itself. As in the case of mixing layers, the Bénard–Kármán vortex street persists in the fully turbulent regime at very high Reynolds numbers but it becomes irregular and displays small scale structures superimposed on large counter-rotating vortices (Roshko, 1961).

2.3 Plane channel flow as a prototype of a viscous instability

The instabilities in mixing layers and wakes of sections 2.1 and 2.2 arise mainly through inviscid processes on a typical time scale* $\tau_i = L/V$. By contrast, the primary instability affecting plane Poiseuille flow takes place on a time scale much longer than $\tau_i = L/V$ (Nishioka *et al.*, 1975) where V is chosen as the center line velocity $U_{\max}$ and L as the channel half-depth (Fig. 2.1d). This flow is chosen here to illustrate the *viscous*

---

* Quantity V is typically the velocity difference ΔU in mixing layers or the uniform upstream velocity U_∞ in wakes and L characterizes the initial vorticity thickness $\delta_\omega(0)$ or the cylinder diameter d.

instability process typically arising in *wall bounded shear flows.* For a thorough review, please consult Bayly *et al.* (1988), Herbert (1988).

Plane Poiseuille flow is generated by maintaining a pressure gradient between the two extremities of a channel. This strictly parallel shear flow occurring between two close parallel plates is an exact solution of the Navier–Stokes equations and it is particularly amenable to a rigorous mathematical treatment, as opposed to spatially developing wall bounded flows such as the Blasius boundary layer (Fig. 2.1a). From an experimental point of view, the situation is paradoxically reversed: boundary layers are easily obtainable while a fully developed plane Poiseuille velocity profile is difficult to achieve. Indeed most of the phenomenology discussed below was first reported in the case of flat-plate boundary layers (Schubauer and Skramstad, 1947; Klebanoff *et al.*, 1962; Saric *et al.*, 1984). Plane Poiseuille flow is governed by a single control parameter, namely the Reynolds number Re $= VL/\nu$. As in the case of free shear layers, it is however essential to specify the nature of incoming disturbances which are observed to drastically affect the transition process.

A thorough experimental investigation of the onset of instability has been performed by Nishioka *et al.* (1975). A fully developed two-dimensional parabolic velocity profile is generated by pushing air through

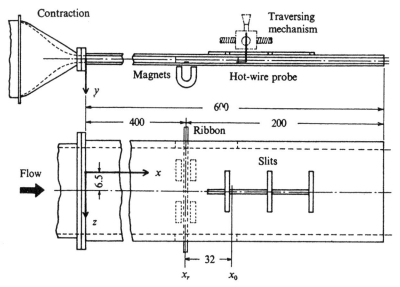

Fig. 2.19.   Experimental set-up for plane channel flow.   Dimensions in centimeters. After Nishioka *et al.* (1975).

a sufficiently long channel as illustrated in the experimental set-up of Fig. 2.19. As in many similar experiments, artificial perturbations are introduced at the entrance of the test section by means of a magnetically-driven vibrating ribbon. Two-dimensional traveling waves, or so-called *Tollmien-Schlichting waves* (Tollmien, 1929; Schlichting, 1933) are then observed in response to a specific frequency and amplitude input with a well-defined phase reference. The background turbulence level is extremely low of the order of 0.05 %. The facility is in fact so quiet that, with forcing turned-off, the flow in the channel remains laminar up to Re = 8,000. Weak natural disturbances are incapable of generating detectable patterns or triggering transition in the test section.

At a fixed Reynolds number above a critical value $\text{Re}_c \simeq 6,000$, only a finite bandwidth of forcing frequencies are observed to be exponentially amplified along the stream while all others are damped or nearly neutral. The typical measured phase velocity varies between $0.2V$ and $0.3V$. Plane Poiseuille flow acts as an *amplifier* of external disturbances in the same manner as free shear layers but the amplification rates are an order of magnitude lower. In the amplified range of forcing frequencies, the cross-stream distribution of streamwise perturbation velocity (Fig. 2.20) is observed to be antisymmetric with respect to the mid-plane of the channel.

Above Re_c, deviations from exponential growth signaling the effects of nonlinearities start to occur when fluctuation amplitudes exceed a few percent of the centerline velocity. Below Re_c, waves of sufficiently small amplitude are exponentially damped for all frequencies but some of them become amplified when their initial amplitude exceeds a threshold value. It is seen in sections 8.2 and 8.3 that this behavior is a manifestation of a *subcritical Hopf bifurcation* whereby nonlinear terms can destabilize a linearly stable flow.

Above Re_c, three-dimensional patterns begin to emerge as spanwise modulations of the wave fronts when two-dimensional waves reach characteristic amplitudes of the order of 1%. The typical scale of spanwise nonuniformities is of the same order of magnitude as their streamwise wavelength. Forcing experiments have been conducted by Thomas and Saric (1981), Kozlov and Ramazanov (1984) to study and control the three-dimensional evolution of Tollmien–Schlichting waves. Three-dimensional perturbations are applied by adding a spanwise row of equally spaced strips of tape at the walls to fix the spanwise wavelength. Two distinct classes of patterns have been documented as a function of forcing ribbon amplitude (Nishioka *et al.*, 1975; Nishioka *et al.*, 1980; Kozlov and Ramazanov, 1984; Ramazanov, 1985).

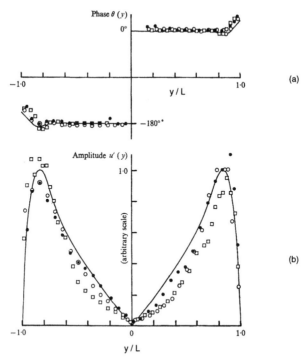

Fig. 2.20. Measured phase (a) and amplitude (b) cross-stream distribution of streamwise perturbation velocities in plane Poiseuille flow. Continuous curves are Itoh's corresponding predictions (1974) for spatially evolving Tollmien–Schlichting waves at $\omega = 0.27$ and Re $= 4,000$ (see section 8.1). Experimental conditions of Nishioka *et al.* (1975): open circles: Re $= 3,000$, $\omega = 0.36$; solid circles: Re $= 4,000$, $\omega = 0.27$; squares: Re $= 6,000$, $\omega = 0.32$.

At high forcing levels (Kozlov and Ramazanov, 1984), one observes a doubly periodic array of Λ-*shaped vortices* (Fig. 2.21) with a streamwise periodicity equal to the initial two-dimensional Tollmien–Schlichting wavelength. This transition route is known as *K-type breakdown*, after the landmark boundary-layer experiments of Klebanoff *et al.* (1962). It is characterized by the appearance of peaks and valleys in the spanwise distribution of rms streamwise velocity fluctuations commonly referred to as "peak-valley splitting".

At low forcing levels (Saric *et al.*, 1984; Ramazanov, 1985), a staggered array of Λ-shaped vortices (Fig. 2.22) is detected, whose streamwise periodicity is equal to *twice* the initial Tollmien–Schlichting wavelength (Fig. 2.22), as confirmed by the existence of a *subharmonic* component in the hotwire signals (Thomas and Saric, 1981). Such a transition route

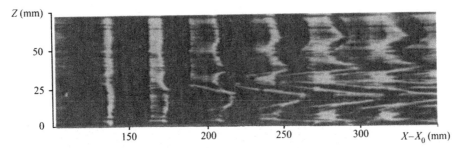

Fig. 2.21. Aligned pattern of Λ-shaped vortices observed in plane channel flow as seen in a plane parallel to the wall. After Kozlov and Ramazanov (1984).

is called *H-type breakdown* after Herbert (1983) who first described it theoretically.

Both these three-dimensional patterns develop over approximately five Tollmien–Schlichting wavelengths, at which stage they experience a catastrophic *breakdown* extending over only one wavelength, characterized, among other features, by the appearance of small scale structures, local high shear layers in the instantaneous velocity profile and spikes in the velocity signal (Fig. 2.23).

Note that forced experiments fail to unveil all possible natural transition scenarios. In the presence of strong localized forcing, *turbulent spots* or Emmons spots are generated which may gradually contaminate the entire flow (Carlson *et al.*, 1982). This so-called *by-pass transition* may occur for Reynolds numbers as low as 1,000. Below this value laminar flow always prevails.

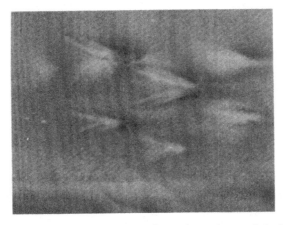

Fig. 2.22. Staggered pattern of Λ-shaped vortices observed in boundary layers as seen in a plane parallel to the wall. After Herbert (1988).

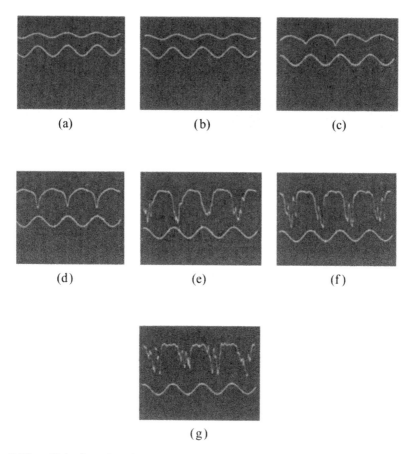

Fig. 2.23. Velocity signals at different stations (a–g) in plane channel flow illustrating the breakdown process. In each plot, upper trace shows u-fluctuations, lower trace the forcing signal. After Nishioka *et al.* (1975).

3 Fundamental concepts

In the previous section, *open flows* were classified according to their various possible dynamical behaviors. Some flows exhibit dynamics analogous to closed systems: the observed patterns are virtually unaffected by the nature of upstream noise. By contrast, others are extremely sensitive to outside perturbations and primarily act as *noise amplifiers*. For instance, the shedding frequency of wakes (section 2.2) is virtually unaffected by low amplitude noise while the downstream evolution of shear layers (section 2.1) crucially depends on the manner in which acoustic waves are converted into vorticity perturbations at the trailing edge of the splitter plate. Stability theory should explain why it is essential to distinguish between open and closed flows and how to determine the influence of an imposed controlled or random forcing on the unsteady evolution of open flows. This section precisely provides such a theoretical description. Rigorous definitions are introduced and instabilities are classified according to their *linear* or *nonlinear*, *spatial* or *temporal*, *primary* or *secondary*, *convective* or *absolute* character (Chandrasekhar, 1961; Drazin and Reid, 1981; Bers, 1983; Landau and Lifshitz, 1987a; Huerre and Monkewitz, 1990; Monkewitz, 1990).

Formal stability definitions are presented in section 3.1. Section 3.2 introduces general linear instability concepts (dispersion relation, impulse response, signaling problem, absolute/convective instability, etc.) in the framework of one-dimensional evolution equations. Nonlinear notions (finite amplitude states, secondary instabilities, supercritical or subcritical bifurcations) are presented in section 3.3.

3.1 Some formal definitions

To start with, one should specify all the field variables $\mathbf{U}_0(\mathbf{x}, t)$ such as velocity, pressure, temperature, necessary to characterize a particular flow called the *basic state*, where $\mathbf{x}$ denotes the position vector and t time. This configuration is the reference flow around which the stability analysis is performed and it satisfies the governing equations of the problem, for instance the Navier–Stokes or Euler equations coupled with appropriate boundary conditions. This definition is rather general since any solution could be chosen as basic state. In practice, one selects a configuration effectively observed in experiments when imposed constraints such as pressure gradients or temperature differences vary within some range. For open flows, examples of such basic states are Poiseuille flow or the Blasius boundary layer. One should not *a priori* associate a particular symmetry to these configurations: basic flows are not bound to be steady or/and parallel. In most theoretical studies, this turns out to be the case: for technical reasons, it is far easier to study fields that are invariant with respect to time translations (*steady flows*) or space translations in one direction (*parallel flows*). Thus, in sections 4 and 7, the instability analysis is restricted to strictly steady and parallel solutions of the Euler or Navier–Stokes equations. This setting, however, constitutes the starting point for a systematic description of slowly spatially developing flows as described in section 5.2 for free shear layers and 6.2 for wakes. Strongly nonparallel basic flows are considered in sections 5.3 and 8.4–5.

It is then natural to ask what is the time evolution of disturbances initially superimposed on the selected basic state. A basic configuration $\mathbf{U}_0(\mathbf{x}, t)$ is said to be *stable in the sense of Lyapunov* when the following holds:

For any ε positive, there exists a positive number $\delta(\varepsilon)$ such that

$$\begin{aligned} &\text{if } \|\mathbf{U}(\mathbf{x},0) - \mathbf{U}_0(\mathbf{x},0)\| \leq \delta(\varepsilon), \\ &\text{then for any } t \geq 0 \quad \|\mathbf{U}(\mathbf{x},t) - \mathbf{U}_0(\mathbf{x},t)\| \leq \varepsilon, \end{aligned} \tag{3.1}$$

where the norm $\|\ldots\|$ within the physical domain V may be defined as

$$\|\mathbf{U}(\mathbf{x},t) - \mathbf{U}_0(\mathbf{x},t)\| \equiv \max\left\{|\mathbf{U}(\mathbf{x},t) - \mathbf{U}_0(\mathbf{x},t)|, \mathbf{x} \in V\right\}. \tag{3.2}$$

In other words, when a system is stable in the sense of Lyapunov, it remains close to the basic state $\mathbf{U}_0(\mathbf{x},t)$ in its latter evolution for sufficiently small initial perturbations. How close it does remain to the basic state depends on the size of the initial perturbations. A basic flow that is unstable in the sense of Lyapunov is not experimentally observable since perturbations drastically alter the flow evolution.

If a system which is stable in the sense of Lyapunov asymptotically returns to the basic state, it is said to be *asymptotically stable*. This statement takes the following mathematical form:

There exists $\delta \geq 0$ such that

$$\begin{aligned} &\lim_{t \to \infty} \|\mathbf{U}(\mathbf{x},t) - \mathbf{U}_0(\mathbf{x},t)\| = 0 \\ &\text{when} \quad \|\mathbf{U}(\mathbf{x},0) - \mathbf{U}_0(\mathbf{x},0)\| \leq \delta. \end{aligned} \tag{3.3}$$

If δ may be chosen arbitrarily, the flow is attracted back to $\mathbf{U}_0(\mathbf{x},t)$ for t large enough, whatever the size of the initial perturbation amplitude. The flow $\mathbf{U}_0(\mathbf{x},t)$ is then said to be *globally stable* in phase space.*

A basic state may be stable in the sense of Lyapunov without being globally stable in phase space. In such instances, a *threshold amplitude* δ_s for the fluctuations may be defined. When the initial perturbation amplitude is less than δ_s, the system converges back to $\mathbf{U}_0(\mathbf{x},t)$. Conversely if the amplitude is greater than δ_s, some disturbances do grow in time and take the system away from $\mathbf{U}_0(\mathbf{x},t)$. Generally the system is brought into a new configuration which can be considered as a new basic state (section 3.3). This threshold behavior shows the central role of finite amplitude effects on the global behavior of the system.

In the context of *linear stability theory*, perturbations are restricted to be infinitesimal. Experimentally, such an approach is pertinent when the noise level is sufficiently low or the forcing is weak. The precise meaning of the statement "sufficiently low" depends on the experiment under consideration. Analytically, this is achieved by linearizing around the basic state $\mathbf{U}_0(\mathbf{x},t)$ to arrive at a new set of equations for the *linear stability problem*. Several powerful results support the consistency of

---

* Global stability in phase space should be distinguished from global stability in physical space as discussed in section 6.2

such an approach: if a system is linearly stable, it is stable in the sense of Lyapunov. However, a linearly stable basic state $U_0(x, t)$ may not be globally stable in phase space since nonlinear terms may destabilize it for sufficiently large amplitudes. Mathematically, one can show through the Hartman–Grobman theorem (Guckenheimer and Holmes, 1983) that, away from bifurcation points, *the linearization procedure faithfully approximates phase space dynamics in the vicinity of the basic state*. The linear approximation is therefore legitimate.

Throughout this review, key concepts are illustrated on the one-dimensional *Ginzburg–Landau* (GL) *equation*

$$\frac{\partial \psi}{\partial t} + U \frac{\partial \psi}{\partial x} = \mu \psi + \frac{\partial^2 \psi}{\partial x^2} - |\psi|^2 \psi, \tag{3.4}$$

where $\psi(x, t)$ is a complex scalar field, μ and U are two real parameters. This amplitude equation which arises in the context of non-equilibrium systems, often describes the dynamics near the onset of linear instability (Newell and Whitehead, 1969; Cross and Hohenberg, 1993). Here, this model is viewed as a "toy" problem replacing the dimensionless *Navier–Stokes equations*

$$\frac{\partial \mathbf{u}}{\partial t} + (\mathbf{U}_0 \cdot \nabla)\mathbf{u} + (\mathbf{u} \cdot \nabla)\mathbf{U}_0 = -\nabla p + \frac{1}{\mathrm{Re}} \nabla^2 \mathbf{u} - (\mathbf{u} \cdot \nabla)\mathbf{u}, \tag{3.5}$$

$$\nabla \cdot \mathbf{u} = 0, \tag{3.6}$$

governing the evolution of disturbances $\mathbf{u}(x, t) \equiv \mathbf{U}(x, t) - \mathbf{U}_0(x, t)$ around $\mathbf{U}_0(x, t)$. In the GL model, the obvious solution $\psi_0(x, t) = 0$ stands for the basic state: it is the analogue of the parabolic velocity profile in the Poiseuille experiment. Similarly, μ is a control parameter analogous to the Reynolds number in the Navier–Stokes equations. Finally, one observes the presence of diffusive and nonlinear terms on the right-hand-side of both equations (3.4) and (3.5).

3.2 Linear instability concepts

In this section, all nonlinear contributions

$$|\psi|^2 \psi \quad \text{or} \quad (\mathbf{u} \cdot \nabla)\mathbf{u} \tag{3.7}$$

are neglected in (3.4) or (3.5). A linear equation results:

$$\frac{\partial \mathbf{u}}{\partial t} - \mathbf{L}\left(\nabla; \mathbf{U}_0(x, t); R\right) \mathbf{u} = 0, \tag{3.8}$$

where the linear operator $\mathbf{L}$ depends on the basic state $\mathbf{U}_0(x, t)$ and R denotes a set of control parameters such as the Reynolds number Re in (3.5). In the case of the GL equation, $L\psi \equiv -U \partial \psi / \partial x + \mu \psi + \partial^2 \psi / \partial x^2$ and R coincides with the set of real parameters μ and U.

Generally the properties of the linear operator $\mathbf{L}$ mirror the symmetries of the governing equations and basic state: when $\mathbf{U}_0(x, y, z, t)$ is steady (parallel), $\mathbf{L}$ is homogeneous in time t (space variable x). Homogeneity with respect to time and space coordinates allows us to make extensive use of Fourier–Laplace transforms to effectively reduce the stability problem (3.8) to an ordinary differential or algebraic equation. Most of our present knowledge resides in situations where such transforms may efficiently be applied, i.e. parallel or/and steady flows. Nothing comparable to Fourier analysis exists to handle the full nonlinear problem which stands in front of us as a formidable task to overcome!

3.2.1　Elementary definitions

In order not to focus on technical issues, the remainder of this section is restricted to a linear one-dimensional analogue of (3.8) that is homogeneous in space and time

$$D\left(-i\partial/\partial x, i\partial/\partial t; R\right)\psi(x, t) = 0. \tag{3.9}$$

For the GL equation, $D \equiv \partial/\partial t - L$. The connection with two- and three-dimensional problems will become clear as real flow problems are examined: similar manipulations can be performed even though the analysis is more involved and generally necessitates numerical computations.

If the basic state is perturbed by applying a prescribed forcing function $S(x, t)$, the perturbation field $\psi(x, t)$ is not freely evolving as in (3.9) but satisfies the forced partial differential equation

$$D(-i\partial/\partial x, i\partial/\partial t; R)\psi(x, t) = S(x, t). \tag{3.10}$$

Forcing $S(x, t)$ initiated at $t = 0$ cannot produce any effect prior to its application. This well-known *causality principle* imposes that the associated physical linear response $\psi(x, t)$ remain zero for $t < 0$:

$$\psi(x, t) = 0 \quad S(x, t) = 0 \quad \text{for} \quad t < 0. \tag{3.11}$$

In the subsequent analysis, three distinct source terms are considered:

$$\text{(a) } S(x, t) = F(x)\delta(t), \qquad \text{(b) } S(x, t) = \delta(x)\delta(t),$$
$$\text{(c) } S(x, t) = \delta(x)H(t)\exp(-i\omega_f t),$$

where δ and H respectively denote the Dirac and Heaviside functions. In case (a), an impulsive "force" $F(x)$ is applied in the entire physical domain of interest at time $t = 0$: this is generally equivalent to solving equation (3.9) as an *initial value problem*.[*] In case (b), the field $\psi(x, t)$ is identical to the *Green function* $G(x, t)$ associated with the linear operator

---

[*] When $D \equiv \partial/\partial t - L$, as in the GL operator (3.4), it amounts to specifying $\psi(x, 0) = F(x)$.

D. Experimentalists will recognize the *impulse response problem*: the basic state is perturbed at a specific space-time location $x = 0$, $t = 0$ and the system is subsequently left to evolve freely. The Green function contains all the information regarding the evolution of linearized disturbances since the response $\psi(x, t)$ to other forcing functions $S(x, t)$ may be expressed as the convolution of $S(x, t)$ and $G(x, t)$. Finally, linear evolution under the source function (c) pertains to the *signaling problem* whereby $\psi(x, t)$ denotes the response to a periodic forcing switched on at $t = 0$ and applied at $x = 0$. Typically, most open flows are experimentally investigated in this configuration: recall for instance that, for mixing layers (section 2.1), a mechanically or acoustically generated sinusoidal signal is frequently imposed to perturb the flow at the trailing edge of the splitter plate.

Three types of responses $\psi(x, t)$ to a localized initial state $\psi(x, 0)$ may occur as depicted in Fig. 3.1. First, the amplitude may asymptotically decay in time in the entire domain (Fig. 3.1a) so that the associated Green function satisfies

$$\lim_{t \to \infty} G(x, t) = 0 \quad \text{along all rays} \quad x/t = \text{const.} \tag{3.12}$$

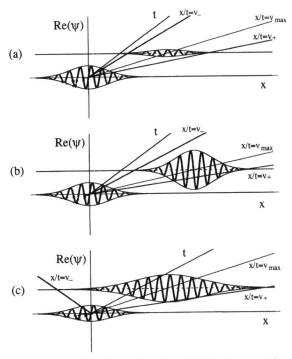

Fig. 3.1. Linear response $\psi(x, t)$ to a localized initial state $\psi(x, 0)$ of (a) stable flow, (b) convectively unstable flow, (c) absolutely unstable flow.

The basic state under consideration is then said to be *linearly stable*. It is called *linearly unstable* otherwise (Fig. 3.1b, c), i.e. when

$$\lim_{t\to\infty} G(x,t) = \infty \quad \text{along at least one ray} \quad x/t = \text{const.} \qquad (3.13)$$

In the latter case, one may further distinguish between two sub-classes of linearly unstable flows. The basic state is *linearly convectively unstable* (Fig. 3.1b) when its impulse response is ultimately advected away from the source location so that:

$$\lim_{t\to\infty} G(x,t) = 0 \quad \text{along the ray} \quad x/t = 0. \qquad (3.14)$$

It is referred to as *linearly absolutely unstable* (Fig. 3.1c) when the impulse response is amplified at the source and gradually contaminates the entire medium:

$$\lim_{t\to\infty} G(x,t) = \infty \quad \text{along the ray} \quad x/t = 0. \qquad (3.15)$$

In other words, in linearly convectively unstable flows, superimposed amplified disturbances are convected away so that the response ultimately decreases at any fixed spatial location. Conversely, for a linearly absolutely unstable flow, some perturbations increase in amplitude at any fixed spatial location. At first sight, this distinction appears to crucially depend on the selected frame of reference. *It is precisely in situations where Galilean invariance is broken that these notions acquire physical significance.* This happens to be the case, either when the flow is continuously forced at a specific spatial location, or when it is weakly spatially developing, or else when no-slip boundary conditions are enforced at the walls as in plane Poiseuille flow. In all such instances, the pertinent reference frame is unambiguously defined.

The objective of this section is to establish criteria capable of discriminating between stable, convectively unstable and absolutely unstable flows in the context of evolution equation (3.10). In a first heuristic approach, particular solutions of (3.9) are sought in the form of *normal modes*

$$\psi(x,t) = A \exp[i(kx - \omega t)]. \qquad (3.16)$$

Such solutions are possible since the linear operator is homogeneous in space and time. The introduction of expression (3.16) into (3.9) results in an algebraic equation for k and ω called the *dispersion relation*. It can readily be obtained by replacing $\partial/\partial t$ by $-i\omega$ and $\partial/\partial x$ by ik in (3.9) and setting it equal to zero to yield

$$D(k, \omega; R) = 0. \qquad (3.17)$$

For the GL equation (3.4), the dispersion relation reads

$$D(k, \omega; \mu, U) \equiv -i\omega + iUk - (\mu - k^2) = 0 .\qquad (3.18)$$

Let us now discuss the nature of the various zeroes of $D(k, \omega; R)$. When the wavenumber k is given real, complex solutions $\omega_j(k; R) \equiv \omega_{j,r}(k; R) + i\omega_{j,i}(k; R)$ of (3.17), where the index j may take several integer values, are called *temporal modes* or *branches*. In the specific case of (3.18), a single temporal mode $\omega(k; U, \mu)$ is found:

$$\omega(k; U, \mu) = Uk + i(\mu - k^2) .\qquad (3.19)$$

In more involved problems such as plane Poiseuille flow, there exists an infinite but discrete number of temporal modes for a single real k. According to (3.16), these freely evolving waves are spatially periodic disturbances of infinite spatial extent which travel with a *phase velocity* $c_{j,r} \equiv \omega_{j,r}(k; R)/k$ and grow or decrease in amplitude with a *temporal growth rate* $\omega_{j,i}(k; R)$. If there exists a particular real wavenumber k_a and index j for which $\omega_{j,i}(k_a; R) > 0$, the basic flow is *linearly unstable* since at least one perturbation is amplified. In the specific case of the GL equation, $\omega_i(k; U, \mu) = \mu - k^2$, and linear instability prevails as soon as $\mu > 0$ (Fig. 3.2).

Conversely, it is shown below that, when all temporal modes are damped, the flow is *linearly stable*. In the case of the GL equation,

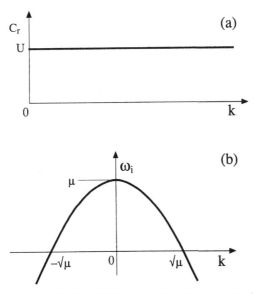

Fig. 3.2. Temporal mode (3.19) of GL equation for $\mu \geq 0$: (a) phase velocity $c_r \equiv \omega_r/k$ versus real wavenumber k; (b) growth rate ω_i versus k. Unstable flow.

linear stability therefore holds when $\mu < 0$ since, in that instance, $\omega_i(k; U, \mu) < 0$ for all k (Fig. 3.3). At this point, it is natural to introduce, in the space spanned by the real wavenumber k and the various control parameters R, the *neutral surface* where the quantity max $\{\omega_{j,i}(k; R)$ for all $j\}$ becomes zero. This surface separates two regions corresponding to linearly damped and amplified perturbations respectively. If only a single control parameter R is allowed to vary, the neutral surface reduces to a *neutral curve* in the k–R plane (Fig. 3.4). The lowest value of R on this curve defines a critical value R_c below which the basic state is linearly stable. The associated wavenumber k_c defines the spatial periodicity of the structure that appears in an infinitely extended system as R crosses R_c. When R is slightly above R_c, a finite bandwidth of wavenumbers around k_c is amplified and all other wavenumbers are damped. In the specific case of the GL equation (3.4), the neutral curve is the parabola defined by $\omega_i(k) \equiv \mu - k^2 = 0$ (Fig. 3.4) and consequently $\mu_c = k_c = 0$.

Zeroes of $D(k, \omega; R)$ may also be sought in terms of the complex wavenumber k when the frequency ω is assumed to be given real. Such solutions $k_j(\omega; R) \equiv k_{j,r}(\omega; R) + ik_{j,i}(\omega; R)$ are called *spatial modes* or *branches*. According to equation (3.16), they represent time-periodic sinusoidal disturbances with a spatially evolving amplitude of *spatial*

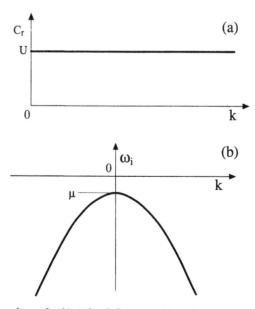

Fig. 3.3. Temporal mode (3.19) of GL equation for $\mu \leq 0$: (a) phase velocity $c_r \equiv \omega_r/k$ versus real wavenumber k; (b) growth rate ω_i versus k. Stable flow.

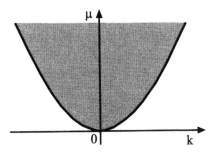

Fig. 3.4. Neutral curve $\omega_i(k) \equiv \mu - k^2 = 0$ of GL equation in $k-\mu$ space. $k_c = 0$, $\mu_c = 0$. Shaded area corresponds to instability.

growth rate $-k_{j,i}(\omega; R)$. In the specific case of the GL equation (3.4), two spatial branches are obtained (Figs. 3.5, 3.6, and later 3.10, 3.11):

$$k^{\pm}(\omega; \mu, U) = \frac{1}{2}\left[-iU \pm \sqrt{4(\mu + i\omega) - U^2}\right]. \tag{3.20}$$

3.2.2 Analysis in complex wavenumber and frequency planes

In order to determine the relevance of temporal modes $\omega_j(k; R)$ and spatial modes $k_j(\omega; R)$, let us now examine how they emerge from the general solution of (3.10) with an arbitrary source function $S(x, t)$. For conciseness, explicit dependence on the set of control parameters R is

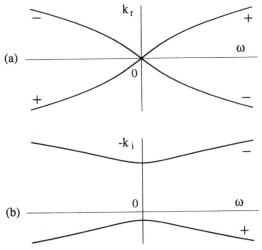

Fig. 3.5. Spatial branches (3.20) of GL equation for $\mu \leq 0$: (a) $k_r^{\pm}(\omega; \mu, U)$ versus real frequency ω; (b) spatial growth rate $-k_i^{\pm}(\omega; \mu, U)$ versus ω.

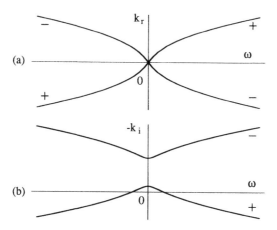

Fig. 3.6.   Spatial branches (3.20) of GL equation for $0 \leq \mu \leq U^2/4$: (a) $k_r^\pm(\omega; \mu, U)$ versus real frequency ω; (b) spatial growth rate $-k_i^\pm(\omega; \mu, U)$ versus ω.

temporarily omitted. The perturbation field $\psi(x, t)$ is expressed as the Fourier superposition

$$\psi(x, t) = \frac{1}{(2\pi)^2} \int_{L_\omega} \int_{F_k} \psi(k, \omega) \exp[i(kx - \omega t)] dk \, d\omega, \qquad (3.21)$$

where the integrations are performed along the path F_k (L_ω) in the complex k-plane (ω-plane), as sketched in Fig. 3.7.

Such contours cannot be chosen arbitrarily since both the convergence of (3.21) and the *causality condition*, $\psi(x, t) = 0$ for $t \leq 0$, should be ensured. A proper choice of contours is a rather delicate matter and the reader is urged to generalize the procedure to more physically relevant situations. The solution $\psi(x, t)$ is assumed to be well-behaved when $x \to \pm\infty$: typically $\psi(x, t)$ is required to decay at least exponentially fast so that the Fourier transform

$$\psi(k, t) = \int_{-\infty}^{+\infty} \psi(x, t) \exp(-ikx) dx \qquad (3.22)$$

is properly defined in a strip of the complex k-plane including the real axis (hatched band in Fig. 3.7b). The contour F_k is then chosen to lie in this particular domain to ensure convergence. To justify the choice of contour L_ω, the growth rates of all possible perturbations for $t \geq 0$ are assumed bounded from above, i.e. there exists a least upper bound γ such that

$$\|\psi(x, t)\| \leq K_0 \exp(\gamma t), \qquad (3.23)$$

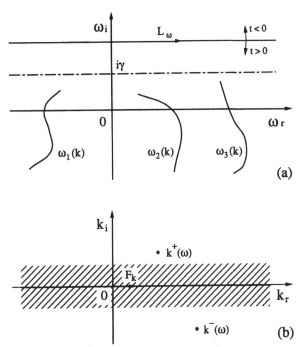

Fig. 3.7. Integration contours (a) in complex ω-plane, (b) in complex k-plane. Curves in (a) represent loci of temporal modes $\omega_j(k)$ as k travels along F_k.

where K_0 is a constant which depends on the particular perturbation considered. Note that γ must be larger or equal to the growth rates of all temporal modes $\omega_j(k)$. When the frequency ω is located above the line $\omega = i\gamma$ in the upper half complex ω-plane (Fig. 3.7a), the integral

$$\psi(x,\omega) = \int_{-\infty}^{+\infty} \psi(x,t)\exp(i\omega t)dt \qquad (3.24)$$

is well-defined at $t = +\infty$ since the integrand decays exponentially for large time. For $t = -\infty$, no convergence problem arises in (3.24) since causality asserts that $\psi(x,t) = 0$ for $t < 0$. Conversely, when $\omega_i < \gamma$, the same integral may diverge for certain classes of perturbations. In particular, if the basic state is unstable, the Fourier transform (3.24) does not exist for real values of ω since the integrand blows up for infinite time. In the following, the particular pair* $C_0 = (F_k, L_\omega)$ in (3.21) is chosen so that the contour F_k lies on the real k-axis and L_ω on a horizontal line in the half plane $\omega_i \geq \gamma$ (Fig. 3.7). *The contour L_ω therefore lies above the loci of all temporal branches $\omega_j(k)$ as k travels along the real k-axis.*

---

* Note that this choice of contour pair is not the only one possible as seen below.

Once these convergence requirements are met, the linear partial differential equation (3.10) reduces, in Fourier space, to the algebraic equation

$$D(k,\omega)\psi(k,\omega) = S(k,\omega),\qquad(3.25)$$

where derivatives with respect to time (space) have been replaced by $-i\omega$ (ik) and $S(k,\omega)$ is the Fourier transform of the source term $S(x,t)$. For an impulsive forcing (case b), $S(k,\omega) = 1$. For more elaborate source functions, it is assumed that convergence problems arising from $S(x,t)$ have been taken care of through a proper choice of contour. In spectral space, one therefore obtains:

$$\psi(k,\omega) = \frac{S(k,\omega)}{D(k,\omega)}.\qquad(3.26)$$

Reverting to physical space *via* (3.21), the linearized perturbation field reads:

$$\psi(x,t) = \frac{1}{(2\pi)^2}\int_{L_\omega}\int_{F_k}\frac{S(k,\omega)}{D(k,\omega)}\exp[i(kx - \omega t)]dk\,d\omega,\qquad(3.27)$$

which formally solves the problem. Unfortunately not much insight is gained from expression (3.27) as it stands. To go any further, one makes the reasonable assumption that the function $D(k,\omega)$ is analytic with respect to k and ω. This is by no means always satisfied and should be checked on a case-by-case basis. For the GL equation, it is easily verified simply by inspection of the polynomial expression (3.18) for $D(k,\omega;\mu,U)$. Branch points are sometimes encountered in realistic dispersion relations but this occurrence is excluded here for simplicity. Returning to the general one-dimensional case, the only possible singularities of the integrand in (3.27) are poles of $\psi(k,\omega)$, i.e. solutions of the dispersion relation $D(k,\omega) = 0$. For the choice of contour pair C_0, F_k lies on the real axis and the only possible singularities of the integrand then coincide with the *temporal modes* $\omega_j(k)$ introduced in the heuristic approach. For the same C_0, it was previously shown that the contour L_ω necessarily lies above the loci of $\omega_j(k)$ as k travels along F_k. This condition is actually sufficient to comply with causality.

Let us now use standard complex variable techniques to evaluate the inverse Fourier transform with respect to ω

$$\psi(k,t) = \frac{1}{2\pi}\int_{L_\omega}\frac{S(k,\omega)}{D(k,\omega)}\exp(-i\omega t)\,d\omega.\qquad(3.28)$$

When $t < 0$, the L_ω contour can be closed at infinity by an upper semi-circle (Fig. 3.7a) since the integrand decays sufficiently fast for the contribution on this additional path to vanish for large radii (Carrier *et al.*, 1966). The L_ω contour being located above all temporal branches, there are no residue contributions and one obtains:

$$\psi(k,t) = 0, \quad t < 0, \tag{3.29}$$

as required by causality. When $t > 0$, the L_ω contour can be closed at infinity by a lower semi-circle (Fig. 3.7a) since the contribution of this additional path vanishes for large radii. If poles are assumed to be all simple then, by the residue theorem (Carrier *et al.*, 1966):

$$\psi(k,t) = -i \sum_j \frac{S(k,\omega_j(k)) \exp(-i\omega_j(k)t)}{\partial D/\partial\omega(k,\omega_j(k))}. \tag{3.30}$$

The explicit solution of the problem thus reads

$$\psi(x,t) = -\frac{i}{2\pi} \sum_j \int_{-\infty}^{+\infty} \frac{S(k,\omega_j(k)) \exp(i(kx - \omega_j(k)t))}{\partial D/\partial\omega(k,\omega_j(k))} \, dk. \tag{3.31}$$

One recognizes in expression (3.31) a wave packet composed of freely evolving temporal modes generated by the source function $S(x,t)$. Note that according to (3.31), the least upper bound γ is indeed given by the maximum temporal growth rate

$$\gamma = \omega_{i,\max} \equiv \max\{\omega_{j,i}(k), \text{ for all real } k \text{ and indices } j\}.$$

The temporal evolution of $\psi(x,t)$, as characterized by equation (3.31), readily provides the following *stability/instability criterion*:

(1) *If $\omega_{i,\max} < 0$, the basic state is linearly stable.* Indeed, if all temporal waves possess a negative growth rate, the integrand in (3.31) decreases exponentially: the basic state is asymptotically stable.

(2) *If $\omega_{i,\max} > 0$, the basic state is unstable.* Indeed when, for some bandwidth of real wavenumbers k, a temporal branch $\omega_j(k)$ lies above the real axis in the complex ω-plane, the integral (3.31) for $\psi(x,t)$ blows up. In the case of the GL equation, recall that instability takes place as μ becomes positive (compare Figs. 3.2 and 3.3).

(3) *If $\omega_{i,\max} = 0$, the basic state is neutrally stable.* In that case, a nonlinear study should be performed to determine the ultimate evolution of infinitesimal perturbations.

Temporal and spatial branches have been defined for real values of k and ω, respectively, i.e. for contours F_k and L_ω lying on the real axis in their respective complex plane. *Generalized temporal (spatial) modes* can also be introduced as solutions of the dispersion relation $D(k,\omega) = 0$ when the contour F_k (L_ω) differs from the real axis. Although a straightforward physical interpretation is missing, these branches are important objects that naturally appear whenever the initial contour pair C_0 is continuously deformed. As a result of analyticity, the contours L_ω and F_k can be legitimately deformed, without violating causality, provided that one avoids crossing the corresponding generalized temporal (spatial) modes. It is essential to mention an important property of the generalized modes. Suppose that, as in the case of C_0, L_ω stands above $\gamma = \omega_{i,\max}$ in the complex ω-plane. Then generalized spatial modes can be partitioned into two disconnected sets $K^+(\omega)$ and $K^-(\omega)$ located respectively above and below the k_r-axis in the complex k-plane (Fig. 3.8a). Spatial modes belonging to these two sets will respectively be denoted $k_j^+(\omega)$ and $k_j^-(\omega)$. Indeed any crossing of the real axis by the generalized spatial branches would imply the existence of a real wavenumber k such that $\omega_j(k)$ belongs to L_ω. This is clearly impossible since L_ω has been chosen to lie above all temporal branches $\omega_j(k)$.

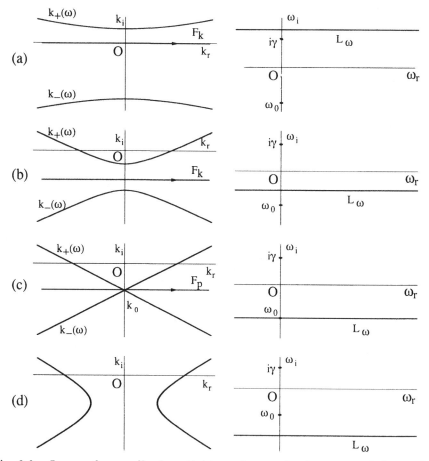

Fig. 3.8.   Locus of generalized spatial branches as the L_ω contour is lowered in the complex ω-plane. Case of the GL equation, $0 \leq \mu \leq U^2/4$. See text for details.

Let us now examine if the partition of generalized spatial branches into two disconnected sets $K^+(\omega)$ and $K^-(\omega)$ is preserved *as the L_ω contour is gradually displaced downward from its C_0 location*. This process is illustrated on Figs. 3.8 and 3.9 for the particular case of the GL equation (3.4) at two parameter settings in the range $0 \leq \mu \leq U^2/4$ and $\mu \geq U^2/4$ respectively. We emphasize that the same scenario is encountered in a wide variety of physical situations.

The initial C_0 configuration is represented on Figs. 3.8a and 3.9a together with the loci K^+ and K^- of the spatial branches. If the partition is preserved (Figs. 3.8b and 3.9b), it is always possible to deform the F_k contour so that it separates the two disconnected sets $K^+(\omega)$ and $K^-(\omega)$. Causality therefore remains enforced and integration of $\psi(k, \omega)$ along the new L_ω and F_k contours yields the same results as integration on the

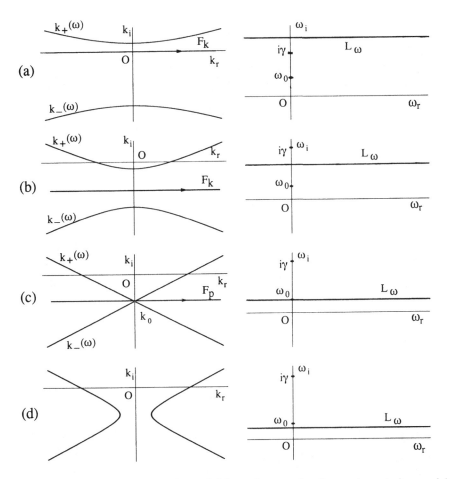

Fig. 3.9.   Locus of generalized spatial branches as the L_ω contour is lowered in the complex ω-plane. Case of the GL equation, $\mu \geq U^2/4$. See text for details.

initial pair C_0. If, in this deformation process, the two sets $K^+(\omega)$ and $K^-(\omega)$ happen to connect at a particular point k_0 in the complex k-plane, *pinching* of the F_k contour by two generalized spatial branches $k_j^+(\omega)$ and $k_l^-(\omega)$ necessarily takes place (Figs. 3.8c, 3.9c). In this instance, any further lowering of L_ω is illegal as any additional deformation of F_k will cross a generalized spatial branch and hence violate causality (Figs. 3.8d, 3.9d).

Let ω_0 denote the value $\omega(k_0)$. Two sub-cases may then be distinguished according to the sign of the elevation $\omega_{0,i}$ of L_ω at pinching. If $\omega_{0,i} < 0$ (Fig. 3.8), spatial branches associated with L_ω lying on the real axis, i.e. real values of ω (Fig. 3.6) do have physical meaning (see the discussion of the signaling problem below). *A contrario*, if $\omega_{0,i} > 0$ (Fig. 3.9), these same spatial branches (Fig. 3.10) cannot be continuously connected to their original C_0 brethren, and the signaling problem will be shown to be ill-defined. In the case of the GL model, k_0 and ω_0 may be explicitly determined by rewriting the generalized spatial branches (3.20) in the form

$$k^\pm(\omega) = k_0 \pm e^{i\pi/4}(\omega - \omega_0)^{1/2},\tag{3.32}$$

where $k_0 = -iU/2$, $\omega_0 = i(\mu - (U^2/4))$, and the branch cut of the square root is chosen to be along the negative imaginary axis in the complex ω-plane. We then note that both generalized spatial branches k^+ and k^- coincide at $\omega = \omega_0$, $k = k_0$, i.e. at the previously identified *pinching point*.

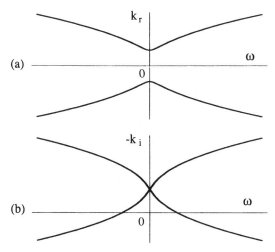

Fig. 3.10. "Unphysical" spatial branches (3.20) of GL equation for $\mu \geq U^2/4$: (a) $k_r(\omega)$ versus real frequency ω; (b) spatial growth rate $-k_i(\omega)$ versus ω.

The two cases $\omega_{0,i} < 0$ and $\omega_{0,i} > 0$ then correspond to the parameter ranges $\mu < U^2/4$ and $\mu > U^2/4$ which were selected in the sketches of Figs. 3.8 and 3.9 respectively. Note that at $\mu = U^2/4$, $\omega_{0,i} = 0$ and the two spatial branches (3.32) reduce to $k^{\pm}(\omega) = -i(U/2) \pm e^{i\pi/4}\omega^{1/2}$. Thus $k^+(\omega)$ and $k^-(\omega)$ display a cusp at $k_0 = -iU/2$ and $\omega_0 = 0$ (Fig. 3.11). This feature may be exploited to track the parameter values for which spatial branches become "unphysical": *the appearance of a cusp in the spatial branches $k^+(\omega)$ and $k^-(\omega)$ at a particular parameter setting serves as a warning signal that the partition into two disconnected sets $K^+(\omega)$ and $K^-(\omega)$ is on the verge of being violated.*

3.2.3 Asymptotic impulse response

This pinching process may be related to the absolute/convective nature of the instability by examining the asymptotic behavior of the impulse response $G(x,t)$ for which, by definition, $S(x,t) = \delta(x)\delta(t)$ and $S(k,\omega) = 1$. According to equation (3.31), $G(x,t)$ reads:

$$G(x,t) = -\frac{i}{2\pi}\int_{F_k} \frac{\exp\left[i(kx - \omega(k)t)\right]}{\partial D/\partial\omega(k,\omega(k))}dk, \qquad (3.33)$$

where, for simplicity, only one generalized temporal branch is considered. Note that the contour F_k does not necessarily coincide with the k_r-axis but may be chosen as discussed previously.

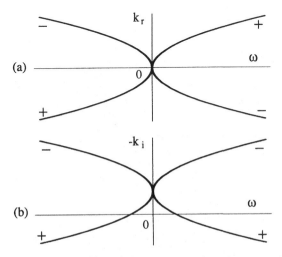

Fig. 3.11. Spatial branches (3.20) of GL equation for $\mu = U^2/4$: (a) $k_r^{\pm}(\omega)$ versus real frequency ω; (b) spatial growth rate $-k_i^{\pm}(\omega)$ versus ω.

Following definitions (3.14–3.15), Green's function $G(x, t)$ is now evaluated at a fixed spatial location x as $t \to \infty$. Let F_k in (3.33) be one of the pinched contours F_p sketched in Figs. 3.8c, 3.9c, 3.12. By definition, the *pinching point* k_0, where two generalized spatial branches meet, is a double zero of the dispersion relation $D(k, \omega) = 0$ at $\omega = \omega_0$, which implies:

$$D(k_0, \omega_0) = 0 \quad \text{and} \quad \frac{\partial D}{\partial k}(k_0, \omega_0) = 0, \tag{3.34}$$

equivalently*

$$\omega_0 = \omega(k_0) \quad \text{and} \quad \frac{\partial \omega}{\partial k}(k_0) = 0. \tag{3.35}$$

Thus k_0 is also a *stationary point* of the function $\omega(k)$. By construction the temporal mode $\omega(k)$ lies below L_ω, defined by $\omega = i\omega_{0,i}$, as k moves along F_p (Fig. 3.12b). It touches L_ω at a single point ω_0. The point k_0 is hence a *global maximum* of the function $\omega_i(k)$ as k travels along F_p. This feature may be exploited to evaluate $G(x, t)$ for large time by an *ad hoc steepest descent method* (Bender and Orszag, 1978).

In this framework, it is important to notice that the integral (3.33) is of the general form

$$G(x, t) = -\frac{i}{2\pi} \int_{F_k} f(k) \exp\left[-i\omega(k)t\right] dk, \tag{3.36}$$

where

$$f(k) \equiv \frac{\exp(ikx)}{\frac{\partial D}{\partial \omega}(k, \omega(k))}. \tag{3.37}$$

The presence of a fast varying complex exponential $\exp[-i\omega(k)t]$ may be used to advantage in order to derive a leading order approximation of $G(x, t)$ as $t \to \infty$, x fixed: in this limit, the magnitude of the integrand in (3.36) is controlled by the exponential factor $\exp[\omega_i(k)t]$, i.e. by the "height" of the surface $\omega_i(k_r, k_i)$ in k_r–k_i–ω_i space (Fig. 3.13). Let us determine the shape of this surface in the vicinity of the stationary point k_0. Around this location, the complex function $\omega(k)$ may be approximated by its Taylor expansion

$$\omega(k) \sim \omega_0 + \frac{1}{2}\frac{\partial^2 \omega}{\partial k^2}(k_0)(k - k_0)^2. \tag{3.38}$$

---

* If one assumes that $\partial D / \partial \omega(k_0, \omega_0) \neq 0$!

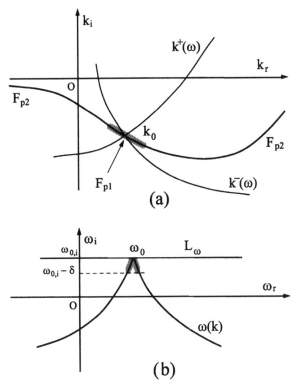

Fig. 3.12. Pinched contour F_p in the complex k-plane (a) and associated locus of temporal modes in the complex ω-plane (b).

It is convenient to set

$$-i\frac{\partial^2\omega}{\partial k^2}(k_0) \equiv a\exp(i\alpha) \quad \text{and} \quad k - k_0 \equiv r\exp(i\theta). \tag{3.39}$$

The Taylor expansion (3.38) then becomes:

$$-i\omega(k) \sim -i\omega_0 + \frac{1}{2}ar^2\left[\cos(\alpha + 2\theta) + i\sin(\alpha + 2\theta)\right]. \tag{3.40}$$

According to (3.40), the so-called lines of constant phase $\omega_r(k) = \omega_{0,r}$ going through the stationary point k_0 are locally given by $\sin(\alpha + 2\theta) = 0$. They are associated with the two mutually orthogonal directions $\theta = -\frac{\alpha}{2} + n\pi$ and $\theta = -\frac{\alpha}{2} + (2n+1)\pi/2$. Along the former direction, $\omega_i(k) \sim \omega_{0,i} + \frac{1}{2}ar^2$, which corresponds to the *steepest ascent path* emerging from k_0. Along the latter, $\omega_i(k) \sim \omega_{0,i} - \frac{1}{2}ar^2$, which corresponds to the *steepest descent path* emerging from k_0. The local shape of the surface $\omega_i(k_r, k_i)$ around k_0 is therefore a saddle as sketched in Fig. 3.13 and k_0 is henceforth referred to as a *saddle point*.

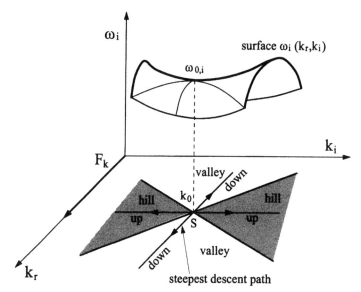

Fig. 3.13　Local topology of surface $\omega_i(k_r, k_i)$ around saddle point k_0.

Let us select as a particular pinched contour in (3.36) the *steepest descent path* of the surface $\omega_i(k_r, k_i)$ emerging from the saddle point k_0. This is always possible since we previously proved F_p to be a descent path [discussion following (3.35)]. In order to isolate the dominant contribution of the saddle point k_0, the steepest descent contour F_p is divided into a segment F_{p1} close to the saddle point k_0 so that $\omega(k)$ satisfies

$$[\omega_{0,i} - \omega_i(k)] \le \delta \,, \tag{3.41}$$

where δ is a small positive number, and the complementary set F_{p2} (Fig. 3.12). On F_{p1}, the function $-i\omega(k)$ may be approximated by its Taylor expansion $-i\omega(k) \sim -i\omega_0 - \frac{1}{2}ar^2$ and the remaining factor $f(k)$ in the integrand of (3.36) may be evaluated at $k = k_0$. Bearing in mind that $k - k_0 \sim ir \exp(-i\alpha/2)$, the contribution of F_{p1} to $G(x,t)$ then reads

$$G_{F_{p1}}(x,t) \sim \frac{1}{2\pi} \frac{\exp[i(k_0 x - \omega_0 t)]}{\frac{\partial D}{\partial \omega}(k_0, \omega_0)} \int_{F_{p1}} \exp\left[-\frac{1}{2}ar^2 t - i\alpha/2\right] dr \,. \tag{3.42}$$

When k spans the complementary set F_{p2}, the generalized temporal branch $\omega(k)$ is, by definition, such that $[\omega_{0,i} - \omega_i(k)] > \delta$: for large time, the F_{p1} contribution (3.42) is therefore dominant compared to the F_{p2} contribution and $G(x,t) \sim G_{F_{p1}}(x,t)$.

In classical steepest descent fashion, the integration contour F_{p1} in (3.42) is now extended to $r = \pm\infty$ since the added contributions remain subdominant compared to (3.42). The Gaussian integral in (3.42) is readily evaluated by introducing the change of variable $s = r\sqrt{at/2}$. Upon making use of the identity

$$\int_{-\infty}^{+\infty} \exp(-s^2)\,ds = \sqrt{\pi}\,, \tag{3.43}$$

and returning to primitive variables via (3.39), the *asymptotic impulse response along the ray* $x/t = 0$ takes the following compact form:

$$G(x,t) \sim \frac{1}{\sqrt{2\pi}} \frac{\exp\left[\frac{i\pi}{4} + i(k_0 x - \omega_0 t)\right]}{\frac{\partial D}{\partial \omega}(k_0, \omega_0)} \left[t \frac{\partial^2 \omega}{\partial k^2}(k_0)\right]^{-1/2}. \tag{3.44}$$

In the laboratory frame, i.e. along the ray $x/t = 0$, the impulse response is dominated by the so-called complex *absolute wavenumber* k_0 and complex *absolute frequency* ω_0 of zero group velocity as defined in (3.34–35). The quantity $\omega_{0,i}$, commonly referred to as the *absolute growth rate*, characterizes the asymptotic growth of disturbances in the laboratory frame. The temporal evolution of $G(x,t)$ in the laboratory frame, as characterized by (3.44), readily provides an *absolute/convective instability criterion*:

(1) *If the basic state is unstable* ($\omega_{i,\max} > 0$) *and* $\omega_{0,i} < 0$, the system goes back to the rest state at any fixed point in the laboratory frame. This apparently paradoxical result is obtained when the packet of unstable waves is washed downstream while increasing in amplitude (Fig. 3.1b). At any fixed station, perturbations grow initially and, as the tail of the wave packet passes by, they ultimately decrease exponentially: according to definition (3.14), the instability is *convective*.* It is shown in section 3.2.4 that spatial modes are then pertinent to describe the response of the flow to external perturbations.

(2) *If the basic state is unstable* ($\omega_{i,\max} > 0$) *and* $\omega_{0,i} > 0$, perturbations exponentially increase in time at any fixed station in the laboratory frame: the instability is then said to be *absolute* (Fig. 3.1c).

Let us now exploit the previous result to determine the nature of the asymptotic impulse response along an arbitrary fixed spatio-temporal ray $x/t = v$, as $t \to \infty$. This is equivalent to analyzing how the impulse response evolves in a reference frame moving at the velocity v. The Green function (3.27) with $S(k, \omega) = 1$ may be expressed in the advected coordinates $x' = x - vt$, $t' = t$ as

$$G(x', t') = \frac{1}{(2\pi)^2} \iint_{C_0} \frac{\exp\left[i(kx' - (\omega - kv)t')\right]}{D(k, \omega)}\,dk\,d\omega\,. \tag{3.45}$$

---

* This term has nothing to do with convection induced by temperature gradients as in Rayleigh-Bénard convection!

Upon introducing the Doppler-shifted frequency $\omega' = \omega - kv$ and $k' = k$, the above integral reads

$$G(x', t') = \frac{1}{(2\pi)^2} \iint_{C_0} \frac{\exp\left[i(k'x' - \omega't')\right]}{D'(k', \omega')} dk' \, d\omega', \qquad (3.46)$$

where the integration contour pair C_0 remains unchanged and D' is related to the original dispersion relation D by

$$D'(k', \omega') = D(k', \omega' + k'v). \qquad (3.47)$$

The asymptotic response at a fixed station in the moving frame may then be directly deduced by applying the previous result. Referring to (3.35), pinching now takes place at (k'_0, ω'_0) such that

$$\omega'_0 = \omega'(k'_0) \quad \text{and} \quad \frac{\partial \omega'}{\partial k'}(k'_0) = 0. \qquad (3.48)$$

In terms of unprimed variables, pinching takes place at $k_* = k'_0$ and $\omega_* = \omega'_0 + k'_0 v$ such that

$$\omega_* = \omega(k_*) \quad \text{and} \quad \frac{\partial \omega}{\partial k}(k_*) = v. \qquad (3.49)$$

Similarly, by applying (3.44) to (3.46) and returning to unprimed variables, one directly obtains the *asymptotic impulse response as* $t \to \infty$ *along the ray* $x/t = v$:

$$G(x, t) \sim \frac{1}{\sqrt{2\pi}} \frac{\exp\left[\frac{i\pi}{4} + i(k_* x - \omega_* t)\right]}{\partial D/\partial \omega(k_*, \omega_*)} \left[t \frac{\partial^2 \omega}{\partial k^2}(k_*)\right]^{-1/2}. \qquad (3.50)$$

According to the above formula, an observer moving at the velocity $x/t = v$ perceives a temporal growth rate equal to $\sigma \equiv (\omega_{*,i} - k_{*,i}v)$, which is by construction less than the maximum temporal growth rate $\omega_{i,\max}$. When $\sigma < 0$ ($\sigma > 0$), perturbations decay (increase) exponentially in time along the ray $x/t = v$. Note that equation (3.49) constitutes an extension of the *group velocity* concept to unstable waves of *complex frequency and wavenumber. Although* $\partial \omega/\partial k$ *is generally complex, it acquires physical significance only for complex wavenumbers* k_* *such that it is real.* In the case of a *single* wave packet, three particular ray velocities may in general be singled out (Fig. 3.14):

1) The real group velocity $v_{\max} = \partial \omega/\partial k(k_{\max})$ at the real wavenumber $k_{\max}$ of highest temporal growth rate $\omega_{i,\max}$. For an observer moving along the ray $x/t = v_{\max}$, the wave packet amplitude is maximum.
2) Two front velocities v_- and v_+ ($v_- < v_{\max} < v_+$) such that $\sigma = 0$. These velocities delineate two moving fronts in the x–t plane within which the wave packet amplitude increases exponentially in time. In

the simplest case of the GL equation, only two such velocities exist. Two distinct behaviors are then possible for the impulse response:

a) Whenever $v_- < 0 < v_+$ (Fig. 3.14a), the system is *absolutely unstable* since the absolute growth rate $\omega_{0,i}$ observed along the ray $v_- < x/t = 0 < v_+$ is necessarily positive.

b) In the opposite case where the front velocities are of the same sign (Fig. 3.14b), e.g. $0 < v_- < v_+$, the system is *convectively unstable*. At a fixed station $x/t = 0$, an observer first perceives a growing wave packet as the first front of velocity v_+ passes by. Once the second front of velocity v_- has reached the same location, the system returns to the rest state with the asymptotic decay rate $\omega_{0,i} < 0$, as expected in a convectively unstable medium.

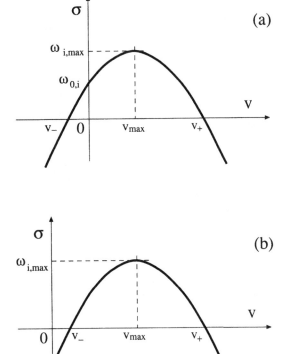

Fig. 3.14.   Temporal growth rate $\sigma \equiv (\omega_{*,i} - k_{*,i}v)$ as a function of observer velocity $x/t = v$ for the GL equation where $v_{max} = U$, $k_{max} = 0$; (a) absolute instability, (b) convective instability.

3.2.4 Signaling problem

Although the impulse response contains all the information required to describe the linear evolution of disturbances, it is worth considering the *signaling problem* associated with a localized harmonic forcing

$$S(x,t) = \delta(x)H(t)\exp(-i\omega_f t). \tag{3.51}$$

The response $\psi(k,\omega)$ in spectral space is readily deduced from (3.26). Upon calculating the Fourier–Laplace transform $S(k,\omega)$ of (3.51), one finds

$$\psi(k,\omega) = \frac{i}{D(k,\omega)(\omega - \omega_f)}. \tag{3.52}$$

Note that $\psi(k,\omega)$ becomes singular not only at zeroes of $D(k,\omega)$ but also at $\omega = \omega_f$. Upon applying the residue theorem in the complex ω-plane (Fig. 3.15a) to evaluate the inverse Fourier transform $\psi(k,t)$ of $\psi(k,\omega)$, one easily obtains:

$$\psi(k,t) = \frac{\exp(-i\omega_f t)}{D(k,\omega_f)} + \frac{\exp(-i\omega(k)t)}{[\omega(k) - \omega_f]\,\partial D/\partial\omega\,[k,\omega(k)]}. \tag{3.53}$$

The inverse Fourier transform $\psi(x,t)$ of $\psi(k,t)$ defined in (3.21) then reads

$$\psi(x,t) = \frac{\exp(-i\omega_f t)}{2\pi}\int_{F_k} \frac{\exp(ikx)}{D(k,\omega_f)}dk$$
$$+ \frac{1}{2\pi}\int_{F_k} \frac{\exp[i(kx - \omega(k)t)]}{[\omega(k) - \omega_f]\,\partial D/\partial\omega\,[k,\omega(k)]}dk. \tag{3.54}$$

According to (3.53), the first integral arises from the residue contribution at $\omega = \omega_f$ in the complex ω-plane and it corresponds to the steady state signal $\psi_{\text{forcing}}(x,t)$ of frequency ω_f. Its integrand is analytic except at points in the complex k-plane where

$$D(k,\omega_f) = 0. \tag{3.55}$$

Such poles precisely coincide with the aforementioned *spatial branches*. To simplify, assume that only two spatial branches $k^+(\omega)$ and $k^-(\omega)$ exist, as in the GL equation. For $x > 0$ ($x < 0$), the contour F_k may be closed by a semi-circle of infinite radius in the upper (lower) half k-plane (Fig. 3.15b) since the latter contribution then vanishes for large radii. The residue theorem may once more be applied to the first integral in (3.54) in the complex k-plane at $k = k^+(\omega_f)$ ($k = k^-(\omega_f)$). The results are summarized in the following formula:

$$\psi_{\text{forcing}}(x,t) = iH(x)\frac{\exp\left[i(k^+(\omega_f)x - \omega_f t)\right]}{\partial D/\partial k\left[k^+(\omega_f), \omega_f\right]}$$

$$- iH(-x)\frac{\exp\left[i(k^-(\omega_f)x - \omega_f t)\right]}{\partial D/\partial k\left[k^-(\omega_f), \omega_f\right]}. \tag{3.56}$$

The computation of the second integral in equation (3.54) is rigorously similar to that of the impulse response case (3.33) and it leads to asymptotic expressions analogous to (3.44) and (3.50). When the instability is *absolute*, this term is predominant since $\omega_{0,i} > 0$: growing switch-on transients invade the entire flow and the periodic forced response (3.56) is overwhelmed and unobservable. When the instability is *convective*, switch-on transients associated with the second integral in (3.54) are advected away and only the steady state response (3.56) remains for large time. The system is therefore controlled by forcing which determines the nature of the observed patterns: it acts as a *noise amplifier*. More explicitly, according to (3.56), downstream of the forcing location, i.e. for $x > 0$, the k^+ branch is observed whereas, for $x \leq 0$, the

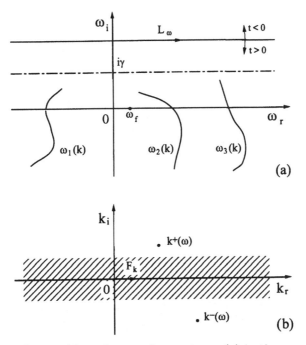

Fig. 3.15. Signaling problem. Integration contours (a) in the complex ω-plane, (b) in the complex k-plane. Curves in (a) represent temporal modes.

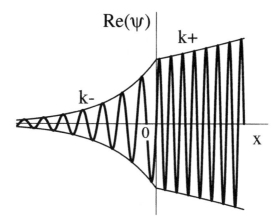

Fig. 3.16. Signaling problem: steady state response in a generic convectively unstable case.

k^- branch shows up (Fig. 3.16). In general, within the range of spatially amplified frequencies, the k^+ branch gives rise to a spatially growing wave downstream of the source and the k^- branch to an attenuated wave upstream.

According to this analysis, it is obvious that the flow behavior as well as its control very much depend on the nature of the instability. *For convective instabilities, spatial branches naturally arise as the solutions of a boundary-value problem for specific forcing amplitudes and frequencies.* For *absolute instabilities*, perturbations grow *in situ* and are unaffected by weak forcing. In that instance, *spatial branches loose their physical meaning* whereas temporal modes are still relevant as solutions of an initial value problem.

3.2.5 Stability criteria

The criteria that determine the nature of the instability are gathered in Table 3.1, in terms of the signs of the maximum temporal growth rate $\omega_{i,\max}$ and *absolute growth rate* $\omega_{0,i}$.

Table 3.1. Stability criteria

$\omega_{i,\max} > 0,$	$\omega_{0,i} > 0$	AU
$\omega_{i,\max} > 0,$	$\omega_{0,i} < 0$	CU
$\omega_{i,\max} < 0,$	$\omega_{0,i} < 0$	S

Let us summarize the various methods that are available to determine k_0, ω_0 as well as the control parameter value for absolute/convective transition.

The first method is purely analytic: one looks for a complex pair (k_0, ω_0) of zero group velocity such that

$$(\partial\omega/\partial k)(k_0; R) = 0, \quad \text{and} \quad \omega_0 = \omega(k_0; R). \tag{3.57}$$

The absolute/convective transition value is then reached when $\omega_{0,i} = 0$ at some parameter value R_t. This criterion as it stands is not entirely satisfactory: spurious values of (k_0, ω_0) may be obtained corresponding to switching between two k^+ or two k^- branches (Huerre, 1987a). *Only the complex wavenumber k_0 of largest growth rate $\omega_{0,i}$ which is a pinching point for two spatial branches k^+ and k^- entirely located, for a high enough L_ω contour, in the upper and lower half complex k-plane needs to be considered.*

The second method is purely geometric and relies on the contour deformation argument sketched in Figs. 3.8, 3.9: the complex pair (k_0, ω_0) coincides with pinching of the generalized spatial branches $k^+(\omega)$ and $k^-(\omega)$ as the contour L_ω is gradually lowered.

Finally, as a third method, any absolute/convective transition characterized by $\omega_{0,i} = 0$ may be directly detected by the appearance of a cusp in the usual spatial branches $k(\omega)$ at a *real* frequency value ω_0, as sketched in Fig. 3.11.

Application of the first method to the dispersion relation (3.19) of the GL equation readily yields

$$\frac{\partial\omega}{\partial k}(k_0; \mu, U) = U - 2ik_0, \tag{3.58}$$

so that

$$k_0 = -iU/2, \quad \omega_0 = i\left[\mu - (U^2/4)\right]. \tag{3.59}$$

Setting $\omega_{0,i} = 0$ leads to the transition value $\mu_t = U^2/4$.

The second method has already been illustrated in Figs. 3.8, 3.9 and the third one is displayed in Fig. 3.11. Note the spatial branch switching evident in Figs. 3.6, 3.10, 3.11 as μ crosses the transition value μ_t from below. The various regimes of the GL equation are summarized on Fig. 3.17 in the space of the control parameters (μ, U).

3.2.6 Transient regime

All the notions previously introduced to define the instability or stability of a given flow, rely on the existence or absence of exponentially growing solutions associated with the temporal $\omega_j(k)$ or spatial $k_j(\omega)$ spectrum.

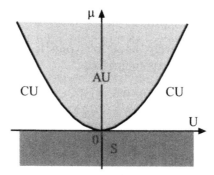

Fig. 3.17. Stability diagram of the GL equation in the U–μ plane. S: stable; CU: convectively unstable; AU: absolutely unstable.

If a basic state is linearly stable in the above sense, can it still sustain transient algebraically growing disturbances? These issues have recently been examined in the context of bounded shear flows, e.g. Poiseuille flow in the linearly stable regime and Couette flow (Reddy *et al.*, 1993). A temporary departure from the expected exponential decay may result from the *non-orthogonality* of the linear eigenfunctions in the sense of a suitably defined scalar product or norm, e.g. the kinetic energy of disturbances. This phenomenon can be understood by examining the influence of initial conditions: assume that two eigenvectors $\mathbf{V}_1$ and $\mathbf{V}_2$ with respective eigenvalues $\sigma_1 = \lambda_1 + i\phi_1$, $\sigma_2 = \lambda_2 + i\phi_2$ are "extremely non-orthogonal," i.e. $(\mathbf{V}_1, \mathbf{V}_2)\,/\,(\|\mathbf{V}_1\|\|\mathbf{V}_2\|) \sim 1$. It is always possible to find initial conditions

$$\mathbf{V}(0) = a(0)\mathbf{V}_1 + b(0)\mathbf{V}_2 + \text{c.c.}\,, \qquad (3.60)$$

where the coefficients $a(0)$ and $b(0)$ are both of order unity but the norm

$$\|\mathbf{V}(t)\|^2 = \|a(t)\|^2\|\mathbf{V}_1\|^2 + \|b(t)\|^2\|\mathbf{V}_2\|^2$$
$$+ a(t)b^*(t)\,(\mathbf{V}_1, \mathbf{V}_2) + a^*(t)b(t)\,(\mathbf{V}_2, \mathbf{V}_1)\,, \qquad (3.61)$$

is nearly zero at $t = 0$ as a result of *destructive* interferences arising from the cross terms $a(0)b^*(0)\,(\mathbf{V}_1, \mathbf{V}_2) + a^*(0)b(0)\,(\mathbf{V}_2, \mathbf{V}_1)$, where $*$ denotes the complex conjugate. In the linearly stable regime, $a(t) = a(0)\exp\left[(\lambda_1 + i\phi_1)t\right]$ and $b(t) = b(0)\exp\left[(\lambda_2 + i\phi_2)t\right]$ are both decaying. However, if the exponential decay is sufficiently weak, the norm itself $\|\mathbf{V}(t)\|$ may experience transient growth due the same cross terms at later times. Indeed $a(t)b^*(t)\,(\mathbf{V}_1, \mathbf{V}_2) + a^*(t)b(t)\,(\mathbf{V}_2, \mathbf{V}_1)$ can become *constructive* in specific ranges of their relative phase $(\phi_2 - \phi_1)\,t$. For plane Poiseuille flow, an "optimum" perturbation is found to reach around forty times its initial amplitude in the stable regime.

3.3 Nonlinear instability concepts

Whenever stable or unstable disturbances reach appreciable amplitudes, the validity of the linearized analysis becomes questionable. From a mathematical standpoint, this stage is reached as soon as nonlinear terms (3.7) significantly modify the temporal evolution predicted by linear theory, e.g. when saturation occurs or higher harmonics are generated. Experimentally, exponential growth rates are no longer observed and finite amplitude effects profoundly alter the primary basic flow and lead to a new topological state that generally possesses different symmetry properties. In this section, notions such as *nonlinear saturation, secondary instability, phase dynamics, supercritical* or *subcritical* behavior are introduced and illustrated in the context of the GL equation. Some of these ideas are later explored and illustrated in the context of mixing layers (section 5.3), wakes (section 6.3), and plane channel flow (section 8.2–5).

We have shown in section 3.2 that the basic state $\psi_0(x,t) = 0$ in the GL equation becomes unstable to a bandwidth of real wavenumbers as soon as $\mu \geq 0$ (Fig. 3.4). In the same parameter range, it can easily be established, by inspection, that a family of *nonlinear Stokes solutions* exists of the form

$$\psi(x,t) = Q \exp\left[ik(x - Ut)\right]$$

$$\text{with} \quad Q = \sqrt{\mu - k^2} \quad \text{and} \quad \mu \geq k^2. \tag{3.62}$$

To each linearly unstable mode within the neutral curve $\mu \geq k^2$ of Fig. 3.4, one may associate a unique *finite amplitude state* (3.62). In order to follow the nonlinear evolution of a monochromatic fluctuation of wavenumber k, space and time variations are decoupled by assuming

$$\psi(x,t) = R(t) \exp\left[ik(x - Ut)\right], \tag{3.63}$$

which reduces the GL equation (3.4) to the so-called *Landau equation*

$$\frac{dR}{dt} = Q^2 R - R^3. \tag{3.64}$$

The solution reads

$$R^2(t) = \frac{Q^2 R^2(0)}{R^2(0) + [Q^2 - R^2(0)]\exp(-2Q^2 t)}. \tag{3.65}$$

For small amplitudes $R/Q \ll 1$, one recovers from equation (3.65) the exponential growth $R(0)\exp(Q^2 t)$ characteristic of the linear regime. In the nonlinear range $R/Q \sim 1$ (Fig. 3.18), the cubic term gradually weakens the growth rate to lead asymptotically to the saturated nonlinear wave (3.62).[*] For a given wavenumber k, *the saturation amplitude Q varies as the square root of the threshold distance $\mu - k^2$* from the neutral point $\mu = k^2$. The system is said to *bifurcate* to the new solution as $\mu > k^2$. The above temporal formulation is valid when the instability of the primary basic state $\psi_0(x, t) = 0$ is absolute, i.e. when $\mu > U^2/4$. If the flow is convectively unstable, $0 < \mu < U^2/4$, this analysis may be replaced by its spatial counterpart: a spatially evolving nonlinear solution must then be connected through a transition region to the spatial modes $k^+(\omega_f)$ and $k^-(\omega_f)$ emerging from the linearized signaling problem examined in section (3.2). This is left to the reader as an exercise.

The saturated Stokes solution (3.62) may itself be regarded as an example of a new basic state $U_0(x, t)$ for the so-called *secondary instability* problem. This study, which generally constitutes a major hurdle, is examined here in the context of *phase dynamics* (Manneville, 1990; Kuramoto, 1984; Fauve in these lectures notes). Such an instability analysis is relevant each time a given solution $U_0(x, t)$ *breaks a continuous symmetry* of the governing equations such as space or time translations. In such an instance, a continuous family of solutions $U(x, t, \theta)$ is generated simply by applying the broken symmetry to the given solution $U_0(x, t)$, where the real parameter θ is a phase discriminating between all these solutions. Consider the specific case of the GL equation: the particular Stokes solution (3.62) spontaneously breaks the continuous space and time invariances $x \rightarrow x + \theta_1$, $t \rightarrow t + \theta_2$. The family of

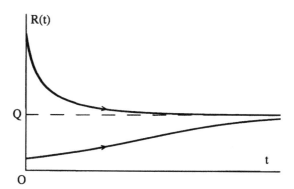

Fig. 3.18. Nonlinear amplitude evolution of $R(t)$ governed by the supercritical Landau equation (3.64) for two typical initial conditions, as given by (3.65) for $\mu > k^2$.

---

[*] For the GL equation, it is possible to follow analytically the growing amplitude of linearly unstable disturbances into the nonlinear regime. This is quite unusual in nonlinear problems and it is due to the form of the nonlinear term in GL which generates no harmonics of the initial single wavenumber k.

solutions $\psi(x,t,\theta) = Q\exp[ik(x-Ut+\theta)]$ is then simply obtained by shifting (3.62) in space $x \to x + \theta_1$ or time $t \to t + \theta_2$.

In the more general setting, solutions $\mathbf{U}(\mathbf{x},t,\theta)$ satisfy the nonlinear governing equations for all *constant* θ. By differentiating the nonlinear equations with respect to θ at θ_0, $(\partial\mathbf{U}/\partial\theta)\,(\mathbf{x},t,\theta_0)$ is readily shown to be a solution of the linear secondary instability problem (3.8) around the basic state $\mathbf{U}_0(\mathbf{x},t) = \mathbf{U}(\mathbf{x},t,\theta_0)$. This perturbation can be viewed as the difference between two neighboring finite amplitude solutions of the form

$$\varepsilon\frac{\partial\mathbf{U}}{\partial\theta}(\mathbf{x},t,\theta_0) \sim \mathbf{U}(\mathbf{x},t,\theta_0+\varepsilon) - \mathbf{U}(\mathbf{x},t,\theta_0). \tag{3.66}$$

If the family of solutions $\mathbf{U}(\mathbf{x},t,\theta)$ are stationary or periodic in time, the disturbance (3.66) cannot grow exponentially in time and it is hence a neutrally stable eigenvector of the secondary instability problem (3.8). The goal of a phase dynamics analysis is precisely to determine whether the finite amplitude state $\mathbf{U}_0(\mathbf{x},t)$ can be destabilized by allowing slow spatio-temporal modulations of the phase θ.

Let us now examine more explicitly the secondary instability problem pertaining to the new basic state specified by the Stokes solution $\psi_0(x,t) = Q\exp[ik(x-Ut)]$ of the GL equation. Fluctuations are taken to be of the form

$$\psi(x,t) = [Q+\rho(x,t)]\exp[ik(x-Ut)+i\theta(x,t)], \tag{3.67}$$

where $\rho(x,t)$ and $\theta(x,t)$ respectively denote the perturbation amplitude and phase. Introduction of this expression into the GL equation (3.4) and linearization around the Stokes solution (3.62) leads to the coupled equations

$$\frac{\partial\rho}{\partial t} = -U\frac{\partial\rho}{\partial x} - 2Q^2\rho - 2kQ\frac{\partial\theta}{\partial x} + \frac{\partial^2\rho}{\partial x^2}, \tag{3.68}$$

$$Q\frac{\partial\theta}{\partial t} = -UQ\frac{\partial\theta}{\partial x} + 2k\frac{\partial\rho}{\partial x} + Q\frac{\partial^2\theta}{\partial x^2}. \tag{3.69}$$

As in the normal mode analysis of section 3.2.1, perturbations are assumed to be of the form* $\theta = A\exp i(\alpha x - \varpi t)$ and $\rho = B\exp i(\alpha x - \varpi t)$. The solution (3.67) may then be interpreted as periodic modulations of the new basic state (3.62) at the two wavenumbers $k + \alpha$ and $k - \alpha$. Upon substituting this ansatz into

---

* Secondary linear instability equations are typically inhomogeneous in x or t. Fortunately, the linear system (3.68–69) remains homogeneous with respect to x and t as a result of the peculiar form of the nonlinear term in the GL equation.

equations (3.68–69), the secondary instability dispersion relation reads:

$$D_s(\alpha, \varpi; \mu, U, k) \equiv (\varpi - \alpha U)^2 + 2i(Q^2 + \alpha^2)(\varpi - \alpha U)$$
$$- 2(\mu - 3k^2)\alpha^2 - \alpha^4 = 0. \tag{3.70}$$

To any wavenumber α, one may associate two temporal branches (Fig. 3.19)

$$\varpi^{\pm}(\alpha) = U\alpha - i(Q^2 + \alpha^2) \pm iQ^2 \left[1 + 2\frac{(1 - D_{\parallel})}{Q^2}\alpha^2\right]^{1/2}. \tag{3.71}$$

The ϖ^- branch admits the low wavenumber expansion

$$\varpi^-(\alpha) \sim U\alpha - i\left[2Q^2 + (1 - D_{\parallel})\alpha^2\right] + \ldots, \tag{3.72}$$

where

$$D_{\parallel} \equiv \frac{\mu - 3k^2}{Q^2}. \tag{3.73}$$

This so-called *amplitude mode* is always damped, even at $\alpha = 0$, and consequently of no significance. In the small wavenumber approximation,

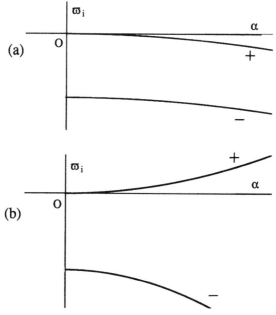

Fig. 3.19. Temporal growth rate ϖ_i of secondary instability modes versus wavenumber α in the GL equation. (a) $D_{\parallel} > 0$, stable regime; (b) $D_{\parallel} < 0$, unstable regime.

the ϖ^+ branch takes the form

$$\varpi^+(\alpha) \sim U\alpha - iD_{\|}\alpha^2 + \dots, \tag{3.74}$$

and the neutrally stable eigenvector defined in (3.66) is therefore recovered at $\alpha = 0$. The behavior of this so-called *phase mode* very much depends on the sign of the diffusion coefficient $D_{\|}$. When $D_{\|} > 0$ (Fig. 3.19a), all perturbations are damped and the Stokes solution (3.62) of wavenumber k is stable. Conversely, when $D_{\|} < 0$ (Fig. 3.19b), the Stokes solution is *modulationally unstable* to a finite bandwidth of wavenumbers α close to $\alpha = 0$ which is typical of the *Eckhaus instability*. According to expression (3.73), a subinterval $k^2 \leq \mu/3$ may be defined for which the corresponding basic state (3.62) is stable with respect to this phase instability (Fig. 3.20).

Within the Eckhaus unstable domain $\mu/3 < k^2 < \mu$, the *convective* or *absolute* nature of the secondary instability may be examined in the same spirit as its primary counterpart (Huerre, 1987b). The pinching points α_0 are given by $(\partial \varpi / \partial \alpha)(\alpha_0) = 0$ and one may calculate the corresponding absolute growth rate $\varpi_{0,i}$ as a function of the basic state parameters μ, U and k. The different regimes pertaining to the primary and secondary instabilities can be summarized as follows [see Huerre (1987b) for details]: when $\mu < U^2/4$ both instabilities are convective. When $\mu > U^2/4$, the primary instability becomes absolute but some secondary Stokes solutions are convectively unstable while others are absolutely unstable. Thus,

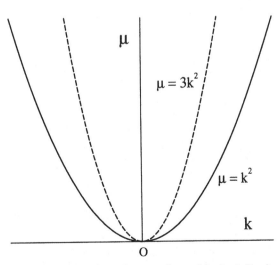

Fig. 3.20. Primary (solid line) and secondary (dashed line) neutral stability curves in k–μ plane.

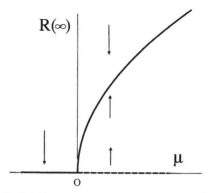

Fig. 3.21.   Supercritical bifurcation diagram in μ–$R(\infty)$ plane for Landau equation (3.64) with stabilizing nonlinear term at wavenumber $k = 0$. Dashed curve indicates unstable solution. Arrows indicate time evolution.

different instabilities coexisting within the same flow need not be of the same nature.

The above analysis has demonstrated that, in the Landau equation (3.64) with stabilizing nonlinear term, the basic state $\psi_0(x, t) = 0$ bifurcates to a family of stable nonlinear solutions $R(\infty) = Q = \sqrt{\mu - k^2}$, as indicated in the *bifurcation diagram* of Fig. 3.21 for $k = 0$. Bifurcations of this type are called *supercritical* (Guckenheimer and Holmes, 1983) since the stable nonlinear bifurcating branch only exists above the linear instability threshold $\mu = 0$.

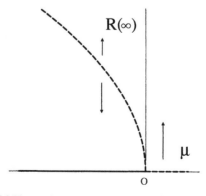

Fig. 3.22.   Subcritical bifurcation diagram in μ–$R(\infty)$ plane for Landau equation (3.77) with destabilizing nonlinear term at wavenumber $k = 0$. Dashed curves indicate unstable solutions. Arrows indicate time evolution.

A distinct behavior is displayed by the modified GL equation

$$\frac{\partial \psi}{\partial t} + U\frac{\partial \psi}{\partial x} = \mu\psi + \frac{\partial^2 \psi}{\partial x^2} + |\psi|^2\psi, \tag{3.75}$$

where the sign of the cubic nonlinearity is now positive. The linear regime in (3.75) is seen to be identical to (3.4). However it is easily checked that, in the linearly stable range $\mu < 0$, a family of nonlinear Stokes solutions (Fig. 3.22) can be obtained in the form

$$\psi(x,t) = Q\exp[ik(x-Ut)] \quad \text{with} \quad Q = \sqrt{-\mu+k^2} \quad \text{and} \quad \mu \leq k^2. \tag{3.76}$$

The nonlinear evolution of an initially monochromatic fluctuation of wavenumber k can be followed by decoupling space and time variations in the same manner as in (3.63). The GL equation (3.75) then reduces to the nonlinear O.D.E.

$$\frac{dR}{dt} = (\mu - k^2)R + R^3, \tag{3.77}$$

the solution of which reads

$$R^2(t) = \frac{Q^2 R^2(0)}{R^2(0) + [Q^2 - R^2(0)]\exp(2Q^2 t)}. \tag{3.78}$$

When the initial amplitude $R(0)$ is less than Q, $Q^2 - R^2(0) > 0$ and the system converges back towards $\psi_0(x,t) = 0$ even though the cubic nonlinear term weakens the rate of exponential decay (Fig. 3.23). Conversely, if the initial amplitude is greater than Q, $Q^2 - R^2(0) < 0$ and fluctuations do grow in time and take the system away from $\psi_0(x,t) = 0$ (Fig. 3.23). In that case, solutions become unbounded in finite time. According to the above discussion, saturated Stokes solutions (3.76) are clearly unstable.

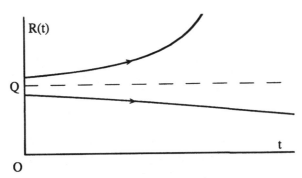

Fig. 3.23. Nonlinear amplitude evolution of $R(t)$ governed by the subcritical Landau equation (3.77) for two typical initial conditions, as given by (3.78) for $\mu < k^2$.

In the Landau equation (3.77) with destabilizing nonlinear term, the basic state $\psi_0(x, t) = 0$ is seen to bifurcate to a family of unstable nonlinear solutions $R(\infty) = Q = \sqrt{-\mu + k^2}$, as indicated in the bifurcation diagram of Fig. 3.22 for $k = 0$. In contrast with the supercritical case (Fig. 3.21), bifurcations of this type are called *subcritical* (Guckenheimer and Holmes, 1983) since the nonlinear bifurcating branch only exists below the linear instability threshold $\mu = 0$. The most important feature of subcritical bifurcations lies in the existence of a *threshold amplitude* $\delta_s = \sqrt{-\mu}$ for $\mu \leq 0$, as introduced in section 3.1. For the subcritical Landau equation (3.77), the basic state $\psi_0(x, t) = 0$ in the linearly stable range, is hence globally unstable in phase space.

4 Inviscid instabilities in parallel flows

The objective of this section is to examine, in a purely *inviscid* context, the general conditions under which *parallel shear flows* may become linearly unstable to infinitesimal disturbances. Classical results are then available as discussed for instance in Drazin and Reid (1981) and Bayly *et al.* (1988). The inviscid framework is adequate to describe the evolution of perturbations in *free shear flows* such as mixing layers, jets and wakes at sufficiently large Reynolds numbers, but it completely fails to account for the primary instability taking place in *bounded shear flows* such as boundary layers, plane Poiseuille flow (section 8) and pipe flow.

Consider the inviscid incompressible flow of a fluid of constant density ρ and let L and V denote the characteristic length and velocity scales of the basic flow under study. Corresponding dynamic pressure scale ρV^2 and inertial time scale $\tau_i = L/V$ may then be defined. With this choice of reference quantities, the evolution of the nondimensional velocity field

$$\mathbf{U}(\mathbf{x}, t) = U(\mathbf{x}, t)\mathbf{e}_x + V(\mathbf{x}, t)\mathbf{e}_y + W(\mathbf{x}, t)\mathbf{e}_z \qquad (4.1)$$

and pressure field $P(\mathbf{x}, t)$ is governed by the *Euler equations*

$$\nabla \cdot \mathbf{U} = 0, \qquad (4.2)$$

$$\frac{\partial \mathbf{U}}{\partial t} + (\mathbf{U} \cdot \nabla)\mathbf{U} = -\nabla P. \qquad (4.3)$$

Note that any unidirectional flow (Fig. 4.1) of the form

$$\mathbf{U}_0(\mathbf{x}, t) = U(y)\mathbf{e}_x, \qquad P_0(\mathbf{x}, t) = P_0 \qquad (4.4a, b)$$

can be chosen as basic state since equations (4.2)–(4.3) are automatically satisfied for arbitrary velocity profiles $U(y)$. The stability of the basic flow is then examined by decomposing total velocity and pressure into basic and perturbation contributions according to

$$\mathbf{U}(\mathbf{x}, t) = U(y)\mathbf{e}_x + \mathbf{u}(\mathbf{x}, t), \qquad P(\mathbf{x}, t) = P_0 + p(\mathbf{x}, t), \qquad (4.5a, b)$$

with

$$\mathbf{u}(\mathbf{x}, t) = u(\mathbf{x}, t)\mathbf{e}_x + v(\mathbf{x}, t)\mathbf{e}_y + w(\mathbf{x}, t)\mathbf{e}_z . \tag{4.6}$$

Upon substituting (4.5a,b) into (4.2), (4.3) and neglecting all contributions quadratic in the fluctuations, one obtains the linearized equations

$$\nabla \cdot \mathbf{u} = 0 , \tag{4.7}$$

$$\left(\frac{\partial}{\partial t} + U(y)\frac{\partial}{\partial x}\right)\mathbf{u} + U'(y)\, v\, \mathbf{e}_x = -\nabla p . \tag{4.8}$$

Since the problem is invariant under arbitrary translations $x \mapsto x+\text{const}$, $z \mapsto z+\text{const}$, $t \mapsto t+\text{const}$, *normal mode* solutions of complex frequency ω and complex wavevector $\mathbf{k} = k_x\mathbf{e}_x + k_z\mathbf{e}_z$ are sought in the form

$$\mathbf{u}(\mathbf{x}, t) = \mathcal{R}e\left\{\hat{\mathbf{u}}(y)\exp[i(k_x x + k_z z - \omega t)]\right\} , \tag{4.9a}$$

$$p(\mathbf{x}, t) = \mathcal{R}e\left\{\hat{p}(y)\exp[i(k_x x + k_z z - \omega t)]\right\} , \tag{4.9b}$$

where $\mathcal{R}e\{\ldots\}$ denotes the real part of $\{\ldots\}$ and the unknown functions $\hat{\mathbf{u}}(y) = \hat{u}(y)\mathbf{e}_x + \hat{v}(y)\mathbf{e}_y + \hat{w}(y)\mathbf{e}_z$, $\hat{p}(y)$ specify the distributions of fluctuations in the cross-stream direction. It is convenient to introduce as an auxiliary quantity the *complex phase velocity*

$$c = \omega/k_x . \tag{4.10}$$

The normal mode decomposition (4.9a,b) then effectively reduces the linearized system (4.7)–(4.8) to the set of ordinary differential equations

$$ik_x\hat{u} + ik_z\hat{w} + \frac{d\hat{v}}{dy} = 0 , \tag{4.11}$$

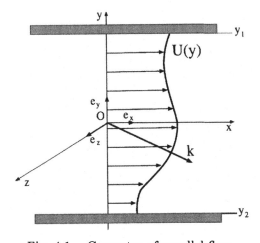

Fig. 4.1. Geometry of parallel flow.

$$ik_x\left[U(y) - c\right]\hat{u} + U'(y)\hat{v} = -ik_x\hat{p}\,, \tag{4.12}$$

$$ik_x\left[U(y) - c\right]\hat{v} = -\frac{d\hat{p}}{dy}\,, \tag{4.13}$$

$$ik_x\left[U(y) - c\right]\hat{w} = -ik_z\hat{p}\,, \tag{4.14}$$

to which may be added, for flows between plane parallel walls (Fig. 4.1), the usual impermeability boundary conditions*

$$\hat{v}(y_1) = \hat{v}(y_2) = 0\,. \tag{4.15a, b}$$

System (4.11)–(4.15) defines an *eigenvalue problem*: nontrivial solutions exist only if the wavevector **k** and frequency ω satisfy a *dispersion relation* which can be formally written

$$D(\mathbf{k}, \omega) = 0\,, \tag{4.16}$$

as in the elementary one-dimensional examples considered in section 3.

4.1 Squire's transformation

The underlying idea is to reduce the three-dimensional linear stability analysis to an equivalent two-dimensional problem through an appropriate change of variables. Following Squire (1933), we introduce tilde variables *via* the transformation

$$\tilde{k}^2 = k_x^2 + k_z^2\,, \qquad \tilde{c} = c\,, \tag{4.17a, b}$$

$$\tilde{k}\tilde{u} = k_x\hat{u} + k_z\hat{w}\,, \qquad \tilde{v} = \hat{v}\,, \qquad \tilde{p}/\tilde{k} = \hat{p}/k_x\,. \tag{4.18a, b, c}$$

The three-dimensional system (4.11)–(4.15) may then be recast, by elementary manipulations, into the equivalent two-dimensional linear stability problem

$$i\tilde{k}\tilde{u} + \frac{d\tilde{v}}{dy} = 0\,, \tag{4.19}$$

$$i\tilde{k}\left[U(y) - \tilde{c}\right]\tilde{u} + U'(y)\tilde{v} = -i\tilde{k}\tilde{p}\,, \tag{4.20}$$

$$i\tilde{k}\left[U(y) - \tilde{c}\right]\tilde{v} = -\frac{d\tilde{p}}{dy}\,, \tag{4.21}$$

$$\tilde{v}(y_1) = \tilde{v}(y_2) = 0\,, \tag{4.22}$$

to which is associated a two-dimensional dispersion relation

$$\tilde{D}(\tilde{k}, \tilde{\omega}) = 0\,, \tag{4.23}$$

---

* These conditions are replaced by exponential decay whenever the boundaries y_1 and y_2 are removed to infinity.

between wavenumber $\tilde{k}$ and frequency $\tilde{\omega} = \tilde{k}\tilde{c}$. Note that, according to (4.17a,b) and (4.10), the frequency $\tilde{\omega}$ can be expressed in terms of **k** and ω as follows :

$$\tilde{\omega} = \tilde{k}\tilde{c} = \tilde{k}(\omega/k_x) = \left(\left(k_x^2 + k_z^2\right)^{1/2} / k_x\right) \omega . \qquad (4.24)$$

Thus, if the two-dimensional dispersion relation (4.23) is known, its three-dimensional counterpart (4.16) can be obtained without any additional calculations as :

$$D(\mathbf{k}, \omega) \equiv \tilde{D}\left[\left(k_x^2 + k_z^2\right)^{1/2} , \left(\left(k_x^2 + k_z^2\right)^{1/2} / k_x\right) \omega\right] = 0 . \qquad (4.25)$$

It can be concluded that the properties of oblique waves $(\mathbf{k}, \omega)$ are readily deduced from those of two-dimensional waves $(\tilde{k}, \tilde{\omega})$ through the *Squire transformation* (4.25). Furthermore, equation (4.24) implies that, to each oblique mode $(\mathbf{k}, \omega)$ of temporal growth rate ω_i is associated a two-dimensional mode $(\tilde{k}, \tilde{\omega})$ of larger temporal growth rate $\tilde{\omega}_i = \sqrt{k_x^2 + k_z^2}\, \omega_i/k_x > \omega_i$. *The wave of maximum growth rate is therefore two-dimensional.* However, this does not imply that all two-dimensional modes are necessarily more amplified than oblique modes. Finally, Squire's transformation (4.25) remains valid when ω and **k** are complex, but it cannot be claimed that the most amplified *spatial* wave is always two-dimensional.

4.2 The two-dimensional stability problem: Rayleigh's equation

We take advantage of the existence of Squire's transformation and restrict, in this section, the presentation to *two-dimensional waves.* Without any possibility of confusion, we make the identifications $k_x \mapsto k, k_z = 0, \hat{w} = 0$ in the linearized system (4.9a,b), (4.10), (4.11)–(4.14). Instead of working with primitive variables, it is more convenient to introduce the streamfunction $\Psi(x, y, t)$ such that the velocity field **U** and vorticity field Ω read:

$$\mathbf{U} = \frac{\partial \Psi}{\partial y}\mathbf{e}_x - \frac{\partial \Psi}{\partial x}\mathbf{e}_y , \qquad \Omega = -\nabla^2 \Psi \mathbf{e}_z . \qquad (4.26a, b)$$

In two-dimensional inviscid flows, the vorticity is conserved along pathlines:

$$\left[\frac{\partial}{\partial t} + (\mathbf{U} \cdot \nabla)\right] \Omega = 0 , \qquad (4.27)$$

which, in terms of Ψ becomes :

$$\left[\frac{\partial}{\partial t} + \frac{\partial \Psi}{\partial y}\frac{\partial}{\partial x} - \frac{\partial \Psi}{\partial x}\frac{\partial}{\partial y}\right] \nabla^2 \Psi = 0 . \qquad (4.28)$$

As previously, the total streamfunction is decomposed into basic and perturbation contributions according to

$$\Psi(x, y, t) = \int U(y)\, dy + \psi(x, y, t)\,. \tag{4.29}$$

For future reference, note that the perturbation velocity field $\mathbf{u} = u\mathbf{e}_x + v\mathbf{e}_y$ is related to ψ in the same way as $\mathbf{U}$ to Ψ in (4.26a). After substitution of (4.29) into (4.28) and linearization around the basic flow $\mathbf{U}_0(y) = U(y)\mathbf{e}_x$, one obtains the equation governing the perturbation streamfunction

$$\left[\frac{\partial}{\partial t} + U(y)\frac{\partial}{\partial x}\right]\nabla^2\psi - U''(y)\frac{\partial\psi}{\partial x} = 0\,. \tag{4.30}$$

In the spirit of section 3.2.1, we seek *normal mode* solutions of the form

$$\psi(x, y, t) = \mathcal{R}e\left\{\phi(y)\exp[i(kx - \omega t)]\right\}, \tag{4.31}$$

where the unknown complex function $\phi(y)$ is related to the perturbation velocity components through the expressions

$$u(x, y, t) = \mathcal{R}e\left\{\phi'(y)\exp[i(kx - \omega t)]\right\}, \tag{4.32a}$$

$$v(x, y, t) = -\mathcal{R}e\left\{ik\phi(y)\exp[i(kx - \omega t)]\right\}. \tag{4.32b}$$

Equation (4.30) then reduces to the *Rayleigh equation*

$$\phi'' - k^2\phi - \frac{U''(y)}{U(y) - c}\phi = 0\,, \tag{4.33}$$

to which may be added the impermeability conditions (4.15a,b), namely

$$\phi(y_1) = \phi(y_2) = 0\,. \tag{4.34}$$

System (4.33)–(4.34) defines an *eigenvalue problem* on the interval $y_1 \le y \le y_2$. In order to assess whether a particular velocity profile $U(y)$ is stable or unstable, it is sufficient to examine the *temporal stability problem*: for a given real wavenumber k, the complex phase velocity c, equivalently the complex frequency $\omega = kc$, and the associated *eigenfunction* $\phi(y)$ are then determined so as to satisfy the Rayleigh equation (4.33) subject to the boundary conditions (4.34). As discussed in section 3.2.1, there may in general be several temporal branches $\omega_j(k)$ which satisfy the dispersion relation $D(k, \omega) = 0$. Furthermore, if c and $\phi(y)$ satisfy (4.33)–(4.34), c^* and $\phi^*(y)$ are also solutions, where a star superscript denotes the complex conjugate. This symmetry property is a consequence of the purely inviscid nature of the present approach: the temporal eigenvalues c are either real or complex conjugate pairs. If they remain real for all k, the basic flow $U(y)$ is neutrally stable. If there exists, for some k, at least one complex conjugate pair (c, c^*) of amplified and decaying waves, the basic flow is unstable. In other words, if one has found a damped wave, the flow is necessarily unstable!

All the notions introduced in section 3 for an arbitrary dispersion relation $D(k, \omega) = 0$ apply. In particular, spatial branches are relevant as solutions of the *signaling problem* only if the basic flow is known to be *convectively unstable*. Furthermore, the determination of the absolute/convective nature of the instability requires a full investigation of the dispersion relation in the complex k- and ω-planes as determined from the Rayleigh eigenvalue problem.

According to standard nomenclature (Bender and Orszag, 1978), the second-order differential equation (4.33) exhibits a *regular singular point* at a pole of the coefficient multiplying $\phi(y)$. In the present context, its location in the complex y-plane is given by

$$U(y_c) = c, \qquad U''(y_c) \neq 0, \qquad \text{(4.35a, b)}$$

and it is commonly referred to as a *critical point* or *critical level*. The behavior of solutions of the Rayleigh equation in the vicinity of y_c is analyzed by resorting to the method of Frobenius (Bender and Orszag, 1978). One independent solution $\phi_1(y)$ is found to be analytic around y_c with the Taylor expansion

$$\phi_1(y) = y - y_c + \frac{U''(y_c)}{2U'(y_c)}(y - y_c)^2 + \cdots, \qquad \text{(4.36a)}$$

while the other displays the following singular structure

$$\phi_2(y) = 1 + \left[\frac{k^2}{2} + \frac{U'''(y_c)}{2U'(y_c)} - \left(\frac{U''(y_c)}{U'(y_c)} \right)^2 \right] (y - y_c)^2 + \cdots$$
$$+ \frac{U''(y_c)}{U'(y_c)} \phi_1(y) \ln(y - y_c). \qquad \text{(4.36b)}$$

The presence of a critical point in the cross-stream distribution of fluctuations is the source of considerable difficulties in many studies of shear flow instabilities both in the linear and nonlinear regimes. As is often the case in fluid dynamics, the presence of a singularity acts as a warning signal that crucial physical effects usually take place in the neighborhood of such points. In the present instance, the logarithmic singularity apparent in (4.36b) may be "smoothed out" by physical phenomena that have been discarded altogether, namely viscous diffusion, nonlinearities, basic flow nonparallelism, transients, etc., or a combination thereof (Maslowe, 1981, 1986; Stewartson, 1981). In these lecture notes, the procedure will be sketched only in the case where viscous diffusion is invoked. It will be shown to lead to a satisfactory interpretation of the singularity appearing in (4.36b). It should be emphasized that the critical point structure makes itself felt in physical space even if it is located well into the complex y-plane and away from the real y-axis. *It cannot simply be argued that critical point dynamics are only relevant when y_c is real.* For neutral modes with c real, one usually distinguishes between *singular neutral modes* for which $U''(y_c) \neq 0$ and *regular neutral modes* for which $U''(y_c) = 0$. In the latter case, y_c coincides with an inflection point of the basic flow and the logarithmic term in (4.36b) disappears so that the inviscid modal structure remains analytic around y_c.

4.3 Rayleigh's inflection point criterion

This classical criterion (Rayleigh, 1880) provides a necessary condition for the inviscid instability of parallel flows $U(y)$. We shall follow Bayly *et al.* (1988) and show that it is a consequence of the conservation of momentum. Temporarily returning to a fully nonlinear three-dimensional framework, the conservative form of Euler's equation (4.3) reads

$$\frac{\partial \mathbf{U}}{\partial t} + \nabla \cdot (\mathbf{U}\,\mathbf{U}) = -\nabla P, \tag{4.37}$$

and its streamwise component can be written as

$$\frac{\partial U}{\partial t} + \frac{\partial U^2}{\partial x} + \frac{\partial UV}{\partial y} + \frac{\partial UW}{\partial z} = -\frac{\partial P}{\partial x}. \tag{4.38}$$

Assume that all flow quantities are spatially periodic of respective wavelength $2\pi/k_x$ and $2\pi/k_z$ in the x and z directions and define the spatial average $\bar{q}(y,t)$ of any quantity $q(\mathbf{x},t)$ as

$$\bar{q}(y,t) \equiv \frac{k_x}{2\pi}\frac{k_z}{2\pi} \int_0^{2\pi/k_x} \int_0^{2\pi/k_z} q(x,y,z,t)\, dx\, dz. \tag{4.39}$$

Upon taking the spatial average of the streamwise momentum equation (4.38) and noting that, by construction, $\overline{\partial q/\partial x} = \overline{\partial q/\partial z} = 0$, one is led to the relation

$$\frac{\partial \bar{U}}{\partial t}(y,t) = \frac{\partial \tau_{xy}}{\partial y}, \tag{4.40}$$

where the *Reynolds stress* component τ_{xy} is defined by

$$\tau_{xy}(y,t) \equiv -\overline{UV}. \tag{4.41}$$

Equation (4.40) is an evolution equation for the streamwise *mean flow* $\bar{U}(y,t)$ under the action of Reynolds stresses generated by the perturbations.

The *mean* flow $\bar{U}(y,t)$ should not, in general, be confused with the *basic* flow $U(y)$. According to (4.40)

$$\frac{d}{dt} \int_{y_1}^{y_2} \bar{U}(y,t)\, dy = \tau_{xy}(y_1,t) - \tau_{xy}(y_2,t) = 0\,, \qquad (4.42)$$

where the second equal sign stems from the fact that, at impermeable walls, $\tau_{xy}(y_1,t) = \tau_{xy}(y_2,t) = 0$. According to equation (4.42), the total mean streamwise momentum is therefore a conserved quantity.

The Reynolds stress τ_{xy} defined in (4.41) can be expressed solely in terms of the perturbation velocity: according to (4.5a), (4.6), $U(\mathbf{x},t) = U(y) + u(\mathbf{x},t)$ and $V(\mathbf{x},t) = v(\mathbf{x},t)$. Furthermore, spatial averaging of the continuity equation (4.7) and enforcement of the boundary conditions $v = 0$ at $y = y_1, y_2$ implies that $\bar{v}(y,t) \equiv 0$. The Reynolds stress (4.41) therefore reduces to

$$\tau_{xy} = -\overline{uv}\,. \qquad (4.43)$$

Returning to the framework of two-dimensional linear theory, $u(x,y,t)$ *and* $v(x,y,t)$ *are related to* ϕ *via* (4.32a,b) *which, when substituted into* (4.43), *leads to the final result*

$$\tau_{xy} = \frac{1}{4}ik\left(\phi\phi'^* - \phi^*\phi'\right)e^{2kc_it}\,. \qquad (4.44)$$

A convenient expression can be obtained for $\partial\tau_{xy}/\partial y$ by differentiating the above formula once. If, in the ensuing relation, the Rayleigh equation is used to express $\phi''(y)$ in terms of $\phi(y)$, one is led to

$$\frac{\partial\tau_{xy}}{\partial y} = \frac{kc_i}{2}\frac{U''(y)|\phi(y)|^2}{[U(y) - c_r]^2 + c_i^2}e^{2kc_it}\,. \qquad (4.45)$$

Integrating (4.45) and using (4.42) therefore leads to the identity

$$\tau_{xy}(y_2,t) - \tau_{xy}(y_1,t) = \frac{kc_i}{2}\int_{y_1}^{y_2}\frac{U''(y)|\phi(y)|^2}{[U(y) - c_r]^2 + c_i^2}\,dy\, e^{2kc_it} = 0\,. \qquad (4.46)$$

The first equality in (4.42) implies that the integral expression in (4.46) is directly proportional to the rate of change of total mean momentum induced by fluctuations. *Identity (4.46) is therefore a statement of mean momentum conservation within the linearized approximation.*

In order for the basic flow $U(y)$ to be unstable, one must necessarily have $c_i \neq 0$, which requires that the integral in (4.46) be zero. This can only occur if $U''(y)$ takes positive and negative values in the interval $y_1 \leq y \leq y_2$ and vanishes at least once, say at some cross-stream station $y = y_s$. *Rayleigh's inflection point criterion* can therefore be stated in the following way : *In order for the basic flow* $U(y)$ *to be unstable, it should have an inflection point say, at* $y = y_s$, *such that* $U''(y_s) = 0$.

This result has had strong implications on the development of the entire subject of hydrodynamic instabilities in shear flows. In effect we can immediately conclude that bounded shear flows without inflection points such as the Blasius boundary layer on a flat plate or plane Poiseuille flow are linearly inviscidly stable. The present linear inviscid formulation cannot be expected to account for the observed unstable evolution of these configurations. Other mechanisms must be invoked : nonlinearities, viscous diffusion, etc.

It should also be noted that for neutral waves ($c_i = 0$), (4.45) and the impermeability conditions at the walls imply that $\tau_{xy} \equiv 0$, $y_1 \leq y \leq y_2$, unless jumps in the cross-stream distribution of Reynolds stresses can be accommodated within the bulk of the flow (see section 7.4).

A physical interpretation of the Rayleigh criterion had in effect been anticipated by Lin (1955) for the case of monotone profiles with no inflection points. The argument relies on the fact that, in two-dimensional inviscid incompressible flow, the vorticity of a fluid element is conserved during the motion [equation (4.27)].

Consider for instance a boundary layer like velocity profile with no inflection point (Fig. 4.2a). The continuous distribution of basic vorticity may conveniently be represented as a discrete collection of vortex filaments. In the undisturbed state, these filaments display the same strength in each plane $y = $ const and increasing strength as the wall is approached (Fig. 4.2b). Assume that two filaments f_1 and f_2 have been interchanged as shown in Fig. 4.2b,c. Filament f_1 now displays a "defect" of vorticity with respect to its neighbors in the same horizontal plane, as indicated by the negative sign and the corresponding sense of rotation around f_1 in Fig. 4.2d. According to Biot and Savart's law, this induces a velocity field (vertical arrows in Fig. 4.2d) which transports weaker vortex filaments from below upstream of f_1 (upwash motion) and stronger vortex filaments from above downstream of f_1 (downwash motion). In turn, these displaced filaments, labeled by $-$ and $+$ signs in Fig. 4.2e, induce on f_1 a downward motion that tends to bring it back to its original position. A similar reasoning can be made to show that the displaced filament f_2 is subjected to an upward velocity which also causes it to return to its initial location. In qualitative agreement with Rayleigh's criterion, we therefore conclude that *boundary layer profiles are inviscidly stable*.

4.4 Fjørtoft's criterion

Whereas Rayleigh's criterion is related to conservation of momentum, Fjørtoft's is associated with conservation of kinetic energy. By suitable rearrangement of the kinetic energy equation derived from the fully

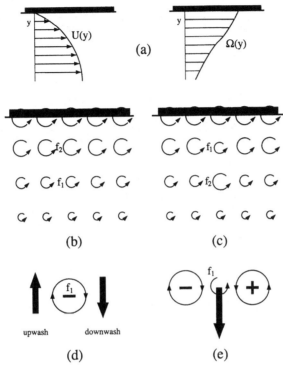

Fig. 4.2. Inviscid stability of boundary layer flows: physical interpretation. The vorticity distribution associated with the boundary layer velocity profile (a) is decomposed into discrete filaments (b). When f_1 and f_2 are interchanged (c), a velocity field (d) is produced around f_1 which alters the vorticity field upstream and downstream of f_1, as indicated by the $+$ and $-$ signs and corresponding oriented circles in (e). This new vorticity field in turn produces a downward velocity on f_1 which tends to bring it back to its original position.

nonlinear Euler's equation (4.37), it is straightforward but tedious to obtain the following exact equation for the perturbation kinetic energy

$$\left[\frac{\partial}{\partial t} + U(y)\frac{\partial}{\partial x}\right]\left(\tfrac{1}{2}\left(u^2 + v^2 + w^2\right)\right)$$
$$+ \nabla \cdot \left[\left(p + \tfrac{1}{2}\left(u^2 + v^2 + w^2\right)\right)\mathbf{u}\right] = -U'(y)uv - U'(y)U(y)v.$$
$$(4.47)$$

Upon taking the spatial average of (4.47) as defined in (4.39), considering that $\overline{\partial q/\partial x} = \overline{\partial q/\partial z} = 0$ for any q and that $\overline{v} \equiv 0$ (see subsection 4.3), one is led to

$$\frac{\partial}{\partial t}\left(\tfrac{1}{2}\left(\overline{u^2} + \overline{v^2} + \overline{w^2}\right)\right) + \frac{\partial}{\partial y}\left[\overline{\left(p + \tfrac{1}{2}\left(u^2 + v^2 + w^2\right)\right)v}\right] = U'(y)\tau_{xy}.$$
$$(4.48)$$

An additional integration in the y direction and application of the boundary conditions $v = 0$ at $y = y_1, y_2$ results in the final form of the equation governing the kinetic energy of the perturbations

$$\frac{d}{dt} \int_{y_1}^{y_2} \tfrac{1}{2} \left(\overline{u^2} + \overline{v^2} + \overline{w^2} \right) dy = \int_{y_1}^{y_2} U'(y) \tau_{xy} \, dy = - \int_{y_1}^{y_2} U(y) \frac{\partial \tau_{xy}}{\partial y} \, dy \, ,$$

$$(4.49)$$

the second identity being obtained by a final integration by parts. This simple balance equation expresses the fact that, in a fully nonlinear context, the perturbation kinetic energy varies in response to the work done by the Reynolds stress τ_{xy} on the basic flow $U(y)$.

In the case of two-dimensional linear theory, the perturbation kinetic energy $(\overline{u^2} + \overline{v^2})/2$ and the Reynolds stress work $-U(y)\partial \tau_{xy}/\partial y$ may readily be evaluated in terms of $\phi(y)$ and its derivative with the help of (4.32a,b) and (4.45). Equation (4.49) then becomes

$$\int_{y_1}^{y_2} \left(|\phi'|^2 + k^2|\phi|^2 \right) dy = - \int_{y_1}^{y_2} \frac{U(y)U''(y)|\phi|^2}{[U(y) - c_r]^2 + c_i^2} \, dy \, .$$

$$(4.50)$$

Assume that the basic velocity profile $U(y)$ satisfies Rayleigh's criterion, namely that it exhibits an inflection point at y_s. Then, from (4.46) we must necessarily have

$$0 = U(y_s) \int_{y_1}^{y_2} \frac{U''(y)|\phi|^2}{[U(y) - c_r]^2 + c_i^2} \, dy \, .$$

$$(4.51)$$

Adding (4.51) to (4.50), one obtains

$$\int_{y_1}^{y_2} \left(|\phi'|^2 + k^2|\phi|^2 \right) dy = - \int_{y_1}^{y_2} \frac{U''(y)[U(y) - U(y_s)]}{[U(y) - c_r]^2 + c_i^2} |\phi|^2 \, dy \, .$$

$$(4.52)$$

Thus the integral appearing on the right-hand side of (4.52) is negative which requires that the quantity $U''(y)[U(y) - U(y_s)]$ be negative in some finite subinterval of $y_1 \leq y \leq y_2$. *Fjørtoft's criterion*, which is a refinement of Rayleigh's, can therefore be stated as follows: *in order for the basic flow $U(y)$ to be unstable, one must necessarily have $U''(y)[U(y) - U(y_s)] < 0$ in some subinterval of the domain of interest $y_1 \leq y \leq y_2$, where y_s is an inflection point such that $U''(y_s) = 0$.*

This result may be expressed in a more palatable way for *monotone velocity profiles with a single inflection point y_s*. In this instance, both factors $U''(y)$ and $[U(y) - U(y_s)]$ may only change sign at $y = y_s$. Thus it is sufficient to examine the sign of these quantities in the vicinity of y_s in order to determine their sign everywhere. In this neighborhood, the function $U''(y)[U(y) - U(y_s)]$ admits the Taylor expansion

$$U''(y)[U(y) - U(y_s)] \sim U'(y_s)U'''(y_s)(y - y_s)^2 + \dots$$

$$(4.53)$$

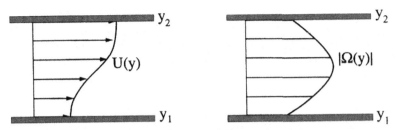

Fig. 4.3. Potentially inviscidly unstable flow according to Fjørtoft's criterion: typical velocity and vorticity profiles.

Fjørtoft's criterion requires that $U'(y_s)U'''(y_s)$ be negative. Since the square of the basic vorticity $\Omega^2(y)$ is approximated by

$$\Omega^2(y) = [U'(y)]^2 \sim \Omega^2(y_s) + U'(y_s)U'''(y_s)(y - y_s)^2 + \dots, \qquad (4.54)$$

this is equivalent to stating that y_s be a maximum of $\Omega^2(y)$. Thus, in the particular case of monotone velocity profiles, a necessary condition for instability is that *the absolute value of the basic vorticity* $|\Omega(y)| \equiv |U'(y)|$ *exhibit a maximum at the inflection point* y_s. This condition is sharper than Rayleigh's criterion: the only potentially inviscidly unstable basic profile corresponds to that sketched in Fig. 4.3. All others (Fig. 4.4a,b,c) are inviscidly stable as a consequence of either Rayleigh's criterion or Fjørtoft's criterion.

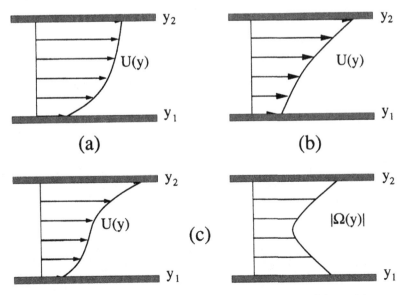

Fig. 4.4. Classes of inviscidly stable velocity profiles. (a) & (b): stable according to Rayleigh's criterion; (c): stable according to Fjørtoft's criterion.

As a word of caution, one should note that the vortex induction mechanism presented in the previous subsection to illustrate Rayleigh's criterion does not apply to Fjørtoft's criterion. One cannot conclude as to the stability or instability of monotone profiles with a vorticity minimum (Fig. 4.4c) or maximum (Fig. 4.3) on the basis of heuristic vorticity dynamics (Orszag and Patera, 1981).

To close this subsection, it is important to mention without proof (see Drazin and Reid, 1981) a general result concerning the location of temporal eigenvalues in the complex c-plane. According to *Howard's semi-circle theorem, amplified or neutral eigenvalues c necessarily lie within a semi-circle centered at $(U_{max} + U_{min})/2$ on the real c axis, of radius $(U_{max} - U_{min})/2$, where U_{max} and U_{min} respectively denote the maximum and minimum values of $U(y)$ in the interval of interest $y_1 \leq y \leq y_2$* (Fig. 4.5). This theorem only holds in the temporal stability problem.

4.5 Jump conditions at an interface. Application to the vortex sheet

Rayleigh's equation (4.33) is only applicable to velocity profiles that are continuous functions of y. It is often convenient, particularly in homework problems (!), to examine, as a first step, the stability of piecewise continuous velocity profiles. In such instances, perturbations on either side of a surface of discontinuity must satisfy the usual continuity of particle displacement and pressure conditions.

If $y = y_0 + \eta(x,t)$ denotes the equation of the interface in the presence of a perturbation $\eta(x,t)$, then, by definition of v in terms of the material

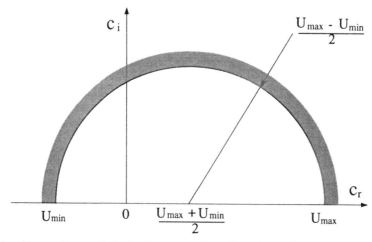

Fig. 4.5. Howard's semi-circle theorem. Complex eigenvalues c with $c_i \geq 0$ lie within the semi-circle.

derivative of η, one has

$$v(x,y,t) = \left[\frac{\partial}{\partial t} + [U(y) + u(x,y,t)]\frac{\partial}{\partial x}\right]\eta \quad \text{on} \quad y = y_0 + \eta(x,t), \quad (4.55)$$

which, in the linearized approximation, reduces to

$$v(x,y_0^{\pm},t) = \left[\frac{\partial}{\partial t} + U(y_0^{\pm})\frac{\partial}{\partial x}\right]\eta, \quad (4.56)$$

on either side of the discontinuity $y = y_0^{\pm}$. Upon assuming that $\eta(x,t) = \mathcal{Re}\left\{\hat{\eta}\,e^{i(kx-\omega t)}\right\}$ and considering the expression (4.32b) for v, equation (4.56) reduces to

$$\phi(y_0^{\pm}) = -\left[U(y_0^{\pm}) - c\right]\hat{\eta}. \quad (4.57)$$

Continuity of particle displacement $\hat{\eta}$ therefore implies that

$$\left[\frac{\phi(y)}{U(y) - c}\right]_{y_0^-}^{y_0^+} = 0, \quad (4.58)$$

where $[\ldots]_{y_0^-}^{y_0^+}$ denotes the jump in the quantity in brackets between y_0^- and y_0^+.

In the same way, one must satisfy the continuity of pressure condition at the interface. Since the basic pressure is uniform [equation (4.4.b)], the linearized approximation reads[*]

$$[\hat{p}(y)]_{y_0^-}^{y_0^+} = 0. \quad (4.59)$$

According to (4.12) and (4.32a,b)

$$\hat{p}(y) = U'(y)\phi - [U(y) - c]\phi', \quad (4.60)$$

and *continuity of pressure* at the interface thus implies that

$$[(U(y) - c)\phi'(y) - U'(y)\phi(y)]_{y_0^-}^{y_0^+} = 0. \quad (4.61)$$

Jump conditions (4.58), (4.61) should be enforced together with Rayleigh's equation and boundary conditions wherever a discontinuity of the basic velocity profile $U(y)$ is present.

As the first and simplest example of unstable flow, consider the unbounded *vortex sheet* (Fig. 4.6a) defined by

$$U(y) = U_1, \quad y > 0; \qquad U(y) = U_2, \quad y < 0, \quad (4.62)$$

---

[*] It is understood that gravity and surface tension have been neglected.

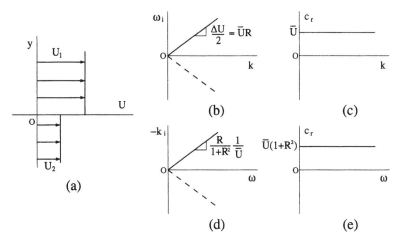

Fig. 4.6. Vortex sheet instability characteristics: (a) basic velocity profile; (b) temporal growth rate ω_i versus real wavenumber k; (c) real phase velocity $c_r \equiv \omega_r/k$ versus k; (d) spatial growth rate $-k_i$ versus real frequency ω, (e) $c_r \equiv \omega/k_r$ versus ω.

where $U_1 > U_2$. We temporarily revert to dimensional variables. Rayleigh's equation (4.33) reduces in both regions to $\phi'' - k^2 \phi = 0$. Upon enforcing exponential decay conditions at $y = \pm\infty$, the solutions in each region read

$$\phi_1(y) = A_1\,e^{-ky}\,, \qquad \phi_2(y) = A_2\,e^{ky}\,, \qquad (4.63a,\,b)$$

where A_1 and A_2 are unknown constants and we have assumed $k_r > 0$. The jump conditions (4.58) and (4.61) lead to the algebraic system

$$(U_1 - c)A_1 + (U_2 - c)A_2 = 0\,, \qquad (4.64a)$$

$$(U_2 - c)A_1 + (U_1 - c)A_2 = 0\,, \qquad (4.64b)$$

which will admit non trivial solutions only when the corresponding determinant is zero. The dispersion relation of the vortex sheet expressed in terms of the complex phase velocity $c \equiv \omega/k$ can therefore be written as

$$(c - U_1)^2 + (c - U_2)^2 = 0\,. \qquad (4.65)$$

Upon introducing the velocity difference $\Delta U \equiv U_1 - U_2$ and the average velocity* between the two streams $\bar{U} \equiv (U_1 + U_2)/2$, the solutions of

---

* not to be confused with the mean velocity $\bar{U}(y, t)$ defined in connection with (4.42).

(4.65) can be written as

$$c \equiv \frac{\omega}{k} = \bar{U} \pm i\frac{\Delta U}{2}.$$

(4.66)

In terms of the dimensionless *velocity ratio* [see section 2.1, equation (2.3)]

$$R \equiv \frac{\Delta U}{2\bar{U}},$$

(4.67)

the complex phase velocity takes the form

$$c \equiv \frac{\omega}{k} = \bar{U}(1 \pm iR).$$

(4.68)

For k real, the temporal growth rate $\omega_i(k) = kc_i$ is proportional to the velocity difference ΔU and to the wavenumber k (Fig. 4.6b). Since this crude model has no intrinsic length scale, there is no cut-off wavenumber and disturbances of all wavelengths are amplified. The vortex sheet model is not expected to offer a realistic representation of shear layers encountered in nature: there is no upper bound to the temporal growth rate and causality in the sense of section 3.2 cannot be enforced. In fact, vortex sheets develop a discontinuity of curvature in finite time. Finally note that the real phase speed of the waves c_r is equal to the average velocity $\bar{U}$ (Fig. 4.6c).

The spatial stability characteristics are immediately deduced from (4.66) as

$$k = \frac{\bar{U} \mp \frac{1}{2}i\Delta U}{\bar{U}^2 + \frac{1}{4}\Delta U^2}\omega,$$

(4.69)

alternatively, in terms of the velocity ratio R :

$$k = \frac{1 \mp iR}{1 + R^2}\frac{\omega}{\bar{U}}.$$

(4.70)

Plots of the spatial growth rate $-k_i$ and phase velocity $c_r \equiv \omega/k_r$ are included for completeness in Figs. 4.6d,e.

As will be seen in section 5.1, the velocity ratio R is an essential parameter in the determination of the absolute/convective nature of the instability and the legitimacy of the spatial approach. R is a measure of the relative importance of the net shear ΔU giving rise to the unstable motion and $\bar{U}$ the advection velocity of the waves. When $|R| \ll 1$, the shear is very weak (Fig. 4.7a); only one stream is present when $|R| = 1$ (Fig. 4.7b) and the vortex sheet separates two counterflowing streams when $|R| > 1$ (Fig. 4.7c).

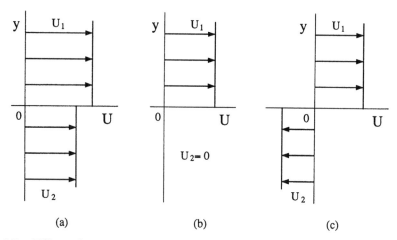

Fig. 4.7. Effect of velocity ratio R on vortex sheet velocity profile. (a) $|R| \ll 1$; (b) $|R| = 1$; (c) $|R| > 1$.

There is in general no simple relationship between temporal and spatial stability results (4.68) and (4.70). However, in the case of small shear $|R| \ll 1$, one obtains, by simple Taylor expansion of (4.70)

$$k \sim (1 \mp iR)\frac{\omega}{\bar{U}} + \mathcal{O}(R^2).\tag{4.71}$$

In this limit, spatial stability characteristics are simply deduced from temporal ones by making the transformation $x \mapsto \bar{U}t$.

Following Batchelor (1967), we give a simple interpretation of the vortex sheet instability mechanism in terms of vorticity dynamics (Fig. 4.8).

The vorticity distribution $\Omega(y)$ of the plane undisturbed vortex sheet is uniform along the sheet (Fig. 4.8a) and given by

$$\Omega(y) = -\Delta U \delta(y),\tag{4.72}$$

where $\delta(y)$ is the usual delta function. If a spatially periodic deformation is applied to the sheet (Fig. 4.8b), a velocity is induced at the crests (troughs) by neighboring troughs (crests) as indicated by the dotted vectors in Fig. 4.8b. Thus net horizontal velocity components are generated along the deformed sheet (solid vectors in Fig. 4.8b), which deplete of vorticity nodes such as B, and gather vorticity towards nodes such as A (Fig. 4.8c). In turn this redistribution of vorticity tends to increase by induction the initial deformation, thereby leading to instability.

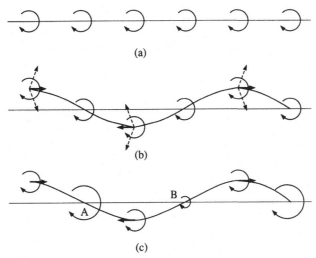

(a)

(b)

(c)

Fig. 4.8. Physical interpretation of vortex sheet instability. Size of depicted vortex lumps is taken to be proportional to their strength. See text for details.

5 The spatial mixing layer

According to the qualitative discussion of section 2.1, spatially-developing mixing layers are dominated by robust quasi-two dimensional *Kelvin–Helmholtz vortices* that evolve on a fast inertial time scale $\tau_i \sim L/\Delta U$, where L is a characteristic initial thickness of the velocity profile immediately downstream of the splitter plate, for instance $\delta_\omega(0)$. These large scale structures are known to be extremely sensitive to external noise. Many of these features can be interpreted within the setting of the inviscid formulation presented in section 4. The spatial mixing layer is chosen here to illustrate the concepts that are applicable to *convective instabilities* driven by an *inviscid* Kelvin-Helmholtz type mechanism.

Section 5.1 summarizes the linear stability characteristics of parallel mixing layers. A WKBJ approximation scheme is then implemented to take into consideration the streamwise variations of the basic velocity profile (section 5.2). We conclude this selective survey of mixing layers with a presentation of the main classes of secondary instabilities supported by two-dimensional Kelvin–Helmholtz vortices (section 5.3). The reader may consult Ho and Huerre (1984) for a comprehensive review of the dynamics of free shear layers from a theoretical point of view.

5.1 Linear instability of parallel mixing layers

The linear stability properties of the crudest model of mixing layer, namely the vortex sheet, have already been derived in subsection 4.5.

To remedy its obvious deficiencies, it appears necessary to explicitly introduce a length scale in the form of a shear layer thickness δ_ω. We temporarily revert to dimensional variables. As an additional example of application of the jump conditions (4.58), (4.61), consider the *broken-line profile* given by

$$U(y) = \begin{cases} U_1, & y > \delta_\omega/2 \\ (U_1 + U_2)/2 + (U_1 - U_2)y/\delta_\omega, & |y| < \delta_\omega/2 \\ U_2, & y < -\delta_\omega/2 \end{cases} \qquad (5.1)$$

and sketched in Fig. 5.1.

As before, Rayleigh's equation (4.33) reduces in each y subdomain to $\phi'' - k^2\phi = 0$. Considering the exponential decay conditions at $y = \pm\infty$, the solutions in the different regions must be of the form

$$\phi_1(y) = A_1 e^{-ky}, \quad y > \delta_\omega/2, \qquad (5.2a)$$

$$\phi_2(y) = B_2 e^{ky}, \quad y < -\delta_\omega/2, \qquad (5.2b)$$

$$\phi_0(y) = A_0 e^{-ky} + B_0 e^{ky}, \quad |y| < \delta_\omega/2, \qquad (5.2c)$$

with A_1, B_2, A_0, B_0 unknown complex constants and $k_r > 0$. Application of the jump conditions (4.58) and (4.61) at $y = \pm\delta_\omega/2$ leads to the algebraic system:

$$A_1 e^{-k\delta_\omega/2} = A_0 e^{-k\delta_\omega/2} + B_0 e^{k\delta_\omega/2}, \qquad (5.3a)$$

$$B_2 e^{-k\delta_\omega/2} = A_0 e^{k\delta_\omega/2} + B_0 e^{-k\delta_\omega/2}, \qquad (5.3b)$$

$$-k(U_1 - c)A_1 e^{-k\delta_\omega/2} = k(U_1 - c)(-A_0 e^{-k\delta_\omega/2} + B_0 e^{k\delta_\omega/2})$$
$$- \frac{\Delta U}{\delta_\omega}(A_0 e^{-k\delta_\omega/2} + B_0 e^{k\delta_\omega/2}), \qquad (5.3c)$$

$$k(U_2 - c)B_2 e^{-k\delta_\omega/2} = k(U_2 - c)(-A_0 e^{k\delta_\omega/2} + B_0 e^{-k\delta_\omega/2})$$
$$- \frac{\Delta U}{\delta_\omega}(A_0 e^{k\delta_\omega/2} + B_0 e^{-k\delta_\omega/2}). \qquad (5.3d)$$

When A_1 and B_2 given by (5.3a,b) are substituted into (5.3c,d), one obtains the reduced system:

$$-\frac{\Delta U}{\delta_\omega}A_0 e^{-k\delta_\omega/2} + \left[2k(U_1 - c) - \frac{\Delta U}{\delta_\omega}\right]B_0 e^{k\delta_\omega/2} = 0, \qquad (5.4a)$$

$$\left[2k(U_2 - c) + \frac{\Delta U}{\delta_\omega}\right]A_0 e^{k\delta_\omega/2} + \frac{\Delta U}{\delta_\omega}B_0 e^{-k\delta_\omega/2} = 0. \qquad (5.4b)$$

Upon setting the corresponding determinant equal to zero and introducing the average velocity $\bar{U}$ such that $U_1 = \bar{U} + \Delta U/2$ and $U_2 = \bar{U} - \Delta U/2$, the following dispersion relation ensues:

$$4(k\delta_\omega)^2(c - \bar{U})^2 - \left[(k\delta_\omega - 1)^2 - e^{-2k\delta_\omega}\right]\Delta U^2 = 0. \qquad (5.5)$$

It is convenient to reason in terms of dimensionless parameters. From here on, we choose $\delta_\omega/2$ as length scale and $\bar{U}$ as velocity scale. Upon making the substitution $k\delta_\omega \mapsto 2k$, $c/\bar{U} \mapsto c$, and introducing the velocity ratio $R = \Delta U/(2\bar{U})$, the dispersion relation may be written in the final form

$$4k^2(c-1)^2 - R^2\left[(2k-1)^2 - e^{-4k}\right] = 0\,. \tag{5.6}$$

When k is given real, there are two *temporal branches*

$$c \equiv \frac{\omega}{k} = 1 \pm \frac{R}{2k}\left[(2k-1)^2 - e^{-4k}\right]^{1/2}\,. \tag{5.7}$$

Let $k_n \sim 0.6392$ denote the wavenumber such that $2k_n - 1 = e^{-2k_n}$. When $k > k_n$, the phase velocities of both branches are purely real and given by (5.7). The disturbances are then neutrally stable (Fig. 5.1b,c). When $k < k_n$, the phase velocities are complex and can be written

$$c = 1 \pm i\frac{R}{2k}\left[e^{-4k} - (2k-1)^2\right]^{1/2}\,. \tag{5.8}$$

As shown in Fig. 5.1b, the broken-line profile is linearly unstable in a finite range of wavenumbers $0 < k < k_n$. The temporal growth rate $\omega_i = kc_i$ reaches a maximum $\omega_{i,max}$ at a well defined wavenumber k_{max} and all unstable waves travel at a phase velocity $c_r = 1$ corresponding to the average velocity $\bar{U}$ between the two streams. The finite thickness of the broken-line model has therefore removed some of the unphysical features of the vortex sheet representation. Note also that in the limit $k \ll 1$, equation (5.8) reduces to the vortex sheet result $c = 1 \pm iR$ obtained in (4.68): instability waves of large wavelength do not "feel" the finite thickness of the layer and they effectively evolve on a vortex sheet with the same ΔU and $\bar{U}$ as the original layer.

Before proceeding to the study of spatial branches, it is first necessary to determine the *absolute/convective nature of the instability* as a function of the parameter R. This issue could not be examined in the vortex sheet problem since this model does not satisfy causality (no finite maximum temporal growth rate $\omega_{i,max}$). Following the methodology summarized in subsection 3.2.5, we need to calculate the complex wavenumber(s) $k_0(R)$ of zero group velocity $(\partial\omega/\partial k)(k_0; R) = 0$. In terms of k and ω, the dispersion relation (5.6) reads:

$$D(k,\omega; R) \equiv 4(\omega - k)^2 - R^2\left[(2k-1)^2 - e^{-4k}\right] = 0\,. \tag{5.9}$$

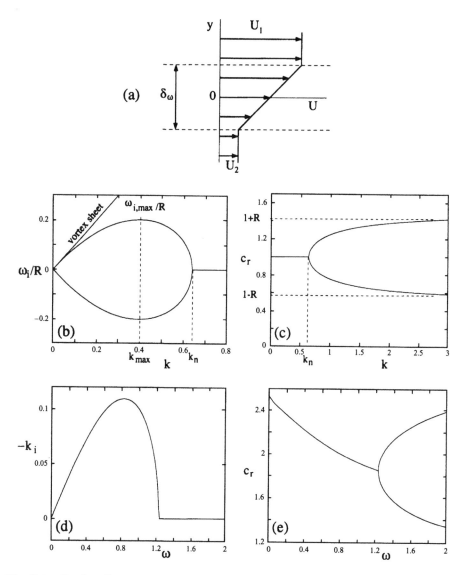

Fig. 5.1. Broken-line profile instability characteristics: (a) basic velocity profile; (b) normalized temporal growth rate ω_i/R versus real wavenumber k; (c) real phase velocity $c_r \equiv \omega_r/k$ versus k; (d) spatial growth rate $-k_i$ versus real frequency ω at $R = 0.5$; (e) $c_r \equiv \omega/k_r$ versus ω at $R = 0.5$. Curves such as (d) and (e) only apply when the flow is convectively unstable, $0 < R < 1$ (courtesy of S. Ortiz, LadHyX, École Polytechnique).

Upon considering ω as a function of k, differentiating (5.9) with respect to k and setting $\partial\omega/\partial k = 0$, we obtain

$$\omega_0 = k_0 - \frac{R^2}{2}\left(2k_0 - 1 + e^{-4k_0}\right). \tag{5.10}$$

When (5.10) is substituted into (5.9), k_0 is found to be a root of

$$\left(2k_0 - 1 + e^{-4k_0}\right)^2 R^2 - \left\{(2k_0 - 1)^2 - e^{-4k_0}\right\} = 0. \tag{5.11}$$

Upon numerically solving (5.11) for k_0 complex and substituting the result into (5.10), one readily obtains the locus of the complex absolute frequency $\omega_0(R)$ as a function of the velocity ratio R: Figure 5.2 indicates that the absolute growth rate $\omega_{0,i}$ becomes positive as soon as $|R|$ exceeds unity.

Thus the broken line profile has been shown to be convectively (absolutely) unstable when $|R| < 1$ ($|R| > 1$). Note that the velocity ratio R can be interpreted in exactly the same way as in the case of the vortex sheet (Fig. 4.7): when $|R| < 1$, both streams are coflowing whereas they are counterflowing when $|R| > 1$. As soon as there is any counterflow, the nature of the instability changes from convective to absolute and the spatial problem is well posed only for coflowing streams ($|R| < 1$). These conclusions are in full agreement with the comprehensive stability

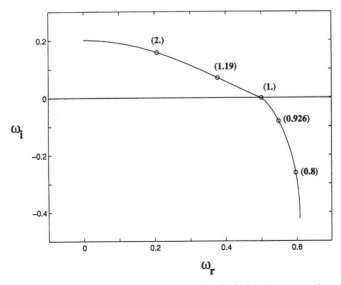

Fig. 5.2. Locus of the absolute frequency $\omega_0(R)$ in the complex ω-plane as a function of the velocity ratio R (in parentheses) for the broken line profile (courtesy of S. Ortiz, LadHyX, École Polytechnique).

analysis of the broken-line profile presented in Balsa (1987, 1988). *Spatial stability* characteristics are displayed in Figs. 5.1d,e for a typical value of the velocity ratio $0 < R < 1$.

The *hyperbolic-tangent mixing layer* has been used by Michalke (1964, 1965) and Monkewitz and Huerre (1982), among others, as a convenient approximation to the continuous mean velocity profiles that are measured in shear layers developing downstream of a splitter plate (section 2.1). In dimensional variables, the basic flow is then taken to be

$$U(y) = \bar{U} + \frac{\Delta U}{2} \tanh\left(\frac{2y}{\delta_\omega}\right), \qquad (5.12)$$

where $\bar{U} = (U_1 + U_2)/2$, $\Delta U = U_1 - U_2$ and δ_ω is the vorticity thickness as sketched in Fig. 5.3a. Upon introducing $\bar{U}$ as velocity scale and $\delta_\omega/2$ as length scale, (5.12) reduces to

$$U(y; R) = 1 + R \tanh y, \qquad (5.13)$$

with $R = \Delta U/(2\bar{U})$ denoting the usual velocity ratio. The dispersion relation $D(k, \omega; R) = 0$ is obtained by solving numerically the Rayleigh equation (4.33) subject to exponential decay boundary conditions at $y = \pm\infty$. By substituting (5.13) into (4.33) the following property is immediately verified: to each eigenvalue $c_1(k)$ associated with the velocity profile $U_1(y) = \tanh y$, corresponds an eigenvalue

$$c(k; R) = 1 + R c_1(k) \qquad (5.14)$$

of $U(y; R) = 1 + R \tanh y$. Equivalently, $\omega(k; R) = k c(k; R)$ can be expressed as

$$\omega(k; R) = k + R \omega_1(k), \qquad (5.15)$$

where $\omega_1(k)$ is a temporal branch of $U_1(y) = \tanh y$. According to (5.14), (5.15), the *temporal stability characteristics* are linear functions of R and it is sufficient to solve the temporal eigenvalue problem for $U_1(y) = \tanh y$. This property is a simple consequence of Galilean invariance. The temporal stability calculations of Michalke (1964) have been summarized on Figs. 5.3b,c. Since $U_1(y)$ is odd, it can be shown by applying the transformation $y \mapsto -y$ and taking the complex conjugate of the Rayleigh equation (4.33) that $\omega_1(k)$ is purely imaginary: $\omega_1(k) = i\omega_{1,i}(k)$. According to (5.15), the phase velocity $c_r \equiv \omega_r/k$ is therefore constant and equal to unity for all k.

As in the case of the broken-line profile, there is a finite range of unstable wavenumbers $0 < k < 1$ and a most amplified wavenumber $k_{\max}$ of growth rate $\omega_{i,\max}$. Note also that in the limit $k \ll 1$, the vortex

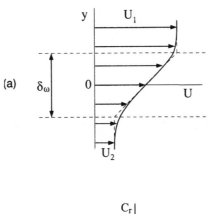

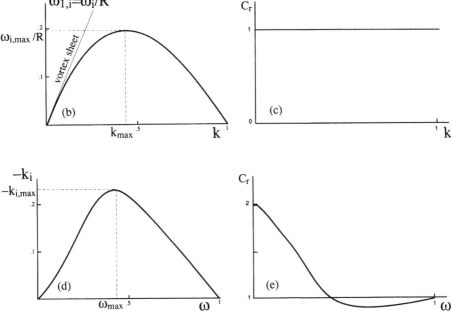

Fig. 5.3. Hyperbolic-tangent mixing layer instability characteristics: (a) basic velocity profile; (b) normalized temporal growth rate ω_i/R versus real wavenumber k; (c) real phase velocity $c_r \equiv \omega_r/k$ versus k; (d) spatial growth rate $-k_i$ of amplified branch $k_1^+(\omega)$ versus real frequency ω at $R = 1$; (e) $c_r \equiv \omega/k_r$ versus ω at $R = 1$. Curves such as (d) and (e) only apply when the flow is convectively unstable, $0 < R < 1.315$. After Michalke (1964, 1965) and Ho and Huerre (1984).

sheet results are recovered. Finally, at the neutral wavenumber $k_n = 1$, the eigenfunction is known analytically as

$$\phi_n(y) = \operatorname{sech} y . \tag{5.16}$$

In this case the critical point $y_c = 0$ satisfying $U(y_c; R) = 1$ coincides with the inflection point, and (5.16) is an example of *regular neutral mode* with no logarithmic singularity at $y_c = 0$ (section 4.2). The cross-stream structure is most easily visualized by representing the total vorticity contours of the flow disturbed by the most amplified temporal mode at k_{max} (Fig. 5.4).

Temporal stability theory is seen to describe the early time evolution of a parallel mixing layer which is perturbed periodically in the streamwise direction at a wavelength $2\pi/k$. This situation is more akin to the *temporal* development of Kelvin–Helmholtz instability waves in counterflow mixing layers produced by tilting a tank than to the *spatial* evolution of the same waves in the mixing layer generated downstream of a splitter plate (section 2.1). It is not surprising that early comparisons between temporal stability predictions and observations were disappointing (Michalke, 1965).

Absolute/convective instability issues have been examined for the hyperbolic-tangent mixing layer by Huerre and Monkewitz (1985). The complex wavenumber $k_0(R)$ and frequency $\omega_0(R)$ of zero group velocity $\partial\omega/\partial k(k_0; R) = 0$ have to be determined numerically from the dispersion relation (5.15). The resulting locus of the absolute frequency $\omega_0(R)$ in the complex ω-plane is shown on Fig. 5.5, corresponding values of

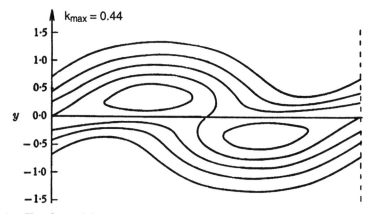

Fig. 5.4. Total vorticity contours of the perturbed hyperbolic tangent velocity profile $U_1(y) = \tanh y$ for the most amplified temporal mode at k_{max}. After Michalke (1964).

the velocity ratio R being indicated in brackets along the curve. When $R < R_t = 1.315$, the absolute growth rate $\omega_{0,i}$ is negative and, according to the notions introduced in section 3.2, the instability is convective.

For sufficiently strong counterflow, namely when $R > R_t = 1.315$, the instability becomes absolute. The spatial stability approach with ω real and k complex is therefore pertinent for mixing layers with co-flowing streams or with very weak counterflow ($R < 1.315$). As emphasized in section 3.2, it should also be checked that the branch point ω_0 corresponds to pinching at k_0 of spatial branches $k^+(\omega; R)$ and $k^-(\omega; R)$ issuing from the upper and lower half k-plane respectively, when ω_i is sufficiently large. It can be shown numerically that the hyperbolic-tangent mixing layer admits three spatial branches $k_1^+(\omega; R)$, $k_2^+(\omega; R)$ and $k_1^-(\omega; R)$. As the contour L_ω is lowered and ω_i decreases, the spatial branch $k_1^+(\omega; R)$ migrates partly into the lower half k-plane to meet $k_1^-(\omega; R)$. When the instability is convective ($R < 1.315$), L_ω can be lowered to coincide with the $\omega_i = 0$ axis without pinching taking place, as shown on Fig. 5.6.

Forcing the mixing layer at a real frequency ω then gives rise to three spatial waves with characteristics displayed on Figs. 5.7a,b. When a time-periodic source is placed within the mixing layer (Fig. 5.8), the damped branches $k_1^-(\omega; R)$ and $k_2^+(\omega; R)$ are respectively located upstream and downstream of the excitation location whereas $k_1^+(\omega; R)$

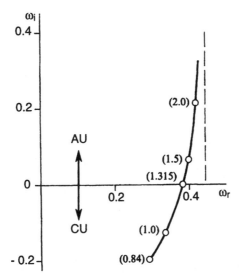

Fig. 5.5. Locus of the complex absolute frequency $\omega_0(R)$ in the complex ω-plane as a function of the velocity ratio R (in brackets) for the hyperbolic-tangent mixing layer. After Huerre and Monkewitz (1985).

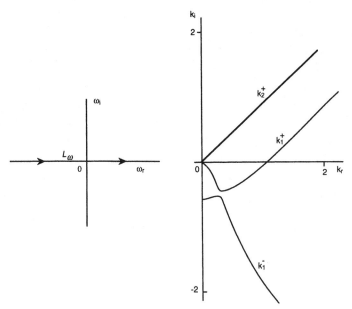

Fig. 5.6.   Locus of spatial branches in complex k-plane as ω moves along L_ω contour coinciding with the $\omega_i = 0$ axis in complex ω-plane.   Case of the convectively unstable hyperbolic tangent mixing layer at $R = 1.3$. After Huerre and Monkewitz (1985).

is spatially growing downstream with an amplification rate $-k_{1,i}^+(\omega; R)$. In the absolutely unstable range $(R > 1.315)$, the pinching of k_1^+ and k_1^- takes place before the L_ω contour reaches the $\omega_i = 0$ axis, i.e. for a value $\omega_{0,i} > 0$.

Finally, note that the convective/absolute transition point can easily be obtained by simply tracking the occurrence of branch switching in the spatial solutions $k_1^+(\omega; R)$ and $k_1^-(\omega; R)$ for ω real, as R crosses R_t from below (dashed curves of Figs. 5.7a,b).

From the above discussion and section 3.2, it can be concluded that convectively unstable mixing layers with $R < 1.315$ are expected to behave as *amplifiers of external noise*, the amplification rate $-k_i(\omega)$ and phase velocity $c_r = \omega/k_r(\omega)$ of resulting streamwise developing perturbations being determined by spatial stability theory as illustrated in Fig. 5.3d,e. It should be emphasized that the spatial point of view was advocated with success by Michalke (1965) to describe the dynamics of coflowing mixing layers, long before absolute/convective instability ideas had been proposed to assess the validity of this approach. The structure of the perturbed flow in the presence of time-periodic forcing applied at $x = 0$ is illustrated by the constant total vorticity contours of Fig. 5.9

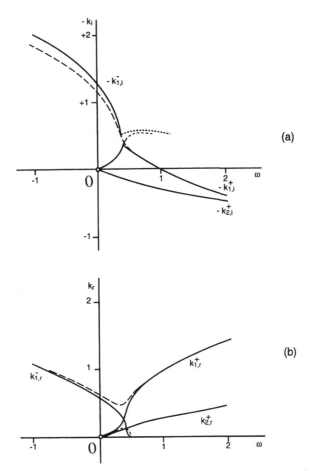

Fig. 5.7. Branch switching as R increases from $R = 1.3$ (solid curves) to $R = 1.4$ (dashed curves) across convective/absolute transition point at $R_t = 1.315$. (a) $-k_i$ versus real ω. (b) k_r versus real ω. After Huerre and Monkewitz (1985).

(Michalke, 1965). This sketch should be compared with its temporal counterpart in Fig. 5.4: the flow is now time-periodic and spatially-developing.

We have seen that, according to (5.15), temporal branch characteristics vary linearly with R and R can be scaled out by plotting ω_i/R as a function of k as illustrated in Fig. 5.3b. The dependence of spatial branches on R is not so trivial: one has to solve for $k(\omega; R)$ implicitly given by

$$\omega = k(\omega; R) + R\,\omega_1[k(\omega; R)]. \tag{5.17}$$

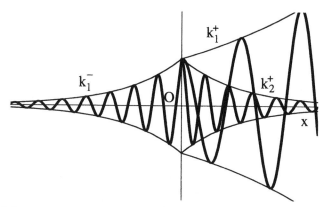

Fig. 5.8. Schematic of relative positions of spatial branches k_1^+, k_2^+ and k_1^- with respect to the location of the periodic source for convectively unstable hyperbolic-tangent mixing layer.

In order to capture the main effect of R, it is tempting to solve (5.17) in the limit $R \ll 1$, when the flow becomes very weakly unstable. Following Monkewitz and Huerre (1982), we set

$$k(\omega; R) \sim k_0(\omega) + R \, k_1(\omega) + \dots \qquad (5.18)$$

Upon substituting (5.18) into (5.17), expanding in powers of R and identifying terms of equal order of magnitude, one readily obtains

$$k_0(\omega) = \omega, \quad k_1(\omega) = -\omega_1(\omega) = -i\omega_{1,i}(\omega), \qquad (5.19\text{a, b})$$

the last equal sign stemming from the observation that $\omega_1(k)$ for $U_1(y) = \tanh y$ is purely imaginary for k real (Fig. 5.3b). Thus, in the limit $R \ll 1$, one has the simple relation

$$k(\omega; R) \sim \omega - iR\,\omega_{1,i}(\omega) + \mathcal{O}(R^2), \qquad (5.20)$$

between the temporal and spatial stability properties. The spatial growth rate $-k_i(\omega; R)$ and phase velocity $c_r \equiv \omega/k_r$ are given by

$$-k_i(\omega; R) \sim R\,\omega_{1,i}(\omega) + \mathcal{O}(R^2), \quad c_r \sim 1 + \mathcal{O}(R^2). \qquad (5.21\text{a, b})$$

This result is equivalent to *Gaster's transformation* (Gaster, 1962): in the limit of small amplification rates, spatially-developing waves are equivalent to temporal waves growing in time at the rate $R\omega_{1,i}(\omega)$ in a reference frame moving at the average velocity of the two streams $c_r = 1$. The result (5.21a,b) seems to apply well beyond its strict range of validity: when numerically determined spatial stability characteristics are represented for different values of R on normalized diagrams $-k_i/R$ and $(c_r - 1)/R^2$ versus ω (Fig. 5.10a,b), all spatial growth rate curves collapse in a thin band around $\omega_{1,i}(\omega)$ and a similar feature prevails for c_r. Relations (5.21a,b) therefore furnish a convenient first-order approximation even for velocity ratios close to $R = 1$.

The predictions of spatial stability theory have been fully confirmed experimentally. A typical experimental configuration has already been presented in section 2.1: the mixing layer produced downstream of the trailing-edge of a splitter plate is forced acoustically or mechanically at a specific frequency ω_f and amplitude. The energy content of the different fundamental, harmonic and subharmonic components is then measured and plotted as a function of streamwise distance x on a semilog plot (Fig. 2.8a, 2.9a). From the local slope of such curves at different streamwise stations, one may retrieve the *measured local growth rate* $-k_i(x; \omega_f)$ of the component ω_f. The streamwise development of the *mean* (averaged over time) velocity profile $\bar{U}(y; x)$ can also be documented as well as the variations of the local thickness $\delta_\omega(x)$ (Fig. 2.8b, 2.9b). To a very good degree of approximation, it is found that $\bar{U}(y; x)$ can effectively be modeled as a self-similar family of hyperbolic-tangent velocity profiles of varying thickness $\delta_\omega(x)$. For a given velocity ratio R, the local growth rates normalized by the local thickness $\delta_\omega(x)$ at the *same* streamwise location can then be positioned on the growth rate diagram of Fig. 5.10a provided that ω_f is also normalized with respect to $\delta_\omega(x)$. This procedure can be repeated at all streamwise stations along the energy curves of Figs. 2.8a, 2.9a. Thus for a given dimensional forcing frequency ω_f, the local dimensionless frequency

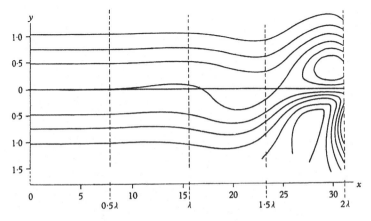

Fig. 5.9. Total vorticity contours of the perturbed hyperbolic-tangent velocity profile for the most amplified spatial mode at $\omega_{\max}$, at a given time t. Velocity ratio is $R = 1$. After Michalke (1965).

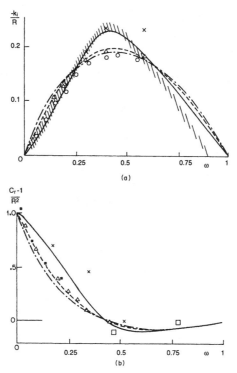

Fig. 5.10. (a) Normalized spatial amplification rates $-k_i/R$ and (b) phase velocities $(c_r - 1)/R^2$ versus frequency ω for hyperbolic-tangent velocity profile at different velocity ratios R (after Monkewitz and Huerre, 1982, and Ho and Huerre, 1984): — $R = 1$; —·— $R = 0.5$; — — — $R \ll 1$, Gaster's transformation (5.21a,b). Experiments: Open squares: $R = 1$ (Sato, 1960); open circles: $R = 1$ (Freymuth, 1966); multiply signs: $R = 0.72$ (Miksad, 1972); hatched band: $R = 1$ (Fiedler *et al.*, 1981); triangles: $R = 0.31$ (Ho and Huang 1982); filled squares: $R = 1$ (Drubka, 1981).

parameter $\omega \equiv \omega_f \delta_\omega(x)/2\bar{U}$ on the horizontal axis of Fig. 5.10a,b can effectively be interpreted as a renormalized streamwise distance. Experimental results obtained by several groups have been collected in such a way on Fig. 5.10a. The agreement between local spatial stability theory and experiments is surprisingly good and this is also confirmed by phase velocity measurements on Fig. 5.10b. Linear theory is *a priori* expected to hold in the initial exponential growth region immediately downstream of the splitter plate, as long as forcing levels are sufficiently low. As a leading order estimate, it is also seen to apply further downstream: agreement appears to persist even close to the neutral point $\omega_f \delta_\omega(x)/2\bar{U} = 1$, where, by definition, the local growth rate is zero and the energy curve of the ω_f component therefore displays a maximum. At such a station, the instability wave has rolled-up into a mature vortex (Ho and Huang, 1982) and nonlinear effects are likely to become significant. Measured cross stream distributions of perturbation quantities are also found to conform with the eigenfunctions of spatial theory. Spatial eigenfunctions differ markedly from their temporal counterparts and this feature further confirms the validity of the spatial point of view (Michalke, 1965). In *convectively unstable mixing layers, the fate of each frequency component fed from upstream can be followed individually in the linear approximation by calculating the complex wavenumber $k^+(\omega; R)$ at each streamwise station as dictated by the shape and scale of the local velocity profile.* The dynamics of the flow is in large measure determined by the composition of external perturbations.

Although the main goal of this section is to illustrate the various methodologies used to describe the response of convectively unstable shear layers, it is worthwhile to briefly digress and discuss the case of counterflow mixing layers (Figs. 2.6d,e). As mentioned in section 2.1, such a configuration is most easily produced by tilting a tube filled with two immiscible fluids of different density or with a continuously stably stratified fluid (Fig. 2.10). According to the results of both broken-line and hyperbolic-tangent profiles, the basic flow is absolutely unstable when $R \gg 1$. Indeed, a temporally growing instability wave is then observed at a well-defined wavenumber close to k_{max}, as one would expect from a

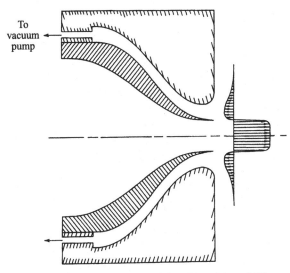

Fig. 5.11. Experimental configuration of Strykowski and Niccum (1991) needed to generate an axisymmetric countercurrent mixing layer.

temporal stability analysis. An ingenious experiment has been devised by Strykowski and Niccum (1991) to study the qualitative changes taking place in a mixing layer as R is increased through the convective/absolute transition value R_t. The classical jet flow issuing from a nozzle of circular cross section takes the form of an axisymmetric mixing layer of small thickness with respect to the jet diameter D. Suction can be applied to an annular outer region to generate a counter current of suitable intensity as sketched in Fig. 5.11.

As the velocity ratio R is increased (Fig. 5.12), the initially broad power spectra that are typical of convectively unstable jets collapse into discrete peaks which signals the spontaneous synchronization of the jet

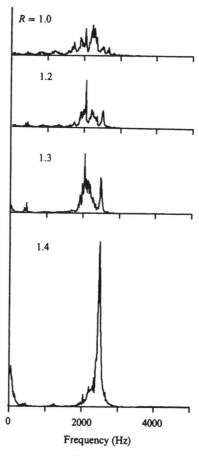

Fig. 5.12. Velocity power spectra (linear scale) measured in the jet shear layer of Fig. 5.11 for different values of the velocity ratio R in counterflow mixing layer experiment of Strykowski and Niccum (1991).

shear layer into a periodic regime of well-defined frequency. According to Fig. 5.13, the amplitude of the limit cycle increases as the square root of the departure from the critical value $R_{G_c} = 1.34$, thereby indicating that the flow has undergone a *supercritical Hopf bifurcation* (see section 3.3).

Note that the critical value R_{G_c} is surprisingly close to the convective/absolute transition value $R_t = 1.315$. In the context of this course, the limit cycle regime is viewed as an example of a *global mode* triggered by the appearance of a sufficiently large region of absolute instability immediately downstream of the nozzle exit. Several spatially developing flows have been shown to undergo such bifurcations, most notorious among them the bluff-body wake as further discussed in section 6. Global mode concepts are presented in detail in section 6.2.

Finally, it is worth pointing out the increasing lack of sensitivity of the axisymmetric mixing layer to external perturbations as R exceeds R_{G_c}. Whereas in the convectively unstable range there is a one to one functional relationship between the forcing amplitude and response, the response becomes independent of forcing level in the limit cycle range (Fig. 5.14).

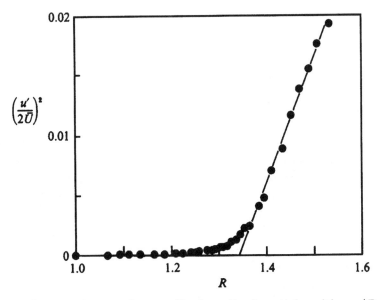

Fig. 5.13. Square of saturation amplitude at fixed spatial position $x/D = 0.26$ versus velocity ratio R in counterflow mixing layer experiments of Strykowski and Niccum (1991). Hopf bifurcation point is extrapolated to be $R_{G_c} = 1.34$.

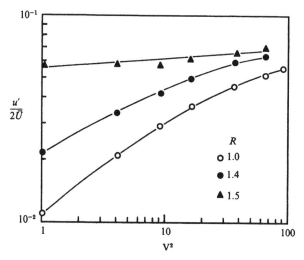

Fig. 5.14.   Response of the axisymmetric mixing layer of Fig. 5.11 to a monochromatic excitation generated by a loudspeaker positioned at $x/D = 10$ and 10 diameters off axis. The forcing frequency coincides with the global oscillation frequency observed as R exceeds 1.4.   The response streamwise perturbation velocity $u'/2\overline{U}$ at $x/d = 0.2$ on the shear layer high-speed side is represented as a function of r.m.s. input speaker power (Strykowski and Niccum, 1991).

5.2 Weakly non parallel WKBJ formulation

The spatial linear stability analysis of parallel mixing layers presented in section 5.1 has led to estimates of the local amplification rates and phase velocities in convectively unstable spatially developing mixing layers subjected to external forcing.   The surprising success of this procedure suggests that the streamwise evolution of the basic flow is effectively slow. The goal of this section is to take advantage of the slow divergence assumption to construct an asymptotic scheme with the locally parallel stability results appearing naturally as the leading-order approximation.   Such formulations have been applied in the past to the development of spatial instability waves in a variety of *convectively unstable slowly diverging shear flows*, for instance boundary layers (Bouthier, 1972), circular jets (Crighton and Gaster, 1976) and forced turbulent mixing layers (Gaster *et al.*, 1985). The main underlying assumption is illustrated in Fig. 5.15: streamwise inhomogeneities of the basic flow are characterized by a length scale $L \sim [(1/\delta_\omega)(d\delta_\omega/dx)]^{-1}$ such that $\delta_\omega \ll L$. The instability wavelength λ being of the same order of magnitude as δ_ω (section 5.1), one also has $\lambda \ll L$: spatial waves effectively "perceive" a locally parallel basic flow as they amplify in the streamwise direction.

The small parameter

$$\varepsilon \sim \frac{\delta_\omega}{L} \sim \frac{\lambda}{L} \ll 1 \qquad (5.22)$$

is therefore introduced as a measure of nonparallel basic flow effects. If $\delta_\omega(0)/2$ and $\overline{U} = (U_1 + U_2)/2$ are chosen as characteristic length and velocity scales, the nondimensional basic velocity profile is formally expressed as $U(y; X)$, where $X = \varepsilon x$ is a slow streamwise coordinate variable. Furthermore, the everywhere convectively unstable basic flow is assumed to be forced at $x = 0$ by a periodic excitation of frequency ω_f.

Following Gaster *et al.* (1985), the flow is taken to be inviscid and governed by the two-dimensional vorticity equation (4.28). The basic flow streamfunction $\Psi_0(y; X)$ should satisfy (4.28): if only $\mathcal{O}(\varepsilon)$ terms are retained, one obtains the governing basic flow equation:

$$\left(\frac{\partial \Psi_0}{\partial y} \frac{\partial}{\partial X} - \frac{\partial \Psi_0}{\partial X} \frac{\partial}{\partial y} \right) \frac{\partial^2 \Psi_0}{\partial y^2} = 0 . \qquad (5.23)$$

Let $U(y; X)$ and $\varepsilon V(y; X)$ denote the associated streamwise and cross-stream velocity components, where

$$U(y; X) = \frac{\partial \Psi_0}{\partial y}, \qquad V(y; X) = -\frac{\partial \Psi_0}{\partial X} \qquad (5.24a, b)$$

are both assumed to be of order unity. Upon integrating relation (5.23) once with respect to y, one recovers the x-component of the Euler

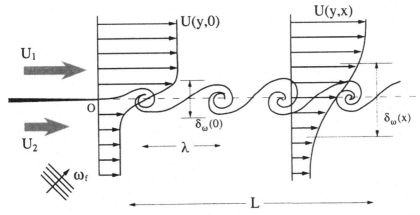

Fig. 5.15. Sketch of slowly-diverging mixing layer considered in WKBJ formulation. The small parameter is $\varepsilon \sim \lambda/L$.

momentum equation

$$\left(U\frac{\partial}{\partial X} + V\frac{\partial}{\partial y}\right)U = -\frac{dP}{dX},\tag{5.25a}$$

to which should be added the continuity equation

$$\frac{\partial U}{\partial X} + \frac{\partial V}{\partial y} = 0.\tag{5.25b}$$

As might have been anticipated, the slowly diverging flow field satisfies the "inviscid boundary layer equations".

The total streamfunction is decomposed into basic and fluctuating contributions according to:

$$\Psi(x, y, t) \sim \Psi_0(y; X) + \psi(x, y, t),\tag{5.26}$$

where $X = \varepsilon x$. Linearizing around the basic flow $\Psi_0(y; X)$ and keeping only terms of order ε then yields the disturbance field equation

$$\begin{aligned}\left[\frac{\partial}{\partial t} + U(y; X)\frac{\partial}{\partial x}\right]\nabla^2\psi - \frac{\partial^2 U}{\partial y^2}\frac{\partial \psi}{\partial x} \\ + \varepsilon\left[\frac{\partial^2 U}{\partial X \partial y}\frac{\partial \psi}{\partial y} + V\frac{\partial}{\partial y}\nabla^2\psi\right] = \mathcal{O}(\varepsilon^2)\end{aligned}\tag{5.27}$$

Note that, to leading order in ε, (5.27) is identical to the linearized partial differential equation (4.30) encountered in the study of strictly parallel flows. The only important caveat is that the streamwise coordinate x appears in the coefficients through the slow dependent variable $X = \varepsilon x$: the locally parallel approximation is therefore recovered at leading order.

In the strictly parallel case, perturbations were sought in terms of spatial normal modes (4.31) with constant wavenumber $k(\omega_f)$. In the present case, the asymptotic response is also assumed to be time-periodic of the form

$$\psi(x, y, t) = \phi(y; X)\exp\left[i\left(\frac{\Theta(X)}{\varepsilon} - \omega_f t\right)\right].\tag{5.28}$$

In the spirit of WKBJ approximations (Bender and Orszag, 1978; Nayfeh, 1973), the perturbation field has been split into a fast varying complex phase function $\Theta(X)/\varepsilon$ and a slowly varying spatial distribution function $\phi(y; X)$. The equation governing $\phi(y; X)$ is obtained by substituting ansatz (5.28) into (5.27). As a result of the assumed form (5.28), this amounts to applying to equation (5.27) the following transformations

$$\frac{\partial}{\partial t} \to -i\omega_f; \quad \frac{\partial}{\partial x} \to i\Theta_X + \varepsilon\frac{\partial}{\partial X}.\tag{5.29a, b}$$

In the process of generating higher order derivative terms, one must pay attention to the fact that multiplication by $i\Theta_X$ and differentiation with respect to X do not commute. For instance, second order derivatives in x should be interpreted as

$$\frac{\partial^2}{\partial x^2} \rightarrow \left(i\Theta_X + \varepsilon\frac{\partial}{\partial X}\right)^2 = -\Theta_X^2 + i\varepsilon\left(\Theta_X\frac{\partial}{\partial X} + \Theta_{XX}\right) + \varepsilon^2\frac{\partial^2}{\partial X^2}. \quad (5.30)$$

The slowly varying spatial distribution function $\phi(y; X)$ is then expanded as

$$\phi(y; X) \sim \phi_1(y; X) + \varepsilon\phi_2(y; X) + \cdots . \quad (5.31)$$

Upon substituting (5.31) into the equation governing $\phi(y; X)$, one is led to a set of successive problems for $\phi_1(y; X)$, $\phi_2(y; X)$, etc.

At $\mathcal{O}(1)$, the governing equation for $\phi_1(y; X)$ is obtained in the form*

$$L[\phi_1] \equiv \phi_1'' - \Theta_X^2\phi_1 - \frac{U''(y; X)}{U(y; X) - \omega_f/\Theta_X}\phi_1 = 0, \quad (5.32)$$

together with the usual exponential decay boundary conditions at $y = \pm\infty$. One recognizes in (5.32) the Rayleigh equation for the basic flow $U(y; X)$ at the forcing frequency ω_f for the unknown wavenumber Θ_X. In order for (5.32) to admit nontrivial solutions, Θ_X must coincide with the local spatial branch $k(X; \omega_f)$ prevailing downstream of the forcing station. In other words $\Theta_X = k(X; \omega_f)$ must satisfy the *local dispersion relation*

$$D[k, \omega_f; X] = 0, \quad (5.33)$$

where the streamwise station X only appears as a parameter. Without loss of generality, the corresponding eigenfunction $\phi_1(y; X)$ may be written as

$$\phi_1(y; X) = A(X)\Phi(y; X, \omega_f), \quad (5.34)$$

where $A(X)$ is an unknown *complex amplitude* and Φ a suitably normalized eigenfunction satisfying for instance $\Phi(0; X, \omega_f) = 1$. Thus Θ_X and its derivatives may be replaced by the wavenumber $k(X; \omega_f)$ and its derivatives in the transformation rules (5.29–30). As an illustration, recall that hyperbolic tangent mixing layers have been shown in section 5.1 to admit three spatial branches k_1^+, k_2^+ and k_1^-. In the convectively unstable range $R < 1.315$, any of these branches may be selected to appear in (5.32,34). However, $k_1^+(X; \omega_f)$ is by far the most interesting to examine since it is the only one that gives rise to spatially amplified waves.

---

* Here primes denote differentiation with respect to y.

At $\mathcal{O}(\varepsilon)$, the following equation for ϕ_2 is obtained:

$$
\begin{aligned}
L\left[\phi_2\right] = \frac{i}{kU - \omega_f} \Bigg[& \left\{2k\omega_f\Phi + U\left(\Phi'' - 3k^2\Phi\right) - U''\Phi\right\}\frac{dA}{dX} \\
& + \left\{\omega_f\left(\Phi\frac{dk}{dX} + 2k\frac{\partial\Phi}{\partial X}\right) + U\left(\frac{\partial\Phi''}{\partial X} - 3k^2\frac{\partial\Phi}{\partial X} - 3k\Phi\frac{dk}{dX}\right)\right. \\
& + \left.\Phi'\frac{\partial U'}{\partial X} - U''\frac{\partial\Phi}{\partial X} + V\left(\Phi''' - k^2\Phi'\right)\right\}A\Bigg] \equiv Q_2(y; X). \quad (5.35)
\end{aligned}
$$

In order for the inhomogeneous equation $L[\phi_2] = Q_2$ to admit solutions, the forcing term $Q_2(y; X)$ should be in the image of the operator L. In other words, $Q_2(y; X)$ *must be orthogonal, with respect to a suitably defined inner product, to the corresponding adjoint eigenfunction* $\tilde{\Phi}(y; X)$ *of the adjoint operator* $\tilde{L}$:

$$
\begin{aligned}
\int_{-\infty}^{+\infty} Q_2(y; X)\tilde{\Phi}(y; X)\,dy = \\
\int_{-\infty}^{+\infty} L[\phi_2]\tilde{\Phi}(y; X)\,dy = \int_{-\infty}^{+\infty} \phi_2\tilde{L}[\tilde{\Phi}]\,dy \equiv 0. \quad (5.36)
\end{aligned}
$$

In the present case, the Rayleigh operator defined by (5.32) is self-adjoint and $\tilde{\Phi} = \Phi$. The above solvability condition yields an *amplitude evolution equation* for $A(X)$ of the form

$$
M(X)\frac{dA}{dX} + N(X)A = 0. \quad (5.37)
$$

According to (5.36), the complex function $M(X)$ (respectively $N(X)$) is given by the y-integral of the product of $\tilde{\Phi} = \Phi$ with the expression multiplying dA/dX (respectively A) in (5.35).

In summary, the WKBJ procedure outlined above has led to an $\mathcal{O}(1)$ approximation of the instability wave field of the form

$$
\psi(x, y, t) \sim A(X)\Phi(y; X; \omega_f)\exp\left[i\left(\frac{1}{\varepsilon}\int_0^X k(X; \omega_f)dX - \omega_f t\right)\right],
$$
$$(5.38)$$

where k is a suitable branch of the local dispersion relation (5.33). The amplitude function $A(X)$ is seen to provide a "smooth" connection between continuous local parallel approximations.

Gaster *et al.* (1985) have conducted a detailed comparison between the predictions of the slowly-diverging flow analysis and experimental observations in forced turbulent mixing layers. As alluded to in section 5.1, the measured *mean* velocity profiles are found to be adequately represented, in nondimensional variables scaled with respect to $\bar{U}$ and $\delta_\omega(0)/2$, by the self-similar family of hyperbolic tangent functions

$$U(y; X) = 1 + R \tanh \left[\frac{y}{\delta(X)} \right] \qquad (5.39)$$

of varying nondimensional thickness $\delta(X) \equiv \delta_w(x)/\delta_w(0)$. Note, however, that this expression is not a solution of the "inviscid boundary layer" equations (5.25), as required in a strict application of the WKBJ formulation. The local spatial instability characteristics may then be recovered by suitably interpreting the parallel flow results in terms of a rescaled local frequency parameter $\omega = \omega_f \delta(X)$ and a rescaled local wavenumber $k = k(X; \omega_f) \delta(X)$ (see section 5.1). Finally the evolution of $A(X)$ is computed via numerical integration of amplitude equation (5.37). The theoretical evolution of the perturbation amplitude resulting from (5.38) is compared with experiments in Fig. 5.16. Slowly-diverging theory results are seen to exceed measured values by a significant margin[*]. This is not too surprising since, in the linear analysis, finite-amplitude effects are not taken into consideration.

It is enlightening to examine the instability analysis of slowly diverging flows from the broader perspective adopted in the presentation of fundamental concepts. In keeping with the philosophy adopted in section 3.2, the hydrodynamic equation (5.27) is hence replaced by the partial differential operator

$$D \left[-i \frac{\partial}{\partial x}, i \frac{\partial}{\partial t}; X \right] \psi + \varepsilon D_\varepsilon \left[-i \frac{\partial}{\partial x}, i \frac{\partial}{\partial t}; X \right] \psi = \mathcal{O}(\varepsilon^2), \qquad (5.40)$$

where D_ε effectively represents nonparallel flow effects associated with the basic flow streamwise derivatives. Note that (5.40) readily follows

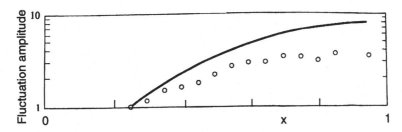

Fig. 5.16. Overall amplification of disturbances with streamwise distance x from the trailing edge of the splitter plate for the forcing Strouhal number $St \equiv f \delta_w(0)/\bar{U} = 0.96$. Continuous line is WKBJ theoretical prediction. After Gaster *et al.* (1985).

---

[*] The more favorable comparison presented in Fig. 5.9 of Ho & Huerre (1984) pertained to a preliminary version of the study by Gaster *et al.* (1985) !

from its parallel flow counterpart $D[-i\partial/\partial x, i\partial/\partial t]\psi = 0$ by "slightly" breaking the invariance with respect to continuous space translations along the streamwise x-direction. Following the previous development, the asymptotic response is taken to be

$$\psi(x,t) = \phi(X) \exp\left\{ i\left[\frac{1}{\varepsilon} \int_0^X k(X;\omega_f)\, dX - \omega_f t \right] \right\}, \qquad (5.41)$$

with

$$\phi(X) \sim \phi_1(X) + \varepsilon\phi_2(X) + \dots. \qquad (5.42)$$

The spatial branch $k(X;\omega_f)$ is a solution of the local dispersion relation $D[k,\omega_f;X] = 0$ associated with (5.40). As in the preceding case, we apply to (5.40) transformation (5.29a,b) with $\Theta_X = k(X;\omega_f)$, whereby $\partial/\partial t \to -i\omega_f$ and $\partial/\partial x \to ik(X,\omega_f) + \varepsilon\partial/\partial X$. The governing equation for ϕ is then found to be

$$D\left[k - i\varepsilon\frac{\partial}{\partial X}, \omega_f; X\right]\phi + \varepsilon D_\varepsilon(k,\omega_f;X)\phi = \mathcal{O}(\varepsilon^2). \qquad (5.43)$$

Bearing in mind that multiplication by $ik(X;\omega_f)$ and differentiation with respect to X do not commute, the local dispersion operator becomes

$$D\left[k - i\varepsilon\frac{\partial}{\partial X}, \omega_f; X\right]\phi \sim D(k,\omega_f;X)\phi \\ - \frac{1}{2}i\varepsilon\left\{ D_k(k,\omega_f;X)\frac{\partial}{\partial X} + \frac{\partial}{\partial X}D_k(k,\omega_f;X) \right\}\phi, \qquad (5.44)$$

i.e.

$$D\left[k - i\varepsilon\frac{\partial}{\partial X}, \omega_f; X\right]\phi \sim D(k,\omega_f;X)\phi \\ - i\varepsilon\left\{ D_k(k,\omega_f;X)\frac{\partial}{\partial X} + \frac{1}{2}D_{kk}(k,\omega_f;X)k_X \right\}\phi. \qquad (5.45)$$

Upon substituting the expansions (5.42) and (5.45) into the governing equation (5.43), one obtains a sequence of problems for ϕ_1 and ϕ_2. The $\mathcal{O}(1)$ problem reads

$$D(k,\omega_f;X)\phi_1(X) = 0. \qquad (5.46)$$

It is identically satisfied provided that k is a spatial branch, and one may set $\phi_1(X) = A(X)$.

The $\mathcal{O}(\varepsilon)$ problem gives rise to

$$D(k,\omega_f;X)\phi_2 = iD_k(k,\omega_f;X)\frac{dA}{dX} \\ + \frac{i}{2}D_{kk}(k,\omega_f;X)k_X A - D_\varepsilon(k,\omega_f;X)A = 0, \qquad (5.47)$$

which yields the governing equation for A. It is convenient to introduce in (5.47) the group velocity $\omega_k = -D_k/D_\omega$ and the expression $\delta\omega \equiv -D_\varepsilon/D_\omega$ so that the final *amplitude equation* reads

$$i\omega_k(k;X)\frac{dA}{dX} - \left\{\delta\omega + \frac{i}{2}\frac{D_{kk}(k,\omega_{\mathrm{f}};X)}{D_\omega(k,\omega_{\mathrm{f}};X)}k_X\right\}A = 0.$$ (5.48)

Equation (5.48) is readily integrated to yield

$$A(X) = A_0\exp\left\{-i\int_0^X\frac{\delta\omega + \frac{i}{2}\{D_{kk}(k,\omega_{\mathrm{f}};X)/D_\omega(k,\omega_{\mathrm{f}};X)\}k_X}{\omega_k(k;X)}dX\right\}.$$ (5.49)

This completes the calculation of the $\mathcal{O}(1)$ approximation to the perturbation field

$$\psi(x,t) \sim A(X)\exp\left\{i\left[\frac{1}{\varepsilon}\int_0^X k(X;\omega_{\mathrm{f}})dX - \omega_{\mathrm{f}}t\right]\right\}.$$ (5.50)

As an additional bonus, we have obtained a convenient form of the coefficients $M(X)$ and $N(X)$ appearing in the original equation (5.37) which only involves derivatives of the local dispersion relation and non parallel flow correction terms k_X and $\delta\omega$. The general operator formulation (5.40) has led to the generic evolution equation (5.48) for the signaling problem in convectively unstable spatially developing flows. This discussion is taken up again in section 6.2 to introduce global modes.

5.3 Secondary instabilities

The instabilities investigated until now have been restricted to parallel or weakly non parallel mixing layers. In the present section, we consider the evolution of three-dimensional perturbations on a periodic alley of two-dimensional large scale vortices which arises as a result of the primary inflectional instability. The methodology used to study such secondary instabilities is discussed at length in sections 8.4–5 where it is applied to the two-dimensional finite amplitude states of plane channel flow. We solely focus here on the results rather than on the method itself. The analysis described below follows the work of Pierrehumbert and Widnall (1982).

In the context of *shear layer secondary instabilities*, the basic flow should represent a two-dimensional *finite amplitude periodic vortex street*. Such a family of solutions has been derived analytically for the steady two-dimensional Euler equations by Stuart (1967).* According to (4.27)

---

* These so-called Stuart vortices are only simple models of the nonlinear states reached by Kelvin-Helmholtz instability waves. The actual state depends on initial conditions and can only be calculated numerically.

with $\frac{\partial}{\partial t} = 0$, the spanwise vorticity $\Omega_0(x, y)$ is then conserved along streamlines. Hence, if $\Psi_0(x, y)$ is the streamfunction, $\Omega_0 \equiv -\nabla^2\Psi_0$ is solely a function of Ψ_0. By assuming a specific form for this function, Stuart (1967) obtained the family of solutions of the Euler equations

$$\Psi_0(x, y) = \tfrac{1}{2}\ln[\cosh(2y) - \rho\cos(2x)], \qquad (5.51)$$

where the vorticity concentration parameter ρ may take continuous values in the range $0 \le \rho \le 1$. Note that (5.51) is written in nondimensional variables by choosing as velocity and length scales half the velocity difference $\Delta U/2$ and λ/π, where λ is the vortex street spatial period. It is understood that the resulting steady *Kelvin cat's eyes* streamline pattern (Fig. 5.17) is viewed in a frame of reference moving with the structures at the average velocity $\bar{U}$. For $\rho = 1$, this state is nothing but a periodic array of point vortices whereas, for $\rho = 0$, it reduces to the familiar parallel velocity profile $U_0(y) = \tanh 2y$. Browand and Weidman (1976) have demonstrated that such a flow with $\rho = 0.25$ approximately displays the same vorticity contours as the fully mature vortices observed in their spatially developing mixing layer experiments. This streamwise periodic non-parallel basic flow is therefore appropriate to study the two main processes which appear in a shear layer beyond roll-up: *vortex pairing*, whereby neighboring vortices amalgamate to form a larger structure, and three-dimensional instabilities which are responsible for the appearance of *streamwise vortices*.

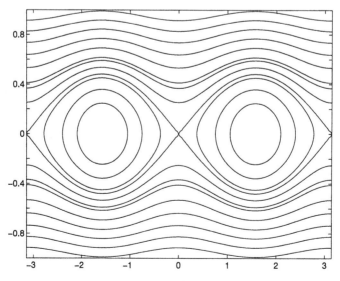

Fig. 5.17.   Typical Kelvin cat's eye streamline pattern of the Stuart vortex solution (5.51) in x–y plane for $\rho = 0.25$.

In this section, it is assumed that the instability is purely inviscid and temporal. An *ad hoc* Taylor hypothesis may then be invoked to connect temporal and spatial evolutions *via* a Galilean transformation of appropriate velocity.

As a consequence of the steady, two-dimensional and periodic nature of (5.51), no explicit dependence on the spanwise variable z and on time t is introduced in the secondary instability operator (3.8). Temporal instability modes are therefore sought in the form (see section 8.5 for details)

$$\mathbf{u}(x,y,z,t) = \mathcal{R}e\left\{\hat{\mathbf{u}}(x,y)\exp\left[i(\alpha x + \beta z - \varpi t)\right]\right\}, \qquad (5.52)$$

where $\varpi \equiv \varpi_r + i\varpi_i$. The quantity ϖ_i measures the unknown temporal growth rate and ϖ_r the unknown frequency shift with respect to the row of large scale vortices. The given parameter β denotes the spanwise wavenumber of fluctuations: it is real since only the temporal stability problem is being investigated. Moreover the eigenfunction $\hat{\mathbf{u}}(x,y)$ is assumed to be streamwise periodic with the same wavelength π as the basic flow (5.51), whereas α ($0 \leq \alpha \leq 1$) designates the given sideband streamwise wavenumber.

As in the primary temporal instability, (5.52) and its analogue for $p(x,y,z,t)$ are substituted into the linearized inviscid form of instability equations (3.5–6), with the basic state $U_0(x,y)$ given by (5.51). Enforcement of the associated boundary conditions then defines an *eigenvalue problem* (see section 8.5 for details) which admits nontrivial solutions if and only if α, β and ϖ satisfy a dispersion relation of the form

$$D(\alpha,\beta,\varpi;\rho) = 0. \qquad (5.53)$$

As summarized in Fig. 5.18, several types of *secondary instability modes* have been identified by Pierrehumbert and Widnall (1982) and Schoppa *et al.* (1995):

fundamental modes ($\alpha \equiv 0$) possess the same streamwise wavelength as the primary vortex row and synchronously propagate ($\varpi_r \equiv 0$) with the basic row of vortices. The associated eigenmodes $\hat{\mathbf{u}}(x,y)$ are such that the spanwise fluctuating vorticity is antisymmetric with respect to the primary vortex center. In the context of mixing layers, such perturbations are commonly referred to as phase-locked *translative instability modes* since they result in spanwise in-phase undulations of adjacent vortices (Fig. 5.18a).

subharmonic modes ($\alpha \equiv 1$) possess *twice* the streamwise wavelength of the primary vortex row and synchronously propagate ($\varpi_r \equiv 0$)

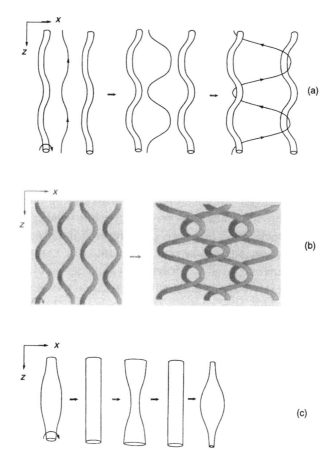

Fig. 5.18.　Flow patterns associated with perturbed Stuart vortices in x–z plane. (a) Translative mode. (b) Helical pairing mode. (c) Bulging mode. After Schoppa *et al.* (1995).

with the basic row of vortices. In the context of mixing layers, such perturbations are commonly referred to as phase-locked *helical pairing instability modes* since, in that instance, adjacent vortices display antisymmetric spanwise undulations (Fig. 5.18b). The resulting spatial pattern is invariant under combined streamwise and spanwise translations of one basic wavelength π and half a spanwise wavelength π/β respectively.

bulging modes $(\alpha \equiv 0)$ possess the same streamwise wavelength as the primary vortex row but, contrary to the translative modes, spanwise vorticity perturbations are symmetric about the vortex center. As shown by Schoppa *et al.* (1995), this core instability is characterized by a spanwise-periodic expansion and contraction of the primary two-dimensional vortex cores (Fig. 5.18c).

In the following, we focus on the subharmonic and fundamental modes of Pierrehumbert and Widnall (1982). The growth rate ϖ_i of the *helical pairing instability* mode is shown in Fig. 5.19 as a function of spanwise wavenumber β. Each curve displays a bandwidth of amplified wavenumbers between $\beta = 0$ corresponding to the maximum growth rate (0.49 for $\rho = 0.25$) and a cutoff β_c ($\beta_c = 2.2$ for $\rho = 0.25$). This class of modes is therefore dominated by purely two-dimensional perturbations ($\beta = 0$). Such fluctuations are associated with the ubiquitous *pairing* process which is largely responsible for the spatial spreading of mixing layers (Fig. 2.3). Furthermore the maximum growth rate at $\beta = 0$ is close to the value 0.5 analytically obtained by Lamb (1995) for a row of point vortices ($\rho = 1$). For finite but small values of β, helical pairing modes can be viewed as giving rise to a localized merging process along the span.

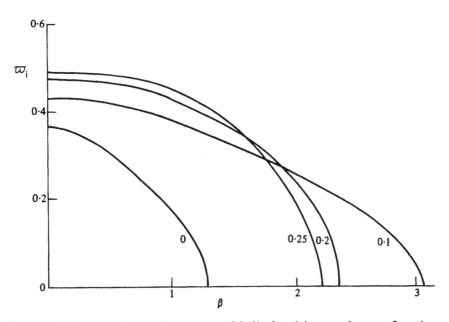

Fig. 5.19. Temporal growth rate ϖ_i of helical pairing mode as a function of spanwise wavenumber β for different values of concentration parameter ρ. After Pierrehumbert and Widnall (1982).

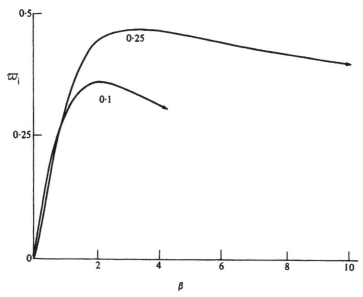

Fig. 5.20.   Temporal growth rate ϖ_i of translative mode as a function of spanwise wavenumber β for different values of concentration parameter ρ. After Pierrehumbert and Widnall (1982).

Translative mode growth rates ϖ_i are represented on Fig. 5.20. They display a large bandwidth of amplified spanwise wavenumbers β with a broad maximum at a spanwise period comparable to the primary Stuart vortex spacing. For $\rho = 0.25$, the preferred spanwise wavelength is 2/3 of the Stuart vortex spacing. Note that, in contrast to the helical pairing instability (Fig. 5.19), translative growth rates ϖ_i tend to zero as $\beta \to 0$ thereby indicating the truly *three-dimensional* nature of the instability. Indeed, the translative mode has been associated with the presence of streamwise streaks (Fig. 2.4) in mixing layer experiments (Bernal and Roshko, 1986; Lasheras and Choi, 1988; Rogers and Moser, 1992).

6 The wake behind a bluff body

According to the qualitative discussion of section 2.2, wake flows past cylindrical bluff bodies experience a well defined *Hopf bifurcation* leading from a stationary state to a temporally periodic regime as the Reynolds number (2.4) based on cylinder diameter d is increased through the value $\mathrm{Re}_{G_c} = 48.5$. In physical space, this transition is marked by the spontaneous appearance of the celebrated *Bénard–Kármán vortex street* composed of counter-rotating spanwise vortices that are shed from the cylinder surface into the wake. Wake mean velocity profiles measured downstream of the cylinder exhibit *two inflection points* (Fig. 2.1c), and, according to the *Rayleigh criterion* of section 4.3, they are likely to be *inviscidly unstable*, just as the spatial mixing layers examined in the previous section. But, in sharp contrast with spatial mixing layers, the dynamics of bluff-body wakes are relatively insensitive to small levels of external noise: the wake instability gives rise to *self-sustained intrinsic oscillations* characterized by a well defined frequency. We shall demonstrate that the intrinsic nature of the instability is intimately related to the appearance of a sufficiently large region of *absolute instability* in the near-wake region. Thus, in these notes, the wake serves as the archetype of inviscidly unstable open flows displaying intrinsic oscillations generated by a sufficiently robust pocket of absolute instability.

The inviscid instability concepts introduced in section 4 are perfectly adequate to describe the properties of parallel wake velocity profiles, as summarized in section 6.1. In order to account for the intrinsic nature of the instability, we choose to regard the Bénard–Kármán vortex street as an extended wave packet or *global mode* that lives on a slowly diverging basic flow and must satisfy specific boundary conditions at infinity in all directions. It is therefore necessary to relate the local instability characteristics at different streamwise stations in order to construct a consistent description of global modes. The WKBJ approximation discussed in section 5.2 naturally provides the necessary framework to achieve this goal. The local instability ideas introduced in section 3.2 then have to be supplemented with *global instability concepts* as developed in section 6.2. It should be emphasized that these notions are relevant not only in wake flows but also in other classes of open flows such as hot or low-density jets, counterflow mixing layers, etc. The intrinsic dynamics of the wake make this open flow an ideal candidate for the kind of *pattern formation* analysis that has been so successful in closed flow systems (Cross and Hohenberg, 1993; Newell *el al.*, 1993). Bénard–Kármán vortex streets have in fact been observed to sustain large-scale three-dimensional patterns which can be successfully modeled in the context of the GL equation, as presented in section 6.3.

Wake dynamics and associated global-mode concepts have been surveyed by Hannemann and Oertel (1990) and Huerre and Monkewitz (1990). Wake patterns are elegantly discussed from an experimentalist's point of view by Williamson (1994, 1996). Current research on wake flow instabilities is compiled in a book of proceedings by Eckelmann *et al.* (1993).

6.1 Linear instability of locally parallel wakes

The inviscid instability characteristics of parallel wake profiles are analyzed by following the same methodology as in the case of mixing layers. Consider, for example, the so-called $\text{sech}^2 y$ wake given, in dimensional variables, by

$$U(y) = U_\infty + (U_c - U_\infty)\,\text{sech}^2 \frac{y}{\delta}\,, \tag{6.1}$$

where U_∞ is the free-stream velocity, U_c the centerline velocity and δ a characteristic thickness (Fig. 6.1). Upon choosing δ and the average velocity $\bar{U} = (U_c + U_\infty)/2$ as reference scales, the nondimensional form of (6.1) reads:

$$U(y; R) = 1 - R + 2R\,\text{sech}^2 y\,, \tag{6.2}$$

with $R \equiv (U_c - U_\infty)/(U_c + U_\infty)$ denoting the *velocity ratio*.

The meaning of R is obvious: wake and jet profiles are obtained for $R < 0$ and $R > 0$ respectively. As R decreases below zero (Fig. 6.2), the velocity defect gradually increases within the wake. In the range $-1 < R < 0$, the wake is co-flowing in the same direction as the free stream (Fig. 6.2a). When $R = -1$, the centerline velocity becomes zero (Fig. 6.2b). Finally, there is a finite region of counterflow as soon as R decreases below -1 (Fig. 6.2c). The dispersion relation $D(k, \omega; R) = 0$ is calculated numerically by solving the Rayleigh equation (4.33) subject to exponential decay boundary conditions at $y = \pm\infty$. As in the case of mixing layers, the functional dependence of D on R can immediately be made explicit: upon substituting (6.2) into (4.33), it is straightforward

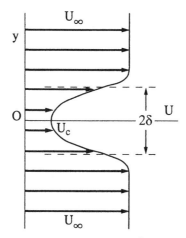

Fig. 6.1. Basic velocity profile of $\text{sech}^2 y$ wake given by (6.1).

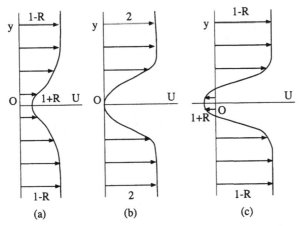

Fig. 6.2. Effect of velocity ratio on basic wake velocity profile (6.2). (a) $-1 < R < 0$; (b) $R = -1$; (c) $R < -1$.

to show that, to each eigenvalue $c_1(k)$ associated with the jet profile $U_1(y) \equiv U(y;1)/2 = \text{sech}^2 y$ corresponds an eigenvalue of the one-parameter family $U(y;R)$

$$c(k;R) = 1 - R + 2Rc_1(k). \qquad (6.3)$$

Equivalently, once the temporal stability properties of $U_1(y)$ have been determined, those of $U(y;R)$ can be retrieved for any velocity ratio R by applying the relation

$$\omega(k;R) = (1-R)k + 2R\omega_1(k), \qquad (6.4)$$

where $\omega_1(k) = kc_1(k)$.

The wake velocity profile $U(y;R)$ is even with two inflection points and admits two unstable temporal branches, namely a *sinuous mode* and a *varicose mode* characterized by even and odd eigenfunctions $\phi(y)$ respectively. The physical meaning of the terms "sinuous" and "varicose" should be made clear: according to (4.32b), the cross-stream perturbation velocity v has the same symmetry properties in y as $\phi(y)$. Thus the sinuous (varicose) mode imparts symmetric (antisymmetric) cross-stream perturbations to the wake which tend to deform it in the manner sketched in Fig. 6.3a,b. The temporal stability calculations of Drazin and Howard (1966) for the jet profile $U_1(y) = \text{sech}^2 y$ have been reproduced and adapted according to the transformation rule (6.4) on Fig. 6.4a,b. The sinuous mode is seen to give rise to the larger growth rates over a wider range of wavenumbers $0 < k < 2$. This prediction is in qualitative agreement with experimental observations: the Bénard–Kármán vortex street (Fig. 2.13) indeed develops from sinuous deformations of the wake

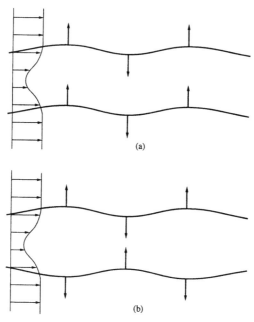

Fig. 6.3. Wake instability modes. (a) Sinuous mode with $v(x, -y, t) = v(x, y, t)$; (b) varicose mode with $v(x, -y, t) = -v(x, y, t)$.

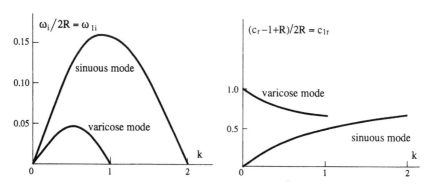

Fig. 6.4.   Temporal instability characteristics of the $\mathrm{sech}^2 y$ wake.   (a) Normalized temporal growth rate $\omega_i/(2R)$ versus real wavenumber k; (b) real phase velocity, suitably normalized, versus k.   After Betchov and Criminale (1966).

and the sinuous mode is also commonly referred to as the Kármán instability.

We have mentioned in section 5.1 that the spatial properties of mixing layers were successfully analyzed and interpreted long before absolute/convective instability issues came to the foreground. In the case of wakes, Betchov and Criminale (1966) and Mattingly and Criminale (1972) realized at an early stage that the spatial instability characteristics of the $\text{sech}^2 y$ velocity profile (6.2) with $R = -1$ (zero centerline velocity) could not as easily be obtained and interpreted: an algebraic branch point singularity appeared in the complex ω-plane that made its presence felt on the real ω-axis. This was indeed a manifestation of *absolute instability* and Koch (1985) seems to have been the first to explicitly recognize the significance of the branch point behavior in terms of absolute/convective instability ideas. More specifically, by applying the absolute/convective criterion of section 3.2.5 to a particular family of quasi-parallel velocity profiles, Koch (1985) demonstrated that the nearfield wake of a blunt rectangular plate indeed presents a pocket of absolute instability. Triantafyllou *et al.* (1986) established, on the basis of numerical simulations, that the circular cylinder wake also displays a region of absolute instability in the Kármán vortex shedding regime. In these notes, we choose to illustrate the essential results by summarizing the elegant and thorough procedure of Monkewitz (1988). The mean flow development of bluff-body wakes is modeled by the two-parameter family of nondimensional velocity profiles

$$U(y; R, N) = 1 - R + 2RU_1(y; N) \,, \tag{6.5}$$

where $R = (U_c - U_\infty)/(U_c + U_\infty)$ is the previously defined velocity ratio and $U_1(y; N)$ is analytically given by

$$U_1(y; N) = \left\{ 1 + \sinh^{2N}[y \sinh^{-1}(1)] \right\}^{-1} \,. \tag{6.6}$$

The effect of the shape-parameter N is illustrated in Fig. 6.5: when $N = \infty$, the velocity profile $U_1(y, \infty)$ is a top-hat jet whereas it reduces to the previously considered $\text{sech}^2 y$ jet in the limit $N = 1$. Each downstream station in the wake is in effect characterized by a particular value of R and N.

Monkewitz (1988) proceeded to determine the viscous dispersion relation

$$D\left[k, \omega; R, N, \widehat{\text{Re}}\right] = 0 \tag{6.7}$$

as a function of R, N and the Reynolds number $\widehat{\text{Re}}$ based on the average velocity $\bar{U}$ and the *wake half width*. The latter Reynolds number should not be confused with the one defined in (2.4). The Orr–Sommerfeld equation (7.16), i.e. the Rayleigh equation with viscous diffusion terms included (see section 7.2), is solved numerically subject to exponential decay boundary conditions at $y = \pm\infty$. As in the case of the $\text{sech}^2 y$ wake,

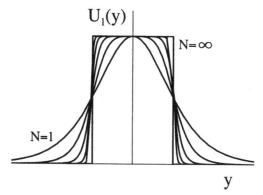

Fig. 6.5. Basic velocity profile $U_1(y; N)$ given by (6.6), for $N = 1, 2, 4, 8, 16, \infty$. After Monkewitz (1988).

there exist two modes of instability, namely a varicose mode and a sinuous Kármán mode. The varicose mode is found to be at most convectively unstable and only the sinuous mode is discussed in the sequel. According to the criterion presented in section 3.2, one seeks to determine the complex wavenumber k_0 of zero group velocity $\partial\omega/\partial k(k_0) = 0$. A sample result is illustrated in Fig. 6.6: contours $\omega_r = $ const and $\omega_i = $ const exhibit a typical saddle-point behavior at a complex wavenumber k_0 as one would expect from the Taylor expansion

$$\omega - \omega_0 \sim \frac{1}{2}\frac{\partial^2\omega}{\partial k^2}(k_0)(k - k_0)^2 \qquad (6.8)$$

around a point of zero group velocity. In this case, the spatial branches $k^+(\omega)$ and $k^-(\omega)$, represented by the $\omega_i = $ const contours, pinch at k_0 as ω_i is decreased from 0.25 to zero. The absolute growth rate $\omega_{0,i}$ is therefore zero and the figure corresponds to parameter values separating domains of absolute and convective instability. This procedure may be repeated to determine the nature of the instability at various points in the space of control parameters.

As an example, the real and imaginary parts of the absolute frequency ω_0 are represented in Fig. 6.7 as a function of $1/N$ for $\widehat{\mathrm{Re}} = 20$ and two values of the velocity ratio $R = -1$, $R = -1.1$. When the centerline velocity is zero ($R = -1$) the family of velocity profiles (6.5) is seen to become absolutely unstable ($\omega_{0,i} > 0$) in the range $1 < N < 3.9$. This feature strikingly illustrates the fact that *a flow may be absolutely unstable even in the absence of counterflow.* However counterflow obviously promotes absolute instability as seen from Fig. 6.7, by comparing the variations of $\omega_{0,i}$ for $R = -1$ and $R = -1.1$. Results of this kind may then be exploited to determine the shape of the marginal surface

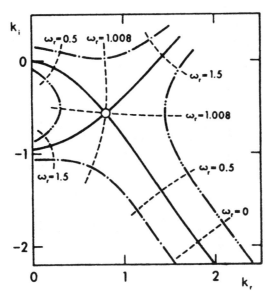

Fig. 6.6. Contour levels $\omega_r = $ const. and $\omega_i = $ const. for profile (6.5) at $R = -1$, $N = 2$, $\widehat{Re} = 11.3$. ——, curve $\omega_i = 0$; —·—·—, $\omega_i = 0.25$; —··—··—, $\omega_i = -0.25$; — — —, curve $\omega_r = $ const.; $\circ$, $k^+ = k^- = k_0$, $\omega_0 = 1.008 + 0i$. After Monkewitz (1988).

$\omega_{0,i} = 0$ separating the absolute and convective instability regions in the space of the control parameter $(R, N, \widehat{Re})$ as shown in Fig. 6.8. At infinite Reynolds number, i.e. in the purely inviscid limit governed by the Rayleigh equation, absolute instability is seen to arise first at $N \sim 2$, for a relatively flat wake (Fig. 6.5), and at a velocity ratio $R = -0.85$ corresponding to a sufficiently "deep" co-flowing wake. Viscous effects tend to push the marginal curve to lower velocity ratios and stronger counterflows.

Through an ingenious and careful analysis of existing experimental data, Monkewitz (1988) successfully established that the family of velocity profiles (6.5) indeed constitutes a satisfactory representation of measured streamwise velocity distributions in the near-wake of circular cylinders as long as the Reynolds number Re based on diameter and free-stream velocity is below 48.5. The downstream development of the wake is effectively assumed to be weakly diverging in the sense of section 5.2 and the "most dangerous" streamwise location, where local absolute instability is just likely to set in, is expected to be at the point of maximum backflow velocity, i.e. minimum velocity ratio R. This station is located approximately one diameter downstream of

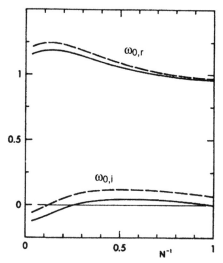

Fig. 6.7. Variations of $\omega_{0,r}$ and $\omega_{0,i}$ as a function of $1/N$ for $\widehat{Re} = 20$. —,
$R = -1$; – – –, $R = -1.1$. After Monkewitz (1988).

the cylinder axis in the same cross-stream section as the vortex centers
within the recirculation bubble (see section 2.2). At this station, the
velocity profile is of the form (6.5) with parameter settings for R, N
and $\widehat{Re}$ that vary with cylinder Reynolds number Re. The sinuous
mode is then found by Monkewitz (1988) to be locally convectively
unstable at this station in the interval $2 < \widehat{Re} < 10$, which corresponds

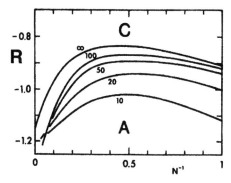

Fig. 6.8. Absolute/convective instability boundary in the R–N^{-1} plane for
the wakes (6.5) at different Reynolds numbers $\widehat{Re}$ indicated near the curves.
The symbols A and C denote the regions of absolute and convective instability
respectively. After Monkewitz (1988).

to 5 < Re < 25. As Reynolds numbers increase through the range $10 < \widehat{\text{Re}} < 16$, i.e. $25 < \text{Re} < 48.5$, the velocity profile becomes absolutely unstable, a pocket of absolute instability gradually increasing in extent within the recirculation bubbles. It can therefore be deduced that the near wake behind a circular cylinder undergoes the following sequence of transitions in its local instability characteristics: *locally stable everywhere, region of local convective instability, finite region of absolute instability embedded within a convectively unstable domain*. It should be emphasized that cylinder wakes are observed to remain stationary in the entire range $0 < \text{Re} < \text{Re}_{\text{G}_c} = 48.5$, with no sign of instability in the absence of forcing (section 2.2). From a qualitative point of view, the transition to an oscillatory regime only takes place when a *sufficiently large region of absolute instability is capable of acting as a continuous source of energy for the self-sustained development of the Bénard–Kármán vortex street*. This scenario has been fully confirmed in numerical experiments of cylinder wakes by Zebib (1987) and Yang and Zebib (1989): the flow is observed to undergo a *Hopf bifurcation* to global self-sustained oscillations only when the region of local absolute instability is sufficiently wide. Wakes and mixing layers are therefore seen to behave in a radically distinct fashion: co-flowing mixing layers are convectively unstable everywhere and, as a result, they primarily act as amplifiers of external noise, their response being described by spatial instability theory; wakes exhibit a finite region of absolute instability which appears to trigger the onset of intrinsically driven oscillations. In order to make further progress, one must step back and extend the parallel flow concepts introduced in section 3.2 to encompass slowly diverging flow situations. This task has already been accomplished for the specific case of convectively unstable spatially developing mixing layers in section 5.2. In the next section, an explicit distinction is made between *local* and *global* instability, for an arbitrary dispersion relation which is slowly varying in space, the ultimate objective being the derivation of *frequency selection criteria* capable of predicting for instance the Strouhal number of the Bénard–Kármán vortex street.

6.2 Global instability concepts for spatially developing flows

The general notions introduced in section 3.2 dealt with steady parallel basic flows that are necessarily invariant with respect to continuous translations in time and along the streamwise x-direction. We now follow the general development succinctly summarized in section 3 of Huerre and Monkewitz (1990). It should be clear that the formulation is applicable to a wide class of spatially evolving flows (mixing layers, jets, wakes, etc.). As in section 5.2, streamwise inhomogeneities of the basic flow are assumed to be characterized by an evolution length scale $L \sim [(1/\delta_\omega)(d\delta_\omega/dx)]^{-1}$, where $\delta_\omega(x)$ is the local vorticity thickness. Let

$$\varepsilon \sim \frac{\lambda}{L} \ll 1 \qquad (6.9)$$

denote the small inhomogeneity parameter of the problem. Upon introducing as reference scales quantities such as $\delta_\omega(0)/2$ and $\bar{U}$, the basic flow takes the form $U(y; X)$, where the slow variable $X = \varepsilon x$ is effectively a local control parameter. Following the general formalism introduced at the end of section 5.2 (equation (5.40)), the full hydrodynamic equations are replaced by the linear partial differential operator

$$D\left[-i\frac{\partial}{\partial x}, i\frac{\partial}{\partial t}; X, R\right]\psi + \varepsilon D_\varepsilon\left[-i\frac{\partial}{\partial x}, i\frac{\partial}{\partial t}; X, R\right]\psi = \mathcal{O}(\varepsilon^2), \qquad (6.10)$$

for the fluctuating field $\psi(x, t)$. The global control parameter R is for instance the velocity ratio in mixing layers or the Reynolds number Re based on cylinder diameter introduced in the previous section.* Note that the *local dispersion relation* is recovered by freezing X and keeping the leading order term only:

$$D(k, \omega; X, R) = 0. \qquad (6.11)$$

Absolute/convective instability concepts (see section 3.2) are readily applicable to (6.11). If the dispersion relation is assumed to admit a single generalized temporal mode $\omega(k; X, R)$, one may introduce a *local absolute frequency* $\omega_0(X, R)$ and wavenumber $k_0(X, R)$, defined by

$$\frac{\partial\omega}{\partial k}(k_0; X, R) = 0, \quad \omega_0(X, R) = \omega(k_0; X, R). \qquad (6.12)$$

Similarly, one may define a local complex frequency $\omega_{max}(X, R)$ and real wavenumber $k_{max}(X, R)$ corresponding to the most amplified temporal disturbance at a given X, such that

$$\frac{\partial\omega_i}{\partial k}(k_{max}; X, R) = 0, \qquad \omega_{max}(X, R) \equiv \omega(k_{max}; X, R). \qquad (6.13)$$

Let $\omega_{i,max}|_{max}$ and $\omega_{0,i}|_{max}$ respectively denote the maximum temporal growth rate and maximum absolute growth rate *over all real X under consideration*. Four broad *classes of spatially developing flows* may then be distinguished according to the nature of the local instability at each streamwise station, as sketched in Fig. 6.9. In the first group (Fig. 6.9a) the flow is *locally stable* uniformly along the streamwise direction so that $\omega_{i,max}|_{max} < 0$ and $\omega_{0,i}|_{max} < 0$. If a region of local convective instability

---

* The parameter R should not necessarily be identified with the velocity ratio.

is present (Fig. 6.9b), whence $\omega_{i,max}|_{max} > 0$ and $\omega_{0,i}|_{max} < 0$, the flow is said to be *locally convectively unstable*. In the third class of flows (Fig. 6.9c), the maximum absolute growth rate $\omega_{0,i}|_{max}$ is close to zero. Finally, in the fourth class of flows (Fig. 6.9d), $\omega_{i,max}|_{max} > 0$, $\omega_{0,i}|_{max} > 0$ and *local absolute instability* prevails over a finite extent in the streamwise direction.

These four classes of spatially developing flows have already been encountered in the various configurations that have been discussed so far. In the counterflow shear layer experiments of Strykowski and Niccum (1991) presented in section 5.1, the flow is locally convectively unstable below a velocity ratio $R_t = 1.315$. As R is increased through R_t the flow becomes marginally absolutely unstable and then locally absolutely unstable. In this case, we have noted that *the onset of discrete global oscillations above R_{G_c} immediately follows the appearance of a finite region of absolute instability.* The spatially developing wake behind a circular cylinder also experiences the four characteristic configurations of Fig. 6.9 as the global control parameter Re is increased through $Re_c = 5$ and $Re_t = 25$, but self sustained oscillations do not appear until $Re_{G_c} = 48.5$. Many other examples can be found in the survey article of Huerre and Monkewitz (1990).

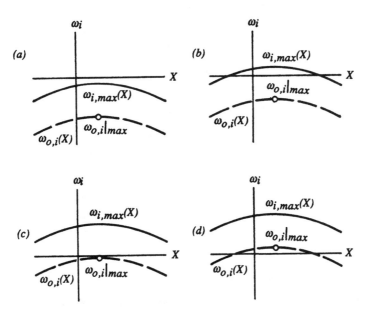

Fig. 6.9. Classes of spatially developing flows according to the nature of the local instability. (a) Uniformly stable $R < R_c$; (b) convectively unstable $R_c < R < R_t$; (c) almost absolutely unstable $R \sim R_t$; (d) pocket of absolute instability $R > R_t$ (Huerre and Monkewitz, 1990).

There remains to establish a precise relationship between the *local* instability properties specified by equation (6.11) at each streamwise station X and any *global* instability that may arise over the entire streamwise extent of the medium. These issues have been examined in various fluid mechanical configurations over the last ten years: the reader may consult the studies of Pierrehumbert (1984), Koch (1985), Chomaz *et al.* (1991) and Hunt and Crighton (1991), among others. In the following discussion, the analysis presented by Monkewitz *et al.* (1993) in the context of the linearized Navier–Stokes equations, is adapted to the one-dimensional operator (6.10).

The local stability concepts introduced in section 3.2 for strictly parallel flows may first be extended to the case of slowly varying media. Let $G(x,t;X)$ denote the global Green function defined by

$$\left\{ D\left[-i\frac{\partial}{\partial x}, i\frac{\partial}{\partial t}; X \right] + \varepsilon D_\varepsilon\left[-i\frac{\partial}{\partial x}, i\frac{\partial}{\partial t}; X \right] \right\} G = \delta(x)\delta(t) . \qquad (6.14)$$

The flow is said to be *globally stable** if

$$\lim_{t\to\infty} G(x,t) = 0 \quad \text{for all } x . \qquad (6.15)$$

It is said to be *globally unstable* if

$$\lim_{t\to\infty} G(x,t) = \infty \quad \text{for some } x . \qquad (6.16)$$

These definitions constitute the analogues of the notions of stability and instability defined in (3.12–13) for strictly parallel flows. In the present WKBJ framework, it is understood that $X = \varepsilon x$. When the flow is globally unstable in the sense of (6.16), the long time behavior of G is expected to be dominated by a linear combination of *global mode* solutions which, for steady nonparallel flows, read

$$\psi(x,t) = \phi(X)\,e^{-i\omega_\mathrm{G}t} . \qquad (6.17)$$

One of the main goals of the global analysis is the determination of the complex *global frequencies* ω_G. If $\omega_{\mathrm{G},i} < 0$ for all ω_G, the medium is globally stable. If some ω_G's are such that $\omega_{\mathrm{G},i} > 0$, the medium is *globally unstable*. For the sake of illustration, consider the specific boundary conditions

$$\lim_{x\to\pm\infty} \psi(x,t) = 0 . \qquad (6.18)$$

The determination of the unknown complex global frequencies ω_G is then tantamount to solving an eigenvalue problem in the streamwise direction X over the interval $-\infty < X < \infty$.

---

* It is recalled that this concept should not be confused with global stability in phase space, see section 3.3.

In order to rapidly generate a *global frequency selection criterion*, it is convenient to derive the WKBJ approximation for the *global Green function* $G(x,t)$. In the same spirit as in section 3.2.2, let us introduce the time Fourier transform $G(x,\omega)$ of $G(x,t)$ defined by

$$G(x,t) = \frac{1}{2\pi} \int_{L_\omega} G(x,\omega)\, e^{-i\omega t}\, d\omega, \qquad (6.19)$$

where the contour L_ω is taken parallel to the real ω-axis and above all singularities in order for the solution to be causal (Fig. 6.10). Following the WKBJ procedure outlined in the latter part of section 5.2 [cf. equation (5.41–42)], we seek a standard WKBJ approximation for $G(x,\omega)$ of the form

$$G^\pm(x,\omega) \sim \left\{ G_1^\pm(X) + \varepsilon G_2^\pm(X) + \cdots \right\} \exp\left[\frac{i}{\varepsilon} \int_0^X k^\pm(X;\omega)\, dX \right].$$
$$(6.20)$$

In the above expansion, the superscripts $+$ and $-$ refer to the downstream ($x > 0$) and upstream ($x < 0$) regions with respect to the source, respectively, the functions $k^\pm(X;\omega)$ denoting the spatial branches, solutions of $D(k,\omega;X) = 0$ which lie in the upper and lower half k-plane respectively, for sufficiently large ω_i (cf. section 3.2). The sequence of equations for $G_1^\pm$, $G_2^\pm$, etc. is identical to the one obtained in the case of the signaling problem in section 5.2. The leading-order approximation G_1 is governed by (5.46) with k, ω_f and ϕ_1 being transposed into $k^\pm$, ω and $G_1^\pm$. The solution is

$$G_1^\pm(X) = A^\pm(X). \qquad (6.21)$$

The second-order term $G_2^\pm(X)$ is governed by (5.47), with k, ω_f, ϕ_2 and A replaced by $k^\pm$, ω, $G_2^\pm(X)$ and $A^\pm$. The resulting amplitude equation (5.48) now reads

$$i\omega_k^\pm \frac{dA^\pm}{dX} - \left\{ \delta\omega^\pm + \frac{i}{2}\frac{D_{kk}^\pm}{D_\omega^\pm} k_X^\pm \right\} A^\pm = 0, \qquad (6.22)$$

where $\omega_k^\pm \equiv \partial\omega/\partial k(k^\pm;X)$, $D_{kk}^\pm \equiv D_{kk}(k^\pm,\omega;X)$, etc. The amplitude functions $A^\pm(X)$ are therefore given by the equivalent of equation (5.49), i.e.

$$A^\pm(X) = A_0^\pm \exp\left\{ -i \int_0^X \frac{\delta\omega^\pm + \frac{i}{2}\left(D_{kk}^\pm/D_\omega^\pm\right) k_X^\pm}{\omega_k^\pm}\, dX \right\}. \qquad (6.23)$$

Finally, by requiring that the Green function be continuous on either side of the source $x = 0$, one immediately obtains $A_0^+ = A_0^-$.

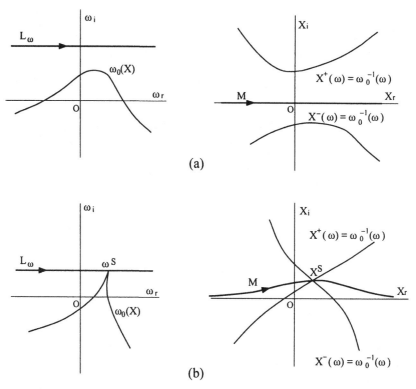

Fig. 6.10.   Global frequency selection criterion.   (a) Integration contour L_ω in complex ω-plane lies above locus of singularity $\omega_0(X)$ as X travels on M; conversely, loci of $X^+(\omega)$ and $X^-(\omega)$ as ω travels on L_ω are well separated on either side of contour M in complex X-plane.   (b) Pinching process as L_ω is gradually lowered; global frequency is given by ω^S.

We are now in a position to determine qualitatively which global frequency will dominate the large-time asymptotics of $G(x,t)$. According to (6.23), the complex amplitude $A^\pm(X)$ exhibits a singularity at complex points X of zero group velocity

$$\omega_k^\pm \equiv \frac{\partial \omega}{\partial k}[k^\pm(\omega; X); X] = 0. \qquad (6.24)$$

Recall, from (6.12), that the absolute frequency $\omega_0(X)$ is precisely defined so that (6.24) is satisfied.   We therefore conclude that, for a fixed frequency ω, the amplitude $A(X)$ and consequently $G(x,\omega)$ becomes singular at points $X = \omega_0^{-1}(\omega)$, where $\omega_0^{-1}(\omega)$ is the reciprocal function of $\omega_0(X)$.   The reasoning may now proceed in analogous fashion to the discussion of absolute and convective instabilities in parallel flows presented in section 3.2. One only needs to replace the complex k-plane by the complex X-plane. The function $\omega_0^{-1}(\omega)$ is in general multiple-valued. For a given contour L_ω in the complex frequency plane, pertaining to the Fourier integral (6.19), the locus of the points $X = \omega_0^{-1}(\omega)$ typically exhibits 2 distinct branches $X^+(\omega)$ and $X^-(\omega)$ that are well separated on either side of the real X-axis, provided that L_ω lies above all singularities (Fig. 6.10a). As L_ω is gradually lowered, the X-contour may be *pinched* at a complex location X^S by the two branches X^+ and X^- (Fig. 6.10b). Correspondingly, a singularity lies on the contour L_ω in the complex ω-plane at the location $\omega^S = \omega_0(X^S)$. It is then no longer possible to lower L_ω any further and one has "reached" the dominant singular contribution to the long-time asymptotics of $G(x,t)$. Typically, the pinching point X^S is a *saddle point* defined by

$$\frac{\partial \omega_0}{\partial X}(X^S) = 0. \tag{6.25}$$

The *global frequency selection criterion* (6.25), as derived by Chomaz *et al.* (1991), may also be expressed in the following alternative form: *knowing the local dispersion relation $\omega(k; X; R)$, the global frequency ω_G is determined by the stationarity conditions*

$$\frac{\partial \omega}{\partial k}(k^S; X^S, R) = 0, \quad \frac{\partial \omega}{\partial X}(k^S; X^S, R) = 0, \quad \omega^S = \omega(k^S; X^S, R). \tag{6.26}$$

Both k^S and X^S have to be taken complex and the global frequency $\omega_G \sim \omega^S$ is seen to be determined by the singularities (6.26) of the local dispersion relation.

Monkewitz *et al.* (1993) have demonstrated that the frequency selection criterion (6.25), equivalently (6.26), may be incorporated into a complete analytical scheme for the determination of the *global mode structure* (6.17). The complex function $\omega_0(X)$ is assumed to display a saddle-point at X^S, as sketched in Fig. 6.11. This structure is entirely consistent with the pinching process illustrated in Fig. 6.10. The function $\omega_{0,i}$ then displays a single maximum on the X_r-axis, as in the various classes of spatially developing flows defined in Fig. 6.9. Eigensolutions of complex frequency ω_G governed by (6.10) and the boundary conditions at infinity (6.18), are sought in the form (6.17). The WKBJ approximation scheme proceeds in the usual manner: one assumes

$$\phi^\pm(X) \sim \left[A^\pm(X) + \varepsilon \phi_2^\pm(X) + \ldots \right] \exp\left\{ \frac{i}{\varepsilon} \int_{X^S}^X k^\pm(X; \omega^S)\, dX \right\}, \tag{6.27}$$

and

$$\omega_G \sim \omega^S + \varepsilon \omega_2 + \cdots. \tag{6.28}$$

The $+$ and $-$ solutions (6.27) are, by construction, subdominant (exponentially small) in the regions $+$ and $-$ of Fig. 6.11 that contain

$X_r = +\infty$ and $X_r = -\infty$ respectively, and become dominant (exponentially large) in the regions $-$ and $+$ respectively. In order to satisfy boundary conditions (6.18) at infinity, the eigenfunction ϕ only consists of ϕ^+ in region $+$ and ϕ^- in region $-$. Again, by duplicating the approach followed in equations (5.41–49), or equivalently in equations (6.20–23), and bearing in mind the expansion (6.28) for ω_G, one is led to the amplitude equations

$$i\omega_k^\pm \frac{dA^\pm}{dX} + \left\{ \omega_2 - \delta\omega^\pm - \frac{i}{2}\frac{D_{kk}^\pm}{D_\omega^\pm}k_X^\pm \right\} A^\pm = 0, \qquad (6.29)$$

where $\omega_k^\pm = \partial\omega/\partial k[k^\pm(\omega^S; X); X]$, $D_{kk}^\pm = D_{kk}[k^\pm(\omega^S; X), \omega^S; X]$, etc. The solution reads

$$A^\pm(X) = A^{S\pm} \exp\left\{ i\int_{X^S}^X \frac{\omega_2 - \delta\omega^\pm - \frac{i}{2}(D_{kk}^\pm/D_\omega^\pm)k_X^\pm}{\omega_k^\pm} dX \right\}, \qquad (6.30)$$

to be compared with (6.23). As expected, this WKBJ approximation breaks down in the vicinity of the point X^S (Fig. 6.11) where $\omega_k^\pm[k^\pm(X^S); X^S] = 0$. This is precisely the region through which the ϕ^+ solution smoothly turns into the ϕ^- solution.

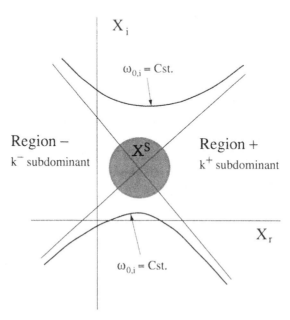

Fig. 6.11. Sketch of different regions in the complex X-plane. The curves represent contours $\omega_{0,i} = $ const. in the vicinity of the saddle point X^S. The shaded area denotes the inner turning-point region of size $\mathcal{O}(\varepsilon^{1/2})$.

To study the inner shaded area of Fig. 6.11, it is necessary to go back to the governing equation (6.10). We make an enlightened guess and assume the size of the inner region to be $\mathcal{O}(\varepsilon^{\frac{1}{2}})$. Upon introducing the inner variable

$$\bar{X} = \frac{X - X^{\mathrm{S}}}{\varepsilon^{1/2}} \tag{6.31}$$

and, following the same steps as in the outer WKBJ regions, the inner global mode structure is assumed to be of the form

$$\bar{\phi}(\bar{X}) \sim \mu(\varepsilon)\left[\bar{A}(\bar{X}) + \varepsilon^{1/2}\bar{\phi}_2(\bar{X}) + \varepsilon\bar{\phi}_3(\bar{X})\right]\exp\left(ik^{\mathrm{S}}\left(x - x^{\mathrm{S}}\right)\right), \tag{6.32}$$

where $\mu(\varepsilon)$ is, at this stage, an unknown gauge function, $\bar{A}(\bar{X})$ is the inner complex amplitude for which we seek a corresponding evolution equation and $x - x^{\mathrm{S}} = \varepsilon^{-1/2}\bar{X}$. In order to generate the governing equation for $\bar{\phi}$, it is then sufficient to apply the transformations

$$\frac{\partial}{\partial t} \to -i\omega_{\mathrm{G}}, \qquad \frac{\partial}{\partial x} \to ik^{\mathrm{S}} + \varepsilon^{1/2}\frac{\partial}{\partial \bar{X}}. \tag{6.33}$$

Bearing in mind expansion (6.28) for ω_{G}, one obtains:

$$D\left(k^{\mathrm{S}}, \omega^{\mathrm{S}}; X^{\mathrm{S}}\right)\bar{\phi} + \varepsilon^{1/2}\left[\bar{X}D_X^{\mathrm{S}} - iD_k^{\mathrm{S}}\frac{\partial}{\partial \bar{X}}\right]\bar{\phi}$$
$$+ \varepsilon\left[\omega_2 D_\omega^{\mathrm{S}} + \frac{1}{2}\bar{X}^2 D_{XX}^{\mathrm{S}} - iD_{kX}^{\mathrm{S}}\bar{X}\frac{\partial}{\partial \bar{X}} - \frac{1}{2}D_{kk}^{\mathrm{S}}\frac{\partial^2}{\partial \bar{X}^2} + D_\varepsilon^{\mathrm{S}}\right]\bar{\phi} = \mathcal{O}(\varepsilon^{3/2}). \tag{6.34}$$

In the above expression, the various operators D_X^{S}, D_k^{S},... are evaluated at $(k^{\mathrm{S}}, \omega^{\mathrm{S}}; X^{\mathrm{S}})$, as in the first term of (6.34). The leading-order equation extracted from (6.34) is identically satisfied since, by construction $D(k^{\mathrm{S}}, \omega^{\mathrm{S}}; X^{\mathrm{S}}) = 0$. According to (6.34), the $\mathcal{O}(\varepsilon^{1/2})$ problem reads:

$$D(k^{\mathrm{S}}, \omega^{\mathrm{S}}; X^{\mathrm{S}})\bar{\phi}_2 = iD_k^{\mathrm{S}}\frac{d\bar{A}}{d\bar{X}} - \bar{X}D_X^{\mathrm{S}}\bar{A}. \tag{6.35}$$

Note, however, that by straightforward differentiation of relation $D(k, \omega(k; X); X) = 0$ with respect to k and X respectively, one has

$$D_k^{\mathrm{S}} + D_\omega^{\mathrm{S}}\omega_k^{\mathrm{S}} = 0, \qquad D_X^{\mathrm{S}} + D_\omega^{\mathrm{S}}\omega_X^{\mathrm{S}} = 0. \tag{6.36a, b}$$

According to (6.26), the point $(k^{\mathrm{S}}, X^{\mathrm{S}})$ is precisely selected so that $\omega_k^{\mathrm{S}} = \omega_X^{\mathrm{S}} = 0$. One therefore has $D_k^{\mathrm{S}} = D_X^{\mathrm{S}} = 0$ and equation (6.35) is identically satisfied.

By identifying terms of $\mathcal{O}(\varepsilon)$ in (6.34), one obtains

$$
\begin{aligned}
D(k^{\mathrm{S}}, \omega^{\mathrm{S}}; X^{\mathrm{S}})\bar{\phi}_3 = {} & \frac{1}{2} D_{kk}^{\mathrm{S}} \frac{d^2 \bar{A}}{d\bar{X}^2} + i D_{kX}^{\mathrm{S}} \bar{X} \frac{d\bar{A}}{d\bar{X}} \\
& - (\omega_2 D_\omega^{\mathrm{S}} + D_\varepsilon^{\mathrm{S}} + \frac{1}{2} D_{XX}^{\mathrm{S}} \bar{X}^2) \bar{A} = 0,
\end{aligned}
\tag{6.37}
$$

which provides the evolution equation for the inner amplitude $\bar{A}$. Its coefficients can more conveniently be expressed in terms of partial derivatives of $\omega(k; X)$. An additional differentiation of (6.36a) and (6.36b) with respect to k and X respectively and evaluation at $k^{\mathrm{S}}, X^{\mathrm{S}}$ leads to the formulas

$$
\omega_{kk}^{\mathrm{S}} = -D_{kk}^{\mathrm{S}}/D_\omega^{\mathrm{S}}, \quad \omega_{kX}^{\mathrm{S}} = -D_{kX}^{\mathrm{S}}/D_\omega^{\mathrm{S}}, \quad \omega_{XX}^{\mathrm{S}} = -D_{XX}^{\mathrm{S}}/D_\omega^{\mathrm{S}}.
\tag{6.38a, b, c}
$$

The inner evolution equation then takes the form

$$
\frac{1}{2}\omega_{kk}^{\mathrm{S}} \frac{d^2 \bar{A}}{d\bar{X}^2} + i\omega_{kX}^{\mathrm{S}} \bar{X} \frac{d\bar{A}}{d\bar{X}} + (\omega_2 - \delta\omega^{\mathrm{S}} - \frac{1}{2}\omega_{XX}^{\mathrm{S}} \bar{X}^2)\bar{A} = 0,
\tag{6.39}
$$

where we recall that $\delta\omega^{\mathrm{S}} = -D_\varepsilon^{\mathrm{S}}/D_\omega^{\mathrm{S}}$. This amplitude equation naturally reflects the structure of the local dispersion relation around $k^{\mathrm{S}}, X^{\mathrm{S}}$. By setting $d/d\bar{X} = i(k - k^{\mathrm{S}})/\varepsilon^{1/2}$ and $\bar{X} = (X - X^{\mathrm{S}})/\varepsilon^{1/2}$ in (6.39), one obtains the Taylor expansion

$$
\begin{aligned}
\varepsilon\omega_2 \sim \omega - \omega^{\mathrm{S}} \sim {} & \varepsilon\,\delta\omega^{\mathrm{S}} + \frac{1}{2}\omega_{kk}^{\mathrm{S}}(k - k^{\mathrm{S}})^2 \\
& + \omega_{kX}^{\mathrm{S}}(k - k^{\mathrm{S}})(X - X^{\mathrm{S}}) + \frac{1}{2}\omega_{XX}^{\mathrm{S}}(X - X^{\mathrm{S}})^2.
\end{aligned}
\tag{6.40}
$$

Note, however, that the constant $\delta\omega^{\mathrm{S}}$ is a genuine "non-parallel" term arising from the $\mathcal{O}(\varepsilon)$ correction D_ε to the local dispersion relation. As a final step, one may express the amplitude equations in terms of absolute instability properties. According to (6.40) and the definitions given in section 3.2, the absolute wavenumber $k_0(X)$ is given by

$$
\frac{\partial\omega}{\partial k} \sim \omega_{kk}^{\mathrm{S}}(k - k^{\mathrm{S}}) + \omega_{kX}^{\mathrm{S}}(X - X^{\mathrm{S}}) = 0,
\tag{6.41}
$$

which yields

$$
k_0(X) \sim k^{\mathrm{S}} + k_{0X}^{\mathrm{S}}(X - X^{\mathrm{S}}), \qquad \text{with} \quad k_{0X}^{\mathrm{S}} = -\omega_{kX}^{\mathrm{S}}/\omega_{kk}^{\mathrm{S}}.
\tag{6.42a, b}
$$

By substituting (6.42a,b) into (6.40), one obtains the following quadratic dependence for the absolute frequency $\omega_0(X)$:

$$
\omega_0(X) \sim \omega^{\mathrm{S}} + \frac{1}{2}\omega_{0XX}^{\mathrm{S}}(X - X^{\mathrm{S}})^2, \qquad \text{with} \quad \omega_{0XX}^{\mathrm{S}} = \omega_{XX}^{\mathrm{S}} - (\omega_{kX}^{\mathrm{S}})^2/\omega_{kk}^{\mathrm{S}}.
\tag{6.43a, b}
$$

We stress that these variations are only valid locally around X^S, in the inner turning-point region. Upon using (6.42b) and (6.43b) into (6.39), the final form of the amplitude equation for $\bar{A}$ is generated:

$$
\frac{1}{2}\omega_{kk}^S \frac{d^2\bar{A}}{d\bar{X}^2} - i\omega_{kk}^S k_{0X}^S \bar{X} \frac{d\bar{A}}{d\bar{X}}
$$
$$
+ \left[\omega_2 - \delta\omega^S - \frac{1}{2}(\omega_{kk}(k_{0X}^S)^2 + \omega_{0XX}^S)\bar{X}^2\right]\bar{A} = 0.
$$
(6.44)

This expression is identical to equation (45) in Huerre and Monkewitz (1990).

When the substitution $-i\omega_2 \mapsto \partial/\partial T_2$ is made in (6.44), one recovers the linearized GL equation with varying coefficients. This is an a posteriori justification for selecting this equation as a model: it arises in a rational approximation scheme of global-mode spatial structure in the vicinity of the turning point X^S. Evolution equation (6.44) may readily be solved by introducing the change of variables

$$
\xi = \left(\frac{4\omega_{0XX}^S}{\omega_{kk}^S}\right)^{1/4} \bar{X}, \qquad \bar{A}(\bar{X}) = \exp\left\{\frac{i}{2}k_{0X}^S\bar{X}^2\right\}\alpha(\xi). \qquad (6.45a, b)
$$

One is led to the standard form of the parabolic cylinder equation:

$$
\frac{d^2\alpha}{d\xi^2} + \left[\frac{\omega_2 - \delta\omega^S + \frac{i}{2}\omega_{kk}^S k_{0X}^S}{(\omega_{kk}^S\omega_{0XX}^S)^{1/2}} - \frac{\xi^2}{4}\right]\alpha = 0. \qquad (6.46)
$$

In anticipation of the matching procedure between the inner region and the outer WKBJ regions $+$ and $-$ (Fig. 6.11), we require that $\alpha(\xi)$ decays exponentially as $|\bar{X}| \to \infty$. According to the classical properties of parabolic cylinder functions (Bender and Orszag, 1978), the frequency correction ω_2 must then take discrete values such that

$$
\frac{\omega_{2n} - \delta\omega^S + \frac{i}{2}\omega_{kk}^S k_{0X}^S}{(\omega_{kk}^S\omega_{0XX}^S)^{1/2}} = n + \frac{1}{2}, \qquad (6.47)
$$

where n is an integer. Under these conditions, the eigenfunction $\alpha(\xi)$ reduces to the Hermite polynomials

$$
\alpha_n(\xi) = e^{-\xi^2/4}\, \mathrm{He}_n(\xi). \qquad (6.48)
$$

Reverting to inner variables, the inner amplitude function reads:

$$
\bar{A}_n(\bar{X}) = \mathrm{He}_n\left[\left(\frac{4\omega_{0XX}^S}{\omega_{kk}^S}\right)^{1/4}\bar{X}\right]\exp\left\{\frac{1}{2}\left[ik_{0X}^S - \left(\frac{\omega_{0XX}^S}{\omega_{kk}^S}\right)^{1/2}\right]\bar{X}^2\right\}.
$$
(6.49)

In summary, the leading-order outer solutions in (6.27) are characterized by the amplitude functions $A^\pm(X)$ given in (6.30). To complete the analysis, both should be matched to the leading-order inner solution in (6.32), where the amplitude function $\bar{A}(\bar{X})$ is of the form (6.49). The outer solutions are first expanded as $|X - X^S| \to 0$ and rewritten in terms of the inner variable $\bar{X}$. To approximate the integrals appearing in (6.27) and (6.30), it is useful to derive from the local dispersion (6.40), *without the correction term* $\varepsilon\,\delta\omega^S$, the following intermediate results:

$$k^\pm(X;\omega^S) \sim k^S + \left[k^S_{0X} + i\left(\frac{\omega^S_{0XX}}{\omega^S_{kk}}\right)^{\frac{1}{2}}\right](X - X^S) + \ldots \quad (6.50a)$$

$$\omega_k[k(X;\omega^S);X] \sim i(\omega^S_{kk}\omega^S_{0XX})^{\frac{1}{2}}(X - X^S) + \ldots \quad (6.50b)$$

$$k^\pm_X(X;\omega^S) \sim k^S_{0X} + i\left(\frac{\omega^S_{0XX}}{\omega^S_{kk}}\right)^{1/2} + \ldots. \quad (6.50c)$$

In view of the eigenvalue relation (6.47), the leading-order expansion of $\phi^\pm(X)$ given in (6.27–6.30) as $|X - X^S| \to 0$ reads

$$\phi^\pm(X) \sim A^{S\pm}(X - X^S)^n$$

$$\times \exp\left\{\frac{1}{2}\left[ik^S_{0X} - (\omega^S_{0XX}/\omega^S_{kk})^{\frac{1}{2}}\right]\left(\frac{X - X^S}{\varepsilon^{1/2}}\right)^2\right\}$$

$$\times \exp\left\{ik^S(x - x^S)\right\}. \quad (6.51)$$

Conversely, the inner solution (6.32), (6.49) is expanded as $|\bar{X}| \to \infty$. Upon using the standard result $\mathrm{He}_n(\xi) \sim \xi^n$, as $\xi \to \infty$, one obtains:

$$\bar{\phi}(\bar{X}) \sim \mu(\varepsilon)\left(\frac{4\omega^S_{0XX}}{\omega^S_{kk}}\right)^{n/4}\bar{X}^n$$

$$\times \exp\left\{\frac{1}{2}\left[ik^S_{0X} - (\omega^S_{0XX}/\omega^S_{kk})^{\frac{1}{2}}\right]\bar{X}^2\right\}$$

$$\times \exp\left\{ik^S(x - x^S)\right\}. \quad (6.52)$$

If both expansions (6.51) and (6.52) are expressed in terms of the same variable, say $\bar{X}$, and required to be identical, the remaining unknown quantities are determined, namely

$$\mu(\varepsilon) = \varepsilon^{n/2}, \qquad A^{S\pm} = \left(4\omega^S_{0XX}/\omega^S_{kk}\right)^{n/4}. \quad (6.53a, b)$$

Bearing in mind the general configuration of the complex X-plane displayed in Fig. 6.11, the above analysis may be summarized as follows: in the $+$ and $-$ regions, the global mode spatial structure is given by the outer WKBJ solutions (6.27), (6.30) and (6.53b). In the inner turning point region close to X_S, it takes the form (6.32), the amplitude function

$\bar{A}(\bar{X})$ being given by the Hermite polynomials (6.49) and the gauge function $\mu(\varepsilon)$ by (6.53a). Furthermore, matching between the + and − outer WKBJ solutions and the turning point solution has led to a quantification condition on the global frequency correction term $\varepsilon\omega_2$, as specified in (6.47). According to (6.25), (6.28) and (6.47), there exists a *discrete infinity of global frequencies*

$$\omega_{Gn} \sim \omega^S + \varepsilon \left[\delta\omega^S - \frac{i}{2}\omega_{kk}^S k_{0X}^S + (\omega_{kk}^S \omega_{0XX}^S)^{1/2}(n+1/2) \right], \qquad (6.54)$$

where n is an integer, and $\omega^S = \omega(k^S; X^S, R)$ is determined by the stationarity conditions (6.26). The above derivation, conducted for the general partial differential operator (6.10), was performed by Chomaz *et al.* (1991) and Le Dizès *et al.* (1996) on the GL equation with spatially varying coefficients and by Monkewitz *et al.* (1993) in the context of the linearized Navier-Stokes equations. If there is at least one global frequency such that $\omega_{Gn,i} > 0$, the medium is globally unstable and supports self-sustained oscillations. Furthermore it can be shown (Chomaz *et al.*, 1991; Le Dizès *et al.*, 1996) that *an upper bound on the global mode growth rates is given by the maximum absolute growth rate* $\omega_{0,i}|_{\max}$ *over all real X.* In other words,

$$\omega_{Gn,i} \leq \omega_{0,i}|_{\max}. \qquad (6.55)$$

In order for a flow to be globally unstable ($\omega_{Gn,i} > 0$), it must necessarily contain a region of absolute instability. It can therefore be concluded that *convectively unstable flows of type (b) or (c) in Fig. 6.9 do not display intrinsic oscillations but behave as noise amplifiers.* This property further confirms the validity of the spatial approach in convectively unstable media. It is also consistent with experimental observations in wakes: well-defined self-sustained oscillations only appear when the absolutely unstable region is sufficiently large as in Fig. 6.9d. The relationship between absolute instability and onset of global oscillations was established in the context of the GL equation with spatially varying coefficients by Chomaz *et al.* (1988).

Experimental and numerical validations of the global frequency selection criterion (6.26), (3.40) are scarce. The numerical simulations of Karniadakis and Triantafyllou (1989), Hannemann and Oertel (1990), and Hammond and Redekopp (1997) indicate that the wake behind a plate of thickness h aligned with the upstream flow undergoes a *Hopf bifurcation* to a global mode as the Reynolds number $\mathrm{Re} \equiv U_\infty h/\nu$ crosses a critical value. Under such conditions, the near wake exhibits a finite region of absolute instability behind the plate trailing edge. Furthermore, according to the recent study of Hammond and Redekopp (1997), application of the *frequency selection criterion* (6.25) to the local viscous dispersion relation of the associated *mean* velocity profile yields a Strouhal number St = 0.101 at Re = 160. This value is in excellent agreement with the observed shedding Strouhal number St = 0.100. Note that the computed global frequency is also very close to the absolute frequency at the transition location between absolute and convective instability as in Koch's (1985) criterion. Schär and Smith (1993) have numerically investigated the flow behind a vertical circular cylinder in a shallow water layer. At a critical value of the Froude number, the wake is observed to undergo a transition to large scale Kármán vortex shedding. *When all nonlinear terms in the numerical code are turned off*, the wake beats at a global frequency $\omega_G \sim 0.17 + 0.045i$. Local stability calculations performed on the unstable *basic* state then indicate the presence of a broad region of absolute instability behind the obstacle. The function $\omega_0(X)$ can be estimated from a parabolic fit near the maximum absolute growth rate $\omega_{0,i}|_{max}$ and application of criterion (6.25) leads to the prediction $\omega_G \sim 0.19 + 0.040i$, which compares very favorably with the computed value. Note, however, that, when all nonlinearities are restored, the observed Strouhal frequency becomes $\omega_G \sim 0.27$, which is noticeably different from the predicted value.

6.3 Phase dynamics of wake patterns

In sections 6.1–2, we have specifically considered the infinite aspect ratio case where the wake dynamics can be considered strictly two-dimensional. However, experimental investigations below Re = 180 (Williamson, 1989), have clearly shown that non-trivial *three-dimensional* phenomena arise because of the finite length of the cylinder (section 2.2). In particular, the puzzling variations of the asymptotic wake frequency found in different experimental facilities have been attributed to oblique shedding induced by different end configurations. It is our purpose here to demonstrate that these finite-end effects can be accounted for in a simple phenomenological approach based on the GL equation. The following presentation is based on the studies of Albarède and Monkewitz (1992) and Albarède and Provansal (1996). For an alternate approach, the reader is referred to Park and Redekopp (1992).

The starting point of the analysis consists in assuming that any *dimensional* physical perturbation field $\psi(\mathbf{x}, t)$ is of the form

$$\psi(\mathbf{x}, t) = \mathcal{Re}\left\{a(z, t)b(x)\phi(y)\exp(i(k_0(x - U_c t)))\right\}, \tag{6.56}$$

where z denotes the spanwise distance along the cylinder axis, k_0 and U_c the wavenumber and advection velocity of the vortical structures.[*] The complex shape function $b(x)$ describes the streamwise modulation of the perturbation amplitude (Albarède and Monkewitz, 1992). The complex amplitude function $a(z,t)$ is assumed to be governed by the one-dimensional GL equation

$$\frac{\partial a}{\partial t} = \sigma a + \gamma \frac{\partial^2 a}{\partial z^2} - b|a|^2 a, \qquad (6.57)$$

where σ, γ and b are complex *dimensional* parameters. Associated boundary conditions at the cylinder ends are taken to be

$$a(\pm \ell/2, t) = 0. \qquad (6.58)$$

It should be emphasized that, at the present time, the ansatz (6.56–58) cannot simply be incorporated within a rational asymptotic global mode analysis of the kind presented in section 6.2. Here, it is assumed that the vortex street can effectively be represented as a wave generated on the cylinder and traveling with the observed convection speed of Bénard–Kármán vortices $U_c = 0.88 U_\infty$. In the purely two-dimensional case $(\partial/\partial z \equiv 0)$, $a(t)$ satisfies the expected Landau equation (2.7) as discussed in section 2.2. In the spirit of Kuramoto (1984), spanwise coupling of two-dimensional hydrodynamic nonlinear oscillators has been introduced by merely adding a diffusive coupling term $\partial^2 a/\partial z^2$. This description, although reminiscent of the amplitude equation approach used to study pattern dynamics in large convection boxes (Newell *et al.*, 1993), is not directly derivable from the Navier–Stokes equations (the latter procedure is outlined in section 8.2). This approach should be viewed as a *phenomenological* model containing the basic ingredients of wake dynamics. In real experiments, end plates, rigid walls or boundary cells located close to the extremities give rise to amplitude nodes and consequently phase discontinuities or vortex dislocations near the cylinder ends. Such phenomena are also likely to be induced by boundary conditions (6.58) in the context of the GL model.

In order to interpret the experimental observations described in section 2.2, it is necessary to relate the observed streakline patterns of Figs. 2.16, 2.18 to the spatio-temporal field (6.56). It is assumed that the spanwise smoke filaments injected at a given streamwise station x_0, y_0 are rapidly concentrated in the vortex cores and passively advected downstream at the convection velocity U_c. This Taylor hypothesis implies that the observed tracer patterns in the x–z plane simply represent

---

[*] The quantity U_c should not be confused with the centerline velocity introduced in section 6.1.

the recording of $a(t, z)b(x_0)\phi(y_0)\exp[i(k_0(x_0 - U_c t))]$ provided that one applies the transformation $t \to -x/U_c$. In other words, flow visualization pictures provide a strip-chart of isophase contours of $a(t, z)\exp(-ik_0 U_c t)$.

By straightforward dimensional considerations, $\sigma d^2/\nu$, γ/ν and b/ν only depend on the Reynolds number Re. It is further assumed that γ/ν and b/ν are constants while $\sigma_r d^2/\nu \approx 0.2(\text{Re} - \text{Re}_c)$ with $\text{Re}_c = 48.5$, as one would expect in a supercritical Hopf bifurcation. The aspect ratio ℓ/d only enters the problem through boundary conditions (6.58). It is convenient to introduce the transformations

$$a(z, t) \mapsto a(z, t)\sqrt{b_r/\sigma_r}\,, \quad t \mapsto t\sigma_r\,, \quad z \mapsto z\sqrt{\sigma_r/\gamma_r}\,, \tag{6.59}$$

whereby the rescaled equations (6.57–58) become without changing notations for $a(z, t)$:

$$\frac{\partial a}{\partial t} = (1 + ic_0)a + (1 + ic_1)\frac{\partial^2 a}{\partial z^2} - (1 + ic_2)|a|^2 a\,, \tag{6.60}$$

$$a(\pm \ell_1/2, t) = 0\,. \tag{6.61}$$

In view of the assumptions made regarding the coefficients of the original equation (6.57), the quantities c_1, c_2 are found to be independent of the Reynolds number. The coefficient c_0 depends on Re but it can be scaled out since it affects the solutions only through an innocuous frequency shift. The single control parameter left is hence the rescaled cylinder length ℓ_1, i.e. the ratio between the dimensional length ℓ and the correlation length $d\sqrt{\gamma_r}/[0.2\nu(\text{Re} - \text{Re}_c)]$.

Most of the experimental observations described in section 2.2 can be captured by considering time-periodic solutions of (6.60–61). The effect of ℓ_1 on the number of excited spanwise modes can be illustrated on the *linearized* version of (6.60–61), where the cubic nonlinear term is momentarily suppressed (Albarède and Provansal, 1996). The general solution can then be written as the infinite sum of modes

$$a(z, t) = \sum_{n=1}^{\infty} \exp(\sigma_n t)\sin[q_n(z + \ell_1/2)]\,. \tag{6.62}$$

The nth linear mode is characterized by a spanwise wavenumber q_n and growth rate σ_n of the form

$$q_n = n\pi/\ell_1\,, \quad \sigma_n = 1 - q_n^2 + i\left(c_0 - c_1 q_n^2\right)\,. \tag{6.63}$$

According to (6.62–63), the basic state $a_0(z, t) = 0$ becomes unstable when the rescaled cylinder length ℓ_1 exceeds the critical value π. Furthermore, in analogy with natural convection (Manneville, 1990), one may identify two distinct regimes corresponding to small and large aspect ratios respectively. For small cylinder lengths, i.e. when ℓ_1 is close to π, only one mode is unstable: the GL partial differential equation effectively reduces to a single ordinary differential equation. *A contrario* when $\ell_1 \gg \pi$, many spanwise modes become unstable and a phase instability analysis can be performed (Kuramoto, 1984) as illustrated below in the context of the fully *nonlinear* problem.

For *small aspect ratios*, i.e. when $\pi \le \ell_1 \le 2\pi$, the severe one-mode truncation

$$a(z,t) = a_1(t) \sin[q_1(z + \ell_1/2)] \tag{6.64}$$

is adequate. In that case, $a_1(t)$ is governed by the *Landau equation*

$$\frac{da_1}{dt} = \sigma_1 a_1 - \tfrac{3}{4}(1 + ic_2)|a_1|^2 a_1, \tag{6.65}$$

not to be confused with the Landau equation obtained by deleting the diffusion term in (6.60). As in section 3.3, the large-time behavior of the general solution reads

$$a(z,t) \sim \sqrt{\tfrac{4}{3}(1 - q_1^2)} \exp\left(it[(c_0 - c_2) - q_1^2(c_1 - c_2)]\right) \sin[q_1(z + \ell_1/2)]. \tag{6.66}$$

The theoretical solution (6.66) qualitatively agrees with experimentally observed wake patterns: when the cylinder length is small enough, the ends directly affect the entire flow field via bending of the vortex lines in a "$\cos(\pi z/\ell_1)$" shape.

For *large aspect ratios*, i.e. when $\ell_1 \gg \pi$, end conditions only indirectly constrain the flow pattern (section 2.2). In the extreme case of infinite cylinder length, the GL equation admits the plane wave solutions

$$a(z,t) = \sqrt{1 - q^2} \exp[i(c_0 - c_2 - q^2(c_1 - c_2))t + iqz], \tag{6.67}$$

of continuous spanwise wavenumber q. Such states are reminiscent of the Stokes solutions (3.62). When $q = 0$, purely two-dimensional shedding is recovered. Plane waves with $q \ne 0$ introduce a new theoretical feature, namely the possibility of oblique shedding.

The purpose of the pattern analysis outlined below is to show (a) how end effects select, among this continuous set, a single spanwise wavenumber q, and (b) how this information is propagated away from the ends into the bulk. A first step consists in rewriting the complex amplitude $a(z,t)$ in terms of a real phase $\theta(z,t)$ and amplitude $R(z,t)$:

$$a(z,t) = R(z,t) \exp[i(c_0 - c_2)t + i\theta(z,t)]. \tag{6.68}$$

Equation (6.60) is then transformed into the two coupled real partial differential equations

$$\frac{\partial R}{\partial t} = R + \frac{\partial^2 R}{\partial z^2} - R\left[\frac{\partial \theta}{\partial z}\right]^2 - R^3 - c_1\left[2\frac{\partial R}{\partial z}\frac{\partial \theta}{\partial z} + R\frac{\partial^2 \theta}{\partial z^2}\right], \tag{6.69}$$

$$R\frac{\partial\theta}{\partial t} = c_2\frac{\partial R}{\partial t} + c_\Delta\left[\frac{\partial^2 R}{\partial z^2} - R\left(\frac{\partial\theta}{\partial z}\right)^2\right] + c_\theta\left[2\frac{\partial R}{\partial z}\frac{\partial\theta}{\partial z} + R\frac{\partial^2\theta}{\partial z^2}\right], (6.70)$$

where $c_\Delta \equiv (c_1 - c_2)$ and $c_\theta \equiv (1 + c_1 c_2)$.

In order to identify the selection mechanism for the spanwise wavenumber q induced by each of the boundaries, let us send to infinity one of the cylinder ends, for instance the one located at $z = \ell_1/2$. Following Albarède and Monkewitz (1992), we then seek a time-harmonic solution tending to a plane wave state (6.67) far away from $z = -\ell_1/2$. More specifically, this solution should satisfy the end condition

$$R(-\ell_1/2, t) = 0, \tag{6.71}$$

and, for large z, tend to the plane wave

$$R(+\infty, t) = \sqrt{1 - q_\infty^2}, \quad \frac{\partial\theta}{\partial z}(+\infty, t) = q_\infty, \tag{6.72}$$

of unknown spanwise wavenumber q_∞. Finally the purely time-harmonic nature of solution (6.68) is enforced through the condition

$$\frac{\partial R}{\partial t}(z, t) \equiv 0, \quad \frac{\partial\theta}{\partial t}(z, t) \equiv -q_\infty^2 c_\Delta. \tag{6.73}$$

System (6.69–73) constitutes a nonlinear eigenvalue problem for q_∞. An exact "*hole solution*" has been obtained by Nozaki and Bekki (1984) in the form

$$R(z) = \sqrt{1 - q_\infty^2}\tanh\left[\eta(z + \ell_1/2)\right], \tag{6.74a}$$

$$\theta(z, t) = -c_\Delta q_\infty^2 t - g\ln\left[2\cosh(\eta(z + \ell_1/2))\right]. \tag{6.74b}$$

The eigenvalue q_∞ and the structural parameters g and η of the "hole solution" are found to be real functions of c_1 and c_2, as specified in Albarède and Monkewitz (1992). This completes the *spanwise wavenumber selection problem associated with end effects*. Note that the parameter η is found to be always positive while q_∞ (resp. g) has a sign identical (resp. opposite) to c_Δ. According to conditions (6.71–73), the solution (6.74) reduces, as $z \to +\infty$, to the plane wave

$$a(z, t) \sim \sqrt{1 - q_\infty^2}\exp[i(c_0 - c_2 - q_\infty^2 c_\Delta)t + iq_\infty(z + \ell_1/2)]. \tag{6.75}$$

of wavenumber q_∞. From the above solution, one can infer that the end condition at $z = -\ell_1/2$ necessarily induces oblique shedding at the angle $\arctan(q_\infty/k_0)$, where k_0 here denotes the nondimensional streamwise wavenumber of the carrier wave. Since c_Δ is experimentally found to be positive (Albarède and Monkewitz, 1992), q_∞ is also positive. Had we retained the end condition at $z = \ell_1/2$ and removed the other end to infinity, we would have found q_∞ to be of the same magnitude but of opposite sign. This is confirmed in the experimental plan view of Fig. 2.16.

Furthermore, the empirical "cosine" law $St_0 = St \cos \theta$ (section 2.2) of Williamson (1989) can partly be justified: for small q_∞ or equivalently for small shedding angle θ, $\cos \theta$ can be approximated by $1 - \frac{1}{2}(q_\infty/k_0)^2$ while, according to (6.56), (6.75), the ratio between the oblique and parallel shedding frequencies is indeed of the form $1 - c_3 q_\infty^2$. The value of c_3 inferred from the coefficients of the GL equation is found to be reasonably close to $1/(2k_0^2)$.

Each extremity $z = \pm \ell_1/2$ therefore specifies a distinct oblique shedding angle. It is now necessary to explain how the bulk flow may accommodate the presence of two distinct shedding angles originating from each end through localized bending of the phase lines as discussed in section 2.2. In the same spirit as in the secondary instability analysis of section 3.3, we examine the fate of fluctuations superimposed on the parallel shedding solution $q = 0$. As in equation (3.67), perturbations are taken to be of the form

$$a(z,t) = [1 + \rho(z,t)] \exp[i(c_0 - c_2)t + i\theta(z,t)]. \qquad (6.76)$$

This decomposition is tantamount to setting $R(z,t) = 1 + \rho(z,t)$ in the exact coupled equations (6.69–70). It is convenient to adopt the *phase dynamics* point of view (Kuramoto, 1984; Manneville, 1990) where three-dimensional effects are assumed to be weak. More specifically, the spanwise wavenumber q, i.e. the phase gradient $q(z,t) = \frac{\partial \theta}{\partial z}$, is taken to be uniformly small. According to the plane wave solutions (6.67) for nonzero q, time derivatives $\frac{\partial}{\partial t}$ and perturbation amplitudes ρ are then expected to be $\mathcal{O}(q^2)$. Upon only retaining terms of $\mathcal{O}(q^2)$ in the phase equation (6.70), one obtains the phase evolution equation

$$\frac{\partial \theta}{\partial t} + c_\Delta \left(\frac{\partial \theta}{\partial z}\right)^2 = c_\theta \frac{\partial^2 \theta}{\partial z^2}. \qquad (6.77)$$

Similarly, equation (6.69) yields, at $\mathcal{O}(q^2)$, an explicit equation for the perturbation amplitude:

$$\rho(z,t) \sim -\frac{1}{2}\left[\left(\frac{\partial \theta}{\partial z}\right)^2 + c_1 \frac{\partial^2 \theta}{\partial z^2}\right], \qquad (6.78)$$

thereby indicating that the amplitude is slaved to the phase. Differentiating (6.77) once with respect to z leads to a *Burgers equation*

$$\frac{\partial q}{\partial t} + 2c_\Delta q \frac{\partial q}{\partial z} = c_\theta \frac{\partial^2 q}{\partial z^2}, \qquad (6.79)$$

for the spanwise wavenumber $q(z,t) = \partial\theta/\partial z$. Such an advection-diffusion equation admits *wavenumber shock* solutions of the form

(Whitham, 1974):

$$q(z,t) = \tfrac{1}{2}(q_{+\infty} + q_{-\infty})$$

$$- \tfrac{1}{2}(q_{-\infty} - q_{+\infty})\tanh\!\left(\frac{c_\Delta(q_{-\infty} - q_{+\infty})}{2c_\theta}[z - c_\Delta(q_{+\infty} + q_{-\infty})t]\right). \quad (6.80)$$

When $z \to -\infty$ ($z \to +\infty$), expression (6.80) tends towards the plane wave (6.67) of spanwise wavenumber $q_{-\infty}$ ($q_{+\infty}$). Such a solution allows a smooth connection between each family of oblique shedding waves via a moving wavenumber shock. Its thickness is proportional to the phase diffusion coefficient c_θ and its celerity is $U_s = c_\Delta(q_{+\infty} + q_{-\infty})$. Since the end selection mechanism prescribes that $q_{+\infty} \le 0$ and $q_{-\infty} \ge 0$, one may also write $U_s = c_\Delta(q_{+\infty} - |q_{-\infty}|)$: regions of larger shedding angle necessarily invade regions of smaller shedding angle as observed experimentally.

The phenomenology described in section 2.2 can now be interpreted as follows. Vortices are initially shed parallel to the cylinder. However, finite size effects produce two oblique shedding waves propagating from each end towards the center via two wavenumber shocks. Since $q_{+\infty} = -q_{-\infty}$, these shocks invade the central parallel shedding region ($q = 0$) with opposite velocity $U_s = \pm c_\Delta q_{+\infty}$. They finally meet at mid-span where they form a single stationary shock ($U_s = 0$) giving rise to a *chevron pattern*. If boundary effects induce asymmetric oblique shedding, i.e. $q_{+\infty} \ne -q_{-\infty}$, the two wavenumber shocks travel at distinct velocities and merge into a single shock with non zero velocity $U_s = c_\Delta(q_{+\infty} + q_{-\infty})$: the wavenumber of larger magnitude imposes its shedding angle, finally leaving a *uniform oblique shedding pattern* over the entire span.[*]

Experimentally observed quasi-periodic dynamics may also be understood in the GL framework (Albarède and Provansal, 1996). Finally, note that three-dimensional intrinsic bulk effects appearing in the range Re = 190–260 (section 2.2) are not accounted for in this phenomenological model. Such intrinsic features necessitate a full *secondary instability* analysis (Henderson and Barkley, 1996) comparable to that performed by Pierrehumbert and Widnall (1982) for mixing layers.

7 Viscous instabilities in parallel flows

According to Rayleigh's criterion, inflectionless velocity profiles that are typical of *bounded shear flows* such as plane Poiseuille flow or the Blasius boundary layer, are linearly inviscidly stable. Yet, as we have seen in section 2, such flows are observed to become unstable. In this section, we

---

[*] In analogy with gas dynamics, Prandtl–Meyer expansion fans have also been predicted in the Burgers equation context and indeed observed (Monkewitz *et al.*, 1996).

demonstrate that, paradoxically, viscous diffusion effects may, in certain cases, be destabilizing over a range of Reynolds numbers and result in the growth of *Tollmien-Schlichting waves*. Here, we reexamine for a viscous fluid the general properties of shear flow instabilities presented in section 4.

Consider the flow of an incompressible fluid of kinematic viscosity ν. As in section 4, we introduce a length scale L and velocity scale V typical of the basic flow under study. In dimensionless variables, the governing Navier–Stokes equations are of the form

$$\nabla \cdot \mathbf{U} = 0, \qquad (7.1)$$

$$\frac{\partial \mathbf{U}}{\partial t} + (\mathbf{U} \cdot \nabla)\mathbf{U} = -\nabla P + \frac{1}{\mathrm{Re}}\nabla^2 \mathbf{U}, \qquad (7.2)$$

where $\mathrm{Re} \equiv VL/\nu$ is the Reynolds number. In contrast with section 4, it is not strictly speaking legitimate to arbitrarily choose any unidirectional flow $U(y)$ as basic state. According to the streamwise component of (7.2), $U(y)$ must necessarily satisfy

$$\frac{1}{\mathrm{Re}}\frac{d^2 U}{dy^2} = -\frac{dP}{dx}. \qquad (7.3)$$

A classical basic state solution of (7.3) is *plane Poiseuille flow* (Fig. 7.1):

$$U(y) = 1 - y^2. \qquad (7.4)$$

Here L is the half-width of the channel and V is the maximum centerline velocity $U_{\max}$ so that the dimensionless pressure gradient is $dP/dx = -2/\mathrm{Re}$. Another solution of interest is plane Couette flow $U(y) = y$ or any combination of these two profiles.

Possible unidirectional steady basic flows are therefore very limited in number. It is common practice among many stability theorists to

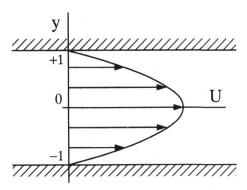

Fig. 7.1. Plane Poiseuille basic flow.

"bend the rules" and to examine the viscous stability of velocity profiles $U(y)$ that do not satisfy (7.3), for instance the Blasius boundary layer on a flat plate, or even continuous mixing-layer, wake or jet profiles that are convenient fits to those measured experimentally. In such cases, it should always be checked *a posteriori* that the evolution time scale τ_{in} of the instability is indeed much shorter than the diffusion time scale $\tau_v \equiv L^2/\nu$. Diffusion of the basic flow is then effectively negligible over the time scales of interest. This condition is easily met for the inviscidly unstable flows of section 4 where $\tau_{\text{in}} \sim L/V = \tau_i$: as long as the Reynolds number $\text{Re} \equiv \tau_v/\tau_i$ is large, viscous diffusion is negligible. For flows with no inflection points that are destabilized by viscosity as we shall see, growth rates are much smaller and one must carefully assess whether indeed $\tau_{\text{in}} \ll \tau_v$.

The usual decomposition

$$\mathbf{U}(\mathbf{x}, t) = U(y)\mathbf{e}_x + \mathbf{u}(\mathbf{x}, t) , \qquad P(\mathbf{x}, t) = P_0(x) + p(\mathbf{x}, t) , \quad \text{(7.5 a, b)}$$

where $U(y)$ and $P_0(x)$ satisfy (7.3), leads to a linearized system of the form (4.7–8) with an additional diffusion term $\text{Re}^{-1} \nabla^2 \mathbf{u}$ on the right-hand side of (4.8). Upon making the normal mode assumption (4.9a,b), the complex phase velocity being defined as in (4.10), one finally obtains a set of ordinary differential equations for the modal shape functions $\hat{u}(y)$, $\hat{v}(y)$, $\hat{w}(y)$ and $\hat{p}(y)$:

$$ik_x\hat{u} + ik_z\hat{w} + \frac{d\hat{v}}{dy} = 0 , \tag{7.6}$$

$$ik_x[U(y) - c]\hat{u} + U'(y)\hat{v} = -ik_x\hat{p} + \frac{1}{\text{Re}}\left(\frac{d^2}{dy^2} - k_x^2 - k_z^2\right)\hat{u} , \tag{7.7}$$

$$ik_x[U(y) - c]\hat{v} = -\frac{d\hat{p}}{dy} + \frac{1}{\text{Re}}\left(\frac{d^2}{dy^2} - k_x^2 - k_z^2\right)\hat{v} , \tag{7.8}$$

$$ik_x[U(y) - c]\hat{w} = -ik_z\hat{p} + \frac{1}{\text{Re}}\left(\frac{d^2}{dy^2} - k_x^2 - k_z^2\right)\hat{w} , \tag{7.9}$$

which should be completed by the no-slip boundary conditions at the parallel walls $y = y_1$ and $y = y_2$:

$$\hat{u}(y_1) = \hat{u}(y_2) = 0 , \quad \hat{v}(y_1) = \hat{v}(y_2) = 0 , \quad \hat{w}(y_1) = \hat{w}(y_2) = 0 . \tag{7.10a, b, c}$$

System (7.6–10) defines an eigenvalue problem with non trivial solutions only if the wavevector $\mathbf{k} = k_x\mathbf{e}_x + k_z\mathbf{e}_z$ and the frequency ω satisfy a *dispersion relation* of the form

$$D(\mathbf{k}, \omega; \text{Re}) = 0 . \tag{7.11}$$

7.1 Squire's transformation

As in section 4.1, the three-dimensional linear stability problem (7.6–10) can be reduced to a two-dimensional one through the transformation to tilde variables (4.17–18), with the additional rescaling of the Reynolds number

$$\tilde{k}\widetilde{\mathrm{Re}} = k_x \,\mathrm{Re} \,. \tag{7.12}$$

By elementary calculations, the tilde variables are found to satisfy an equivalent two-dimensional problem of the form (4.19–22), provided one adds appropriate viscous diffusion terms $\widetilde{\mathrm{Re}}^{-1}(d^2/dy^2 - \tilde{k}^2)\tilde{u}$ and $\widetilde{\mathrm{Re}}^{-1}(d^2/dy^2 - \tilde{k}^2)\tilde{v}$ to the right-hand sides of (4.20) and (4.21) respectively, and the no-slip conditions $\tilde{u}(y_1) = \tilde{u}(y_2) = 0$ to the impermeability conditions (4.22). The two-dimensional dispersion relation

$$\tilde{D}(\tilde{k}, \tilde{\omega}; \widetilde{\mathrm{Re}}) = 0 \,, \tag{7.13}$$

is therefore effectively recovered. By following the same argument as the one leading to (4.25), this result implies that the three-dimensional dispersion relation can readily be inferred from the two-dimensional dispersion relation through the transformation

$$D(\mathbf{k}, \omega; \mathrm{Re}) \equiv \tilde{D}\left[(k_x^2 + k_z^2)^{1/2}, \frac{(k_x^2 + k_z^2)^{1/2}}{k_x}\omega; \frac{k_x}{(k_x^2 + k_z^2)^{1/2}}\mathrm{Re}\right] = 0 \,. \tag{7.14}$$

To each oblique mode $(\mathbf{k}, \omega)$ of temporal growth rate ω_i, at Reynolds number Re, corresponds a two-dimensional mode $(\tilde{k}, \tilde{\omega})$ of larger growth rate $\tilde{\omega}_i = \omega_i\sqrt{k_x^2 + k_z^2}/k_x$, at a *lower* Reynolds number $\widetilde{\mathrm{Re}} = \mathrm{Re}\,k_x/\sqrt{k_x^2 + k_z^2}$.

Squire's transformation (7.14) has important consequences. Assume that there exists a critical value Re_c of the Reynolds number above which a given basic flow is unstable. Then, *the wave that first becomes unstable at the critical Reynolds number* Re_c *is necessarily two-dimensional.* If this were not true, one could find *via* Squire's transformation an amplified two-dimensional wave at $\mathrm{Re} < \mathrm{Re}_c$, which contradicts the definition of Re_c. Finally, note that, at a given Reynolds number $\mathrm{Re} > \mathrm{Re}_c$, the most amplified wave is not necessarily two-dimensional.

7.2 The two-dimensional stability problem: the Orr–Sommerfeld equation

In the same way as in section 4.2, one introduces a two-dimensional streamfunction $\Psi(x, y, t)$ defined by (4.26a,b) and satisfying the vorticity equation

$$\left[\frac{\partial}{\partial t} + \frac{\partial \Psi}{\partial y}\frac{\partial}{\partial x} - \frac{\partial \Psi}{\partial x}\frac{\partial}{\partial y}\right]\nabla^2 \Psi = \frac{1}{\mathrm{Re}}\nabla^4 \Psi, \qquad (7.15)$$

to be compared with the inviscid version (4.28). When the basic flow is perturbed according to (4.29), the perturbation streamfunction $\psi(x, y, t)$ is found to be governed, in the linearized approximation, by (4.30) with an additional diffusion term $\mathrm{Re}^{-1}\nabla^4\psi$ on the right-hand side. The normal mode decomposition (4.31) finally leads to the *Orr–Sommerfeld equation*

$$[U(y) - c][\phi'' - k^2\phi] - U''(y)\phi = \frac{1}{ik\,\mathrm{Re}}\left(\frac{d^2}{dy^2} - k^2\right)^2\phi, \qquad (7.16)$$

to which are added the no-slip boundary conditions (7.10a,b), namely

$$\phi(y_1) = \phi(y_2) = 0, \qquad \phi'(y_1) = \phi'(y_2) = 0. \qquad (7.17\mathrm{a, b})$$

The system (7.16–17) defines a challenging eigenvalue problem which in principle can be solved to yield the two-dimensional dispersion relation $D(k, \omega; \mathrm{Re}) = 0$. Note that, in contrast to the Rayleigh equation, the Orr–Sommerfeld equation is dissipative: temporal eigenvalues no longer appear as complex conjugate pairs. Thus, in the present setting, the basic flow $U(y)$ may either be stable ($c_i < 0$ for all k and temporal branches), neutrally stable (there exists at least one branch and one value of k such that $c_i = 0$, otherwise $c_i(k) < 0$), or unstable (there exists at least one branch and a range of k such that $c_i(k) > 0$).

The Rayleigh equation (4.33) is formally recovered in the limit $\mathrm{Re} \to \infty$, but the order of the governing equation is reduced from four to two, thereby indicating that this limit process is *singular*: in the same way as in classical singular perturbation problems (Bender and Orszag, 1978), there are four boundary conditions (7.17) to be simultaneously enforced with a linear combination of only two independent solutions. Nonuniformities are in fact present at the walls y_1 and y_2 and at *critical points* y_c defined by $U(y_c) = c$. In thin layers around these locations, viscous diffusion terms of the Orr–Sommerfeld equation must partly be retained, as outlined in section 7.4.

7.3 A first look at the instability mechanism: the energy equation

At a superficial level of understanding, it appears overly optimistic to explain the observed instability of bounded shear flows without inflection points by taking into account viscosity. If a flow is inviscidly stable, the inclusion of viscous diffusion will only lead to the attenuation of waves that are neutrally stable on purely inviscid grounds. To dispel this paradox, we appeal to the evolution equation for the perturbation kinetic energy.

By performing elementary but tedious manipulations on the Navier–Stokes equation (7.2), one is led to the exact viscous analogue of the inviscid energy equation (4.47)

$$\left[\frac{\partial}{\partial t} + U(y)\frac{\partial}{\partial x}\right]\left(\frac{1}{2}\left(u^2 + v^2 + w^2\right)\right) + \boldsymbol{\nabla}\cdot\left[\left(p + \frac{1}{2}\left(u^2 + v^2 + w^2\right)\right)\mathbf{u}\right] =$$
$$- U'(y)uv - U'(y)U(y)v + \frac{1}{\mathrm{Re}}\left[\boldsymbol{\nabla}\cdot(\mathbf{u}\times\boldsymbol{\omega}) - \boldsymbol{\omega}\cdot\boldsymbol{\omega}\right]$$
$$(7.18)$$

where $\boldsymbol{\omega} = \boldsymbol{\nabla}\times\mathbf{u}$ is the perturbation vorticity. Upon spatially averaging (7.18) in the x–z plane according to definition (4.39), considering that $\overline{\partial q/\partial x} = \overline{\partial q/\partial z} = 0$ for any q and that $\overline{v} = 0$ (see subsection 4.3), one is led to:

$$\frac{\partial}{\partial t}\left(\frac{1}{2}\left(\overline{u^2} + \overline{v^2} + \overline{w^2}\right)\right) + \frac{\partial}{\partial y}\left[\overline{\left(p + \frac{1}{2}\left(u^2 + v^2 + w^2\right)\right)v}\right] = U'(y)\tau_{xy}$$
$$+ \frac{1}{\mathrm{Re}}\frac{\partial}{\partial y}\left(\overline{\omega_x w - \omega_z u}\right) - \frac{1}{\mathrm{Re}}\overline{\boldsymbol{\omega}\cdot\boldsymbol{\omega}} .$$
$$(7.19)$$

Recall that τ_{xy} denotes the Reynolds stress defined in (4.43). A final integration in the y direction and application of the no-slip boundary conditions $u = v = w = 0$ at $y = y_1$, y_2 results in the exact equation governing the kinetic energy of the perturbations

$$\frac{d}{dt}\int_{y_1}^{y_2}\frac{1}{2}\left(\overline{u^2} + \overline{v^2} + \overline{w^2}\right)dy = \int_{y_1}^{y_2}U'(y)\tau_{xy}\,dy - \frac{1}{\mathrm{Re}}\int_{y_1}^{y_2}\overline{\boldsymbol{\omega}\cdot\boldsymbol{\omega}}\,dy , \quad (7.20)$$

to be compared with the inviscid form (4.49).

The temporal variations of the perturbation kinetic energy are seen to involve a delicate balance between the production term, represented by the work of Reynolds stress on $U(y)$ (first term on the right-hand side of (7.20)), and viscous dissipation (second positive definite term on the right-hand side of (7.20)). At first sight, viscosity has only a stabilizing influence. But this reasoning fails to consider the subtle effects that viscosity may have on the production term. Let us take a closer look at the *Reynolds stress* $\tau_{xy} \equiv -\overline{uv}$, in the context of temporal theory. The velocity components $u(\mathbf{x}, t)$ and $v(\mathbf{x}, t)$ admit normal mode decompositions of the form (4.9a), which lead to the expression

$$\tau_{xy} \equiv -\overline{uv} = -\frac{1}{2}|\hat{u}(y)||\hat{v}(y)|\cos[\varphi_u(y) - \varphi_v(y)]\,\mathrm{e}^{2kc_it} , \quad (7.21)$$

where

$$\hat{u}(y) \equiv |\hat{u}(y)|\,\mathrm{e}^{i\varphi_u(y)} , \qquad \hat{v}(y) \equiv |\hat{v}(y)|\,\mathrm{e}^{i\varphi_v(y)} . \quad (7.22\mathrm{a, b})$$

The Reynolds stress is seen to be controlled by the phase difference $\varphi_u(y) - \varphi_v(y)$. If, within the inviscid formulation, a basic flow $U(y)$ is neutrally stable (c real), then, according to a result obtained at the end of section 4.3, $\tau_{xy} \equiv 0$ and the production term vanishes. From (7.21), this implies that u and v are in quadrature: $\varphi_u(y) - \varphi_v(y) = \pm\pi/2$. However, in a certain intermediate range of large but finite Reynolds numbers, viscous diffusion may slightly upset the phase difference $\varphi_u(y) - \varphi_v(y)$ away from perfect quadrature so as to generate a *Reynolds stress* large enough to overcome the stabilizing effect of dissipation. Naturally, if the Reynolds number is too small, it is expected that the flow will return to stability as a consequence of the large amount of dissipation. The cross-stream structure is investigated in more detail in the next section, to demonstrate that this mechanism indeed takes place.

7.4 Heuristic analysis of the structure of two-dimensional Tollmien–Schlichting waves

The determination of approximations to the solutions of the Orr–Sommerfeld problem (7.16–17) in the limit of large Reynolds numbers constitutes an intricate problem. It involves delicate asymptotic analyses that have been conceived over the years by several generations of fluid dynamicists, among them Heisenberg (1924), Tollmien (1947) and Schlichting (1933). A concise presentation of the main ideas can be found in the book by Lin (1955). For a more up-to-date and in-depth account of *high Reynolds number asymptotics* of the Orr–Sommerfeld equation, the reader is referred to chapters 4 and 5 of Drazin and Reid (1981) and to the surveys of Maslowe (1981, 1986) and Stewartson (1981). We proceed in two stages: first, we present an elementary and heuristic derivation of the eigenfunction structure [equations (7.23–58)], the main objective being the identification of the driving mechanism giving rise to the viscous instability; secondly, we outline the calculation of the resulting eigenvalue c and neutral curve [equations (7.59–68)].

Plane Poiseuille flow is used here as an illustrative example but a very similar structure prevails in other bounded shear flows such as the Blasius layer on a flat plate. We first note that, the velocity profile (7.4) being even, eigensolutions neatly separate into symmetric modes ($\phi(y)$ even, $\hat{u}(y)$ odd, $\hat{v}(y)$ even) and antisymmetric modes ($\phi(y)$ odd, $\hat{u}(y)$ even, $\hat{v}(y)$ odd). Numerical eigenvalue calculations have indicated that the only unstable mode is symmetric. In the sequel, we seek neutral symmetric solutions (c real) of the Orr–Sommerfeld equation (7.16) on the interval $y_1 \leq y \leq y_2$, where $y_1 = 0$ temporarily denotes the lower wall and $y_2 = 1$ the plane of symmetry (Fig. 7.2). The no-slip and symmetry conditions

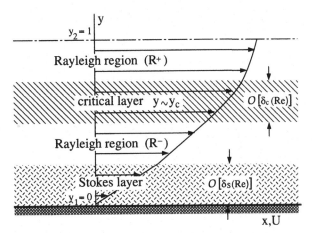

Fig. 7.2. High Reynolds number structure of even eigenfunctions for plane Poiseuille flow. The flow is decomposed into two Rayleigh regions (R+) and (R−), a viscous critical layer around y_c and a Stokes layer close to the lower wall $y_1 = 0$.

are expressed as

$$\phi(y_1) = \phi'(y_1) = 0\,, \quad \phi'(y_2) = \phi'''(y_2) = 0\,. \qquad (7.23a, b)$$

In the limit of large Reynolds numbers, the inviscid Rayleigh equation is formally recovered at leading order, but we anticipate that viscous diffusion will come into play to smooth the critical point singularity and to enforce the no-slip condition at the wall. We therefore postulate the existence of four distinct subdomains in the $x - y$ plane (Fig. 7.2): two Rayleigh regions (R+) and (R−) located on either side of $y = y_c$, a *viscous critical layer* of unknown thickness $\delta_c(\mathrm{Re}) \ll 1$ around $y = y_c$ and a *Stokes boundary layer* of unknown thickness $\delta_S(\mathrm{Re}) \ll 1$ lining the lower wall $y = y_1$.

In the Rayleigh regions (R+) and (R−), the leading-order approximation $\phi_{R,1}^{\pm}(y)$, assumed to be of order unity, is governed by the *Rayleigh equation* in the limit $\mathrm{Re} \gg 1$. The general solution of the Rayleigh equation is written as a linear combination of the two independent solutions $\phi_1(y)$ and $\phi_2(y)$ defined in (4.36) by their Frobenius expansions around y_c:

$$\phi_{R,1}^{\pm}(y) = a_R^{\pm}\phi_1(y) + b_R^{\pm}\phi_2(y)\,, \qquad (7.24)$$

equivalently

$$\phi_{R,1}^{\pm}(y) \sim a_R^{\pm}\left[y - y_c + \frac{U''(y_c)}{2U'(y_c)}(y - y_c)^2 + \ldots \right]$$

$$+ b_R^\pm \left[1 + \left\{ \frac{k^2}{2} + \frac{U'''(y_c)}{2U'(y_c)} - \left(\frac{U''(y_c)}{U'(y_c)} \right)^2 \right\} (y - y_c)^2 + \dots \right.$$

$$\left. + \frac{U''(y_c)}{U'(y_c)} \{y - y_c + \dots\} \ln(y - y_c) \right] . \tag{7.25}$$

The superscripts $\pm$ indicate that the multiplicative constants $a_R^\pm$ and $b_R^\pm$ may take distinct values in regions (R+) and (R−) and the branch selected for the logarithmic function is temporarily left unspecified. According to (4.45), the *Reynolds stress* is constant for *inviscid neutral waves* and we have at leading order

$$\tau_{xy,R}^\pm \sim \tau^\pm . \tag{7.26}$$

To determine the structure of the *Stokes layer* indicated by an S subscript, it is first necessary to evaluate its thickness $\delta_S(\mathrm{Re})$. The change of variable

$$y = \delta_S(\mathrm{Re})\xi, \quad \delta_S(\mathrm{Re}) \ll 1, \quad \xi = \mathcal{O}(1), \tag{7.27}$$

and the basic flow expansion $U(y) \sim U'(0)\, y + \dots$ lead to the following estimates for the various terms of the Orr–Sommerfeld equation (7.16)

$$\underline{-c \left(\frac{1}{\delta_S^2} \frac{d^2}{d\xi^2} - k^2 \right) \phi_S} - U''(0)\phi_S \sim$$

$$\frac{1}{ik\,\mathrm{Re}} \left(\underline{\frac{1}{\delta_S^4} \frac{d^4}{d\xi^4}} - 2\frac{k^2}{\delta_S^2} \frac{d^2}{d\xi^2} + k^4 \right) \phi_S . \tag{7.28}$$

Dominant balance arguments suggest that the largest inviscid term (underlined) on the left-hand side be of the same order of magnitude as the largest diffusive term (underlined) on the right-hand side, which yields the estimate

$$\delta_S(\mathrm{Re}) = \frac{1}{(k\,\mathrm{Re})^{1/2}} . \tag{7.29}$$

In the Stokes layer, the leading-order equation is obtained by substituting (7.29) into (7.28) and retaining dominant terms only:

$$\frac{d^4\phi_S}{d\xi^4} + ic\frac{d^2\phi_S}{d\xi^2} = 0, \tag{7.30}$$

to which should be added the no-slip conditions (7.23a):

$$\phi_S(0) = \phi_S'(0) = 0. \tag{7.31}$$

The general solution of (7.30) is a linear combination of the four independent solutions 1, ξ, $e^{-(-ic)^{1/2}\xi}$ and $e^{(-ic)^{1/2}\xi}$, where the principal branch of the square root $z^{1/2}$ is selected with argument in the range

$-\pi < \arg(z) \leq \pi$. The last exponentially growing solution must then be excluded in anticipation of matching requirements with the Rayleigh region (R−). Upon enforcing the boundary conditions (7.31), one is left with a ξ-structure of the form

$$\phi_S(\xi) \sim b_S \left[1 - (-ic)^{1/2}\xi - e^{-(-ic)^{1/2}\xi} \right] , \qquad (7.32)$$

where the constant b_S is to be determined by matching with the Taylor expansion of the outer solution close to the wall.

Let us tentatively assume that, at higher order, the outer expansion in the (R+) and the (R−) regions proceeds as:

$$\phi_R^{\pm}(y) \sim \phi_{R,1}^{\pm}(y) + \frac{1}{(k\,\mathrm{Re})^{1/2}}\phi_{R,2}^{\pm}(y) + \cdots . \qquad (7.33)$$

It is readily found by substitution of (7.33) into the Orr–Sommerfeld equation (7.16) that $\phi_{R,2}^{\pm}(y)$ also satisfies the Rayleigh equation. The Taylor expansion of the outer solution (7.33) close to the wall $y = 0$ reads in inner variable ξ

$$\phi_R^-(y) \sim \phi_{R,1}^-(0) + \frac{1}{(k\,\mathrm{Re})^{1/2}}\left[\phi_{R,2}^-(0) + \phi_{R,1}^{-\prime}(0)\xi \right] + \cdots . \qquad (7.34)$$

Leading-order matching between (7.32) and (7.34) gives the identity $b_S \left[1 - (-ic)^{1/2}\xi \right] = \phi_{R,1}^-(0)$. This can only be satisfied for all ξ if $b_S = 0$: one therefore recovers the expected impermeability condition

$$\phi_{R,1}^-(0) = 0 . \qquad (7.35)$$

Such a result implies that the inner expansion in the Stokes region should start at order $1/(k\,\mathrm{Re})^{1/2}$:

$$\phi_S(\xi) \sim \frac{1}{(k\,\mathrm{Re})^{1/2}}b_{S,1}\left[1 - (-ic)^{1/2}\xi - e^{-(-ic)^{1/2}\xi} \right] . \qquad (7.36)$$

Invoking the impermeability condition $\phi_{R,1}^-(0) = 0$, leading-order matching between (7.36) and (7.34) then yields :

$$b_{S,1}\left[1 - (-ic)^{1/2}\xi \right] = \phi_{R,2}^-(0) + \phi_{R,1}^{-\prime}(0)\xi . \qquad (7.37)$$

This is satisfied for all ξ provided that

$$\phi_{R,2}^-(0) = b_{S,1} , \qquad (7.38a)$$

$$\phi_{R,1}^{-\prime}(0) = -(-ic)^{1/2}b_{S,1} , \qquad (7.38b)$$

To the distribution (7.36) are associated velocity distributions $\hat{u}(y)$ and $\hat{v}(y)$ that are not in quadrature within the Stokes layer. The Reynolds stress is therefore non zero and it can be determined from (4.44) by evaluating $\phi_S(y)$ according to (7.36) and setting $d/dy = (k\,\mathrm{Re})^{1/2}d/d\xi$. After some elementary algebra, one finds

$$\tau_{xy,\mathrm{S}}(\xi) \sim \tfrac{1}{2}k\sqrt{c/2}(k\,\mathrm{Re})^{-1/2}|b_{\mathrm{S},1}|^2 \left[1 - 2\,e^{-\xi\sqrt{c/2}}\cos\left(\xi\sqrt{c/2}\right)\right.$$
$$\left. - \sqrt{2c}\,\xi\,e^{-\xi\sqrt{c/2}}\sin\left(\xi\sqrt{c/2}\right) + e^{-\xi\sqrt{2c}}\right]. \tag{7.39}$$

When $\xi \to \infty$, the Reynolds stress reaches a constant level that matches with the value in the (R−) region:

$$\tau_{xy,\mathrm{S}}(\infty) = \tfrac{1}{2}k\sqrt{c/2}(k\,\mathrm{Re})^{-1/2}|b_{\mathrm{S},1}|^2 = \tau^-. \tag{7.40}$$

Note that, as a result of the conditions (7.38a,b), the limiting value (7.40) is indeed compatible with the value prevailing in the (R−) region $\tau^- = \left(ik/4(k\,\mathrm{Re})^{1/2}\right)\left[\phi_{\mathrm{R},2}^-(0)\phi_{\mathrm{R},1}^{-\,\prime*}(0) - \phi_{\mathrm{R},2}^{-*}(0)\phi_{\mathrm{R},1}^{-\,\prime}(0)\right]$, as calculated by substituting the two-term Rayleigh expansion (7.33) into (4.44) and appealing to the impermeability condition $\phi_{\mathrm{R},1}^-(0) = 0$.

The inner structure of the *critical layer* is somewhat more involved. As in the previous case, we first determine its characteristic thickness $\delta_c(\mathrm{Re})$ by performing the change of variable

$$y - y_c = \delta_c(\mathrm{Re})\eta, \quad \delta_c(\mathrm{Re}) \ll 1, \quad \eta = \mathcal{O}(1). \tag{7.41}$$

Considering that around y_c, $U(y) \sim U'(y_c)(y - y_c)$ and $U''(y) \sim U''(y_c)$, the following estimates of the various terms in the Orr–Sommerfeld equation (7.16) are obtained:

$$\underline{(U'(y_c)\,\delta_c\eta)\left(\frac{1}{\delta_c^2}\frac{d^2}{d\eta^2} - k^2\right)\phi_c} - U''(y_c)\phi_c$$
$$\sim \underline{\frac{1}{ik\,\mathrm{Re}}\left(\frac{1}{\delta_c^4}\frac{d^4}{d\eta^4} - 2\frac{k^2}{\delta_c^2}\frac{d^2}{d\eta^2} + k^4\right)\phi_c}. \tag{7.42}$$

Dominant balance arguments lead to the requirement that the two underlined terms in the above equation be comparable. Thus the critical layer thickness is found to be

$$\delta_c(\mathrm{Re}) = \frac{1}{(k\,\mathrm{Re})^{1/3}}. \tag{7.43}$$

Setting

$$\phi_c(\eta) \sim \phi_{c,1}(\eta) + \frac{1}{(k\,\mathrm{Re})^{1/3}}\phi_{c,2}(\eta) + \cdots, \tag{7.44}$$

one finds at leading order

$$\left[\frac{d^4}{d\eta^4} - iU'(y_c)\eta\frac{d^2}{d\eta^2}\right]\phi_{c,1} = 0. \qquad (7.45)$$

The general solution of (7.45) is a linear combination of four independent solutions η, 1, $\phi_c^{(3)}(\eta)$ and $\phi_c^{(4)}(\eta)$. As discussed in detail in Eagles (1969) and Drazin and Reid (1981), the viscous-dominated solutions $\phi_c^{(3)}(\eta)$ and $\phi_c^{(4)}(\eta)$ may be expressed as successive double integrals of Hankel functions. For the present purpose, it is sufficient to isolate their behavior for large $|\eta|$.

The asymptotic expansion of $\phi_c^{(3)}(\eta)$ and $\phi_c^{(4)}(\eta)$ displays the *Stokes phenomenon* (Bender and Orszag, 1978): it is dependent on the direction along which η tends to infinity. The complex η-plane (Fig. 7.3) is divided into three sectors S_1, S_2 and S_3 of angle $2\pi/3$ bounded by Stokes lines (Bender and Orszag, 1978) emanating from $\eta = 0$ (equivalently $y = y_c$). It is found that $\phi_c^{(3)}(\eta)$ $[\phi_c^{(4)}(\eta)]$ is exponentially small in the sector S_1 $[S_2]$ whereas it is exponentially large in the two remaining sectors S_2 and S_3 $[S_1$ and $S_3]$. This behavior is incompatible with matching requirements between the critical layer solution $\phi_{c,1}(\eta)$ and the leading-order Rayleigh solution (7.25) in the (R+) and (R−) regions on the real η-axis. We therefore anticipate that $\phi_{c,1}(\eta)$ only involves a linear combination of the first two solutions, i.e.

$$\phi_{c,1}(\eta) = a_c\eta + b_c. \qquad (7.46)$$

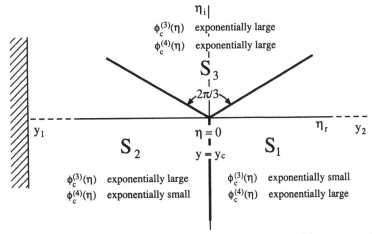

Fig. 7.3. Asymptotic behavior of independent solutions $\phi_c^{(3)}(\eta)$ and $\phi_c^{(4)}(\eta)$ in complex η-plane for large $|\eta|$, $U'(y_c) > 0$: Stokes phenomenon.

Matching (7.46) and (7.25) to order unity requires that $a_c(k\,\mathrm{Re})^{1/3}(y - y_c) + b_c = b_R^{\pm}$, which leads to the relations

$$b_c = b_R^+ = b_R^- \equiv b_R,\tag{7.47}$$

$$a_c = 0.\tag{7.48}$$

The constant $b_R^{\pm}$ is seen to be continuous across the critical point which implies that

$$\phi_{R,1}^+(y_c) = \phi_{R,1}^-(y_c) \equiv \phi_{R,1}(y_c),\tag{7.49}$$

equivalently, at leading order,

$$\hat{v}_R^+(y_c) = \hat{v}_R^-(y_c) \equiv \hat{v}_R(y_c).\tag{7.50}$$

In order to reach a conclusion regarding the connection formula relating a_R^+ to a_R^-, it is necessary to pursue the critical layer solution to $\mathcal{O}(k\,\mathrm{Re})^{-1/3}$ included. At the next order, $\phi_{c,2}(\eta)$ is given by

$$\left[\frac{d^4}{d\eta^4} - iU'(y_c)\eta\frac{d^2}{d\eta^2}\right]\phi_{c,2} = -iU''(y_c)b_R.\tag{7.51}$$

As shown in Drazin and Reid (1981), a particular solution $\phi_{c,2}(\eta)$ of (7.51) can be constructed which smoothes the logarithmic singularity at y_c. As $|\eta| \to 0$ (Fig. 7.4), this solution exhibits in the sectors S_1 and S_2 an "inviscid-like" behavior fully compatible with the Frobenius solutions (7.25).

However in the sector S_3, it is dominated by viscosity and becomes exponentially large, a feature which is incompatible with the necessarily inviscid solutions of the Rayleigh equation. We merely state the main

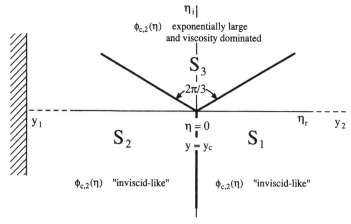

Fig. 7.4. Asymptotic behavior of particular solution $\phi_{c,2}(\eta)$ of equation (7.51) in complex η-plane for large $|\eta|$, $U'(y_c) > 0$.

result: upon matching $\phi_c(\eta) \sim \phi_{c,1}(\eta) + (k\,\mathrm{Re})^{-1/3}\phi_{c,2}(\eta)$ and $\phi_{R,1}(y)$ to order $(k\,\mathrm{Re})^{-1/3}$ included along the real y axis, it is found that $a_R^{\pm}$ remains continuous across the critical layer,

$$a_R^+ = a_R^- \equiv a_R\,, \tag{7.52}$$

provided that the logarithmic function appearing in (7.25) is interpreted in the following manner:

$$\ln(y - y_c) = \begin{cases} \ln|y - y_c| & y > y_c \\ \ln|y - y_c| - i\pi\,\mathrm{sgn}\left\{U'(y_c)\right\} & y < y_c \end{cases} \tag{7.53}$$

When $U'(y_c) > 0$, the branch cut of the logarithm must therefore be chosen to lie in the upper half-plane as shown in Fig. 7.5.

This important conclusion holds in general, independently of the specific flow under consideration. Thus, *in order to ensure that the logarithmic singularity at the critical point on the real y axis can be smoothed by viscosity, it is sufficient to integrate the Rayleigh equation along a path that is deformed into the lower complex y-plane when $U'(y_c) > 0$, so as to avoid y_c, as sketched in Fig. 7.5. The phase shift applied to the logarithm is then $-i\pi$ as dictated by (7.53).* In this fashion, one may avoid altogether the complexities associated with the viscous critical layer structure.

Note that the *normal velocity remains continuous across y_c* [see (7.50)], but that *the streamwise component is discontinuous.* By applying (7.53) to (7.25) and taking into consideration the continuity of a_R and b_R across

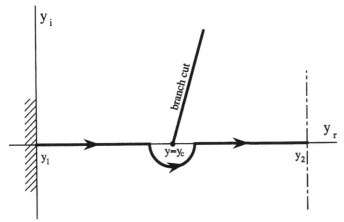

Fig. 7.5. Contour deformation rule in the complex y-plane to integrate the Rayleigh equation around a critical point on the real y-axis, when $U'(y_c) > 0$.

y_c, one readily obtains

$$\phi_{R,1}^{+}{}'(y_c) - \phi_{R,1}^{-}{}'(y_c) = i\pi \frac{U''(y_c)}{|U'(y_c)|}\phi_{R,1}(y_c)\,, \tag{7.54}$$

or in terms of velocity components

$$\hat{u}_R(y_c^{+}) - \hat{u}_R(y_c^{-}) = -\frac{\pi}{k}\frac{U''(y_c)}{|U'(y_c)|}\hat{v}_R(y_c)\,. \tag{7.55}$$

Correspondingly, the Reynolds stress $\tau_{xy,R}$ exhibits a *jump* across y_c that can be evaluated by substituting (7.54) into (4.44). One finds:

$$\tau^{+} - \tau^{-} = \frac{\pi k}{2}\frac{U''(y_c)}{|U'(y_c)|}|\phi_{R,1}(y_c)|^2\,. \tag{7.56}$$

The above results regarding the Stokes and critical layers have been derived *independently of the particular flow under consideration*. We now proceed to exploit our findings for the symmetric modes of plane Poiseuille flow. All conditions have been enforced except those concerning the even nature of the perturbations. The symmetry conditions (7.23b) together with (4.44) imply that the constant Reynolds stress in Rayleigh region (R+) is necessarily zero:

$$\tau^{+} = 0\,. \tag{7.57}$$

Two distinct expressions have therefore been derived for the Reynolds stress τ^{-} in the Rayleigh region (R−): the first one results from the jump

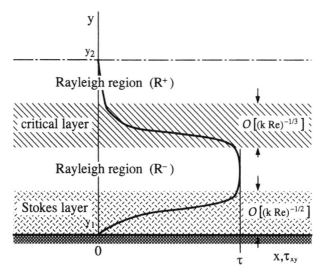

Fig. 7.6. Reynolds stress distribution of even neutral modes on the upper branch of the plane Poiseuille flow neutral stability curve.

condition (7.56) across the critical point y_c, where τ^+ is set equal to zero. The second expression comes from the asymptotic value (7.40) at the outer edge of the Stokes layer. The eigenvalue c must be chosen in such a way that these two values coincide:

$$\tau^- = -\frac{\pi k}{2}\frac{U''(y_c)}{|U'(y_c)|}|\phi_{R,1}(y_c)|^2 = \tfrac{1}{2}k\sqrt{c/2}(k\,\mathrm{Re})^{-1/2}|b_{S,1}|^2\,. \tag{7.58}$$

Thus, the Reynolds stress distribution corresponding to an even neutral mode takes the form sketched in Fig. 7.6: *a finite Reynolds stress is created in the Stokes layer as a result of the phase difference between $\hat{u}$ and $\hat{v}$* generated by viscous diffusion. Within the Rayleigh region (R−), the Reynolds stress remains constant and returns to zero through the critical layer in order to satisfy the symmetry conditions on the plane $y = 0$. In this fashion, a positive production term is induced by a small but finite viscosity, which, for neutral waves, exactly balances the viscous dissipation term in the energy equation (7.20). Although the present considerations strictly pertain to neutral waves, one may easily envision eigenmodes where the *induced Reynolds stress exceeds viscous dissipation, thereby leading to instability*: the potentially destabilizing role of viscosity has therefore been demonstrated.

To complete the reasoning, the derivation of the real eigenvalue c is now briefly outlined. The more physically inclined reader may, without harm, proceed to the next section. Following Tollmien (1947), the wavenumber is assumed to be small and even solutions of the Rayleigh equation are sought in powers of k^2:

$$\phi_{R,1}(y) \sim \phi_\alpha(y) + k^2\phi_\beta(y) + \dots \tag{7.59}$$

By substitution into the Rayleigh equation (4.33), the following sequence of problems is generated:

$$\phi_\alpha'' - \frac{U''(y)}{U(y) - c}\phi_\alpha = 0\,, \tag{7.60}$$

$$\phi_\beta'' - \frac{U''(y)}{U(y) - c}\phi_\beta = \phi_\alpha\,. \tag{7.61}$$

We are only interested here in even solutions of (7.60,61) which can be expressed in the form

$$\phi_\alpha(y) = U(y) - c\,, \tag{7.62a}$$

$$\phi_\beta(y) = [U(y) - c]\int_{y_2}^y \frac{ds}{[U(s) - c]^2}\int_{y_2}^s [U(t) - c]^2 dt\,. \tag{7.62b}$$

The sole remaining issue is the determination of the coefficients a_R and b_R multiplying the Frobenius solutions in (7.25) where $\phi_{R,1}(y)$ is the particular even solution specified above. Tedious but straightforward manipulations [see, for instance, Eagles (1969)] indicate that, in the limit $y_c \to 0$, $k \to 0$,

$$a_R \sim U'(0), \quad b_R \sim \frac{1}{U'(0)} \int_{y_1}^{y_2} U^2(y)\, dy \cdot k^2. \tag{7.63a, b}$$

The eigenvalue is then conveniently obtained as a solution of the impermeability condition (7.35) and the matching conditions (7.38b) and (7.58). In view of (7.24,25) and the continuity of $a_R^\pm$ and $b_R^\pm$ across the critical layer, they can be expressed as

$$a_R \phi_1(0) + b_R \phi_2(0) = 0, \tag{7.64a}$$

$$a_R \phi_1'(0) + b_R \phi_2'(0) = -(-ic)^{1/2} b_{S,1}, \tag{7.64b}$$

$$-\frac{\pi k}{2} \frac{U_c''}{|U_c'|} |b_R|^2 = \tfrac{1}{2} k \sqrt{c/2} (k\,\mathrm{Re})^{-1/2} |b_{S,1}|^2. \tag{7.64c}$$

To leading order in $y_c \ll 1$, expressions (7.25) for $\phi_1(y)$ and $\phi_2(y)$ yield

$$\phi_1(0) \sim -y_c, \quad \phi_2(0) \sim 1, \tag{7.65a, b}$$

$$\phi_1'(0) \sim 1, \quad \phi_2'(0) \sim \mathcal{O}(1). \tag{7.66a, b}$$

Bearing in mind the estimates (7.63a,b) for a_R and b_R, the first two matching conditions (7.64a,b) lead to the following expressions for the location y_c of the critical point and for $b_{S,1}$:

$$y_c \sim \frac{1}{U'^2(0)} \int_{y_1}^{y_2} U^2(y)\, dy \cdot k^2, \tag{7.67a}$$

$$b_{S,1} \sim -\frac{U'(0)}{\sqrt{-ic}}. \tag{7.67b}$$

Use of these relations in (7.64c), together with the phase velocity estimate $c = U(y_c) \sim U'(0) y_c$, finally result in the relationship between Reynolds number and wavenumber

$$\mathrm{Re} \sim \frac{1}{2\pi^2} \frac{[U'(0)]^{11}}{[U''(0)]^2} \left[\int_{y_1}^{y_2} U^2(y)\, dy \right]^{-5} k^{-11}. \tag{7.68}$$

The above relation provides an asymptotic estimate of the *upper branch of the neutral curve* in the Re–k plane. In terms of the original small parameter $(k\,\mathrm{Re})^{-1}$, the critical point is located at a distance $\mathcal{O}[(k\,\mathrm{Re})^{-1/5}]$ from the lower wall [see equation (7.67a)]. Under these circumstances, the critical layer and the Stokes layer are indeed distinct, as postulated initially (Figs. 7.2 and 7.6). However, the assumed cross-stream structure has failed to capture all the even neutral modes in the limit $k \ll 1$ and $\mathrm{Re} \gg 1$. There also exists a *lower branch* to the neutral curve for which the critical point y_c is so close to the wall that the critical layer and wall layer strongly overlap. Under these conditions, the present analysis fails and one must adopt a distinct scaling procedure which will not be presented here. The above calculations have illustrated the complexity of the asymptotic structure of viscous instabilities at large Reynolds numbers. The main conclusion is that, in an intermediate range of Reynolds numbers, viscous diffusion can serve to extract energy from the basic flow and even result in its destabilization.

8 Plane channel flow

Plane channel flow will serve here as the archetype of bounded shear flows with no inflection points which, according to the results of sections 4 and 7, can only be linearly destabilized by weak viscous diffusion. This study will offer the opportunity to introduce and implement various methods that could also be used in other open shear flow configurations. Following the philosophy outlined in section 3, the procedure is decomposed into successive steps. The main features of the primary viscous instability are summarized in section 8.1, thereby completing the presentation of section 7.4. The weakly nonlinear evolution of unstable finite amplitude disturbances is then examined close to the critical Reynolds number by resorting to the *method of multiple scales* (section 8.2). The perturbation amplitude is shown to be governed by the same GL equation that appears prominently elsewhere in this volume. Nonlinear effects are however found to be destabilizing below onset and stable finite-amplitude two-dimensional states must be determined numerically. Primary perturbations evolve into finite-amplitude waves over long viscous time scales τ_v and this behavior cannot account for the rapid transition to turbulence which is observed experimentally. Combined analytical-numerical studies have demonstrated that finite-amplitude two-dimensional waves (section 8.3) are themselves subject to a powerful *inviscid instability* arising from the presence of elliptically shaped vortical regions (section 8.4). This mechanism is responsible for the development of several types of secondary three-dimensional instabilities growing over fast inertial time scales τ_i (section 8.5). In preparing this section, we have been inspired by the lucid reviews of Bayly *et al.* (1988) and Herbert (1988) to which we refer the reader for additional information.

8.1 Primary linear instability

The large Reynolds number analysis outlined in section 7.4 can be considerably generalized (Drazin and Reid, 1981) to arrive at an excellent approximation for the two-dimensional neutral curve $\omega_i(k; \mathrm{Re}) = 0$. A lower neutral branch $\mathrm{Re} \propto k^{-7}$, for which the critical layer and Stokes layer become undistinguishable, can also be identified, as sketched in Fig. 8.1.

Efficient numerical algorithms (Orszag, 1971; Mack, 1976) have been developed to accurately solve the eigenvalue problem associated with the Orr–Sommerfeld equation and to determine the dispersion relation $D(k,\omega;\mathrm{Re}) = 0$ either in the temporal case or the spatial case. The neutral curve in the Re–k plane is found to have the shape illustrated on Fig. 8.1. When Re $\gg$ 1, plane Poiseuille flow is only unstable to long wavelength disturbances and the upper and lower branches are asymptotically recovered as Re $\to \infty$. Note that, in this limit, the unstable band of wavenumbers shrinks to zero, as required by the fact that the flow is inviscidly stable. Below the critical value $\mathrm{Re}_c \equiv 5,772$ corresponding to $k_c = 1.02$, plane Poiseuille flow becomes linearly stable since $\omega_i < 0$ for all k. Thus the shape of the neutral curve corroborates the intuitive arguments of section 7: viscous diffusion can indeed be destabilizing in an intermediate range of Re. However, if Re is too low, the flow is stabilized by viscosity.

These results should briefly be compared with those for inflectional shear flows such as the hyperbolic tangent mixing layer analyzed in section 5. Asymptotic (Tatsumi and Gotoh, 1960; Tatsumi *et al.*, 1964) and numerical (Betchov and Szewczyk, 1963) solutions of the Orr–Sommerfeld equation on the infinite interval for the basic flow $U_1(y) = \tanh y$ have indicated that the neutral curve is of the form sketched in Fig. 8.2. As the Reynolds number becomes infinite, the inviscidly unstable band of wavenumbers $0 \le k \le 1$ obtained by Michalke (1964) is recovered (section 5). When Re decreases, viscous diffusion is found to continuously reduce the unstable range of wavenumbers as well

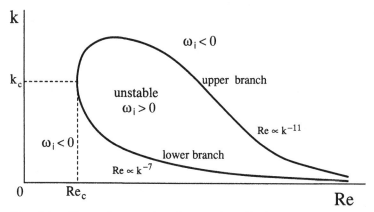

Fig. 8.1.   Neutral curve for plane Poiseuille flow in Re–k plane.

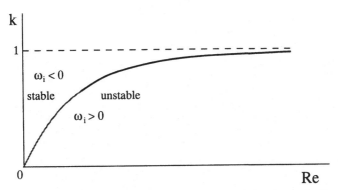

Fig. 8.2. Neutral curve for hyperbolic tangent mixing layer in Re–k plane.

as the temporal growth rates. The nominal critical Reynolds number is in this case zero but this feature should be viewed with extreme caution: the basic flow $U_1(y) = \tanh y$, which does not satisfy the Navier–Stokes equations, diffuses over viscous time scales $\tau_v = L^2/\nu$. When $Re = \mathcal{O}(1)$, the instability evolution time scale τ_{in} becomes of the same order of magnitude as τ_v, which invalidates the steady basic flow assumption. The shapes of the neutral curves illustrated on Figs. 8.1 and 8.2 are typical of flows that are subjected to viscous and inviscid instability mechanisms respectively.

A representative eigenfunction for a temporally growing mode and the associated *Reynolds stress* distribution are shown in Figs. 8.3 and 8.4. The latter should be compared with the result of the asymptotic analysis in section 7.4.

The absolute frequency $\omega_0 \equiv \omega(k_0)$ of zero group velocity (section 3.2) has been numerically determined from the two-dimensional dispersion relation $D(k, \omega; Re) = 0$ by Deissler (1987). In a wide range of Reynolds numbers, the absolute growth rate $\omega_{0,i}$ is negative and it may be concluded that *plane Poiseuille flow is convectively unstable*. It is therefore expected to be sensitive to external noise, the streamwise evolution of imposed perturbations being described by spatial instability concepts. The Blasius boundary layer is also locally convectively unstable as vividly demonstrated theoretically and experimentally by Gaster (1975) and Gaster and Grant (1975): the three-dimensional impulse response is observed to be advected in the downstream direction and its early development closely follows the predictions of a numerically synthesized Green's function.

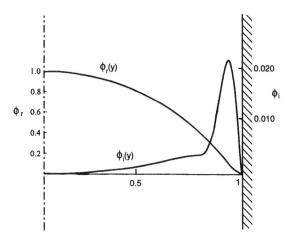

Fig. 8.3.   Even eigenfunction at $k = 1$ and $\mathrm{Re} = 10^4$ for the eigenvalue $c = 0.23 + 0.0037i$ in plane Poiseuille flow.   After Drazin and Reid (1981), original results of Thomas (1953).

These theoretical results have conclusively been verified experimentally in the case of plane Poiseuille flow by Nishioka *et al.* (1975). In particular, the amplitude and phase distributions of Tollmien–Schlichting waves are seen to closely follow the spatial stability analysis of Itoh (1974), as shown on Fig. 2.20, and measured spatial growth rates $-k_i$ are consistent with theoretically computed eigenvalues (Fig. 8.5). Furthermore, channel flow is found to be linearly stable or unstable in distinct regions of the Re–k_r plane, in satisfactory agreement with the shape of the calculated neutral curve (Fig. 8.6).

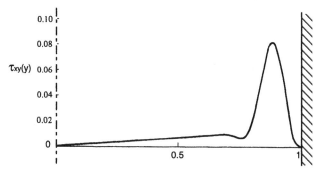

Fig. 8.4.   Reynolds stress distribution $\tau_{xy} \propto (\phi\phi'^{*} - \phi^{*}\phi')i/4$ calculated according to formula (4.44) for eigenfunction in Fig. 8.3.   After Drazin and Reid (1981) and Stuart (1963).

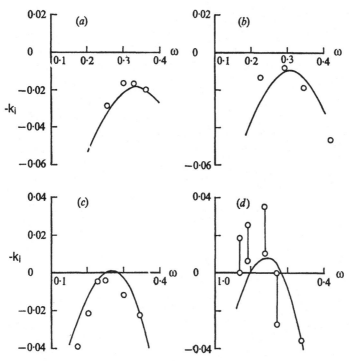

Fig. 8.5. Spatial amplification rates $-k_i$ versus frequency ω (Nishioka *et al.*, 1975). $-$: Itoh's theoretical prediction (1974); o: experiments. (a): Re $= 3,000$. (b): Re $= 4,000$. (c): Re $= 6,000$; experiments: Re $= 5,700$. (d): Re $= 8,000$; experiments: Re $= 7,000$.

This comparison only holds in the early linear stages of the evolution, i.e. when streamwise velocity fluctuations are less than 1% of the center line velocity. This takes place before three-dimensional processes and catastrophic breakdown* develop which rapidly lead to turbulence.

Linear theory has been successful inasmuch as it has led (a) to the identification of a linear viscous instability mechanism strong enough to destabilize plane Poiseuille flow over long time scales ω_i^{-1} beyond a critical value of the Reynolds number Re$_c$ = 5,772 and (b) to the prediction of the evolution of predominantly two-dimensional waves in carefully conducted forced experiments at low turbulence level close to this instability threshold (section 2.4). However the linear framework is incapable of explaining experimental observations (Carlson *et al.*, 1982; Nishioka *et al.*, 1975; Nishioka and Asai, 1985) which indicate that, with a *high level of incoming perturbations*, transition may take place

---

* Breakdown commonly occurs when streamwise velocity fluctuations reach 25% of the center line velocity and it is characterized by high-frequency oscillations.

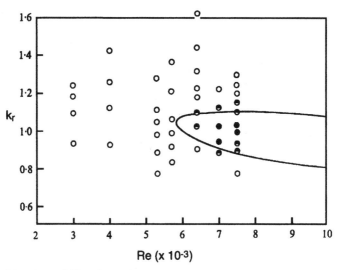

Fig. 8.6. Linear stability boundary in Re–k_r plane. —: Itoh's neutral curve (1974). Nishioka *et al.*'s results (1975): open circles: damped; half-filled circles: nearly neutral; full circles: amplified.

for Reynolds numbers as low as 1,000. Furthermore, in uncontrolled experiments or natural conditions, most growing disturbances are *three-dimensional structures* evolving over inviscid time scales $\tau_i = L/V$ that are considerably shorter than those typically encountered for *Tollmien–Schlichting waves*. We must therefore appeal to other mechanisms in order to bridge the gap between experimental evidence and the predictions of linear instability theory. One must enter the realm of nonlinear formulations.

8.2 Weakly nonlinear analysis

The most natural way to investigate the effect of nonlinearities analytically is to generate, in a systematic manner, nonlinear corrections to linear stability results, assuming that fluctuations remain small, say $\mathcal{O}(\varepsilon)$. Thus, within a two-dimensional setting, we postulate the existence of an expansion in powers of ε

$$\Psi(x,y,t) \sim \Psi_0(y) + \varepsilon\psi_1(x,y,t) + \varepsilon^2\psi_2(x,y,t) + \dots, \qquad (8.1)$$

where $\Psi_0(y) \equiv \int U(y)\,dy$ characterizes the basic state and $\psi_1(x,y,t)$ small perturbations of the form (4.31) given by linear theory. Such an approach constitutes the starting point of the weakly nonlinear theories pioneered by Malkus and Veronis (1958) in the context of Rayleigh–Bénard convection and by Stuart (1960) and Watson (1960) in the context

of plane Poiseuille flow. Possible first-order fluctuations ψ_1 *a priori*
encompass all points in the Re–k plane of Fig. 8.7.

Ideally, one would prefer to determine the "nonlinear fate" of the most
amplified Tollmien–Schlichting wave for a fixed supercritical Reynolds
number Re > Re_c. Nonlinear self-interactions of the linearized
disturbance (4.31) will then generate in the higher-order approximations
ψ_2, ψ_3, etc., higher harmonics containing terms of the form $e^{n\omega_i t}$. The
expansion (8.1) will therefore become disordered over very short time
scales and this approach is then doomed to fail. One must be content
with exploring regions of the k–Re plane where the linear growth rate
$\omega_i(k; \mathrm{Re})$ is small enough that it can be approximated by its Taylor
expansion, i.e. regions made up of thin bands centered around the neutral
curve $\omega_i(k; \mathrm{Re}) = 0$ in Fig. 8.7. If Re is set at an arbitrary supercritical
value, the weakly nonlinear formulation will then obviously miss the most
amplified wave $k_{\max}$ in the middle of the unstable band of wavenumbers.
In order to capture all unstable components, one must further restrict
the analysis to the vicinity of the critical condition[*] (Re_c, k_c). In the
sequel the weakly nonlinear formulation of Stewartson and Stuart (1971)
is outlined for wave packets "centered" around criticality Re_c, k_c. The
method of multiple scales (Bender and Orszag, 1978) is used to generate
an amplitude evolution equation valid close to Re_c, k_c.

As in many other asymptotic modeling approaches, it is a definite plus
to know what to expect! The temporal mode $\omega(k; Re)$ that becomes

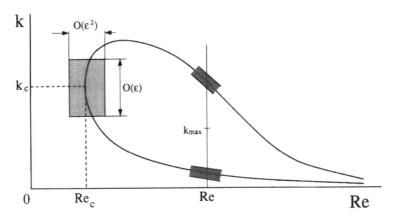

Fig. 8.7. Sketch of regions in Re–k plane where weakly nonlinear theory can be
applied.

---

[*] Note that a weakly nonlinear analysis will successfully determine the nature of the
bifurcation at that particular point on the neutral curve.

unstable at k_c, Re_c necessarily admits a Taylor expansion of the form

$$\omega - \omega_c \sim \omega_k^c(k - k_c) + \tfrac{1}{2}\omega_{kk}^c(k - k_c)^2 + \omega_{\mathrm{Re}}^c(\mathrm{Re} - \mathrm{Re}_c) + \dots, \qquad (8.2)$$

where $\omega_c \equiv \omega(k_c; \mathrm{Re}_c)$ is the real frequency at onset, $\omega_k^c \equiv \partial\omega/\partial k(k_c; \mathrm{Re}_c)$ is the corresponding real group velocity and the remaining coefficients ω_{kk}^c and ω_{Re}^c are complex. This expansion is consistent with the parabolic shape of the neutral curve $\omega_i(k; \mathrm{Re}) = 0$ around k_c, Re_c. In all modesty, the leading-order perturbation is assumed to be

$$\psi(x, y, t) \sim \mathcal{R}e\left\{a(x, t)\phi_c(y)\exp[i(k_c x - \omega_c t)]\right\}. \qquad (8.3)$$

Expression (8.3) represents a wave packet made up of carrier *Tollmien–Schlichting waves* at the critical wavenumber with an unknown complex amplitude $a(x, t)$ of order ε. The amplitude evolution equation for $a(x, t)$ should be such that the dominant linear terms represented in Fourier space by (8.2) balance the dominant nonlinear term. To determine the form of the latter, we note that the quadratic nonlinearity in the vorticity equation (7.15), equivalently the Navier–Stokes equation (7.2), induces harmonics at successive orders in ε which take the following form

$$\begin{aligned}
\mathcal{O}\left(\varepsilon^2\right): &\quad a^2\exp[2i(k_c x - \omega_c t)] + \text{c.c.}, &\quad |a|^2, \\
\mathcal{O}\left(\varepsilon^3\right): &\quad a^3\exp[3i(k_c x - \omega_c t)] + \text{c.c.}, &\quad |a|^2 a\,\exp[i(k_c x - \omega_c t)] + \text{c.c.}.
\end{aligned}$$

The carrier wave at k_c, ω_c is seen to be reproduced at $\mathcal{O}(\varepsilon^3)$ through a component of amplitude $|a|^2 a$. Furthermore, the Taylor approximation (8.2) to the linear dispersion relation can be transformed back into physical space by applying the rules $(\omega - \omega_c) \mapsto i\partial/\partial t$, $(k - k_c) \mapsto -i\partial/\partial x$. When the resulting linear operator is balanced with the dominant nonlinearity $|a|^2 a$, one obtains the general form of the anticipated *amplitude evolution equation* for $a(x, t)$:

$$\frac{\partial a}{\partial t} + \omega_k^c\frac{\partial a}{\partial x} - \frac{1}{2}i\omega_{kk}^c\frac{\partial^2 a}{\partial x^2} + i\omega_{\mathrm{Re}}^c(\mathrm{Re} - \mathrm{Re}_c)a = -b|a|^2 a, \qquad (8.4)$$

where b is an unknown complex number, commonly referred to as the Landau constant in the fluid mechanics literature. Not surprisingly, the GL evolution equation is recovered. All linear terms in (8.4) are naturally known from linear theory and the objective of the multiple scale procedure is to calculate the Landau constant. Let us introduce the following explicit scaling relationships:

$$a(x, t) = \varepsilon A(X, T_1, T_2), \qquad \mathrm{Re} - \mathrm{Re}_c = \Delta\delta(\varepsilon), \qquad (8.5\text{a, b})$$
$$X = \mu(\varepsilon)x, \qquad T_1 = \nu_1(\varepsilon)t, \qquad T_2 = \nu_2(\varepsilon)t. \qquad (8.5\text{c, d, e})$$

The *slow space and time scales* X, T_1, T_2 and the departure from criticality $\mathrm{Re} - \mathrm{Re}_c$ are characterized by unknown gauge functions $\mu(\varepsilon)$,

$\nu_1(\varepsilon)$, $\nu_2(\varepsilon)$ and $\delta(\varepsilon)$. Under the change of variables (8.5), evolution equation (8.4) becomes

$$\nu_1 \frac{\partial A}{\partial T_1} + \nu_2 \frac{\partial A}{\partial T_2} + \mu \omega_k^c \frac{\partial A}{\partial X} - \frac{\mu^2}{2} i \omega_{kk}^c \frac{\partial^2 A}{\partial X^2} + i \delta \Delta \omega_{\mathrm{Re}}^c A = -\varepsilon^2 b |A|^2 A \,. \quad (8.6)$$

It is impossible to find gauge functions such that all terms in the above equation become of the same order of magnitude. The leading-order time variation $\nu_1 \partial A/\partial T_1$ is tentatively chosen to balance the group velocity advection term $\mu \omega_k^c \partial A/\partial X$ whereas the second-order time variation $\nu_2 \partial A/\partial T_2$ is required to balance all other remaining terms including the $\mathcal{O}(\varepsilon^2)$ nonlinearity. The following estimates for the gauge functions are then obtained

$$\nu_1(\varepsilon) = \varepsilon \,, \quad \nu_2(\varepsilon) = \varepsilon^2 \,, \quad \mu(\varepsilon) = \varepsilon \,, \quad \delta(\varepsilon) = \varepsilon^2 \,, \quad (8.7\mathrm{a,b,c,d})$$

and (8.6) separates into two distinct evolution equations

$$\frac{\partial A}{\partial T_1} + \omega_k^c \frac{\partial A}{\partial X} = 0 \,, \quad (8.8\mathrm{a})$$

$$\frac{\partial A}{\partial T_2} - \frac{1}{2} i \omega_{kk}^c \frac{\partial^2 A}{\partial X^2} + i \Delta \omega_{\mathrm{Re}}^c A = -b |A|^2 A \,. \quad (8.8\mathrm{b})$$

Alternatively, the time variable T_1 can be absorbed into a moving space variable

$$X' = X - \omega_k^c T_1 = \varepsilon (x - \omega_k^c t) \,, \quad (8.9)$$

to arrive at a single *amplitude evolution equation* for $A(X', T)$:

$$\frac{\partial A}{\partial T_2} - \frac{1}{2} i \omega_{kk}^c \frac{\partial^2 A}{\partial X'^2} + i \Delta \omega_{\mathrm{Re}}^c A = -b |A|^2 A \,. \quad (8.10)$$

The above reasoning constitutes by far the most delicate step in the implementation of the method. In order to confirm the validity of the result (8.10), it is necessary to carry out the painful multiple scale expansion explicitly. In view of (8.5), (8.7), (8.9) and (8.10), the following slow scales and supercriticality parameter are introduced:

$$T_2 = \varepsilon^2 t \,, \quad X' = \varepsilon (x - \omega_k^c t) \,, \quad \mathrm{Re} - \mathrm{Re}_c = \Delta \varepsilon^2 \,. \quad (8.11)$$

The starting procedure is the vorticity equation (7.15) where we make the substitutions

$$\frac{\partial}{\partial t} \mapsto \frac{\partial}{\partial t} - \varepsilon \omega_k^c \frac{\partial}{\partial X'} + \varepsilon^2 \frac{\partial}{\partial T_2} \,, \quad \frac{\partial}{\partial x} \mapsto \frac{\partial}{\partial x} + \varepsilon \frac{\partial}{\partial X'} \,. \quad (8.12)$$

Expansion (8.1) for the total streamfunction remains valid but the various perturbations ψ_1, ψ_2, ψ_3, etc. are now functions of the slow scales X', T as well as the fast scales x, y, t. Upon substituting (8.1), (8.11) and (8.12) into the vorticity equation (7.15), one obtains a set of governing equations for ψ_1, ψ_2, ψ_3, etc. that can be solved sequentially. At $\mathcal{O}(\varepsilon)$, the linearized problem is recovered in the form

$$\mathcal{L}[\psi_1] \equiv \left(\frac{\partial}{\partial t} + U(y)\frac{\partial}{\partial x}\right)\nabla^2\psi_1$$

$$- U''(y)\frac{\partial\psi_1}{\partial x} - \frac{1}{\mathrm{Re_c}}\nabla^4\psi_1 = 0, \tag{8.13a}$$

$$\frac{\partial\psi_1}{\partial x} = \frac{\partial\psi_1}{\partial y} = 0 \qquad \text{at} \quad y = \pm 1, \tag{8.13b}$$

where we have reverted to a coordinate system centered on the midplane of the channel as in Fig. 7.1. We choose as solution a scaled version of (8.3), namely

$$\psi_1(x, y, t; X', T_2) = \mathcal{R}e\left\{A(X', T_2)\phi_c(y)\exp[ik_c(x - ct)]\right\}. \tag{8.14}$$

The function $\phi_c(y)$ is the even neutral eigenfunction of the Orr–Sommerfeld eigenvalue problem at $\mathrm{Re} = \mathrm{Re_c}$ given by

$$\mathrm{L}[\phi_c] \equiv [U(y) - c][\phi_c'' - k_c^2\phi_c] - U''(y)\phi_c$$

$$- \frac{1}{ik_c\,\mathrm{Re_c}}\left(\frac{d^2}{dy^2} - k_c^2\right)^2\phi_c = 0, \tag{8.15a}$$

$$\phi_c(1) = \phi_c'(1) = 0, \quad \phi_c'(0) = \phi_c'''(0). \tag{8.15b}$$

At $\mathcal{O}(\varepsilon^2)$, one obtains:

$$\mathcal{L}[\psi_2] = -(U - \omega_k^c)\frac{\partial\nabla^2\psi_1}{\partial X'} - 2\left(\frac{\partial}{\partial t} + U\frac{\partial}{\partial x}\right)\frac{\partial^2\psi_1}{\partial x\partial X'}$$

$$+ U''\frac{\partial\psi_1}{\partial X'} + \frac{4}{\mathrm{Re_c}}\frac{\partial^2\nabla^2\psi_1}{\partial x\partial X'} - \left(\frac{\partial\psi_1}{\partial y}\frac{\partial}{\partial x} - \frac{\partial\psi_1}{\partial x}\frac{\partial}{\partial y}\right)\nabla^2\psi_1, \tag{8.16a}$$

$$\frac{\partial\psi_2}{\partial x} = \frac{\partial\psi_2}{\partial y} = 0, \qquad \text{at} \quad y = \pm 1. \tag{8.16b}$$

In view of the forcing term in (8.16a), we seek a solution of the form

$$\psi_2(x, y, t; X', T_2) = \phi_2^{(0)}(y; X', T_2)$$

$$+ \mathcal{R}e\left\{\phi_2^{(1)}(y; X', T_2)\exp[ik_c(x - ct)]\right\}$$

$$+ \mathcal{R}e\left\{\phi_2^{(2)}(y; X', T_2)\exp[2ik_c(x - ct)]\right\}, \tag{8.17}$$

the subscript denoting the order of the approximation and the superscript the index of the harmonic component. The $\mathcal{O}(\varepsilon^2)$ streamfunction is seen to include a mean flow distortion at $k = 0$ and a second harmonic at $k = 2k_c$, both induced by the quadratic nonlinearities appearing in the forcing term of (8.16a), as well as a modification of the fundamental at

$k = k_c$. Substitution of (8.17) into (8.16a) first leads to the *mean flow distortion equation*

$$\frac{1}{\mathrm{Re}_c} \frac{\partial^4 \phi_2^{(0)}}{\partial y^4} = -\frac{1}{4} i k_c |A|^2 (\phi_c \phi_c'^* - \phi_c^* \phi_c')'' . \tag{8.18}$$

Bearing in mind relation (4.44) for the Reynolds stress τ_{xy}, equation (8.18) can be integrated once to read

$$\frac{1}{\mathrm{Re}_c} \frac{\partial^3 \phi_2^{(0)}}{\partial y^3} = -\frac{\partial \tau_{xy}^c}{\partial y} + \frac{\partial P_2^{(0)}}{\partial X'} , \tag{8.19}$$

where the "constant" of integration has been expressed as a modified pressure gradient $\partial P_2^{(0)}/\partial X'$. Equation (8.19) is readily interpreted as a streamwise momentum balance for the mean flow distortion $U_2^{(0)} \equiv \partial \phi_2^{(0)}/\partial y$, generated by the Reynolds stress τ_{xy}^c. *If the externally imposed pressure gradient is assumed to remain unchanged in the presence of perturbations* (see section 8.3 for a full discussion), one may take $\partial P_2^{(0)}/\partial X' \equiv 0$. Upon integrating (8.19) twice and applying the no-slip conditions $\partial \phi_2^{(0)}/\partial y = 0$ at $y = \pm 1$, one obtains the final expression for the *streamwise mean flow distortion* as

$$U_2^{(0)} = -\mathrm{Re}_c \int_1^y \tau_{xy} \, dy = -\frac{1}{4} i k_c \mathrm{Re}_c |A|^2 \int_1^y (\phi_c \phi_c'^* - \phi_c^* \phi_c') \, dy . \tag{8.20}$$

The *modification of the fundamental* $\phi_2^{(1)}$ is governed by

$$\mathrm{L}[\phi_2^{(1)}] = \frac{i}{k_c} \frac{\partial A}{\partial X'} \left[(U - \omega_k^c)(\phi_c'' - k_c^2 \phi_c) - 2k_c^2 \phi_c (U - c) \right.$$

$$\left. + U'' \phi_c - \frac{4 i k_c}{\mathrm{Re}_c} (\phi_c'' - k_c^2 \phi_c) \right]$$

$$= Q_2^{(1)}(y; X', T_2) , \tag{8.21a}$$

with the boundary conditions

$$\phi_2^{(1)}(1) = \phi_2^{(1)'}(1) = 0 , \qquad \phi_2^{(1)'}(0) = \phi_2^{(1)'''}(0) = 0 . \tag{8.21b}$$

System (8.21) admits no solutions unless the forcing term $Q_2^{(1)}(y; X', T_2)$ is in the image of the operator L. This only occurs when $Q_2^{(1)}(y; X', T_2)$ is orthogonal to the adjoint eigenfunction $\tilde{\phi}_c(y)$ defined by the adjoint eigenvalue problem (Berger, 1977)

$$\tilde{\mathrm{L}}[\tilde{\phi}_c] \equiv (U - c) \left(\tilde{\phi}_c'' - k_c^2 \tilde{\phi}_c \right) + 2U' \tilde{\phi}_c$$

$$- \frac{1}{i k_c \mathrm{Re}_c} \left(\frac{d^2}{dy^2} - k_c^2 \right)^2 \tilde{\phi}_c = 0 , \tag{8.22a}$$

$$\tilde{\phi}_c(1) = \tilde{\phi}'_c(1) = 0, \quad \tilde{\phi}'_c(0) = \tilde{\phi}'''_c(0) = 0. \tag{8.22b}$$

Thus the following *orthogonality* or *compatibility condition* must be enforced:

$$\int_{-1}^{+1} Q_2^{(1)}(y; X', T_2) \tilde{\phi}_c(y) \, dy = \int_{-1}^{+1} L[\phi_2^{(1)}] \tilde{\phi}_c(y) \, dy$$

$$= \int_{-1}^{+1} \phi_2^{(1)}(y) \tilde{L}\left[\tilde{\phi}_c\right] dy = 0. \tag{8.23}$$

From the form of the forcing term in (8.21a), this constraint *de facto* provides a means of calculating the group velocity ω_k^c as the ratio of two integrals involving the eigenfunction $\phi_c(y)$ and its adjoint $\tilde{\phi}_c(y)$. Under this condition, (8.21) has infinitely many solutions

$$\phi_2^{(1)}(y; X', T_2) = \frac{\partial A}{\partial X'} \phi_2^{(1)}(y) + A_2(X', T_2) \phi_c(y), \tag{8.24}$$

where $\phi_2^{(1)}(y)$ satisfies (8.21) with the factor $\partial A/\partial X'$ suppressed in $Q_2^{(1)}$. Finally, the *second harmonic* $\phi_2^{(2)}$ is governed by

$$(U - c)(\phi_2^{(2)''} - 4k_c^2 \phi_2^{(2)}) - U'' \phi_2^{(2)}$$

$$- \frac{1}{2ik_c \, \mathrm{Re}_c} \left(\frac{d^2}{dy^2} - 4k_c^2\right)^2 \phi_2^{(2)}$$

$$= -\frac{1}{8} A^2 (\phi'_c \phi''_c - \phi_c \phi'''_c), \tag{8.25a}$$

with

$$\phi_2^{(2)} = \phi_2^{(2)'} = 0 \qquad \text{at} \quad y = \pm 1. \tag{8.25b}$$

The solution is

$$\phi_2^{(2)}(y; X', T_2) = A^2 \phi_2^{(2)}(y), \tag{8.26}$$

where $\phi_2^{(2)}(y)$ satisfies (8.25) with the factor A^2 suppressed in the forcing term.

In order to arrive at the amplitude equation for $A(X', T_2)$, it is sufficient to examine the *resonant part* of the $\mathcal{O}(\varepsilon^3)$ solution *which reproduces the wavenumber* k_c. The modification $\phi_3^{(1)}$ to the fundamental satisfies a system of the form

$$L\left[\phi_3^{(1)}\right] = Q_3^{(1)}(y; X', T_2) \tag{8.27a}$$

with

$$\phi_3^{(1)}(1) = \phi_3^{(1)'}(1) = 0, \qquad \phi_3^{(1)'}(1) = \phi_3^{(1)'''}(1) = 0. \tag{8.27b}$$

The forcing term $Q_3^{(1)}$ explicitly reads:

$$
Q_3^{(1)}(y; X', T_2) \equiv \frac{i}{k_c}\left(\phi_c'' - k_c^2\phi_c\right)\frac{\partial A}{\partial T_2}
$$

$$
+ \left\{ -(U - c)\left(\phi_c + 2ik_c\phi_2^{(1)}\right) \right.
$$

$$
+ \frac{i}{k_c}(U - \omega_k^c)\left(\phi_2^{(1)''} - k_c^2\phi_2^{(1)} - 2k_c^2\phi_c\right) - \frac{iU''}{k_c}\phi_2^{(1)}
$$

$$
\left. - \frac{2i}{k_c\,\mathrm{Re}_c}\left[\phi_c'' - 3k_c^2\phi_c - 2ik_c\left(\phi_2^{(1)''} - k_c^2\phi_2^{(1)}\right)\right]\right\}\frac{\partial^2 A}{\partial X'^2}
$$

$$
+ \frac{i}{k_c\,\mathrm{Re}_c}\left(\frac{d^2}{dy^2} - k_c^2\right)^2\phi_c\Delta A
$$

$$
+ \left\{ \phi_2^{(2)'}\left(\phi_c''^* - k_c^2\phi_c^*\right) + 2\phi_2^{(2)}\left(\phi_c''^* - k_c^2\phi_c^*\right)\right.
$$

$$
- 2\phi_c'^*\left(\phi_2^{(2)''} - 4k_c^2\phi_2^{(2)}\right) - \phi_c^*\left(\phi_2^{(2)'''} - 4k_c^2\phi_2^{(2)'}\right)
$$

$$
\left. - \phi_2^{(0)'}\left(\phi_c'' - k_c^2\phi_c\right) + \phi_2^{(0)'''}\phi_c\right\}\frac{|A|^2 A}{4}. \tag{8.28}
$$

As before, the compatibility condition, i.e. the orthogonality condition

$$
\int_{-1}^{+1} Q_3^{(1)}(y; X', T_2)\tilde{\phi}_c(y)\,dy = 0 \tag{8.29}
$$

involves multiplying (8.28) by $\tilde{\phi}_c(y)$ and integrating across the channel. Upon comparing the result with the formal amplitude equation (8.10), one readily obtains explicit relations for ω_{kk}^c, ω_{Re}^c and b in terms of ratios of integrals involving the cross-stream distribution functions $\phi_c(y)$, $\tilde{\phi}_c(y)$, $\phi_2^{(0)}(y)$, $\phi_2^{(1)}(y)$ and $\phi_2^{(2)}(y)$.

The GL equation (8.4), equivalently (8.8) or (8.10), has repeatedly been encountered in these notes as a simple model of nonlinear pattern evolution (sections 3 and 6). It also arises in a wide variety of fluid mechanical contexts (Rayleigh–Bénard convection between differentially heated horizontal plates, Taylor–Couette flow between rotating cylinders, etc.) as an asymptotically valid evolution equation close to the onset of unstable motion (Cross and Hohenberg, 1993). In all such cases, the *multiple scale* formalism outlined above provides a systematic method of derivation. For plane Poiseuille flow, all the complex coefficients ω_k^c, ω_{kk}^c, ω_{Re}^c and b have to be evaluated numerically. According to the computations of Stewartson and Stuart (1971), the Landau constant b is then found to have a *negative real part*. To interpret this result, we examine the purely temporal development of a single wavenumber $k_c + q$

by setting $a(x,t) = a(t)\, e^{iqx}$ in the wave packet decomposition (8.3). Upon substituting this expression into (8.10) the GL equation reduces to the following *Landau equation* for $a(t)$:

$$\frac{da}{dt} = \sigma a - b|a|^2 a,\qquad (8.30)$$

the complex coefficient σ taking the form

$$\sigma = -i\left[\omega_k^c q + \frac{1}{2}\omega_{kk}^c q^2 + \omega_{\mathrm{Re}}^c(\mathrm{Re} - \mathrm{Re_c})\right],\qquad (8.31a)$$

equivalently, to the present level of accuracy,

$$\sigma \sim -i\left[\omega(k_c + q;\mathrm{Re}) - \omega(k_c;\mathrm{Re_c})\right].\qquad (8.31b)$$

As seen in section 3.3, the Landau equation (8.30) with complex coefficients governs the nonlinear evolution of fluctuations close to a *Hopf bifurcation*, i.e. a transition from a stationary state to a limit cycle (see sections 2.2 and 6). However, by contrast with wakes, one numerically finds that $b_r < 0$. The bifurcation is therefore *subcritical* and the bifurcating finite-amplitude branch is *unstable* (section 3.3, Figs. 3.22,23). Consequently plane Poiseuille flow exhibits a threshold-like behavior outside the linear neutral curve in the Re–k plane. For small enough fluctuations, the laminar parabolic velocity profile remains stable, in agreement with the linearized approach. But, *for fluctuations exceeding the amplitude of the unstable limit cycle, trajectories in phase space diverge and move to a finite-amplitude attractor which is beyond the scope of the weakly nonlinear formulation.* The present results bring the hope of bridging the gap between the predictions of linear theory and the observed transitional values Re $\sim 1,000$ *provided that one searches for finite-amplitude bifurcated branches numerically.*

8.3 Finite-amplitude two-dimensional vortical states

The analysis carried out in section 8.2 relies on a perturbation expansion of physical quantities in powers of Re – Re$_c$. The results are hence only valid when the expansion parameter is small, i.e. for Reynolds numbers close to Re$_c$ = 5,772, and they fail to account for the transitions observed in uncontrolled experiments. Fortunately recent progress in computing facilities coupled with the use of the implicit function theorem (Keller, 1977) and classical bifurcation theory (Guckenheimer and Holmes, 1983) has alleviated the need for amplitude expansions restricted to a neighborhood of onset. New sets of finite-amplitude solutions can be determined, which hopefully provide some clues as to the overall dynamics of the flow. In particular, fully nonlinear periodic solutions of the Navier–Stokes equations have been found to persist at

Reynolds numbers far below Re_c (Zahn *et al.*, 1974; Bayly *et al.*, 1988; Herbert, 1988). These finite-amplitude states characteristically take the form of *Kelvin cat's eyes* in which the vorticity of the basic parallel flow has rolled up into a periodic array of elliptically shaped structures, as in the Stuart vortices considered in section 5.3 (Fig. 5.17). In a way, if the basic steady parallel flow corresponds to a fixed point for ordinary differential equations, these traveling vortical structures are analogous to periodic orbits in the same phase space. Recall that, in dynamical systems theory, one studies fixed points and their respective stability and then looks for periodic solutions, which arise from these fixed points *via* primary bifurcations, or for homoclinic orbits which are linked to them. In analogous fashion, we have now completed the study of the instability of steady parallel flows and we proceed to investigate the nature and properties of *two-dimensional periodic traveling waves** with the following motivations in mind. Firstly, it is known from dynamical systems theory (Guckenheimer and Holmes, 1983) that periodic orbits, whether stable or unstable, set specific topological constraints on allowable phase space trajectories. Secondly, it is argued in section 8.4 that these two-dimensional nonlinear vortical states allow an efficient transfer of energy to smaller scales *via* an instability mechanism relying on the elliptical nature of the vortex cores.

In the same way as in sections 4.2 and 7.2, one introduces a total streamfunction Ψ which satisfies the vorticity equation (7.15). As in equation (4.29), this nondimensional streamfunction is decomposed into two parts: a *basic flow* contribution $\int_0^y U(y)\,dy$ associated with the parabolic Poiseuille velocity profile $U(y) = 1 - y^2$ and a *finite-amplitude perturbation* $\psi(\mathbf{x}, t)$. Upon substituting (4.29) into the full vorticity equation (7.45), the governing equation for $\psi(\mathbf{x}, t)$ is found to be

$$
\left(\frac{\partial}{\partial t} + U(y) \frac{\partial}{\partial x} \right) \nabla^2 \psi - U''(y) \frac{\partial \psi}{\partial x}
$$
$$
+ \left(\frac{\partial \psi}{\partial y} \frac{\partial}{\partial x} - \frac{\partial \psi}{\partial x} \frac{\partial}{\partial y} \right) \nabla^2 \psi = \frac{1}{\text{Re}} \nabla^4 \psi, \tag{8.32a}
$$

with the usual no-slip conditions at the walls

$$
\frac{\partial \psi}{\partial x}(x, \pm 1, t) = 0, \qquad \frac{\partial \psi}{\partial y}(x, \pm 1, t) = 0, \tag{8.32b, c}
$$

and the periodicity condition

$$
\psi(x, y, t) = \psi(x + 2\pi/k, y, t). \tag{8.32d}
$$

---

* For simplicity we disregard here homoclinic orbits (Balmforth, 1995), though in a recent investigation, Cherhabili and Ehrenstein (1995) have computed their counterpart for a specific shear flow in the form of stationary localized solutions.

Recall that all the variables have been made nondimensional with respect to the half-width L of the channel and the velocity scale V so that $\mathrm{Re} = VL/\nu$. The problem is not completely defined as it stands: a transition experiment is typically carried out by requiring that the volume flow rate *or* the externally imposed mean pressure gradient be kept constant as perturbations evolve from the laminar Poiseuille solution $U(y)$.

At *constant volume flow rate*, the volumetric flux must be maintained to the nondimensional value $4/3$ imposed by $U(y)$, namely the total streamwise velocity $U(\mathbf{x}, t)$ must satisfy

$$\int_{-1}^{+1} U(\mathbf{x}, t) = \int_{-1}^{+1} U(y)\, dy = \frac{4}{3}. \tag{8.33}$$

Since $U(\mathbf{x}, t) = U(y) + \partial\psi/\partial y$, this condition translates into

$$\psi(x, +1, t) - \psi(x, -1, t) = 0. \tag{8.34}$$

Considering that ψ is defined up to an arbitrary additive constant, one may always impose that $\psi(x, -1, t) = 0$. In the end, the original no-slip conditions (8.32b,c) and the constant flow rate condition (8.34) translate into the equivalent relations

$$\psi(x, \pm 1, t) = 0, \qquad \frac{\partial\psi}{\partial y}(x, \pm 1, t) = 0. \tag{8.35a, b}$$

The set of equations (8.32a, 8.35a,b) then completes the formulation of the finite amplitude periodic problem for the *constant flow rate* case. According to (8.33), the velocity scale V is in this instance defined in terms of the imposed *dimensional* volume flow rate Q by the relation $Q = \frac{4}{3}VL$, so that the Reynolds number is effectively $\mathrm{Re} = 3Q/4\nu$.

Alternatively, one may impose a *constant mean pressure gradient*. In view of the periodicity condition (8.32d), let $\bar{q}(y, t) \equiv (k/2\pi)\int_{-1}^{+1} q(x, y, t)\,dx$ denote the mean (or spatial average) of any quantity $q(\mathbf{x}, t)$ over one wavelength $2\pi/k$. The *constant mean pressure gradient condition then effectively states that, even in the presence of finite amplitude perturbations, $-\partial P/\partial x$ is kept at the value needed to drive the basic Poiseuille flow $U(y) = 1 - y^2$*. In other words

$$\frac{\overline{\partial P}}{\partial x} = \frac{1}{\mathrm{Re}}\frac{d^2 U}{dy^2} = -\frac{2}{\mathrm{Re}}. \tag{8.36}$$

In order to translate this constraint in terms of $\psi(\mathbf{x}, t)$, it is appropriate to consider the mean streamwise momentum equation. In the same spirit as in the derivation of the Rayleigh criterion (section 4.3), the streamwise momentum equation for $U(x, y, t)$

$$\frac{\partial U}{\partial t} + \frac{\partial U^2}{\partial x} + \frac{\partial UV}{\partial y} = -\frac{\partial P}{\partial x} + \frac{1}{Re} \left(\frac{\partial^2}{\partial x^2} + \frac{\partial^2}{\partial y^2} \right) U \qquad (8.37)$$

is spatially averaged and integrated over the channel width. Upon exploiting the periodicity and the no-slip conditions (see section 4.3), one is immediately led to the general momentum balance equation

$$\frac{d}{dt} \int_{-1}^{+1} \bar{U}(y, t) dy = -2 \frac{\overline{dP}}{dx} + \frac{1}{Re} \left[\frac{\partial \bar{U}}{\partial y}(1, t) - \frac{\partial \bar{U}}{\partial y}(-1, t) \right]. \qquad (8.38)$$

Note that, in contrast to the analogous inviscid relation (4.42), the mean pressure gradient $\overline{dP/dx}$ is non-zero and there is viscous momentum transfer at the walls. When the specific value (8.36) is substituted into (8.38) and the total streamwise velocity is decomposed into $U(\mathbf{x}, t) = U(y) + \partial \psi / \partial y$, the momentum balance equation (8.38) translates into the following constraint

$$\frac{d}{dt} \left[\bar{\psi}(1, t) - \bar{\psi}(-1, t) \right] = \frac{1}{Re} \left[\frac{\partial^2 \bar{\psi}}{\partial y^2}(1, t) - \frac{\partial^2 \bar{\psi}}{\partial y^2}(-1, t) \right]. \qquad (8.39)$$

The set of equations (8.32a–d), (8.39) completes the formulation of the finite amplitude periodic problem for the *constant mean pressure gradient* case. As before, one may also assume, without loss of generality, that $\psi(x, -1, t) = 0$. According to (8.36), the velocity scale is, in this instance, defined in terms of the imposed *dimensional* (here indicated by a * superscript) mean pressure gradient $-\overline{dP/dx}^*$ by the relation $-\overline{dP/dx}^* = -(\rho V^2/L) \overline{dP/dx} = 2\mu V/L^2$. Thus in the constant mean pressure gradient case, $V = (L^2/2\mu)(-\overline{dP/dx}^*)$ and the Reynolds number is effectively defined as $Re = VL/\nu = \rho L^3 (-\overline{dP/dx}^*)/(2\mu^2)$. Hereafter, *we shall solely consider constant mean pressure gradient situations.*[*] Note that this condition was precisely enforced in the weakly nonlinear analysis presented in section 8.2 [see discussion following equation (8.19)].

From now on, we restrict our attention to periodic *traveling wave* solutions of the form

$$\psi(x, y, t) = \psi'(x - ct, y) \qquad (8.40)$$

---

[*] It should be made clear that the choice of velocity scale V measuring the strength of the basic flow is somewhat arbitrary, whether we consider the constant flow rate or constant mean pressure gradient case. It is here defined so that the resulting basic parabolic velocity profile is the one which would prevail at the same volume flow rate or mean pressure gradient in the absence of perturbations.

of *unknown real phase speed* c. Let us introduce the co-moving streamwise variable $\xi = x - ct$. Then one must set $\partial \psi / \partial t = -c \partial \psi' / \partial \xi$ in (8.32a). In a reference frame moving with the wave speed c, the flow is effectively *steady* and the streamfunction $\psi'(\xi, y)$ satisfies the spatial partial differential equation:

$$(U(y) - c) \frac{\partial}{\partial \xi} \nabla^2 \psi' - U''(y) \frac{\partial \psi'}{\partial \xi} + \left(\frac{\partial \psi'}{\partial y} \frac{\partial}{\partial \xi} - \frac{\partial \psi'}{\partial \xi} \frac{\partial}{\partial y} \right) \nabla^2 \psi' = \frac{1}{\text{Re}} \nabla^4 \psi'.$$

$$(8.41)$$

The periodicity and the no-slip conditions (8.32b–d) remain formally unchanged provided that $\psi(x, y, t)$ is replaced by $\psi'(\xi, y)$. Finally, the spatial average in the constant mean pressure gradient condition (8.39) is now taken over $2\pi/k$ with respect to the variable ξ. Since, in the co-moving frame, the left-hand side of (8.39) becomes $-c \frac{\partial}{\partial \xi} [\bar{\psi}'(+1) - \bar{\psi}'(-1)] = 0$, the mean pressure gradient constraint simply takes the form

$$\frac{\partial^2 \bar{\psi}'}{\partial y^2}(+1) - \frac{\partial^2 \bar{\psi}'}{\partial y^2}(-1) = 0.$$

$$(8.42)$$

The phase speed c is seen to be the solution of a *nonlinear eigenvalue problem* for the steady periodic solutions $\psi'(\xi, y)$ in the co-moving frame.

It is important to emphasize that, *in the case of traveling waves of the form (8.40), the constant volume flow rate condition and the constant mean pressure gradient condition are strictly equivalent*. Indeed, the constant mean pressure gradient condition (8.42) in the co-moving frame for $\psi'(\xi, y)$ implies the same condition for $\psi(x, y, t)$ in the laboratory frame. Hence the time derivative of the volume flux is zero in (8.39) and the volume flow rate is effectively constant. For more general unsteady solutions the two conditions may not be equivalent.

The periodic function $\psi'(\xi, y)$ admits the Fourier series expansion

$$\psi'(\xi, y) = \sum_{n=-\infty}^{+\infty} \psi'_n(y) \exp(ink\xi).$$

$$(8.43)$$

Since this physical quantity is necessarily real, its Fourier components satisfy the relation $\psi'^*_n(y) = \psi'_{-n}(y)$. Introducing equation (8.43) into the governing equation (8.41) and boundary conditions (8.32b–d,8.42), one obtains an infinite set of nonlinear ordinary differential equations for the Fourier modes $\psi'_n(y)$ with n positive. The linear part of this system is nothing else but the Orr–Sommerfeld operator written for each Fourier mode. Needless to say, these coupled equations are impossible to solve analytically. To go any further, one then leaves the shores of exact analytical solutions and sails towards the beauties and dangers of numerical computations. In particular the infinite series (8.43) is truncated at order N to read

$$\psi'(\xi, y) = \sum_{n=-N}^{+N} \psi'_n(y) \exp(ink\xi) . \qquad (8.44)$$

The infinite system is then reduced to a set of $N + 1$ coupled nonlinear ordinary differential equations. At this stage, one approximates the modes $\psi'_n(y)$ by a combination of M Chebyshev polynomials $T_m(y)$:

$$\psi'_n(y) = \sum_{m=0}^{M} a_{nm} T_m(y) . \qquad (8.45)$$

In order to ensure convergence, the number of Chebyshev polynomials required is at least $M = 30$, while the number of Fourier modes is generally chosen to be $N = 3$ or $N = 4$ (Ehrenstein and Koch, 1991). For convergence-related problems such as aliasing, choice of proper collocation points or number of spectral modes, the reader is referred to Canuto *et al.* (1988). For a given value of Re and k, the above discretization process leads to a purely *algebraic* nonlinear eigenvalue problem symbolically written as:

$$F(a_{nm}, c; \mathrm{Re}, k) = 0 , \qquad (8.46)$$

where the $(N + 1) \times (M + 1)$ Chebyshev coefficients a_{nm} are unknown and the real phase velocity c is the eigenvalue. For a given Reynolds number and wavenumber close to the linear neutral stability boundary, the weakly nonlinear theory of section 8.2 provides a nontrivial solution of (8.46). The main idea is then to reconstruct from this particular solution the whole set $F = 0$, or at least a large subset of solutions by gradually changing the value of the Reynolds number or wavenumber. The procedure is as follows: if a solution $a_{nm}^{(0)}$, $c^{(0)}$ of (8.46) is known for a given pair Re_0, k_0, a guess value of the true finite-amplitude solution $a_{nm}^{(0)} + \delta a_{nm}$, $c^{(0)} + \delta c$ for the nearby values $\mathrm{Re}_0 + \delta \mathrm{Re}$, $k_0 + \delta k$ is found by solving the linear system:

$$\left(\frac{\partial F}{\partial a_{nm}}\right) \delta a_{nm} + \left(\frac{\partial F}{\partial c}\right) \delta c + \left(\frac{\partial F}{\partial k}\right) \delta k + \left(\frac{\partial F}{\partial \mathrm{Re}}\right) \delta \mathrm{Re} = 0 \qquad (8.47)$$

where the partial derivatives are evaluated at $(a_{nm}^{(0)}, c^{(0)}; \mathrm{Re}_0, k_0)$. Note that relation (8.47) is obtained by differentiating (8.46) along the branch of nonlinear solutions. The implicit function theorem (Keller, 1977) then asserts that, if the Jacobian matrix $(\partial F/\partial a_{nm})(a_{nm}^{(0)}, c^{(0)}; \mathrm{Re}_0, k_0)$ is invertible, the solution $a_{nm}^{(0)} + \delta a_{nm}$, $c^{(0)} + \delta c$ of (8.47) constitutes a good approximation of the true finite-amplitude state given by (8.46) for $\mathrm{Re}_0 + \delta \mathrm{Re}$, $k_0 + \delta k$. If the Jacobian matrix is non-invertible, it is nevertheless possible to adapt this method as described in Keller (1977). This guess value is then introduced into a Newton iteration scheme (Press *et al.*, 1992) to rapidly converge towards the true solution satisfying $F = 0$. In spite of severe truncations in the number of Fourier and Chebyshev modes, this analysis removes the small amplitude hypothesis and allows to get away from the linear neutral stability curve. Finally note that, for numerically robust cases, one may even choose as starting solution the linear critical mode at $\mathrm{Re_c}$, k_c with an *a priori* small amplitude instead of the true weakly nonlinear state.

For given k and Re, the above procedure leads to the determination of c and all the a_{nm}'s. If ε characterizes the overall amplitude of such nonlinear solutions, e.g. the total perturbation energy, this parameter and $\omega = kc$ are then effectively determined by two nonlinear relations of the form

$$\omega = \Omega(k; \mathrm{Re}), \qquad \varepsilon = \mathcal{E}(k; \mathrm{Re}). \qquad (8.48a, b)$$

The second relation is represented in k–Re–ε parameter space by a paraboloid surface (Fig. 8.8) intersecting the plane $\varepsilon = 0$ along the neutral stability boundary.

For wavenumbers close to unity, the intersection of the same surface with a plane of constant wavenumber lies along a curve (Fig. 8.9) which possesses a turning point for a Reynolds number well below the linear

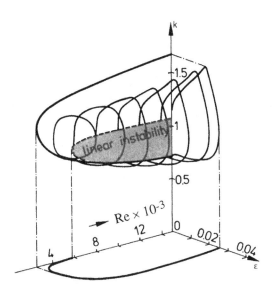

Fig. 8.8.   Locus of finite-amplitude traveling wave solutions ($F = 0$) in Re–ε–k space for plane channel flow. After Ehrenstein and Koch (1991).

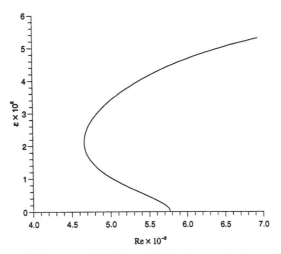

Fig. 8.9. Locus of finite-amplitude traveling wave solutions $(F = 0)$ in Re–ε plane for a constant wavenumber $k = 1.02$ in plane channel flow (courtesy of U. Ehrenstein, Laboratoire de Mécanique de Lille, France).

critical Reynolds number $\mathrm{Re_c}$. Finally, in a plane $\mathrm{Re} = \mathrm{const.}$, one gets isolated curves (Fig. 8.10) which shrink to a point at $\mathrm{Re} = 2,900$.

These features confirm the highly *subcritical* nature of this instability: well below the linear threshold, for Reynolds numbers as low as 2,900,

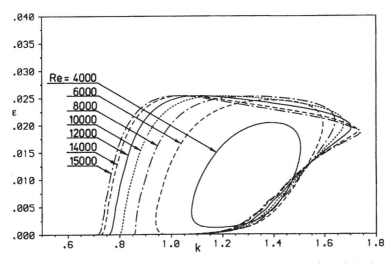

Fig. 8.10. Locus of finite-amplitude traveling wave solutions (F=0) in k–ε plane for constant Reynolds numbers in plane channel flow. After Ehrenstein and Koch (1991).

the laminar stable plane Poiseuille solution $U(y)$ coexists with highly nonlinear states. The lower branch of nonlinear solutions below the turning point in Fig. 8.9 is unstable. It is the continuation of the unstable bifurcating branch obtained in the weakly nonlinear analysis of section 8.2. The upper branch above the turning point in Fig. 8.9 is unstable close to the turning point but becomes stable further on in a finite Re-interval.

Two-dimensional direct numerical simulations (Jimenez, 1990) demonstrate that the system jumps from laminar Poiseuille flow $U(y)$ to the upper *stable* traveling wave branch, when perturbed with a high enough amplitude. The required *threshold amplitude*, typical of *subcritical instabilities*, coincides with the *unstable* finite amplitude lower branch. Since finite amplitude waves only exist down to Re $= 2,900$ (Fig. 8.8), this analysis does not completely explain the lowest experimental values Re $\sim 1,000$ for which transition may take place. One partly accounts for such a discrepancy by seeking more general solutions. Indeed, *three-dimensional traveling waves* (Ehrenstein and Koch, 1991) or two-dimensional *quasi-equilibria* (Orszag and Patera, 1983), i.e. slowly decaying traveling waves, have been shown to exist at even lower values of the Reynolds number.

Experimental observations indicate that viscosity does not play a prominent role in the temporal evolution. The spatial structure of

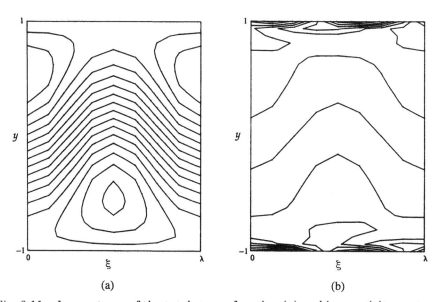

(a)   (b)

Fig. 8.11. Iso-contours of the total streamfunction (a) and iso-vorticity contours (b) for finite amplitude waves in plane channel flow, in a reference frame moving with the wave. After Orszag and Patera (1983).

the finite amplitude waves confirms this assertion: apart from the wall regions, vorticity contours and streamlines of the total velocity field are almost identical when viewed in a frame of reference moving with the wave (Fig. 8.11). This is typical of steady flows governed by quasi-inviscid processes. To demonstrate this property, note that, apart from inner regions such as the vicinity of walls, the total streamfunction $\Psi(\xi, y)$ in a steady nearly inviscid flow satisfies

$$\left(\frac{\partial \Psi}{\partial y}\frac{\partial}{\partial \xi} - \frac{\partial \Psi}{\partial \xi}\frac{\partial}{\partial y}\right)\nabla^2\Psi \approx \mathcal{O}(1/\operatorname{Re}). \tag{8.49}$$

Equation (8.49) states that the vorticity field $\Omega = -\nabla^2\Psi$ is related to the streamfunction by $\Omega = G(\Psi) + \mathcal{O}(1/\operatorname{Re})$, where G is an undetermined function. *Iso-vorticity lines and streamlines are then almost identical.* An example of such a class of flows is given by the two-dimensional family of Stuart vortices (see section 5.3, Fig. 5.17) which has proved to be convenient in the study of secondary instabilities in mixing layers (Pierrehumbert and Widnall, 1982). The respective role of viscous and nonlinear processes is clearly put in evidence in the direct two-dimensional numerical simulations of Orszag and Patera (1980) and Jimenez (1990): for a wide class of initial conditions, solutions are attracted to the neighborhood of the traveling wave states. The flow field first adapts to align vorticity contours and streamlines as required by the time-dependent inviscid dominant balance

$$\left(\frac{\partial \Psi}{\partial y}\frac{\partial}{\partial \xi} - \frac{\partial \Psi}{\partial \xi}\frac{\partial}{\partial y}\right)\nabla^2\Psi \approx -\frac{\partial \nabla^2\Psi}{\partial t}. \tag{8.50}$$

This process takes place on a convective time scale $\tau_i = L/V$ which corresponds to the rapid temporal evolution seen in the experiments. Once this relation is established within $\mathcal{O}(1/\operatorname{Re})$, the solution closely approaches an exact equilibrium state on the viscous time scale $\tau_v = L^2/\nu$. These traveling waves exhibit further bifurcations displaying features analogous to the classical ejections, sweeps and bursting events observed in turbulent boundary layers (Robinson, 1991). Large-scale intermittency of the type encountered in wall turbulence (Jimenez, 1990) has also been identified.

Critical minds will and should question the relevance of *two-dimensional* amplitude states and *two-dimensional* numerical simulations to describe real *three-dimensional* turbulence. For instance, *by-pass modes* of transition to turbulence (Morkovin, 1988) show no sign of such two-dimensional saturated states. However, some aspects of transition that are hidden in three-dimensional simulations may appear more clearly in a two-dimensional study. As discussed in section 8.4, the significance of such two-dimensional patterns is fundamental when studying the three-dimensional energy transfer mechanisms from the basic state into small scales. These processes depend on the existence of vortical structures analogous to those computed here for channel flow. The persistence of *quasi-equilibrium vortical states* during a sufficiently long period of time appears to be a prerequisite for transition. As argued by Saffman (1983), such structures may constrain the global dynamics even though they do not show up in experiments.

8.4 Universal elliptical instability

It is highly desirable to connect two well-documented features of turbulent or pre-turbulent shear flows: organized large scale vortices and three-dimensional small scale structures. This section examines how three-dimensional effects are intimately related to an instability process which simply requires the existence of two-dimensional *elliptical vortex* structures. The analysis is based on vorticity dynamics and it is not limited to the plane channel flow configuration. For this specific case, the theory explains how two-dimensional finite amplitude states are necessary to overcome the energy barrier that prevents weak perturbations from extracting energy directly from the basic shear below $\mathrm{Re}_c = 5,772$. Two main differences exist between the primary instability of section 8.1 and its secondary counterpart examined in the present section. Viscous diffusion is central to the primary mechanism while it plays no role in the development of *secondary instabilities*, a property which is more in line with experimental observations of *natural transition*. Furthermore, whereas iso-vorticity contours are straight lines for the unidirectional Poiseuille flow solution, they become elliptically shaped for the vortical states (Fig. 8.11) arising from the primary bifurcations discussed in section 8.3.

In the following discussion adapted from Orszag and Patera (1983), viscosity is therefore neglected and the Euler equations are linearized around a given two-dimensional basic flow $\mathbf{U}_0(x, y)$ with vorticity $\Omega_0(x, y)\mathbf{e}_z$.[*] Three-dimensional vorticity and velocity perturbations ω and $\mathbf{u}$ satisfy the equivalent of the Rayleigh equation for the primary instability, namely,

$$\frac{\partial \omega}{\partial t} + \{(\mathbf{u} \cdot \nabla)(\Omega_0 \mathbf{e}_z) - (\Omega_0 \mathbf{e}_z \cdot \nabla)\mathbf{u}\} + \{(\mathbf{U}_0 \cdot \nabla)\omega - (\omega \cdot \nabla)\mathbf{U}_0\} = 0. \tag{8.51}$$

---

[*] If the basic flow under consideration coincides with the finite amplitude traveling waves identified in section 8.3, it is understood that the analysis is conducted in the co-moving ξ–y frame.

Besides the advection term already present in the Rayleigh equation, two new effects are included that radically alter the dynamics: vortex tilting and vortex stretching. Such processes, often considered as fundamental hallmarks of three-dimensional turbulence, are accounted for in equation (8.51) by the two sets of bracketed terms.

The first group is related to *advection, tilting and stretching of the basic vorticity $\Omega_0 e_z$ by the perturbation velocity* **u**. These effects are discussed by examining the reduced equation

$$\frac{\partial \omega}{\partial t} + \left\{ (\mathbf{u} \cdot \boldsymbol{\nabla})(\Omega_0 \mathbf{e}_z) - \Omega_0 \frac{\partial \mathbf{u}}{\partial z} \right\} = \frac{\partial \omega}{\partial t} - \boldsymbol{\nabla} \times \{\mathbf{u} \times (\Omega_0 \mathbf{e}_z)\} = 0 \,. \quad (8.52)$$

The projection of equation (8.52) on the x–y plane, shows that the perturbation vorticity component $\omega_\perp$ in that plane is governed by

$$\frac{\partial \omega_\perp}{\partial t} = \Omega_0(x, y) \frac{\partial \mathbf{u}_\perp}{\partial z} \,. \quad (8.53)$$

The perturbation vorticity $\omega_\perp$ is therefore permanently generated by tilting of the basic vorticity $\Omega_0 \mathbf{e}_z$ induced by the velocity perturbations $\mathbf{u}_\perp$. As the following perturbation kinetic energy argument indicates, these mechanisms alone are incapable of giving rise to an instability. A straightforward integration of (8.52) leads to

$$\frac{\partial \mathbf{u}}{\partial t} = \mathbf{u} \times (\Omega_0(x, y) \mathbf{e}_z) + \boldsymbol{\nabla} \phi \,, \quad (8.54)$$

where ϕ is an arbitrary scalar field. Upon integrating the scalar product of (8.54) with **u** over the entire physical domain and making use of incompressibility , one obtains the perturbation kinetic energy equation

$$\frac{\partial}{\partial t} \cdot \int \frac{1}{2} \mathbf{u} \cdot \mathbf{u} \, d\tau = \int \mathbf{u} \cdot (\mathbf{u} \times (\Omega_0(x, y) \mathbf{e}_z)) \, d\tau + \int \boldsymbol{\nabla} \cdot (\phi \mathbf{u}) \, d\tau \,. \quad (8.55)$$

The first term on the right hand side of equation (8.55) is obviously zero, whereas the second can be rewritten as a surface integral by the divergence theorem. Since boundary contributions generally vanish as a result of periodicity or finite extent, the perturbation kinetic energy is conserved and *no instabilities arise if advection, tilting and stretching of the basic vorticity $\Omega_0 \mathbf{e}_z$ by the perturbation velocity* **u** *are the only mechanisms taken into consideration*.

The second group of bracketed terms in (8.51) has to do with *advection, tilting and stretching of the perturbation vorticity ω by the basic flow $\mathbf{U}_0$ in the x–y plane*. The governing equation (8.51) is then reduced to:

$$\frac{\partial \omega}{\partial t} + \{(\mathbf{U}_0 \cdot \boldsymbol{\nabla}) \omega - (\omega \cdot \boldsymbol{\nabla}) \mathbf{U}_0\} = 0 \,. \quad (8.56)$$

Since the vectors $\boldsymbol{\omega}$ and $\mathbf{U}_0$ are both solenoidal, i.e. divergence-free, this relation may also be written as

$$\frac{\partial \boldsymbol{\omega}}{\partial t} + \boldsymbol{\nabla} \times \{\boldsymbol{\omega} \times \mathbf{U}_0\} = 0 . \tag{8.57}$$

The projection of (8.56) along the z-axis indicates that the z-component of the vorticity perturbation acts as a passive scalar convected by the basic flow:

$$\frac{\partial \omega_z}{\partial t} + (\mathbf{U}_0 \cdot \boldsymbol{\nabla})\omega_z = 0 . \tag{8.58}$$

This component is hence incapable of producing an instability. Furthermore it is shown in the Appendix that the perturbation vorticity $\boldsymbol{\omega}_\perp$ in the x–y plane may be decomposed into

$$\boldsymbol{\omega}_\perp = \Lambda_1 \mathbf{U}_0 + \boldsymbol{\nabla} \times (\Lambda_2 \mathbf{e}_z) = \Lambda_1 \mathbf{U}_0 + \boldsymbol{\nabla}\Lambda_2 \times \mathbf{e}_z , \tag{8.59}$$

where Λ_1 and Λ_2 are both passive scalars advected by the basic flow. Since Λ_1 and Λ_2 are passive scalars, $\boldsymbol{\omega}_\perp$ may only grow through the second gradient term in (8.59). More specifically, gradients normal to the basic flow streamlines may experience transient growth as a result of velocity differences between neighboring streamlines. When this takes place, the increase is thus algebraic. The second term in (8.59) is then dominant and tends to align the vorticity $\boldsymbol{\omega}_\perp$ along the streamlines. As a consequence, advection, tilting and stretching of the perturbation vorticity $\boldsymbol{\omega}$ by the basic flow $\mathbf{U}_0$ in the x–y plane cannot give rise to exponential growth.

When both groups of terms are present, as in (8.51), vorticity filaments are constantly redirected at an angle with respect to the streamlines as observed in the numerical computations of Orszag and Patera (1983). Thus, instability is a two-fold process: the first group of terms permanently generates vorticity in the x–y plane and the second group provides growth.

The direct transfer of energy from a basic two-dimensional flow to three-dimensional perturbations has also been examined by Cambon *et al.* (1985), Bayly (1986), Craik and Criminale (1986), Pierrehumbert (1986), Landman and Saffman (1987) and Waleffe (1990). In particular, the above analyses demonstrate the existence of a *universal instability mechanism* capable of producing arbitrarily small scales, whenever the basic streamlines are closed and elliptically shaped. Such a secondary instability is hence active in a wide class of shear flows giving rise to two-dimensional vortices such as Kelvin–Helmholtz vortices in mixing layers (Fig. 2.2b) and finite amplitude Tollmien–Schlichting waves in plane channel flow (Fig. 8.11).

We follow here Waleffe's treatment. Within vortex cores, the basic streamfunction locally reads, in dimensional variables

$$\Psi_0(x,y) = -\tfrac{1}{2}\left((\Omega/2 + \varepsilon)x^2 + (\Omega/2 - \varepsilon)y^2\right), \tag{8.60}$$

where Ω and ε are two independent positive or negative parameters. Streamlines are therefore ellipses of eccentricity

$$E = \sqrt{(\Omega/2 + \varepsilon)/(\Omega/2 - \varepsilon)}, \tag{8.61}$$

and the basic vorticity reduces to the uniform field

$$\boldsymbol{\Omega}_0 \equiv -\nabla^2\Psi_0(x,y)\,\mathbf{e}_z = \Omega\,\mathbf{e}_z. \tag{8.62}$$

Note that such a velocity field (Fig. 8.12), though it describes only locally real vortices, is an exact solution of the Navier–Stokes equations in an infinite physical domain. Furthermore, it can be viewed as the superposition of a pure rotation and a pure strain:

$$\mathbf{U}_0(\mathbf{x}) = (\Omega/2)\mathbf{e}_z \times \mathbf{x} + \varepsilon D.\mathbf{x}, \tag{8.63}$$

where

$$D \equiv \begin{pmatrix} 0 & 1 & 0 \\ 1 & 0 & 0 \\ 0 & 0 & 0 \end{pmatrix}. \tag{8.64}$$

It is enlightening to consider first the case $\varepsilon = 0$, where the flow (8.60) reduces to a *pure solid body rotation* with circular streamlines and uniform rotation around the z-axis of angular velocity $\Omega/2$. When perturbed, such a basic flow is known to sustain neutrally stable *inertial waves* (Whitham, 1974).

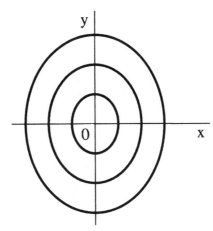

Fig. 8.12 Elliptically shaped streamlines of the basic flow (8.60).

In a reference frame rotating with angular velocity $\Omega/2$, the basic flow (8.60) with $\varepsilon = 0$ reduces to the rest state and linearization of the dimensional Euler equations leads to the homogeneous set of equations

$$\frac{\partial \mathbf{u}}{\partial t} + \boldsymbol{\Omega}_0 \times \mathbf{u} = -\frac{1}{\rho} \, \boldsymbol{\nabla} \, p \,, \tag{8.65}$$

$$\boldsymbol{\nabla} \cdot \mathbf{u} = 0. \tag{8.66}$$

Propagating plane wave solutions of *constant* wavevector $\mathbf{k}$ and frequency ω are sought in the form

$$\mathbf{u}(\mathbf{x}, t) = \mathcal{R}e \left\{ \tilde{\mathbf{u}} \exp\left[i(\mathbf{k} \cdot \mathbf{x} - \omega t) \right] \right\} , \tag{8.67}$$

where $\tilde{\mathbf{u}}$ is an arbitrary constant complex vector. *In the non-rotating frame*, the wavevector $\mathbf{k}$ is no longer constant but evolves in time according to the law

$$\mathbf{k}(t) = R_{Oz}(\Omega t/2)\mathbf{k}(0) \,, \tag{8.68}$$

where $R_{Oz}(\Omega t/2)$ denotes the rotation matrix of instantaneous angle $\Omega t/2$ around the z-axis:

$$R_{Oz}(\Omega t/2) = \begin{pmatrix} \cos(\Omega t/2) & -\sin(\Omega t/2) & 0 \\ \sin(\Omega t/2) & \cos(\Omega t/2) & 0 \\ 0 & 0 & 1 \end{pmatrix} . \tag{8.69}$$

Thus, the wavevector $\mathbf{k}(t)$ remains of constant amplitude but spins with the fluid around the z-axis so that, *in the non-rotating frame*, the plane wave solution (8.67) becomes

$$\mathbf{u}(\mathbf{x}, t) = \mathcal{R}e \left\{ \tilde{\mathbf{u}} \exp\left[i(\mathbf{k}(t) \cdot \mathbf{x} - \omega t) \right] \right\} . \tag{8.70}$$

Let us determine the dispersion relation in the rotating frame. Upon substituting (8.67) into equations (8.65-66), one obtains:

$$-i\omega \mathbf{u} + \boldsymbol{\Omega}_0 \times \mathbf{u} = -\frac{i\mathbf{k}}{\rho} p \,, \tag{8.71}$$

$$\mathbf{k} \cdot \mathbf{u} = 0 \,. \tag{8.72}$$

The incompressibility condition (8.72) requires that waves be of transverse type, i.e. that the induced velocity field be confined to a plane perpendicular to the wavevector $\mathbf{k}$. Let u_1, u_2 denote the two complex components of $\mathbf{u}$ in a plane perpendicular to $\mathbf{k}$. Upon taking the scalar product of equation (8.71) with $\mathbf{u}$, one is immediately led to

$$\mathbf{u} \cdot \mathbf{u} \equiv u_1^2 + u_2^2 = 0 \,. \tag{8.73}$$

These two velocity components are therefore in phase quadrature: $u_1 = \pm i \, u_2$. Finally the vector product of equation (8.71) with $\mathbf{k}$ leads to

$$i\omega \mathbf{k} \times \mathbf{u} = (\boldsymbol{\Omega}_0 \cdot \mathbf{k})\mathbf{u} \,, \tag{8.74}$$

where one has used the fact that $\mathbf{k} \times (\mathbf{\Omega}_0 \times \mathbf{u}) = -(\mathbf{\Omega}_0 \cdot \mathbf{k})\mathbf{u}$. Upon writing (8.74) in component form in a plane perpendicular to the wavevector $\mathbf{k}$, the vector equation reduces to

$$-i\omega k u_2 = (\mathbf{\Omega}_0 \cdot \mathbf{k})u_1 , \tag{8.75a}$$

$$i\omega k u_1 = (\mathbf{\Omega}_0 \cdot \mathbf{k})u_2 . \tag{8.75b}$$

This algebraic system for the components u_1 and u_2 will admit non-trivial solutions if and only if the *dispersion relation for inertial waves*

$$\omega(\mathbf{k}) = \pm\frac{\mathbf{\Omega}_0 \cdot \mathbf{k}}{|\mathbf{k}|} \tag{8.76}$$

is satisfied. If θ denotes the angle* between $\mathbf{\Omega}_0$ and $\mathbf{k}$, the above relation takes the form

$$\omega(\theta) = \pm\Omega \cos\theta . \tag{8.77}$$

Inertial waves of given frequency ω are seen to exist if and only if $\omega < \Omega$: propagating waves are then generated only when the wavevector $\mathbf{k}$ is located on the cone of half-angle $\arccos(\omega/\Omega)$ (Fig. 8.13). To each such frequency corresponds a specific wavevector angle given by (8.77) but the modulus of $\mathbf{k}$ remains unspecified: relation (8.76) is scale invariant. Arbitrarily small scales can therefore be sustained by a smooth basic flow in solid body rotation, at least until viscous effects become predominant.

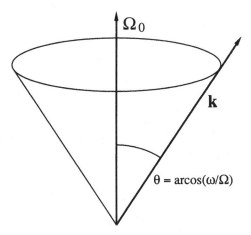

Fig. 8.13.   Cone of allowable wavevectors for neutral inertial waves of given frequency ω supported by a solid body rotation of constant vorticity $\mathbf{\Omega}_0$.

---

* Its value does not depend upon the choice of reference frame, whether rotating or non-rotating.

Following Waleffe (1990), the above *inviscid* analysis for a solid body rotation $E = 1$, $\varepsilon = 0$ may be extended to the *viscous* case for an elliptically-shaped basic flow (8.60) of arbitrary eccentricity $E \neq 1$, $\varepsilon \neq 0$. The entire investigation is conducted in the non-rotating frame.

In the usual way, let

$$\mathbf{U}(\mathbf{x}, t) = \mathbf{U}_0(x, y) + \mathbf{u}(\mathbf{x}, t), \tag{8.78a}$$

$$P(\mathbf{x}, t) = P_0(x, y) + p(\mathbf{x}, t), \tag{8.78b}$$

where

$$\mathbf{U}_0(x, y) = -\left(\frac{\Omega}{2} - \varepsilon\right) y \, \mathbf{e}_x + \left(\frac{\Omega}{2} + \varepsilon\right) x \, \mathbf{e}_y \tag{8.79}$$

denotes the basic velocity field associated with the elliptical vortex streamfunction (8.60) and $P_0(x, y)$ the corresponding pressure field. The three-dimensional perturbation velocity vector $\mathbf{u}(\mathbf{x}, t)$ is written in component form as in (4.6), and decomposition (8.78a,b) is then substituted into the Navier–Stokes equations (7.1–2) written in dimensional form. Upon linearizing around the elliptical vortex state $\mathbf{U}_0(x, y)$, $P_0(x, y)$, one is led to the following system:

$$\frac{\partial u}{\partial x} + \frac{\partial v}{\partial y} + \frac{\partial w}{\partial z} = 0 \tag{8.80a}$$

$$\frac{Du}{Dt} - \left(\frac{\Omega}{2} - \varepsilon\right) u = -\frac{1}{\rho}\frac{\partial p}{\partial x} + \nu \nabla^2 u, \tag{8.80b}$$

$$\frac{Dv}{Dt} + \left(\frac{\Omega}{2} + \varepsilon\right) v = -\frac{1}{\rho}\frac{\partial p}{\partial y} + \nu \nabla^2 v, \tag{8.80c}$$

$$\frac{Dw}{Dt} = -\frac{1}{\rho}\frac{\partial p}{\partial z} + \nu \nabla^2 w, \tag{8.80d}$$

where D/Dt is the convective derivative

$$\frac{D}{Dt} \equiv \frac{\partial}{\partial t} + \mathbf{U}_0 \cdot \nabla \ . \tag{8.80e}$$

Straightforward but painful algebraic manipulations are necessary in order to replace the linearized problem (8.80) by a single partial differential equation for the velocity component $w(\mathbf{x}, t)$. The unmotivated reader may directly proceed to the final result (8.85).

We first note the following useful relations between spatial and convective derivatives as readily obtained from (8.80e):

$$\frac{\partial}{\partial x}\frac{D}{Dt} = \frac{D}{Dt}\frac{\partial}{\partial x} + \left(\frac{\Omega}{2} + \varepsilon\right)\frac{\partial}{\partial y}, \tag{8.81a}$$

$$\frac{\partial}{\partial y}\frac{D}{Dt} = \frac{D}{Dt}\frac{\partial}{\partial y} - \left(\frac{\Omega}{2} - \varepsilon\right)\frac{\partial}{\partial x}, \tag{8.81b}$$

Differentiating (8.80b) and (8.80c) with respect to y and x respectively, making use of (8.81b) and (8.81a) and invoking the continuity equation (8.80a) leads to the useful equations

$$\left(\frac{D}{Dt} - \nu\nabla^2\right)\frac{\partial u}{\partial y} + \left(\frac{\Omega}{2} - \varepsilon\right)\frac{\partial w}{\partial z} = -\frac{1}{\rho}\frac{\partial^2 p}{\partial x \partial y}, \qquad (8.82a)$$

$$\left(\frac{D}{Dt} - \nu\nabla^2\right)\frac{\partial v}{\partial x} - \left(\frac{\Omega}{2} + \varepsilon\right)\frac{\partial w}{\partial z} = -\frac{1}{\rho}\frac{\partial^2 p}{\partial x \partial y}. \qquad (8.82b)$$

Differentiating (8.80b), (8.80c), and (8.80d) with respect to x, y, and z respectively, making use of (8.81a) and (8.81b) and adding all resulting expressions leads to the Poisson equation:

$$\frac{1}{\rho}\nabla^2 p = \Omega\left(\frac{\partial v}{\partial x} - \frac{\partial u}{\partial y}\right) - 2\varepsilon\left(\frac{\partial v}{\partial x} + \frac{\partial u}{\partial y}\right). \qquad (8.83)$$

Taking the difference between (8.82b) and (8.82a) and the sum of these same relations yields

$$\left(\frac{D}{Dt} - \nu\nabla^2\right)\left(\frac{\partial v}{\partial x} - \frac{\partial u}{\partial y}\right) = \Omega\frac{\partial w}{\partial z}, \qquad (8.84a)$$

$$\left(\frac{D}{Dt} - \nu\nabla^2\right)\left(\frac{\partial v}{\partial x} + \frac{\partial u}{\partial y}\right) = 2\varepsilon\frac{\partial w}{\partial z} - \frac{2}{\rho}\frac{\partial^2 p}{\partial x \partial y}. \qquad (8.84b)$$

By applying the Laplacian operator to (8.80d), substituting (8.83) into the resulting equation, applying the operator $\left(\frac{D}{Dt} - \nu\nabla^2\right)$ once and making use of (8.84a,b) and (8.80d), one finally obtains the long sought after partial differential equation for $w(\mathbf{x}, t)$:

$$\left(\frac{D}{Dt} - \nu\nabla^2\right)\nabla^2\left(\frac{D}{Dt} - \nu\nabla^2\right)w$$

$$+ (\Omega^2 - 4\varepsilon^2)\frac{\partial^2 w}{\partial z^2} - 4\varepsilon\frac{\partial^2}{\partial x \partial y}\left(\frac{D}{Dt} - \nu\nabla^2\right)w = 0. \qquad (8.85)$$

The linear operator in (8.85) is $\mathbf{x}$-dependent through the operator D/Dt and cannot be efficiently studied by usual Fourier transform techniques. Inspired by the study of inertial waves [see equation (8.70)], we try the following *ansatz* for the solutions of (8.85):

$$w(\mathbf{x}, t) = \mathcal{Re}\left\{\tilde{w}(t)\exp\left[i\mathbf{k}(t)\cdot\mathbf{x} - \nu\int_0^t k^2\, d\tau\right]\right\}, \qquad (8.86)$$

where $k(t)$ denotes the magnitude of the wavevector $\mathbf{k}(t)$. Note that the last exponential decay factor anticipates on the damping effect induced by viscous dissipation, and that any temporal growth is contained in $\tilde{w}(t)$. Inserting this expression into (8.85), it is possible to remove the terms which explicitly depend on the position vector $\mathbf{x}$ if the time evolution of the wavevector $\mathbf{k}(t) = k_x(t)\mathbf{e}_x + k_y(t)\mathbf{e}_y + k_z(t)\mathbf{e}_z$ is appropriately chosen such that

$$\frac{D}{Dt}\left[\exp(i\mathbf{k}(t)\cdot\mathbf{x})\right] = 0\,. \tag{8.87}$$

If the linear terms in respectively x, y, and z, are required to be zero in (8.87), one finally obtains the evolution equation for $\mathbf{k}$ in component form:

$$\frac{dk_x}{dt} = -\left[(\Omega/2) + \varepsilon\right]k_y\,, \tag{8.88a}$$

$$\frac{dk_y}{dt} = \left[(\Omega/2) - \varepsilon\right]k_x\,, \tag{8.88b}$$

$$\frac{dk_z}{dt} = 0\,. \tag{8.88c}$$

Under these conditions, equation (8.85) magically becomes separable and reduces to an ordinary differential equation for $\tilde{w}(t)$:

$$k^2\frac{d^2(k^2\tilde{w})}{dt^2} + \left[\Omega^2 k_z^2 - 4\varepsilon^2 k^2 - 2\Omega\varepsilon(k_y^2 - k_x^2)\right]k^2\tilde{w} = 0\,. \tag{8.89}$$

Note again that the action of viscosity ν solely appears in the last term of (8.86). The analysis has therefore succeeded in completely decoupling the *inviscid* dynamics governed by (8.88–89) from *viscous diffusion* effects. Inviscid aspects are now investigated in detail.

The wavevector equations (8.88) readily integrate into

$$k_x(t) = \frac{k_0}{E}\cos\left[\Omega_0(t - t_0)\right]\,, \tag{8.90a}$$

$$k_y(t) = k_0\sin\left[\Omega_0(t - t_0)\right]\,, \tag{8.90b}$$

$$k_z(t) = q_0\sqrt{(E^2 + 1)/2E^2}\,, \tag{8.90c}$$

where the time shift t_0 and the "rescaled" horizontal and vertical wavenumbers k_0, q_0 are integration constants and

$$\Omega_0 \equiv \left[(\Omega/2)^2 - \varepsilon^2\right]^{1/2}\,. \tag{8.91}$$

Recall that E is the eccentricity defined in (8.61). As in the case of inertial waves ($\varepsilon = 0$, $E = 1$), the wavevector $\mathbf{k}(t)$ spins around the z-axis with angular velocity Ω_0 given by (8.91), but its tip follows an ellipse, the major axis of which is perpendicular to that of the basic streamlines* (Fig. 8.14). Finally, by substituting expressions (8.90a,b,c) into equation

---

* In that case, the angle θ between Ω_0 and $\mathbf{k}$ is time-dependent.

(8.89), one obtains the ordinary differential equation for $\tilde{w}$

$$[1 - a\cos(2\Omega_0(t - t_0))]\frac{d^2(k^2\tilde{w})}{dt^2} + \Omega_0^2[c - 4a\cos(2\Omega_0(t - t_0))]k^2\tilde{w} = 0,$$
(8.92)

where the quantities a and c are real constants given by

$$a \equiv -\frac{2\varepsilon}{\Omega}\frac{k_0^2}{k_0^2 + q_0^2}; \quad c \equiv \frac{4q_0^2}{k_0^2 + q_0^2}.$$
(8.93a, b)

Equation (8.92) is a homogeneous second-order linear ordinary differential equation where time explicitly appears through a *periodic forcing* of period π/Ω_0. It falls within the class of *Hill's equations* which commonly appear in the study of *parametrically excited systems* (Magnus and Winkler, 1966; Nayfeh and Mook, 1979):

$$\frac{d^2u}{dt^2} + p(t)u = 0,$$
(8.94)

where the real forcing function $p(t)$ is T-periodic. In particular, when $a \ll 1$, equation (8.92) reduces at leading order to the *Mathieu equation*

$$\frac{d^2(k^2\tilde{w})}{dt^2} + \Omega_0{}^2[c - a(4 - c)\cos(2\Omega_0(t - t_0))]k^2\tilde{w} = 0$$
(8.95)

governing the oscillations of a *periodically forced pendulum* (Nayfeh and Mook, 1979) or the Faraday instability arising at the free surface of a periodically forced liquid-filled container (see chapter 4).

The general solutions $u(t)$ of a second-order linear homogeneous equation is the sum of two independent complex solutions $u_1(t)$, $u_2(t)$.[*] When the coefficients of this equation, e.g. $p(t)$ in (8.94), are periodic, the general *Floquet theory* (Nayfeh and Mook, 1979) asserts that the fundamental set $u_1(t)$, $u_2(t)$ may be chosen either as

$$u_1(t) = \hat{u}_1(t)\exp(-i\omega_1 t), \quad u_2(t) = \hat{u}_2(t)\exp(-i\omega_2 t),$$
(8.96)

or, exceptionally as

$$u_1(t) = \hat{u}_1(t)\exp(-i\omega t),$$
$$u_2(t) = \left(\hat{u}_2(t) + \frac{t}{T\exp(-i\omega T)}\hat{u}_1(t)\right)\exp(-i\omega t),$$
(8.97)

where $\hat{u}_1(t)$, $\hat{u}_2(t)$ are T-periodic complex functions and ω_1, ω_2, ω are complex *characteristic frequencies* defined within an additive constant

---

[*] These have nothing to do with the u_1 and u_2 components introduced in (8.73–75)!

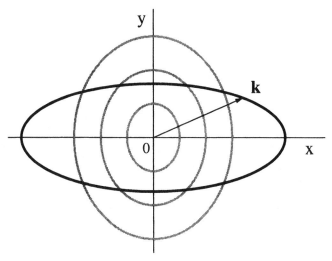

Fig. 8.14.   Wavevector evolution $\mathbf{k}(t)$ projected on x–y plane for instability waves in elliptical flow (8.60) of arbitrary eccentricity E. Gray curves correspond to the basic flow streamlines and black curve to locus of wavevector $\mathbf{k}(t)$.

$2n\pi/T$ where n is an integer. In the generic case (8.96), the fundamental set is such that

$$u_1(t+T) = \mu_1 u_1(t), \qquad u_2(t+T) = \mu_2 u_2(t), \qquad (8.98)$$

where $\mu_1 \equiv \exp(-i\omega_1 T)$, $\mu_2 \equiv \exp(-i\omega_2 T)$ are the so-called *Floquet multipliers*. According to standard Floquet procedure, they are uniquely determined by constructing the 2×2 *fundamental matrix* of solutions of (8.94)

$$M(t) = \begin{pmatrix} u_1(t) & du_1/dt(t) \\ u_2(t) & du_2/dt(t) \end{pmatrix} \qquad (8.99)$$

with initial conditions $M(0) = I$, and then by computing the eigenvalues of the fundamental matrix $M(T)$ which then yield μ_1 and μ_2.

In the case of Hill's equation (8.94), it is straightforward to show (Nayfeh and Mook, 1979) that $\mu_1 \mu_2 = 1$.[*] Since the function $p(t)$ is real, the latter condition implies (see Fig. 8.17a,b) that either (a) μ_1 and μ_2 are complex conjugate numbers lying on the unit circle, or (b) μ_1 and μ_2 are real with $\mu_2 = 1/\mu_1$. The transition between the two situations occurs when $\mu_1 = \mu_2 = \pm 1$.

---

[*] One first infers from (8.94) that the Wronskian $u_1(t)\frac{du_2}{dt}(t) - u_2(t)\frac{du_1}{dt}(t) = \text{const.}$ and one then exploits relations (8.98).

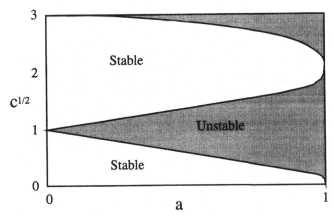

Fig. 8.15. Stability diagram of elliptical vortex (8.60) in a–$\sqrt{c}$ plane for equation (8.92). After Waleffe (1990).

For the equation of interest (8.92), case (b) occurs in the shaded domains of a–$\sqrt{c}$ parameter space, as sketched in Fig. 8.15. According to (8.96–98), this generically corresponds to unbounded oscillatory solutions with an exponentially increasing amplitude: such features are characteristic of the *unstable domain*.

Case (a) which prevails in unshaded areas of Fig. 8.15, defines the stable regime since solutions (8.96) are then bounded aperiodic functions with two fundamental frequencies: the forcing frequency $2\pi/T$ and $\omega_1 = i\ln(\mu_1)/T$. Finally, note that the neutral stability boundary is generally associated with *fundamental solutions* of period T ($\mu_1 = \mu_2 = 1$) or *subharmonic solutions* of period $2T$ ($\mu_1 = \mu_2 = -1$).

Figure 8.15 is reminiscent of the resonance tongue displayed in the forced pendulum case. Indeed, when $a \ll 1$, (8.92) becomes Mathieu's equation (8.95) with a small forcing amplitude $a(4-c)$. In the absence of forcing ($a = 0$), the natural frequency of the oscillator is seen to be $\Omega_0\sqrt{c}$. It is known (Nayfeh and Mook, 1979) that the solutions of (8.95) are unstable when the ratio $\sqrt{c}/2$ of the natural frequency $\Omega_0\sqrt{c}$ to the forcing frequency $2\Omega_0$ is almost equal to an integer (in which case one finds $\mu_1 = \mu_2 = 1$, i.e. *fundamental solutions*) or half an integer (in which case on finds $\mu_1 = \mu_2 = -1$, i.e. *subharmonic solutions*). In Fig. 8.15, the tongues issuing from $\sqrt{c} = 1$ and $\sqrt{c} = 3$ are hence associated with a *subharmonic instability*.[*]

---

[*] Note that the tongue emerging from $\sqrt{c} = 2$ does not appear in Fig. 8.15 because the forcing amplitude $a(4-c)$ in (8.95) precisely vanishes for this value.

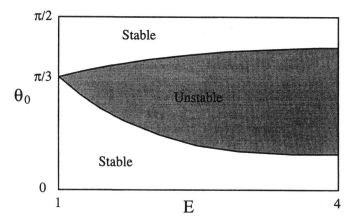

Fig. 8.16. Stability diagram of elliptical vortex in E–θ_0 plane, where E denotes the eccentricity (8.61) and θ_0 is a measure of the angle between the wavevector **k** and the vortex axis (equation (8.100)). After Bayly (1986).

These properties may now be expressed in terms of the elliptical vortex parameters (Fig. 8.16). According to (8.93a,b), a and c are related to the eccentricity E and the angle

$$\theta_0 \equiv \arccos\left(\frac{q_0}{\sqrt{k_0^2 + q_0^2}}\right), \tag{8.100}$$

which is a measure of the varying angle between the wavevector $\mathbf{k}(t)$ and the vortex z-axis [see equations (8.90a,b,c)]. The pure solid body rotation case ($\varepsilon = 0$, $E = 1$) thus corresponds to $a = 0$, i.e. to the vertical axis in Fig. 8.15 and it has previously been shown to give rise to *neutral inertial waves* governed by the dispersion relation (8.76). As soon as the basic state differs from a pure solid body rotation ($\varepsilon \neq 0$, $E \neq 1$), a is non zero: according to Fig. 8.15, inertial waves of angle θ_0 close to θ_c for which $c = 1$ are then unstable. That such an angle θ_c does exist, is proven as follows: when $a = 0$, the coefficients μ_1, μ_2 are easily computed from the dispersion relation (8.77) for neutral inertial waves with $T = 2\pi/\omega$:

$$\mu_1 \equiv \exp(-i\omega T) = \exp(-2i\pi \cos\theta), \tag{8.101a}$$
$$\mu_2 \equiv \exp(+i\omega T) = \exp(+2i\pi \cos\theta). \tag{8.101b}$$

The Floquet multipliers μ_1 and μ_2 pertaining to different values of $\sqrt{c}$ therefore lie on the unit circle as shown on Fig. 8.17a. When a becomes positive, μ_1 and μ_2 gradually migrate from their locations (8.101a,b). In particular, as soon as $a \neq 0$, the *double* Floquet multiplier $\mu = -1$ at

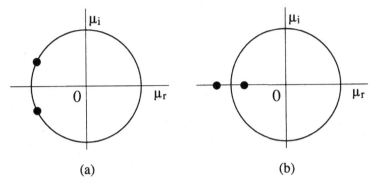

Fig. 8.17. Floquet multipliers μ_1, μ_2 in the complex μ–plane for equation (8.92): (a) stable configuration; (b) unstable configuration.

$a = 0$, $\sqrt{c} = 1$ (Fig. 8.15) spawns a pair of real eigenvalues characteristic of the unstable regime (case (b)) as sketched on Fig. 8.17b. According to (8.101a,b), this bifurcation first takes place at the critical angle θ_c such that

$$\mu = -1, \tag{8.102}$$

i.e.

$$\cos\theta_c = 1/2 \quad \text{i.e.} \quad \theta_c = \pi/3. \tag{8.103a, b}$$

Note that, in the limit $a \ll 1$ ($E \approx 1$), the forcing frequency $2\Omega_0$ appearing in the coefficients of (8.95) is equal to Ω. Inertial waves of frequency equal to the angular velocity $\Omega/2$ of the circular vortex, are therefore destabilized via a subharmonic instability produced by the imposed ellipticity.

As seen from Fig. 8.16, ellipticity destabilizes a bandwidth of wavevector angles around $\theta_c = \pi/3$. Furthermore, as in the case of inertial waves, a rescaling of the wavevector leaves expressions (8.93a,b) unchanged: the inviscid perturbation dynamics are independent of the modulus of $\mathbf{k}(t)$ and there exists a continuous family of self-similar solutions with identical dynamical behavior. Such an inviscid mechanism is therefore capable of amplifying a large spectrum of scales directly from a smooth basic flow. According to equation (8.86), viscous effects do not affect the overall picture but introduce a cut-off wavenumber, a most amplified wavenumber,[*] and its corresponding growth rate σ_{max} at a given ellipticity (Landman and Saffman, 1987).

---

[*] However it predicts a most amplified wavenumber $k = 0$ for which the local approximation (8.60) is no longer valid: finite size core effects should hence be taken into account.

This analysis is particularly interesting in the context of transition
to turbulence since it leads to a three-dimensional *universal* short-wave
instability for elliptically shaped two-dimensional vortices, independently
of the detailed vortex structure. Indeed, the flow field (8.60) is only
pertinent within the vortex core where the vorticity is supposed to be
uniform and streamlines are ellipses of constant eccentricity. Away from
this region of characteristic size L, the flow rapidly becomes potential
but this outer region does not affect the evolution of small scales as long
as their wavenumber k is such that $kL \gg 1$. Corresponding eigenmodes
are then localized in the core and rapidly decay to zero away from it:
it is then legitimate to approximate the steady vortex by the assumed
elliptical vortex (8.60) without modifying the small scale dynamics.

Direct numerical simulations of transition in the elliptical vortex
(Pierrehumbert, 1986), in plane channel flow (Orszag and Patera, 1983)
or mixing layers (Rogers and Moser, 1992), and secondary instability
analyses of mixing layers (Pierrehumbert and Widnall, 1982) fully confirm
the universality of the elliptical instability mechanism, as evidenced in
Fig. 8.18. In *bounded shear flows*, the *primary instability* mechanism has
been shown in sections 7 and 8.1 to be of *viscous* origin while in *free
shear flows* it is *inviscidly* driven (sections 4 and 5). At the *secondary
instability* stage, both classes of shear flows find themselves on the same
boat: closed streamline patterns directly produce a large bandwidth of
spatial scales within an inertial time scale τ_i.

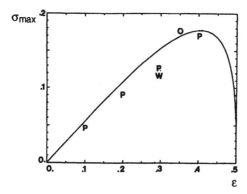

Fig. 8.18. Nondimensional growth rate σ_{max} as a function of eccentricity
parameter ε. Solid curve is from elliptical instability theory (Landman and
Saffman, 1987).   Points P are numerical results of Pierrehumbert (1986)
pertaining to the elliptical vortex (8.60), point O is the numerical result of Orszag
and Patera (1983) for plane channel flow, point W is the secondary instability
result for mixing layers (Pierrehumbert and Widnall, 1982). After Bayly *et al.*
(1988).

8.5 Fundamental and subharmonic secondary instability routes

The instability mechanism investigated in the previous section is independent of the detailed form of the basic flow but it is restricted to small scales. In the present section, we examine the evolution of arbitrary scales in the specific context of secondary instabilities in plane channel flow, i.e. *three-dimensional instabilities of finite amplitude Tollmien–Schlichting waves of given wavenumber k* computed in section 8.3. It is our purpose to demonstrate that, in the case of streamwise periodic basic flows with wavelength $2\pi/k$, a modified normal mode analysis may still be performed (Bayly *et al*, 1988; Herbert, 1988).

In the context of *secondary instabilities*, the basic flow is the combination of plane Poiseuille flow and two-dimensional saturated finite amplitude Tollmien–Schlichting waves ψ' of wavenumber k (see (8.43)). *Viewed in a reference frame moving at their phase velocity c*, this basic flow

$$\mathbf{U}_0(\mathbf{x}; k, \mathrm{Re}) = \left[1 - y^2 + \frac{\partial \psi'(\xi, y)}{\partial y}\right] \mathbf{e}_x - \frac{\partial \psi'(\xi, y)}{\partial \xi} \mathbf{e}_y, \qquad (8.104)$$

is steady and two-dimensional. As a consequence, no explicit dependence on the spanwise variable z and on time t is introduced in the secondary instability operator

$$\frac{\partial \mathbf{u}}{\partial t} + (\mathbf{U}_0 \cdot \nabla)\mathbf{u} + (\mathbf{u} \cdot \nabla)\mathbf{U}_0 = -\nabla p + \frac{1}{\mathrm{Re}} \nabla^2 \mathbf{u}, \qquad (8.105)$$

$$\nabla \cdot \mathbf{u} = 0. \qquad (8.106)$$

In that frame, instability modes may therefore be sought in the form

$$\mathbf{u}(\xi, y, z, t) = \mathcal{R}e \left\{ \tilde{\mathbf{u}}(\xi, y) \exp\left[i(\beta z - \varpi t)\right] \right\}, \qquad (8.107)$$

where $\varpi \equiv \varpi_r + i\varpi_i$ is the complex frequency and β the real spanwise wavenumber. Note that ϖ_i and ϖ_r respectively represent the temporal growth rate and the frequency shift with respect to the Tollmien–Schlichting frequency $\omega = kc$. The form of $\tilde{\mathbf{u}}(\xi, y)$ is specified by Floquet theory (Nayfeh and Mook, 1979): the linear operator (8.105–106) being periodic in ξ of period $2\pi/k$, $\tilde{\mathbf{u}}(\xi, y)$ may be expressed as

$$\tilde{\mathbf{u}}(\xi, y) = \hat{\mathbf{u}}(\xi, y) \exp(i\alpha\xi), \qquad (8.108)$$

where the eigenfunction $\hat{\mathbf{u}}(\xi, y)$ is assumed to be periodic in ξ of period $2\pi/k$. The quantity $\alpha \equiv \alpha_r + i\alpha_i$ is the complex streamwise wavenumber, $-\alpha_i$ and α_r respectively representing the spatial growth rate and the sideband wavenumber with respect to the Tollmien–Schlichting wavenumber k. Note that, without loss of generality, the range of α_r may be confined to*

$$|\alpha_r| \leq k/2 . \tag{8.109}$$

Introduction of (8.107–108) into the secondary instability equations (8.105–106) and associated no-slip boundary conditions defines an eigenvalue problem which admits nontrivial solutions if and only if α, β and ϖ satisfy a *dispersion relation* of the form

$$D(\alpha, \beta, \varpi; k, \mathrm{Re}) = 0 . \tag{8.110}$$

Note that this dispersion relation depends on the amplitude of the periodic basic flow *via* its streamwise wavenumber k and the Reynolds number Re as seen from Fig. 8.8. In the same fashion as for the primary instability problem, two alternate points of view may be considered: in the *temporal* framework, the wavevector of components (α, β) is given real and the dispersion relation (8.110) is solved for the complex frequency ϖ. If $\varpi_r = 0$, perturbation modes synchronously propagate with the basic flow field: they are said to be *phase locked* with the Tollmien–Schlichting waves. In the *spatial* setting, external forcing is imposed *in the laboratory x–y frame* at a given real frequency ϖ_{f}. The reader may verify that the dispersion relation in that frame is obtained by substituting in (8.110) the Doppler shift $\varpi = \varpi_{\mathrm{f}} - \alpha c$ so that

$$D(\alpha, \beta, \varpi_{\mathrm{f}} - \alpha c; k, \mathrm{Re}) = 0 . \tag{8.111}$$

Dispersion relation (8.111) is then solved for complex α at given real spanwise wavenumber β and forcing frequency ϖ_{f}. The spatial growth rate in the laboratory frame is given by $-\alpha_i$ and the detuning in wavenumber with respect to k by α_r.

To the present day, most investigations have addressed the temporal instability problem, the properties of which are discussed below. The streamwise periodic temporal eigenfunction $\hat{\mathbf{u}}(\xi, y)$ is decomposed into Fourier series so that three-dimensional modes (8.107–108) become:

$$\mathbf{u}(\xi, y, z, t) = \mathcal{R}e \left\{ \exp\left[i(\alpha\xi + \beta z - \varpi t)\right] \sum_{m=-\infty}^{+\infty} \hat{\mathbf{u}}_m(y) \exp(imk\xi) \right\} . \tag{8.112}$$

In order to determine the complex frequency ϖ, the functions $\hat{\mathbf{u}}_m(y)$ are decomposed, as in (8.45), into a combination of Chebyshev polynomials which, after substitution into the secondary instability equations (8.105–106) and appropriate truncation, leads to an algebraic eigenvalue problem for ϖ. This numerically yields the dispersion relation (8.110) (Orszag and

---

* If α_r lies outside this range, $\hat{\mathbf{u}}(\xi, y)$ may be suitably renormalized by a multiplicative factor of the form $\exp(ink\xi)$ where n is an integer.

Patera, 1983; Herbert, 1983; Herbert *et al.*, 1986). These computations confirm the *inviscid* character of the three-dimensional instability: at a fixed Tollmien–Schlichting frequency and spanwise wavenumber, the growth rate indeed saturates for increasing Reynolds number.

As in section 5.3, eigenmodes may be classified according to their symmetry properties specified by the value of $\alpha_r = \alpha$.

When $\alpha = 0$, the overall flow —Poiseuille parabolic velocity profile + Tollmien–Schlichting wave + three-dimensional perturbations— is streamwise (spanwise) periodic of wavelength $2\pi/k$ $(2\pi/\beta)$. This *fundamental mode* (Fig. 8.19a) is the analogue of the translative mode in mixing layers (section 5.3, Fig. 5.18a).

When $\alpha = \pm k/2$, the overall flow is still spanwise periodic of period $2\pi/\beta$ but its streamwise wavelength $4\pi/k$ is now twice that of the Tollmien–Schlichting wave. Furthermore, according to (8.112), it is invariant with respect to the translation $(\xi, z) \mapsto (\xi + 2\pi/k, z + \pi/\beta)$. This *subharmonic mode* (Fig. 8.19b) belongs to the same class as the

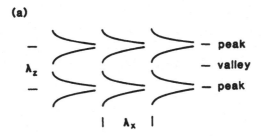

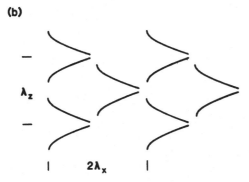

Fig. 8.19. Typical streakline patterns in x–z plane associated with secondary instabilities in plane channel flow. (a) Fundamental mode with aligned amplitude maxima. (b) Subharmonic mode with staggered amplitude maxima. After Herbert (1983).

helical pairing instability mode discussed in the context of mixing layers (section 5.3, Fig. 5.18b).

Finally, when α is neither 0 nor $\pm k/2$, one obtains the so-called *combination modes*. Since real disturbances require two complex conjugate components of opposite sideband wavenumber α, solution (8.112) then contains wavenumber pairs $mk \pm \alpha$ with m a positive integer, the sum of which matches the Tollmien–Schlichting wavenumber k and its harmonics mk. Concomitantly, the sum of associated frequencies $\omega \pm \varpi_r$ matches the Tollmien–Schlichting frequency ω. Such modes also exist in the mixing layer case though we did not allude to them in section 5.3.

The *subharmonic mode* characteristics are summarized in Figs. 8.20–21. According to Fig. 8.20, the growth rate curve displays a large bandwidth of amplified spanwise wavenumbers β with a broad maximum at a value comparable to the Tollmien–Schlichting wavenumber k. Since the destabilizing agent is inviscid, computed growth rates are considerably larger than for the primary viscous instability giving birth to Tollmien–Schlichting waves. Furthermore, note the distinction with the corresponding curve for the secondary helical pairing instability in mixing layers (Fig. 5.19): whereas the two-dimensional pairing mode $\beta = 0$ is the most unstable for shear layers, it is damped for plane channel flow. Finally, the frequency shift ϖ_r almost vanishes over a wide extent of the unstable bandwidth, indicating that disturbances are almost phase locked. Figure 8.21 demonstrates the sensitivity of the growth rate ϖ_i to the Tollmien–Schlichting amplitude level ε: there exists a *critical threshold amplitude* ε_s above which the subharmonic mode is

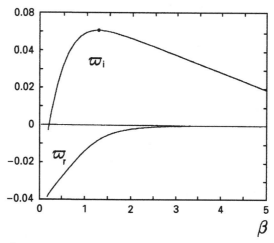

Fig. 8.20. Growth rate ϖ_i and frequency shift ϖ_r for the subharmonic instability mode as a function of spanwise wavenumber β in plane channel flow at $k = 1.12$, $\mathrm{Re} = 5,000$. After Herbert (1983).

destabilized.* A representative eigenfunction shape displays a maximum near the critical point of the Tollmien–Schlichting wave as sketched in Figs. 8.22. The resulting disturbance field in the x–z plane near the wall takes the form of a *staggered streakline pattern* (Fig. 8.19b). This solution is consistent with experimental findings (Saric *et al.*, 1984; Ramazanov, 1985): at low forcing levels, Tollmien–Schlichting waves evolve into an ordered pattern of staggered Λ-shaped vortices, the well known *H-type vortices* discussed in section 2.3 (Fig. 2.22). Corresponding frequency spectra exhibit peaks at odd multiples of the subharmonic frequency.

The *fundamental mode* characteristics are similar to those of the subharmonic: eigenvalues ϖ are of the same order of magnitude except that the most unstable fundamental mode is exactly phase locked ($\varpi_r = 0$). Furthermore, the growth rate curve displays a large bandwidth of amplified spanwise wavenumbers β with a broad maximum at a value comparable to the Tollmien–Schlichting wavenumber k. The fundamental mode has been experimentally identified in plane channel flow by Kozlov and Ramazanov (1984) and in boundary layers by Klebanoff *et al.* (1962): in both instances, Tollmien–Schlichting waves evolve into an ordered *pattern of aligned Λ-shaped structures* known as *K-type vortices*, with the same streamwise periodicity as the underlying Tollmien–Schlichting wave. Note that, according to (8.112), the fundamental mode ($\alpha = 0$) is

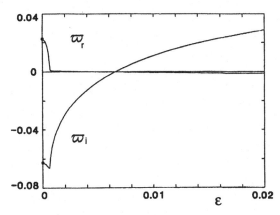

Fig. 8.21. Growth rate ϖ_i and frequency shift ϖ_r for the subharmonic instability mode as a function of Tollmien–Schlichting amplitude level ε in plane channel flow at $k = 1.02$, $\beta = 2$. After Herbert (1983).

---

* This is not the case for boundary layers where the triad resonance mechanism proposed by Craik (1971) is active at small amplitude levels. Symmetry considerations preclude the existence of such a resonance in plane Poiseuille flow.

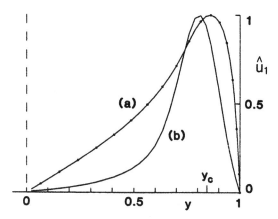

Fig. 8.22. Normalized eigenfunction $\hat{u}_1(y)$ of subharmonic mode (b) in plane channel flow at $k = 1.12$, $\beta = 2$, Re $= 5,000$. Normalized eigenfunction of corresponding two-dimensional Tollmien–Schlichting wave (a) at the same k and Re. The critical layer of the Tollmien–Schlichting wave is located at y_c. After Herbert (1983).

characterized by the presence of a spanwise periodic mean flow distortion of the form $\hat{u}_0(y) \exp[i(\beta z - \varpi t)]$. This feature is entirely consistent with experimental findings: alternating peaks and valleys are observed in the mean streamwise velocity distributions along the span (Klebanoff *et al.*, 1962).

One major characteristic of the linear stability analysis is the sensitivity with respect to the Tollmien–Schlichting wave amplitude level. *For low amplitudes, the subharmonic instability is predominant while for higher amplitudes fundamental modes exhibit slightly higher growth rates.* Different routes are hence favored according to the amplitude level ε reached by Tollmien–Schlichting waves: a) If ε is less than a threshold value ε_s no secondary instability is observed; b) subharmonic modes are dominant when $\varepsilon_s \leq \varepsilon \leq \varepsilon_t$, with ε_t typically less than 1 percent of the centerline velocity; c) above the second threshold ε_t, subharmonic and fundamental modes exhibit growth rates of the same order of magnitude and the observed transition scenario very much depends on the nature of upstream perturbations. In natural transition experiments, combination modes, where time signals display beating modulations, are more likely to occur than in controlled experiments where fundamental and subharmonic modes seem to dominate the dynamics.

Finally, one should bear in mind that, on account of their inviscid origin, secondary instability growth rates are considerably larger than the their primary Tollmien–Schlichting counterparts. As a result, the above analysis remains pertinent below Re $= 2,900$ where no self-sustaining two-dimensional finite-amplitude states exist (Fig. 8.8). Indeed, below Re $= 2,900$, finite amplitude Tollmien–Schlichting waves decay slowly enough that they may be regarded as quasi-stationary on the evolution time scale of the associated three-dimensional instability modes. Such a quasi-steady assumption is apparently valid down to Re $\sim 1,000$ and it may account for the common occurrence of natural transition at such low values of the Reynolds number. Another possible scenario for the natural transition has been suggested by Ehrenstein and Koch (1991) who demonstrated the existence of three-dimensional finite amplitude waves below Re $= 2,900$.

Appendix

The objective of this appendix is to prove that the second group of terms in (8.51) cannot give rise to an instability on its own. Since ω_z has been shown in (8.58) to be a passive scalar, we focus here on the perturbation vorticity component $\omega_\perp$ in the x–y plane. This quantity may be decomposed into

$$\omega_\perp = \Lambda_1 \mathbf{U}_0 + \tilde{\omega}\,, \tag{A1}$$

where $\tilde{\omega}$ is in the x–y plane and Λ_1 is assumed to be a passive scalar chosen so that $\tilde{\omega}$ is divergenceless at $t = 0$. Bearing in mind that $\nabla \cdot \omega = 0$, this condition takes the form

$$\frac{\partial \Lambda_1}{\partial t} = -(\mathbf{U}_0 \cdot \nabla)\Lambda_1 = -\nabla \cdot \omega_\perp = \frac{\partial \omega_z}{\partial z} \quad \text{at} \quad t = 0\,. \tag{A2}$$

Considering that Λ_1 and ω_z are both passive scalars, $\partial \Lambda_1 / \partial t$ and $\partial \omega_z / \partial z$ are also passive scalars since $\mathbf{U}_0$ is independent of z and t. Condition (A2) states that these two quantities are equal at time $t = 0$. They are therefore equal for all time:

$$\frac{\partial \Lambda_1}{\partial t} = \frac{\partial \omega_z}{\partial z}\,. \tag{A3}$$

Upon substitution of decomposition (A1) into (8.57) and making use of equations (8.58), (A2–3) and the standard identity for $\nabla \times (\mathbf{A} \times \mathbf{B})$, one obtains:

$$\frac{\partial \tilde{\omega}}{\partial t} + \nabla \times [\tilde{\omega} \times \mathbf{U}_0] = 0\,. \tag{A4}$$

Since $\tilde{\omega}$ is chosen to be initially divergenceless, it remains so thereafter in accordance with the divergence of (A4). This vector contained in the x–y plane can hence be written as

$$\tilde{\omega} = \nabla \times (\Lambda_2 \mathbf{e}_z)\,, \tag{A5}$$

where Λ_2 is a scalar function. Upon substituting expression (A5) into equation (A4) and using standard identities for $\nabla \times (\psi \mathbf{A})$ and $\nabla \times (\mathbf{A} \times \mathbf{B})$, one easily obtains:

$$\nabla \left[\left(\frac{\partial \Lambda_2}{\partial t} + (\mathbf{U}_0 \cdot \nabla) \Lambda_2 \right) \right] \times \mathbf{e}_z = 0 . \tag{A6}$$

Integration of equation (A6) finally leads to

$$\frac{\partial}{\partial t} \Lambda_2 + \mathbf{U}_0 \cdot \nabla \Lambda_2 = -\frac{\partial}{\partial t} \theta(z, t) , \tag{A7}$$

where $\theta(z, t)$ is an arbitrary scalar field. The latter function can be absorbed into Λ_2 by a simple redefinition $\Lambda_2 \to \Lambda_2 + \theta(z, t)$, a change which does not affect the vorticity field $\tilde{\omega}$. The function Λ_2 is therefore a passive scalar.

Acknowledgements

The authors are grateful beyond all exponential orders to Paul Manneville, Claude Godrèche, and the co-authors of this volume for their infinite patience while these notes were being written. They also wish to thank Ivan Delbende, Thérèse Lescuyer, Sabine Ortiz, and Pierre-Yves Lagrée for their kind and generous help.

References

Acheson, D.J. (1990). *Elementary Fluid Dynamics* (Clarendon Press, Oxford).

Albarède, P., Monkewitz, P.A. (1992). *Phys. Fluids* **A4**, 744–756.

Albarède, P., Provansal, M. (1996). *J. Fluid Mech.* **291**, 191–222.

Andereck, C.D., Liu, S.S., Swinney, H.L. (1986). *J. Fluid Mech.* **164**, 155–183.

Balmforth, N.J. (1995). *Ann. Rev. Fluid Mech.* **27**, 335–373.

Balsa, T.F (1987). *J. Fluid Mech.* **174**, 553–563.

Balsa, T.F (1988). *J. Fluid Mech.* **187**, 155–177.

Batchelor, G. K. (1967). *An Introduction to Fluid Dynamics* (Cambridge University Press, Cambridge).

Bayly, B.J. (1986). *Phys. Rev. Lett.* **57**, 2160–2163.

Bayly, B.J., Orszag, S.A., Herbert T. (1988). *Ann. Rev. Fluid Mech.* **20**, 359–391.

Bender, C.M., Orszag, S.A. (1978). *Advanced Mathematical Methods for Scientists and Engineers* (McGraw-Hill, Singapore).

Bergé, P., Pomeau, Y., Vidal, Ch. (1984). *Order within Chaos* (Wiley, New York).

Berger, M. (1977). *Nonlinearity and Functional Analysis* (Academic Press, New York).

Bernal, L.P., Roshko, A. (1986). *J. Fluid Mech.* **170**, 499–526.

Bers, A. (1983). In *Handbook of Plasma Physics*, M.N. Rosenbluth, R.Z. Sagdeev, eds. (North-Holland, Amsterdam).

Betchov, R., Criminale, W.O. (1966). *Phys. Fluids* **9**, 359–362.

Betchov, R., Criminale, W.O. (1967). *Stability of Parallel Flows* (Academic Press, New York).

Betchov, R., Szewczyk, A. (1963). *Phys. Fluids* **6**, 1391–1396.

Bonnet, J.P., Glauser, M.N. (1993). *Eddy Structure Identification in Free Turbulent Shear Flows* (Kluwer, Dordrecht).

Bouthier, M. (1972). *J. Méc.* **11**, 599–621.

Briggs, R.J. (1964). *Electron-Stream Interaction with Plasmas* (MIT Press, Cambridge Mass.).

Browand, F.K., Weidman, P.D. (1976). *J. Fluid Mech.* **76**, 127–144.

Brown, G.L., Roshko, A. (1974). *J. Fluid Mech.* **64**, 775–816.

Buell, J.C., Huerre, P. (1988). Inflow/outflow boundary conditions and the global dynamics of spatial mixing layers, in *Proc. NASA Ames/Stanford Center for Turbulence Res. Summer Program*, Rep. CTR-S88, 19–27.

Busse, F.H. (1981). Transition to turbulence in Rayleigh–Bénard convection, in Swinney and Gollub (1981).

Cambon, C., Teissedre, C., Jeandel, D. (1985). *J. Méc. Théor. Appl.* **4**, 629–657.

Canuto, C., Hussaini, M.Y., Quarteroni, A., Zang, T.A. (1988). *Spectral Methods in Fluid Dynamics* (Springer-Verlag, Berlin).

Carlson, D.R., Widnall, S.E., Peeters, M.F. (1982). *J. Fluid Mech.* **121**, 487–505.

Carrier, G.F., Krook, M., Pearson, C.E. (1966). *Functions of a Complex Variable* (McGraw-Hill, New York).

Chandrasekhar, S. (1961). *Hydrodynamic and Hydromagnetic Stability* (Clarendon Press, Oxford).

Cherhabili, A., Ehrenstein, U. (1995). *Eur. J. Mech. B/Fluids* **14**, 6, 677–696.

Chomaz, J.M., Huerre, P., Redekopp, L.G. (1988). *Phys. Rev. Lett.* **60**, 25–28.

Chomaz, J.M., Huerre, P., Redekopp, L.G. (1991). *Stud. Appl. Math.* **84**, 119–144.

Chossat, P., Iooss, G. (1994). *The Couette-Taylor Problem* (Springer-Verlag, Berlin).

Corcos, G.M., Sherman, S.J. (1984). *J. Fluid Mech.* **139**, 29–65.

Craik, A.D.D. (1971). *J. Fluid Mech.* **50**, 393–413.

Craik, A.D.D. (1985). *Wave Interactions and Fluid Flows* (Cambridge University Press, Cambridge).

Craik, A.D.D., Criminale, W.O. (1986). *Proc. R. Soc. London* **A406**, 13–26.

Crighton, D.G., Gaster, M. (1976). *J. Fluid Mech.* **77**, 397–413.

Crighton, D.G, Dowling, A.P., Ffowcs Williams, J.E., Heckl, M., Leppington, F.G. (1992). *Modern Methods in Analytical Acoustics* (Springer-Verlag, Berlin).

Cross, M.C., Hohenberg, P.C. (1993). *Rev. Mod. Phys.* **65**, 851–1112.

Crow, S.C., Champagne, F.H. (1971). *J. Fluid Mech.* **48**, 547–591.

Davis, S.H. (1978). *Ann. Rev. Fluid Mech.* **8**, 57–74.

Deissler, R.J. (1987). *Phys. Fluids* **30**, 2303–2305.

Di Prima, R.C., Swinney, H.L. (1981). Instabilities and transition in flow between concentric rotating cylinders, in Swinney and Gollub (1981).

Dirac, P.A.M. (1984). *The Principles of Quantum Mechanics*, 4th ed. (Oxford Science publications, Oxford).

Drazin, P.G., Howard, L.N. (1966). *Adv. in Appl. Mech.* **7**, 1–89 (Academic Press, New York).

Drazin, P.G., Reid, W.H. (1981). *Hydrodynamic Stability* (Cambridge University Press, Cambridge).

Drubka, R.E. (1981). *Instabilities in Near Field of Turbulent Jets and their Dependence on Initial Conditions and Reynolds Number.* Ph.D. Thesis, Ill. Inst. Technol., Chicago.

Eagles, P.M. (1969). *Quart. J. Mech. Appl. Math.* **22**, 129–182.

Eckelmann, H., Graham, J.M.R., Huerre, P., Monkewitz, P.A., eds. (1993). *Bluff-body Wakes, Dynamics and Instabilities* (Springer-Verlag, Berlin).

Ehrenstein, U., Koch, W. (1991). *J. Fluid Mech.* **228**, 111–148.

Fenstermacher, P.R., Swinney, H.L., Gollub, J.P. (1979). *J. Fluid Mech.* **94**, 103–128.

Freymuth, P. (1966). *J. Fluid Mech.* **25**, 683–704.

Fiedler, H.E., Dziomba, B., Mensing, P., Rosgen, T. (1981). Initiation and global consequences of coherent structures in turbulent shear flows, in *The Role of Coherent Structures in Modelling Turbulence and Mixing*, Lecture Notes in Physics **136**, J. Jimenez, ed. (Springer-Verlag, New York).

Frisch, U. (1995). *Turbulence : The Legacy of A.N. Kolmogorov* (Cambridge University Press, Cambridge).

Gaster, M. (1962). *J. Fluid Mech.* **14**, 222–224.

Gaster, M. (1975). *Proc. R. Soc. London* **A347**, 271–289.

Gaster, M., Grant I. (1975). *Proc. R. Soc. London* **A347**, 253–269.

Gaster, M., Kit, E., Wygnanski, I. (1985). *J. Fluid. Mech.* **150**, 23–39.

Goldstein, M.E. (1983). *J. Fluid. Mech.* **127**, 59–81.

Guckenheimer, J., Holmes, P. (1983). *Nonlinear Oscillations, Dynamical Systems and Bifurcations of Vector Fields* (Springer-Verlag, Berlin).

Guyon, E., Hulin, J-P., Petit L. (1991). *Hydrodynamique physique* (InterEdition CNRS, Paris).

Hammache, M., Gharib, M. (1989). *Phys. Fluids* **A1**, 1611.

Hammond, D.A., Redekopp, L.G. (1997). *J. Fluid Mech.* **331**, 231–260.

Hannemann, K., Oertel, H.Jr (1990). *Ann. Rev. Fluid Mech.*, **22**, 539–564.

Heisenberg, W. (1924). *Ann. Phys. Leipzig* **74**, 577–627.

Henderson, R., Barkley, D. (1996). *Phys. Fluids* **8**, 1683–1685.

Herbert, T. (1983). *Phys. Fluids* **26**, 871–74.

Herbert, T., Bertolotti, F.P., Santos, G.R. (1986). Floquet analysis of secondary instability in shear flows, in *Stability of Time-Dependent and Spatially*

Varying Flows, D.L. Dwoyer, M.Y. Hussaini, eds. (Springer-Verlag, New York).

Herbert, T. (1988). *Ann. Rev. Fluid Mech.* **20**, 487–526.

Ho, C.M., Huang, L.S. (1982). *J. Fluid. Mech.* **119**, 443–473.

Ho, C.M., Huerre, P. (1984). *Ann. Rev. Fluid Mech.* **16**, 365–424.

Huerre, P. (1987a). Spatio-temporal instabilities in closed and open flows, in *Instabilities and Nonequilibrium Structures*, E.Tirapegui, D. Villaroel, eds. (Reidel, Dordrecht).

Huerre, P. (1987b). On the absolute-convective nature of primary and secondary instabilities, in *Propagation in Systems far from Equilibrium*, J.E. Wesfreid, H.R. Brand, P. Manneville, G. Albinet, N. Boccara, eds. (Springer-Verlag, Berlin).

Huerre, P., Monkewitz, P.A. (1985), *J. Fluid Mech.* **159**, 151–168.

Huerre, P., Monkewitz, P.A. (1990). *Ann. Rev. Fluid Mech.* **22**, 357–473.

Hunt, R.E., Crighton, D.G. (1991). *Proc. R. Soc. Lond.* **A435**, 109–128.

Iooss, G., Adelmeyer, M. (1992). *Topics in Bifurcation Theory and Applications*, Advanced Series in Nonlinear Dynamics (World Scientific, Singapore).

Itoh, N. (1974), *Trans. Japan Soc. Aero. Space Sci.* **17**, 65–75.

Jimenez, J. (1990). *J. Fluid Mech.* **218**, 265–297.

Karniadakis, G.E., Triantafyllou, G.S. (1989). *J. Fluid. Mech.* **199**, 441–469.

Keller, H.B. (1977). Numerical solutions of bifurcation and nonlinear eigenvalue problems, in *Applications of Bifurcation Theory*, P.H. Rabinowitz, ed. (Academic Press, New York).

Kerschen, E. (1989). Boundary layer receptivity, *AIAA paper* **89-1109**.

Klebanoff, P.S., Tidstrom, K.D., Sargent, L.M. (1962). *J. Fluid. Mech.* **12**, 1–34.

Koch, W. (1985). *J. Sound Vib.* **99**, 53–83.

Koochesfahani, M.M., Dimotakis, P.E. (1986). *J. Fluid Mech.* **170**, 83–112.

Kozlov, V.V, Ramazanov, M.P. (1984). *J. Fluid. Mech.* **147**, 149–157.

Kundu, P.K. (1990). *Fluid Mechanics* (Academic Press, London).

Kuramoto Y. (1984). *Chemical Oscillations, Waves, and Turbulence* (Springer-Verlag, Berlin).

Lamb, H. (1995). *Hydrodynamics*, 6th. Edition (Cambridge University Press, Cambridge).

Landau, L.D., Lifshitz, E.M. (1987a). *Fluid Mechanics* (Pergamon, Oxford).

Landau, L.D., Lifshitz, E.M. (1987b). *Kinetic Theory* (Pergamon, Oxford).

Landman, M.J., Saffman, P.G. (1987). *Phys. Fluids* **30**, 2339–2342.

Lasheras, J.C., Choi, H. (1988) *J. Fluid Mech.* **189**, 53–86.

Le Dizès, S., Huerre, P., Chomaz, J.M., Monkewitz, P.A. (1996). *Phil. Trans. R. Soc. Lond.* **A354**, 169–212.

Lesieur, M. (1991). *Turbulence in Fluids.* 2nd ed. (Kluwer, Dordrecht).

Lin, C.C. (1955). *The Theory of Hydrodynamic Stability* (Cambridge University Press, Cambridge).

Lindgren, E.R. (1969). *Phys. Fluids* **12**, 418–425.

Lyapunov, A. (1988). *Problème Général de la Stabilité du Mouvement* (Gabay, Paris).

Mack, L.M. (1976). *J. Fluid Mech.* **73**, 497–520.

Magnus, W., Winkler, S. (1966). *Hill's Equation* (Wiley, New York).

Malkus, W.V.R., Veronis, V. (1958). *J. Fluid Mech.* **4**, 225–260.

Manneville, P. (1990). *Dissipative Structures and Weak Turbulence* (Academic Press, New York).

Maslowe, S.A. (1981). Shear flow instabilities and transition, in Swinney and Gollub (1981).

Maslowe, S.A. (1986). *Ann. Rev. Fluid Mech.* **18**, 405–32.

Mathis, C., Provansal, M., Boyer, L. (1984). *J. Physique Lett.* **45**, L 483–491.

Mattingly, G.E., Criminale, W.O. (1972). *J. Fluid Mech.* **51**, 233–272.

Michalke, A. (1964). *J. Fluid Mech.* **19**, 543–556.

Michalke, A. (1965). *J. Fluid Mech.* **23**, 521–544.

Miksad, R.W. (1972). *J. Fluid Mech.* **56**, 695–719.

Monin, A.S., Yaglom, A.M. (1975). *Statistical Fluid Mechanics* (MIT Press, Cambridge Mass.).

Monkewitz, P.A. (1988). *Phys. Fluids* **31**, 999–1006.

Monkewitz, P.A. (1990). *Eur. J. Mech. B/Fluids* **9**, 395–413.

Monkewitz, P.A., Huerre, P. (1982). *Phys. Fluids* **25**, 1137–1143.

Monkewitz, P.A., Huerre, P., Chomaz, J.M. (1993). *J. Fluid Mech.*, **251**, 1–20.

Monkewitz, P.A., Williamson, C.H.K., Miller, G.D. (1996). *Phys. Fluids* **8**, 91–96.

Morkovin, M.V. (1988). Recent insights into instability and transition to turbulence in open flow systems, *AIAA paper* **88-3675**.

Nayfeh, A.H. (1973). *Perturbation Methods* (Wiley, New York).

Nayfeh, A.H., Mook, D.T. (1979). *Nonlinear Oscillations* (Wiley, New York).

Newell, A.C., Whitehead, J.A. (1969). *J. Fluid Mech.* **38**, 279–303.

Newell, A.C., Passot, T., Lega, J. (1993). *Ann. Rev. Fluid Mech.* **25**, 399–453.

Nishioka, M., Asai, M. (1985). *J. Fluid Mech.* **150**, 441–450.

Nishioka, M., Asai, M., Iida, S. (1980). Wall phenomena in the final stage of transition to turbulence, in *Laminar–Turbulent Transition*, R. Eppler and H. Fasel, eds. (Springer-Verlag, Berlin).

Nishioka, M., Iida, S., Ichikawa, Y. (1975). *J. Fluid Mech.* **72**, 731–751.

Nozaki, K., Bekki, N. (1984). *J. Phys. Soc. Jpn.* **53**, 1581.

Orszag, S.A. (1971). *J. Fluid Mech.* **50**, 689–703.

Orszag, S.A., Patera, A.T. (1980). *Phys. Rev. Lett.* **45** 989–992.

Orszag, S.A., Patera, A.T. (1981). Hydrodynamic stability of shear flows, in *Chaotic Behaviour of Deterministic systems*, G. Iooss, R.H.G. Helleman, R. Stora, eds. (North Holland, Amsterdam).

Orszag, S.A., Patera, A.T. (1983). *J. Fluid Mech.* **128**, 347–385.

Park, D.S., Redekopp, L.G. (1992). *Phys. Fluids* **A4**, 1–10.

Patel, V.C., Head, M.R. (1969). *J. Fluid Mech.* **38**, 181–201.

Perry, A.E., Chong, M.S., Lim, R.T. (1982). *J. Fluid Mech.* **116**, 77–90.

Pierrehumbert, R.T. (1984). *J. Atmos. Sci.* **41**, 2141–2162.

Pierrehumbert, R.T. (1986). *Phys.Rev.Lett.***57**, 2157–2159.

Pierrehumbert, R.T., Widnall, S.E. (1982). *J. Fluid Mech.* **114**, 59–82.

Pouliquen, O., Chomaz, J.M., Huerre, P. (1994). *J. Fluid Mech.* **266**, 277–302.

Press, W.H., Teukolsky, S.A., Vetterling, W.T., Flannery, B.P. (1992). *Numerical Recipes* (Cambridge University Press, Cambridge).

Provansal, M., Mathis, C., Boyer, L. (1987). *J. Fluid Mech.* **182**, 1–22.

Ramazanov, M.P. (1985). Development of finite-amplitude disturbances in Poiseuille flow, in *Laminar-Turbulent Transition*, V.V. Kozlov, ed. (Springer-Verlag, Berlin).

Rayleigh, Lord (1880). *Proc. London Math. Soc.* **11**, 57–70.

Reddy, S.C., Schmid, P.J., Henningson, D.S. (1993). *SIAM J. Appl. Math.*, **53**, 1, 15–47.

Reed, H.L., Saric, W.S., Arnal, D. (1996). *Ann. Rev. Fluid Mech.* **28**, 389–428.

Reynolds, O. (1883). *Phil. Trans. R. Soc. Lond.* **174**, 935–982.

Robinson, S.K. (1991). *Ann. Rev. Fluid Mech.* **23**, 601–39 .

Rogers, M.M., Moser, R.D. (1992). *J. Fluid Mech.* **243**, 183–226.

Roshko, A. (1961). *J. Fluid Mech.* **10**, 345–356.

Saffman, P.G. (1983). *Ann. N.Y. Acad. Sci.* **404**, 12–24.

Saffman, P.G. (1992). *Vorticity Dynamics* (Cambridge University Press, Cambridge).

Saric, W.S., Kozlov, V.V., Levchenko Y.Ya. (1984). *AIAA paper* **84–0007**.

Sato, H. (1960). *J. Fluid Mech.* **7**, 53–80.

Schär, C., Smith, R.B. (1993). *J. Atm. Sci.* **50**, 1401–1412.

Schlichting, H. (1933). *Nachr. Ges. Wiss. Göttingen, Math. Phys. Kl.*, 181–208.

Schlichting, H. (1979). *Boundary-Layer Theory*, 7th ed. (McGraw-Hill, New York).

Schoppa, W., Hussain, F., Metcalfe, R.W. (1995). *J. Fluid Mech.* **298**, 23–80.

Schubauer, G.B., Skramstad, H.K. (1947). *J. Res. Nat. Bur. Stand.* **38**, 251–292.

Squire, H.B. (1933). *Proc. R. Soc. London* **A142**, 621–628.

Stewartson, K. (1981). *IMA J. Appl. Math.* **27**, 133–175.

Stewartson, K., Stuart, J.T. (1971). *J. Fluid Mech.* **48**, 529–545.

Strykowski, P.J., Niccum, D.L. (1991). *J. Fluid Mech.* **227**, 309–343.

Stuart, J.T. (1960). *J. Fluid Mech.* **9**, 353–370.

Stuart, J.T. (1963). Hydrodynamic Stability, in *Laminar Boundary Layers*, L. Rosenhead, ed. (Clarendon Press, Oxford).

Stuart, J.T. (1967). *J. Fluid Mech.* **29**, 417–440.

Swinney, H.L., Gollub, J.P., eds. (1981). *Hydrodynamic Instabilities and the Transition to Turbulence* (Springer-Verlag, Berlin).

Tatsumi, T., Gotoh, K. (1960). *J. Fluid Mech.* **7**, 433–441.

Tatsumi, T., Gotoh, K., Ayukawa, K. (1964). *J. Phys. Soc. Japan* **19**, 1966–1980.

Taylor, G.I. (1923). *Phil. Trans. R. Soc. Lond.* **A223**, 289–343.

Tennekes, H., Lumley, J.L. (1972). *A First Course in Turbulence* (MIT Press, Cambridge Mass.).

Thomas, L.H. (1953). *Phys. Rev.* **91**, 780–783.

Thomas, A.W.S., Saric W.S. (1981). *Bull. Am. Phys. Soc.* **26**, 1252.

Thorpe, S.A. (1971). *J. Fluid Mech.* **46**, 299–320.

Tollmien, W. (1929). *Nachr. Ges. Wiss. Göttingen, Math. Phys. Kl.*, 21–44.

Tollmien, W. (1947). *Z. Angew. Math. Mech.* **25/27**, 33–50/70–83.

Triantafyllou, G.S., Triantafyllou, M.S., Chrysostomidis, C. (1986). *J. Fluid Mech.* **170**, 461–477.

Tritton, D.J. (1988). *Physical Fluid Dynamics*, 2nd edition (Clarendon Press, Oxford).

Van Dyke, M. (1982). *An Album of Fluid Motion* (The Parabolic Press, Stanford).

Waleffe, F. (1990). *Phys. Fluids* **A2**, 76–80.

Watson, J.T.(1960). *J. Fluid Mech.* **9**, 371–389.

Whitham, G.B. (1974). *Linear and Nonlinear Waves* (Wiley, New York).

Williamson, C. H. K. (1988). *Phys. Fluids* **31**, 2742.

Williamson, C. H. K. (1989). *J. Fluid Mech.* **206**, 579–627.

Williamson, C. H. K. (1994). Vortex dynamics in the wake of a cylinder, in *Fluid Vortices*, S. Green, ed. (Kluwer Academic, London).

Williamson, C. H. K. (1996). *Ann. Rev. Fluid Mech.* **28**, 477–539.

Winant, C.D., Browand, P.K. (1974). *J. Fluid Mech.* **63**, 237–255.

Wygnanski, I.J., Champagne, F.H. (1973). *J. Fluid Mech.* **59**, 281–336.

Yang, X., Zebib, A. (1989). *Phys. Fluids* **A1**, 689–696.

Zahn, J.P., Toomre, J., Spiegel, E.A., Gough, D.O. (1974). *J. Fluid Mech.* **64**, 319–345.

Zebib, A. (1987). *J. Eng. Math.* **21**, 155–165.

3

Asymptotic techniques in nonlinear problems: some illustrative examples

Vincent Hakim

Introduction

Nonlinear dynamics is a rich subject, still only partly understood. Progress is being made in different ways through experiments, computer simulations and theoretical analysis. In this latter category, perturbation theory and asymptotic methods are classical but useful techniques. The aim is to introduce the reader to the nuts and bolts of their use. These notes correspond to an introductory course given at the Beg-Rohu Summer School in the Spring of 1991. It was for graduate students and young post-docs who were interested in theoretical computations in hydrodynamics. The emphasis is put on practical applications rather than on theory. We present, as completely as possible, several calculations on some examples that we find interesting both from a physical and mathematical standpoint. For the expert reader, we should apologize because many subtleties are left aside in these lectures. However, more thorough treatments are referred to as we proceed. Finally, this is an appropriate place to acknowledge my debt to Yves Pomeau from whom I have learned almost everything that I know in this field. I would also like to thank Claude Godrèche for giving me the opportunity to present these lectures in Beg-Rohu and even for convincing me of writing them up, all the "students" for their interest, for catamaran sailing and for sharing unexpected salted baths and crêpes which made teaching in Beg-Rohu an enjoyable experience. Finally, I am grateful to Angélique Manchon for patiently typing these notes, and to Paul Manneville for carefully editing them.

1 Boundary layers and matched asymptotic expansions

In this section, we begin our exploration of the oddities that can appear when one tries to understand a phenomenon using perturbation theory. Therefore, we assume that there is a small parameter ϵ in the problem under consideration. Then, one can try to find its solution f as a formal power series in ϵ

$$f = f_0 + \epsilon f_1 + \ldots + \epsilon^n f_n + \ldots \tag{1.1}$$

In the simplest cases, it is possible to find $f_0, \ldots f_n$, which satisfy the boundary conditions. One is usually happy to compute the first few terms of the series and assume that they are sufficient to understand the small ϵ behavior of the actual solution (see, however, section 5, for examples where this does not work). Another possibility is that one cannot satisfy the required conditions for the actual solution using the perturbation series. This very often appears as the impossibility of finding f_0 with the right boundary conditions at a limiting point. Typically, this happens when the small parameter stands in front of the highest derivative. Setting it to zero reduces the order of the equation and there is no longer enough freedom to satisfy all the boundary conditions.

The main idea behind the analysis of this case is that rapid variation of a function in the neighborhood of some point can make derivatives much larger than the function itself and that this can compensate for the explicit appearance of the small parameter in front of the high derivative. The strategy is to write a new perturbation expansion in the zone where the function varies rapidly. The two developments are then matched in order to find a global approximation of the actual solution.

The historical example of this type of matched asymptotic expansions comes, of course, from Prandtl's description of solutions of Navier–Stokes equation at high Reynolds number (Prandtl, 1905). In this case, the small parameter is $1/\mathrm{Re}$. When it is set to zero the Navier–Stokes equation reduces to the Euler equation and it usually becomes impossible to have a zero tangential velocity of the fluid on solid boundaries (no-slip boundary conditions) while maintaining desired characteristics of the flow. Prandtl reasoned that there is a boundary layer where the tangential velocity varies rapidly from a non zero to a zero tangential velocity and he obtained simplified equations that describe the flow in the boundary layer to lowest order.

In this section, we first describe the appearance of a boundary layer in the very simple setting of a linear equation which illustrates clearly the basic mechanism. Then we describe an interesting application of boundary layer analysis to coating films due to Landau and Levich. Of course many other examples exist and many subtleties have been left aside in this short presentation. The interested reader who wants to get a more thorough view of this subject should consult the general references (Bender and Orszag, 1978; Lagerström, 1988; Nayfeh, 1973; Van Dyke, 1975). Eckhaus (1979) attempts to put the method on rigorous foundations.

1.1 An elementary example of a boundary layer; inner and outer expansions

In order to demonstrate simply the existence of boundary layers we consider the elementary case of a linear equation (Bender and Orszag, 1978):

$$\epsilon \frac{d^2 X}{dt^2} + t\frac{dX}{dt} - tX = 0, \qquad \epsilon \ll 1. \tag{1.2}$$

We consider the motion that starts from the origin at $t = 0$ and reaches $X = e$ at $t = 1$, i.e., with the boundary conditions:

$$X(0) = 0, \qquad X(1) = e. \tag{1.3}$$

Since ϵ is a small parameter, we can hope to analyze this problem using perturbation theory. Let us first proceed straightforwardly. We search for X as a perturbation series

$$X(t) = X_0(t) + \epsilon X_1(t) + \cdots \tag{1.4}$$

The first approximation verifies:

$$\frac{dX_0}{dt} - X_0 = 0. \tag{1.5}$$

The general solution of this equation is:

$$X_0(t) = A\,e^t. \tag{1.6}$$

Therefore it appears impossible to impose both boundary conditions (1.3). The source of the difficulty is that near $t = 0$ the neglected term $(\epsilon d^2 X_0/dt^2)$ in the equation is not negligible as compared to the two terms kept which vanish at zero. This will cause a significant departure of the actual solution from $X_0(t)$ in the neighborhood of $t = 0$. Elsewhere, the series (1.4), the so-called outer expansion in this context, is a fair approximation of the actual solution. Imposing $X(1) = e$ in equation (1.5) we obtain its first term:

$$X_{\text{out},0} = e^t. \tag{1.7}$$

Let us analyze more precisely under which conditions the term $\epsilon d^2 X/dt^2$ is comparable to the other terms in the equation. A convenient way to do that is to remember that the derivative is given by the variation of the function divided by the variation of its argument. Moreover, near the origin, since $X(0) = 0$, the variation of the function is the function itself. So the comparison of the first two terms $\epsilon d^2 X/dt^2$ and $t\,dX/dt$ in equation (1.2) can be written:

$$\epsilon \frac{X}{t^2} \sim t\frac{X}{t}, \qquad \text{i.e.,} \qquad t \sim \sqrt{\epsilon}. \tag{1.8}$$

Therefore these two terms are comparable in magnitude if the t-scale of variation of the function X is $\sqrt{\epsilon}$ in the neighborhood of $t = 0$. If this is the case, then the last term $(-tX)$ in equation (1.2) is of order $\sqrt{\epsilon} X$ for t of order $\sqrt{\epsilon}$ and is negligible with respect to the first two.

The comparison of the first and last terms in equation (1.2) gives:

$$\epsilon \frac{X}{t^2} \sim tX , \qquad \text{i.e.,} \qquad t \sim \epsilon^{1/3} . \tag{1.9}$$

These two terms are comparable if the t-scale of variation of the function X is $\epsilon^{1/3}$ in the neighborhood of $t = 0$. If this is the case, then the second term is of order X much larger than the two terms retained for the comparison of order $\epsilon^{1/3} X$. Clearly only the first possibility is consistent. So, in order to analyze the behavior of X in this ϵ-dependent neighborhood of $t = 0$, the so-called boundary layer, we define $t = \sqrt{\epsilon}\, \tau$ and we rewrite (1.2) as

$$\frac{d^2 X}{d\tau^2} + \tau \frac{dX}{d\tau} - \sqrt{\epsilon}\, \tau X = 0 . \tag{1.10}$$

Therefore in the boundary layer we can perform a different perturbation expansion, the so-called inner expansion:

$$X(\tau) = X_{\text{in},0}(\tau) + \ldots \tag{1.11}$$

with

$$\frac{d^2 X_{\text{in},0}}{d\tau^2} + \tau \frac{dX_{\text{in},0}}{d\tau} = 0 , \qquad X_{\text{in},0}(0) = 0 , \tag{1.12}$$

The solution of this inner equation is readily obtained:

$$X_{\text{in},0} = B \int_0^\tau \exp(-u^2/2) \, du . \tag{1.13}$$

In order to get a uniform approximation, to the actual solution X of (1.2) and (1.3), it remains to match the inner and outer expansions. This is done by requiring that the asymptotic behavior of the inner solution around $\tau = \infty$ coincides with the behavior near zero of the outer expansion. Here, to lowest order, this reduces to equating their limiting values:

$$\lim_{t \to 0} X_{\text{out},0} = \lim_{\tau \to \infty} X_{\text{in},0} . \tag{1.14}$$

This gives $B = \sqrt{2/\pi}$.

Finally a uniform approximation of the solution is obtained by adding the inner and outer solutions and subtracting their common limiting behavior which would be counted twice otherwise. In our simple example, this gives:

$$X_{\text{uniform},0} = e^t + \sqrt{\frac{2}{\pi}} \int_0^{t/\sqrt{\epsilon}} \exp(-u^2/2)\, du - 1. \tag{1.15}$$

The different functions are plotted and compared in Fig. 1.1.

It is worth summarizing the general strategy that we have applied in this simple example.

i) We have first written a straightforward perturbation series. This has given a fair approximation of the function except in the neighborhood of t = 0. For this reason, this is called an outer expansion. In particular one has:

$$\lim_{\epsilon \to 0} X(t, \epsilon) = X_{\text{out},0}(t) \qquad \text{for} \quad t \neq 0. \tag{1.16}$$

ii) Near t = 0 we have determined a zone of rapid ϵ-dependent variation of the function: the inner region or boundary layer. In this region a different inner expansion has been performed.

iii) Finally, in order to get a uniform approximation to the function sought, the inner and outer expansions have been matched in a common region of validity, fixing the arbitrariness in one development by using information coming from the other one.

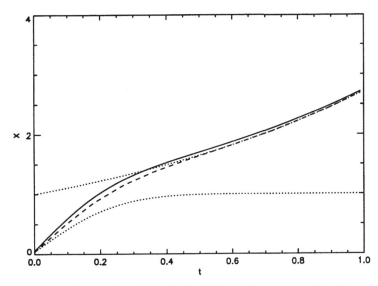

Fig. 1.1. The exact solution of (1.2) (solid line) is compared to the outer and inner solutions, (1.7) and (1.13) respectively (dotted lines), and to the uniform approximation (1.15) (dashed line) for $\epsilon = 0.04$.

In the following subsection, we describe a much more interesting physical example of this general strategy. Before doing that we would like to make a last general remark. In the above example, we have performed both developments to their lowest order. It is, of course, possible to push each expansion to a higher order and then to apply the same matching strategy in order to get a better uniform approximation. The interested reader should consult the general references cited at the beginning of this section for more information on higher order expansions. However, one should be aware that higher order matched asymptotic expansions tend to get complicated and new powers (or even logarithms) of the small parameter can appear. An example (Bender and Orszag, 1978) is given by the period $T(\epsilon)$ of the limit cycle of the Van der Pol equation:

$$\epsilon \frac{d^2 X}{dt^2} - (1 - X^2)\frac{dX}{dt} + X = 0. \tag{1.17}$$

The development of $T(\epsilon)$ as ϵ tends to zero has been analyzed by Haag (1944) and Dorodnitsyn (1947). Its first terms are:

$$T(\epsilon) = 3 - 2\ln(2) - 3\alpha\epsilon^{2/3} + \frac{1}{6}\epsilon \ln(\epsilon) + \ldots \tag{1.18}$$

where α is the first zero of the Airy function Ai, $\alpha = -2.338\ldots$.

1.2 Landau and Levich coating flow problem

Let us suppose that one takes out a solid plate from a liquid bath. What is the thickness of the wetting film on the solid substrate as a function of the pulling velocity? This question is of some importance in the coating and photographic film industries (for a review see Ruschak, 1985). An answer was found by Landau and Levich (1942) in the low velocity limit (unfortunately not the most relevant one for industrial purposes). It is a beautiful example of the use of matched asymptotic expansions that has been adapted to similar situations and with various degrees of formalization by different authors (Bretherton, 1961; Park and Homsy, 1984; Dombre and Hakim, 1987).

1.2.1 Physical setting and dimensional arguments

The model problem that we considered is sketched in Fig. 1.2. A flat plate is pulled vertically out of a liquid bath at a velocity V_0. The liquid is supposed to wet the solid plate perfectly (for a generalization to partial wetting, see de Gennes, 1986). A liquid film of width w_0 coats the plate. Before turning to its precise dependence on the pulling velocity, let us make some general dimensional considerations.

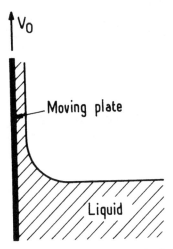

Fig. 1.2. A solid plate is drawn vertically out of a liquid bath. It is coated by a liquid film of width w_0.

There are five dimensional parameters that enter into this physical problem. These are the pulling velocity V_0, the acceleration due to gravity g and three parameters characterizing the fluid: its density ρ, its viscosity μ and the surface tension γ between the liquid and the air. Their physical dimensions are:

$$[V_0] = \frac{L}{T}, \quad [g] = \frac{L}{T^2}, \quad [\rho] = \frac{M}{L^3}, \quad [\mu] = \frac{M}{LT}, \quad [\gamma] = \frac{M}{T^2}. \quad (1.19)$$

Since only three fundamental quantities M, L, T enter in (1.19), out of these five dimensional parameters only three are independent and two dimensionless parameters can be formed. Lengths are conveniently made dimensionless by introducing the capillary length $a = (\gamma/\rho g)^{1/2}$. The two dimensionless numbers are conventionally chosen to be:

1) the capillary number $\mathrm{Ca} = \mu V_0/\gamma$ which compares the strength of viscous forces to the capillary forces,
2) the Reynolds number $\mathrm{Re} = \rho V_0 a/\mu$ which compares the inertial forces to the viscous forces.

From general arguments (see, e.g., Barenblatt, 1979), the dimensionless wetting film width w_0/a can only depend on the two dimensionless parameters:

$$w_0 = a\, f(\mathrm{Ca}, \mathrm{Re})\,. \quad (1.20)$$

Landau and Levich have analyzed the limit of small Reynolds and capillary numbers by performing a double expansion in powers of Re and

Ca. While the Reynolds number can be set to zero to lowest order, this cannot be done for the capillary number since f tends to zero with Ca. The aim of the analysis is therefore to determine the behavior of $f(\text{Ca}, 0)$ (1.20) as Ca tends to zero.

In this limit, the free surface $z(x)$ of the liquid can be separated into two regions (see Fig. 1.3):

1) In region I the liquid interface is well approximated by the static meniscus and it is intuitively clear that it tends towards it as the pulling velocity tends to zero.
2) On the contrary in region II the liquid interface is very different from the static meniscus. It is described by the behavior of $z(x)$ in the neighborhood of $x = 0$. This region shrinks to zero as the pulling velocity tends to zero.

One clearly recognizes in this description the occurrence of a boundary layer near $x = 0$, region I being the outer one and region II the inner one.

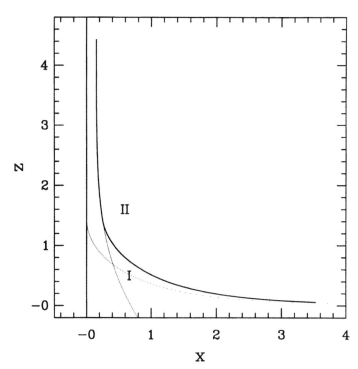

Fig. 1.3. Different regions of the boundary layer analysis. The static meniscus and the inner solution are shown as dotted lines. The uniform approximation is the solid line.

Before writing detailed expansions it is worth summarizing the dominant physical mechanism in both regions.

In region I gravity is mainly competing with capillarity in an almost hydrostatic situation. The hydrostatic variation of pressure between the free surface of the liquid and the top of region I is simply $\rho g a$ where a is the characteristic size of the static meniscus. It can also be estimated using Laplace law to be γ/a (the pressure variation in the air is negligible). Equating the two expressions simply gives that the capillary length is the characteristic size of the static meniscus.

In region II capillarity is competing with viscous forces. At the top of region II, the liquid layer has reached a constant width w_0 and a uniform velocity V_0 equal to the velocity of the moving plate. On the contrary in the lower part of region II, the velocity of the fluid is much smaller and the width of the layer is larger. This is the result of the pressure difference between the top and bottom of region II. Let us denote by ℓ the characteristic length over which the pressure is changing in region II. Then the pressure drop is related to the magnitude of the velocity by (as in Poiseuille flow):

$$\frac{\Delta P}{\ell} \sim \frac{\mu V_0}{w_0^2} . \tag{1.21}$$

The pressure variation between B and C is of capillary origin and related to the curvature change of the interface. This can be estimated as w_0/ℓ^2 and therefore the corresponding change of pressure is:

$$\Delta P \sim \gamma \frac{w_0}{\ell^2} . \tag{1.22}$$

Comparing equations (1.21) and (1.22), one obtains the relation between the characteristic length of variation in the vertical direction in region II and the width of the coating:

$$\ell \sim w_0 \, \mathrm{Ca}^{-1/3} . \tag{1.23}$$

The analysis is then concluded by matching regions I and II, that is by requiring that the values of physical quantities in the upper part of region I or in the lower part of region II are identical. For example, doing this for the curvature of the interface gives:

$$\frac{1}{a} = \frac{w_0}{\ell^2} . \tag{1.24}$$

Taking into account relation (1.23) between ℓ and w_0, one finds the dependence of the coating-film width on pulling velocity:

$$w_0 \sim a \, \mathrm{Ca}^{2/3} . \tag{1.25}$$

Now that we have outlined the argument, we describe it in more detail. The outcome of this more detailed analysis simply replaces the order of magnitude relation (1.25) by the precise estimate:

$$w_0 = 0.94 \, a \, \mathrm{Ca}^{2/3} + \mathcal{O}(\mathrm{Ca}) . \tag{1.26}$$

1.2.2 Outer problem: the static meniscus

To lowest order in region I, we can entirely neglect the motion of the solid plate. In other words the problem has a non trivial limit in this region for $\mathrm{Ca} = 0$. It reduces simply to the determination of the shape of a meniscus that wets perfectly a solid wall. Taking the origin of height and pressure at the horizontal liquid surface, the pressure at height z is $-\rho g z$ from elementary hydrostatics. On the other hand, from Laplace law it should be equal to γ/R, where R is surface radius of curvature at height z and is negative for a liquid on the concave side of the surface as in Fig. 1.3. Equating the two expressions gives the equation of the static meniscus shape:

$$-\gamma \frac{d^2 z/dx^2}{(1 + (dz/dx)^2)^{3/2}} = -\rho g z . \tag{1.27}$$

This can be integrated once after multiplication by dz/dx on both sides

$$\frac{1}{(1 + (dz/dx)^2)^{1/2}} - 1 = -\frac{z^2}{2a^2} . \tag{1.28}$$

As will be seen below, matching with the inner equation will only be possible if the outer equation has a vertical slope at $x = 0$, i.e., if the static meniscus joins tangentially the solid plate. This gives, for the meniscus height, from (1.28)

$$z_0 = \sqrt{2} \, a \tag{1.29}$$

and for the curvature of the liquid surface at this point:

$$\frac{1}{|R|} = \frac{\sqrt{2}}{a} = \frac{d^2 x_{\mathrm{out}}}{dz^2} . \tag{1.30}$$

1.2.3 The inner region

In this region (region II of Fig. 1.3) a different expansion is performed. To lowest order, we show that the Navier–Stokes equation can be treated within the lubrication approximation which is much easier to handle.

  First let us recall that we are considering the lowest order in Reynolds number. Therefore the Navier–Stokes equation reduces to the Stokes equation:

$$\nabla^2 \mathbf{v} = \frac{1}{\mu}\nabla P - \frac{g\rho}{\mu}. \tag{1.31}$$

The second simplification comes from the fact that, in the inner region, physical quantities vary on a much larger scale in the z direction than in the x direction, as explained in the previous paragraph, see (1.23). Let us denote by ϵ the ratio of these two scales. Our aim is to obtain systematic expansions in powers of ϵ. It is convenient to introduce dimensionless quantities of order unity. The previous argument suggests to take $\epsilon = \mathrm{Ca}^{1/3}$ and to introduce the dimensionless lengths X and Z and the dimensionless pressure Π and velocity $\mathbf{V}$

$$x = a\epsilon^2 X, \quad z = a\epsilon Z, \quad P = \frac{\gamma}{a}\Pi, \quad \mathbf{v} = \frac{\gamma}{\mu}\epsilon^3 \mathbf{V} = V_0 \mathbf{V}. \tag{1.32}$$

We can now expand everything in powers of ϵ. For example, the incompressibility equation becomes:

$$\frac{\partial V_x}{\partial X} + \epsilon\frac{\partial V_z}{\partial Z} = 0. \tag{1.33}$$

This shows that V_x is smaller by a power of ϵ than V_z in the inner region. Then, the x-component of (1.31) shows that, to lowest order, the pressure is independent of the X-coordinate:

$$\frac{\partial \Pi^{(0)}}{\partial X} = 0, \quad \text{i.e.,} \quad \Pi(X, Z) = \Pi^{(0)}(Z) + \epsilon\Pi^{(1)}(X, Z) + \dots \tag{1.34}$$

Therefore, the z-component of (1.31) is simply, to lowest order:

$$\frac{\partial^2 V_z}{\partial X^2} = \frac{d\Pi^{(0)}}{dZ}. \tag{1.35}$$

Let us remark here that, to this order, the draining of the coating film by gravity is neglected. The boundary conditions on v_z are: a) it should be equal to the velocity of the moving plate at $x = 0$, and b) the stress parallel to the surface should vanish at the free surface of the liquid layer,

$$V_z^{(0)} = 1 \quad \text{at} \quad X = 0, \qquad \frac{\partial V_z}{\partial X} = 0 \quad \text{at} \quad X = W(Z). \tag{1.36}$$

Equation (1.35) and the boundary conditions (1.36) determine $V_z^{(0)}$ as a function of the pressure field and the layer width $W(Z)$

$$V_z^{(0)} = \frac{1}{2}\frac{d\Pi^{(0)}}{dZ}X(X - 2W(Z)) + 1. \tag{1.37}$$

The pressure itself is given by Laplace Law as a function of the curvature of the free surface. To lowest order, this gives:

$$\Pi^{(0)} = -\frac{d^2 W}{dZ^2}.$$
(1.38)

Therefore all quantities in the inner region are determined as a function of the surface shape. In order to obtain it and conclude the computation to lowest order, the simplest way is to write the conservation of liquid flux. The liquid flux at height Z is obtained by integrating (1.37) from $X = 0$ to $X = W(Z)$. It should be equal to the flux at $Z \gg 1$ where the liquid layer has a constant width and a uniform velocity equal to the velocity of the moving plate. Thus, one obtains:

$$\frac{W^3}{3}\frac{d^3 W}{dZ^3} + W = W_0.$$
(1.39)

This is the equation of the surface shape in the internal region. Defining $W = \omega W_0$ and $Z = 3^{-1/3} W_0 \xi$, it takes the classical form:

$$\omega^3 \frac{d^3 \omega}{d\xi^3} = 1 - \omega.$$
(1.40)

1.2.4   Matching between the inner and outer regions

In order to obtain a complete description of physical quantities it remains to match the outer and inner expansions. As explained in section 1.1, this is done by requiring that the asymptotic behavior in the inner region coincides with the limiting behavior in the outer region as $X \to 0$.

   Let us first see that the inner equation (1.40) determines an essentially unique shape. As ξ tends to $+\infty$ the width $W(Z)$ of the liquid layer should tend toward the constant width W_0. This means that $\omega(x)$ should tend toward 1 as $\xi \to +\infty$. The approach toward this limiting value can be analyzed by linearizing equation (1.40):

$$\frac{d^3 \eta}{d\xi^3} = -\eta \qquad \text{with} \qquad \omega = 1 + \eta, \qquad \eta \ll 1.$$
(1.41)

The general solution of this linear equation is a linear combination of three eigenfunctions:

$$\eta = \sum_{j=1}^{} 3A_j\, e^{\lambda_j \xi} \qquad \text{with} \qquad \lambda_1 = -1,\ \lambda_2 = e^{i\pi/3},\ \lambda_3 = e^{-i\pi/3}.$$

   Requiring that ω tends toward 1 as ξ tends toward $+\infty$ imposes that the coefficients A_2 and A_3 of the diverging modes vanish. This imposes two conditions on the third order equation (1.40). The only remaining degree of freedom, which is reflected in the arbitrariness of the coefficient A_1, clearly reduces to translate the solution of (1.40). Thus (1.40) essentially defines a unique shape, once its translation invariance is taken care of. As ξ tends toward $-\infty$, ω becomes large and its asymptotic behavior is:

$$\omega = \frac{A}{2}\xi^2 + B\xi + C + \frac{2}{3A^2\xi} + \cdots \tag{1.42}$$

The constant A is independent of the choice of origin on the ξ axis. It is found to be approximately 0.643 by a numerical integration of (1.40). B can be set to zero by an appropriate choice of origin and the value of C can then be determined ($C \simeq 2.8$).

   The inner and outer region can be matched smoothly if physical quantities have the same value when computed in the $\omega \to \infty$ asymptotics of the inner region or the $x \to 0$ limiting behavior of the outer region. For example the limiting curvature of the outer free surface is given by (1.30). When expressed with the inner coordinates $z = 3^{1/3}\epsilon aW_0\xi$, $x = a\epsilon^2 W_0\omega$, this is

$$\left.\frac{d^2\omega}{d\xi^2}\right|_{\text{out}} = 3^{-2/3}\sqrt{2}\,W_0. \tag{1.43}$$

This should be equal to A which is the inner estimate of the same quantity. Therefore W_0 is determined to be:

$$W_0 = A\,\frac{3^{2/3}}{\sqrt{2}} = 0.946. \tag{1.44}$$

Finally, this gives the expression (1.26) for the width of the wetting film.

1.2.5   Experimental measurement of the wetting film

It is interesting to compare the previous analytical estimate of the coating film to experimental determinations. Experimentally, it seems easier to measure coating film on fibers than on plates. For a fiber of radius large compared to the width of the coating film, the analytic computation is very similar to the previous one (Deryaguin, 1943). The only difference is that the fiber radius b is small compared to the capillary length ($a \sim$ 0.3 cm). Hydrostatic effects can be neglected in the outer region and Laplace law then implies that the two radii of curvature of the surface are equal and opposite. The inner equation should thus be matched with an outer curvature equal to $1/b$ instead of $\sqrt{2}/a$. The resulting predicted coating width is:

$$W_0 = 1.34\,b\,\text{Ca}^{2/3}. \tag{1.45}$$

The principle of an experiment by de Ryck and Quéré (1996) which tests this relationship is sketched in Fig. 1.4. A thin tungsten or nickel wire ($b = 63.5\ \mu\text{m}$) is drawn at constant velocity through a drop of silicon oil.

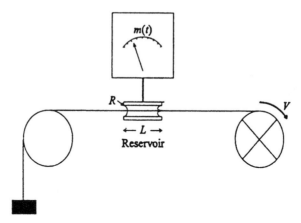

Fig. 1.4.   Sketch of the experiment by de Ryck and Quéré (1996).

The width of the wetting film on the wire is inferred from the decrease of the drop weight with time.

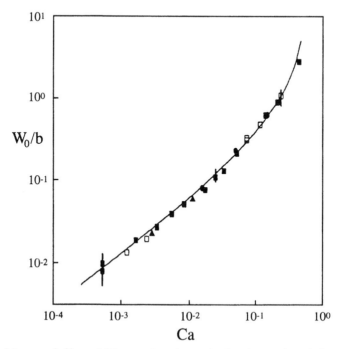

Fig. 1.5.   Measured film width *vs.* drawing velocity (reproduced from de Ryck and Quéré (1996). Results obtained with different silicon oils (symbols) are compared to eq. (1.46) (continuous line).

Representative results are shown in Fig. 1.5 and are compared with a phenomenological refinement of (1.45) due to White and Tallmadge (1966)

$$W_0 = \frac{1.34b\,\mathrm{Ca}^{2/3}}{1 - 1.34\,\mathrm{Ca}^{2/3}}. \tag{1.46}$$

The different causes (gravity, inertia,...) of departure from the asymptotic law (1.45) for large velocities are discussed by de Ryck and Quéré (1996). The asymptotic law (1.45) also breaks down for very small capillary numbers ($\mathrm{Ca} < 10^{-6}$). In this regime, the width of the wetting film saturates to a constant value as reported in an earlier experiment by Quéré *et al.* (1989). This is attributed to Van der Waals forces which become dominant for very small film thickness (de Gennes, 1986).

2 Multiscale analysis and envelope equations

In this section, we continue the exploration of perturbation expansions. We want to illustrate the phenomenon of resonance between different orders of perturbation theory and the simultaneous appearance of secular terms. These terms usually grow with time or space. As a consequence straightforward perturbation expansions break down for large values of the time or space variable in the sense that higher order terms then become larger than lower order terms. This pathology was first analyzed in the context of celestial mechanics and was found to derive from the variation of a planet's basic orbit characteristics on a time scale that is long as compared to the orbiting period. Many methods have been invented to deal with these phenomena, like the time-stretching of Lindstedt and Poincaré and averaging which goes back at least to Gauss and has been strongly advocated by Bogoliubov, Krylov and Mitropolsky (Bogoliubov and Mitropolsky, 1961) (for a review see Sanders and Verhulst, 1985). The multiscale method is a neat technique that seems to have been first introduced by Cole around 1960. It consists in postulating a priori that the functions vary on two (or more) time or spatial scales. Suppressing secular terms fixes the ensuing arbitrariness. The method is very convenient to analyze weakly nonlinear phenomena and, in particular, to determine if nonlinearities saturate or reinforce a weak instability. We first explain the method with an elementary example, the computation of the dependence of pendulum oscillation frequency on its oscillation amplitude for weak amplitude.

One of our main motivations for introducing the multiscale method is that it is widely used for obtaining amplitude equations which describe the slow dynamics of the large scale modulation of a basic predetermined structure (Segel, 1969; Newell and Whitehead, 1969). This is illustrated in subsection 2.2 by deriving the well-known result that the nonlinear Schrödinger equation is the envelope equation for a periodic water wave of small amplitude on deep water (Hasimoto and Ono, 1972). The computation is slightly more cumbersome than for model equations but we hope that this is more than compensated by the physical interest of the result. Subsection 2.3 is devoted to a more general discussion of amplitude equations. The reader will find more applications of the use of amplitude equations in Chapter 4.

2.1 The period of the pendulum by the multiscale method

We consider a simple pendulum

$$\frac{d^2x}{dt^2} + \sin(x) = 0 \,. \tag{2.1}$$

This problem is of course completely integrable but, here, we want to illustrate the phenomenon of resonance between different orders of perturbation theory and the consequent nonuniformity of perturbation theory at large time. We therefore suppose that the oscillation amplitude is small. Formally, we write $x = \sqrt{\epsilon}A(t)$ and develop $A(t)$ as a power series in ϵ

$$A(t) = A_0(t) + \epsilon A_1(t) + \dots \tag{2.2}$$

At first order, we obtain harmonic oscillations

$$\frac{d^2A_0}{dt^2} + A_0 = 0 \,, \qquad \text{i.e.,} \qquad A_0 = a_0 \cos(t + \phi) \,. \tag{2.3}$$

At next order, we get

$$\frac{d^2A_1}{dt^2} + A_1 = \frac{1}{6}A_0^3 = \frac{a_0^3}{24}\left[3\cos(t+\phi) + \cos\left(3(t+\phi)\right)\right] \,. \tag{2.4}$$

The two terms on the r.h.s. give two contributions of different types in A_1. The second one gives $(-a_0^3/196)\cos\left(3(t+\phi)\right)$ which is bounded at all times. On the contrary, the first one gives $(a_0^3/16)\,t\sin(t)$ which grows linearly in time. Therefore for time of order $1/\epsilon$, A_1 becomes comparable to A_0 and the perturbative treatment loses its coherence. The formal growth of the orbit amplitude with time is obviously a spurious effect here since it is clear that energy is conserved by equation (2.1). This spurious effect appears because the forcing by products of lower order terms of the perturbation series (first term on the r.h.s. of (2.4)) is resonant with the free oscillations described by the linear operator on the l.h.s. of (2.4). In order to turn this pathology into a benefit, it is helpful to remark that any function that varies on a time scale of $1/\epsilon$ will produce growing terms since we have expanded everything in powers of ϵ

$$f(\epsilon t) = f(0) + \epsilon t f'(0) + \dots \tag{2.5}$$

The idea of the multiscale method is to postulate a priori that there are two (or more) different scales in the problem and therefore, that some functions depend on t only through the product ϵt. In order to avoid expanding slow variations as in (2.5) one introduces explicitly the slow variable:

$$T = \epsilon t. \tag{2.6}$$

Then slow variations can be separated from fast ones by making the substitution:

$$\frac{d}{dt} = \frac{\partial}{\partial t} + \epsilon \frac{\partial}{\partial T} \tag{2.7}$$

as if t and T were independent variables. Let us see how this works in our elementary example. Instead of (2.1) one has:

$$\left(\frac{\partial^2}{\partial t^2} + 2\epsilon \frac{\partial^2}{\partial t \partial T} + \epsilon^2 \frac{\partial^2}{\partial T^2} \right) A + \frac{1}{\sqrt{\epsilon}} \sin\left(\sqrt{\epsilon}\, A \right) = 0 \tag{2.8}$$

and the perturbation for A is assumed to be of the form:

$$A(t) = A_0(t, T) + \epsilon A_1(t, T) + \dots \tag{2.9}$$

Substituting (2.9) into (2.8) one gets back (2.3) at first order except that total derivatives are replaced by partial t-derivatives. The solution is the same as before, see (2.3), but now a_0 and ϕ are functions of the slow variable T instead of being constants. At next order, one obtains:

$$\frac{\partial^2 A_1}{\partial t^2} + A_1 = \frac{1}{6} A_0^3(t, T) - 2 \frac{\partial^2}{\partial t \partial T} A_0. \tag{2.10}$$

The last term is the new one. When the r.h.s. is written explicitly one obtains

$$\frac{a_0^3}{24} \left[3\cos(t + \phi) + \cos\left(3(t + \phi) \right) \right] + 2\frac{da_0}{dt} \sin(t + \phi) + 2a_0 \frac{d\phi}{dT} \cos(t + \phi).$$

The new arbitrariness in $a_0(T)$ and $\phi(T)$ is fixed by requiring that secular terms do not appear in the solution to (2.10). One gets:

$$\frac{da_0}{dT} = 0,$$
$$\frac{d\phi}{dT} = -\frac{3}{16} a_0^2. \tag{2.11}$$

We have therefore obtained the dependence of the pendulum period on this amplitude for small amplitude of oscillation,

$$x = \sqrt{\epsilon}a_0 \cos(t + \phi_0\epsilon - \tfrac{3}{16}\epsilon a_0^2 t + \ldots) + \ldots \qquad (2.12)$$

Another interesting elementary example of application of the multiscale method is the computation of the size of the limit cycle of a Van der Pol oscillator:

$$\ddot{x} + \epsilon\left(x^2 - 1\right)\dot{x} + x = 0 \qquad \text{for} \qquad \epsilon \ll 1. \qquad (2.13)$$

The zeroth order solution is

$$x_0 = A\cos(t + \phi) \qquad (2.14)$$

with an arbitrary amplitude A. Introducing a slow time scale as in (2.6), the cancellation of secular terms at first order gives the evolution of the amplitude

$$2\frac{dA}{dT} = A - \frac{A^3}{4}. \qquad (2.15)$$

This shows that the amplitude of the limit cycle is equal to 2 and that (2.15) describes the approach to the limit cycle on a time scale that is long compared to the oscillation frequency. At next order, the change of the period of the limit cycle appears as

$$\omega = 1 - \frac{1}{16}\epsilon^2. \qquad (2.16)$$

This computation is left as an exercise to the reader. It is described in more detail in Chapter 4.

2.2 The nonlinear Schrödinger equation as an envelope equation for small amplitude gravity waves in deep water

2.2.1   A short introduction to water waves

The problem of water waves has played a central role in the development of nonlinear physics, with Stokes's discovery that the dispersion relation involves the amplitude for periodic nonlinear wave trains, Russell's first observation of a solitary wave, the derivation of model equations by Boussinesq, Rayleigh, Korteweg–de Vries... Moreover, the observation of water waves is a common but fascinating experience. A good introduction to the subject can be found in Stoker (1957) and Whitham (1974).

We consider the problem in the simplified setting of a two-dimensional flow of an inviscid incompressible liquid. We assume that the flow is irrotational and introduce the potential ϕ for the fluid velocity $\mathbf{v}$

$$\mathbf{v} = \nabla\phi. \qquad (2.17)$$

The incompressibility of the fluid implies that ϕ satisfies the Laplace equation

$$\text{div } \mathbf{v} = 0, \qquad \nabla^2 \phi = 0. \tag{2.18}$$

The potential ϕ is determined by two boundary conditions. The first one is of dynamic nature. If surface tension is neglected, the pressure in the liquid and the pressure in the air should be equal at the surface. The pressure in the liquid is determined by Bernouilli's equation, as can be seen by substituting (2.17) into Euler's equation. Assuming a constant air pressure (i.e., neglecting air density) one obtains at the surface

$$\left[\frac{\partial \phi}{\partial t} + \frac{1}{2} \left(\nabla \phi \right)^2 + gy \right]_{\text{surface}} = 0 \tag{2.19}$$

where y denotes the vertical coordinate and all space independent constants have been cancelled by a suitable redefinition of the potential. There is also a second boundary condition at the surface of purely kinematic nature. By the very definition of the surface, the normal velocity of the surface should be equal to the normal velocity of the fluid at the surface:

$$\mathbf{N} \cdot \mathbf{v}_{\text{int}} = \mathbf{N} \cdot \nabla \phi. \tag{2.20}$$

In the case of a layer of liquid of finite depth, one gets another kinematic boundary condition from the bottom impermeability. For simplicity, we will only consider the case of waves in deep water and we require simply that the fluid velocity vanishes as $y \to -\infty$.

At this stage, some readers may feel that our system of equations is overdetermined since we have imposed two boundary conditions on the Laplace equation (2.18). This is not the case, because we are considering a free-boundary problem. The surface of the liquid and therefore the shape of the region where one should solve the Laplace equation is not fixed in advance but it is itself determined in the course of solving the equation. A good way to get an intuitive feeling may be to say that (2.18) together with (2.19) determines ϕ at time t as a function of the surface and the field at time $t - \Delta t$. Then equation (2.20) can be solved as an equation of motion for the fluid surface and gives the new position of the surface at time t.

In order to analyze these equations perturbatively, it is convenient to rewrite them in a less compact but more explicit fashion. We denote by $\eta(x, t)$ the vertical position of the water surface at position x and time t. Then the surface normal $\mathbf{N}$ and surface velocity $\mathbf{v}_{\text{int}}$ are:

$$\mathbf{N} \propto \left(-\frac{\partial \eta}{\partial x}, 1 \right), \qquad \mathbf{v}_{\text{int}} = \left(0, \frac{\partial \eta}{\partial t} \right). \tag{2.21}$$

The kinematic boundary condition (2.20) is thus

$$\frac{\partial \eta}{\partial t} - \frac{\partial \phi}{\partial y} = -\frac{\partial \eta}{\partial x} \frac{\partial \phi}{\partial x}\bigg|_{y=\eta(x)} \tag{2.22}$$

where we have explicitly indicated that this relation should be satisfied at the liquid surface. With the same notation, the other boundary condition (2.19) reads:

$$\frac{\partial \phi}{\partial t} + g\,\eta = -\frac{1}{2}\left[\left(\frac{\partial \phi}{\partial x}\right)^2 + \left(\frac{\partial \phi}{\partial y}\right)^2\right]\bigg|_{y=\eta(x)}. \tag{2.23}$$

2.2.2 Weakly nonlinear expansion of periodic waves and secular terms

We study water waves as a perturbation of the simplest case, a liquid at rest with a horizontal surface. This case is described with a suitable choice of vertical coordinates by

$$\phi_0(x,t) = 0, \qquad \eta_0(x,t) = 0. \tag{2.24}$$

For a slight perturbation, ϕ and η will be small and the nonlinear terms in the l.h.s. of equations (2.22) and (2.23) can be neglected at first order. Moreover, the boundary conditions can be evaluated at $y = 0$ instead of $y = \eta(x)$. One should therefore find a Laplacian field ϕ_1 satisfying the two boundary conditions:

$$\frac{\partial \eta_1}{\partial t} - \frac{\partial \phi_1}{\partial y} = 0,$$
$$\frac{\partial \phi_1}{\partial t} + g\,\eta_1 = 0. \tag{2.25}$$

The equations being linear, the motion of an arbitrary surface can be determined from the superposition of the motions of its Fourier components. Therefore, we consider η_1 in the form:

$$\eta_1 = \mathcal{Re}\left\{A_1\,e^{i(kx-\omega t)}\right\}. \tag{2.26}$$

The form of the associated periodic Laplacian field which vanishes as $y \to -\infty$ is:

$$\phi = \mathcal{Re}\left\{F_1\,e^{i(kx-\omega t)}\right\}e^{ky}, \qquad k > 0. \tag{2.27}$$

Substitution of (2.26) and (2.27) in (2.25) gives the homogeneous linear system

$$i\omega A_1 + kF_1 = 0,$$
$$gA_1 - i\omega F_1 = 0. \tag{2.28}$$

A non trivial solution of this homogeneous linear system exists only if its determinant is zero. That is, the frequency is related to the wave number by

$$\omega^2 - kg = 0. \tag{2.29}$$

In this case, the liquid flow is related to the wave shape by

$$F_1 = -i\frac{\omega}{k}A_1. \tag{2.30}$$

We have thus recovered the well-know linear dispersion relation of gravity waves. Now, we want to analyze the influence of the neglected nonlinear terms and therefore consider η_1 and ϕ_1 as first order terms in a systematic expansion of a nonlinear periodic wave, solution of the full equations (2.18), (2.22–23), as Stokes first did in 1847. At this stage it is worth clarifying what the actual expansion parameter is. The acceleration due to gravity g appears explicitly in the equation. If we admit, moreover, that for a nonlinear periodic wavetrain we can impose arbitrarily the wave height a and the wave length $2\pi/k$ then the solution is determined by three dimensional parameters. Out of these, only one dimensionless parameter can be formed, the so-called "wave steepness" ak. Therefore, the weakly nonlinear expansion is really an expansion in powers of the wave steepness. It can be explicitly checked that the wave steepness is the dimensionless parameter which stands in front of the nonlinear terms in (2.22), (2.23) when dimensionless variables are introduced by setting $x = \tilde{x}/k$, $y = \tilde{y}/k$, $t = \tau/\sqrt{kg}$, $\eta = a\tilde{\eta}$ and $\phi = a\sqrt{g/k}\,\tilde{\phi}$. Without finding it necessary to do so, we can now proceed and compute the nonlinear corrections to η_1 and ϕ_1. We formally expand ϕ and η as:

$$\eta = \epsilon\eta_1 + \epsilon^2\eta_2 + \epsilon^3\eta_3 + \cdots,$$
$$\phi = \epsilon\phi_1 + \epsilon^2\phi_2 + \epsilon^3\phi_3 + \cdots, \tag{2.31}$$

where we have introduced a formal expansion parameter ϵ which can be set to one at the end of the computation to remind us that terms of order n are proportional to $(ak)^n$, as pointed out above. Substituting (2.31) in equations (2.22, 2.23) one gets at second order

$$\frac{\partial\eta_2}{\partial t} - \frac{\partial\phi_2}{\partial y} = -\frac{\partial\eta_1}{\partial x}\frac{\partial\phi_1}{\partial x} + \frac{\partial^2\phi_1}{\partial y^2}\eta_1\bigg|_{y=0}$$

$$g\eta_2 + \frac{\partial\phi_2}{\partial t} = -\frac{1}{2}\left[\left(\frac{\partial\phi_1}{\partial x}\right)^2 + \left(\frac{\partial\phi_1}{\partial y}\right)^2\right] - \frac{\partial^2\phi_1}{\partial y\partial t}\eta_1\bigg|_{y=0} \tag{2.32}$$

The last terms on the r.h.s. of these equations come from the fact that the boundary conditions are evaluated at $y = 0$ instead of $y = \eta(x)$.

Using the expressions (2.26, 2.27) and of ϕ_1 and η_1, the r.h.s. of (2.32) can be computed and one obtains

$$\left[\frac{\partial \eta_2}{\partial t} - \frac{\partial \phi_2}{\partial y}\right]_{y=0} = \mathcal{R}e\left\{-i\omega k A_1^2 e^{2i(kx-\omega t)}\right\},$$

$$g\eta_2 + \frac{\partial \phi_2}{\partial t} = \mathcal{R}e\left\{\tfrac{1}{2}\omega^2 A_1^2 e^{2i(kx-\omega t)}\right\}.$$

(2.33)

One can seek for η_2 and ϕ_2 in the form

$$\eta_2 = \mathcal{R}e\left\{A_2 e^{2i(kx-\omega t)}\right\}, \qquad \phi_2 = \mathcal{R}e\left\{F_2 e^{2i(kx-\omega t)}\right\} e^{2ky}. \qquad (2.34)$$

A_2 and F_2 are easily determined by substitution of (2.34) into (2.33)

$$A_2 = \tfrac{1}{2}k A_1^2, \qquad F_2 = 0. \qquad (2.35)$$

Therefore, nothing special happens at second order. Things change at third order. The equations for ϕ_3 and η_3 are found as before by substitution of the formal expansions (2.31) into equations (2.22, 2.23). One gets (with $\phi_2 = 0$):

$$\frac{\partial \eta_3}{\partial t} - \frac{\partial \phi_3}{\partial y} = -\frac{\partial \eta_2}{\partial x}\frac{\partial \phi_2}{\partial x} - \frac{\partial \eta_1}{\partial x}\frac{\partial^2 \phi_1}{\partial x \partial y},$$

$$+ \frac{\partial^2 \phi_1}{\partial y^2}\eta_2 + \frac{1}{2}\frac{\partial^3 \phi_1}{\partial y^3}\eta_1^2$$

$$g\eta_3 + \frac{\partial \phi_3}{\partial t} = -\frac{1}{2}\left[2\frac{\partial \phi_1}{\partial x}\frac{\partial^2 \phi_2}{\partial x \partial y} + 2\frac{\partial \phi_1}{\partial y}\frac{\partial^2 \phi_1}{\partial y^2}\eta_1\right]$$

$$- \frac{\partial^2 \phi_1}{\partial t \partial y}\eta_2 - \frac{1}{2}\frac{\partial^3 \phi_1}{\partial t \partial^2 y}\eta_1^2.$$

(2.36)

All terms on the r.h.s. except the first one in the first equation come from evaluating the boundary conditions at $y = 0$ instead of $y = \eta(x)$. Evaluation of the r.h.s. proceeds as before, by substituting the previously found expressions for ϕ_1, η_1 and η_2 (2.30,2.35). After multiplication of the exponentials, one obtains terms proportional to $A_1^2 A_1^*$ and A_1^3 and their complex conjugates:

$$\frac{\partial \eta_3}{\partial t} - \frac{\partial \phi_3}{\partial y} = \frac{5}{8}\omega k^2 \mathcal{R}e\left\{-iA_1^2 A_1^* e^{i(kx-\omega t)}\right\}$$

$$+ \frac{9}{8}\omega k^2 \mathcal{R}e\left\{-iA_1^3 e^{3i(kx-\omega t)}\right\},$$

$$g\eta_3 + \frac{\partial \phi_3}{\partial t} = -\frac{3}{8}\omega^2 k \mathcal{R}e\left\{A_1^2 A_1^* e^{i(kx-\omega t)}\right\}$$

$$+ \frac{3}{8}\omega^2 k \mathcal{R}e\left\{A_1^3 e^{3i(kx-\omega t)}\right\}.$$

(2.37)

The terms proportional to $\exp(3i(kx - \omega t))$ on the l.h.s. do not pose any special problems and give contributions in ϕ_3 and η_3 which are uniformly small compared to previous terms in the expansion. On the contrary terms proportional to $\exp(i(kx - \omega t))$ have the same functional form as first order terms and are a potential problem. If one tries to determine their contribution to ϕ_3 and η_3 under the same form as above, see (2.26–27), the linear inhomogeneous system has a determinant equal to zero owing to (2.29) and can only be solved if the inhomogeneous term satisfies a particular relation. In general, the presence of this resonant inhomogeneous term produces contributions in η_3 and ϕ_3 which grow linearly as compared to η_1 or ϕ_1, in the time or spatial domain. An alternative way to determine the resonant contribution is to combine equations (2.37) into a single equation. Taking the time derivative of the second one and subtracting g times the first one, one gets:

$$\frac{\partial^2 \phi_3}{\partial t^2} + g \frac{\partial \phi_3}{\partial y} = \omega^3 k \, \mathcal{Re} \left\{ i A_1^2 A_1^* \, e^{i(kx - \omega t)} \right\}$$
$$+ \frac{3}{4} \omega^3 k \, \mathcal{Re} \left\{ i A_1^3 \, e^{3i(kx - \omega t)} \right\} . \tag{2.38}$$

If one looks for a solution ϕ_3 which is uniformly small in space with respect to ϕ_1, then the contribution to ϕ_3 of the resonant first term is:

$$[\phi_3]_{\mathrm{r}} = \tfrac{1}{2} \omega^2 \, k \, t A_1^2 A_1^* \, e^{i(kx - \omega t)} \tag{2.39}$$

which grows linearly in time and become larger than the first order contribution. Therefore, the expansion breaks down in the long-time limit.

2.2.3   Stokes's nonlinear dispersion relation

Stokes's idea was that the appearance of the secular term (2.39) was simply a reflection of a small frequency change due to nonlinear interactions. The difference between the unperturbed and perturbed frequency gives rise to a long-time scale and we can exploit Stokes's idea by using the multiscale method. We introduce a long-time scale $T = \epsilon^2 t$ and write ϕ and η as if they depended separately on t and T

$$\eta(x, t) = \epsilon \eta_1(x, t, T) + \epsilon^2 \eta_2(x, t, T) + \cdots ,$$
$$\phi(x, y, t) = \epsilon \phi_1(x, y, t, T) + \epsilon^2 \phi_2(x, y, t, T) + \cdots . \tag{2.40}$$

Here, it is to be understood that one should make the substitution

$$\frac{\partial}{\partial t} \longrightarrow \frac{\partial}{\partial t} + \epsilon^2 \frac{\partial}{\partial T} \tag{2.41}$$

in every time differentiation. At first and second order, nothing changes except that all constants appearing now depend on the long time. The interesting result appears when we collect the third order terms. Supplementary terms appear from differentiation with respect to the slow variable:

$$\frac{\partial \eta_3}{\partial t} - \frac{\partial \phi_3}{\partial y} = \left[\quad \right]_{\text{old}} - \mathcal{R}e \left\{ \frac{\partial A_1}{\partial T} \, e^{i(kx-\omega t)} \right\},$$

$$g \, \eta_3 + \frac{\partial \phi_3}{\partial y} = \left[\quad \right]_{\text{old}} - \mathcal{R}e \left\{ \frac{\partial F_1}{\partial T} \, e^{i(kx-\omega t)} \right\}. \tag{2.42}$$

As before, the two boundary conditions can be combined into a single one for ϕ_3, cf. (2.38):

$$\frac{\partial^2 \phi_3}{\partial t^2} + g \frac{\partial \phi_3}{\partial y} = \omega^3 k \, \mathcal{R}e \left\{ i A_1^2 A_1^* \, e^{i(kx-\omega t)} \right\}$$

$$+ 2 \frac{\omega^2}{k} \mathcal{R}e \left\{ \frac{\partial A_1}{\partial T} \, e^{i(kx-\omega t)} \right\} \tag{2.43}$$

$$+ \text{ non-secular terms},$$

where we have used the relation (2.30) between F_1 and A_1 and the linear dispersion relation. To avoid linearly growing terms in ϕ_3, we can now set the coefficients of $\exp(i(kx - \omega t))$ and $\exp(-i(kx - \omega t))$ separately to zero. This is a constraint that must be satisfied at all values of T

$$\frac{\partial A_1}{\partial T} = -i \tfrac{1}{2} k^2 \omega A_1^2 A_1^*. \tag{2.44}$$

As anticipated A_1 evolves on a time scale which is longer than the basic time scale $2\pi/\omega$ by the inverse square wave steepness. This is Stokes's celebrated result for the nonlinear dispersion relation

$$\omega_{\text{NL}} = \sqrt{kg} \left(1 + \tfrac{1}{2} k^2 |A_1|^2 + \ldots \right). \tag{2.45}$$

2.2.4 Evolution of slowly modulated wavetrains

In the previous subsection, we considered a perfectly periodic wavetrain. It is also very interesting to consider the evolution of wave packets and the interaction between dispersion and nonlinearity. A wave packet centered around wave number k_0 can be represented as:

$$w(x,t) = A(x,t) \, e^{i(k_0 x - \omega_0 t)}. \tag{2.46}$$

We want to determine the time evolution of $A(x,t)$ when $A(x,0)$ is a slowly varying function of x. For convenience, we have explicitly extracted the linear pulsation $\omega_0 = \sqrt{k_0 g}$.

We begin with a simple approach and first consider the linear evolution of $A(x,t)$. We expand $A(x,0)$ in Fourier series and write the initial condition as

$$w(x,0) = \sum_k A_k\, e^{i(k_0+k)x},$$ (2.47)

where k is small compared to k_0. The form of the wave packet at all times is given by the linear dispersion relation

$$w(x,t) = \sum_k A_k\, e^{i[(k_0+k)x - \omega(k_0+k)t]}.$$ (2.48)

Since k is small compared to k_0, we can expand the dispersion relation in the neighborhood of k_0

$$\omega(k_0 + k) = \omega_0 + k\omega'(k_0) + \tfrac{1}{2}k^2\omega''(k_0) + \ldots$$ (2.49)

Substituting this relation into (2.48) and comparing with (2.46), one obtains the linear evolution of $A(x,t)$

$$A(x,t) = \sum_k A_k\, e^{i\left[kx - \left(\omega'(k_0)k + \frac{k^2}{2}\omega''(k_0)+\ldots\right)t\right]}.$$ (2.50)

We thus find that $A(x,t)$ satisfies the equation

$$\frac{\partial A}{\partial t} + \omega'(k_0)\frac{\partial A}{\partial x} = +\frac{i}{2}\omega''(k_0)\frac{\partial^2 A}{\partial x^2} + \ldots$$ (2.51)

For a slowly varying amplitude ($A(x) = f(\epsilon x)$), this equation reduces at first order to its first two terms, i.e., one obtains the well-known result that the wave packet moves at the group velocity $\omega'(k_0)$.

At next order, in a reference frame which moves at the group velocity, one sees the dispersion of the wave packet

$$\frac{\partial A}{\partial t} = \frac{i}{2}\omega''(k_0)\frac{\partial^2 A}{\partial x^2}.$$ (2.52)

The combined effects of dispersion and nonlinearity are simply described by adding the r.h.s. of (2.44) and (2.52). One obtains the so-called "nonlinear Schrödinger" equation:

$$\frac{\partial A}{\partial t} = \frac{i}{2}\omega''(k_0)\frac{\partial^2 A}{\partial x^2} - \frac{i}{2}k_0^2\omega_0^2|A|^2 A.$$ (2.53)

This is a well-known equation which appears in a variety of contexts and which possesses many interesting properties, as we will shortly discuss in the next section. Before doing that we want to show briefly how (2.53) is more formally derived by the multiscale method.

Again, the strategy is to separate explicitly the slow variations from the fast ones. We therefore introduce "long" space variables X and Y, and two slow time variables T_1 and T_2 since dispersion occurs on a longer time scale than motion of the wave packet, as seen above:

$$X = \epsilon x, \qquad Y = \epsilon y, \qquad T_1 = \epsilon t, \qquad \text{and} \quad T_2 = \epsilon^2 t. \qquad (2.54)$$

For simplicity, we focus on the velocity potential ϕ and therefore combine the two boundary conditions (2.22), (2.23) into a single one for ϕ, as explained above:

$$\frac{\partial^2 \phi}{\partial t^2} + g\frac{\partial \phi}{\partial y} = g\frac{\partial \eta}{\partial x}\frac{\partial \phi}{\partial x} - \frac{\partial^2 \phi}{\partial t \partial y}\frac{\partial \eta}{\partial t} - \frac{\partial \phi}{\partial x}\frac{\partial^2 \phi}{\partial x \partial t}$$
$$- \frac{\partial \phi}{\partial y}\frac{\partial^2 \phi}{\partial y \partial t} - \frac{\partial \phi}{\partial x}\frac{\partial^2 \phi}{\partial x \partial y}\frac{\partial \eta}{\partial t} - \frac{\partial \phi}{\partial y}\frac{\partial^2 \phi}{\partial y^2}\frac{\partial \eta}{\partial t}\Bigg|_{y=\eta(x)}, \qquad (2.55)$$

where η should be understood as being expressed as a function of ϕ using boundary condition (2.23)

$$\eta = -\frac{1}{g}\left[\frac{\partial \phi}{\partial t} + \frac{1}{2}\left(\left(\frac{\partial \phi}{\partial x}\right)^2 + \left(\frac{\partial \phi}{\partial y}\right)^2\right)\right]_{\eta(x)}. \qquad (2.56)$$

In the Laplace equation and equations (2.55–56) we then make the substitutions

$$\frac{\partial}{\partial x} \longrightarrow \frac{\partial}{\partial x} + \epsilon\frac{\partial}{\partial X}, \qquad \frac{\partial}{\partial y} \longrightarrow \frac{\partial}{\partial y} + \epsilon\frac{\partial}{\partial Y},$$
$$\frac{\partial}{\partial t} \longrightarrow \frac{\partial}{\partial t} + \epsilon\frac{\partial}{\partial T_1} + \epsilon^2\frac{\partial}{\partial T_2}. \qquad (2.57)$$

At first order, one obtains again (2.30) but now A_1 and F_1 depend on the slow variables

$$F_1(X, 0, T_1, T_2) = -i\frac{\omega}{k_0}A_1(X, T_1, T_2). \qquad (2.58)$$

At second order, supplementary terms appear in the Laplace equation

$$\nabla^2 \phi_2 = \left[-2\frac{\partial^2 \phi_1}{\partial X \partial x} - 2\frac{\partial^2 \phi_1}{\partial Y \partial x}\right]$$
$$= -2k_0\, \mathcal{R}e\left\{\left(i\frac{\partial F_1}{\partial X} + \frac{\partial F_1}{\partial Y}\right)e^{i(k_0 x - \omega_0 t)}e^{k_0 y}\right\}. \qquad (2.59)$$

Secular terms do not appear in ϕ_2 if

$$\frac{\partial F_1}{\partial Y} = -i\frac{\partial F_1}{\partial X}, \qquad (2.60)$$

that is if F is a function of $X - iY$. New terms also appear in the boundary condition for ϕ_2

$$\frac{\partial^2 \phi_2}{\partial t^2} + g\frac{\partial \phi_2}{\partial y} = [\quad]_{old} - 2\frac{\partial^2 \phi_1}{\partial t \partial T_1} - g\frac{\partial \phi_1}{\partial Y}$$

$$= [\quad]_{old} - \mathcal{Re}\left\{\left[-2i\omega_0 \frac{\partial F_1}{\partial T_1} + g\frac{\partial F_1}{\partial Y}\right] e^{i(k_0 x - \omega_0 t)}\right\}.$$
(2.61)

Secular terms do not appear only if the new contribution is set to zero. This gives the result that the wave packet propagates at the group velocity

$$\frac{\partial A_1}{\partial T_1} + \frac{g}{2\omega_0}\frac{\partial A_1}{\partial X}\bigg|_{Y=0} = 0$$
(2.62)

where we have utilized relation (2.60) between X and Y derivatives of F_1 and replaced $F_1|_{Y=0}$ by A_1, cf. (2.58).

At third order, no secular terms appear in the Laplace equation since F_1 satisfies (2.60) and $\phi_2 = 0$,

$$\nabla^2 \phi_3 = -\left(\frac{\partial^2}{\partial X^2} + \frac{\partial^2}{\partial Y^2}\right)\phi_1 = 0.$$
(2.63)

New contributions appear in the boundary condition for ϕ_3

$$\frac{\partial^2 \phi_3}{\partial t^2} + g\frac{\partial \phi_3}{\partial y} = [\quad]_{old} - 2\frac{\partial^2 \phi_1}{\partial t \partial T_2} - \frac{\partial^2 \phi_1}{\partial T_1^2} + \text{non-secular terms.}$$
(2.64)

We have only explicitly written the new secular contributions.

Secular contributions are of two types:

1) They involve a single exponential and differentiation with respect to the slow variables. The only such terms are the second and third term on the r.h.s. of (2.64).
2) They involve the product of three exponentials and cannot involve differentiation with respect to the slow variables. These terms have already been computed (r.h.s. of (2.38). They are denoted by $[\quad]_{old}$ in the above equation. Setting the secular contribution to zero gives

$$i\omega^3 k A_1^2 A_1^* + 2i\omega\frac{\partial F_1}{\partial T_2} - 2\frac{\partial^2 F_1}{\partial T_1^2} = 0.$$
(2.65)

After replacing F_1 by A_1 using (2.58) and replacing T_1-derivatives by X-derivatives using (2.62), one recovers the nonlinear Schrödinger equation:

$$\frac{\partial A_1}{\partial T_2} = -i\frac{\omega_0}{8k_0^2}\frac{\partial^2 A_1}{\partial X^2} - i\frac{\omega_0 k_0^2}{2}A_1^2 A_1^*.$$
(2.66)

2.2.5   Miscellaneous remarks

In the previous subsection, we have described Stokes's result for weakly nonlinear periodic wavetrains and a generalization for slowly modulated wavetrains. Some remarks may be needed to put them in a proper perspective (see Stoker, 1957; Benjamin, 1974; Whitham, 1974).

Stokes's investigation started a mathematical controversy about the actual existence of nonlinear periodic gravity waves with a continuum ratio of amplitude over wavelength. This culminated in Levi-Civita's (1925) rigorous constructive proof of existence of periodic waves of small amplitude. The restriction to small amplitude was then removed in the fifties when it was shown that for every value of the wave maximum slope between 0 and 1/3 (i.e., for $\sup |\partial\eta/\partial X| \in]0, \frac{1}{3}[$) a steady progressive wave train exists.*

After these difficult mathematical proofs, Benjamin and Feir's (1967) discovery came as a bad surprise. They showed that the strictly periodic wavetrains are unstable. This is very simply derived, at least for weak amplitude, by using the nonlinear Schrödinger equation. Writing it in the form:

$$\frac{\partial A}{\partial t} = i\gamma \frac{\partial^2 A}{\partial x^2} - i\beta A^2 A^*, \qquad (2.67)$$

one easily checks that the solution

$$A = a \exp(-i\beta a^2 t) \qquad (2.68)$$

is unstable to x dependent disturbances when $\beta\gamma < 0$ (see Chapter 4).

The later development of the instability is to break the wavetrain into a number of pulses which are described by solitary waves of the nonlinear Schrödinger equation. These solitary waves have been called *solitons* because they show many remarkable properties associated with the complete integrability of the NLS equation. This means that the integration of this one-dimensional nonlinear equation can be reduced to the solution of linear equations. This beautiful mathematical development is well explained in a number of recent books (for example Ablowitz and Clarkson, 1991; Novikov *et al.*, 1984) to which we refer the interested reader.

The nonlinear Schrödinger equation appears in different physical contexts, like plasma physics and nonlinear optics, since it simply describes the interaction of dispersion and weak nonlinearity, as shown by the simple but general derivation above. Due to its use in nonlinear optics, the

---

* The value 1/3 can be understood from another Stokes's result which states that when a sharp crest is attained in a steady state profile the angle there must be 120°. This is simply derived by a local analysis of the equation in the neighborhood of the wave crest (see Whitham, 1974)

case $\beta\gamma < 0$ is sometimes called "focusing" in the literature ("defocusing" for $\beta\gamma > 0$).

In higher dimensions, the NLS equation produces singularities in finite time in the focusing case. This is shown by an ingenious argument (Zakharov, 1972). Normalizing the equation by taking $\gamma = 1$, $\beta = -1$, two conserved quantities are:

$$I_1 = \int |A|^2 \, d^d\mathbf{r} \,, \tag{2.69}$$

$$I_2 = \int \left(|\nabla A|^2 - \tfrac{1}{2}|A|^4 \right) d^d\mathbf{r} \,. \tag{2.70}$$

Then, computing the evolution of

$$C = \int \mathbf{r}^2 |A|^2 \, d^d\mathbf{r} \tag{2.71}$$

after some calculations one finds:

$$\frac{d^2 C}{dt^2} = 8\, I_2 + 4\left(1 - \frac{d}{2}\right) \int |A|^4 d^d\mathbf{r} \,. \tag{2.72}$$

Therefore for a negative I_2 and for $d \geq 2$, the r.h.s. of (2.72) is negative and C goes to zero at some finite time t_0. Since A cannot vanish everywhere by conservation of I_1, it is clear that a singularity should be produced for $t \leq t_0$. The formation of these singularities has been the subject of recent analyses (for example Fraiman, 1985; Le Mesurier *et al.*, 1988; Dyachenko *et al.*, 1992; Merle, 1992).

2.3 Amplitude equation from a more general viewpoint

In the previous subsection, we have derived the NLS equation as an envelope equation for gravity waves. It is a special case of an amplitude equation for a conservative system. In the more common case where dissipation exists, the usual amplitude equation is the so-called "complex Ginzburg–Landau" equation (Newell, 1974):

$$\frac{\partial A}{\partial t} = \mu A + \alpha |A|^2 A + \beta \frac{\partial^2 A}{\partial x^2} \tag{2.73}$$

where μ, α and β are complex coefficients which are not purely imaginary. Their physical meaning is as follows. A tends to increase or decrease depending on the balance between dissipation and injected energy. For $\mathcal{R}e\{\mu\} < 0$ the trivial solution $A = 0$ is stable. On the contrary for $\mathcal{R}e\{\mu\} > 0$ a small A tends to grow. The real part of β reflects the dependence of the instability growth rate on wavelength while the imaginary part comes from dispersion as above. The imaginary part of α describes the variation of the basic frequency with the amplitude of A, while its real part reflects the dependence of dissipation on the amplitude of A.

If $\mathcal{R}e\,\{\alpha\} < 0$, dissipation increases when $|A|$ grows and nonlinearities tend to saturate the instability. This is referred to as a "normal" or "supercritical" instability. Near the instability threshold $(0 < \mu \ll 1)$, the restabilized value of $|A|$ is small and one can perform controlled weakly nonlinear expansions. A typical example is given by Rayleigh–Bénard convection. In the opposite case, $\mathcal{R}e\,\{\alpha\} > 0$ nonlinearities decrease the dissipation for small A amplitude and $|A|$ saturates for values of order unity which are not controlled by the distance to threshold. Weakly nonlinear expansions are therefore much less informative since high order terms are as important as low order ones. This is commonly referred to as a "subcritical" instability or as an "inverted" bifurcation in the literature. Examples include plane Poiseuille flow (Stewartson and Stuart, 1971) and directional solidification in metals (Wollkind and Segel, 1970; Müller-Krumbhaar and Kurz, 1990).

Amplitude equations seem to have been used in three rather distinct manners in the literature. Our derivation of the nonlinear Schrödinger equation is an illustration of the first one. One starts with well defined equations and derives amplitude equations using a weakly nonlinear expansion. This is how envelope equations have been introduced for Rayleigh–Bénard convection (Segel, 1969; Newell and Whitehead, 1969) and it has since been useful in the study of many different instabilities. It is worth pointing out that the rigorous mathematical derivation of amplitude equations is a subject of current interest (see, for example, Kirchgässner, 1988; Collet and Eckmann, 1990; Craig *et al.*, 1992). A second direction that has been followed, is to classify the possible modes of instabilities and the different types of amplitude equations. Usually, this is done by using the various possible symmetries of the system. Of course, the particular coefficients characterizing a definite physical situation cannot be obtained but similar phenomena appearing in different systems can be described in a synthetic way. This is well illustrated by the work of Coullet and Iooss (1990) on possible instabilities of one-dimensional cellular patterns and a nice application to traveling waves is given by Coullet *et al.* (1989). This strategy is described in detail in Chapter 4.

Finally, amplitude equations are used in a much more radical way in the literature, in a spirit close to the original Landau proposal and to the use of Ginzburg–Landau free energies in the description of phase transitions. The idea is that simple amplitude equations describe the dynamics of the system order parameter. For example $A = 0$ would characterize a laminar flow and $A = 1$ a turbulent one. One imagines that it would, in principle, be possible to obtain a coarsened description involving A

only. One then argues that simple equation like (2.74) are rich enough to describe the main phenomena. This is analogous to the description of transitions to chaos using one-dimensional maps although the case is much weaker at present. Examples of this strategy are provided by Pomeau (1986) or Coullet and Lega (1988) and an application to spiral turbulence is described in Andereck *et al.* (1989). This viewpoint gives much interest to the analysis of simple structures in Ginzburg–Landau equations, like fronts and localized states which are the subject of the forthcoming section.

3 Fronts and localized states

In this chapter, we use the simplified setting of Ginzburg–Landau like equations to investigate the spatial coexistence of different states. We begin by considering a system which possesses two linearly stable homogeneous states. We discuss their relative stability by considering a situation where the system is in the different stable states at different spatial locations. Then, an interface or front separates the two states. We show that this front generally moves so that one state, which is by definition the more stable, invades the other one. We then discuss the interaction between two fronts and find that in some cases stable localized inclusion of the less stable phase can exist in the more stable phase.

After treating this bistable situation which usually occurs in the vicinity of a subcritical bifurcation, we turn to the case of a normal bifurcation. Specifically, we consider the invasion of an unstable state by a stable state. This turns out to be a subtle problem, the basic features of which we discuss along simple lines.

3.1 Front between linearly stable states

3.1.1 Front solution for the real Ginzburg–Landau equation and mechanical analogy

As a model of a spatially extended bistable system, we study the Ginzburg–Landau equation:

$$\frac{\partial A}{\partial t} = \frac{\partial^2 A}{\partial x^2} - \frac{\partial V(A)}{\partial A}, \tag{3.1}$$

where V is a real function with two local minima, say at $A = A_1$ and $A = A_2$ (see Fig. 3.1) which are the two homogeneous stable states of the system. We suppose that the system is in the $A = A_2$ state at $x = -\infty$ and in the A_1 state at $x = +\infty$. Clearly there is a transition region between these two states. We assume that after some transients, it relaxes to a steadily propagating front solution that we try to determine. So we look for a solution of (3.1) of the form $A(x,t) = A(x - ct)$ such that

$A \to A_2$ for $x \to -\infty$ and $A \to A_1$ for $x \to +\infty$. Substituting in (3.1), we obtain:

$$\frac{d^2 A}{dx^2} = -c\frac{dA}{dx} + \frac{\partial V(A)}{\partial A}. \tag{3.2}$$

This is conveniently interpreted as the equation of motion of a particle in the inverted potential $-V(A)$, where x plays the role of time (see Fig. 3.2) and c the role of friction. A front solution is given by a trajectory which starts from the top of the hill $A = A_2$ at $x \to -\infty$ and reaches the top of the hill $A = A_1$ at $x \to +\infty$. It is intuitively clear that such a trajectory exists only for a particular value of the "friction" c. If c is too small, then not enough energy is dissipated during the motion from $A = A_2$ to $A = A_1$ and the particle continues its motion past $A = A_1$. On the other hand, if c is too large, the particle never reaches $A = A_1$ and ends its motion at the bottom of the valley between $A = A_1$ and $A = A_2$. The critical friction which corresponds to the front velocity can be computed perturbatively if the height difference between the two hills is small, i.e., for $|V(A_2) - V(A_1)| \ll 1$. Indeed, for a potential such that $V_0(A_2) = V_0(A_1)$, energy conservation shows that $c = 0$. The corresponding trajectory $A_0(x - x_0)$ is straightforwardly obtained by quadrature since it satisfies:

$$\tfrac{1}{2}(dA/dx)^2 = V_0(A) - V_0(A_1). \tag{3.3}$$

The front solution $A(x) = A_0(x) + A^{(1)}(x) + \ldots$ for a potential $V(A) = V_0(A) + \delta V(A)$ which is slightly different from $V_0(A)$, can be obtained perturbatively from $A_0(x)$. The linear correction to $A_0(x)$ is given by

$$\frac{d^2 A^{(1)}}{dx^2} - \frac{d^2 V}{dA^2}(A_0(x))A^{(1)} = \frac{d}{dA}\delta V(A_0(x)) - c\frac{dA_0}{dx}. \tag{3.4}$$

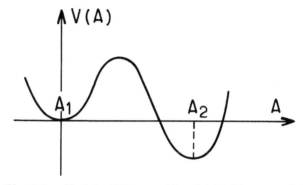

Fig. 3.1.   Sketch of the considered bistable potential.

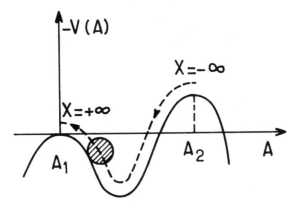

Fig. 3.2. Motion of a particle in x-time corresponding to a front between the two phases.

Invariance by translation implies that dA_0/dx is a zero mode of the linear hermitian operator of the l.h.s. of (3.4). Secular contributions do not appear in A_1 only if the r.h.s. is orthogonal to this zero mode

$$\int_{-\infty}^{+\infty} dx \, \frac{dA_0}{dx} \left(\frac{d}{dA} \delta V \left(A_0(x) \right) - c \frac{dA_0}{dx} \right) = 0 \, . \tag{3.5}$$

This determines the front velocity c as:

$$c = \left(\delta V \left[A_1 \right] - \delta V \left[A_2 \right] \right) \Big/ \int \left(dA_0/dx \right)^2 dx \, , \tag{3.6}$$

which simply means, in the mechanical analogy, that the energy dissipated by friction should be equal to the potential energy difference between the tops of the two hills.

An illustrative example is given by the family of potentials:

$$V_\mu(A) = \tfrac{1}{2} \mu A^2 - \tfrac{1}{4} \beta A^4 + \tfrac{1}{6} \gamma A^6 \tag{3.7}$$

which has a local minimum for $A_1 = 0$ if μ is positive ($\beta > 0$, $\gamma > 0$). This is the only minimum if $\mu > \beta^2/4\gamma$. For $\mu < \beta^2/4\gamma$ two non-trivial symmetric minima, A_2 and $-A_2$, appear which become deeper than the minimum at $A = 0$ for $0 < \mu < \mu_c = 3\beta^2/16\gamma$. For $\mu = \mu_c$, the three minima are equal and the corresponding resting front $A_0(x - x_0)$ joining $A_2 = A_0^*$ at $x = -\infty$ to $A = 0$ at $x = +\infty$ is given by

$$A_0^2 \left(x - x_0 \right) = \frac{(A_0^*)^2}{2} \left[1 - \tanh \left(\frac{x - x_0}{\xi} \right) \right] \tag{3.8}$$

with

$$(A_0^*)^2 = \frac{3\beta}{4\gamma} \, , \qquad \xi = \frac{4}{\beta} \sqrt{\frac{\gamma}{3}} \, . \tag{3.9}$$

For a slight departure $\delta\mu$ of μ from μ_c the front moves at a speed c equal to

$$c = -2\delta\mu\,\xi\,. \qquad (3.10)$$

3.1.2   Localized states and front interactions

Since fronts exist for equation (3.1), it is interesting to discuss their interactions. We begin by discussing the existence of a localized bubble of one state into the other one which can be viewed as a bound state of two fronts. Using the mechanical analogy described before, it is simple to see that one can obtain inclusion of the stabler state into the less stable one. Indeed, let us consider the case of Fig. 3.2 and release the particle on top of the A_1 hill at $x = -\infty$. Then for $c = 0$, the motion is conservative. The particle will slide down the A_1 hill and then climb up the A_2 hill but not up to the top. When the initial releasing height is reached the velocity changes sign and the particle returns to its initial position. For a small height difference between the two hills, the particle spends a long time near $A = A_2$ before returning to A_1, and this motion clearly describes an inclusion of the A_2 state into the A_1 one. However, in the simple case of (3.1) all such localized states are unstable. The reader familiar with first order transitions will have recognized that our localized state is nothing other than a critical nucleation droplet. If it is made slightly larger then the A_2 phase invades the A_1 phase surrounding it, while a droplet smaller than the critical one shrinks and disappears. A more mathematical way of seeing this is to analyze the linear stability of a localized state $A_{\mathrm{loc}}(x)$. The equation of an eigenmode $e^{\sigma t}\,A_1(x)$ is obtained by linearization of (3.1)

$$-\sigma A_1(x) = -\frac{d^2 A_1}{dx^2} + \frac{\partial^2 V}{\partial A^2}\,[A_{\mathrm{loc}}(x)]\,A_1\,. \qquad (3.11)$$

The linear operator on the r.h.s. is a Schrödinger operator describing a one-dimensional motion in the potential $V''[A_{\mathrm{loc}}(x)]$. Moreover, $dA_{\mathrm{loc}}(x)/dx$ is a zero-mode of this Hamiltonian (from translation invariance) with a node (since $A_{\mathrm{loc}}(x)$ which goes from A_1 at $x = -\infty$ to A_1 at $x = +\infty$ must have an extremum). Therefore, the nodeless ground-state wave function has a strictly lower and therefore negative energy. In other words, there exists a solution of (3.11) for a positive value of σ and the localized state is indeed unstable.

An instructive way of reaching the same conclusion, which is more easily applied to other situations than this clever classical argument, is to compute directly the interaction between two widely separated fronts. We first analyze the case where the two minima have the same height, i.e., $V(A_1) = V(A_2)$ and we consider two widely separated fronts located at $x_1(t)$ and $x_2(t)$ $(x_1 < x_2)$. We want to compute the time evolution of $x_1(t)$ and $x_2(t)$ which we anticipate to be very slow. Therefore we treat it as a perturbation. In the regions near $x_1(t)$ and $x_2(t)$ at zeroth order the solution is a motionless front, $A_0(x - x_1(t))$ near $x_1(t)$ and $A_0(x_2(t) - x)$ near $x_2(t)$ where

$$\frac{d^2 A_0}{dx^2} = \frac{\partial V}{\partial A}, \qquad A_0(x) \to A_{1,2} \qquad \text{as} \qquad x \to -\infty, +\infty. \qquad (3.12)$$

In the intermediate region between the two fronts A is close to A_2 and (3.1) can be linearized

$$\frac{d^2 A_{\text{int}}}{dx^2} = V''(A_2)[A_{\text{int}}(x) - A_2]. \qquad (3.13)$$

Therefore

$$A_{\text{int}}(x) - A_2 = -\alpha \exp\left[\sqrt{V''(A_2)}\left(x - \frac{x_1 + x_2}{2}\right)\right] \\
- \beta \exp\left[\sqrt{V''(A_2)}\left(\frac{x_1 + x_2}{2} - x\right)\right] \qquad (3.14)$$

where α and β are two constants that should be determined. In order to have a uniformly valid first approximation it remains to match $A_0(x - x_1(t))$ and $A_0(x_2(t) - x)$ to $A_{\text{int}}(x)$. When $x \to \infty$ the asymptotic behavior of $A_0(x)$ is

$$A_0(x) - A_2 \approx -c \exp\left[-\sqrt{V''(A_2)}\, x\right]. \qquad (3.15)$$

The constant c is of course well defined only after we have fixed the translation invariance of (3.12). It is in fact necessary to define precisely what we mean when we say that a front is standing at x_1 or x_2. For definiteness, we arbitrarily define the front positions as the locations x where $A(x) = \frac{1}{2}(A_1 + A_2)$. Therefore $A_0(x)$, (3.12), is precisely determined by requiring:

$$A_0(0) = \frac{1}{2}(A_1 + A_2). \qquad (3.16)$$

The constant c is now well defined and can be computed by integrating the motionless front equation (3.12). Matching of $A_0(x - x_1(t))$ and $A_{\text{int}}(x)$ for $x_1(t) \ll x \ll \frac{1}{2}(x_1 + x_2)$ gives:

$$\beta = c \exp\left[\sqrt{V''(A_2)}\frac{1}{2}(x_1 + x_2)\right]. \qquad (3.17)$$

In a similar way matching to $A_0(x_2(t) - x)$ gives the same equation (3.17) with α replacing β. Using the previously described mechanical analogy, it is clear that the particle has lost some energy during its motion from A_1 at $x = -\infty$ to $A_2 - (\alpha + \beta)$ at $x = \frac{1}{2}(x_1 + x_2)$ where its velocity vanishes. At next order this implies the existence of a small front velocity which gives a friction term. The energy dissipated by friction is simply computed at first order by using the above constructed first order motion and should be equal at this order to the potential energy loss:

$$\dot{x}_1(t) \int_{-\infty}^{+\infty} \left(\frac{dA_0}{dx} (x - x_1(t)) \right)^2 = +V(A_2 - \alpha - \beta) - V(A_1)$$

$$= +\tfrac{1}{2} V''(A_2)(\alpha + \beta)^2. \tag{3.18}$$

The integration in the l.h.s. has been extended to $x = +\infty$ since this only changes its value by exponentially small terms in $(x_2 - x_1)$ which are of higher orders. Finally, using (3.17) we obtain the time evolution of $x_1(t)$

$$\dot{x}_1(t) = \frac{2V''(A_2)\,c^2}{\int_{-\infty}^{+\infty} (dA_0/dx)^2\,dx} \exp\left[\sqrt{V''(A_2)}\,(x_1 - x_2) \right]. \tag{3.19}$$

A similar calculation would, of course, give the same equation for $-x_2(t)$. Equation (3.19) can be interpreted as an attractive interaction between the two fronts since without the existence of another front at x_2, the front at x_1 would not move. In order to compute meaningfully the prefactor in (3.19), it has been found necessary to precisely define the meaning of front positions, since changes of order one in $x_1(t)$ and $x_2(t)$ affect the prefactor by quantities of order one. However, it is worth noting that the front velocities $x_1(t)$, $x_2(t)$ are independent of this arbitrary choice as they should be, since a redefinition would affect both the value of $x_1 - x_2$ and c in (3.19). Several other derivations of (3.19) exist in the literature (see for example Kawasaki and Ohta, 1982).[*] A rigorous mathematical proof can be found in Carr and Pego (1989).

Now that we have obtained an attractive interaction between widely separated fronts, the existence and instability of a localized state can be understood more intuitively. Let us suppose that $V(A_1)$ and $V(A_2)$ are slightly different so that, for example, $V(A_2)$ is slightly lower than $V(A_1)$, $V(A_2) \leq V(A_1)$. In this case, a single front moves so that the A_2 phase invades the A_1 phase, as described previously, (3.6). Let us consider the case of two fronts located at x_1 and x_2 so that $A(x) \simeq A_2$ for $x_1 < x < x_2$ and $A(x) \simeq A_1$ for x outside the interval $[x_1, x_2]$. For widely separated fronts and small potential differences, the motion of each front is simply given by the superposition of equations (3.10) and (3.19).

---

[*] While (3.19) agrees with the result of Carr and Pego (1989) (attributed by them to Neu, unpublished), our prefactor of the exponential term seems to differ from the one found by Kawasaki and Ohta (1982) by a factor of two.

For the x_1 front, it reads

$$\dot{x}_1(t) = k_1 \left[V(A_2) - V(A_1)\right] + k_2 \exp\left[\sqrt{V''(A_2)}\,(x_1 - x_2)\right] \qquad (3.20)$$

where the expression for the constants k_1 and k_2 is given in equations (3.6) and (3.10). So each front motion is driven by two forces:

1) one constant force which tends to push the fronts apart and which is independent of the front position (the first term on the l.h.s of (3.20)),
2) one attractive interaction between the two fronts which decreases to zero with their separation.

It is therefore clear that if the repulsive force is not too strong it can be exactly counterbalanced by the attractive force between the fronts for one well defined value of their separation given by

$$|x_1 - x_2| = -\frac{1}{\sqrt{V''(A_2)}} \log\left[\frac{k_1}{k_2}\left(V(A_1) - V(A_2)\right)\right]. \qquad (3.21)$$

This position is unstable since the repulsive interaction wins over the decreasing attractive interaction if the separation between the two fronts is increased and the contrary happens if the separation is reduced, so that in any case the perturbation tends to increase. So, after a long detour, we have again reached the conclusion that localized states of (3.1) are unstable. The advantage of this perturbative approach is that it can be applied without much difficulty to more complicated cases as we show in the next subsection. In conclusion, let us note that the above calculation can be generalized to a case with many fronts. The equivalent of (3.19) shows in this case, that each front motion is driven by exponential interactions with its two neighbors. When two fronts become close enough they merge and disappear. The analysis of coarsening in such a situation as time proceeds, has given rise to many interesting statistical studies (see for example Ohta *et al.*, 1982; Mazenko, 1990; Derrida *et al.*, 1990).

3.1.3 The complex Ginzburg–Landau case

Localized structures are often observed in nonequilibrium systems which possess two stable homogeneous states as for example in Poiseuille flow at moderately high Reynolds number where laminar and turbulent flows can coexist. Two types of such structures can be distinguished, there are stable localized "pulses" which travel without changing their size and expanding domains which grow at a constant speed. While the latter case can be accounted for, at least at a phenomenological level, by the simple description of the previous section, this is not the case for localized pulses which we found to be unstable. It therefore came as a nice surprise when the observation of stable localized pulses was reported (Thual and

Fauve, 1988) in a numerical simulation of the complex Ginzburg–Landau equation

$$\frac{\partial A}{\partial t} = \alpha \frac{\partial^2 A}{\partial x^2} - \mu A + \beta |A|^2 A - \gamma |A|^4 A \,, \qquad (3.22)$$

where α, β, γ are complex and* $\mu > 0$, $\mathcal{R}e\{\beta\} > 0$, $\mathcal{R}e\{\gamma\} > 0$, so as to be in a subcritical case. The stable pulses exist for values of μ within a finite band and have the form:

$$A(x,t) = R(x) \exp\left[i\left(\Omega t + \phi(x)\right)\right]. \qquad (3.23)$$

In order to explain the pulse stability, two different limits of (3.22) have been analyzed:

1) Equation (3.22) can be viewed as a perturbation of the nonlinear Schrödinger equation, (2.67), in the limit where $\mathcal{I}m\{\alpha\}$ and $\mathcal{I}m\{\beta\}$ are of order one and all the other coefficients are small. Among the continuum family of motionless solitons only a finite number survive the perturbation and give rise to pulses of (3.22). This analysis, which uses the well-studied perturbation of integrable equations (see, for example the review of Kivshar and Malomed, 1989) is described in Chapter 4.
2) Equation (3.22) can be viewed as a perturbation of the real equation (3.1), (3.7) of the previous section in the limit where all coefficients have small imaginary parts. This is the path which we follow in order to illustrate how a stable pulse emerges from an analysis of fronts and their interactions (Hakim *et al.*, 1990; Hakim and Pomeau, 1991; Malomed and Nepomnyashchy, 1990).

Thus, we suppose that all imaginary coefficients in (3.22) are small and scale the x-coordinate so that $\alpha_i = \mathcal{R}e\{\alpha\} = 1$. We first consider a generalization $A(x,t)$ of the real front solution (3.2) to the complex case in the form

$$A(x,t) = R\left(x - x_0\right) \exp\left[i\left(\Omega t + \phi\left(x - x_0\right)\right)\right]. \qquad (3.24)$$

Substituting this ansatz into (3.22) we obtain:

$$R_{xx} = -\frac{\partial V_{\mu_c}}{\partial R} = -x_{0,t}\,R + \delta\mu\,R + \frac{S^2}{R^3} + \alpha_i \frac{S_x}{R}\,, \qquad (3.25)$$

$$S_x = -x_{0,t}S + \Omega\,R^2 - \beta_i R^4 + \gamma_i R^6 - \alpha_i R R_{xx}\,. \qquad (3.26)$$

We have defined $S = R^2 \phi_x$ and denoted by $\delta\mu$ the departure of μ from the value μ_c $(\delta\mu = \mu - \mu_c)$ where the potential (3.7) (corresponding to

---

* μ can be chosen real without loss of generality by replacing $A(t)$ by $A(t)\exp(-\mathcal{I}m\{\mu\}\,t)$.

the real part of the coefficients in (3.22)) has standing front solutions. Now, we expand (3.25, 26) both in $\delta\mu$ and α_i, β_i, γ_i which we suppose all small. It turns out that both Ω and the front velocity are then small so that (3.26) shows that S is small. Therefore at zeroth order (3.25) just reduces to the real standing front equation (3.8)

$$R_0(x) = A_0(x) \,. \tag{3.27}$$

Equation (3.26) can then be used to compute S to first order. This determines the phase variation of a complex front

$$S^{(1)}(x) = -\int_x^{+\infty} \left[\Omega R_0^2 - \beta_i R_0^4 + \gamma_i R_0^6 - \alpha_i R_0 R_{0,xx} \right] dx \,. \tag{3.28}$$

Convergence of this integral at $x = -\infty$ fixes Ω to be

$$\Omega = \beta_i \left(A_0^* \right)^2 - \gamma_i \left(A_0^* \right)^4 \tag{3.29}$$

where A_0^* is the asymptotic $x = +\infty$ value of $A_0(x)$ (given by (3.9)). S_1 being determined, the front velocity is obtained as in the real case by requiring that the r.h.s. of (3.25) be orthogonal to the zero mode of the linearized operator of the l.h.s.

$$x_{0,t} \int_{-\infty}^{+\infty} \left(\frac{dR_0}{dx} \right)^2 = \frac{\delta\mu}{2} \left(R_0^* \right)^2 + \int_{-\infty}^{+\infty} dx\, F_S^1 \frac{R_0}{dx} \tag{3.30}$$

where F_S^1 is due to the phase variation:

$$F_S^1 = \frac{S^{(1)2}}{R_0^3} + \alpha_i \frac{S_x^{(1)}}{R_0} \,. \tag{3.31}$$

This formula is completely analogous to (3.6). The second term simply gives a shift μ_F of μ of the locus of motionless fronts (which is proportional to the magnitude of the imaginary part of the coefficients):

$$x_{0,t} = -2\xi \left(\mu - \mu_F \right) \,. \tag{3.32}$$

The interaction between two fronts can be computed perturbatively for widely separated fronts as in section 3.1.2. The outcome of the calculation is that, contrary to the case studied in section 3.1.2, due to the phase contributions, the interaction is no longer monotonously decreasing with distance. In a situation like the one depicted in Fig. 3.3 with a bifurcated medium of size L in an otherwise trivial ($A = 0$) medium, the length L is found to evolve according to (for $L \gg 1$)

$$\frac{dL}{dt} = -2\xi \left(\mu - \mu_F \right) - a_0 \exp\left(-2L/\xi \right) + a_1/L \tag{3.33}$$

where a_0 is a positive constant of order one and a_1 a positive constant of order ϵ^2, ϵ denoting the magnitude of α_i, β_i, and γ_i in (3.22). The last two terms on the r.h.s. of (3.33) are the interaction terms between the two fronts and describe a short range attraction and a longer range repulsion.* For μ slightly bigger than μ_F a stable pulse exists (together with a smaller unstable pulse). This can be intuitively understood as follows. For a large enough L, interactions do not matter so a large pulse shrinks until the repulsive interaction is felt. If $\mu - \mu_F$ is not too large, the shrinking stops at $L = L_2$. This is a stable position since for a pulse of size L slightly below L_2 the repulsive interaction is slightly larger and L grows toward L_2. Finally if L is sufficiently small $(L < L_1)$ the short range attractive interaction overcomes the weak repulsive interaction and the pulse shrinks and disappears.

3.1.4   Reaction–diffusion models

a) *The excitable medium case*

An interesting variation on equation (3.1) is obtained when another field is present. A particularly remarkable example is given by excitable media. The best known phenomenon in this context is the propagation of an action potential along a neuron axon, but traveling waves in chemical reactions are also very striking. A simple model is obtained by modifying (3.1) as follows

$$\frac{\partial A}{\partial t} = \frac{\partial^2 A}{\partial x^2} + A - A^3 - B \,, \tag{3.34a}$$

$$\frac{\partial B}{\partial t} = \epsilon \left(A - A_1 \right) . \tag{3.34b}$$

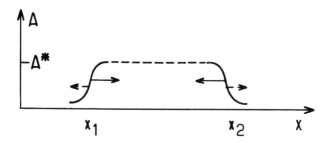

Fig. 3.3.   A localized phase of state A^* in an otherwise trivial $A = 0$ medium described by the interaction of two fronts.

---

* In (3.33) the term a_1/L is the behavior of the repulsive part for L sufficiently large. Of course, for small L, this term is simply of order ϵ^2 and there is no divergence.

This is simply (3.1) with $V(A) = -\frac{1}{2}A^2 + \frac{1}{4}A^4$ coupled to a slowly varying ($\epsilon \ll 1$) field. Let us first consider homogeneous states. The system's rest state is $A^* = A_1$, $B^* = A_1 - A_1^3$. Its stability can, of course, be determined by linearizing (3.34) around (A^*, B^*). However, it is more readily understood pictorially by looking at the phase space (A, B).

On Fig. 3.4, it is seen that in the small ϵ limit the resting state is unstable for $|A_0| < 1/\sqrt{3}$ so that the whole medium oscillates. Hereafter, we consider only the so-called excitable case where the resting state is linearly stable $\left(|A_0| > 1/\sqrt{3}\right)$. For $\epsilon \ll 1$, B changes much more slowly than A and it is enlightening to first analyze (3.34a) with B constant. This is simply (3.1) with $V_B(A) = -\frac{1}{2}A^2 + \frac{1}{4}A^4 - BA$. For definiteness let us suppose that $-1 < A_1 < -1/\sqrt{3}$ so that $V_{B*}(A)$ is a bistable potential. Let us consider a front of (3.34a) between $A = A_1$ and the other stable state A_2 of the potential. With the chosen value of A_1, B being fixed at B^*, the A_2-state is more stable than the A_1-state and it invades the A_1-state as described in section 3.1.1. How does the dynamics of B, (3.34b), change the picture? At any point, the front passage results in a jump of

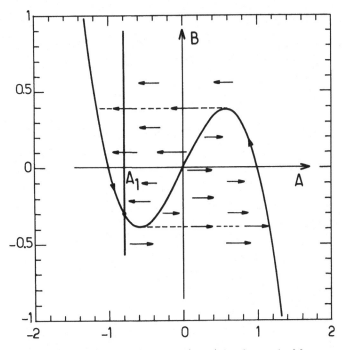

Fig. 3.4. Flow diagram for the equations (3.34) in the excitable case. The fixed point A_1, B_1 is stable.

A from A_1 to A_2 ($> A_1$) in a time of order one,[*] too small for a noticeable change of B. However, after the passage of the front, A is different from A_1 and B increases slowly from (3.34b). This continues until B reaches the value $2/3\sqrt{3}$ for which $V_B(A)$ is no longer a bistable potential so that the A_2-state of (3.34a) disappears. At this point A returns suddenly to the A_1-branch creating a second front behind the leading front. With the symmetric potential chosen in (3.34) the front velocity (in absolute value) depends only on the magnitude of $|B|$ and it is an increasing function of $|B|$. The second front which appears in a region where $B = 2/3\sqrt{3}$ moves initially more rapidly than the initial front which propagates with $B = B^*$. So the second front begins by catching up the leading front, but in so doing moves to a region with lower B values. The distance between the two fronts remains constant when the B value at the second front position has reached $|B^*|$. In conclusion, systems described by (3.34) possess one-dimensional traveling pulses which can be simply described and understood as bound states of two fronts.

In a two-dimensional geometry, as for a chemical reaction in a shallow Petri dish, the pulse just described corresponds to planar waves. It is of course also possible to have circular waves. More interestingly, planar and circular waves with free tips can be created (for example by mixing locally the reagents). Because of the tip tangential motion, these evolve naturally into remarkable spirals (see Fig. 3.5) as can be seen from elementary cellular automata models. The analysis of these spirals and their dynamics is a subject of much current interest (Tyson and Keener, 1988; Winfree, 1991).

b) *A model with long range inhibition*

In the previous activator-inhibitor model (3.34) the diffusion of B was neglected, but similar effects would have been obtained by choosing it to be comparable to the diffusion of A. However, things change, if B diffuses much more rapidly than A. In this case, a localized motionless structure may exist (Gierer and Meinhardt, 1972; Koga and Kuramoto, 1980; Ermentrout *et al.*, 1984)[†]. A simple model of this effect reads as:

$$\frac{\partial A}{\partial t} = \frac{\partial^2 A}{\partial x^2} + A - A^3 - B\,, \qquad (3.35a)$$

---

[*] This is true as long as the front velocity is not very small, i.e., if A_0^* is not very close to -1.

[†] I am grateful to S. Fauve for pointing out the paper of Koga and Kuramoto (1980). The analysis outlined below was developed during discussions with R. Goldstein and S. Leibler (1990, unpublished).

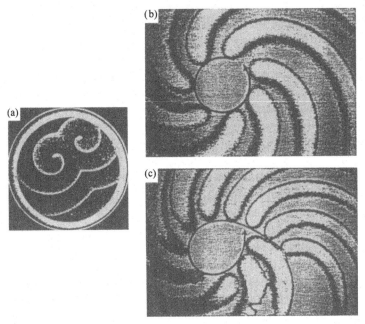

Fig. 3.5. Spirals in an excitable chemical medium (reproduced from Skinner and Swinney, 1991). (a) A general view of the Petri dish; (b) a spiral tip at successive times in a regime of simple rotation; (c) a spiral tip in the "meandering" regime.

$$\frac{\partial B}{\partial t} = \ell_0^2 \frac{\partial^2 B}{\partial x^2} + A - A_1 - \gamma B\,, \qquad \ell_0 \gg 1\,, \qquad (3.35b)$$

where A_1 and γ are chosen so that the system has only one stable homogeneous state (A^*, B^*) as before, with $B^* < 0$. In order to understand the possible existence of a localized state, let us first consider (3.35a) with B constant. For $B = B^*$, a state with two fronts with $A = A_1(B) < 0$ outside and $A = A_2(B) > 0$ inside grows if the fronts are sufficiently separated so that their attractive interaction is not too strong. This is true as long as $B < 0$ since in this case the state $A_2(B)$ is more stable than $A_1(B)$. Now, we can analyze the role of (3.35b). Since $\ell_0 \gg 1$, the B field varies on a much longer scale than the A field. The B field is produced by the bifurcated state between the two fronts and its value at the front location increases with the size of the bifurcated state. For two fronts separated by a distance small compared to ℓ_0, B is everywhere close to B^* and the two fronts move away. However, if they are sufficiently separated there is a large region where B is produced and the local B value can be positive at the position of the fronts (and in between). Since B does not vary much on the scale of the front width, the dynamics of each front is given by equation (3.35a) with a constant B value equal to the local value of B. If it is positive, the two fronts move toward each other. Therefore, when the separation between the two fronts is large they move toward each other, and a stable localized state can exist in this model. In the parameter range where this stable localized state exists a smaller unstable localized state due to the direct short range front interaction also exists, very analogously to the CGLE case (section 3.1.3). Similar localized states also exist in higher dimensions (Ohta *et al.*, 1989).

3.2 Invasion of an unstable state by a stable state

Here we consider a front between a stable and an unstable state. The observation of the single propagating front is of course much more difficult to perform than in the previous case, since the unstable state is very sensitive to any disturbance. This has nonetheless been achieved in several experiments (Ahlers and Cannell, 1983; Fineberg and Steinberg, 1987; Limat *et al.*, 1992). In Fig. 3.6, reproduced from Fineberg and Steinberg (1987), one sees a convective state which invades the unstable conductive state in a Rayleigh–Bénard experiment. The front moves at a constant velocity which is reproducible from experiment to experiment and depends on material and control parameters. The theory of this phenomenon reveals some subtleties and surprises, the main one being that a continuum set of front velocities and shapes exist. The question is therefore to find out what selects the physical solutions. We will consider a simple example and try to explain the main physical ideas. However, we will not do justice to any of the rigorous mathematical developments. The interested reader is referred to the original articles (Kolmogorov, Petrovskii, Piscounoff, 1937; Aronson and Weinberger, 1978; Bramson, 1983) and to the book by Collet and Eckmann (1990).

3.2.1 Continuum of front velocities in the Fisher–KPP equation

The invasion of an unstable state by a stable state seems to have been first considered by Fisher in the thirties as a problem in population dynamics. It was then carefully analyzed by Kolmogorov, Petrovskii and Piscounoff (1937) on the example of the equation

$$\frac{\partial A}{\partial t} = \frac{\partial^2 A}{\partial x^2} + f(A), \qquad f'(0) = 1. \tag{3.36}$$

For concreteness, we choose $f(A) = A - A^2$ as an example. Here the state $A = 0$ is unstable and the state $A = 1$ is stable. We thus consider a situation such that $A(x) \to 1$ for $x \to -\infty$ and $A(x) \to 0$ at $x \to +\infty$. Let us first search for a steady propagating front as we did in section 3.1.

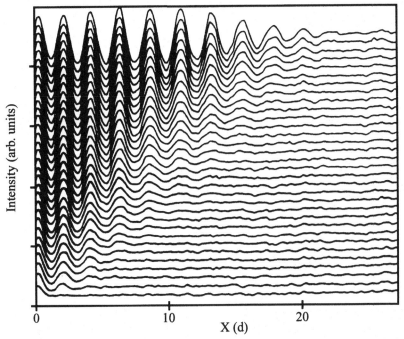

Fig. 3.6. Invasion of the unstable conductive state by the stable convective state in a Rayleigh–Bénard experiment (reproduced from Fineberg and Steinberg, 1987).

Substituting $A(x, t)$ by $A(x - ct)$ in (3.36), we obtain

$$\frac{d^2 A}{dx^2} = -c\frac{dA}{dx} - \frac{\partial U}{\partial A}, \qquad \text{with} \quad U(A) = \tfrac{1}{2}A^2 - \tfrac{1}{3}A^3. \qquad (3.37)$$

As before, this can be interpreted as a motion of a particle in the potential $U(A)$, the role of time being played by the x coordinate (Fig. 3.7).

The front is represented by a particle starting at $x = -\infty$ on the top of the hill at $A = 1$ and arriving in the bottom of the valley at $A = 0$ at "time" $x = +\infty$. How should the "friction" c be chosen so that such a particle motion is possible. The intuitively clear but surprising answer is that any non zero (and positive) friction is adequate: the particle will dissipate its energy and at "time" $x = +\infty$ it will rest on the bottom of the valley. So, to any positive velocity corresponds a front solution. We are thus faced with a selection problem: what is the velocity of the physical propagating front? Firstly, it is possible to eliminate all solutions where A does not remain positive because it is not difficult to show that they are all unstable. Using the mechanical analogy, it is clear that this happens for sufficiently small friction (i.e., velocity) since in this case the particle oscillates around the bottom of the well before coming to rest at

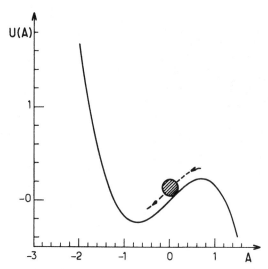

Fig. 3.7. Motion of a particle corresponding to a front solution.

$A = 0$. So the friction must be larger than a critical friction. In order to see it mathematically, we can determine the behavior of (3.36) near $A = 0$ where it reduces to:

$$\frac{d^2 A}{dx^2} = -c\frac{dA}{dx} - A.$$
(3.38)

Therefore, when $x \to \infty$, the asymptotic behavior of A is a linear super-position of two exponentials $\exp(-qx)$ with:

$$q^2 - cq + 1 = 0 \qquad \text{or} \qquad c = q + 1/q.$$
(3.39)

For $c < 2$ this equation has two complex roots with positive real part. So A oscillates before reaching 0 as stated. On the contrary, for $c > 2$ the two roots of (3.39) are real and positive so these solutions are not obviously forbidden. The velocity is plotted against q in Fig. 3.8.

At this stage two remarks can be made:

1) In general the branch with $q > 1$ (dashed in Fig. 3.8) does not corre-spond to the convergence rate of a front. The reason is simply that the large x asymptotic behavior of a solution of (3.37) moving at $c > 2$ is a superposition of the two possible asymptotic modes. Therefore its convergence rate toward zero is controlled by the slowest $q < 1$ converging mode making the $q > 1$ mode unobservable.

2) The fact that c behaves like $1/q$ for small q is easily interpreted. This is the behavior that would be obtained for (3.36) if the diffusive term was absent. So this has nothing to do with a real propagation. It is

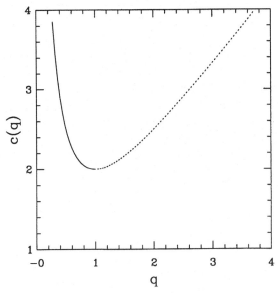

Fig. 3.8. Front velocity $c(q)$ as a function of the front asymptotics.

simply the independent growth of A at different points set up by the initial condition.

This last remark makes it clear that the velocity of the invading front depends on the initial condition and that it can be as fast as one wants if the initial condition decreases sufficiently slowly at $x \to +\infty$. The physical question is therefore to determine the front velocity for an initial condition that decreases sufficiently rapidly as $x \to +\infty$. In the case of the second order nonlinear diffusion equation the problem has been rigorously solved in the above-cited mathematical references. The result is that the propagation speed is the smallest one for which $A(x)$ remains everywhere positive. The main mathematical tool is a comparison theorem which states that given two solutions $A(x,t)$ and $B(x,t)$ of (3.36), if A is everywhere larger than B at time 0 $(A(x,0) > B(x,0))$ then this remains true at all time $(A(x,t) > B(x,t)$ for all $t)$. Then a sufficiently localized initial condition can be bounded above at time 0 by the slowest front solution and therefore cannot propagate faster than it. Instead of detailing these interesting mathematical proofs, we explain an heuristic argument (Ben-Jacob *et al.*, 1985) of wider applicability.

3.2.2 The criterion of linear marginal stability

We consider the fate of a small perturbation when viewed from a referential moving at velocity c. The idea is that the leading edge of the front should be a small stationary perturbation of the unstable state in the front referential. A small perturbation can be written as:

$$W(x,t) = \int dk\, A(k) \exp[ikx - w(k)t] , \qquad (3.40)$$

where $A(k)$ is a slowly varying function of k, in order to have a localized spatial disturbance. In the coordinate $x' = x + ct$ of a referential moving at speed c, W is given by:

$$W(x',t) = \int dk\, A(k) \exp[ik(x' + ct) - w(k)t] . \qquad (3.41)$$

For large t the behavior of W can be estimated by the saddle point method

$$W(x',t) \sim A(\bar{k}) \exp(i\bar{k}x') \exp[-\sigma(\bar{k})t] \qquad (3.42)$$

where $\sigma(k)$ is defined by

$$\sigma(k) = w(k) - ikc \qquad (3.43)$$

and $\bar{k}$ is determined by the phase stationary condition:

$$\left.\frac{dw}{dk}\right|_{k=\bar{k}} = ic . \qquad (3.44)$$

Now, as explained in Chapter 2, two physically different kinds of instability can be obtained:

1) For $\mathcal{R}e\{\sigma(\bar{k})\} > 0$ the perturbation disappears at any fixed position in the moving referential. It grows but is convected away sufficiently rapidly. This is a "convective instability" in the moving referential.
2) For $\mathcal{R}e\{\sigma(\bar{k})\} < 0$ the perturbation grows at fixed position in the moving referential. It is an "absolute instability."

Ben-Jacob *et al.* (1985) argue self-consistently that since the leading edge of the front is a stationary perturbation in the front referential, the front should move at a velocity such that $\mathcal{R}e\{\sigma(\bar{k})\} = 0$. This is the marginal stability criterion for selecting the front velocity. Let us show how this works in the case of (3.36). Substituting $A(x,t) = A_k \exp[ikx - w(k)t]$ in (3.36) and linearizing, we obtain:

$$-w(k) = 1 - k^2 . \qquad (3.45)$$

Replacing this in (3.44) we obtain $\bar{k} = ic/2$ and $\sigma(\bar{k}) = \frac{1}{4}c^2 - 1$. The marginal stability condition, $\mathcal{R}e\{\sigma(\bar{k})\} = 0$, therefore gives $c = 2$ in agreement with the mathematical results cited before.

As we are going to discuss in the next section, this ingenious self-consistent marginal stability criterion does not always work because it does not take into account all the instability modes around the nonlinear front solution (for a discussion see Van Saarloos, 1989). Nonetheless, since it is based on a simple linear stability analysis, it can be applied to a large range of phenomena. For example it is not difficult to extend it to the complex amplitude equation (Ben-Jacob *et al.*, 1985):

$$\tau \frac{\partial A}{\partial t} = \xi^2 \frac{\partial^2 A}{\partial x^2} + \epsilon A - g|A|^2 A \tag{3.46}$$

In this case, the prediction for the moving front velocity is $V = 2\xi\sqrt{\epsilon}/\tau$. This has been verified experimentally in the beautiful experiment of Fineberg and Steinberg (1987) mentioned previously from which Fig. 3.9 is extracted. A comparison with a numerical solution of the equation can be found in Lücke *et al.* (1987).

3.2.3   Nonlinear selection of the front velocity

In the previous sections, we tried to make plausible the result that the propagating front created by a sufficiently localized initial condition has the smallest speed c^* among all the positive front solutions. The linear analysis (3.39) shows that all fronts for which $c \leq 2$ have an oscillating leading edge and therefore cannot be positive. So c^* is certainly greater than 2. We showed that for $c = 2$ the leading edge is marginally stable

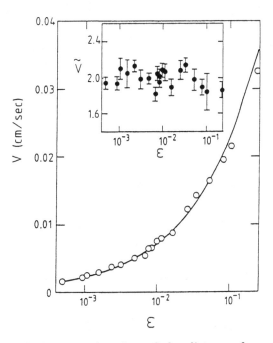

Fig. 3.9.   Front velocity as a function of the distance from threshold in the experiment of Fineberg and Steinberg. The solid line is the marginal stability prediction for (3.46) (reproduced from Fineberg and Steinberg, 1987).

and argued that this was therefore a good candidate for the selected velocity. For some choices of $f(A)$ this reasoning can go wrong, because the leading edge of the $c = 2$ front solution of (3.36) is negative and the corresponding solution is unstable. In these cases, the selected velocity c^* is strictly greater than $c = 2$. We consider an interesting example (Ben-Jacob *et al.*, 1985) where c^* can be determined analytically. A slightly more complicated form for $f(A)$ is chosen:

$$f(A) = A(1 - A)(1 + \alpha A). \tag{3.47}$$

where α is a parameter. The corresponding equation for a front moving at speed c is:

$$\frac{d^2 A}{dx^2} = -c\frac{dA}{dx} - A(1 - A)(1 + \alpha A). \tag{3.48}$$

We want to determine the asymptotic behavior at $x = +\infty$, $A(x) = k_S \exp[-q_S x] + k_F \exp[-q_F x]$ of the solution of (3.48) which starts at[*] $x = -\infty$ from $A = 1$. More precisely the sign of the coefficient k_S of the slowest decreasing mode is important: if k_S is negative then we know that the corresponding A is not an allowed solution. Unfortunately, in general we do not know how to solve (3.48). For large c, it is intuitively clear that A remains positive since it will creep toward zero at $x \to +\infty$. If the leading part of the solution becomes negative for $c \geq 2$ this should happens by a change of sign of k_S. Therefore precisely at this point, the asymptotic of the front solution is given only by the single fast exponential decrease. This gives the idea of searching this particular front solution as a solution of a first order differential equation (a more general application of this strategy is given by Van Saarloos and Hohenberg, 1992):

$$\frac{dA}{dx} = g(A), \tag{3.49}$$

where g is an unknown function still to be determined. The advantage of (3.49) over (3.48) is that it can be integrated without difficulty. In order to find $g(A)$, we substitute (3.49) into (3.48) and obtain:

$$(g'(A) + c)g(A) = -A(1 - A)(1 + \alpha A). \tag{3.50}$$

We can try a polynomial solution for $g(A)$. Matching degrees on both sides shows that g should be of second degree. Moreover since we look

---

[*] This solution is unique once translation invariance is taken care of. This is clear from the mechanical analogy. A more mathematical argument is that linearization around $A = 1$ gives a converging and a diverging mode at $x = -\infty$. Once the coefficient of the diverging mode is set to zero, the only arbitrariness is the choice of the coefficient of the converging mode which gives simply a translation of the solution.

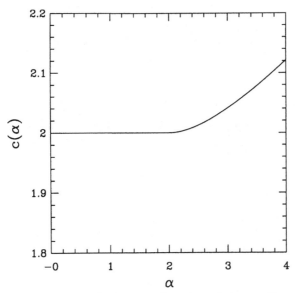

Fig. 3.10. Plot of the front velocity as a function of the nonlinearity strength α for (3.48). For $\alpha > 2$ the velocity is larger than the constant value 2 predicted by the linear marginal stability criterion.

for a front solution, (3.49) shows that $g(A)$ should vanish for $A = 0$ or 1 and:

$$g(A) = \beta \left(A^2 - A \right). \tag{3.51}$$

Then (3.50) gives the two equations $2\beta^2 = \alpha$, $\beta(c - \beta) = 1$. Therefore, for each choice of the function $f(A)$, a particular front $A^*(x)$ moving at velocity $c^* = \sqrt{\alpha/2} + \sqrt{2/\alpha}$ is obtained

$$A^*(x) = \left[1 + e^{\sqrt{\alpha/2}(x+x_0)} \right]^{-1}. \tag{3.52}$$

$A^*(x)$ is a special solution of (3.48) whose asymptotic approach toward zero is controlled by one exponential (and its harmonics) instead of two in the general case. Moreover for $\alpha > 2$, this asymptotic approach belongs to the non generic fast branch (dashed in Fig. 3.8) which means that $k_S(c)$, the coefficient of the slow mode, is exactly zero at c^*. If we assume that $k_S(c)$ has only one simple non degenerate zero, we can conclude that c^* is precisely the point where $k_S(c)$ changes sign, being positive for $c \geq c^*$ and negative for $c \leq c^*$. Therefore for $\alpha > 2$ the slowest velocity of positive front solutions is $c^* = \sqrt{\alpha/2} + \sqrt{2/\alpha}$ and it is strictly greater than 2 (Fig. 3.10). This departure from the marginal velocity criterion as nonlinearities are increased is often observed (for a recent example see Pellegrini and Jullien, 1991).

4 Exponentially small effects and complex-plane boundary layers

4.1 Introduction

For usual physical problems, it is generally enough to compute the first few terms of a perturbation series if a perturbative approach is worth following at all. In sections 2 and 3, we have seen how difficulties such as the appearance of boundary layers or of secular terms are generally revealed by low order analyses. In this section, we are going to describe some counter-examples to this common wisdom, which have recently appeared in a variety of contexts. The general feature of these problems is that the perturbation series to any finite order is well-behaved but fails to reveal the main interesting phenomena although, in some cases, the complete solution of the problem is already close to the zeroth order solution of the perturbation series. The common underlying reason is that the key effect is exponentially small, i.e., behaves like $\exp(-c/\epsilon)$ in the small perturbation parameter ϵ. It is therefore invisible to any finite order in ϵ and can be rather subtle to determine. As a concrete example in hydrodynamics we describe the viscous finger puzzle.

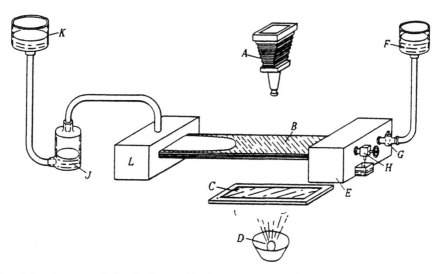

Fig. 4.1. Sketch of the Saffman–Taylor experiment (reproduced from Saffman and Taylor, 1958).

At the end of the fifties, Saffman and Taylor studied the dynamics of the air/oil interface when air is blown into the oil-filled space between two close glass plates and pushes out the oil (see Fig. 4.1).

The use of this device was motivated by the fact that the relation between pressure and velocity in lubricating flows is analogous to the macroscopic Darcy's description of flow through porous media, as pointed out by Hele-Shaw. As shown in Fig. 4.2, when air pushes the oil, a planar air/oil interface is unstable. In the asymptotic nonlinear regime a long finger of air advances through the oil. This is qualitatively similar to the observation that a large proportion of mineral oil remains in porous rocks after attempts at extraction by water injections. Now, for the puzzle. By using the lubrication approximation to reduce the problem to two dimensions and a clever change of variables, Saffman and Taylor succeeded in finding steady shapes of moving air/oil interfaces (Saffman and Taylor, 1958). But they found too many! They obtained a continuous family of moving fingers which could have any width between zero and the whole cell width. The experimental shape was found to occupy approximately half the channel width for sufficiently high air injection speed and it was fitted rather well by the corresponding analytic solution. The puzzle was to find why this particular shape was singled out of the continuum of analytic solutions.

Indeed, these solutions are exact solutions of reduced two-dimensional equations, and not complete solutions of full Navier–Stokes equations. The neglected features, the most important one being surface tension between air and oil and the three-dimensional nature of the flow near the interface could presumably be taken into account by perturbation since the analytically determined shape already agreed quite satisfactorily with the experimental one. But, when this was attempted, it appeared that all the analytic solutions could be perturbed without encountering any feature which would have indicated a particular role for one member of the family. The puzzle solution was finally found when it was realized that the selection was coming from exponentially small effects that were invisible at any finite order of perturbation theory.

We are going to discuss first the simple example of an ordinary differential equation where such exponentially small effects change completely the nature of the solution. This is used to explain in detail how these effects can be computed by using a kind of boundary layer analysis in the complex plane. We will then come back to the more realistic example of the viscous fingers. In the last part of this section, we discuss several other physical examples, where exponentially small effects have been found to play a key role.

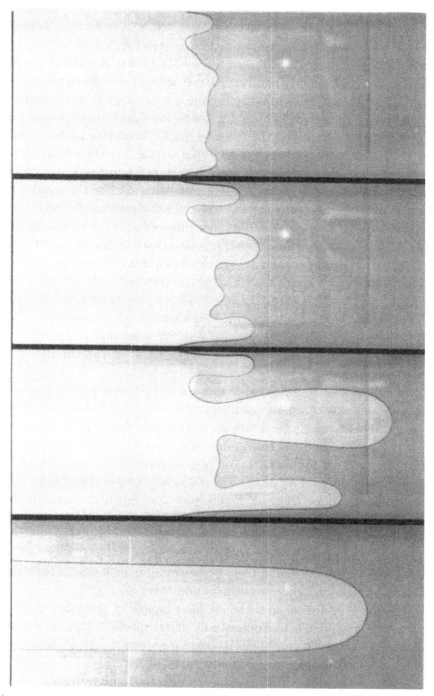

Fig. 4.2. Development of the instability of the air/oil interface in a Hele-Shaw cell viewed from above. In the end state, a single finger of air advances through the cell (Courtesy of Y. Couder and M. Rabaud).

4.2 The example of the geometric model of interface motion

4.2.1 The geometric model

The geometric model is a simple phenomenological model of interface dynamics (Brower *et al.*, 1984). It mimics realistic situations like the solid/liquid interface in dendritic growth or the air/oil interface in viscous fingering in that outward-pointing parts tend to move faster than flat ones. But, it is much simpler because non-local interactions between different parts of the interface are entirely neglected. Specifically, one postulates that the normal velocity v_n is proportional to the curvature κ of the interface. One adds a term which stabilizes the interface at short distances and plays the role of surface tension in more realistic descriptions. In two dimensions, the interface is simply a line, the motion of which is given by:

$$v_n = V_0 \left[\kappa \ell_0 + \ell_0^3 \frac{\partial^2 \kappa}{\partial s^2} \right] \tag{4.1}$$

where ℓ_0 and V_0 are reference length and velocity and s is the curvilinear abscissa along the interface. The simple question that we are going to analyze is the existence of steady shapes moving at constant velocity V that have the form of a symmetric needle along the velocity axis (Fig. 4.3).

If θ denotes the angle between the velocity and the local normal to the interface then:

$$v_n = V \cos \theta , \qquad \kappa = \frac{d\theta}{ds} . \tag{4.2}$$

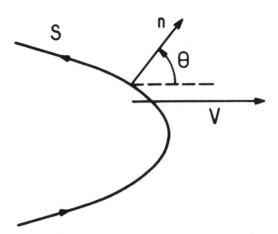

Fig. 4.3. A needle shaped geometric interface moving at velocity V.

Therefore the equation satisfied by the interface profile is:

$$\epsilon^2 \frac{d^3\theta}{dz^3} + \frac{d\theta}{dz} = \cos\theta \qquad (4.3)$$

where we have introduced the dimensionless parameter $\epsilon = V/V_0$ and rescaled the curvilinear abscissa $s = z\ell_0/\epsilon$. The question of the existence of a steady symmetric shape can now be phased rather simply. "Does equation (4.3) have an odd solution such that $\theta(\pm\infty) = \pm\pi/2$?"

4.2.2   A perturbative approach

We are going to analyze the above-mentioned question in the small ϵ limit where a perturbative approach can be tried (a review of the mathematical literature can be found in Segur, 1991). In order to do that we search for θ in the form:

$$\theta = \theta_0 + \epsilon^2\theta_1 + \ldots + \epsilon^{2n}\theta_n + \ldots \qquad (4.4)$$

When this form is substituted in (4.3), one obtains successive equations for $\theta_0, \theta_1, \ldots, \theta_n$.

At lowest order, one gets:

$$\frac{d\theta_0}{dz} = \cos\theta_0 . \qquad (4.5)$$

This is easily solved and one obtains for θ_0

$$\theta_0 = -\frac{\pi}{2} + 2\arctan\left(e^z\right), \quad \sin\theta_0 = \tanh z, \quad \cos\theta_0 = 1/\cosh z. \quad (4.6)$$

θ_0 is an odd function that satisfies the required boundary conditions $\theta_0(\pm\infty) = \pm\pi/2$. Similarly at next order, one gets:

$$\frac{d\theta_1}{dz} + \theta_1 \sin\theta_0 = -\frac{d^3\theta_0}{dz^3}. \qquad (4.7)$$

Using (4.6) this first order equation is also easily solved and gives:

$$\theta_1 = -\frac{z}{\cosh z} + 2\frac{\sinh z}{\cosh^2 z}. \qquad (4.8)$$

Therefore θ_1 is also an odd function which satisfies the right boundary conditions $\theta_1(\pm\infty) = 0$. Therefore, up to this order of perturbation theory nothing strange* happens and it seems that we have determined

---

* The careful reader will notice, as G. Joulin did during the oral presentation of this lecture in Beg-Rohu, that the first term on the r.h.s. of (4.8) looks secular. This is indeed true and it happens because the third order equation (4.3) and the first order equations (4.5–7) have different convergence rates at infinity. This can be handled by introducing boundary layers at $\pm\infty$ and by matching them to the present (outer) expansion. This purely local matching does not shed light on the difficulty encountered by the perturbation expansion

an approximation for the steady-state shape, we were looking for. In fact, the same is true to all-orders in perturbation theory, as we now argue.

The equation which determines θ_n is:

$$\frac{d\theta_n}{dz} + \theta_n \sin \theta_0 = P_n(\theta_1, \ldots, \theta_{n-1}) \sin \theta_0$$
$$+ Q_n(\theta_1, \ldots, \theta_{n-1}) \cos \theta_0 - \frac{d^3\theta_{n-1}}{dz^3}. \tag{4.9}$$

P_n and Q_n are polynomials in $\theta_1, \ldots, \theta_{n-1}$ that come from the coefficient of ϵ^{2n} in the expansion of $\cos(\theta_0 + \epsilon^2\theta_1 + \ldots)$. Since $\cos(\theta_0 + \epsilon^2\theta_1 + \ldots)$ is invariant when all θ_n are replaced by their opposite ($\theta_n \to -\theta_n$), this is also true for all coefficients of the expansion. Since under this change $\sin(\theta_0) \to -\sin(\theta_0)$ and $\cos(\theta_0) \to \cos(\theta_0)$, one has:

$$P_n(-\theta_1, \ldots, -\theta_{n-1}) = -P_n(\theta_1, \ldots, \theta_{n-1}),$$
$$Q_n(-\theta_1, \ldots, -\theta_{n-1}) = Q_n(\theta_1, \ldots, \theta_{n-1}). \tag{4.10}$$

Now, we can reason by induction and suppose that $\theta_1, \ldots, \theta_{n-1}$ are odd functions that vanish rapidly at $\pm\infty$ (like polynomials multiplied by $\exp(-|z|)$. Then, it follows from (4.10) and (4.6) that the r.h.s. of (4.9) is an even function. Moreover, the linear operator $L(= \partial_z + \tanh(z))$ on the l.h.s. of (4.9) anticommutes with parity. Therefore, if $\theta_n(z)$ is a solution of equation (4.9), $\theta_n(z) - \theta_n(-z)$ is also a solution, and (4.9) has an odd solution. This solution also vanishes rapidly at infinity since the r.h.s. does, by induction, and L has no diverging mode at infinity.

The conclusion of this perturbative approach seems to be that a symmetric needle shape exists since we have found an expansion to every order in ϵ. Nonetheless, we will now see that this is very misleading in this problem.

4.2.3 A simple shooting argument

We are now going to present a simple argument (Kessler *et al.*, 1985; Ben-Jacob *et al.*, 1984) which strongly suggests that no steady needle solution exists, in contradiction to what was indicated by the perturbative approach of subsection 4.2.2.

Let us linearize (4.3) around $z = -\infty$, $\theta = -\frac{1}{2}\pi + \phi$. Then one obtains to linear order in ϕ

$$\epsilon^2 \frac{d^3\phi}{dz^3} + \frac{d\phi}{dz} = \phi. \tag{4.11}$$

---

which comes from connecting the behaviors at $-\infty$ and $+\infty$. This local matching has been, for instance, performed by McLean and Saffman (1981) in the viscous fingering context but this did not give any hint on the width selection mechanism.

ϕ is a linear combination of three exponential modes $\exp(\lambda z)$ where λ is a solution of

$$\epsilon^2 \lambda^3 + \lambda - 1 = 0. \qquad (4.12)$$

The three solutions of this algebraic equation are:

$$\lambda_1 = 1 + \mathcal{O}(\epsilon), \qquad \lambda_\pm = \pm \frac{i}{\epsilon} - \frac{1}{2} + \mathcal{O}(\epsilon). \qquad (4.13)$$

There are two divergent modes and one convergent mode around $z = -\infty$. The same result is also obtained by linearization around $z = +\infty$. However (4.3) is a third order equation and its solution only depends on three arbitrary constants. If we start the integration at $z = -\infty$, two constants are fixed by the cancellation of the two diverging modes at $z = -\infty$ and, one constant is missing to kill the two diverging modes at $z = +\infty$. In a similar way, once the two diverging modes at $z = -\infty$ have been set to zero, one can use the last constant to choose $\theta(0) = 0$. There is then no reason for $\theta''(0)$ to vanish. Therefore, this shooting argument suggests that there are no symmetric needle shapes (except maybe at some discrete value of ϵ *).

The two approaches appear to lead to different conclusions because it turns out that $\theta''(0)$ is exponentially small in ϵ in the small ϵ limit

$$\left. \frac{d^2\theta}{dz^2} \right|_{z=0} \sim \frac{2\Gamma}{\epsilon^{5/2}} \exp\left(-\frac{\pi}{2\epsilon}\right). \qquad (4.14)$$

This finite value is zero to all orders in perturbation theory. Therefore, the main physical conclusion that no steady-state exists is missed by the perturbation approach of section 4.2.2.

4.2.4   Boundary layers in the complex plane

How can this shortcoming of perturbation theory in the small ϵ limit be overcome? As we have seen in section 1 on boundary layers, perturbative approaches can fail when there is a small parameter in front of the highest derivatives. This happens when the function varies sufficiently rapidly in the neighborhood of some point so that derivatives are so much larger than the function that this compensates for the explicit appearance of the small parameter in front of them. Here, the needle shape (4.6) varies on a scale large compared to ϵ when z is real so no problem appears. However, this is no longer true in the complex plane in the neighborhood of one of the singularities of θ_0. There θ_0 varies rapidly and $\epsilon^2 \theta'''(z)$ is not negligible compared to $\theta'(z)$. Another way to see that the perturbative

---

* That this does not happen has been proved by Amick and McLeod (1990).

approach is in trouble around one singularity z_s of θ_0 is to notice that
the higher is n, the stronger is the divergence of θ_n. As shown below:

$$\theta_n \sim \frac{i\,a_n}{(z - z_s)^{2n}}. \tag{4.15}$$

So, in a neighborhood of size ϵ around z_s, the contribution $\epsilon^{2k}\theta_k$ is of the
same order in ϵ for all k. The idea of Kruskal and Segur (1991) is to define
a new perturbation expansion in this inner region around a singularity
where effects invisible within the usual perturbation expansion grow and
become visible. This is in spirit similar to the usual boundary layer
analysis explained in section 1, but here the boundary layer is located in
the complex plane instead of being in the real physical plane.

In order to follow this strategy, we begin by analyzing the perturbation
expansion (4.4) in the neighborhood of one of its singularities. Firstly,
since (4.9) is linear, the singularities of θ_n are located where the coefficients of the equation are singular. Therefore, it is clear by induction
that the singularities of $\theta_1, \ldots, \theta_n$ are located at the same position as
those of θ_0, i.e., at $z = i(\pi/2 + k\pi)$, $k \in \mathbb{Z}$. We analyze the behavior of
the perturbation series around $z = i\pi/2$ since the singularities closest to
the real axis will turn out to be the most important ones.

We first show that the strength of the divergence of the terms of the
perturbation expansion (4.9) increase with their order as stated in (4.15).
From the explicit expression of θ_1, (4.8), one obtains that its behavior near
$i\pi/2$ is

$$\theta_1 \sim \frac{-2i}{(z - i\pi/2)^2}. \tag{4.16}$$

So that $a_1 = -2$. Let us obtain, by induction, that if (4.15) is true for
$1 \le k \le n - 1$ it is also true for θ_n. θ_n is given by (4.9)

$$\frac{d\theta_n}{dz} + \theta_n \sin\theta_0 = P_n\,(\theta_1, \ldots, \theta_{n-1})\,\sin\theta_0$$

$$+ Q_n\,(\theta_1, \ldots, \theta_{n-1})\,\cos\theta_0 - \frac{d^3\theta_{n-1}}{dz^3}, \tag{4.17}$$

where P_n and Q_n are polynomials in $\theta_1, \ldots, \theta_{n-1}$ that are the coefficients of order ϵ^{2n} of $\sin\theta_0$ in the expansion of $\cos\,(\theta_0 + \epsilon^2\theta_1 \ldots)$. By
the induction hypothesis the power of the strongest divergence of θ_k for
$i \le k \le n - 1$ is equal to $2k$ which coincides with the power of its order in ϵ. Therefore the strongest divergence of P_n and Q_n is also equal
to their order in ϵ and they diverge near $i\pi/2$ like $(z - i\pi/2)^{-2n}$. Since
$\sin\theta_0$ and $\cos\theta_0$ diverge like $(z - i\pi/2)^{-1}$ the l.h.s. of (4.17) diverges
like $(z - i\pi/2)^{-(2n+1)}$ near $i\pi/2$. By comparing with the r.h.s. of (4.17)
one concludes that θ_n diverges like $(z - i\pi/2)^{-2n}$ near $i\pi/2$ as stated in

(4.15). The consequence is that when $i\pi/2$ is approached at distances of order ϵ all the terms $\epsilon^{2n}\theta_{2n}$ of the perturbation expansion (4.9) have the same magnitude. It is therefore necessary to define a new perturbation expansion in this neighborhood of $\pi/2$, the inner region. In order to do so let us define a rescaled variable w to measure the distance from $i\pi/2$

$$z - i\frac{\pi}{2} = \epsilon w .\tag{4.18}$$

Then formally in the inner region, $\theta(z)$ is equal to

$$\theta(z) = -\frac{\pi}{2} - i\log\left(-\frac{w\epsilon}{2}\right) + \sum_{n=1}^{\infty}\frac{ia_n}{w^{2n}} + \mathcal{O}(\epsilon) ,\tag{4.19}$$

where the most diverging term of each order of the regular perturbation equation (4.9) appears in the sum. In order to define this sum more precisely, it is convenient to introduce a new function ϕ by

$$\theta\left(i\frac{\pi}{2} + \epsilon w\right) \equiv -\frac{\pi}{2} - i\log\frac{\epsilon}{2} + i\phi(w) .\tag{4.20}$$

The equation obeyed by ϕ is simply obtained by substituting the definition (4.20) in the original equation (4.3) for θ

$$\frac{d^3\phi}{dw^3} + \frac{d\phi}{dw} = e^{\phi} - \left(\frac{\epsilon}{2}\right)^2 e^{-\phi} .\tag{4.21}$$

This is simply a rewriting of (4.3) but it permits to obtain a regular perturbation expansion in the neighborhood of $z = i\pi/2$ by expanding ϕ in powers of ϵ. At zeroth order one obtains:

$$\frac{d^3\phi_0}{dw^3} + \frac{d\phi_0}{dw} = e^{\phi_0} .\tag{4.22}$$

Therefore, one way to give a meaning to the sum in (4.19) is to solve the inner problem (4.22) with the boundary condition

$$\phi_0(w) \sim \log(-w) , \qquad w \to -\infty ,\tag{4.23}$$

so that the inner and outer expansion can be matched when w tends to $-\infty$.

4.2.5 Solution of the inner problem and non existence of the symmetric needle

How can the solution of the inner problem give some information about the existence of the symmetric needle? For a symmetric needle, $\theta(z)$ is purely imaginary on the imaginary axis since, in this case, $\theta(z)$ is an odd real function for z real. So around $z = 0$ it looks like

$$\theta(z) = t_1 z + t_3 z^3 + t_5 z^5 + \dots\tag{4.24}$$

where the coefficients t_n are real. For a purely imaginary z, θ is then purely imaginary. So, given the relation (4.20), $\phi(w) + i\pi/2$ is purely real on the lower w imaginary axis ($w = -i\rho$, $\rho \geq 0$) if a symmetric needle exists. The strategy consists therefore in solving the inner equation (4.22) with the boundary condition (4.23) to compute the imaginary part of $\phi(-i\rho)$. The alternative is therefore:

$$\mathcal{I}m\left\{\phi_0(-i\rho) + \pi/2\right\} \begin{cases} = 0 & \text{no problem at this order,} \\ \neq 0 & \theta_0 \text{ cannot be odd.} \end{cases}$$

What is the expected form of this quantity when ρ is large? Starting from the asymptotic behavior (4.23) of $\phi_0(w)$, the asymptotic series for the large w behavior of $\phi_0(w)$ can be generated

$$\phi_0 = -\log(-w) - \frac{2}{w^2} - \frac{50}{3w^4} + \ldots \tag{4.25}$$

The general term of this series is of course a_n/w^{2n}, (4.19), since the inner problem was defined so as to sum the most divergent terms near $z = i\pi/2$ of the perturbation expansion (4.9). Indeed $\mathcal{I}m\left\{a_n/(i\rho)^{2n}\right\} = 0$ for any n and each term in the asymptotics gives a zero contribution to the imaginary part of $\phi_0 + \pi/2$. This again comes from the fact that the perturbation expansion shows no sign of the non-existence of the symmetric needle. But now, it is also possible to solve (4.22) to obtain the complete asymptotic behavior of ϕ_0. One can then see if there exist terms which contribute to $\mathcal{I}m\left\{\phi_0(-i\rho) + \pi/2\right\}$ but are invisible in perturbation theory because they are smaller than $|w|^{-n}$ for any n when $|w| \to \infty$ with $-\pi \leq \arg w \leq -\pi/2$. What can be the functional form of these terms? Let us add a potential small correction $\eta(w)$ to the asymptotic development (4.25). By substitution in (4.22), one can determine the equation which η satisfies to linear order:

$$\frac{d^3\eta}{dw^3} + \frac{d\eta}{dw} = \eta \exp\left[-\log(-w) + \ldots\right] \approx -\frac{\eta}{w}. \tag{4.26}$$

The asymptotic expansion has been truncated to its first term since this is sufficient to determine the leading asymptotic behavior of η for large $|w|$. It can be searched under the form

$$\eta(w) = \gamma(w)\, e^{\lambda w}. \tag{4.27}$$

At leading order, one obtains:

$$\lambda^3 + \lambda = 0, \quad \text{i.e.,} \quad \lambda = 0, \quad +i, \quad -i. \tag{4.28}$$

At next order the leading behavior of the prefactor is determined by:

$$\left(3\lambda^2 + 1\right)\gamma'(w) = -\frac{\gamma(w)}{w}. \tag{4.29}$$

Therefore, the general asymptotic behavior of a solution of (4.26) is a linear superposition of $1/w$, $w^{1/2} \exp(-iw)$, and $w^{1/2} \exp(+iw)$. Clearly, only $w^{1/2} \exp(iw)$ is smaller than every term in the asymptotic expansion (4.25) when $|w| \to \infty$ with $-\pi \le \arg w < -\pi/2$. It is therefore the only possible correction term. If it does not vanish for some "miraculous" reason, one expects that the asymptotic behavior of $\mathcal{I}m\,\{\phi + \pi/2\}$ (or of the real part of θ) on the lower imaginary w axis is given by:

$$- \mathcal{R}e \left\{ \theta \left(i\frac{\pi}{2} - i\epsilon\rho \right) \right\} = \mathcal{I}m \left\{ \phi_0(-i\rho) + \frac{\pi}{2} \right\} \sim \Gamma\sqrt{\rho}\exp(-\rho). \quad (4.30)$$

The determination of the constant Γ should be performed numerically (see section 4.2.7 for an alternative strategy) by solving the inner problem (4.22). One obtains $\Gamma \approx 2.11$ (Kruskal and Segur, 1991). Therefore, the imaginary part of $\mathcal{I}m\,\{\phi_0(w) + \pi/2\}$ does not vanish on the lower imaginary w-axis and one can conclude that there is no even solution of equation (4.3). At this stage, it is instructive to translate back (4.30) into information on $\theta(z)$ for real values of z.

4.2.6 Matching between the inner and outer region and exponentially small terms

Each term of the asymptotic series (4.25) can be matched to a term of the regular perturbation expansion (4.9). In the same way, it is possible to match the correction term $\eta(w)$ (4.27–30) with a correction term to the regular perturbation expansion (4.9). The information gained by going into the complex plane and solving the inner problem can then be given a precise meaning on the real axis.

In order to obtain the functional form of potential corrections $C(z)$ to the regular perturbation expansion, we use a strategy similar to the previously developed one. By substitution in (4.3) and linearization around the regular perturbation expansion, one obtains to linear order:

$$\epsilon^2 \frac{d^3C}{dz^3} + \frac{dC}{dz} = -\sin\left(\theta_0 + \epsilon^2\theta_1 + \dots\right) C. \quad (4.31)$$

Now, C can be searched for under the WKB form

$$C(z) = g(z) \exp\left[\frac{f(z)}{\epsilon}\right]. \quad (4.32)$$

At lowest order one obtains:

$$(f'(z))^3 + f'(z) = 0 \quad (4.33)$$

with the three possible solutions $f(z) = \text{cst.}$, $iz + \text{cst.}$, or $-iz + \text{cst.}$.

In order to find the lowest order approximation for the prefactor g, it is sufficient to keep only the first term θ_0 of the asymptotic series for θ

in (4.31). One then finds

$$\left(3f'^2(z) + 1\right) g'(z) = - \tanh(z)\, g(z) \,. \tag{4.34}$$

At this order, the general form of $C(z)$ is a linear combination of the three different solutions. Remembering that θ is a real function and that it vanishes at $z = 0$ by definition of the problem, one obtains:

$$C(z) = g\left(\frac{2}{\cosh(z)} - \sqrt{\cosh(z)}\,\exp(iz/\epsilon) - \sqrt{\cosh(z)}\,\exp(-iz/\epsilon)\right) \,. \tag{4.35}$$

In order to determine the three arbitrary constants, (4.35) should be matched to the asymptotics of the inner problem (4.30) in the neighborhood of $i\pi/2$. Replacing z by $i\pi/2 - i\rho\epsilon$, one obtains that (4.35) reduces to (4.30) for

$$g = +\frac{\Gamma}{\sqrt{\epsilon}}\exp\left(-\frac{\pi}{2\epsilon}\right) \,. \tag{4.36}$$

This implies of course that $\theta''(0)$ is given by (4.14) and therefore that θ is not an odd function.

We have thus shown how exponentially small terms can be obtained. In the next section, we will give several recent physical examples where these effects play a crucial role. In some cases, it is important to compute precisely the magnitude of the exponential terms and the prefactor Γ. This can be done by a pure numerical solution of the inner problem as originally performed for the geometric model by Kruskal and Segur (1991). It is also possible to compute them from a more thorough study of the asymptotic expansion (4.25) which also sheds some light on their origin. This method is shortly described in the next subsection.

4.2.7 Asymptotics of the inner problem *via* Borel summation

In the previous subsection, we have argued that the asymptotics of the inner problem on the lower w-axis are given by (4.30) but the constant Γ remained to be estimated. As explained previously, one way to calculate it is to solve numerically the inner problem. In this section, an alternative way of estimating Γ is explained (Combescot *et al.*, 1988; Hakim, 1991).

ϕ_0 can be thought of as a generating function for the coefficients a_n in (4.15) since in the asymptotic w-region one has the relation (4.25)

$$\phi_0(w) + \log(-w) = \sum_{n=1}^{\infty} a_n w^{-2n} \,. \tag{4.37}$$

The main idea is that the inner problem gives a definite way of summing the series on the l.h.s. of (4.37) but that one can try to sum this series more directly. A study of the recurrence relation for the a_n shows that they grow very quickly as

$$a_n \sim \kappa \, \frac{(-4)^n n!^2}{2n}, \tag{4.38}$$

where κ is a constant estimated below, see (4.46). So the series (4.37) is badly divergent. In a sense, this is fortunate since we are trying to get an imaginary result by summing a real series. One classic way to give a meaning to such a divergent series is to use Borel summation (Borel, 1928; Dingle, 1973). We define $B(x)$ by

$$B(x) = \sum_{n=1}^{\infty} \frac{a_n}{(2n)!} x^n. \tag{4.39}$$

The advantage is that $B(x)$ is a convergent series for sufficiently small x since the asymptotic behavior of its coefficients is:

$$\frac{a_n}{(2n)!} \sim \frac{\kappa(-1)^n}{2} \sqrt{\frac{\pi}{n}}. \tag{4.40}$$

$B(x)$ is therefore a convergent series for $|x| < 1$ and it defines an analytic function there. If we analytically continue $B(x)$, (4.40) shows that the closest singularity to zero is located at $x = -1$. The behavior of the singular part $B_{\mathrm{s}}(x)$ of $B(x)$ near $x = -1$ is given by

$$B_{\mathrm{s}} \sim \kappa \frac{\pi}{2} (1 + x)^{-1/2}. \tag{4.41}$$

In order to go backward from $B(x)$ to $\phi(w) + \log(-w)$, we define

$$\Psi(w) = \int_0^{\infty} dt \, \mathrm{e}^{-t} B\left(t^2/w^2\right). \tag{4.42}$$

We then identify $\Psi(w)$ and $\phi_0(w) + \log(-w)$ which satisfy the same differential equation and have the same asymptotic series on the real negative axis. From this explicit representation for Ψ, we can find the asymptotic behavior of its imaginary part on the negative imaginary z axis. We analytically continue $\Psi(w)$ from the negative real axis to the negative imaginary axis (staying along the way in the lower half complex w plane). In this way, we obtain on the negative imaginary z axis:

$$\Psi(w) = \int_0^{|w|} dt \, \mathrm{e}^{-t} B\left(t^2/w^2\right) + \int_{|w|}^{\infty} dt \, \mathrm{e}^{-t} B\left(t^2/w^2\right). \tag{4.43}$$

The integral has been split into two parts. In the first integral the integration is along the real t axis from 0 to $|w|$ and the argument of B goes from 0 to -1. Therefore B is purely real there and the first integral does not contribute to the imaginary part of Ψ. In the second integral the integration contour goes from $|w|$ to $+\infty$ along a contour that is determined by the singularities of $B(x)$ outside the unit circle. Moreover the asymptotic behavior of the second integral is dominated by the neighborhood* of $t = |w|$ so that it is possible to replace $B(x)$ by its singular behavior $B_s(x)$ near $x = -1$. One obtains therefore

$$\mathcal{I}m\,\{\Psi(w)\} = \frac{\kappa\pi}{2} \int_{|w|}^{\infty} dt\, e^{-t} \left(\frac{t^2}{|w|^2} - 1\right)^{-1/2} \sim \frac{\kappa\pi^{3/2}}{2\sqrt{2}} \sqrt{|w|}\, e^{-|w|}.$$

(4.44)

This is exactly the asymptotic form obtained in (4.30). Moreover, Γ is now related to the asymptotic behavior of the a_n by

$$2\Gamma = \kappa\frac{\pi^{3/2}}{\sqrt{2}}.$$

(4.45)

A computation of the first few a_n gives approximations to κ using (4.38). From a_n, $n = 1,\ldots,8$ one obtains

$$\kappa_1 = 1, \quad \kappa_2 = 1.041\ldots, \quad \kappa_3 = 1.059\ldots, \quad \kappa_4 = 1.067\ldots,$$
$$\kappa_6 = 1.072\ldots, \quad \kappa_8 = 1.074\ldots$$

(4.46)

in agreement with the numerical estimate for Γ given previously. In this section, we have obtained information on the solution of the differential equation (4.22) by Borel-summing a formally divergent series. It should be noted that this strategy has recently been the subject of systematic mathematical analyses by Ecalle and others (see, for example, Candelpergher *et al.* (1993) and references therein).

4.3 The viscous finger puzzle

4.3.1 Analytic solutions vs. experiments and numerics

As explained in the introduction, when two-fluid displacement is studied in a linear Hele-Shaw cell, in the unstable case where the less viscous fluid (e.g., air) pushes the more viscous one (e.g., oil), after a transient regime, a large single bubble or "finger" of air advances through the cell (Fig. 4.2). What is the shape of this bubble? For closely spaced glass plates, the problem can be addressed within a simplified two-dimensional formulation (Saffman and Taylor, 1958). The oil flow can be approximated by a plane Poiseuille flow in the vertical direction, the characteristics of which slowly vary in the horizontal direction. So locally, the velocity $\mathbf{V}(x,y)$

---

* This implicitly assumes that $B(x)$ has no other singularities in the lower complex plane above the curve $\rho^{1/2} \sin(\theta/2)$ (in polar coordinates). A finer analysis of the perturbation series is needed to prove it.

averaged along the vertical is related to the horizontal pressure gradient

$$\mathbf{V}(x,y) = -\frac{b^2}{12\mu}\,\boldsymbol{\nabla}_{x,y}\,P\,. \tag{4.47}$$

where μ is the dynamic oil viscosity and b the spacing between the glass plates. The incompressibility of oil then, requires that the pressure be a two-dimensional harmonic function

$$\nabla^2 P = 0 \tag{4.48}$$

which is determined by the boundary conditions on the wall of the cell and on the advancing finger. These are of two kinds. Firstly, conditions on the pressure gradient arising from requirements on the local flux of oil, i.e., that it should be zero through the lateral walls of the cell and the kinematic requirement that the normal interface velocity $\mathbf{V_n}$ be the same as that of the oil just in front of it:

$$V_{\mathrm{n}}(s) = -\frac{b^2}{12\mu}\mathbf{n}\cdot\boldsymbol{\nabla}\,P(x(s),y(s)) \tag{4.49}$$

where $(x(s),y(s))$ is an interface point and $\mathbf{n}$ the normal to the interface at this point. In writing (4.49), we have supposed for simplicity that no oil goes through the interface, which can appear obvious at first sight. However, the real physics is three-dimensional and oil films remain on the upper and lower glass plates. Oil films of constant width, can simply be taken into account by an effective proportionality coefficient in (4.49) but their velocity dependence (see section 1.2) complicates the matter (for more details see Reinelt, 1987; Tanveer, 1990). In addition to boundary conditions on the pressure gradient, there is also an interface boundary condition on the pressure itself. Saffman and Taylor (1958) chose to neglect surface tension between air and oil and pressure variation in the (low-viscosity) air. In this case, one has on the interface:

$$P = P_{\mathrm{air}}\,, \tag{4.50}$$

where P_{air} is the constant injected air pressure. Given (4.47-4.50), can the steady propagating shapes be determined?

Although it seems a formidable task, this can be achieved in two dimensions by cleverly using the relation between harmonic functions and analytic functions. The main idea is first that, because of (4.48), P is the real part of a complex velocity potential which is an analytic function of $x + iy$. Secondly, to search $x + iy$ as a function of the velocity potential and not the contrary, i.e., to use the so-called hodograph method, because this transforms the original free boundary problem into a fixed boundary problem. Using this strategy, Saffman and Taylor (1958) have

obtained steady propagating solutions symmetric around the channel centerline. Denoting by w the complex velocity potential, $w = \phi + i\psi$ with
$\phi = -(b^2/12\mu)\,(P - P_{\text{air}})$, they obtained

$$z = x + iy = \frac{w}{V} + \frac{2a}{\pi}(1 - \lambda) \log \left[\frac{1}{2} \left(1 + \exp \left(-\frac{\pi w}{Va} \right) \right) \right] \qquad (4.51)$$

where a is the cell half width, V is the oil velocity far in front of the
channel which is determined by the injection conditions and λ is an undetermined parameter varying between 0 and 1. The equation of the
interface $x_0(y)$ is obtained by putting $\phi = 0$ in (4.51) which gives, after
elimination of ψ

$$x_0(y) = \frac{1 - \lambda}{\pi} a \log \left[\frac{1}{2} \left(1 + \cos \left(\frac{\pi y}{\lambda a} \right) \right) \right]. \qquad (4.52)$$

Equation (4.52) represents not a single shape but rather a one parameter
family of shapes depending upon the value of the undetermined constant
λ, the width of the advancing finger. The experimental data at moderately high velocity is well-fitted by the analytic solution (4.52) when λ is
chosen to be $1/2$. Why is this particular analytic solution selected?

Since there appears to be no immediate explanation, one can try to
reintroduce some of the physical effects that have been neglected in the
simple model (4.47)-(4.52). As already pointed out, one such effect is surface tension. McLean and Saffman (1981) proposed to test its relevance
in the framework of a two-dimensional model by modifying (4.50) into:

$$P(s) = P_{\text{air}} - \frac{T}{R(x(s), y(s))}, \qquad (4.53)$$

where R is the two-dimensional lateral curvature of the interface at the
point $(x(s), y(s))$. This is simply Laplace law where one assumes a constant transverse curvature, the (much larger) effect of which is included
in a redefinition of P_{air}. Unfortunately, no analytic solution is known[*]
when (4.53) is imposed instead of (4.50) and one has to resort to numerical analysis. The influence of surface tension is measured by a capillary
number $Ca = \sqrt{\mu V/T}$. The numerical solutions (McLean and Saffman,
1981; Vanden-Broeck, 1983) show that the continuum of zero-surface tension shapes does not exist as soon as Ca^{-1} is non zero. It is instead replaced by a discrete number of steady-state solutions, the relative widths
$(\lambda_n(Ca), n = 0, 1, 2,...)$ of which tend to $1/2$ when Ca^{-1} tends to zero in
qualitative agreement with the experimental results (Fig. 4.4). Moreover,
only the narrowest shapes $\lambda_0(Ca)$ are linearly stable and of immediate
relevance for the dynamics. Therefore, the modification (4.53) explains

---

[*] See however Kadanoff (1990) for a suggestion that the problem may still be integrable.

qualitatively the experimental findings. It can further be refined to include the three-dimensional effects alluded to before (Reinelt, 1987). The mathematical problem arises when one tries to understand the numerical findings. Since for small Ca^{-1}, the shapes are close to the $Ca^{-1} = 0$, $\lambda = 1/2$ shape, it would be expected that a perturbative treatment would explain its particular role among the continuum of $Ca^{-1} = 0$ shapes. However, when this is attempted, each $Ca^{-1} = 0$ shape seems to give rise to a well defined perturbation series and the mechanism responsible for the selection of the particular $\lambda = 1/2$ shape appears mysterious. The solution of this puzzle is that exponentially small terms prevent the existence of solutions which seems to exist in perturbation theory, as in the simpler geometric model example. In the viscous finger case, the analytic zeroth order solutions depend on the parameter λ. Qualitatively, what happens is that the analog of Γ in (4.30), the magnitude of the exponential term, is a function of λ instead of being a constant as in the geometric model case. Therefore, for special values of λ, the zeroes of $\Gamma(\lambda)$, exponentially small terms vanish and steady-state fingers exist. Selection of dendrite shape and velocity in crystal growth is another similar case where exponentially small effects manifest themselves by selecting a solution out of a continuum and thus play a major role (for a review, see Kessler, Koplik and Levine, 1988, or Brener and Melnikov, 1991).

4.3.2 Asymptotics beyond all orders

Our aim is now to show how the function $\Gamma(\lambda)$ can be determined. That is, we would like to obtain the analog for viscous fingering of the inner

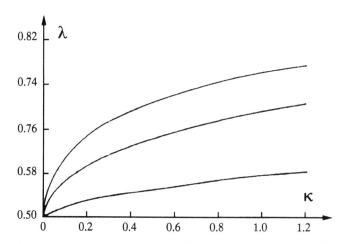

Fig. 4.4. Selected finger widths λ *vs.* capillary number (reprinted from Vanden-Broeck, 1983).

problem (4.22) (Combescot *et al.*, 1988; Ben Amar and Pomeau, 1986). Firstly, it is conceptually convenient to write a closed equation for the interface shape. One fairly general method is to use Green functions. In the channel geometry, this is the function $G(r', r)$ which satisfies:

$$\nabla_{\mathbf{r}'}^2 G(\mathbf{r}', \mathbf{r}) = \delta^2 (\mathbf{r}' - \mathbf{r}) \tag{4.54}$$

with the boundary conditions $\partial_n G(\mathbf{r}', \mathbf{r}) = 0$ on the channel walls and $G(\mathbf{r}', \mathbf{r}) \to 0$ as $\mathbf{r}$ tends to infinity in the part of the channel in front of the finger. Then, Green's identity determines the velocity potential ϕ at any point p in the viscous fluid given both its value $\phi(y')$ and the value of its normal gradients $\mathbf{n} \cdot \nabla|_{\mathbf{r}(y')}$ at the interface points $\mathbf{r}(y') = (x(y'), y')$

$$\phi(p) = \int \frac{dy'}{\cos\theta(y')} \left[G(\mathbf{r}(y'), p) \, \mathbf{n} \cdot \nabla \, \phi|_{\mathbf{r}(y')} \right. $$
$$ \left. -\phi(\mathbf{r}(y')) \, \mathbf{n} \cdot \nabla_{\mathbf{r}'} G(\mathbf{r}(y'), p) \right], \tag{4.55}$$

where the integration is along the whole finger interface parametrized by its abscissa y'. Letting p tend to an interface point $\mathbf{r}(s)$ and using the boundary conditions (4.49), (4.53) the equation for the interface shape is obtained

$$\frac{b^2 T}{12\mu a^2} \frac{1}{2R(y)} = \int_{-\lambda}^{\lambda} \frac{dy'}{\cos\theta(y')} \left[U \cos\theta(y') \, G(\mathbf{r}(y'), \mathbf{r}(y)) \right. $$
$$ \left. -\frac{b^2 T}{12\mu a^2} \frac{1}{R(y')} \mathbf{n} \cdot \nabla_{\mathbf{r}'} G(\mathbf{r}(y'), \mathbf{r}(y)) \right], \tag{4.56}$$

where U is the finger velocity (equal to V/λ where V is the previously defined fluid velocity at infinity) and the coordinates have been made dimensionless by taking the channel half width as unit length. The regular perturbation series can in principle be obtained from (4.56) by treating $\sigma = (b^2 T)/(12\mu a^2 U)$ as a small parameter and expanding $x(y)$ around the zero surface tension solution (4.52) (although (4.56) is not the most convenient expression for this purpose)

$$x(y) = x_0(y) + \sigma x_1(y) + \ldots + \sigma^n x_n(y) + \ldots \tag{4.57}$$

Here, we simply assume that this perturbation series is well behaved and can be obtained starting from $x_0(y)$ for any λ, so that the regular perturbation series does not give any indication of selection for real y. However, as in the simpler geometric model, each $x_n(y)$ is singular for complex values of y determined by the zeroth order $x_0(y)$. In the neighborhood of such a singularity different orders of perturbation theory give terms of the same magnitude and exponentially small contributions on the real axis can be computed by summing the main contributions of each order.

Since the curvature $1/R(y)$ can be written as

$$\frac{1}{R(y)} = -\frac{d}{dy}\left(\frac{dx/dy}{[1 + (dx/dy)^2]^{1/2}}\right), \tag{4.58}$$

singularities of the x_n are located at the zeroes of $1 + (dx_0/dy)^2$. Using (4.52), this is found to be $1 + ((1 - \lambda)/\lambda)^2 \tan^2(\pi y/2\lambda)$. The pattern of zeroes of this function is qualitatively different depending upon whether $\lambda/(1 - \lambda)$ is larger or smaller than 1, that is λ larger or smaller than the particular value $1/2$, an indication that we follow the right track. In order to analyze (4.56) in the neighborhood of a curvature singularity, we analytically continue (4.56) to the complex plane, i.e., in each member we go from real values of y to complex values of y, which we denote by ξ, assuming that $x(\xi)$ is an analytic function (the integration variable y' remains real of course). This being done, the integral term should be taken care of. In a neighborhood of a curvature singularity, we assume a priori, that the sum $F(y)$ of the contributions of the x_ns, $n \geq 1$ (cf. (4.57)) is small compared to $x_0(y)$ so that we can estimate the integral by linearizing it (of course, the assumption has to be checked afterwards for self consistency). This being assumed, (4.56) is given at dominant order by:

$$\frac{\sigma}{2}\frac{d}{d\xi}\left[\frac{-dx_0/d\xi}{\left(1 + (dx_0/d\xi)^2 + 2(dx_0/d\xi)(dF/d\xi)\right)^{1/2}}\right]$$

$$= F(\xi)\int_{-\lambda}^{\lambda} dy' \frac{\partial G}{\partial x}(\mathbf{r}_0(y'), \mathbf{r}_0(\xi)), \tag{4.59}$$

where $\partial G/\partial x$ is meant to be the x-component of the Green function gradient with respect to the second variable. A simple way to compute the integral on the l.h.s. for real values y of x is to express it in terms of the velocity field of the zeroth order solution. Letting $x = y + i\,0^+$, one obtains:

$$\int_{-\lambda}^{\lambda} dy' \frac{\partial G}{\partial x}(\mathbf{r}_0(y'), \mathbf{r}_0(y + i\,0^+))$$

$$= \mathcal{P}\int_{-\lambda}^{\lambda} dy' \frac{\partial G}{\partial x}(\mathbf{r}_0(y'), \mathbf{r}_0(y)) - \frac{i}{2}\frac{dx_0/dy}{1 + (dx_0/dy)^2}, \tag{4.60}$$

where the local behavior of the Green function has been used, $G(\mathbf{r}', \mathbf{r}) \sim \log|\mathbf{r} - \mathbf{r}'|/2\pi$. Using (4.55), the integral on the r.h.s. of (4.60) is seen to be simply given by the velocity field ϕ_0,

$$2U\,\mathcal{P}\int_{-\lambda}^{\lambda} dy' \frac{\partial G}{\partial x}(\mathbf{r}_0(y'), \mathbf{r}_0(y)) = \frac{\partial \phi_0}{\partial x} = \frac{U}{1 + (dx_0/dy)^2}, \tag{4.61}$$

where the last equality comes from (4.51) or directly from the boundary conditions on the finger ($\phi = 0$, $\partial_n \phi = U \cos \theta$). So, the l.h.s. of (4.59) is simply $F(y)/[2(1 + i\, dx_0/dy)]$ and in the upper ξ plane (4.59) reads as

$$\frac{\sigma}{2} \frac{d}{d\xi} \left\{ \frac{\frac{1-\lambda}{\lambda} \tan\left(\frac{\pi\xi}{2\lambda}\right)}{\left[1 + \left(\frac{1-\lambda}{\lambda}\right)^2 \tan^2\left(\frac{\pi\xi}{2\lambda}\right) - 2\left(\frac{1-\lambda}{\lambda}\right)\tan\left(\frac{\pi\xi}{2\lambda}\right) F'(\xi)\right]^{1/2}} \right\}$$

$$= \frac{F(\xi)}{2\left[1 - i\frac{1-\lambda}{\lambda}\tan\left(\frac{\pi\xi}{2\lambda}\right)\right]}. \tag{4.62}$$

It is convenient to use, instead of ξ, the variable $\tan(\pi\xi/2\lambda)$ for which the curvature singularities are located in the neighborhood of $+i$ in the upper complex plane when λ is close to $1/2$. Expanding $\tan(\pi\xi/2\lambda) = i(1 - \tau/2 + \dots)$, $\lambda = 1/2 + \alpha/8 + \dots$, and replacing ξ-derivatives by τ-derivatives ($d/d\xi = 2\pi i\tau\, d/d\tau$), we obtain in the neighborhood of $\xi = i$

$$\frac{\sigma}{2} 2\pi i\tau \frac{d}{d\tau} \frac{i}{[\alpha + \tau + 4\pi\tau\, dF/d\tau]^{1/2}} = \frac{F(\tau)}{4}. \tag{4.63}$$

The orders of magnitude of the different quantities are $\alpha \sim \tau \sim F$ which comes from balancing the three terms on the l.h.s., and $F \sim \sigma/\sqrt{\alpha}$ from the comparison of the two sides of (4.63). From these scalings, it follows that the inner problem depends on the dimensionless parameter $\sigma/\alpha^{3/2}$, so that λ, the relative finger width, tends to $1/2$ as $U^{-2/3}$ for large finger velocities. A nicer looking form for this problem is obtained by introducing the function Q

$$Q/\sqrt{\alpha} = [\alpha + \tau + 4\pi\tau\, dF/d\tau]^{-1/2}, \tag{4.64}$$

so, that (4.63) reads, when differentiated once:

$$\frac{1}{a} x \frac{d}{dx} x \frac{d}{dx} Q + \frac{1}{Q^2} = \pm 1 + x^2, \tag{4.65}$$

where $a^{-1} = 4\pi^2\sigma/|\alpha|^{3/2}$, the $+$ $(-)$ sign obtains for $\lambda > 1/2$ ($\lambda < 1/2$) and we have introduced the variable x with $\tau = |\alpha|x^2$ so as to make exact contact with the original notations of Combescot et al. (1988).[*] Going back to the definition of the different variables, the symmetric finger existence condition is that Q be strictly real on the positive real axis starting from $Q \sim 1/x$ for $\mathcal{I}m\{x\} \to -\infty$. The qualitative difference between the position of curvature singularities is most easily seen on (4.65) in the

---

[*] It is worth pointing out that in the dendritic growth literature (e.g., Ben Amar and Pomeau, 1986; Brener and Melnikov, 1991) the form (4.63) is used most often.

large a limit. For $\lambda > 1/2$ (i.e., the $+$ sign), at zeroth order Q has two singularities at $\pm i$. On the contrary, the singularities are located at ± 1 for $\lambda < 1/2$ (the case of a $-$ sign in (4.65)). In the large a limit, (4.65) itself can be treated by the method of section 4.2. We leave a simple exercise to the reader to show that in a region of extension $a^{-2/7}$ around each singularity (4.65) gives rise to a reduced inner problem (Combescot *et al.*, 1988)

$$\frac{d^2G}{dr^2} + \frac{1}{G^2} = r \qquad \text{with} \qquad G = \left(\frac{a}{4}\right)^{1/7} Q \qquad (4.66)$$

where r denotes the (rescaled) distance to the singularity. For $\lambda < 1/2$ the magnitudes of the exponential terms arising from these singularities are different and the singularity at $x = -1$ can be effectively neglected at dominant order. The situation is thus very similar to the one obtained in the simple geometric model. On the contrary, for $\lambda > 1/2$, the singularities $x = \pm i$ are located in symmetric positions relative to the x axis. The exponential small term on the real x axis is thus a superposition of the contributions of the two singularities and its magnitude oscillates with a. It vanishes for a equal to a_n, with the large n estimate

$$a_n \approx 2\left(n + \tfrac{4}{7}\right)^2, \qquad n \in \mathbb{N}. \qquad (4.67)$$

Even for small n, (4.67) is a good approximation of the exact values obtained by a numerical solution of (4.65)

$$a_0 = 0.7368\,, \quad a_1 = 5.067\,, \quad a_2 = 13.36\,. \qquad (4.68)$$

We have thus shown how the method of section 4.2 can be used to obtain the selected zero-surface tension steady-state solutions. By an extension of this analysis, it can be shown that only the lowest branch of solution is linearly stable (Tanveer, 1987; Bensimon *et al.*, 1987) and therefore observable in experiments.

4.3.3 Self-dilating fingers and merging of selected solutions

An interesting variation of the Saffman–Taylor experiment is to inject air at the apex of a wedge-shaped Hele-Shaw cell instead of a linear cell (Thomé *et al.*, 1989). Instead of a finger moving at constant velocity, a finger growing self-similarly in time is observed (see Fig. 4.5). The dimensionless relative width λ is replaced by the ratio of the angular width of the finger to the angle of the wedge and is a function of the angle of the cell and the injection conditions.

   This experiment can be analyzed along the previous lines. Using the two-dimensional effective equations (4.47), (4.48), (4.53), one search for a

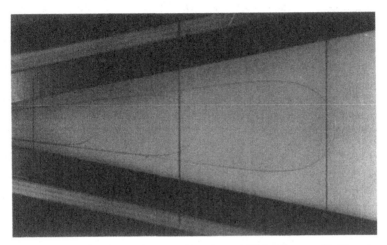

Fig. 4.5. A self-similar finger in a wedge-shaped Hele-Shaw cell (Courtesy of Y. Couder and M. Rabaud).

one-dimensional interface growing self-similarly in time. That is, in polar coordinates,

$$r(\theta, t) = f(t)r(\theta) \,. \tag{4.69}$$

This hypothesis for the interface shape time dependence fixes the time dependence of the velocity potential since it is completely determined by the Neumann boundary condition, (4.49)

$$\phi(\mathbf{r}, t) = \dot{f}(t)f(t)\phi\left(\mathbf{r}/f(t), 0\right) \,. \tag{4.70}$$

Since the curvature varies in time like the inverse of the interface coordinates, the boundary condition (4.53) can only be satisfied at all times if $f^2\dot{f}$ is a time independent constant

$$f(t) = (3At + 1)^{1/3} \,.$$

This implies that strictly self-dilating fingers are possible only if the injected flux of air is decreased in time as $t^{-1/3}$. We will suppose this for the theoretical analysis. Although this was not the case in the experiments of Thomé *et al.* (1989), the observed self-similarity can be attributed to the weak λ dependence on the injected flux.

A continuum of zero surface tension solutions were found for the particular angle $\theta = \pi/2$ by Thomé *et al.* (1989) and have since then been obtained for other angles (Ben Amar, 1991; Tu, 1991). In order to investigate selection out of this continuum, surface tension (i.e., (4.53)) should be taken into account. One can follow closely the previous analysis by introducing a conformal transformation which maps the wedge geometry into a linear channel (see Fig. 4.6):

$$x = \frac{1}{\theta} \log \left(\frac{r}{r_0} \right), \quad y = \frac{\theta}{\theta_0}. \tag{4.71}$$

In these coordinates (4.55) is valid. One needs only to evaluate $\mathbf{n} \cdot \boldsymbol{\nabla} \phi$ and ϕ on the interface in the (x, y) coordinate system in order to obtain the analog of (4.56)

$$\mathbf{n} \cdot \boldsymbol{\nabla}_{x,y} \, \phi = \theta_0 r_0^2 \exp \left(2\theta_0 x \right) \cos \psi \,, \tag{4.72}$$

$$\phi_{\text{int}} = \frac{b^2 T}{12\mu A r_0 \theta_0} \frac{\exp \left(-\theta_0 x \right)}{\left[1 + (dx/dy)^2 \right]^{3/2}}$$
$$\left[-\frac{d^2 x}{dy^2} + \theta_0 \left(1 + (dx/dy)^2 \right) \right], \tag{4.73}$$

where we have denoted by ψ the angle between the interface normal and the x axis (in order to avoid confusion with the coordinate θ). (4.73) is simply the expression of interface real curvature in the coordinate (x, y), (4.71). The dimensionless parameter σ which measures the strength of capillary forces reads:

$$\sigma = \frac{b^2 T}{12\mu A r_0^3 \theta_0^2}. \tag{4.74}$$

When the analog of (4.56) is solved numerically a result similar to the one obtained in the linear case is found. Surface tension selects a discrete family of solutions out of the continuum of analytic zero surface tension solutions. But, there is a surprise (Ben Amar *et al*, 1991): instead of

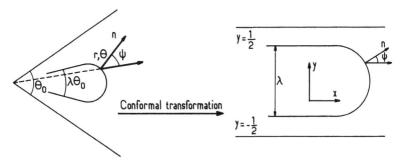

Fig. 4.6.   Corresponding points in the physical plane and in the (x, y) coordinates.

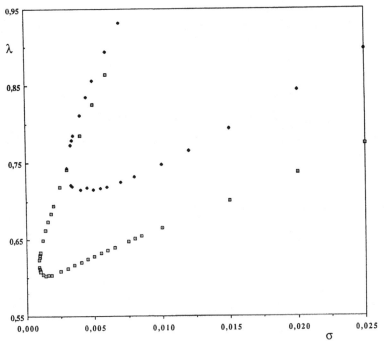

Fig. 4.7. Curves of the first two selected λs for wedge-shape angles $\theta = 15°$ (dotted squares) and $\theta = 45°$ (black diamonds). The two curves merge for finite σ, (4.74), contrary to the linear case (Fig. 4.4). A similar phenomenon is observed for higher branches (reproduced from Ben Amar *et al.*, 1991).

existing all the way up to $\sigma = 0$, the selected states merge two by two at finite surface tension (see Fig. 4.7).

This new feature can be qualitatively understood from the selection theory described previously. When one computes the curvature singularity in the complex plane, one finds three relevant singularities. In addition to the previous two, there is a supplementary one right on the imaginary x axis. For fixed cell angle, as λ is decreased from 1 to 0, this supplementary singularity becomes dominant at $\lambda = \lambda_c$. The magnitude of exponential terms on the imaginary axis is given by the superposition of the contributions of these different singularities. It goes smoothly from an oscillating function of $1/\sigma$ with a discrete number of zeroes to a monotonously decreasing one with no zero. In the process, the zeroes (i.e., the selected states) have to merge two by two (see Fig. 4.8). This competition between different singularities can be described quantitatively by a single inner problem in the small θ_0 limit where all three singularities are close enough (Combescot, 1992). One obtains instead of (4.65)

$$\gamma^3 x \frac{d}{dx}\left(x \frac{dQ}{dx}\right) + \frac{1}{Q^2} = x^2 + 1 + \delta \log(x),\qquad (4.75)$$

where γ plays the role of a in (4.65) and δ is a rescaled cell tip angle.

4.4 Miscellaneous examples of exponential asymptotics in physical problems

In the previous sections, we have shown that exponentially small effects can prevent the existence of solutions which would appear to exist in conventional perturbation theory. This is representative of a class of phenomena where exponentially small effects are important because they manifest themselves through a "selection mechanism": the problem has no solution except for special parameter values. Dendritic growth and viscous fingering are nice physical examples of this selection by exponentially small terms. An important role is also played by exponentially small terms when they control qualitatively new physical effects. In this class, we recall the classic analysis of the reflection of a high energy particle on a low energy potential barrier and summarize the more recent computations of the decays of ϕ^4-breathers and solitary water waves in

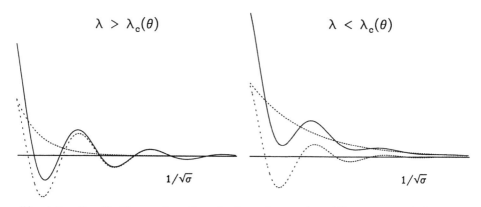

Fig. 4.8. Qualitative explanation for branch merging. The exponentially small term is given as a sum of two contributions: 1) an oscillating one coming from complex conjugate singularities in the x plane as in the linear case; 2) a non zero one coming from the singularity on the real x-axis. For $\lambda > \lambda_c(\theta)$, the side singularities are dominant and the exponentially small contribution oscillates as a function of $1/$ Ca. It has zeroes (the selected states) as in the linear case. On the contrary, for $\lambda < \lambda_c(\theta)$ the axis singularity is dominant and the exponentially small contribution never vanishes. When λ is varied from one case to the other, the function deforms smoothly and the zeroes have to merge two by two to disappear in the process.

the presence of small surface tension, and the appearance of chaos in a rapidly forced pendulum due to separatrix splitting.

4.4.1 Linear examples: above barrier reflection in the semiclassical limit and conservation of adiabatic invariants

The problem that we consider is the motion of an incident particle of energy E over a potential barrier $V(x)$ of height smaller than E. Classically of course the particle is not reflected by the barrier. Quantum mechanically, this is no longer true and the amplitude for reflection of the particle does not vanish in general. The Schrödinger equation reads

$$\epsilon^2 \frac{d^2\psi}{dx^2} + p^2(x)\psi = 0, \quad \epsilon^2 = \frac{\hbar^2}{2mE}, \quad p(x) = \sqrt{1 - \frac{V(x)}{E}}. \quad (4.76)$$

One can try to estimate the reflection coefficient in the case of a heavy particle where the WKB approximation can be used. However, in this limit the reflection coefficient is exponentially small when the potential is an analytic function and the correct solution of this seemingly simple exercise is more subtle than would appear at first thought. It was obtained in Pokrovskii and Khalatnikov (1961) by inner-outer matching in the complex plane in one of the first uses of the technique to compute an exponentially small effect. The simplification as compared with the example of section 4.2 is due to the linearity of the problem which makes the inner equation analytically soluble.

Before showing it, it is worth noticing that the problem is equivalent to the classical question of estimating the precision of adiabatic invariant conservation for a classical harmonic oscillator. By introducing $t = x/\epsilon$ and $w(\epsilon t) = p(x)$, equation (4.76) is seen to describe a harmonic oscillator, the frequency of which is slowly varied from w_- at $t = -\infty$ to w_+ at $t = +\infty$

$$\ddot{q} + w^2(\epsilon t)q = 0. \quad (4.77)$$

As is well known, the action $J = H/w$ remains constant when the variation is infinitely slow. For finite ϵ, this adiabatic invariant is conserved to all orders in ϵ and its variation is exponentially small (see Arnold, 1988, and references therein). The calculation of this variation is equivalent to finding the amplitude of the reflected wave in the quantum mechanics problem (Dykne, 1960). In this case, one has the two asymptotic forms of the wave function:

$$\psi_- = \frac{1}{T} e^{ip_-x/\epsilon} + \frac{R}{T} e^{-ip_-x/\epsilon}, \quad p_- > 0 \text{ as } x \to -\infty, \quad (4.78)$$

$$\psi_+ = e^{ip_+x/\epsilon}, \quad p_+ > 0 \text{ as } x \to +\infty, \quad (4.79)$$

and conservation of the probability implies that the in-going and out-going current are equal:

$$p_- = p_-|R|^2 + p_+|T|^2. \qquad (4.80)$$

For the harmonic oscillator this shows that an oscillation $A\cos(w_+t+\phi)$, at $t = +\infty$, corresponds, at $t = -\infty$, to the oscillations

$$q = A\,\mathcal{R}e\left\{\frac{e^{i\phi}}{T}\left(e^{iw_-t}+R\,e^{-iw_-t}\right)\right\} = A\,\mathcal{R}e\left\{\left(\frac{e^{i\phi}}{T}+\frac{e^{-i\phi}}{T^*}R^*\right)e^{iw_-t}\right\}.$$

$$(4.81)$$

The associated actions are $J_+ = \frac{1}{2}A^2w_+$ and

$$J_- = \frac{1}{2}\frac{A^2w_-}{|T|^2}\left|1+e^{-2i\phi}\frac{T}{T^*}R^*\right|^2$$

$$= \frac{1}{2}\frac{A^2w_-}{|T|^2}\left[1+|R|^2+2\,\mathcal{R}e\left\{e^{-2i\phi}\frac{T}{T^*}R^*\right\}\right].$$

$$(4.82)$$

Therefore using (4.80), one can write the variation

$$J_+ - J_- = -\frac{A^2w_-}{|T|^2}\left[|R|^2+\mathcal{R}e\left\{e^{2i\phi}\frac{T^*}{T}R\right\}\right]. \qquad (4.83)$$

This exact expression shows as desired that the variation of the action is related to the scattering amplitudes of the quantum mechanics problem and that it is as exponentially small as the reflection coefficient in this case. Moreover, the second term shows that it depends on the phase of the oscillation which is a kind of pinning phenomenon of the rapid oscillation on the slow frequency variation.

How can the exponentially small reflection be computed? Firstly it is natural to search for $\psi(x)$ solution of (4.76) under the WKB form

$$\psi(x) = g(x)\exp i\left(\frac{f(x)}{\epsilon}\right), \qquad (4.84)$$

which leads to

$$f'(x) = \pm p(x), \qquad \frac{d}{dx}\left(g\sqrt{f'}\right) = \frac{i\,\epsilon}{2}\frac{g''(x)}{\sqrt{f'}}. \qquad (4.85)$$

If we choose the $+$ sign in (4.55) and impose $g(\infty) = 1$, we can try to solve perturbatively the equation for g. This gives:

$$g_0 = \sqrt{\frac{p_+}{p(x)}}, \qquad g_1 = \frac{-i}{2\sqrt{p}}\int_x^{+\infty}\frac{g_0''(u)}{\sqrt{p(u)}}\,du. \qquad (4.86)$$

At any order one obtains functions which tend to constant values at $x = -\infty$. Therefore, perturbatively, one does not obtain any reflected wave. The trouble is of course that we have treated perturbatively the highest derivative term. For real values of x, this is not a problem since $p(x)$ smoothly varies and does not vanish (the particle energy is higher than the barrier height). On the contrary, this is no longer legitimate in the complex plane near a singularity of g_n. Let us simply consider what happens near a simple root x_0 of $p^2(x)$, where

$$p^2(x) \sim c(x - x_0) . \tag{4.87}$$

The size of the inner region is such that all most divergent terms are of the same order. In other words, x should be scaled near x_0 in such a way that ϵ disappears from equation (4.85) or (4.76) to leading order. Thus, one finds that the inner region size is of order $\epsilon^{2/3}$ and we introduce w such that

$$x = x_0 + \epsilon^{2/3} c^{-1/3} w . \tag{4.88}$$

Instead of (4.85), it is simpler to study directly (4.76), which reduces to leading order to the Airy equation

$$\frac{d^2\psi}{dw^2} + w\psi = 0 . \tag{4.89}$$

Matching between the inner solution of (4.89) and the outer functions (4.84–86) is conveniently done on the "Stokes" line where the imaginary part of f is constant so that the out-going and in-going waves remain of the same magnitude. We denote by L_+ the Stokes line which starts from x_0 along the positive w axis and reaches the x-real line at $+\infty$ (see Fig. 4.9). On L_+, the absence of an in-going wave imposes that ψ is equal to:

$$\psi(w) = K \, \mathrm{Ai}\left(e^{-i\pi/3} w\right) \sim \frac{K}{2\sqrt{\pi}} e^{i\pi/12} \frac{1}{w^{1/4}} \exp\left(\tfrac{2}{3} i w^{3/2}\right) , \tag{4.90}$$

where K is a constant. The behavior of ψ at $x = -\infty$ is determined by going to L_-, the Stokes line which starts from x_0 at angle $-2\pi/3$ from L_+ and reaches the real x axis at $x = -\infty$. Using the properties of Airy functions (see for example Bender and Orszag, p.130), the asymptotic behavior of ψ on L_- is found to be

$$\psi(w) \sim \frac{K}{2\sqrt{\pi}} e^{i\pi/12} \left[\frac{1}{w^{1/4}} \exp\left(\tfrac{2}{3} i w^{3/2}\right) - \frac{i}{w^{1/4}} \exp\left(-\tfrac{2}{3} i w^{3/2}\right) \right] . \tag{4.91}$$

It remains to match (4.91) to the (outer) solution with the appropriate asymptotic behavior (4.78) on L_-. From (4.84) and (4.85), it is given by

$$\psi = \frac{1}{T} \sqrt{\frac{p_-}{p(x)}} \left[\exp\left(i \frac{f(x)}{\epsilon}\right) + R \exp\left(-i \frac{f(x)}{\epsilon}\right) \right] \tag{4.92}$$

$$\text{with} \quad f(x) = p_- x + \int_{-\infty}^{x} du \, [p(u) - p_-] .$$

Matching (4.92) and (4.91) in the asymptotics of the inner region finally determines R

$$R = -i \exp\left(+2i\frac{f(x_0)}{\epsilon}\right). \qquad (4.93)$$

This is indeed exponentially small since $\mathcal{I}m\{f(x_0)\}$ is positive. It is equal to $d_0\, p_-$ where d_0 is the distance between L_- and the real axis when $\mathcal{R}e\{x\} \to -\infty$ (see Fig. 4.9). The different magnitudes of the incoming and reflected waves simply comes from the fact that the Stokes line L_-, where they are of similar amplitude, is a distance d_0 away from the real axis in the complex plane.

This analysis of the linear problem (4.76) is essentially the one presented in Pokrovskii and Khalatnikov (1961). The problem has since been analyzed with more concern for mathematical rigor by various authors. The interested reader will find appropriate references in the reviews by Arnold *et al.* (1988), Fedoryouk (1983), Meyer (1980).

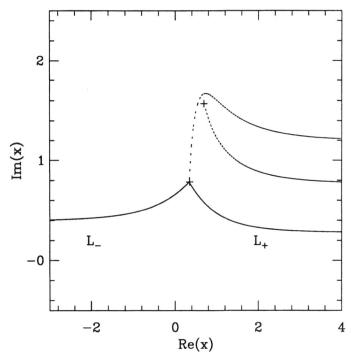

Fig. 4.9. Stokes lines corresponding to equation (4.76) for $p^2(x) = \left[(e^x - 1)^2 + 1\right] / \left[e^{2x} + 4\right]$. The locations of the two singularities closest to the real axis are marked by plus signs. The lower one is a zero and the upper one a pole.

4.4.2 Exponentially small decay of solitary waves

The methodology of section 4.2 has proved useful in analyzing the existence of solitary waves. We briefly describe here two such examples, the cases of the breather solutions in ϕ^4 theory and the problem of solitary water waves in the presence of surface tension.

a) *Existence of ϕ^4 breather*

Nonlinear Klein–Gordon equations arise in various physical contexts in nonlinear optics, condensed matter physics and high-energy physics. In one dimension, they read

$$\frac{\partial^2 \phi}{\partial t^2} = \frac{\partial^2 \phi}{\partial x^2} - V'(\phi). \tag{4.94}$$

When $V(\phi)$ is a multistable potential, stable solutions connecting different minima exist (these are the fronts of section 3 also called kinks, domain walls, topological solitons,...). A particularly interesting example is obtained by taking $V(u) = -\cos u$. The obtained so-called "Sine-Gordon" equation is completely integrable (see Novikov *et al.*, 1984). Besides the above-mentioned topological solitons, it admits remarkable exact solutions called breathers which are localized in space and periodic in time. They can be viewed as bound states of two solitons. The question then arises of the existence of similar breather solutions in non exactly integrable theories. It has been analyzed in the usual ϕ^4 model where $V(u)$ is a symmetric double-well potential

$$V(\phi) = -\tfrac{1}{2}\phi^2 + \tfrac{1}{4}\phi^4. \tag{4.95}$$

Spatially uniform solution of small amplitude oscillates with a pulsation $w = \sqrt{2}$ around one of the minima of the potential. For w slightly less than $\sqrt{2}$, it appears possible to obtain small amplitude breather solutions in perturbation (Dashen *et al.*, 1975). One expands around one of the potential minima by setting $\phi = -1 + u$. Then a periodic solution can be searched for in the form of a Fourier representation

$$u(x,t) = a_0(x) + \sum_{n=1}^{\infty} (a_n(x) \exp(inwt) + \text{c.c.}). \tag{4.96}$$

For small values of $\epsilon = (2 - w^2)^{1/2}$, $\xi = \epsilon x$ is the appropriate spatial scale for the expansion. Coefficients $a_n(\xi)$ start at order ϵ^n (for $n > 0$) and can be determined recursively by substituting (4.96) in (4.94). For example $a_0(\xi)$ is found to be

$$a_0(\xi) = \frac{\epsilon^2}{2\cosh^2(\xi)} + \epsilon^4 \left(\frac{85}{36} \frac{1}{\cosh^2(\xi)} - \frac{395}{144} \frac{1}{\cosh^4(\xi)} \right) + \mathcal{O}\left(\epsilon^6\right). \tag{4.97}$$

Order by order the $a_n \to 0$ as $x \to \pm\infty$ and there is no apparent problem with the expansion. However, as in the simpler equations, counting of modes at infinity makes breather existence unlikely (see Eleonsky, 1991).

Since all a_n are given as power series in $\epsilon/\cosh(\xi)$, the perturbation expansion breaks down near zeroes of $\cosh(\xi)$. Following the methodology described previously, this can be used to show that terms beyond all orders prevent the existence of true breathers and that approximate breathers radiate away their energy at an exponentially small rate (Segur and Kruskal, 1987). The long lifetime of breather-like excitations explains their observation in numerical computations.

b) *Solitary gravity-capillary water waves*

The important role of water waves in the development of nonlinear physics was already mentioned in section 2 when we described Stokes's and related results for water waves in deep water. Here, we will briefly discuss another historically very important facet of this subject, the existence of solitary water waves in shallow water. The basic equations describing the problem are (2.17-20).

In order to distinguish the different regimes it is helpful to introduce the two dimensionless numbers α and β:

$$\alpha = \frac{\text{wave height}}{\text{water depth}}, \qquad \beta = \left[\frac{\text{water depth}}{\text{characteristic wavelength}}\right]^2 . \qquad (4.98)$$

The perturbation technique of section 2 is essentially valid for α small and β arbitrary but much larger than α. For α of order one and β small, one can derive the shallow water equations (see for example Whitham, 1974) where dispersive effects are neglected. Dispersive effects compete with weak nonlinear effects for α and β small and of the same order of magnitude ($\alpha \ll 1$, $\beta \ll 1$, $\alpha \sim \beta$). In this interesting regime of small amplitude and long wavelength, solitary water waves can propagate without deformation of shapes, as first reported by Scott Russell around 1840. This can be described using a remarkable equation (below) derived by Korteweg and de Vries, following earlier work of Boussinesq and Rayleigh (for some historical information and references see Miles, 1981). The KdV equation has since played a key role in the discovery of complete integrability and the development of inverse scattering theory (see Ablowitz and Segur, 1981). The question we want to consider here is the effect on solitary waves of terms neglected in the KdV equation (Hunter and Vanden-Broeck, 1983; Hunter and Scheurle, 1988; Pomeau *et al.*, 1988; Boyd, 1991).

The KdV equation can be formally derived using the multiple-scale technique explained in section 2 (see e.g., Ablowitz and Segur, 1981). Here, we use a lighter heuristic approach to motivate the result. We want to include the effect of surface tension which was neglected in section 2. The only modification, due to Laplace law, is to add to the r.h.s. (2.19) the term $T/\rho R(x)$, where T is surface tension and R the local interface radius of curvature. Linearizing the equation as in (2.25), the dispersion relation for a water wave moving with angular frequency ω and wave number k is found to be

$$\omega = \frac{c_0}{h} \left[\left(1 + \tau k^2 h^2 \right) kh \tanh(kh) \right]^{1/2}. \qquad (4.99)$$

h is the depth of the water layer and we have introduced the velocity c_0 and Bond number τ

$$c_0 = \sqrt{gh}, \qquad \tau = T/\rho gh^2. \qquad (4.100)$$

For shallow water ($\beta = kh \ll 1$), (4.99) can be expanded around $kh = 0$

$$\omega = c_0 k \left(1 + \frac{1}{2} \left(\tau - \frac{1}{3} \right) k^2 h^2 + \frac{1}{4} \left(\frac{19}{90} - \frac{\tau}{3} - \frac{\tau^2}{2} \right) k^4 h^4 + \dots \right). \qquad (4.101)$$

Therefore, the linear equation describing the propagation of a long wave is given by:

$$\frac{\partial \eta}{\partial t} = -c_0 \frac{\partial^2 \eta}{\partial x^2} + \frac{c_0}{2} \left(\tau - \frac{1}{3} \right) h^2 \frac{\partial^3 \eta}{\partial x^3} - \frac{c_0}{4} \left(\frac{19}{90} - \frac{\tau}{3} - \frac{\tau^2}{2} \right) h^4 \frac{\partial^5 \eta}{\partial x^5} + \dots \qquad (4.102)$$

Since this conservative problem is invariant under the simultaneous change $x \to -x$ and $t \to -t$, the first term which is added on the r.h.s. of (4.102) in a weakly nonlinear expansion should be proportional to $\eta \partial \eta / \partial x$. The actual coefficient is found to be $3c_0/2h$ (see e.g., Hunter and Vanden-Broeck, 1983). In a reference frame moving at velocity c_0, the equation describes the competition between dispersive effects and nonlinearity. As stated above, the last term in (4.102) and other nonlinear terms besides $\eta \partial \eta / \partial x$ are multiplied by small coefficients when α and β (cf. (4.98)) are small and of the same order and are usually neglected. This is of course not correct if $\tau = 1/3$ where the third order derivative term disappears. More generally, it is interesting to study the effect of the higher derivative terms. A simple model equation is obtained by making (4.102) dimensionless for $\tau < 1/3$

$$\frac{\partial \eta}{\partial t} + \eta \frac{\partial \eta}{\partial t} + \frac{\partial^3 \eta}{\partial x^3} + \epsilon^2 \frac{\partial^5 \eta}{\partial x^5} = 0, \qquad (4.103)$$

where ϵ^2 is of order α. For $\epsilon = 0$, this is the classical KdV equation which possess elevation solitary waves[*]

$$\eta(x,t) = \frac{3}{\cosh^2\left(\frac{\sqrt{c}}{2}(x - ct)\right)}. \qquad (4.104)$$

For $\epsilon \ll 1$ the fifth derivative in (4.103) is a singular perturbation of the KdV equation. As in the previous problem, its effect on solitary waves can be guessed by a simple counting of modes around $|x| \to \infty$. When a constant velocity solution $\eta(x - ct)$ is substituted in (4.103), the resulting equation can be integrated once. Linearizing the obtained fourth-order equation around $\eta = 0$, one finds four modes: one convergent, one divergent and two purely oscillatory. Imposing that the solution tends to zero at $x = -\infty$ fixes three constants and the last one simply reflects translation invariance of the solution. No freedom remains to determine the asymptotic behavior at $+\infty$. So, oscillatory tails at infinity should be admitted. Their magnitude in the small ϵ limit can be computed by using the analysis of section 4.2 (Pomeau *et al.*, 1988). It starts from a KdV soliton at zeroth order and proceeds very similarly as for the geometric model. The obtained asymptotic estimates have been confirmed in a direct numerical study of (4.103) (Boyd, 1991). Hunter and Vanden-Broeck (1983) have numerically integrated the fully non local water wave equation in the long wavelength-shallow water limit. They have also obtained traveling waves with oscillatory tails and their existence has been recently rigorously demonstrated (see Beale, 1991). The Kruskal–Segur method remains to be applied to this nonlocal problem to compute the magnitude of the oscillatory tail and explain the numerical data of Vanden-Broeck (1991). This can presumably be done along lines similar to the analysis of the viscous finger puzzle (section 4.3.2).

4.4.3 Pinning of fronts and splitting of separatrices

Exponentially small effects appear in two classic phenomena which seem unrelated at first sight. The first one concerns the motion of fronts, as studied in section 3.1, when the underlying medium is discrete or has a periodic spatial modulation. The second one is the phenomenon of bifurcation of separatrices in forced mechanical systems which has far-reaching consequences. It was discovered by Poincaré (1893) about a century ago together with the simultaneous appearance of extremely complicated orbits in phase space, i.e., "chaotic motion."

---

[*] For $\tau > 1/3$ (i.e., for a water depth of less than about 0.5 cm) the sign of the third derivative term in (4.103) is changed and these turn into depression solitary waves.

Let us first consider front motion in a bistable medium using a simple Ginzburg–Landau like equation as in section 3 (cf. (3.1)).

$$\frac{\partial A}{\partial t} = \frac{\partial^2 A}{\partial x^2} - V'_\mu(A) \,. \tag{4.105}$$

Then a front is motionless only for isolated values of the control parameter μ. Such a front obeys the time-independent version of (4.105)

$$\frac{\partial^2 A_c}{\partial x^2} - V'_\mu(A_c) = 0 \,. \tag{4.106}$$

The corresponding particle trajectory in the phase (A, A_x) can link one stable state $(A_1, 0)$ to the other $(A_2, 0)$ only if $V_\mu(A_1) = V_\mu(A_2)$ and this happens only for special values μ_c of μ. This particular (heteroclinic) trajectory separating oscillating from unbounded motion exist at $\mu = \mu_c$ but disappears as soon as μ is varied (Fig. 4.10): fronts move as soon as μ differs from μ_c.

Let us now consider a discrete analog of (4.105) which may be more appropriate for several experimental situations where A is only defined in discrete cells (e.g., Willaime *et al.*, 1991)

$$\frac{\partial A_i}{\partial t} = K \left(A_{i+1} - 2A_i + A_{i-1} \right) - V'_\mu(A_i) \,. \tag{4.107}$$

In this case, there exists a continuum interval of values of μ for which fronts do not move. This is the phenomenon of front pinning (Cahn,

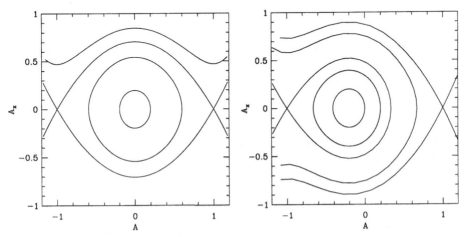

Fig. 4.10. Left: a standing front for the continuum Ginzburg–Landau equation (4.105) corresponds to a heteroclinic trajectory which joins the two stable states. Right: this particular trajectory disappears as soon as the potential is deformed.

1960). A motionless front solution in this model is a set of points which satisfy:

$$K\left(A_{n+1} - 2A_n + A_{n-1}\right) = V'_\mu\left(A_n\right), \qquad n \in \mathbb{Z}, \qquad (4.108)$$

together with the boundary conditions $A_n \rightarrow A_1, A_2$ when $n \rightarrow -\infty, +\infty$. By introducing a discrete analog of momentum $P_{n+1} = (A_{n+1} - A_n)/\sqrt{K}$, a two-dimensional mapping of the (A, P) plane is obtained

$$A_{n+1} = A_n + \frac{1}{\sqrt{K}} P_{n+1}, \qquad (4.109a)$$

$$P_{n+1} = P_n + \frac{1}{\sqrt{K}} V'_\mu\left(A_n\right). \qquad (4.109b)$$

This kind of mapping has been thoroughly analyzed in the dynamical systems literature. For $V'_\mu\left(A_n\right) = \sin\left(q_n\right)$, it is referred to as the "standard map" (Chirikov, 1979). The mapping (4.109) has two hyperbolic fixed points. As said previously, in the continuous case for $\mu = \mu_c$, the unstable manifold of $(A_1, 0)$ coincides with the stable manifold of $(A_2, 0)$ and forms a unique separatrix. However in the discrete case of (4.109) the two curves no longer coincide (Fig. 4.11a). The unique separatrix splits into two transversally intersecting invariant manifolds in the phase space (A, P) as discovered by Poincaré (1893) (see also Holmes *et al.*, 1988). For dynamical systems theory, this is important since the existence of transversal (heteroclinic) intersection points produces extremely complicated behavior in phase space. In the more limited context of front motion, this is also important because a front linking the stable state $(A_1, 0)$ to the stable state $(A_2, 0)$ is included in the set of intersection points (Hakim and Mallick, 1993). Indeed, the set of points (A_n, P_n) which constitutes such a front satisfies the two conditions:

1) $(A_n, P_n) \rightarrow (A_1, 0)$ as $n \rightarrow -\infty$, i.e., it belongs to the unstable manifold of $(A_1, 0)$;
2) $(A_n, P_n) \rightarrow (A_2, 0)$ as $n \rightarrow +\infty$, i.e., it belongs to the stable manifold of $(A_2, 0)$.

The difference between the discrete and continuous cases come from the fact that the transversal intersection points cannot disappear as soon as the potential is deformed (Fig. 4.11b,c). So, motionless fronts still exist when μ varies in a small range around μ_c, i.e., in the discrete case, fronts are "pinned" by the underlying lattice and do not move as soon as μ is varied as in the continuous case (Cahn, 1960). The calculation of the pinning interval around μ_c brings us back to the main theme of

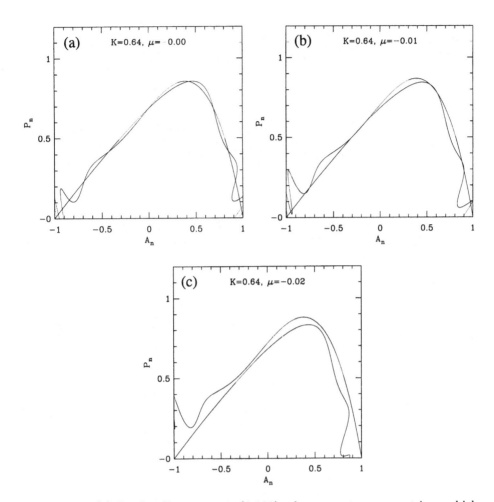

Fig. 4.11. (a) In the discrete case (4.109), there are two separatrices which have transversal intersection points. The angle of intersection is exponentially small in K. The motionless front trajectory is represented by a subset of these intersection points. (b) When the potential is deformed the intersection points do not disappear immediately and a motionless front exists in a finite interval around μ_c. (c) Finally, the intersection points disappear for sufficient potential deformation.

this section. Like the intersection angle between the separatrices of a rapidly forced system, it is exponentially small in the large K limit and the method of matched asymptotic expansions in the complex plane can be used to compute it (Lazutkin *et al.*, 1989; Hakim and Mallick, 1993). In order to see clearly how this technique can be applied to the present problem, it is useful to write a parametric equation for the separatrices in the form (from equation (4.109))

$$P(x) = \sqrt{K} \left[A(x) - A\left(x - 1/\sqrt{K}\right) \right], \tag{4.110a}$$

$$K \left[A\left(x + 1/\sqrt{K}\right) - 2A(x) + A\left(x - 1/\sqrt{K}\right) \right] = V'_\mu(A(x)). \tag{4.110b}$$

By using Taylor's expansion, these are seen to be the equations of the continuous front perturbed by higher derivatives, e.g., (4.110b) can be written as a perturbation of (4.106)

$$\frac{d^2 A}{dx^2} = V'_\mu(A) - 2 \sum_{n=2}^{\infty} \frac{1}{(2n)! K^{n-1}} \frac{d^{2n} A}{dx^{2n}}. \tag{4.111}$$

The matching technique of section 4.2 is applied around the singularity of the continuous front solution $A_c(x)$, (4.106), when continued to the complex x-plane. For example in the standard mapping case, the following asymptotic formula is obtained for the angle ϕ between the two separatrices:

$$\phi \sim \pi \theta_1 K^{3/2} \exp\left(-\pi^2 \sqrt{K}\right), \qquad \theta_1 = 1118.8 \ldots \tag{4.112}$$

As usual, the argument of the exponential simply reflects the location of the singularities of $A_c(x)$ in the complex x-plane. On the contrary, the estimation of the prefactor requires a resummation of perturbation theory to all orders, as done by solving a nonlinear inner problem. It is not correctly obtained by a simple first order perturbation theory calculation, like the classical Melnikov formula.

References

Ablowitz M.J., Clarkson P.A. (1991). *Solitons, Nonlinear Evolution Equations and Inverse Scattering*, London Math. Soc. Lecture note series 149 (Cambridge University Press).

Ablowitz M.J., Segur H. (1981). *Solitons and the Inverse Scattering Transform* (SIAM, Philadelphia).

Ahlers G., Cannell D. (1983). *Phys. Rev. Lett.* **50**, 383.

Amick C.J., McLeod J.B. (1990). *Arch. Rat. Mech. Anal* **109**, 139.

Andereck C., Hegseth J., Hayot F., Pomeau Y. (1989). *Phys. Rev. Lett.* **62**, 257.

Arnold V.I., ed. (1988). *Dynamical systems III* (Springer-Verlag, Berlin).

Aronson D., Weinberger H. (1978). *Adv. Math.* **30**, 33.

Barenblatt G.I. (1979). *Similarity, Self-similarity, and Intermediate Asymptotics* (Consultant Bureau, New York).

Beale J.T. (1991). *Comm. Pure Appl. Math.* **44**, 211.

Ben Amar M. (1991). *Phys. Rev. A* **43**, 5726.

Ben Amar M., Pomeau Y. (1986). *Europhys. Lett.* **2**, 307.

Ben Amar M., Hakim V., Mashaal M., Couder Y. (1991). *Phys. Fluids A* **3**, 1687.

Bender C., Orszag S.A. (1978). *Advanced Mathematical Methods for Scientists and Engineers* (McGraw-Hill).

Ben-Jacob E., Brand H., Dee G., Kramer L., Langer J.S. (1985). *Physica D* **14**, 348-364.

Ben-Jacob E., Goldenfeld N.D., Kotliar G., Langer J.S. (1984). *Phys. Rev. Lett.* **53**, 2110.

Benjamin T.B. (1974). *Lectures in Appl. Math.* **15**, 3.

Benjamin T.B., Feir J.E. (1967). *J. Fluid Mech.* **27**, 417.

Bensimon D., Pelcé P., Shraiman B.I. (1987). *J. Physique* **48**, 2081.

Berry M., Howls C.J. (1990). *Proc. Roy. Soc. (London) A* **430**, 653.

Berry M. (1991). in Segur *et al.* (1991).

Bogoliubov N.N., Mitropolsky Y.A. (1961). *Asymptotic Methods in the Theory of Nonlinear Oscillations* (Gordon and Breach, New York).

Borel E. (1928). *Leçons sur les Séries Divergentes* (Gauthier-Villars, Paris).

Boyd J.P. (1991). *Physica D* **48**, 129.

Bramson M. (1983). *Memoirs of the AMS* **44**, 185.

Brener E., Melnikov V.I., (1991). *Adv. Phys.* **40**, 53-97.

Bretherton F.P. (1961). *J. Fluid Mech.* **10**, 166.

Brower R., Kessler D., Koplik J., Levine H. (1984). *Phys. Rev. A* **29**, 1335.

Cahn J.W. (1960). *Act. Metall.* **8**, 554.

Candelpergher B., Nosmas J.C., Pham F. (1993). "Approches de la résurgence," *Actualités Mathématiques* (Hermann, Paris).

Carr J., Pego R.L. (1989). *Comm. Pure Appl. Math.* **42**, 523.

Chirikov B.V. (1979). *Phys. Rep.* **52**, 263.

Collet P., Eckmann J.P. (1990). *Instabilities and Fronts in Extended Systems* (Princeton University Press, Princeton).

Combescot R. (1992). *Phys. Rev. A* **45**, 873.

Combescot R., Hakim V., Dombre T., Pumir A., Pomeau Y. (1988). *Phys. Rev. A* **37**, 1270-1283.

Coullet P., Goldstein R.E., Gunaratne (1989). *Phys. Rev. Lett.* **63**, 1954.

Coullet P., Iooss G. (1990). *Phys. Rev. Lett.* **64**, 866.

Coullet P., Lega J. (1988). *Europhys. Lett.* **7**, 511.

Craig W., Sulem C., Sulem P.L. (1992). *Nonlinearity* **5**, 497.

Dashen R. F., Hasslacher B., Neveu A. (1975). *Phys. Rev. D* **11**, 3424.

de Gennes P.G. (1986). *Colloid Polym. Sci.* **264**, 463.

de Ryck A., Quéré D. (1996). *J. Fluid Mech.* **311**, 219.

Derrida B., Godrèche C., Yekutieli I. (1990). *Europhys. Lett.* **12**, 385.

Deryaguin B.V. (1943). *Dokl. Akad. Nauk. SSSR* **39**, 11.

Dingle R.B. (1973). *Asymptotic Expansions: their Derivation and Interpretation* (Academic Press, New York).

Dombre T., Hakim V. (1987). *Phys. Rev. A* **36**, 2811.

Dorodnitsyn A.A. (1947). *Inst. Mech. of the Acad. of Sci. of the USSR*, Vol. XI.

Dyachenko S., Newell A.C., Pushkarev A., Zakharov V.E. (1992) *Physica D* **57**, 96.

Dykne A.M. (1960). *Sov. Phys. JETP* **11**, 411.

Eckhaus W. (1979). *Asymptotic Analysis of Singular Perturbations* (North Holland Publ. Co, Amsterdam).

Eleonsky V. (1991). in Segur *et al.* (1991).

Ermentrout G.B., Hastings S.P., Troy W.C. (1984). *SIAM J. Appl. Math.* **44** 1133.

Fauve S., Thual O. (1990). *Phys. Rev. Lett.* **64**, 282.

Fedoryouk M. (1983). *Méthodes Asymptotiques pour les Équations Différentielles Ordinaires* (Mir, Moscow).

Fineberg J., Steinberg V. (1987). *Phys. Rev. Lett.* **58**, 1332.

Fraiman G.M. (1985). *Sov. Phys. JETP* **61** 228.

Gierer A., Meinhardt H. (1972). *Kybernetik* **12**, 30.

Haag J. (1944). *Ann. Sci. Ecole Normale Supérieure* **61**, 10.

Hakim V. (1991). in Segur, Tanveer, Levine (1991)

Hakim V., Jakobsen P., Pomeau Y. (1990). *Europhys. Lett.* **11**, 19.

Hakim V., Mallick K. (1993). *Nonlinearity* **6**, 57, and unpublished.

Hakim V., Pomeau Y. (1991). *Eur. J. Mech. B, Fluids* **10** n°2-Suppl, 137.

Hasimoto H., Ono H. (1972). *J. Phys. Soc. Japan* **33**, 805.

Holmes P., Marsden J., Scheurle J. (1988). *Contemp. Math.* **81**, 213.

Hunter J.K., Scheurle J. (1988). *Physica D* **32**, 253.

Hunter J.K., Vanden-Broeck J.M. (1983). *J. Fluid Mech.* **134**, 205.

Kadanoff L. (1990). *Phys. Rev. Lett.* **65**, 2986.

Kawasaki K., Ohta T. (1982). *Physica A* **116**, 573.

Kessler D.A., Koplik J., Levine H. (1985). *Phys. Rev. A* **31**, 1712.

Kessler D.A., Koplik J., Levine H. (1988). *Adv. Phys.* **37**, 255.

Kirchgässner K. (1988). *Adv. Appl. Mech.* **26**, 135.

Kivshar Y.S., Malomed B.A. (1989). *Rev. Mod. Phys.* **61**, 763.

Koga S., Kuramoto Y. (1980). *Prog. Theor. Phys.* **63**, 106.

Kolmogorov A.N., Petrovskii I.G., Piscounoff N.S. (1937). *Bulletin de l'Université d'État à Moscou, Série Internationale* **A1**, 1-25; reprinted in Pelcé (1988).

Kruskal M., Segur H. (1991). *Stud. Appl. Math.* **85**, 129.

Lagerström P.A. (1988). *Matched Asymptotic Expansions* (Springer-Verlag, New York).

Landau L., Levich B. (1942). *Acta Physicochimie URSS* **17**, 42; reprinted in Pelcé (1988).

Landau L., Lifshitz E. (1987). *Fluid Mechanics* (Pergamon Press, Oxford).

Lazutkin V.F., Schahmannski I.G., Tabanov M.B. (1989). *Physica D* **40**, 235.

Le Mesurier B.J., Papanicolaou G.C., Sulem C., Sulem P.L. (1988). *Physica D* **32**, 210.

Levi-Civita T. (1925). *Math. Ann.* **93**, 264.

Limat L., Jenffer P., Dagens B., Touron E., Fermigier M., Wesfreid J.E. (1992). *Physica D* **61**, 166.

Lücke M., Mihelcic M., Kowalski B. (1987). *Phys. Rev. A* **35**, 4001.

Malomed B.A., Nepomnyashchy A.A. (1991). *Phys. Rev. A* **42**, 6009.

Mazenko G.F.(1990). *Phys. Rev. B* **42**, 4487.

McLean J.W., Saffman P.G. (1981). *J. Fluid Mech.* **102**, 455.

Merle F. (1992). *Comm. Pure. Appl. Math.* **45**, 203.

Meyer R.E. (1980). *SIAM Rev.* **22**, 213.

Miles J.W. (1981). *J. Fluid Mech.* **106**, 131.

Müller-Krumbhaar H., Kurz W. (1991). *Phase Transformation in Materials*, P. Haasen, ed. (VCH-Verlag, Weinheim).

Nayfeh A.H. (1973). *Perturbation Methods* (Wiley, New York).

Newell A.C., Whitehead J.A. (1969). *J. Fluid Mech.* **38**, 279.

Newell A.C. (1974). *Lectures in Appl. Math.* **15**, 157.

Novikov S., Manakov S.V., Pitaevskii L.P., Zakharov V.E. (1984). *Theory of Solitons* (Plenum Press, New York).

Ohta T., Jasnow D., Kawasaki K. (1982). *Phys. Rev. Lett.* **49**, 1223.

Ohta T., Mimura M., Kobayashi R. (1989). *Physica D* **34**, 115.

Park C.W., Homsy G.M. (1984). *J. Fluid Mech.* **139**, 291.

Pelcé P., ed. (1988). *Dynamics of Curved Fronts* (Academic Press, New York).

Pellegrini Y., Jullien R. (1991). *Phys. Rev. A* **44**, 920.

Poincaré H. (1893). *Les Méthodes Nouvelles de la Mécanique Céleste* Vol. 2 (Gauthier-Villars, Paris)

Pokrovskii V.L., Khalatnikov I.M. (1961). *Sov. Phys. JETP* **13**, 1207.

Pomeau Y. (1986). *Physica D* **23**, 3.

Pomeau Y., Ramani A., Grammaticos B. (1988). *Physica D* **31**, 127.

Prandtl L. (1905). *III Inter. Math. Kongress Heidelberg*, Teubner Leipzig, 484.

Quéré D., Di Meglio J.M., Brochard-Wyart F. (1989). *Europhys. Lett.* **10**, 335.

Reinelt D.A. (1987). *Phys. Fluids* **30**, 2617.

Ruschak K.J. (1985). *Ann. Rev. Fluid Mech.* **17**, 65.

Saffman P.G., Taylor G.I. (1958). *Proc. Roy. Soc. (London) A* **245**, 312.

Sanders J.A., Verhulst F. (1985). *Averaging Methods in Nonlinear Dynamics* (Springer-Verlag, New York).

Segel L. (1969). *J. Fluid Mech.* **38**, 203.

Segur H., Kruskal M.D. (1987). *Phys. Rev. Lett.* **58**, 747.

Segur H., Tanveer S., Levine H., eds. (1991). *Asymptotics Beyond All Orders* (Plenum Press, New York).

Segur H. (1991). in Segur *et al.* (1991).

Skinner G.S., Swinney H.L. (1991). *Physica D* **48**, 1.

Stewartson J., Stuart J.T. (1971). *J. Fluid Mech.* **48**, 529.

Stoker J.J. (1957). *Water Waves* (Interscience, New York).

Stokes G.G. (1847). *Camb. Trans.* **8**, 441.

Tanveer S. (1987). *Phys. Fluid* **30**, 1589.

Tanveer S. (1990). *Proc. Roy. Soc. (London) A* **428**, 511.

Taylor G.I. (1961). *J. Fluid Mech.* **10**, 161.

Thomé H., Rabaud M., Hakim V., Couder Y. (1989). *Phys. Fluids A* **1**, 224.

Thual O., Fauve S. (1988). *J. Physique (Paris)* **49**, 1829.

Tu Y. (1991). *Phys. Rev. A* **44**, 1203.

Tyson J.J., Keener J.P. (1988). *Physica D* **32**, 327.

Van Dyke M. (1975). *Perturbation Methods in Fluid Mechanics* (Parabolic Press, Stanford).

Van Saarloos W. (1989). *Phys. Rev. A* **39**, 6367.

Van Saarloos W., Hohenberg P.C. (1992), *Physica D* **56**, 303.

Vanden-Broeck J. M. (1983). *Phys. Fluids* **26**, 2033.

Vanden-Broeck J. M. (1991). *Phys. Fluids A* **3**, 2659.

White D.A., Tallmadge J.A. (1966). *AIChE J.* **12**, 333.

Whitham G. B. (1974). *Linear and Nonlinear Waves* (Wiley, New York).

Willaime H., Cardoso O., Tabeling P. (1991). Eur. J. Mech. B Fluids Suppl. **10**, 165; *Phys. Rev. Lett.* **67** 3247.

Winfree A.T. (1991). *Chaos* **1**, 303.

Wollkind D., Segel L. (1970). *Philos. Trans. R. Soc. London* **268**, 351.

Zakharov V.E. (1972). *Sov. Phys. JETP* **35**, 908.

4

Pattern forming instabilities

Stephan Fauve

1 Introduction

Instabilities in nonlinear systems driven far from equilibrium often consist of a transition from a motionless state to one varying periodically in space or time. Various examples, widely studied in the past, are Rayleigh–Bénard convection, Couette–Taylor flow, waves in shear flows, instabilities of liquid crystals, oscillatory chemical reactions,... The appearance of periodic structures in these systems driven externally by a forcing homogeneous in space or constant in time, corresponds to a bifurcation, characterized by one or several modes that become unstable as a *control parameter* is varied. Linear stability analysis of the basic state gives the critical value of the control parameter for the primary instability onset, the nature of the most unstable modes and their growth rate above criticality. Many examples have been studied for a long time and can be found for instance in the books of Chandrasekhar (1961) or Drazin and Reid (1981). However, linear stability analysis does not describe the saturation mechanism of the primary instability, and thus a nonlinear analysis should be performed to determine the selected pattern, its dynamics and in particular the secondary instabilities that occur as the control parameter is increased above criticality. Before considering these problems, we present some examples of the characteristic phenomena that occur above a pattern-forming instability onset.

1.1 Example: the Faraday instability

As a first example, consider a cylindrical vessel containing a liquid and its vapor (or any other gas), vertically vibrated at frequency ω_e (see Fig. 1.1). It was discovered by Faraday (1831) that when the vibration amplitude exceeds a critical value, the flat liquid-gas interface becomes unstable to standing waves. Figures 1.2 and 1.3 display the surface waves excited by vertical shaking, seen from above. The axisymmetric standing wave of Fig. 1.2 is observed as a transient at instability onset; the shape of this

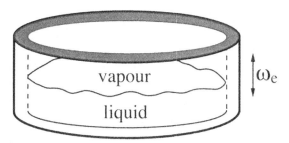

Fig. 1.1. Sketch of the apparatus for the Faraday instability.

mode is due to the circular geometry of the vessel. This is the most unstable linear eigenmode of the flat interface. However, this axisymmetric pattern is not nonlinearly stable; in a large enough container, nonlinear interactions select a one-dimensional standing wave pattern (see Fig. 1.3) if the viscosity of the fluid is large enough. Thus, from this first example, one observes that the patterns ultimately selected by the instability do not generally correspond to the most unstable linear mode. In large containers, nonlinear effects are strong enough to overcome boundary effects that trigger the axisymmetric pattern.

When the viscosity of the fluid is small, a square pattern is observed above instability onset (see Fig. 1.4). We can think of the square pattern

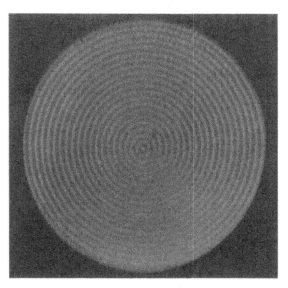

Fig. 1.2. Axisymmetric standing waves observed at onset on the interface between liquid CO_2 and its vapor submitted to vertical vibrations.

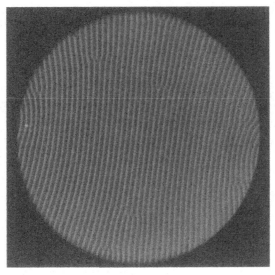

Fig. 1.3. Standing wave pattern at the interface between liquid CO_2 and its vapor. Note the defects in the wave pattern.

as the result of two standing waves of equal amplitudes with perpendicular wave-vectors. When the viscosity is increased, there is a critical point for which the nonlinear interaction between the intersecting waves changes so that they cannot both remain stable, and one of them vanishes. We then

Fig. 1.4. Square pattern at the interface between liquid CO_2 and its vapor.

Fig. 1.5.   Line-defects in surface waves on vertically-shaken mercury. The regions on opposite sides of the lines are out of phase with one another.

get a one-dimensional standing wave pattern (see Fig. 1.3). Thus nonlinear effects not only saturate the growth of the linearly unstable modes but they also act as a selection mechanism for the pattern. In Fig. 1.3 we see that some of the wave-crests do not reach all the way across the vessel. The endpoint of a crest is called a *defect*. If one integrates the phase gradient along a closed curve around one of these defects, the integral comes to $\pm n2\pi$ (n integer) instead of 0, because one passes more waves on one side of the defect than on the other. At the defect itself, the amplitude of the wave vanishes, so that at this point the phase is undefined. Defects are widely observed when a periodic pattern undergoes a secondary instability. In the above example, they nucleate or annihilate by pair or at the lateral boundary and during their life-time, move in the underlying periodic pattern. Their dynamics plays an important role in the transition to spatiotemporal disorder. Their shape follows the pattern symmetries. For instance, defects of the square standing wave pattern consist of lines instead of points; they separate two regions of the wave-field that oscillate out of phase (see Fig. 1.5). Secondary instabilities of periodic patterns do not always generate defects. An important class of secondary instabilities consists of long-wavelength modulations of the primary pattern. The surface wave pattern in an elongated rectangular geometry or in an annular container, i.e., when a one-dimensional wave is forced by the boundary conditions, exhibits a secondary instability consisting of a long-wavelength spatiotemporal modulation of the primary pattern wavelength (see Fig. 1.6).

1.2 Analogy with phase transitions: amplitude equations

The different phenomena described above are not particular to parametrically generated surface waves; long-wavelength instabilities or dynamics of defects are widely observed above the onset of most pattern-forming instabilities (Wesfreid and Zaleski, 1984; Wesfreid *et al.*, 1988). An obvious unifying description consists of looking for evolution equations for the amplitude and the phase, i.e., for the complex amplitude of the periodic pattern generated by the primary instability. Indeed, at the primary instability onset, the critical modes have, by definition, a vanishing growth rate; we will show that adiabatic elimination of all the other (faster) modes leads to nonlinear partial differential equations that govern the amplitude of the critical modes, and so describes the slow modulations in space or time of the periodic structure envelope: these are the *amplitude equations*. We will see that the form of the amplitude equations can be derived simply from symmetry considerations, and that the underlying detailed equations are needed only to evaluate parameters in the amplitude equations. Another possible way is to use experiments to determine the parameters in an amplitude equation, because we can know the form of the equation without knowing all the details of the

Fig. 1.6. Snapshot of the long-wavelength modulation of the basic standing wave in an annular geometry. The basic standing wave consists of 21 wavelengths and the wavelength of the modulation is equal to the perimeter of the annulus.

microscopic dynamics. This is similar to the description of a fluid flow using the Navier–Stokes equation, the form of which follows conservation laws; the coefficients, viscosity for instance, depend on the microscopic dynamics and might be computed using the Boltzmann equation; it is, however, simpler and more reliable to use the experimentally measured coefficients.

We will mainly be studying amplitude equations, rather than the microscopic governing equations of the underlying systems. We take this approach because similar patterns are observed in a wide range of systems, and their behavior is a result of the broken symmetries at the primary instability onset rather than of the microscopic dynamics. Systems with different microscopic descriptions frequently exhibit, on a macroscopic level, similar patterns which are governed by the same amplitude equation. The situation is analogous to the one we encounter in phase transitions in condensed-matter physics where the behavior of the order parameter is governed by symmetries and does not depend on the "chemical details" of the system. The close analogy between instabilities in nonlinear systems driven far from equilibrium and phase transitions is now well documented experimentally as well as theoretically. This idea was fathered by Landau (1941), and developed by several people in the context of hydrodynamics, electric circuits, nonlinear optics and chemical instabilities. In this context, the complex amplitude of the periodic pattern plays the role of an order parameter and characterizes the broken symmetries at instability onset. Amplitude equations are analogous to the Ginzburg–Landau description of phase transitions.

1.3 Long-wavelength neutral modes: phase dynamics

The second area that we will study is the disorganization of the primary pattern through secondary instabilities. When the primary instability saturates nonlinearly and gives rise to a finite amplitude periodic pattern, only its phase remains neutral in the long-wavelength limit. Indeed, a spatially uniform modification of the phase corresponds to a shift of the periodic pattern, and thus is neutral because of translational invariance in space. Likewise, other broken symmetries, translational invariance in time at the onset of an oscillatory instability, Galilean invariance at the onset of a pattern-forming instability, ... may generate long-wavelength neutral modes, i.e., modes that are neither dissipated nor amplified at zero wavenumber. These modes are analogous to Goldstone modes in particle physics or condensed-matter physics, and often lead to secondary instabilities of the primary pattern. Because their growth rate vanishes in the long-wavelength limit, we eliminate adiabatically the other fast modes in order to obtain evolution equations that describe pattern dynamics through its slowly varying phases. Thus, contrary to the situation at instability onset, the pattern amplitude is no longer a neutral mode above criticality; for a perfectly periodic pattern, it saturates at a finite value. However, phase instabilities, that usually occur at small wavenumber do not always saturate in this long wavelength limit; they often cascade to short scales, leading to defect nucleation in the primary pattern. Although non-neutral, the pattern amplitude locally vanishes, thus breaking the long-wavelength approximation. A consistent description of this type of pattern dynamics is still an open problem.

1.4 Localized nonlinear structures

Shock waves or solitons are well known examples of nonlinear localized structures. As was said above, defects of periodic patterns are another class of localized structures. Although incompletely, we will define the main characteristics of these objects, try to classify them and to understand their dynamics.

Another type of localized structure is well known in fluid dynamics and is illustrated by plate 109 in Van Dyke (1982). It is a turbulent spot, i.e., a region of turbulent flow advected in a laminar flow. This type of structure is widely observed in pipe flows or boundary layers. The turbulent region can expand when moving, but there are also situations where the spot keeps a nearly constant size. Similarly, pattern forming instabilities can display localized structures consisting of a region in the bifurcated state surrounded by the basic state. This occurs when there is a parameter range in which the system has two stable states of different form; then one might observe both of them in separate regions. Thermal convection of a binary fluid mixture in an annulus displays such localized patterns (Kolodner *et al.*, 1988). An earlier example is the localized standing wave observed in Faraday instability (Wu, Keolian and Rudnick, 1984). A two-dimensional example of these localized structures, observed in a Faraday experiment with two forcing frequencies, is displayed in Fig. 1.7. "Bubbles" of standing waves are surrounded by regions with flat surfaces. In the limit of small dissipation, we will show that these localized structures trace back to the solitary waves of conservative systems.

This chapter is organized as follows. In section 2 we recall elementary results on nonlinear oscillators and describe the multiple-scale perturbation method. We then consider nonlinear waves in dispersive media in section 3. We describe Rayleigh–Bénard convection as a canonical example of pattern forming instability (section 4), and we then give, in section 5, a catalogue of simple bifurcations to spatiotemporal patterns. We consider secondary instabilities of cellular flows in sections 6 and 7. Finally,

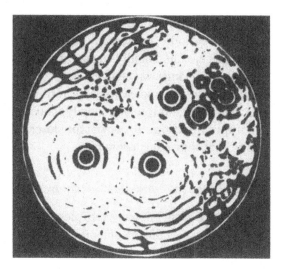

Fig. 1.7.   Localized standing waves on the surface of a vertically-shaken fluid.

we give a few examples of nonlinear localized structures in section 8. The references contain most of the material used for this chapter but are far from being exhaustive. Several recently published books might give a different approach about pattern forming instabilities: Grindrod (1991), Manneville (1990), Rabinovich and Trubetskov (1989), Sagdeev, Usikov and Zaslavsky (1988). These notes were written in 1991 for the Beg-Rohu school in Condensed Matter Physics and for the Woods Hole Oceanographic Institution Summer Program in Fluid Dynamics. I acknowledge Claude Godrèche, Edward Spiegel and William Young for having motivated this work.

2 Nonlinear oscillators

In this section we study dissipative nonlinear oscillators as the simplest examples of temporal patterns, i.e., periodic limit cycles, quasiperiodic regimes and frequency locking phenomena. Two different amplification mechanisms are considered to balance dissipation and sustain an oscillatory regime:

- "negative dissipation," for which the system is autonomous,
- parametric forcing, where an external time dependent perturbation is applied.

Our objective is to find an amplitude equation for the amplitude and phase of the oscillation. We show that its form depends on the amplification mechanism, because the broken symmetries at the oscillatory instability onset are different, but that universal behavior of the oscillation amplitude and the frequency exists in both types of oscillator. We next consider a negative dissipation oscillator with an external time-dependent forcing, and study frequency-locking phenomena.

2.1 Van der Pol oscillator

We begin with an autonomous system with negative dissipation. The canonical example is the Van der Pol oscillator, introduced in electronics a long time ago (Van der Pol, 1934). The governing equation is

$$\ddot{u} - 2\lambda\dot{u} + u^2\dot{u} + \omega_0^2 u = 0. \tag{2.1}$$

Here, the dissipation is negative if $\lambda > 0$ and the nonlinear term causes the system to saturate when $u^2 \sim \lambda$.

2.1.1 Global or linear stability

Some insight into the behavior of this equation can be obtained by studying the total energy of the system, which is here a simple example of Lyapunov functional. This is a useful first step before attempting a more detailed analysis. The sum of potential and kinetic energies is

$$E = \omega_0^2 \frac{u^2}{2} + \frac{\dot{u}^2}{2} \tag{2.2}$$

and the time derivative of the energy is

$$\dot{E} = \omega_0^2 u\dot{u} + \dot{u}\ddot{u} = (2\lambda - u^2)\dot{u}^2. \tag{2.3}$$

Thus if $\lambda < 0$ (linear damping), $\dot{E} < 0$. Moreover, there is a lower bound, $E = 0$, which occurs if and only if $u = \dot{u} = 0$, and consequently the motionless state is globally stable.

We are interested in the case $\lambda > 0$. A linear stability analysis about $u = 0$ is done by taking $u \propto \exp(\eta t)$, and if $\mathcal{R}e(\eta) > 0$ the system is unstable. Substituting into the Van der Pol equation and linearizing gives the characteristic polynomial

$$\eta^2 - 2\lambda\eta + \omega_0^2 = 0 \tag{2.4}$$

or $\eta = \lambda \pm \sqrt{\lambda^2 - \omega_0^2}$. If we let $\lambda = \mu\epsilon$ where ϵ is small, $\eta \simeq \mu\epsilon \pm i\omega_0^2$ and the system becomes unstable as λ changes sign; this is a Hopf bifurcation and the system's behavior changes from a damped oscillation to an amplified oscillation as λ increases. Near the bifurcation point, the growth rate is small and the time scale of the growth is $T \sim 1/\epsilon$.

2.1.2 Nonlinear effects

The linear analysis predicts exponential growth, but one expects the nonlinear term to saturate the instability when $u^2 \sim \epsilon$. We can see this by studying the energy balance of the system. We suppose that there is a harmonic limit cycle when λ is small and so we are close to the bifurcation, and we also presume that the amplitude of the oscillation varies on a slow time scale, thus

$$u \approx a(t) \cos \omega_0 t,$$
$$\dot{u} \approx -\omega_0 a(t) \sin \omega_0 t. \tag{2.5}$$

The energy balance over one cycle requires

$$\frac{1}{T} \int_0^T \dot{E} dt = 0, \tag{2.6}$$

thus

$$2\lambda \overline{\dot{u}^2} - \overline{u^2 \dot{u}^2} = 0, \tag{2.7}$$

where the overbar denotes averaging over one cycle. Using the above expressions for u, we find $a^2 = 8\lambda = 8\mu\epsilon$, and so the oscillation amplitude above criticality is proportional to the square-root of the distance to criticality,

$$u \sim \sqrt{\epsilon}. \tag{2.8}$$

A second role of the nonlinearity is the production of higher harmonics and the subsequent shift of the fundamental frequency. We show next that these effects are connected. If the governing equation is multiplied by u and averaged over one cycle, the terms involving $\dot{u}$ all average to zero and we are left with the virial equation

$$\overline{\dot{u}^2} = \omega_0^2 \overline{u^2}. \tag{2.9}$$

Note that both the nonlinear and dissipative terms have averaged to zero. However, the effect of nonlinearity still enters, since we assume that u contains some higher harmonic components:

$$u = \sum_{n=1}^{\infty} a_n \cos (n\omega t + \phi_n). \tag{2.10}$$

When this is substituted into the virial equation, we obtain

$$\sum n^2 \omega^2 a_n^2 = \omega_0^2 \sum a_n^2. \tag{2.11}$$

and this can be written as

$$\frac{\omega^2 - \omega_0^2}{\omega_0^2} = -\frac{\sum (n^2 - 1) a_n^2}{\sum n^2 a_n^2}. \tag{2.12}$$

Assuming that the successive harmonics amplitude decreases fast enough with increasing n, the frequency shift, $\Delta\omega \equiv \omega - \omega_0$, is to leading orders,

$$\frac{\Delta\omega}{\omega_0} \simeq -\frac{3}{2}\frac{a_2^2}{a_1^2} - 4\frac{a_3^2}{a_1^2} + \dots \tag{2.13}$$

Thus the shift of the fundamental frequency is related to the existence of higher order harmonics. For the Van der Pol equation, there is no quadratic nonlinearity and $a_2 = 0$; thus there is no frequency correction to leading order. Note however that the relationship between the fundamental frequency shift and the harmonics amplitude, is in general not so simple as the one above. For instance, replace $\omega_0^2 u$ by $\omega_0^2 \sin u$ in the Van der Pol equation and try the same exercise.

2.1.3 Amplitude equations: the multiple-scale method

In cases where the amplitude varies much more slowly than the underlying oscillation, we can use the disparity in temporal scale to obtain a simplified description of amplitude variations independently of the faster time scale (Bender and Orszag, 1978; Nayfeh, 1973). This is sketched schematically in Fig. 2.1. The oscillation has time scale t. We introduce a second time variable, $T = \epsilon t$, to parameterize the slow variation in amplitude; indeed, in the vicinity of instability onset, $\lambda = \mu\epsilon$, the instability growth-rate scales like ϵ.

We have already seen that the saturation amplitude for u scales like $\sqrt{\epsilon}$, so we take

$$u(t) = \sqrt{\epsilon}\,\tilde{u}(t, T)\,. \tag{2.14}$$

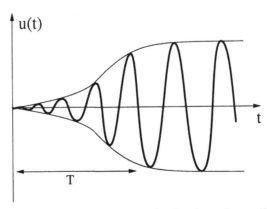

Fig. 2.1. Amplitude varying much more slowly than the oscillation gives multiple time scales.

Using

$$\frac{d}{dt} = \frac{\partial}{\partial t} + \epsilon \frac{\partial}{\partial T} \tag{2.15}$$

we rewrite the Van der Pol equation:

$$\frac{\partial^2 \tilde{u}}{\partial t^2} + 2\epsilon \frac{\partial^2 \tilde{u}}{\partial t \partial T} + \epsilon^2 \frac{\partial^2 \tilde{u}}{\partial T^2} + \epsilon(\tilde{u}^2 - 2\mu)\left(\frac{\partial \tilde{u}}{\partial t} + \epsilon \frac{\partial \tilde{u}}{\partial T}\right) + \omega_0^2 \tilde{u} = 0. \tag{2.16}$$

We can now consider a perturbation expansion

$$\tilde{u} = \tilde{u}_0 + \epsilon \tilde{u}_1 + \epsilon^2 \tilde{u}_2 + \ldots, \tag{2.17}$$

and collect terms in increasing powers of ϵ. To leading order we have a simple harmonic oscillator:

$$\mathcal{L}.\tilde{u}_0 \equiv \left(\frac{\partial^2}{\partial t^2} + \omega_0^2\right)\tilde{u}_0 = 0, \tag{2.18}$$

thus

$$\tilde{u}_0 = A(T) \exp(i\omega_0 t) + \bar{A}(T) \exp(-i\omega_0 t), \tag{2.19}$$

where $A(T)$ is the slowly-varying amplitude of the oscillation. Substituting (2.19) into the ϵ^1 equation,

$$\mathcal{L}.\tilde{u}_1 = -2\frac{\partial^2 \tilde{u}_0}{\partial t \partial T} - \tilde{u}_0^2 \frac{\partial \tilde{u}_0}{\partial t} + 2\mu \frac{\partial \tilde{u}_0}{\partial t}, \tag{2.20}$$

yields

$$\begin{aligned}
\mathcal{L}.\tilde{u}_1 = {} & 2i\omega_0 \left[\mu A(T) - \frac{dA}{dT}\right] \exp i\omega_0 t \\
& - i\omega_0 \left(|A|^2 A \exp i\omega_0 t + A^3 \exp 3i\omega_0 t\right) + \text{c.c.}
\end{aligned} \tag{2.21}$$

where "c.c." refers to the complex conjugate of the preceding expression.

In order to avoid secular growth of $\tilde{u}_1$, it is necessary to set the resonant terms i.e., the terms in $\exp(i\omega_0 t)$, to zero. Doing so yields

$$\frac{dA}{dT} = \mu A - \frac{1}{2}|A|^2 A. \tag{2.22}$$

Taking $A = R \exp(i\theta)$, factoring out $\exp(i\theta)$ and separating real and imaginary parts, gives

$$\begin{aligned}
\frac{dR}{dT} &= \left(\mu - \frac{1}{2}R^2\right)R, \\
\frac{d\theta}{dT} &= 0.
\end{aligned} \tag{2.23}$$

The amplitude R approaches $\sqrt{2\mu}$ when $\mu > 0$ and the phase θ does not vary with the slow time scale at all. This second feature is not generic for a Hopf bifurcation and traces back to the absence of frequency shift to leading order. As a simple exercise, you can again consider the Van der Pol equation with $\omega_0^2 \sin u$ instead of $\omega_0^2 u$, and show that a term proportional to R^2 occurs the θ equation.

Generically, the amplitude equation for a Hopf bifurcation is of the form

$$\frac{dA}{dT} = \mu A - \beta |A|^2 A \qquad (2.24)$$

with $\beta = \beta_r + i\beta_i$ some complex number. Then, the equations for R and θ become

$$\begin{aligned}
\frac{dR}{dT} &= (\mu - \beta_r R^2) R\,, \\
\frac{d\theta}{dT} &= -\beta_i R^2\,.
\end{aligned} \qquad (2.25)$$

If $\beta_r > 0$, the leading order nonlinearity saturates the instability and the bifurcation is said to be "direct" or "supercritical" (see Fig. 2.2). Above instability onset, the oscillation amplitude is proportional to $\sqrt{\mu}$. Otherwise, the bifurcation is "inverse" or "subcritical." In the supercritical situation, the phase increases linearly in time, indicating the fundamental frequency shift proportional to μ. Oscillatory instabilities observed experimentally often display this characteristic behavior for the oscillation amplitude and frequency slightly above criticality. The corresponding measurements are a useful check that the system undergoes a Hopf bifurcation. Note however, that the frequency ω_0 itself usually depends on the experimental control parameter (proportional to μ); this dependence is generally linear to leading order and adds to the one due to β_i. This makes β_i difficult to measure precisely.

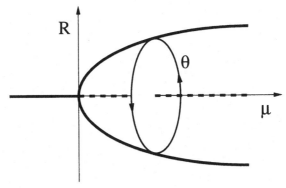

Fig. 2.2. Diagram of the supercritical Hopf bifurcation.

2.1.4 Symmetry arguments

We show next that the form of the evolution equation for the oscillation complex amplitude A is determined by symmetry constraints. It is clear that the original Van der Pol equation (2.1) describes an autonomous system, i.e., it is invariant under translation in time and is unaffected by a change of variables $t \rightarrow t + \theta$. This does not mean that any solution should be translationally invariant in time; any oscillatory solution, and in particular the expansion we used for u

$$u = \sqrt{\epsilon} \left[A(T) \exp(i\omega_0 t) + \bar{A}(T) \exp(-i\omega_0 t) + \epsilon u_1 + \dots \right], \qquad (2.26)$$

obviously breaks translational invariance in time. However, the ensemble of possible solutions should have this invariance. In other words, any image of an oscillatory solution under the transformation corresponding to the broken symmetry must be a solution of the autonomous system. Thus, if $u(t)$ is an oscillatory solution, so is $u(t + t_0)$. This obvious requirement is enough to determine the form of the amplitude equation, if one assumes, and this is a crucial assumption, that dA/dT can be expanded in powers of A and $\bar{A}$ in the vicinity of instability onset. Making the transformation $t \rightarrow t + \theta/\omega_0$, one gets

$$u(t + \theta/\omega_0) = \sqrt{\epsilon} \left[A(T) \exp(i\omega_0 t) \exp(i\theta) + \text{c.c.} + \epsilon u_1 + \dots \right]. \qquad (2.27)$$

The dynamics of the transformed u is to be the same as the original u, therefore the dynamics of A should be invariant under the transformation,

$$A(T) \rightarrow A(T) \exp(i\theta).$$

This transformation selects the combinations of A and $\bar{A}$ that can appear in the amplitude equation: only those which also transform with a factor of $\exp i\theta$ will do. Consider all the possibilities up to cubic terms:

$$\underline{A \rightarrow Ae^{i\theta}} \qquad \bar{A} \rightarrow \bar{A}e^{-i\theta}$$

$$A^2 \rightarrow A^2 e^{2i\theta} \qquad A\bar{A} \rightarrow A\bar{A} \qquad \bar{A}^2 \rightarrow \bar{A}^2 e^{-2i\theta}$$

$$A^3 \rightarrow A^3 e^{3i\theta} \quad \underline{A^2\bar{A} \rightarrow A^2\bar{A}\,e^{i\theta}} \quad A\bar{A}^2 \rightarrow A\bar{A}^2 e^{-i\theta} \quad \bar{A}^3 \rightarrow \bar{A}^3 e^{-3i\theta}$$

Only the underlined terms scale appropriately, so they are the only ones which can appear in the amplitude equation. Hence, we could have deduced the form of the amplitude equation simply from the symmetry of the original equation and the solution to the zeroth-order equation that fixes the broken symmetry.

Note that we considered t and T as independent variables and did not change $A(T) \rightarrow A(T + \epsilon \theta/\omega_0)$. This is only correct to leading order in (2.27).

Another remark concerns the locality assumption which is made when one looks for dA/dT as an expansion in powers of A and $\bar{A}$. A term of the form

$$A(T) \int_{-\infty}^{T} A(T')\bar{A}(T')K(T - T')\,dT', \qquad (2.28)$$

which includes memory effects, would be perfectly valid in the amplitude equation since it satisfies the symmetry requirement. Try to work out a simple oscillator example with memory effects.

Finally, note that not all the symmetries of the problem give constraints on the form of the amplitude equation. The symmetries that have to be considered are the ones that are broken by the linearly unstable solution. The symmetry requirement for the ensemble of possible bifurcated solutions then constrains the form of the amplitude equation. A symmetry of the original problem that is not broken by the linearly unstable mode gives no constraint on the amplitude equation. Amplitude equations can also have symmetries that are not forced by the original equation. As we will see when studying frequency-locking phenomena, the form of the linearly unstable mode is of crucial importance; an amplitude equation is meaningless without the expression of the original field as a function of the amplitude.

2.2 Parametric oscillators

We will now discuss the bifurcation structure of parametric oscillators. Parametric amplification occurs widely in physical situations. Examples include Langmuir waves in plasmas, spin waves in ferromagnets, surface waves on a ferrofluid in a time-dependent magnetic field, or on a liquid dielectric in an alternting electric field. As shown in the first section, parametric amplification of surface waves on a horizontal layer of fluid vertically vibrated, is a simple experimental model of pattern dynamics. Let us mention also that parametric amplifiers were widely used in electronics. We study here the simplest example which is a pendulum whose support is vibrated. The pendulum angle $u(t)$ from the vertical axis, is governed by the damped *Mathieu equation*

$$\ddot{u} + 2\lambda\dot{u} + \omega_0^2(1 + f\sin\omega_e t)\sin u = 0, \qquad (2.29)$$

where λ is the damping, ω_0 is the natural frequency, f is the forcing amplitude and ω_e is the external forcing frequency. The resonance characteristics of the Mathieu equation are well known to occur whenever $n\omega_e = 2\omega_0$ and are shown in Fig. 2.3.

We now examine the strongest resonance $\omega_e = 2\omega_0$ in detail. We obtain the amplitude equation by the method of multiple scales. The

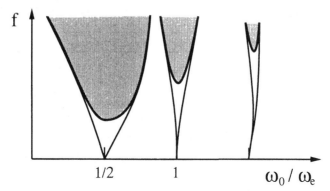

f

1/2   1 ω_0 / ω_e

Fig. 2.3.   Resonance tongues of the Mathieu equation (unstable regions are shaded).

three parameters in the equation are damping, forcing and detuning $\delta = (\omega_0^2/\omega_e^2) - \frac{1}{4}$. To bring in all the effects at the same order we use the scaling $\delta = \epsilon\Delta$, $\lambda = \epsilon\Lambda$, $f = \epsilon F$ and let

$$u(t) = \sqrt{\epsilon}\,[u_0(t,T) + \epsilon u_1(t,T) + \ldots]\,, \qquad (2.30)$$

where the slow time is $T = \epsilon t$. After substituting these in the equation and collecting terms we obtain at the zeroth order

$$\mathcal{L}.u_0 \equiv \left(\frac{d^2}{dt^2} + \omega_0^2\right)u_0 = 0\,, \qquad (2.31)$$

which has the solution

$$u_0 = A(T)\exp(i\omega_0 t) + \bar{A}(T)\exp(-i\omega_0 t)\,. \qquad (2.32)$$

The next order (ϵ^1) problem is

$$\mathcal{L}.u_1 = -2\frac{\partial^2 u_0}{\partial t\partial T} - 2\Lambda\frac{\partial u_0}{\partial t} - \omega_0^2 F u_0 \sin\omega_e t + \frac{1}{6}\omega_0^2 u_0^3\,. \qquad (2.33)$$

Using the fact that $\omega_e = 2\omega_0 - 4\epsilon\Delta\omega_0 + \mathcal{O}(\epsilon^2)$ the solvability condition to eliminate resonant terms is

$$\frac{dA}{dT} = -\Lambda A + \frac{\omega_0 F}{4}\bar{A}\exp(-4i\Delta\omega_0 T) - \frac{\omega_0}{4}iA^2\bar{A}\,. \qquad (2.34)$$

By moving to a frame of reference rotating with $\omega_e/2$ (instead of ω_0) with the transformation $A = B\exp(-2i\Delta\omega_0 T)$ we obtain the autonomous amplitude equation

$$\frac{dB}{dT} = (-\Lambda + i\nu)B + \mu\bar{B} + i\beta|B|^2 B \qquad (2.35)$$

where $\nu = 2\Delta\omega_0$, $\mu = \omega_0 F/4$ and $\beta = -\omega_0/4$.

The form of this equation could have been guessed by using symmetry arguments, the relevant symmetry of the Mathieu equation being $t \rightarrow t + 2\pi/\omega_e$, which restricts terms in the amplitude equation to be invariant under $B \rightarrow -B$. This symmetry is a much weaker restriction than the one for the Van der Pol equation ($A \rightarrow A\exp i\theta$). The term proportional to iB in the amplitude equation corresponds to a rotation of B at constant velocity in the complex plane and thus to a detuning. In other words, $\nu \neq 0$ indicates that $\omega_e/2$ and ω_0 are slightly different, thus that the forcing frequency is not exactly at parametric resonance. Moreover, collecting all the terms with pure imaginary coefficients, $i(\nu + \beta|B|^2)B$, shows that $\beta|B|^2$ is a nonlinear detuning. It is associated with the amplitude dependence of the oscillator frequency, and this nonlinear effect is the one that saturates the instability, by shifting the oscillator away from parametric resonance. This is to be contrasted to the Van der Pol oscillator where the instability is saturated by nonlinear damping. The term $\mu\bar{B}$, that breaks rotational invariance in the complex plane of the amplitude equation, is precisely the one that results from the parametric forcing.

Let us now study the linear stability of the solution $u = 0$. Writing $B = X + iY$ and inserting a mode proportional to $\exp \eta T$ we obtain the following quadratic equation for the eigenvalues:

$$\eta^2 + 2\Lambda\eta + (\Lambda^2 - \mu^2 + \nu^2) = 0. \qquad (2.36)$$

Since the damping, Λ, is positive, we see that there is no Hopf bifurcation contrary to the Van der Pol case. There is a stationary bifurcation at a threshold forcing amplitude $\mu_c = \sqrt{\Lambda^2 + \nu^2}$. This is shown in Fig. 2.4 which reproduces the $2:1$ resonance curve of Fig. 2.3.

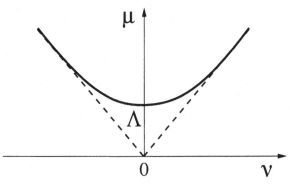

Fig. 2.4. Linear stability in parameter space of $u = 0$ with $2:1$ forcing.

The nonlinear stability is only slightly more complicated. We write $B = R \exp i\theta$ and obtain the equations

$$\frac{dR}{d\tau} = (-\Lambda + \mu \cos 2\theta)R,$$

$$\frac{d\theta}{d\tau} = \nu - \mu \sin 2\theta + \beta R^2. \tag{2.37}$$

To find the stationary solutions we set the right hand side of the equations to zero. Defining the finite amplitude stationary solution to be R_0 we obtain

$$\beta R_0^2 = -\nu \pm \sqrt{\mu^2 - \Lambda^2}. \tag{2.38}$$

Without loss of generality we take $\beta > 0$ (otherwise we consider the complex conjugate equation). For real solutions we need $\mu > \Lambda$. Then if $\nu > 0$ only the positive sign is valid and there is one solution for the amplitude, i.e., two solutions with different phases (labeled by 2×1 in Fig. 2.5). If $\nu < 0$, then for $\mu < \sqrt{\nu^2 + \Lambda^2}$ we can have four solutions (2×2). Figure 2.5 shows these different regions.

The behavior is made clearer in the bifurcation diagrams. For $\nu > 0$ (see Fig. 2.6a) we have a supercritical bifurcation at $\mu_c = \sqrt{\nu^2 + \Lambda^2}$. For $\nu < 0$ (Fig. 2.6b) we have a subcritical bifurcation, which is why we have 2×2 non-zero solutions. The point $\nu = 0$, $\mu = \Lambda$ is a tricritical point (in the language of phase transitions). As usual in subcritical bifurcations one of the solutions is unstable (shown by the dashed branch in the diagram). The stability of the branches can be derived by perturbing the finite amplitude solutions. If we write $R = R_0 + r$ and $\theta = \theta_0 + \phi$, then the eigenvalue σ of the perturbation satisfies

$$\sigma^2 + 2\Lambda\sigma + 4\beta R_0^2(\nu + \beta R_0^2) = 0. \tag{2.39}$$

Thus, the bifurcated solutions are stable if $\nu + \beta R_0^2 > 0$.

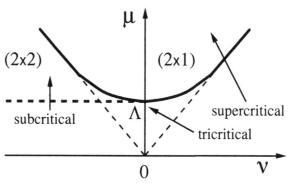

Fig. 2.5. Nonlinear solutions in parameter space with $2 : 1$ forcing.

Scaling behaviors above criticality display an interesting feature: for the tricritical point at $\nu = 0$, if one writes $\mu = \mu_c + \epsilon$, the amplitude scales as $R_0 \sim \epsilon^{1/4}$; for the supercritical case $R_0 \sim \epsilon^{1/2}$ as for the Hopf bifurcation. For $\nu \simeq 0$, one expects a cross-over between the two behaviors. The bifurcation diagram of the parametric oscillator is richer than the one of the Hopf bifurcation. This is because parametric forcing involves two control parameters, the forcing amplitude and its frequency, instead of one for the Hopf bifurcation.

The analysis we have performed above is valid in the limit of small dissipation; harmonic oscillations are almost neutral for the unforced oscillator, and a small external driving generates a nearly sinusoidal limit cycle which can be computed perturbatively. When the dissipation is large, one obviously needs a larger driving to generate the instability, but in addition the unstable eigenmode is no longer a nearly sinusoidal one. The parametrically driven damped pendulum (2.29) displays a strong relaxation type behavior at instability onset when λ becomes comparable to ω_0.

2.3 Frequency locking

Two independent oscillators have generically incommensurate frequencies. In simple words, this means that if one sets up two clocks, even as similar as possible, they oscillate at slightly different frequencies and thus finally indicate a different time if one waits long enough. It has been known since Huygens that a coupling, even very small, can lock the phases of the oscillators i.e., force them to oscillate at the same frequency, or more generally with commensurate frequencies. In the phase-locking process, the system thus bifurcates from a quasiperiodic to a periodic regime. A similar situation exists in crystallography for spatial patterns, known as the commensurate–incommensurate transition.

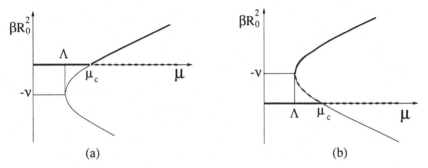

Fig. 2.6. Bifurcation diagram for the parametric oscillator: a) $\nu > 0$; b) $\nu < 0$.

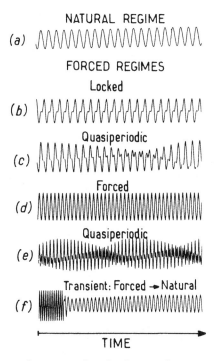

NATURAL REGIME

(a)

FORCED REGIMES

Locked

(b)

Quasiperiodic

(c)

Forced

(d)

Quasiperiodic

(e)

Transient: Forced → Natural

(f)

TIME

Fig. 2.7. Temperature time records of a layer of mercury heated from below, for different forcing amplitude and frequency, see Chiffaudel and Fauve (1987) for details.

Let us first show experimental results in Rayleigh–Bénard convection (Chiffaudel and Fauve, 1987). Convective rolls in a horizontal layer of mercury heated from below become unstable to an oscillatory motion as the temperature difference across the layer is increased above a critical value. The mercury temperature thus oscillates at a "natural" frequency $\omega_0/2\pi$ (see Fig. 2.7a). We apply an external periodic forcing by rotating the mercury layer about its vertical axis, with a sinusoidal angular velocity of frequency $\omega_e/2\pi$. Figure 2.7 displays the different flow regimes when the external frequency is about twice the natural one. One can observe locked (b) or quasiperiodic regimes (c, e). These regimes are located in the experimental parameter space displayed in Fig. 2.8. At small forcing amplitude the locked regime is observed within a tongue (the "Arnold tongue") that begins at twice the natural frequency for vanishing external forcing amplitude. When the detuning is increased there is a transition from the locked to the quasiperiodic regime. For large forcing amplitude the natural oscillation is completely inhibited and the temperature oscillates at the forcing frequency ("forced" regime).

The transitions between these regimes can be modeled by considering the simpler system of a Van der Pol oscillator that is externally forced. The governing equation for this system is as considered in section 2, except for an additional forcing term $\sqrt{\epsilon}f\cos\omega_e t$. Assuming that the dissipation is small ($\lambda = \mu\epsilon$) and rescaling u by $\sqrt{\epsilon}$ we have

$$\ddot{u} + \omega_0^2 u = \epsilon(2\mu - u^2)\dot{u} + f\cos(\omega_e t). \tag{2.40}$$

The method of multiple time scales is used to determine the amplitude equation for this system in precisely the same manner as for the previous systems.

We first consider non-resonant forcing (i.e., ω_e and ω_0 incommensurable). The leading order term in u is

$$u_0 = A(T)\exp(i\omega_0 t) + \text{c.c.} + \frac{f}{\omega_0^2 - \omega_e^2}\cos(\omega_e t), \tag{2.41}$$

where the last term is due to the external forcing. At the next order we have

$$\mathcal{L}.u_1 = (2\mu - u_0^2)\frac{\partial u_0}{\partial t} - 2\frac{\partial^2 u_0}{\partial t\partial T}. \tag{2.42}$$

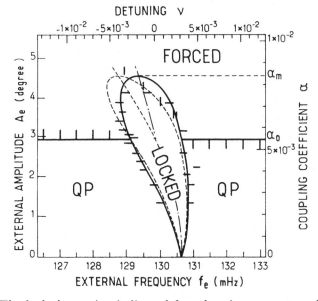

Fig. 2.8. The locked, quasiperiodic and forced regimes as external forcing amplitude and frequency are varied, see Chiffaudel and Fauve (1987) for details.

The solvability condition then gives

$$\frac{dA}{dT} = \xi A - \frac{1}{2}|A|^2 A,\tag{2.43}$$

where

$$\xi = \mu - \frac{1}{4}\left(\frac{f}{\omega_0^2 - \omega_e^2}\right)^2.$$

This amplitude equation is the same as in the unforced case except for the form of the coefficients (when $f = 0$, $\xi = \mu$). The additional forcing term shifts the onset of instability to larger μ by reducing ξ (see Fig. 2.9). One knows other examples of stabilization by applying a periodic forcing; for instance, the unstable up-position of a pendulum can be stabilized by vibrating the point of support.

Note that the symmetry, $t \to t + 2\pi n/\omega_e$ (n is an integer), imposes the invariance $A \to A \exp\left(2\pi i n \omega_0/\omega_e\right)$, on the terms in the amplitude equation. For ω_0 and ω_e incommensurate, this is a constraint as strong as the rotation in the complex plane, $A \to A \exp i\theta$. Thus, the form of the amplitude equation is the same as in the unforced case, and phase-locking terms that break the $A \to A \exp i\theta$ invariance, cannot be found at any order in the amplitude equation. This is due to the form of the leading order solution, u_0.

Indeed, let us consider a small amplitude forcing with $\omega_e = \omega_0 + \epsilon\sigma$ and $f = \epsilon F$. As the forcing is of order ϵ there is no term due to the forcing in u_0, and the leading order solution is

$$u_0 = A(T) \exp i\omega_0 t + \text{c.c.} .\tag{2.44}$$

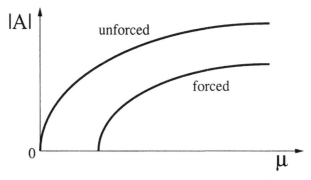

Fig. 2.9. Variation of amplitude $|A|$ with μ for unforced (a) and forced (b) Van der Pol oscillators.

However, the additional term $F \cos(\omega_0 t + \sigma t)$ at the next order changes the solvability condition so that

$$\frac{dA}{dT} = \mu A - \frac{1}{2}|A|^2 A - \frac{iF}{4\omega_0} \exp i\sigma T. \tag{2.45}$$

Writing $A = B \exp i\sigma T$, gives the amplitude equation

$$\frac{dB}{dT} = (\mu - i\sigma)B - \frac{1}{2}|B|^2 B - \frac{iF}{4\omega_0}. \tag{2.46}$$

This transformation amounts to writing

$$u_0 = B(T) \exp i\omega_e t + \text{c.c.}, \tag{2.47}$$

and thus to looking for an amplitude equation in the "reference frame" of the external oscillator. For this new choice of u_0, the symmetry $t \to t + 2\pi n/\omega_e$ only requires $B \to B$. There is no constraint on the amplitude equation, and indeed the rotation symmetry in the complex plane is broken to leading order by the forcing through the constant term $iF/4\omega_0$.

This simple example shows how important is the choice of the leading order solution, $u_0(t)$. The second choice is called "resonant forcing" although with a non-zero detuning σ one can describe a quasiperiodic regime. However, to leading order the response is assumed to be at the forcing frequency. It is the correct choice if one wants to describe frequency-locking phenomena.

Let us now generalize to the case: $\omega_e = n\omega_0/p$ (n, p are integers), where the system is invariant under discrete translation in time $t \to t + 2\pi/\omega_e$. The equation for the amplitude B of

$$u_0 = B(T) \exp \left(i\frac{p}{n}\omega_e t \right) + \text{c.c.}, \tag{2.48}$$

must be invariant to the rotation $B \to B \exp(2i\pi p/n)$. Therefore an additional term $\bar{B}^{n-1}$ related to the forcing is allowed, and the amplitude equation is of the form

$$\frac{dB}{dT} = (\mu + i\nu)B - \beta|B|^2 B - \alpha\bar{B}^{n-1}. \tag{2.49}$$

The coefficient ν represents the detuning, while α is related to the forcing. When $n = 1, 2, 3$ and 4 the forcing term is of at least the same order as the β term, and these are known as strong resonances.

The three different regimes correspond to:

1) $B = 0$, the forced regime,
2) $B = \text{constant}$, the locked regime,
3) B time dependent, the quasiperiodic regime.

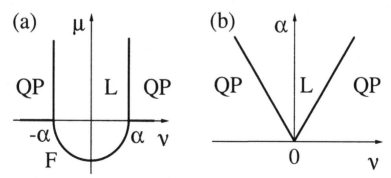

Fig. 2.10. The three regimes for a forced Van der Pol oscillator. There is a stationary bifurcation between the forced (F) and locked (L) regimes, a Hopf bifurcation between the forced and quasiperiodic (QP) regimes, and a saddle-node bifurcation between the quasiperiodic and locked regimes. There are codimension-two points at $(\pm \mu, 0)$.

We consider the specific case $n/p = 2$. For simplicity we choose β to be real and equal to unity. The linear stability of the the forced regime $B = 0$ is exactly the same as that of the parametric oscillator considered earlier. It is stable when $\mu < 0$ and $\alpha^2 < \mu^2 + \nu^2$. The boundary at $\alpha^2 = \mu^2 + \nu^2$ corresponds to a stationary bifurcation (in the reference frame of the external oscillator), while the boundary at $\mu = 0$ is a Hopf bifurcation provided that $\nu^2 > \alpha^2$, and corresponds to the boundary between the forced and quasiperiodic regimes. Substituting $B = R \exp i\theta$ into the amplitude equation we have that the locked regime ($d\theta/dt = 0$) occurs when $\sin 2\theta = \nu/\alpha$, hence when $|\nu| < |\alpha|$. When this inequality is not satisfied, there exists no constant non-zero solution for B and the system bifurcates to the quasiperiodic regime through a saddle-node bifurcation. Figure 2.10 shows the three time-forced regimes on the (α, ν) plane.

Finally, let us remember that interacting oscillators can display complex chaotic dynamics (see for instance, Arnold, 1983 or Guckenheimer and Holmes, 1984).

3 Nonlinear waves in dispersive media

In this section we consider the propagation of a quasi-monochromatic wave and study the dynamics of dispersion and nonlinearity (Whitham, 1974). To wit, the objective is to find an evolution equation for the slowly varying amplitude and phase of the wave (Newell, 1985). Using this amplitude equation we can then study the long-wavelength stability of periodic waves, and look for solitary wave-trains.

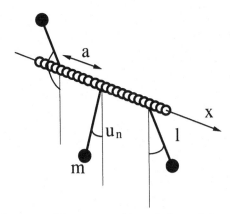

Fig. 3.1. The array of coupled pendulums.

We begin with a simple example, namely the array of pendulums shown in Fig. 3.1. Each pendulum oscillates in the plane perpendicular to the axis of the array and is coupled to its neighbors by torsion springs.

The equation governing the angle from the vertical, $u_n(t)$, of the nth pendulum is,

$$m\,l^2\,\frac{d^2 u_n}{dt^2} = -m\,g\,l\,\sin u_n + C\left(u_{n-1} - 2u_n + u_{n+1}\right), \qquad (3.1)$$

where m is the mass of a pendulum, l is its length and C is the spring torsion constant. We want to investigate phenomena on a length-scale $\lambda \gg a$, where a is the distance between two pendulums. In this case we can take the continuous limit of the above equation, and after rescaling time and space we obtain:

$$\frac{\partial^2 u}{\partial t^2} = -\sin u + \frac{\partial^2 u}{\partial x^2}. \qquad (3.2)$$

This is the Sine–Gordon equation, which is also found in nonlinear optics, where it models the propagation of pulses in resonant media, in condensed-matter physics where it describes charge-density waves in periodic pinning potentials or propagation along Josephson transmission lines, and in field theory where it was used to describe elementary particles. However, it is also a long-wavelength approximation of our array of pendulums and it will be helpful to keep this example in mind to understand the results of this section and the different approximation levels.

3.1 Evolution of a wave-packet

Consider a wave-packet which is peaked around $k = k_0$, and can be written

$$u(x,t) = \int_{-\infty}^{\infty} F(k) \, \exp i[\omega(k)t - kx] \, dk \,. \tag{3.3}$$

Linearizing (3.2) about $u = 0$ and substituting for u using (3.3) gives the dispersion relation,

$$\omega^2 = 1 + k^2 \tag{3.4}$$

shown in Fig. 3.2.

The group velocity, $U(k)$, is given by

$$U = \frac{d\omega}{dk} = \frac{k}{\omega} = \frac{k}{\sqrt{1+k^2}} \,. \tag{3.5}$$

Notice that

$$\frac{d^2\omega}{dk^2} = \frac{\omega - U k}{\omega^2} = \frac{1 - U^2}{\omega} \neq 0 \,,$$

so that the medium is dispersive.

The largest contribution to the integral in equation (3.3) will come from k in the neighborhood of k_0 and we make the approximation

$$u(x,t) \simeq \exp i[\omega(k_0)t - k_0 x] \int_{-\infty}^{\infty} F(k_0 + K) \exp i[\Omega(K)t - Kx] dK \tag{3.6}$$

where

$$\Omega(K) = U_0 K + \frac{\omega_0'' K^2}{2} + \cdots, \tag{3.7}$$

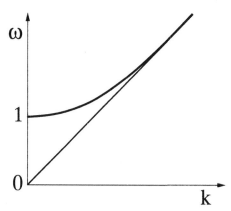

Fig. 3.2.   The dispersion relation for the wave described by equation (3.2).

with

$$U_0 = U(k_0) \,,$$

$$\omega_0'' = \frac{d^2\omega}{dk^2}(k_0) \,.$$

This gives us

$$u(x,t) \simeq A(x,t) \exp i(\omega_0 t - k_0 x) + \text{c.c.} \tag{3.8}$$

where the envelope is

$$A(x,t) = \int_{-\infty}^{\infty} \hat{A}(K,t) \exp(-iKx) \, dK \,, \tag{3.9}$$

with

$$\hat{A}(K,t) = F(k_0 + K) \exp i\Omega(K)t \,. \tag{3.10}$$

Now using (3.7) gives

$$i\Omega\hat{A} = iU_0 K\hat{A} + \frac{i\omega_0'' K^2 \hat{A}}{2} + \cdots$$

Taking the inverse Fourier transform we see that this corresponds to

$$\frac{\partial A}{\partial t} = -U_0 \frac{\partial A}{\partial x} - \frac{i\omega_0''}{2} \frac{\partial^2 A}{\partial x^2} \,. \tag{3.11}$$

Note that A is slowly varying in space compared to $2\pi/k_0$ since the wave packet is peaked around k_0. Thus it is clear from (3.11) that A is also slowly varying in time compared to $2\pi/\omega_0$. Equation (3.11) describes the amplitude of a slowly varying wave. The first term on the right hand side represents the propagation of amplitude perturbations at the group velocity, and can be removed if we transform to a reference frame moving at U_0. We then get a Schrödinger equation

$$\frac{\partial A}{\partial t} = i\alpha \frac{\partial^2 A}{\partial x^2} \,, \tag{3.12}$$

where $\alpha = -\omega_0''/2$ represents dispersion. It is easily seen that $\xi = x^2/4i\alpha t$ is a similarity variable and that $\int |A|^2 \, dx$ is conserved. Thus, a self-similar solution is,

$$A \propto (4i\alpha t)^{-1/2} f(\xi) \,. \tag{3.13}$$

In particular, with an initial condition of the form

$$A(x,0) \propto \exp(-x^2/x_0^2) \,, \tag{3.14}$$

we have

$$A(x,t) \propto (x_0^2 + 4i\alpha t)^{-1/2} \exp[-x^2/(x_0^2 + 4i\alpha t)] \,, \tag{3.15}$$

and we get a well-known result: dispersion causes the amplitude of the wave-packet to decrease as $t^{-1/2}$. This is valid for large t and x_0, since A is assumed to vary slowly in time and space. Notice that here the approximation is included at the stage of formulating the amplitude equation.

The next step is to include nonlinear terms in the evolution equation of the amplitude A. Assuming that these terms are monomials in A and $\bar{A}$, we proceed using symmetry arguments in a similar way as we did for nonlinear oscillators. The Sine-Gordon equation (3.1) is invariant to translations in both time and space,

$$t \to t + \theta, \qquad x \to x + \phi,$$

and from (3.4) we see that this corresponds to

$$A \to A \exp i\psi,$$

where ψ can vary through all real values. Considering all possible nonlinear terms, we find that the lowest order term with the right transformation property is $|A|^2 A$. So we can write

$$\frac{\partial A}{\partial t} = -U_0 \frac{\partial A}{\partial x} - i \frac{\omega_0''}{2} \frac{\partial^2 A}{\partial x^2} + \beta |A|^2 A. \tag{3.16}$$

There are two further symmetries: time reversal and space reflection. In the general case these can be applied separately, but we have taken the particular form of u given by (3.8) that consists only of waves propagating to the right, and this constrains us to applying both transformations together (see below for the general case). Applying the symmetries together implies the invariance of the amplitude equation under the transformation

$$t \to -t, \qquad x \to -x, \qquad A \to \bar{A},$$

and applying this to (3.16) gives

$$-\frac{\partial \bar{A}}{\partial t} = U_0 \frac{\partial \bar{A}}{\partial x} + i\alpha \frac{\partial^2 \bar{A}}{\partial x^2} + \beta |A|^2 \bar{A}.$$

However the complex conjugate of (3.16) is

$$\frac{\partial \bar{A}}{\partial t} = -U_0 \frac{\partial \bar{A}}{\partial x} - i\alpha \frac{\partial^2 \bar{A}}{\partial x^2} + \bar{\beta} |A|^2 \bar{A}$$

Hence $\bar{\beta} = -\beta$, and β is pure imaginary, so we can replace β with $-i\beta$ in (3.16). If we also transform to a frame moving with the group velocity, U_0, we obtain

$$\frac{\partial A}{\partial t} = i\alpha \frac{\partial^2 A}{\partial x^2} - i\beta |A|^2 A \tag{3.17}$$

This is the nonlinear Schrödinger equation. It shows that the dynamics of the wave-packet consists of a balance between dispersion, $i\alpha A_{xx}$, and nonlinearity, $-i\beta|A|^2A$, that traces back in this problem to the amplitude dependence of the frequency of each oscillator. These points are illustrated by the particular solution

$$A = Q \exp i(\Omega t - qx),$$
$$\Omega = -\alpha q^2 - \beta Q^2,$$
(3.18)

which corresponds to shifting $\omega_0 \to \omega_0 + \Omega$, $k_0 \to k_0 + q$. Thus, nonlinearity and dispersion act in antagonistic ways if $\alpha\beta < 0$.

We now derive the nonlinear Schrödinger equation from the Sine-Gordon equation using a multiple-scale expansion. Considering an initial condition which is a slowly modulated wave in space, we take as a small parameter the typical modulation wavenumber compared to the wavenumber of the carrier wave, thus

$$\frac{\partial}{\partial x} \to \frac{\partial}{\partial x} + \epsilon \frac{\partial}{\partial X}.$$

As discussed above, we expect two characteristic time scales, one corresponding to the propagation of the wave envelope at the group velocity, and the other to the dispersion of the wave-packet, thus

$$\frac{\partial}{\partial t} \to \frac{\partial}{\partial t} + \epsilon \frac{\partial}{\partial T_1} + \epsilon^2 \frac{\partial}{\partial T_2}.$$

We need now to scale the oscillation amplitude $u(x, t, X, T_1, T_2)$ to be able to handle the nonlinear term of (3.2) perturbatively. There does not exist a correct scaling versus a wrong one. If the amplitude is scaled too small, we get, to leading orders, an amplitude equation with only linear terms, which is correct; if it is scaled too large, we get nonlinear terms at a lower order than dispersion, which is also true if the amplitude is large. One generally considers that the most interesting situation consists of having both effects, nonlinearity and dispersion, at the same order in the amplitude equation; this fixes the scale for the amplitude, and

$$u(x, t) = \epsilon \left[A(X, T_1, T_2) \exp i(\omega_0 t - k_0 x) + \text{c.c.} + \epsilon u_1 + \dots \right].$$
(3.19)

At $\mathcal{O}(\epsilon)$, the solvability condition is

$$\frac{\partial A}{\partial T_1} = -U_0 \frac{\partial A}{\partial X},$$
(3.20)

which leads us to take

$$A(X, T_1, T_2) = A(X - U_0 T_1, T_2).$$
(3.21)

Then at $\mathcal{O}(\epsilon^2)$, the solvability condition gives

$$\frac{\partial A}{\partial T_2} = i\alpha \frac{\partial^2 A}{\partial X^2} - i\beta|A|^2A,$$
(3.22)

with $\alpha = -\omega_0''/2$ and $\beta = 1/4\omega_0$. Note that $\alpha\beta < 0$ so that dispersion and nonlinearity are antagonistic. We have recovered the nonlinear Schrödinger equation (3.17). We could have taken two slow length-scales and one slow time scale, and this would have resulted in a different form of the nonlinear Schrödinger equation which is widely used in nonlinear optics.

To deal with the general case we must consider both left and right-propagating waves, and begin the expansion with

$$u = \epsilon\left[A\exp i(\omega_0 t - k_0 x) + B\exp i(\omega_0 t + k_0 x) + \text{c.c.} + \ldots\right]. \quad (3.23)$$

This leads to the two coupled amplitude equations

$$
\begin{aligned}
\frac{\partial A}{\partial t} &= -U_0\frac{\partial A}{\partial x} - i\frac{\omega_0''}{2}\frac{\partial^2 A}{\partial x^2} - i\beta|A|^2 A - i\gamma|B|^2 A, \\
\frac{\partial B}{\partial t} &= U_0\frac{\partial B}{\partial x} - i\frac{\omega_0''}{2}\frac{\partial^2 B}{\partial x^2} - i\gamma|A|^2 B - i\beta|B|^2 B.
\end{aligned}
\quad (3.24)
$$

Translational invariances in time and space constrain the form of the leading order nonlinear terms. Space reflection symmetry implies the invariance under the transformation

$$x \to -x, \qquad A \to B, \qquad B \to A,$$

and shows that the coefficients of the similar nonlinear terms should be the same in both equations. Time reversal symmetry implies

$$t \to -t, \qquad A \to \bar{B}, \qquad B \to \bar{A},$$

so that the coefficients of the nonlinear terms are pure imaginary. Note that one can check that the coefficient of the propagative term is real whereas that of the dispersive term is pure imaginary.

A slight problem arises if one tries to get these coupled equations with a multiple-scale expansion. Indeed, one cannot remove both propagative terms by transforming to a frame moving at the group velocity. It is straightforward to change the scaling of the amplitude in order to bring nonlinear terms at the same order as propagative ones, but dispersive terms are smaller, and one should not, in principle, keep them. This is obviously a bad choice since one does not expect dispersion to become negligible because of the presence of counter-propagating waves. The problem here is that one small adjustable parameter is not enough to balance all the relevant terms allowed by symmetries. One way out is to look for another small parameter, here obviously the group velocity, however this restricts our study to the case of small carrier wave frequency. Another way is to keep the dispersive terms; but should we scale amplitude as for the nonlinear Schrödinger equation or in order to bring leading order nonlinear terms with propagative terms, and then keep higher order nonlinearities at the order of dispersive terms? The two different scalings might be relevant, one when the counter-propagating wave-packets are far apart, the other when they collide. There is perhaps no rigorous way to describe that situation with amplitude equations.

The nonlinear Schrödinger equation is the generic evolution equation that governs the complex amplitude of a nonlinear wave in dispersive media (Newell, 1985). It has been widely used to describe surface waves and light propagation in optical fibers. It should be modified in the vicinity of a caustic where the group velocity is stationary and correspondingly dispersive effects are small.

It also occurs that the wave amplitude is coupled to a mean field. This is a general situation when there exists a neutral mode at zero wave-number and we will discuss that later; as a simple example we can derive the nonlinear Schrödinger equation for the envelope of a quasi-monochromatic wave governed by the Korteweg–de Vries equation,

$$\frac{\partial u}{\partial t} + u\frac{\partial u}{\partial x} + \frac{\partial^3 u}{\partial x^3} = 0 \,. \tag{3.25}$$

Writing $u(x,t)$ in the form (3.19), we get at $\mathcal{O}(\epsilon)$ the solvability condition (3.20) with $U_0 = -3k_0{}^2$ and A is given by (3.21). Then

$$u_1(x, X, t, T_1, T_2) = B(X, T_1, T_2) + \frac{A^2}{6k_0{}^2}\exp i(\omega_0 t - k_0 x) + \text{c.c.} \,. \tag{3.26}$$

The solvability conditions at $\mathcal{O}(\epsilon^2)$ give

$$\begin{aligned}
\frac{\partial B}{\partial T_1} &= -\frac{\partial}{\partial X}|A|^2 = -\frac{1}{U_0}\frac{\partial}{\partial T_1}|A|^2 \,, \\
\frac{\partial A}{\partial T_2} &= 3ik_0\frac{\partial^2 A}{\partial X^2} + \frac{i}{6k_0}(|A|^2 A + 6k_0{}^2 AB) \,.
\end{aligned} \tag{3.27}$$

B can be eliminated and we obtain the nonlinear Schrödinger equation (3.22) with $\alpha = -\omega_0''/2 = 3k_0{}^2$ and $\beta = 1/6k_0$. However, in general the mean field cannot be eliminated; here it just modifies the coefficient of the nonlinear term of (3.22).

Finally, note that when a conservative system undergoes a dispersive instability, such as the Kelvin–Helmholtz instability, the amplitude of the unstable waves is not governed by the nonlinear Schrödinger equation (see section 5).

3.2 The side-band or Benjamin–Feir instability

We now use the nonlinear Schrödinger equation to study the stability of a quasi-monochromatic wave. The original motivation was to understand the instability of Stokes waves. When a wave train of surface gravity waves is generated with a paddle oscillating at constant frequency, one observes that if the fluid layer is deep enough compared to the wavelength, the quasi-monochromatic wave is unstable and breaks into a series of pulses (Fig. 3.3).

Return to the nonlinear Schrödinger equation (3.17), and consider the particular solution

$$A_0 = Q \exp(i\Omega t), \qquad \Omega = -\beta Q^2, \tag{3.28}$$

that represents a quasi-monochromatic wave of amplitude Q, wavenumber k_0 and frequency $\omega_0 + \Omega$. If we perturb A_0 slightly, so that

$$A = [Q + r(x,t)] \exp i[\Omega t + \theta(x,t)], \tag{3.29}$$

we obtain after separating real and imaginary parts,

$$\frac{\partial r}{\partial t} = -\alpha(Q+r)\frac{\partial^2 \theta}{\partial x^2} - 2\alpha\frac{\partial r}{\partial x}\frac{\partial \theta}{\partial x},$$

$$Q\frac{\partial \theta}{\partial t} = -\beta r\left(2Q^2 + 3Qr + r^2\right) + \alpha\frac{\partial^2 r}{\partial x^2} - \alpha(Q+r)\left(\frac{\partial \theta}{\partial x}\right)^2 - r\frac{\partial \theta}{\partial t}.$$
$$\tag{3.30}$$

Linearizing and taking $(r, \theta) \propto \exp(\eta t - iKx)$, one finds the dispersion relation

$$\eta^2 = -[2\alpha\beta Q^2 K^2 + \alpha^2 K^4]. \tag{3.31}$$

η^2 is always negative for $\alpha\beta > 0$, but has a positive region in K for $\alpha\beta < 0$ as shown in Fig. 3.4.

Thus if $\alpha\beta > 0$ then η is pure imaginary and the quasi-monochromatic wave (3.8) is a stable solution. On the other hand if $\alpha\beta < 0$ then in the long wavelength region, η has both a negative and a positive root. When η is positive, there is an instability, the Benjamin–Feir or side-band instability. It has the name "side-band" because if one takes a band of frequencies centered on ω_0 as shown in Fig. 3.5, the interaction of one

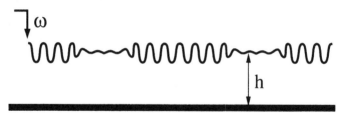

Fig. 3.3. A deep layer of water, forced by an oscillating paddle, exhibits the side-band instability (Benjamin and Feir, 1967; Lake *et al.*, 1977; Melville, 1982).

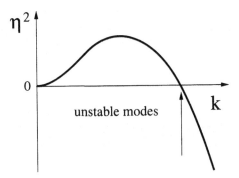

Fig. 3.4. The dispersion relation for the perturbation to A_0.

side-mode with the second harmonic is resonant with the other side-mode, causing it to be amplified, i.e., $2\omega_0 - (\omega_0 - \Omega) = \omega_0 + \Omega$ (Stuart and DiPrima, 1978). As an exercise, write the perturbation to A_0 as the sum of two side-band modes and see how their coupling generates the instability.

Another way to understand this instability is to linearize (3.31) in the form

$$\frac{\partial^2 \theta}{\partial t^2} = 2\alpha\beta Q^2 \frac{\partial^2 \theta}{\partial x^2} - \alpha^2 \frac{\partial^4 \theta}{\partial x^4}. \tag{3.32}$$

So if $\alpha\beta < 0$, the phase of the wave obeys an unstable propagation equation (the propagation velocity is imaginary).

In the stable case it might be interesting to find the higher order terms of (3.32). This is simple only if one considers slowly varying propagative solutions of (3.30) in the form

$$r(\xi, \tau), \qquad \theta = \theta(\xi, \tau), \tag{3.33}$$

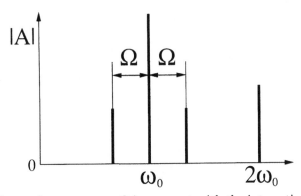

Fig. 3.5. The mode at $\omega = \omega_0 - \Omega$ is resonant with the interaction of the modes at $\omega_0 + \Omega$ and $2\omega_0$.

with $\xi = \epsilon(x - ct)$, $c = Q\sqrt{2\alpha\beta}$, $\tau = \epsilon^3 t$. Expanding r and θ

$$r = \epsilon^2 r_0 + \epsilon^4 r_1 + \dots ,$$
$$\theta = \epsilon\theta_0 + \epsilon^3\theta_1 + \dots , \tag{3.34}$$

we obtain

$$\frac{\partial r_0}{\partial t} + 6\sqrt{2\alpha\beta}\, r_0\frac{\partial r_0}{\partial \xi} - \frac{\alpha}{Q}\sqrt{\frac{\alpha}{2\beta}}\frac{\partial^3 r_0}{\partial \xi^3} = 0 , \tag{3.35}$$

which is the Korteweg–de Vries equation. It has well-known solitary wave solutions consisting of a region with a non zero r, or correspondingly a region with a non-zero phase gradient, thus a localized region with a different local wavenumber for the wave train. We consider, in the next section, these localized structures as solitary wave solutions of the nonlinear Schrödinger equation.

First, try this exercise. If we were now to consider a slowly modulated wave solution of (3.35), we would find that its slowly varying amplitude A_1 satisfies the nonlinear Schrödinger equation with coefficients depending on those of the nonlinear Schrödinger equation we start from at the beginning of this section. Derive the mapping between the old and new coefficients. Is there a fixed point? If yes, what would this mean? Find other similar examples using symmetry arguments to guess the form of the successive equations.

3.3 Solitary waves

Nonlinear wave equations sometimes have solitary wave solutions, which have locally distributed amplitudes and propagate without changing their profiles, because of the balance between nonlinearity and dispersion. In this section, we look for solitary wave solutions of the nonlinear Schrödinger equation.

3.3.1 Solitary wave solutions in the Benjamin–Feir unstable regime

Firstly, we solve the nonlinear Schrödinger equation to get solitary wave solutions in the Benjamin–Feir unstable case, i.e., $\alpha\beta < 0$. For simplicity, we select the parameters to be $\alpha = 1$ and $\beta = -2$ with appropriate scalings of space and amplitude. Then the nonlinear Schrödinger equation is

$$\frac{\partial A}{\partial t} = i\frac{\partial^2 A}{\partial x^2} + 2i|A|^2 A . \tag{3.36}$$

We assume the form of the solution is $A_s(x,t) = R(x)\exp(i\Omega t)$. Substituting this into (3.36), we get

$$\frac{d^2 R}{dx^2} = -\frac{dV}{dR}$$

$$V(R) \equiv -\tfrac{1}{2}\Omega R^2 + \tfrac{1}{2}R^4 . \tag{3.37}$$

Figure 3.6 shows the profile of the potential $V(R)$. Equation (3.37) corresponds to the equation of motion of a particle in the potential $V(R)$ by considering x as time and R as the position of the particle. Multiplying by dR/dx and integrating, we have

$$\frac{1}{2}\left(\frac{dR}{dx}\right)^2 + V(R) = E , \tag{3.38}$$

where E is a constant. In the case of $E < 0$, the solution of (3.37) corresponds to the periodic motion between $R_1 < R < R_2$ in Fig. 3.6. Thus, we get periodic solutions with respect to x as shown in Fig. 3.7. These solutions are called cnoidal waves. In the case of $E = 0$, the solution corresponds to the motion of the particle which starts with $R = 0$ at $x \to -\infty$, reaches $R = R_0$ and returns to $R = 0$ as $x \to \infty$. Therefore, we get a solitary wave solution whose profile tends to zero as $x \to \pm\infty$. There is a simple analytic form for this special case:

$$A_s = \sqrt{\Omega} \operatorname{sech}\left(\sqrt{\Omega}x\right)\exp\left(i\Omega t\right) . \tag{3.39}$$

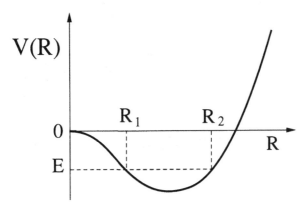

Fig. 3.6.   The potential $V(R)$.

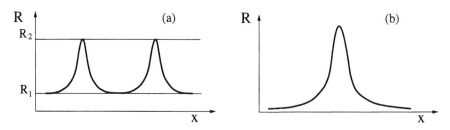

Fig. 3.7. a) Cnoidal wave ($E < 0$). b) Solitary wave ($E = 0$).

3.3.2 Symmetries and other solitary wave solutions

By symmetry arguments, we can derive other solitary waves from one simple solitary wave. The nonlinear Schrödinger equation has the following symmetry property,

$$
\begin{cases}
x \longrightarrow \lambda x, \\
t \longrightarrow \lambda^2 t, \\
A \longrightarrow \lambda^{-1} A
\end{cases}
\tag{3.40}
$$

Applying this symmetry to the simple solitary wave solution

$$ A_{s1} = \text{sech}(x)\,\exp(it), $$

we have

$$ A_{s2} = \sqrt{\Omega}\,\text{sech}(\sqrt{\Omega}x)\exp(i\Omega t). \tag{3.41} $$

Further, using another symmetry

$$
\begin{cases}
x \longrightarrow x + vt, \\
A \longrightarrow A \exp\left(-i\frac{v}{2}x + i\frac{v^2}{4}t\right),
\end{cases}
\tag{3.42}
$$

we get a further solitary wave solution,

$$ A_{s3} = \sqrt{\Omega}\,\text{sech}[\sqrt{\Omega}(x + vt)]\exp i \left[\frac{1}{2}vx + (\Omega - \frac{v^2}{4})t\right]. \tag{3.43} $$

One can also consider translational invariance in space, thus replacing x by $x - x_0$, and rotational invariance in the complex plane that leads to an arbitrary phase factor in A. The important point to notice is that continuous families of solutions are associated with the invariance properties of the evolution equation. We will use this later to study the dynamics of localized structures.

3.3.3 Solitary wave solutions in the Benjamin–Feir stable regime

Next we consider solitary wave solutions in the Benjamin–Feir stable case, i.e., $\alpha\beta > 0$. For simplicity, we select the parameters to be $\alpha = 1$ and $\beta = 2$.

$$\frac{\partial A}{\partial t} = i\frac{\partial^2 A}{\partial x^2} - 2i|A|^2 A. \tag{3.44}$$

We assume the form of the solution is $A_s = R(x)\exp\left[i\Omega t + \theta(x)\right]$. Substituting it into (3.44), we get

$$\Omega R = -2R^3 + \frac{d^2 R}{dx^2} - R\left(\frac{d\theta}{dx}\right)^2,$$

$$0 = -2\frac{dR}{dx}\frac{d\theta}{dx} - R\frac{d^2\theta}{dx^2} = -\frac{d}{dx}\left(R^2\frac{d\theta}{dx}\right). \tag{3.45}$$

We get $h \equiv R^2\theta_x = $ constant, and substituting into (3.45),

$$\frac{d^2 R}{dx^2} = -\frac{dV}{dR},$$

$$V(R) = \frac{1}{2}\left(-\Omega R^2 - R^4 + \frac{h^2}{R^2}\right), \tag{3.46}$$

In the same manner as for the Benjamin–Feir unstable case, we can find the solitary wave solution by selecting the homoclinic orbit of the potential $V(R)$ (see Fig. 3.8). We obtain

$$R^2 = R_0^2 - \frac{a^2}{\cosh^2 ax}, \tag{3.47}$$

where $R = R_0$ gives the maximum of $V(R)$, and the parameters are

$$\begin{cases} \Omega = -3R_0^2 + a^2, \\ h^2 = R_0^4(R_0^2 - a^2). \end{cases}$$

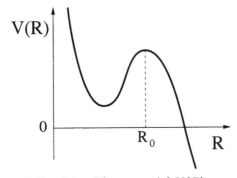

Fig. 3.8.   The potential $V(R)$.

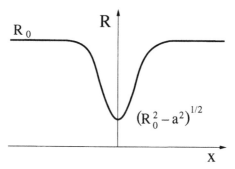

Fig. 3.9. A envelope-hole soliton or dark soliton.

This solution is called an "envelope hole soliton," or a "dark soliton" in optics since it consists of a region with a smaller oscillation amplitude (Fig. 3.9). Note that the local wavenumber is changed according to the relation $R^2\theta_x = $ constant. In particular, when we choose $h = 0, a = R_0$, we have

$$R^2 = R_0^2 \tanh^2 R_0 x \,. \tag{3.48}$$

This represents a non-oscillating location that separates two regions which oscillate out of phase.

4 Cellular instabilities, a canonical example: Rayleigh–Bénard convection

Various fluid flows display instabilities that generate cellular structures. Examples are the Couette–Taylor flow, Rayleigh–Bénard convection, the Faraday instability and many shear flow experiments. The fluid is sometimes assumed to be inviscid but we will first consider situations where dissipation cannot be neglected. These "dissipative instabilities" are described by simpler amplitude equations, in particular when they generate stationary patterns, such as convective rolls for instance. A canonical example of stationary cellular instability is Rayleigh–Bénard convection that we will study in this section.

4.1 Rayleigh–Bénard convection

4.1.1 Convection in the Rayleigh–Bénard geometry

Thermal convection occurs widely in geophysical and astrophysical flows: in the earth mantle, it is responsible for the motion of tectonic plates; in the earth core, it generates the earth magnetic field by a dynamo effect; in the sun or other stars, it is the advection mechanism for the energy generated in the core. Thermal convection has also been studied extensively in laboratory experiments, both for engineering purposes and also as one of the simplest examples of hydrodynamic instability that displays pattern formation and transition to turbulence. There exist many reviews about thermal convection, for instance, Spiegel (1971, 1972), Palm (1975), Normand *et al.* (1977), Busse (1978, 1981); the reader may also look at the book by Gershuni and Zhukovitskii (1976).

Convection in the Rayleigh–Bénard geometry is achieved by uniformly heating from below a horizontal layer of fluid (Fig. 4.1). For small temperature gradients, the fluid remains in a stable heat-conducting state, with a linear temperature profile and no fluid motion. However, if the fluid has a negative thermal expansion coefficient, $-\alpha$, the thermal gradient generates a density stratification with cold heavy fluid above warm light fluid. For sufficiently large temperature differences, the resulting buoyancy force overcomes dissipative effects due to viscosity and heat diffusivity, causing less dense warmer fluid to rise and cooler fluid to sink. With appropriate boundary conditions, periodic parallel convection rolls result from the circulation of the fluid.

At the onset of convection, any field $u(x, y, z, t)$ representing the state of the system, i.e., one of the velocity components, the temperature fluctuation from the heat-conduction profile or the pressure fluctuation from hydrostatic equilibrium, takes the form

$$u(x, y, z, t) = \epsilon \left[A(X, Y, T) \, \exp\left(ik_c x\right) + \text{c.c.} \right] f(z) + \dots$$

This represents periodic convection rolls perpendicular to the x-axis with a slowly varying complex amplitude $A(X, Y, T)$. The modulus of A accounts for the convection amplitude whereas the phase of A is related to the local wavenumber difference from its critical value, k_c. The vertical structure of the convection mode is described by $f(z)$ and depends on the boundary conditions at the lower and upper plates (see below).

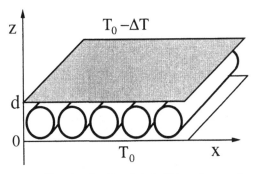

Fig. 4.1. Sketch of a Rayleigh–Bénard experiment.

Our objective is to find the amplitude equation, i.e., the evolution equation for A, and to use it to describe pattern dynamics in the vicinity of convection onset.

4.1.2 The Boussinesq approximation

The Boussinesq approximation is reasonably valid in usual experimental situations. In this approximation, the fluid behaves as though it were incompressible, the density varying only as a consequence of changes in temperature; the density variation about its mean value is taken into account only in the buoyancy force term; the mechanical dissipation rate is neglected in the heat equation, and the fluid parameters, viscosity, heat diffusivity and heat capacity are assumed to be constant. These *a priori* physical assumptions can be replaced by a rigorous asymptotic expansion of the conservation equations of mass, momentum and energy (see the review by Malkus, 1964); the resulting Boussinesq equations are

$$\nabla \cdot \mathbf{v} = 0\,, \tag{4.1}$$

$$\rho_0 \left[\frac{\partial \mathbf{v}}{\partial t} + (\mathbf{v} \cdot \nabla)\mathbf{v} \right] = -\nabla p + \rho_0 \nu \nabla^2 \mathbf{v} - \rho(T)g\hat{\mathbf{z}}\,, \tag{4.2}$$

$$\frac{\partial T}{\partial t} + \mathbf{v} \cdot \nabla T = \kappa \nabla^2 T\,, \tag{4.3}$$

where ν is the fluid kinematic viscosity and κ is the heat diffusivity; ρ_0 is the fluid density at a reference temperature and T is the temperature difference from that reference. Thus,

$$\rho(T) \simeq \rho_0(1 - \alpha T)\,. \tag{4.4}$$

Defining θ as the temperature fluctuation from the heat-conducting profile,

$$T = T_0 - \frac{\Delta T}{d} z + \theta, \tag{4.5}$$

where ΔT is the temperature difference across the layer of height d, and using d, d^2/κ and ΔT as scales for length, time and temperature, one gets

$$\nabla \cdot \mathbf{v} = 0\,, \tag{4.6}$$

$$\frac{\partial \mathbf{v}}{\partial t} + (\mathbf{v} \cdot \nabla)\mathbf{v} = -\nabla \pi + P\nabla^2 \mathbf{v} + RP\theta\hat{\mathbf{z}}\,, \tag{4.7}$$

$$\frac{\partial \theta}{\partial t} + \mathbf{v} \cdot \nabla \theta = \mathbf{v} \cdot \hat{\mathbf{z}} + \nabla^2 \theta\,, \tag{4.8}$$

where $P = \nu/\kappa$ is the Prandtl number and $R = g\alpha\Delta T d^3/\nu\kappa$ is the Rayleigh number. These two dimensionless numbers, together with the boundary conditions, characterize the convection problem in the Boussinesq approximation. Let us mention that the small or large Prandtl number or large Rayleigh number limits of equations (4.6, 4.7, 4.8) are usually considered without any caution in the literature, although these limits might invalidate the Boussinesq approximation.

The Prandtl number is the ratio of the time scales of the two diffusive processes involved in convection, heat diffusion and momentum diffusion. Depending on the microscopic mechanisms of transport, the Prandtl number varies on many orders of magnitude in different convective flows of interest (Fig. 4.2).

The Rayleigh number is proportional to the temperature difference across the fluid layer, and relates the strength of the driving mechanism to dissipative processes. It is the control parameter in a convection experiment.

4.1.3 Boundary conditions

We need now to specify the boundary conditions. We consider a fluid layer of infinite horizontal extent or periodic lateral boundary conditions. At the upper and lower boundaries, the temperature and the heat flux are assumed to be continuous. There exist two simple limit situations:

a) boundaries with high heat conductivity

$$\theta|_B = 0, \tag{4.9}$$

b) insulating boundaries

$$\frac{\partial\theta}{\partial z}\Big|_B = 0. \tag{4.10}$$

Depending on the nature of the boundaries, the boundary condition for velocity can be either "no-slip" or "stress-free". If the boundary is a rigid plate, the "no-slip" boundary condition is applicable for viscous fluids, i.e.,

$$\mathbf{v}|_B = 0.$$

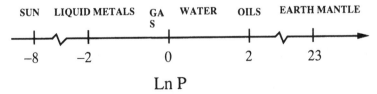

Fig. 4.2.   Typical values of the Prandtl number.

We separate the velocity into horizontal and vertical components, $\mathbf{v} = \mathbf{v}_h + w\hat{\mathbf{z}}$, where $\mathbf{v}_h$ is the horizontal velocity, and w is the z component. Since $\mathbf{v}_h$ must vanish identically at $z = z_B$, its horizontal derivatives also must vanish at $z = z_B$. Using $\nabla \cdot \mathbf{v} = 0$, we have

$$w|_B = 0 \,,$$
$$\left.\frac{\partial w}{\partial z}\right|_B = 0 \,. \tag{4.11}$$

If the boundary is an interface with another fluid or a free surface open to the air, boundary conditions have to account for the continuity of both the normal velocity w and of the tangential stress in the plane of the interface. Assuming that surface tension effects are not involved and that the interface remains flat, then we have the following "stress-free" boundary conditions:

$$w|_B = 0 \,, \qquad \left.\frac{\partial \mathbf{v}_h}{\partial z}\right|_B = 0 \,.$$

Again using the incompressibility condition $\nabla \cdot \mathbf{v} = 0$, we get

$$w|_B = 0 \,,$$
$$\left.\frac{\partial^2 w}{\partial z^2}\right|_B = 0 \,. \tag{4.12}$$

One example of a "stress-free" experimental boundary condition consists of a convection layer of oil sandwiched between layers of mercury and gaseous helium (Goldstein and Graham, 1969). The "stress-free" boundary conditions can be applied since the viscosity of oil is much larger than that of mercury or helium, and the temperature fluctuation should be zero at the oil–mercury interface, while the fluctuation heat flux should be zero at the oil–helium interface.

Finally, let us mention the crucial effect of temperature boundary conditions on the convective regime observed at onset. As buoyancy is the driving mechanism, the length-scale of the convection pattern is primarily fixed by the characteristic scale for the temperature disturbances. In the case of boundaries with a high heat-conductivity compared to the one of the fluid, the temperature disturbances should vanish on the boundaries and the relevant length-scale is the height of the layer d. For insulating boundaries, the isotherms can penetrate into the boundaries and the temperature can vary on a very large length-scale compared to d. The pattern wavelength goes to infinity in the insulating limit, i.e., there is only one roll in the fluid container (see Fig. 4.3).

(a)   (b)

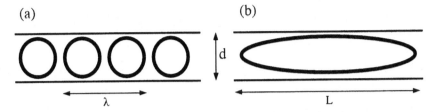

Fig. 4.3. Convective regime at onset a) with boundaries of high heat conductivity, b) with insulating boundaries.

4.2 Linear stability analysis

The linear stability of the motionless heat-conducting state can be studied analytically using the Boussinesq equations (4.6, 4.7, 4.8) with stress-free (4.12) and perfectly conducting (4.9) boundary conditions.

We first eliminate the pressure field by applying the operators $\nabla \times (\cdot)$ and $\nabla \times \nabla \times (\cdot)$ to the momentum equation (4.7) and we get the evolution equations for the vertical vorticity, ζ, and the vertical velocity, w, by projecting on the vertical axis:

$$\frac{\partial \zeta}{\partial t} + \hat{\mathbf{z}} \cdot \nabla \times [(\mathbf{v} \cdot \nabla)\mathbf{v}] = P\nabla^2 \zeta, \tag{4.13}$$

$$\frac{\partial}{\partial t}\nabla^2 w - \hat{\mathbf{z}} \cdot \nabla \times \nabla \times [(\mathbf{v} \cdot \nabla)\mathbf{v}] = P\nabla^4 w + RP\nabla_h^2 \theta, \tag{4.14}$$

where ∇_h^2 stands for the Laplacian operator in the horizontal plane. Note that, at the linear stage, the vertical vorticity decouples and obeys a diffusion equation. Thus, the vertical vorticity modes can be ignored in the linear stability analysis; however, they should be kept in the study of finite amplitude convection since they are nonlinearly coupled to the linear convection modes. Neglecting the nonlinear terms in equations (4.14) and (4.8) yields the coupled linearized system for w and θ:

$$\frac{\partial}{\partial t}\nabla^2 w = P\nabla^4 w + RP\nabla_h^2 \theta,$$
$$\frac{\partial \theta}{\partial t} = w + \nabla^2 \theta. \tag{4.15}$$

From the requirement of spatial periodicity in the horizontal plane, we consider a normal mode of the disturbances (w, θ) under the form

$$w(x, y, z, t) = W(z) \exp[i\mathbf{k} \cdot \mathbf{r} + \sigma t],$$
$$\theta(x, y, z, t) = \Theta(z) \exp[i\mathbf{k} \cdot \mathbf{r} + \sigma t], \tag{4.16}$$

where $\mathbf{r}$ is the position vector in the horizontal plane, and $\mathbf{k}$ is the pattern (horizontal) wave-vector. Boundary conditions (4.9, 4.12) together with equations (4.15) require that W and all its even derivatives vanish for $z = 0$ and $z = 1$. It follows that

$$W(z) = W_0 \sin(n\pi z), \qquad (n = 0, 1, \ldots). \tag{4.17}$$

Using (4.15), (4.16) and (4.17) we obtain the dispersion relation for the growth rate σ of the normal mode k

$$\sigma^2 + q_n^2(1 + P)\sigma + \left(Pq_n^4 - \frac{RPk^2}{q_n^2}\right) = 0, \tag{4.18}$$

where $q_n^2 = k^2 + n^2\pi^2$.

A stationary instability occurs when the constant term in σ of the dispersion relation vanishes and becomes negative. Thus, as the Rayleigh number is increased, a mode with $n = 1$ bifurcates first for $R = R_c(k)$ with

$$R_c(k) = \frac{(\pi^2 + k^2)^3}{k^2}. \tag{4.19}$$

This defines the marginal stability curve on which a mode with $n = 1$ and horizontal wavenumber k has a zero growth rate (Fig. 4.4). The critical Rayleigh number R_c and the critical wavenumber k_c at convection onset correspond to the minimum of the marginal stability curve (4.19),

$$R_c = \frac{27\pi^4}{4}, \qquad k_c = \frac{\pi}{\sqrt{2}}.$$

These critical values depend on the boundary conditions; in particular, as said above, k_c vanishes in the limit of thermally insulating upper and lower boundaries.

Slightly above criticality we expand the positive solution σ_+ of the dispersion relation (4.18) and get the growth rate of the unstable modes

$$\sigma_+(R, k) \simeq (\pi^2 + k_c^2)\frac{P}{1 + P}\frac{R - R_c(k)}{R_c(k)},$$

which is proportional to the distance to criticality. Using

$$R_c(k) = R_c + \frac{(k - k_c)^2}{2}\left(\frac{\partial^2 R_c}{\partial k^2}\right)_c + \cdots,$$

we get to leading order in $R - R_c$ and $k - k_c$,

$$\sigma_+(k, R) = (\pi^2 + k_c^2)\frac{P}{1 + P}\frac{(R - R_c)}{R_c} - \alpha(k - k_c)^2 + \cdots, \tag{4.20}$$

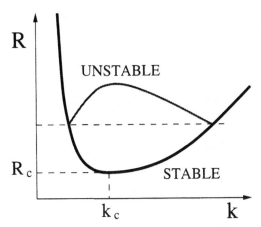

Fig. 4.4. The marginal stability curve $R_c(k)$; the curve in gray represents the growth rate of the unstable modes for $R > R_c$.

with

$$\alpha = (\pi^2 + k_c^2)\,\frac{P}{1+P}\,\frac{1}{2R_c}\,\frac{\partial^2 R_c}{\partial k^2}\bigg|_c. \qquad (4.21)$$

Note that $\sigma_+(R,k)$ involves a term proportional to $(R - R_c)(k - k_c)$; as the marginal stability curve is locally a parabola close to its minimum, we have $(k - k_c) \leq (R - R_c)^{1/2}$, and this term is of higher order in (4.20). Thus, for R larger than R_c, there exists a band of unstable modes with growth rates determined by equation (4.20) (see Fig. 4.4).

Linear analysis gives the critical Rayleigh number R_c for instability onset and determines the modulus k_c of the critical wave-vector $\mathbf{k}$ of the unstable modes. The direction of $\mathbf{k}$ is arbitrary; this orientational degeneracy is obviously related to the isotropy in the horizontal plane. There is also a translational degeneracy which is related to the translational invariance of the layer of infinite horizontal extent. These degeneracies do not result from the linear approximation but from the symmetries of the Rayleigh–Bénard geometry; thus, they will subsist in the nonlinear analysis. On the other hand, there is a pattern degeneracy that results from the linear approximation; indeed, any superposition of normal modes

$$w(\mathbf{r}, z) = \sum_p c_p \exp(i\mathbf{k}_p \cdot \mathbf{r})\,W(z), \qquad (4.22)$$

with $|\mathbf{k}_p| = k_c$ and where the c_p's are constant coefficients, is a solution of the linear problem with a zero growth rate at criticality. In order to represent a real field w, we must impose the conditions, $c_{-p} = \bar{c}_p$ and

$\mathbf{k}_{-p} = -\mathbf{k}_p$, but the number of non zero c_p's, i.e., the shape of the pattern, and their modulus, i.e., the amplitude of the convection velocity, remain undeterminate. Three basic examples of cellular pattern described by (4.22), that involve respectively one, two and three wave-vectors, are sketched in Fig. 4.5. Nonlinear interactions between the modes with different wave-vectors generally select one pattern at instability onset and determine the amplitude above criticality. However, it sometimes happens that no stationary pattern exists even immediately above a stationary instability onset; the nonlinear regime is then time-periodic or chaotic.

Another problem results from the existence of a continuous band of unstable modes above criticality as described by equation (4.20). Linear analysis only determines the one with the highest growth rate, but the wavenumber selected by nonlinear interactions may correspond to a different one. The interaction of two (or several) modes within the unstable band gives rise to a spatial modulation of the periodic pattern on a large length-scale compared to the pattern wavelength. The inverse of this length-scale is of order $k - k_c$, thus within a multiple-scale expansion procedure, it corresponds to a "slow scale" X such that

$$X = (R - R_c)^{1/2}x. \tag{4.23}$$

Let us recall, that close to the instability onset, the slow time scale T that corresponds to the vanishing growth rate of the unstable mode (4.20) is

$$T = (R - R_c)t. \tag{4.24}$$

4.3 Nonlinear saturation of the critical modes

4.3.1 Nonlinear saturation of a roll pattern: the Landau equation

We first show how nonlinear terms saturate the amplitude of the convection velocity of a roll pattern above R_c. We use a stream function,

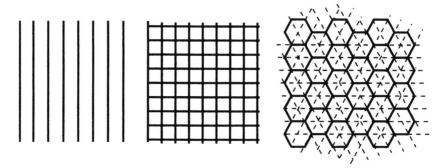

Fig. 4.5. Rolls, squares, and hexagons.

$\psi(x, z, t)$, and write the velocity field as

$$\mathbf{v} = (-\partial_z\psi, 0, \partial_x\psi).$$

Equations (4.7, 4.8) become

$$\frac{\partial}{\partial t}\nabla^2\psi + J(\psi, \nabla^2\psi) = P\nabla^4\psi + RP\frac{\partial\theta}{\partial x}, \tag{4.25}$$

$$\frac{\partial\theta}{\partial t} + J(\psi, \theta) = \frac{\partial\psi}{\partial x} + \nabla^2\theta, \tag{4.26}$$

where J is the Jacobian,

$$J(f, g) \equiv \frac{\partial f}{\partial x}\frac{\partial g}{\partial z} - \frac{\partial f}{\partial z}\frac{\partial g}{\partial x}.$$

Assuming stress-free perfectly conducting boundary conditions at $z = 0, 1$, the marginal mode that describes rolls perpendicular to the x-axis for $R = R_c$ reads

$$\psi(x, z) = [A \exp(ik_c x) + \text{c.c.}] \sin\pi z,$$
$$\theta(x, z) = \frac{ik_c}{k_c^2 + \pi^2}[A \exp(ik_c x) - \text{c.c.}] \sin\pi z. \tag{4.27}$$

The problem is to determine how the convection amplitude, $|A|$, saturates above criticality because of nonlinear interactions. This has been studied by Gorkov (1957) and Malkus and Veronis (1958) with a Poincaré–Lindstedt expansion. We will use a multiple-scale expansion, which is only slightly different, in order to keep the time-dependence of A. We do not consider a possible modulation of A on a slow length-scale in this section (see section 4.3.3). We expand ψ and θ

$$\psi = \sum_{n=1} \epsilon^n\psi_n,$$
$$\theta = \sum_{n=1} \epsilon^n\theta_n, \tag{4.28}$$

where ϵ is a small parameter related to the distance to criticality by

$$R = R_c + \sum_{n=1} \epsilon^n R_n. \tag{4.29}$$

The Boussinesq equations are symmetric under a mid-plane reflection, $z \to -z$, coupled with a temperature inversion, $\theta \to -\theta$, which corresponds to $\psi \to -\psi$ and $\theta \to -\theta$; thus to leading order, $R_1 = 0$, and

$$R - R_c \simeq \epsilon^2 R_2. \tag{4.30}$$

Using the result of the linear theory for the growth rate of the unstable mode, we obtain for the convection mode time scale

$$\partial_t = \epsilon^2\partial_T.$$

To leading order in ϵ, equations (4.25) and (4.26) give the linear problem, and the solutions for ψ_1 and θ_1 are given by the linear modes (4.27). To the next order we get,

$$\psi_2 = 0 \, ,$$

$$\theta_2 = -\frac{k_c^2}{2\pi(k_c^2 + \pi^2)}\sin 2\pi z \, , \tag{4.31}$$

that describes how the temperature advection nonlinear term deforms the vertical temperature profile. The solvability condition at the next order gives the evolution equation for A,

$$\frac{dA}{dT} = \mu A - \beta|A|^2 A \, , \tag{4.32}$$

with

$$\mu = (\pi^2 + k_c^2)\frac{P}{1+P}\frac{(R - R_c)}{\epsilon^2 R_c} \, ,$$

$$\beta = \frac{k_c^2}{2}\frac{P}{1+P} \, .$$

Note that, since $R - R_c$ is of order ϵ^2, all the terms of the amplitude equation (4.32) are of order one. However, we can easily write this equation using original unscaled variables for time, amplitude and distance to criticality, $R - R_c$; we can check on the unscaled form that all terms are of order $(R - R_c)^{3/2}$.

As previously shown for nonlinear oscillators, the form of the amplitude equation is here also determined by symmetry constraints. Translational invariance in the horizontal plane implies that, if $(\psi_0(x, z), \theta_0(x, z))$ represents a roll solution, $(\psi_0(x + x_0, z), \theta_0(x + x_0, z))$ is another solution; this amounts to shifting the rolls in the horizontal plane, or to changing the origin of the x-axis. From equation (4.27) this transformation corresponds to a rotation in the complex plane for A, and the amplitude equation should be invariant under the transformation

$$A \to A \exp i\phi \, .$$

As shown for nonlinear oscillators, the only allowed nonlinear term up to third order in amplitude is thus $|A|^2 A$, and the amplitude equation is of the form (4.32), where β is *a priori* a complex number.

However, there is here an additional symmetry, space-reflection: $x \to -x$. From equation (4.27) this transformation corresponds to

$$A \to \bar{A}$$

for the complex amplitude. Taking the complex conjugate of the amplitude equation, applying the reflection transformation and comparing to the original amplitude equation, gives $\beta = \bar{\beta}$, thus β real. The form of the amplitude equation (4.32) is determined by symmetry constraints. The perturbative calculation starting from the Boussinesq equations is only useful to get the sign of β and shows that the bifurcation is supercritical. This can also be shown using variational methods (Sorokin, 1953), and for a large variety of boundary conditions, the motionless state is globally stable below R_c in the Boussinesq approximation. Above R_c, the convection velocity amplitude increases continuously from zero and scales as $(R - R_c)^{1/2}$. It is the order parameter of the transition, the corresponding broken symmetry being translational invariance in space. Note finally that equation (4.32) can be written in a variational form

$$\frac{dA}{dT} = -\frac{\partial V}{\partial \bar{A}},$$
$$V(A, \bar{A}) = -\mu A \bar{A} + \tfrac{1}{2}\beta A^2 \bar{A}^2, \qquad (4.33)$$

where $V(A, \bar{A})$ is the "Landau free-energy" in the vicinity of the transition.

4.3.2 Pattern selection

We now consider the problem of pattern selection via nonlinear interactions. As was said above, in the slightly supercritical range, any superposition (4.22) of marginal modes has the same growth rate. Let us consider two examples, squares and hexagons.

For squares, we have

$$w(x, y, t) = \epsilon \left([A_1 \exp(ik_c x) + \text{c.c.}] + [A_2 \exp(ik_c y) + \text{c.c.}] \right) \sin \pi z + \ldots, \qquad (4.34)$$

where $A_1(T)$ and $A_2(T)$ are the complex amplitudes of the two sets of perpendicular rolls. Using symmetry considerations, the amplitude equations read

$$\frac{dA_1}{dT} = \mu A_1 - \left[\beta |A_1|^2 + \gamma |A_2|^2 \right] A_1,$$
$$\frac{dA_2}{dT} = \mu A_2 - \left[\gamma |A_1|^2 + \beta |A_2|^2 \right] A_2. \qquad (4.35)$$

It is an easy exercise to show that for $\mu > 0$, stationary squares ($|A_1| = |A_2|$) are stable when $|\gamma| < \beta$, i.e., when the cross-coupling nonlinear term is small enough so that the two sets of rolls weakly interact; when their interaction is too strong, more precisely when, $\gamma > \beta$, one of the two sets of rolls nonlinearly damps out the other, and rolls are the stable nonlinear state. This is the situation for Boussinesq convection with stress-free perfectly heat-conducting boundary conditions. On the contrary, with insulating boundaries, squares are observed.

For hexagons, we have

$$w(x, y, t) = \epsilon \sum_{p=1}^{3} [A_p \exp(i\mathbf{k}_p \cdot \mathbf{r}) + \text{c.c.}] \sin \pi z + \dots, \qquad (4.36)$$

with $|\mathbf{k}_p| = k_c$ and $\mathbf{k}_1 + \mathbf{k}_2 + \mathbf{k}_3 = 0$, and where the $A_p(T)$'s are the complex amplitudes of the three sets of rolls. Using symmetry considerations, the amplitude equations read

$$\frac{dA_l}{dT} = \mu A_l - \left[\beta |A_l|^2 + \delta(|A_m|^2 + |A_n|^2)\right] A_l. \qquad (4.37)$$

Note that a term proportional to $\bar{A}_m \bar{A}_n$ in the evolution equation for A_l respects the translational and reflection $(x \to -x)$ symmetries, but is forbidden here because of the additional Boussinesq symmetry $(z \to -z,$ $w \to -w, \theta \to -\theta)$. A possible exercise at this stage is to determine the stability domains of rolls, squares and hexagons as a function of the real coupling constants β, γ, δ and μ. Show also that the square–hexagons transition is "first order" and relate that to a symmetry argument.

The general problem of pattern selection is much more difficult to solve and has been studied by Schlüter *et al.* (1965) for the case of rigid perfectly conducting boundary conditions. They have found that rolls are the only stable stationary pattern just above the onset of convection. Using a similar analysis, Riahi (1983) has found stationary squares in the case of thermally insulating rigid boundaries.

4.3.3 Slowly varying amplitude of a roll pattern: the Ginzburg–Landau equation

We now consider the problem of the existence of a band of unstable modes above R_c. For simplicity we assume stress-free perfectly heat conducting boundaries, so that the pattern consists of parallel rolls. To take into account the modes around k_c, we consider a wave packet

$$\psi(x, z) = \epsilon [A(X, Y, T) \exp(ik_c x) + \text{c.c.}] \sin \pi z + \dots, \qquad (4.38)$$

where $A(X, Y, T)$ represents the slowly varying envelope of the roll pattern. We first consider modulations only along the x-axis, thus $A = A(X, T)$. We have

$$A(X, T) = \int \hat{A}(K, \Sigma) \exp(\Sigma T + iKX) \delta[\Sigma - \Sigma(\mu, K)] \, dK \, d\Sigma, \qquad (4.39)$$

where

$$\Sigma(\mu, K) = \mu - \alpha K^2 + \dots, \qquad (4.40)$$

is the dispersion relation (4.20) in terms of scaled (order one) variables, i.e., $K = (k - k_c)/\epsilon$. The Fourier–Laplace transform of the dispersion relation (4.40) gives the linear part of the evolution equation for the amplitude $A(X, T)$

$$\frac{\partial A}{\partial T} = \mu A + \alpha \frac{\partial^2 A}{\partial X^2} \tag{4.41}$$

Taking into account the leading order nonlinear term, we get

$$\frac{\partial A}{\partial T} = \mu A + \alpha \frac{\partial^2 A}{\partial X^2} - \beta |A|^2 A. \tag{4.42}$$

We now consider modulations also along the rolls axis, thus $A = A(X, Y, T)$. We have

$$\mathbf{k} = (k_c + \delta k_x)\, \hat{\mathbf{x}} + (\delta k_y)\, \hat{\mathbf{y}},$$

and the generalization of the dispersion relation (4.20) that respects rotational invariance in the horizontal plane, is

$$\sigma(k, R) = (\pi^2 + k_c^2) \frac{P}{1+P} \frac{(R - R_c)}{R_c} - \xi_0^2 \left(k^2 - k_c^2\right)^2 + \dots \tag{4.43}$$

with $4k_c^2 \xi_0^2 = \alpha$. We have

$$\left(k^2 - k_c^2\right)^2 = \left(2k_c \delta k_x + (\delta k_x)^2 + (\delta k_y)^2\right).$$

Thus, the relevant scalings are

$$\delta k_x = \epsilon K_x, \qquad \delta k_y = \epsilon^{1/2} K_y.$$

Slower y-modulations do not affect the amplitude equation to leading order, whereas modes corresponding to modulations on shorter scales are too strongly damped to be marginal. In terms of scaled variables, the dispersion relation reads

$$\Sigma = \mu - \alpha \left[K_x + \frac{1}{2k_c} K_y^2\right]^2 + \dots \tag{4.44}$$

Taking its Fourier–Laplace transform and adding the leading order nonlinear term, gives

$$\frac{\partial A}{\partial T} = \mu A + \alpha \left(\frac{\partial}{\partial X} - \frac{i}{2k_c} \frac{\partial^2}{\partial Y^2}\right)^2 A - \beta |A|^2 A. \tag{4.45}$$

In terms of unscaled variables, all the terms of (4.45) are of order ϵ^3. This equation has been obtained by Newell and Whitehead (1969) and Segel (1969) using a multiple-scale expansion both in space and time. Note that partial derivatives in X and Y are not involved similarly because the roll pattern breaks the rotational invariance in the horizontal plane.

We can write the amplitude equation (4.45) in variational form,

$$\frac{\partial A}{\partial T} = -\frac{\delta \mathcal{L}}{\delta \bar{A}}$$

(4.46)

where

$$\mathcal{L}[A] = \int \left[-\mu |A|^2 + \frac{\beta}{2}|A|^4 + \alpha \left| \left(\frac{\partial}{\partial X} - \frac{i}{2k_c}\frac{\partial^2}{\partial Y^2} \right) A \right|^2 \right] dX\, dY \,,$$

(4.47)

is analogous to a "Ginzburg–Landau" free-energy.

In the following sections we will study pattern dynamics governed by this "Ginzburg–Landau" equation, also named in the context of convection, the Newell–Whitehead–Segel equation. In the case of stress-free boundary conditions, this equation is incorrect because it does not take into account the nonlinear interaction with vertical vorticity modes (Zippelius and Siggia, 1982). We will discuss this effect later. With different coefficients than the ones derived above, the Ginzburg–Landau equation is correct to leading order for the description of slowly modulated roll patterns in convection with rigid thermally conducting boundaries. More generally, it describes slowly modulated one-dimensional patterns that occur via a stationary bifurcation in a dissipative system, invariant under translations, rotations and space reflections in the horizontal plane, when no other marginal mode than the roll-mode at wavenumber k_c is involved.

5 Amplitude equations in dissipative systems

We consider a dissipative system governed by a nonlinear partial differential equation

$$\frac{\partial \mathbf{U}}{\partial t} = L_\mu(\nabla) \cdot \mathbf{U} + N(\nabla, \mathbf{U}) \,,$$

(5.1)

where $\mathbf{U}(\mathbf{r}, t)$ represents a set of scalar or pseudo-scalar fields. The system is driven externally by a control parameter μ constant in space and time (except in section 5.3 where we consider parametric instabilities). Its basic state is thus homogeneous in space and constant in time, and corresponds to $\mathbf{U} = 0$, say. $L_\mu(\nabla)$ is a linear operator which involves spatial-derivatives, and $N(\nabla, \mathbf{U})$ represents nonlinear terms. We assume that equation (5.1) is invariant under continuous translations in space and time (except in section 5.3 where it is invariant under continuous translations in space and discrete translations in time). We also assume in some cases, space-reflection symmetry or invariance under Galilean transformations.

For a critical value of μ, the basic state, $\mathbf{U} = 0$, loses its stability. The linear stability analysis consists of solving the eigenvalue problem

$$L_\mu(\nabla) \cdot \mathbf{U} = \eta \mathbf{U}. \tag{5.2}$$

The basic state is stable when all the eigenvalues η have a negative real part. The instability onset, or the bifurcation of $\mathbf{U} = 0$, is characterized by one or several eigenvalues with a zero real part. The corresponding eigenfunctions, $\mathbf{U}_{i,\mathbf{k}}(\mathbf{r}, t)$, are the critical modes, and characterize the temporal or spatial pattern that sets in at the instability onset, and breaks spontaneously some of the invariances listed above. In the vicinity of the instability onset, the amplitudes A_i of the critical modes vary on a time scale much slower than that of the other modes, and thus contain all the information about the asymptotic time-dependence of $\mathbf{U}$. More precisely, the amplitude of the non-critical (damped) modes does not vanish only because they are forced by the critical (slightly unstable) modes through nonlinear interactions; thus, they follow adiabatically critical modes; adiabatic elimination of fast modes leads to amplitude equations. We observed in the previous sections that symmetry constraints determine their form. We give below a catalogue of the most frequent situations (Newell, 1974, Fauve, 1985).

5.1 Stationary instability

In the previous section we studied the example of Rayleigh–Bénard convection. A stationary instability corresponds to a marginal mode with a real growth-rate. We shall discuss the case when the critical wave number, k_c, is non-zero; situations where the critical wave number is vanishingly small will be considered in section 5.5.

5.1.1 One-dimensional pattern

Let us first consider the situation where the nonlinear terms select a one-dimensional "roll pattern." In the vicinity of the instability onset, $r \simeq r_c$, the growth-rate is a real function of k with a maximum around k_c (see Fig. 5.1)

$$\sigma = r - r_c + \frac{1}{2}\left(\frac{\partial^2 \sigma}{\partial k^2}\right)_c (k - k_c)^2 + \dots \tag{5.3}$$

We take the distance to criticality, $r - r_c$, of order ϵ^2 ($\epsilon \ll 1$), accordingly the instability growth-rate is of order ϵ^2; this is the time scale for the slow critical modes. Correspondingly, there is a large spatial scale generated by mode interaction in the unstable wavenumber band; as $\sigma(k)$ is locally a parabola, $k - k_c \leq \mathcal{O}(\epsilon)$, and the large spatial scale corresponds to

$k - k_c = \epsilon K$. Thus, the scaled dispersion relation is

$$\Sigma(K) = \mu - \alpha K^2 + \ldots, \tag{5.4}$$

where $\Sigma = \sigma/\epsilon^2$, $\mu = (r - r_c)/\epsilon^2$, $\alpha = -\frac{1}{2}(\partial^2 \sigma/\partial k^2)_c$, and K, are all of order one.

We first consider modulations of the pattern only along the x-axis, and write

$$\mathbf{U}(x,t) = \epsilon\,[A(X,T)\,\exp(ik_c x) + \text{c.c.}]\,\tilde{\mathbf{U}}_{k_c} + \ldots, \tag{5.5}$$

where $\tilde{\mathbf{U}}_{k_c}\exp(ik_c x)$ is the critical mode. As shown in previous sections, the linear evolution equation for A is the Fourier–Laplace transform of (5.4), and the leading order nonlinear term compatible with translational invariance in space is $|A|^2 A$. Moreover, its coefficient β is real if the system is invariant under space-reflection symmetry, $x \to -x$. Thus, to leading order, A obeys the Ginzburg–Landau equation

$$\frac{\partial A}{\partial T} = \mu A + \alpha \frac{\partial^2 A}{\partial X^2} - \beta |A|^2 A\,. \tag{5.6}$$

The nonlinear term saturates the instability growth if $\beta > 0$, and the bifurcation is supercritical. In terms of the original scaled variables, the critical mode amplitude scales like ϵ, i.e., like $(r-r_c)^{1/2}$. The slow spatial scale can be understood as a coherence length and diverges at criticality like $(r - r_c)^{-1/2}$.

At this stage it is important to discuss finite size effects due to lateral boundary conditions. If the dissipative system under consideration consists, for instance, of a fluid in a container of horizontal size L, the expansion (5.5) does not satisfy, in general, lateral boundary conditions and should be replaced by an expansion on the linear modes of the container. However, (5.5) is roughly valid in the bulk of the container if L is

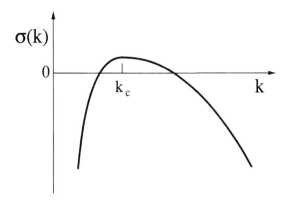

Fig. 5.1.   Growth-rate versus wave number for a stationary instability at $k = k_c$.

large compared to the coherence length $2\pi\sqrt{\alpha/(r - r_c)}$. This is of course never true at the instability onset but might be satisfied slightly above the onset if α is small enough, i.e., if the instability mechanism is not too selective in wavenumber. Then, the pattern, roughly described by (5.5) in the bulk of the container, matches the lateral boundary conditions within a few coherence lengths. In other words, in order to neglect finite size effects, the width of the growth-rate curve should be large compared to the interval in k imposed by the quantization due to the lateral boundaries. Note that it is often claimed in the literature that a pattern with many wavelengths within the container is an "extended system" involving many degrees of freedom. This is not correct in general, except if L is also large compared to the coherence length. This discussion also applies to oscillatory instabilities. An interesting example is the laser instability; although the length of the cavity is huge compared to the optical wavelength, the selectivity of the instability mechanism is so high that a laser very often involves only few unstable modes.

If $\beta < 0$, higher order nonlinear terms should be taken into account, and a simple model for a subcritical bifurcation is

$$\frac{\partial A}{\partial T} = \mu A + \alpha \frac{\partial^2 A}{\partial X^2} - \beta|A|^2 A - \gamma|A|^4 A, \qquad (5.7)$$

with $\gamma > 0$. However, in terms of scaled variables, β should be small in order to get the two nonlinear terms at the same order; thus (5.7) is asymptotically correct only in the vicinity of a tricritical point. Like the Ginzburg–Landau equation, (5.7) can be put in variational form

$$\frac{\partial A}{\partial T} = -\frac{\delta\mathcal{L}}{\delta\bar{A}} \qquad (5.8)$$

where

$$\mathcal{L}[A] = \int \left[\alpha\left|\frac{\partial A}{\partial X}\right|^2 - V(A, \bar{A})\right] dX, \qquad (5.9)$$

$$V(A, \bar{A}) = \mu|A|^2 - \frac{\beta}{2}|A|^4 - \frac{\gamma}{3}|A|^6.$$

$\mathcal{L}[A]$ is a Lyapunov functional; indeed, multiplying (5.7) by $\partial\bar{A}/\partial T$, and adding to the complex conjugate expression, yields

$$\frac{d}{dt}\mathcal{L}[A] = -2\int\left|\frac{\partial A}{\partial T}\right|^2 dX < 0. \qquad (5.10)$$

Thus $\mathcal{L}[A]$ is a decreasing function, bounded from below for the constant A's that are the maxima of $V(A, \bar{A})$. For a supercritical bifurcation ($\beta > 0$, fifth order term neglected), the uniform state $A = 0$ is globally stable for $\mu < 0$, whereas perfectly periodic patterns corresponding to

$A = \sqrt{\mu/\beta} \exp(i\phi)$, with constant ϕ, are globally stable for $\mu > 0$. The degeneracy in ϕ is obviously related to translational invariance in space, but a more subtle effect that involves the phase might happen when the system is not invariant under reflection symmetry, $x \to -x$.

When the system is not invariant under reflection symmetry, $x \to -x$, the coefficient β is in general a complex number. Writing $A(X, T) = R(X, T) \exp(i\phi(X, T))$, equation (5.6) yields

$$\frac{\partial R}{\partial T} = \left[\mu - \alpha \left(\frac{\partial \phi}{\partial X} \right)^2 \right] R + \alpha \frac{\partial^2 R}{\partial X^2} - \beta_r R^3 ,$$

$$R \frac{\partial \phi}{\partial T} = 2 \frac{\partial R}{\partial X} \frac{\partial \phi}{\partial X} + R \frac{\partial^2 \phi}{\partial X^2} - \beta_i R^3 . \tag{5.11}$$

Consequently, a homogeneous roll pattern of amplitude R_0 has its phase that linearly increases in time

$$\phi_0 = -\beta_i R_0^2 T.$$

Thus, from (5.5), one observes that this "stationary" instability gives rise to a traveling pattern

$$\mathbf{U}_0(x, t) = \epsilon \left[R_0 \exp \left(i \left(k_c x - \epsilon^2 \beta_i R_0^2 t \right) \right) + \text{c.c.} \right] \tilde{\mathbf{U}}_{k_c} + \dots$$

due to the externally broken reflection symmetry. We will study a similar effect when the reflection symmetry is spontaneously broken; a secondary instability then generates a drifting pattern from a stationary one (see section 7).

Note that the absence of a term proportional to $\partial A/\partial X$ in (5.6) is not related to reflection symmetry. Indeed, a term of the form $i\partial A/\partial X$ is compatible with the $x \to -x$ symmetry, and is present as soon as one expands $\mathbf{U}$ in (5.5) at $k \neq k_c$. This corresponds to the change of variable

$$A = B \exp(iqX)$$

that gives

$$\frac{\partial B}{\partial T} = (\mu - \alpha q^2)B + 2iq\alpha \frac{\partial B}{\partial X} + \alpha \frac{\partial^2 B}{\partial X^2} - \beta |B|^2 B , \tag{5.12}$$

thus showing that the growth-rate of the mode $k = k_c + \epsilon q$ is $\mu - \alpha q^2$, i.e., in scaled terms, $r - r_c - \alpha(k - k_c)^2$, in agreement with the dispersion relation. The absence of a term in $i\partial A/\partial X$ in (5.6) is thus only related to the fact that the first unstable mode k_c is the one with the maximum growth-rate $\sigma(k)$.

A simple model, that mimics the formation of a one-dimensional pattern is the Swift–Hohenberg equation for a field $u(x, y, t)$

$$\frac{\partial u}{\partial t} = \left[r - r_c - (k_c^2 + \Delta)^2 \right] u - u^3 . \tag{5.13}$$

Consider a one-dimensional field in space, $u(x,t)$, and using a multiple-scale expansion technique, derive the Ginzburg–Landau equation in the vicinity of the instability onset, $r - r_c \simeq \mu\epsilon^2$. Try the same exercise with the model

$$\frac{\partial u}{\partial t} = \left[r - r_c - (k_c{}^2 + \Delta)^2 \right] u - au^2 - u\frac{\partial u}{\partial x}, \tag{5.14}$$

which is not invariant under the $x \to -x$ reflection symmetry, and show that β is complex. Find the nature of the bifurcation as a function of a, and derive (5.7) in the vicinity of the tricritical point.

Finally, let us consider a situation where the spatial phase is quenched. This occurs in the convection problem for instance, if one takes stress-free boundary conditions also at the lateral boundaries, $x = 0$ and $x = 2n\pi/k_c$, where n is an integer. Then, the phase of the pattern is fixed,

$$\mathbf{U}(\mathbf{x}, \mathbf{t}) = \epsilon R(X, T) \sin(k_c x) \, \tilde{\mathbf{U}}_{k_c} + \dots, \tag{5.15}$$

and its amplitude is governed by a real Ginzburg–Landau equation

$$\frac{\partial R}{\partial T} = \mu R + \alpha \frac{\partial^2 R}{\partial X^2} - \beta R^3. \tag{5.16}$$

The broken symmetry at the instability onset is not translational invariance, which is here externally broken because of the lateral boundary conditions, but the $R \to -R$ symmetry of Boussinesq convection.

5.1.2 Two-dimensional modulations of a one-dimensional roll pattern

We now consider the dynamics of two-dimensional modulations of a one-dimensional roll pattern, parallel to the x-axis. In isotropic systems, the growth-rate depends on $\mathbf{k}^2$ and is maximum for $\mathbf{k}^2 = k_c{}^2$; thus, for $|\mathbf{k}| \simeq k_c$

$$\sigma(\mathbf{k}) = r - r_c - \xi_0{}^2(\mathbf{k}^2 - k_c{}^2)^2 + \dots, \tag{5.17}$$

and as shown in section 4, the amplitude equation reads

$$\frac{\partial A}{\partial T} = \mu A + \alpha \left(\frac{\partial}{\partial X} - \frac{i}{2k_c}\frac{\partial^2}{\partial Y^2} \right)^2 A - \beta |A|^2 A, \tag{5.18}$$

or in variational form

$$\frac{\partial A}{\partial T} = -\frac{\delta \mathcal{L}}{\delta \bar{A}} \tag{5.19}$$

with

$$\mathcal{L}[A] = \int \left[-\mu|A|^2 + \frac{\beta}{2}|A|^4 + \alpha \left| \left(\frac{\partial}{\partial X} - \frac{i}{2k_c}\frac{\partial^2}{\partial Y^2} \right) A \right|^2 \right] dX \, dY. \tag{5.20}$$

For anisotropic systems, the growth-rate is maximum for $|\mathbf{k}| = k_c$ with $\mathbf{k}$ along a preferred axis, x say. Thus

$$\sigma(\mathbf{k}) = r - r_c - \alpha(k_x - k_c)^2 + \alpha' k_y{}^2 + \dots , \qquad (5.21)$$

and

$$\Sigma(\mathbf{K}) = \mu - \alpha K_X{}^2 + \alpha' K_Y{}^2 + \dots , \qquad (5.22)$$

so the Ginzburg–Landau equation takes the form

$$\frac{\partial A}{\partial T} = \mu A + \alpha \frac{\partial^2 A}{\partial X^2} + \alpha' \frac{\partial^2 A}{\partial Y^2} - \beta |A|^2 A , \qquad (5.23)$$

5.1.3 Two-dimensional patterns

We have already considered two-dimensional patterns in the previous section. Let us take the example of hexagons or more precisely of patterns with three basic wave-vectors, $\mathbf{k}_1, \mathbf{k}_2, \mathbf{k}_3$, such that $|\mathbf{k}_p| = k_c$ and $\mathbf{k}_1 + \mathbf{k}_2 + \mathbf{k}_3 = 0$,

$$\mathbf{U}(x, y, t) = \epsilon \sum_{p=1}^{3} [A_p \exp(i\mathbf{k}_p \cdot \mathbf{r}) + \text{c.c.}] \, \tilde{\mathbf{U}}_{k_c} + \dots , \qquad (5.24)$$

without the $\mathbf{U} \to -\mathbf{U}$ invariance. Using symmetry considerations, the amplitude equations read

$$\frac{dA_l}{dT} = \mu A_l + \rho \bar{A}_m \bar{A}_n - \left[\beta |A_l|^2 + \delta(|A_m|^2 + |A_n|^2) \right] A_l , \qquad (5.25)$$

Note that this equation is asymptotically valid only if the $\mathbf{U} \to -\mathbf{U}$ symmetry is slightly broken, so that the quadratic and cubic nonlinearities are obtained at the same order. The quadratic nonlinearities correspond to the resonant triad interaction $\mathbf{k}_1 + \mathbf{k}_2 + \mathbf{k}_3 = 0$. Although one can change the phase ϕ_p of each wave by shifting the origin in the horizontal plane along $\mathbf{k}_p$, the above relation implies

$$\phi_1 + \phi_2 + \phi_3 = \Phi = \text{constant} . \qquad (5.26)$$

Φ determines the shape of the pattern (hexagons, triangles,...). Using translational and reflection symmetry, one can restrict Φ to the interval $[0, \pi/2]$. Note that the leading order amplitude equations (5.25) select $\Phi = 0$. In the presence of the $\mathbf{U} \to -\mathbf{U}$ symmetry, $\rho = 0$, but hexagons might be selected by third order nonlinearities. In that case, Φ is arbitrary; this is analogous to a phason mode in condensed-matter physics.

Periodic tilings of the plane consist either of rhombs, hexagons or triangles (see for instance, Chandrasekhar, 1961). It has been known since the discovery of quasicrystals that there exist structures with orientational order and no translational order. It has been shown recently that patterns with quasicrystalline order can be generated by nonlinear interaction between parametrically driven instability modes, whatever the shape of the fluid container (Edwards and Fauve, 1992, 1993).

5.2 Oscillatory instability

We now consider situations where the instability growth-rate has an imaginary part $\omega(k)$; thus, the unstable mode has an oscillatory behavior with a pulsation $\omega(k_0)$ at onset. We begin with the case $k_0 = 0$.

5.2.1 Oscillatory instability at zero wavenumber

When the real part of the growth-rate is maximum at zero wavenumber, we have

$$\eta(k) = \sigma(k) + i\omega(k) \tag{5.27}$$

with

$$\sigma(k) = r - r_{\text{c}} + \frac{1}{2}\left(\frac{\partial^2 \sigma}{\partial k^2}\right)_0 k^2 + \cdots$$
$$\omega(k) = \omega_0 + \left(\frac{\partial \omega}{\partial k}\right)_0 k + \frac{1}{2}\left(\frac{\partial^2 \omega}{\partial k^2}\right)_0 k^2 + \cdots \tag{5.28}$$

In the vicinity of the instability onset, $r - r_{\text{c}} = \mu\epsilon^2$,

$$\mathbf{U}(\mathbf{x}, \mathbf{t}) = \epsilon\left[A(X, T)\exp(i\omega_0 t) + \text{c.c.}\right]\tilde{\mathbf{U}}_0 + \cdots, \tag{5.29}$$

where $\tilde{\mathbf{U}}_0 \exp(i\omega_0 t)$ is the critical mode. As previously observed, the Fourier–Laplace transform of the dispersion relation gives the linear part of the amplitude equation, and translational invariance in time determines the form of the leading order nonlinear term, $|A|^2 A$. We get

$$\frac{\partial A}{\partial T} = \mu A - c\frac{\partial A}{\partial X} + \alpha\frac{\partial^2 A}{\partial X^2} - \beta|A|^2 A, \tag{5.30}$$

where $c = (\partial \omega/\partial k)_0$ is the group velocity, $\alpha_r = -\frac{1}{2}(\partial^2 \sigma/\partial k^2)_0$ is related to the diffusion of space-dependent perturbations ($\alpha_r > 0$), $\alpha_i = -\frac{1}{2}(\partial^2 \omega/\partial k^2)_0$ corresponds to the dispersion, β_r is the nonlinear dissipation, and β_i is related to the nonlinear amplitude-dependence of the frequency. Note that, as for the nonlinear Schrödinger equation, one needs two slow time scales T_1 and T_2 when deriving (5.30) with an asymptotic expansion. Transforming to the reference frame moving at the group velocity yields

$$\frac{\partial A}{\partial T} = \mu A + \alpha\frac{\partial^2 A}{\partial X^2} - \beta|A|^2 A. \tag{5.31}$$

Equation (5.31) is a Ginzburg–Landau equation with complex coefficients α and β. This is a crucial difference from (5.6) for stationary cellular instabilities, that involves real coefficients. Indeed, no variational formulation is known for (5.31). Thus, A does not evolve in order to minimize a functional $\mathcal{L}[A]$ as for (5.6), but displays in some parameter range, periodic or even chaotic behaviors in space and time. Equation (5.31) has two simple limits, a "variational" one for α and β real

$$\frac{\partial A}{\partial T} = -\frac{\delta \mathcal{L}}{\delta \bar{A}},$$

$$\frac{d}{dt}\mathcal{L}[A] = -2 \int \left|\frac{\partial A}{\partial T}\right|^2 dX < 0,$$

and a conservative limit (the nonlinear Schrödinger equation), for α and β pure imaginary,

$$\frac{\partial A}{\partial T} = -\frac{\delta \mathcal{L}}{\delta \bar{A}},$$

$$\frac{d}{dt}\mathcal{L}[A] = 0.$$

One method is to investigate (5.31) perturbatively, starting from one of these limit situations (section 8).

5.2.2 Oscillatory instability at finite wavenumber

Hydrodynamic instabilities often lead to time-dependent cellular patterns. The Couette–Taylor flow between concentric cylinders (DiPrima and Swinney, 1981), thermal convection in the presence of a salinity gradient or in binary fluid mixtures (Turner, 1973), thermal convection in a layer of fluid rotating about a vertical axis (Chandrasekhar, 1961), display a Hopf bifurcation at a finite wavenumber $k_0 \neq 0$. The growth-rate of the marginal modes reads

$$\eta(k) = \sigma(k) + i\omega(k) \tag{5.32}$$

with

$$\sigma(k) = r - r_{\rm c} + \frac{1}{2}\left(\frac{\partial^2 \sigma}{\partial k^2}\right)_0 (k - k_0)^2 + \dots$$

$$\omega(k) = \omega_0 + \left(\frac{\partial \omega}{\partial k}\right)_0 (k - k_0) + \frac{1}{2}\left(\frac{\partial^2 \omega}{\partial k^2}\right)_0 (k - k_0)^2 + \dots \tag{5.33}$$

In the vicinity of the instability onset, $r - r_{\rm c} = \mu\epsilon^2$,

$$\mathbf{U}(x, t) = \epsilon\left[A(X, T)\,\exp(i\omega_0 t - k_0 x) + {\rm c.c.}\right]\tilde{\mathbf{U}}_{-k_0}$$
$$+ \epsilon\left[B(X, T)\,\exp(i\omega_0 t + k_0 x) + {\rm c.c.}\right]\tilde{\mathbf{U}}_{k_0} + \dots, \tag{5.34}$$

where A and B are the complex amplitudes of the waves propagating to the right and to the left, and obey the following amplitude equations

$$\frac{\partial A}{\partial T} = \mu A - c\frac{\partial A}{\partial X} + \alpha\frac{\partial^2 A}{\partial X^2} - (\beta|A|^2 + \gamma|B|^2)A \,,$$
$$\frac{\partial B}{\partial T} = \mu B + c\frac{\partial B}{\partial X} + \alpha\frac{\partial^2 B}{\partial X^2} - (\gamma|A|^2 + \beta|B|^2)B \,.$$

(5.35)

Equations (5.35) are invariant under the transformations

$$A \to A \exp(-i\phi)\,, \qquad B \to B \exp(i\phi)\,,$$
$$A \to A \exp(i\theta)\,, \qquad B \to B \exp(i\theta)\,,$$

that reflect translational invariance in space and time, and under the transformation

$$X \to -X\,, \qquad A \to B\,, \qquad B \to A\,,$$

that traces back to space-reflection symmetry, $x \to -x$. Note that terms of the form $i\partial A/\partial X$, $i\partial B/\partial X$, satisfy symmetry requirements but are not involved because σ is maximum for $k = k_0$. Let us also remark that the group velocity c should be small in order to get all the terms of (5.35) at the same order of an asymptotic expansion (see the discussion about counter-propagating waves in section 3).

Restricting the discussion to spatially homogeneous solutions, it is easy to check that (5.35) describe either:

a) propagating waves ($|A| \neq 0$, $|B| = 0$, or $|A| = 0$, $|B| \neq 0$), which are stable if $\gamma_r > \beta_r > 0$,
b) standing waves ($|A| = |B| \neq 0$), which are stable if $\beta_r > |\gamma_r|$.

When $\beta_r < 0$ or $\gamma_r < -\beta_r$, the bifurcation is subcritical.

5.3 Parametric instability

Parametric instabilities in spatially extended systems also generate waves. Let us consider, for example, the one-dimensional array of coupled pendulums, already studied in section 3, with an additional damping and a parametric forcing. In the long-wavelength limit, the governing equation is

$$\frac{\partial^2 u}{\partial t^2} + 2\lambda\frac{\partial u}{\partial t} + (1 + f\sin\omega_e t)\sin u = \frac{\partial^2 u}{\partial x^2} \,.$$

(5.36)

In the limit of small dissipation ($\lambda \simeq 0$), the dispersion relation of the unforced array of pendulums is

$$\omega_0{}^2(k) = 1 + k^2.$$

(5.37)

Waves of the form $\exp\left(i(\omega_1 t - k_1 x)\right)$ or $\exp\left(i(\omega_2 t + k_2 x)\right)$ are nearly marginal and can be parametrically amplified. Momentum conservation implies $k_1 = k_2 \equiv k$, whereas energy conservation gives $\omega_1 + \omega_2 = n\omega_e$, where n is an integer. We get from (5.37)

$$\omega_1 = \omega_2 = \omega_0(k) = \frac{n\omega_e}{2}.$$

Thus, the system selects its wavenumber in order to satisfy the parametric resonance condition. In the presence of a small dissipation, the resonance $n = 1$ is the strongest and the selected wavenumber k is such that

$$\omega_e \equiv 2\,\omega = 2\,\omega_0(k).$$

In the vicinity of the instability onset,

$$
\begin{aligned}
u(x,t) = \epsilon[A(X,T)\,&\exp i(\omega t - kx) \\
+ B(X,T)\,&\exp i(\omega t + kx) + \text{c.c.}\,] + \dots .
\end{aligned}
\tag{5.38}
$$

The symmetry requirements are

a) continuous translational invariance in space, that implies the invariance under the transformation

$$A \to A\,\exp(-i\phi), \qquad B \to B\,\exp(i\phi),$$

b) discrete translational invariance in time, $t \to t + 2\pi/\omega_e$, that implies

$$A \to -A, \qquad B \to -B,$$

c) space-reflection symmetry, $x \to -x$, that implies

$$X \to -X, \qquad A \to B, \qquad B \to A.$$

All the terms of (5.35) respect these requirements, but additional terms are allowed due to the less restrictive requirement about translational invariance in time (discrete instead of continuous). To leading order, two additional terms, proportional to the forcing, are allowed, $F\bar{A}$ and $F\bar{B}$. We get

$$
\begin{aligned}
\frac{\partial A}{\partial T} &= (-\Lambda + i\nu)A + F\bar{B} - c\frac{\partial A}{\partial X} + \alpha\frac{\partial^2 A}{\partial X^2} - (\beta|A|^2 + \gamma|B|^2)A \\
\frac{\partial B}{\partial T} &= (-\Lambda + i\nu)B + F\bar{A} + c\frac{\partial B}{\partial X} + \alpha\frac{\partial^2 B}{\partial X^2} - (\gamma|A|^2 + \beta|B|^2)B,
\end{aligned}
\tag{5.39}
$$

where Λ is proportional to the dissipation and ν is proportional to the detuning from parametric resonance. Equations (5.39) show that the right-going wave is forced by the left-going wave, and vice-versa, so that only standing waves are parametrically generated in the vicinity of the instability onset. In other words, the phase of the parametric response being quenched by the external forcing in time, the only possibility at onset is a standing wave.

Note that this result is obvious if we do not consider the zero dissipation limit. With finite dissipation, propagative waves are not neutral modes. From Floquet theory, one has

$$u(x,t) = \epsilon\,[W(X,T)\,\exp(ikx) + \text{c.c.}]\,\chi(t)\,\exp(i\sigma t) + \dots\,,$$

where $\chi(t)$ is a time-periodic function with period $2\pi/\omega_e$, and where $i\sigma$ is the Floquet exponent ($\sigma = \omega_e/2$ for the subharmonic response). Then, symmetry requirements show that $W(X,T)$ obeys a Ginzburg–Landau equation. As an exercise, find this equation from (5.39) by adiabatic elimination of the damped mode in the vicinity of the instability onset. The only advantage of the analysis in the zero dissipation limit is that (5.39) also describes secondary instabilities of a parametrically generated wave as the forcing is increased (see section 7).

5.4 Neutral modes at zero wavenumber. Systems with Galilean invariance

Symmetry properties or conservation laws often imply the existence of neutral modes at zero wavenumber. Consider the following model of one-dimensional stationary cellular instability

$$\frac{\partial u}{\partial t} = \frac{\partial^2 u}{\partial x^2} + r\frac{\partial^4 u}{\partial x^4} + \frac{\partial^6 u}{\partial x^6} - u\frac{\partial u}{\partial x}. \tag{5.40}$$

The growth-rate $\sigma(k)$ for a perturbation of wavenumber k around the $u = 0$ solution reads

$$\sigma(k) = -k^2(1 - rk^2 + k^4)$$

and is displayed in Fig. 5.2.

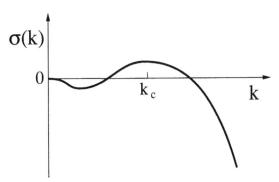

Fig. 5.2.   Growth-rate versus wave number for the model (5.40).

For $r \simeq r_c = 2$, the nul state undergoes a stationary instability to a cellular structure of wavenumber k_c. The local behavior of the growth-rate around k_c is thus similar to that of the Swift–Hohenberg model (5.13). However, an important difference is that (5.40) can be written in a conservative form

$$\frac{\partial u}{\partial t} = \frac{\partial}{\partial x}\left(\frac{\partial u}{\partial x} + r\frac{\partial^3 u}{\partial x^3} + \frac{\partial^5 u}{\partial x^5} - \frac{1}{2}u^2\right). \qquad (5.41)$$

If u is considered as a velocity field along the x-axis, this traces back to the Galilean invariance of (5.40), i.e., the invariance under the transformation

$$x \to x - vt, \quad u \to u + v.$$

This implies the existence of marginal modes at zero wavenumber, as shown by the dispersion relation of Fig. 5.2. In the vicinity of the instability onset, the amplitude of these slow modes should be taken into account, since it is generally coupled to the amplitude of the critical modes at k_c. Thus, we write

$$u(x,t) = \epsilon\,[A(X,T)\,\exp(ix) + \text{c.c.} + B(X,T)] + \dots,$$

and get coupled evolution equations for A and B,

$$\frac{\partial A}{\partial T} = \mu A + 4\frac{\partial^2 A}{\partial X^2} - \frac{1}{36}|A|^2 A - iAB, \qquad (5.42)$$

$$\frac{\partial B}{\partial T} = \frac{\partial^2 B}{\partial X^2} - \frac{\partial}{\partial X}|A|^2. \qquad (5.43)$$

Note that all the terms in (5.42) and (5.43) cannot be obtained at the same order of an asymptotic expansion, and one should again use two different time scales. If u is considered as a velocity field, B is a large-scale flow that advects the cellular pattern, thus shifting its phase through the term iAB in (5.42); in turn, the amplitude inhomogeneities of the cellular pattern generate the large-scale flow B in (5.43). Note that u is a "compressible velocity field," but a similar effect occurs with Boussinesq convection rolls when two-dimensional perturbations are taken into account. The important point to remember is that the amplitude of neutral modes couples to that of the critical modes, and modifies the dynamics described by the Ginzburg–Landau equation. This is the case of vertical vorticity modes in Rayleigh–Bénard convection with stress-free boundary conditions (Zippelius and Siggia, 1982).

5.5 Conserved order parameter

Another class of stationary cellular instabilities where additional symmetries modify the form of the evolution equation for the amplitude of the critical modes, is when the amplitude obeys a conservation law. This situation is similar to a phase transition with a conserved order parameter, spinodal decomposition for instance (Langer, 1975). Let us consider again the example of Rayleigh–Bénard convection, but this time with upper and lower perfectly insulating boundaries. The Boussinesq equations are

$$\nabla \cdot \mathbf{v} = 0 \tag{5.44}$$

$$\frac{\partial \mathbf{v}}{\partial t} + (\mathbf{v} \cdot \nabla)\mathbf{v} = -\nabla \pi + P\nabla^2 \mathbf{v} + RP\theta \hat{\mathbf{z}} \tag{5.45}$$

$$\frac{\partial \theta}{\partial t} + \mathbf{v} \cdot \nabla \theta = \mathbf{v} \cdot \hat{\mathbf{z}} + \nabla^2 \theta, \tag{5.46}$$

with

$$\frac{\partial \theta}{\partial z} = 0, \tag{5.47}$$

$$\mathbf{v} = 0, \tag{5.48}$$

at the boundaries, $z = 0$ and $z = 1$. We consider rigid boundaries in order to avoid coupling with a neutral mean flow at instability onset. We first consider a solution of equations (5.44–5.48),

$$\mathbf{U}_0 = [\mathbf{v}_0(\mathbf{r}, t), \theta_0(\mathbf{r}, t), \pi_0(\mathbf{r}, t)],$$

and note that

$$\mathbf{U}_\Theta = [\mathbf{v}_0, \theta_0 + \Theta(t), \pi_0 + RP\Theta(t)z],$$

is also a solution if

$$\frac{\partial \Theta}{\partial t} = 0.$$

In other words, $D\mathbf{U} = (0, 1, RPz)$ is an eigenmode with a zero eigenvalue. Thus, there is a neutral mode at zero wavenumber, that traces back to the existence of the above continuous family of solutions. To investigate the stability of the static state, $\mathbf{U} = 0$, we consider

$$\mathbf{U} = \Theta(X, Y, T) D\mathbf{U} + \dots, \tag{5.49}$$

where $\Theta(X, Y, T)$ is slowly varying in space and time. When R is small, the space-dependent buoyancy in equation (5.45) is small and generates a small velocity field; thus, the coupling with velocity in equation (5.46) is weak, and Θ is damped because of thermal diffusion. The growth-rate is negative and the static state is stable in the long wavelength limit. However, when R increases above a critical value R_c, the effective diffusivity in the equation for Θ changes sign due to the coupling with the velocity field that enhances the temperature perturbation Θ; we get a stationary instability at zero wavenumber (see Fig. 5.3).

The evolution equation for $\Theta(X, Y, T)$ can be obtained with a multiple-scale expansion (Chapman and Proctor, 1980). We wish to point out here that it should have the form of a conservation equation. Indeed, using the boundary conditions, the vertical average of (5.46) reads

$$\frac{\partial}{\partial t} \int_0^1 \theta \, dz + \nabla_{\mathrm{h}} \cdot \left[\int_0^1 [(\theta - z)\mathbf{v}_{\mathrm{h}} - \nabla_{\mathrm{h}}\theta] \right] = 0 . \qquad (5.50)$$

In the vicinity of the instability onset, $R - R_{\mathrm{c}} = \epsilon^2 R_{\mathrm{c}} r$, the evolution equation for Θ is at leading order

$$\frac{\partial \Theta}{\partial T} = -r\nabla_{\mathrm{h}}^2 \Theta - \kappa\nabla_{\mathrm{h}}^4 \Theta + \gamma\nabla_{\mathrm{h}} \cdot \left[(\nabla_{\mathrm{h}}\Theta)^2 \nabla_{\mathrm{h}}\Theta \right] . \qquad (5.51)$$

The absence of terms that involve explicitly Θ is related to the freedom of constant shift in temperature that we have with the Boussinesq equations with insulating boundary conditions. The even number of space-derivatives is due to x and y-reflection symmetries in the horizontal plane. The absence of quadratic nonlinearities traces back to the Boussinesq symmetry. We will find a lot of similar examples of stationary instability at zero wavenumber, when studying phase instabilities of cellular patterns (see section 6).

5.6 Conservative systems and dispersive instabilities

Another additional symmetry that affects the form of the amplitude equation is the time reversal symmetry of conservative systems. A well-known example is that of two inviscid fluid layers of different densities, possibly moving at different velocities, in which Kelvin–Helmholtz or Rayleigh–Taylor instabilities can occur.

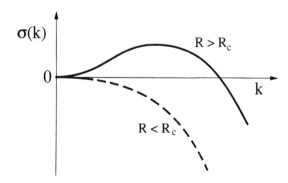

Fig. 5.3.   Dispersion relation for fixed-flux convection.

We restrict our attention to the Kelvin–Helmholtz instability. A layer of density ρ' moves at velocity U' above a layer of density ρ, $(\rho > \rho')$, moving at velocity U. The stability analysis is governed by the following dispersion relation (Chandrasekhar, 1961)

$$\frac{\eta}{k} = -i\frac{\rho U + \rho' U'}{\rho + \rho'} \pm i \left[\frac{g}{k}\frac{\rho - \rho'}{\rho + \rho'} - \frac{\rho\rho'}{(\rho + \rho')^2}(U - U')^2 + \frac{T}{\rho + \rho'}k\right]^{\frac{1}{2}} \quad (5.52)$$

where k is the wavenumber and T is the surface tension. Instability occurs when the term in the square brackets is negative, a result of the shear being large enough. The first term in the square brackets shows that the gravitational restoring force stabilizes low wavenumbers, whereas the last term shows that surface tension stabilizes high wavenumbers. The net effect is that in the presence of both a density step and surface tension, there is a critical velocity difference $(\Delta U^2)_c$ for the onset of instability, at which a critical wavenumber k_c becomes unstable first; at slightly higher values of ΔU^2 there is a narrow band of unstable wavenumbers. This looks very similar to Rayleigh–Bénard convection, although the growth-rate for the Kelvin–Helmholtz instability has an imaginary part that corresponds to the frequency of the unstable waves, but the main difference is related to the conservative nature of the present problem. At the onset of the Kelvin–Helmholtz instability, four pure imaginary eigenvalues collide in pairs and give rise to four complex eigenvalues, $(\pm\eta, \pm\bar{\eta})$, whereas for an instability in a dissipative system, a real eigenvalue, or pairs of complex-conjugate eigenvalues cross the imaginary axis. The former situation is a dispersive instability, and is related to time-reversal symmetry; indeed, this implies that, if η is an eigenvalue, then $-\eta$ is another eigenvalue. As the original problem involves real quantities, $\pm\bar{\eta}$ are also eigenvalues, and a conservative system involves either pure imaginary eigenvalues (stable range), or complex eigenvalues $(\pm\eta, \pm\bar{\eta})$ (unstable range).

In the unstable range, the amplitude equation is also modified by the additional time-reversal symmetry constraint. Let us consider for instance the following conservative model of Swift–Hohenberg type:

$$\frac{\partial^2 u}{\partial t^2} = \left[\mu - \left(1 + \frac{\partial^2}{\partial x^2}\right)^2\right]u + u\frac{\partial u}{\partial x} \quad (5.53)$$

When $\mu \simeq 0$, the system is unstable about $k_c = 1$ and the linear growth-rate around k_c is

$$\eta^2 \simeq \mu - \alpha(k - k_c)^2.$$

Using the scalings

$$\frac{d}{dx} = \frac{\partial}{\partial x} + \sqrt{\mu}\frac{\partial}{\partial X}$$

$$\frac{\partial}{\partial t} = \sqrt{\mu}\frac{\partial}{\partial T},$$

shows that the amplitude equation is

$$\frac{\partial^2 A}{\partial T^2} = \mu A + \alpha\frac{\partial^2 A}{\partial X^2} - \beta|A|^2 A.$$

This is a nonlinear Klein–Gordon equation.

6 Secondary instabilities of cellular flows: Eckhaus and zigzag instabilities

6.1 Broken symmetries and neutral modes

We have so far considered the onset of cellular structures as a control parameter is changed. We now consider the instabilities of such structures which occur as the parameter is changed further. We call these *secondary instabilities* of the system. In many cases, a secondary instability arises from a *neutral mode* associated with a symmetry of the governing equations broken by the primary instability.

For example, consider convection in a container of infinite extent in the horizontal plane, or with periodic boundary conditions. The onset of convection breaks translational symmetry in the direction perpendicular to the rolls, the x-axis, say. However, one can imagine pushing the rolls along the x-axis without any expenditure of energy, since this only amounts to a shift of the x-axis origin. This translation is a neutral mode. In other words, one roll-solution breaks translational invariance, but the ensemble of all the possible roll-solutions should be invariant according to the Curie principle. Thus, as we noticed earlier, the phase of the periodic structure above the instability onset is arbitrary. Changing the phase amounts to moving along the orbit of all the possible roll-solutions. This "motion" does not require any energy and is a neutral mode of the periodic structure, i.e., has a zero growth-rate.

Let us illustrate this concept with Swift–Hohenberg-type models. We first consider

$$\frac{\partial u}{\partial t} = \left[\mu - \left(1 + \frac{\partial^2}{\partial x^2}\right)^2\right]u - u^3 \tag{6.1}$$

Suppose there exists a periodic solution $u_0(x)$ to the full nonlinear equation (6.1), that is,

$$L \cdot u_0 = u_0{}^3,$$

where $L \equiv [\mu - (1 + \partial^2/\partial x^2)^2]$. Taking derivatives yields

$$L \cdot \left(\frac{du_0}{dx}\right) = 3u_0{}^2 \frac{du_0}{dx} . \tag{6.2}$$

Now, we write

$$u = u_0(x) + \epsilon\, v(x, t)$$

in order to investigate the linear stability of u_0, and we get from equation (6.1)

$$\frac{\partial v}{\partial t} = L \cdot v - 3u_0{}^2\, v + \mathcal{O}(\epsilon^2) \tag{6.2'}$$

Comparison of equation (6.2) with equation (6.2′) shows that

$$v = \frac{du_0}{dx}$$

is an eigenmode with a zero eigenvalue. A perturbation v proportional to this eigenmode corresponds to a translation $u_0(x + \epsilon)$. This can be shown differently by looking for a solution in the form

$$u(x, t) = u_0\,[x + \phi(t)] \tag{6.3}$$

where u_0 is our steady-state solution, and $\phi(t)$ is a time dependent phase. Substituting this into equation (6.1) yields

$$u_0' \frac{d\phi}{dt} = L \cdot u_0 - u_0{}^3$$

where u_0' is the derivative of u_0 with respect to its argument. We get

$$\frac{d\phi}{dt} = 0 , \tag{6.4}$$

which confirms that the translational perturbation is a neutral mode of equation (6.1).

Translational invariance is of course not the only possible broken-symmetry at the onset of a pattern-forming instability. Consider for instance the model (5.40) of section 5

$$\frac{\partial u}{\partial t} = \frac{\partial^2 u}{\partial x^2} + r\frac{\partial^4 u}{\partial x^4} + \frac{\partial^6 u}{\partial x^6} - u\frac{\partial u}{\partial x} , \tag{6.5}$$

which is Galilean invariant, i.e., invariant under the transformation

$$x \to x - ct , \qquad u \to u + c$$

We consider a periodic solution $u_0(x)$ of equation (6.5) and look for a perturbation in the form

$$u = u_0[x + \phi(t)] - \psi(t). \tag{6.6}$$

We get from (6.5)

$$\frac{d\phi}{dt} u_0' - \frac{d\psi}{dt} - \psi u_0' = 0,$$

thus,

$$\frac{d\phi}{dt} = \psi$$

$$\frac{d\psi}{dt} = 0.$$

(6.7)

Two neutral modes are now involved, which trace back to the broken translational and Galilean invariances. The form of the coupling between ϕ and ψ can be understood as follows: if one pushes the pattern $u_0(x)$ at a constant velocity ψ along the x-axis, one observes a spatial phase ϕ that increases linearly in time.

6.2 Phase dynamics

We could have imposed perturbations of the form (6.3) but with the phase ϕ slowly varying both in time and space, $\phi(X, T)$, where $X = \epsilon x$ and $T = \epsilon^\tau t$ are slow variables (for a diffusive behavior for instance, $\tau = 2$). More generally, a long wavelength perturbation of the perfectly periodic pattern $u_0(x)$ can be written in the form

$$u(x, t) = u_0 [x + \phi(X, T)] + u_\perp (X, T),$$

(6.8)

where

$$u_0 [x + \phi(X, T)] - u_0 \simeq \phi(X, T) \frac{du_0}{dx}$$

represents perturbations along the orbit of the symmetry group corresponding to translational invariance, i.e., phase perturbations, and $u_\perp$ consists of perturbations transverse to the group orbit, i.e., amplitude perturbations. A constant phase perturbation is neutral, and correspondingly weakly damped in the long wavelength limit, whereas amplitude perturbations decay on a faster time scale when the perfectly periodic pattern is stable. Adiabatic elimination of the amplitude modes leads to an evolution equation for the phase, that can be derived assuming that an expansion in powers of the gradients of the phase is valid,

$$\frac{\partial \phi}{\partial T} = \mathcal{F} \left[\frac{\partial \phi}{\partial X}, \frac{\partial^2 \phi}{\partial X^2}, \cdots \right].$$

(6.9)

The linear part of this equation is the Fourier–Laplace transform of the dispersion relation for the growth-rate of the phase modes. The growth-rate is zero at zero wavenumber due to translational invariance. If it is negative in the long wavenumber limit, the phase perturbation is damped;

however, due to the coupling with amplitude modes, the growth-rate may be positive in the long wavelength limit, thus describing a phase instability of the perfectly periodic pattern (Pomeau and Manneville, 1979; Kuramoto, 1984).

In the case of model (6.5), long wavelength perturbations should be considered in the form

$$u(x,t) = u_0 \left[x + \phi(X,T) \right] - \psi(X,T) + u_\perp(X,T). \qquad (6.10)$$

This leads to phase equations of the form

$$\frac{\partial \phi}{\partial T} = \psi + \mathcal{F} \left[\frac{\partial \phi}{\partial X}, \frac{\partial^2 \phi}{\partial X^2}, \cdots \right]$$

$$\frac{\partial \psi}{\partial T} = \mathcal{G} \left[\frac{\partial \phi}{\partial X}, \frac{\partial^2 \phi}{\partial X^2}, \cdots \right]$$

$$(6.11)$$

Thus the phase dynamics is second order in time, and propagative modes may result from the coupling between the phases associated with broken translational and Galilean invariances (Coullet and Fauve, 1985).

Before we pursue phase dynamics further, it is worth pointing out that it is only useful for secondary instabilities with wavenumbers much smaller than those of the primary instability. In other words, we can only investigate variations of the underlying pattern which are much larger in scale than the basic pattern wavelength. Furthermore, we had earlier said that amplitude and phase ought to vary together, but here we have considered phase variations in isolation. This turns out to be reasonable when the secondary instability is well separated from the onset of the primary instability. In this case, the amplitude follows adiabatically the phase gradients, so no additional equation for the amplitude is needed. However, in the vicinity of the primary instability, coefficients in the phase equation diverge, and we must use the full amplitude equation instead.

Other symmetries besides translation and Galilean invariance can be found in the basic equations, but all symmetries do not, in general, lead to slowly-varying local dynamics, and so are not amenable to the method of phase dynamics. For example, rotational and dilatational invariances, regardless how small the rate of transformation, both result in arbitrarily large effects at sites sufficiently distant from the center, and do not lead, in general, to additional phase modes.

6.3 Eckhaus instability

6.3.1 Compression mode of a one-dimensional pattern

We considered stationary cellular instability earlier, and derived the amplitude equation, that reads in the supercritical case with appropriate amplitude and x scales,

$$\frac{\partial A}{\partial T} = \mu A + \frac{\partial^2 A}{\partial X^2} - |A|^2 A. \tag{6.12}$$

Equation (6.12) has stationary solutions

$$A_q = Q \exp(iqX) \qquad \text{where} \qquad Q^2 = \mu - q^2, \tag{6.13}$$

that represent all the possible roll-solutions which correspond to the unstable band of modes above the instability onset (Fig. 6.1).

We investigate the stability of these stationary patterns by perturbing their amplitude and phase, thus writing

$$A = [Q + r(X,T)] \exp i[qX + \phi(X,T)] \tag{6.14}$$

in equation (6.12). Expanding and separating real and imaginary parts yields

$$\frac{\partial r}{\partial T} = \mu[Q + r] + \frac{\partial^2 r}{\partial X^2} - [Q + r]\left[q + \frac{\partial \phi}{\partial X}\right]^2 - [Q + r]^3$$

$$[Q + r]\frac{\partial \phi}{\partial T} = 2\frac{\partial r}{\partial X}\left[q + \frac{\partial \phi}{\partial X}\right] + [Q + r]\frac{\partial^2 \phi}{\partial X^2}$$

If we linearize in r and ϕ, and use $Q^2 = \mu - q^2$, we obtain

$$\begin{aligned}
\frac{\partial r}{\partial T} &\simeq -2Q^2 r + \frac{\partial^2 r}{\partial X^2} - 2Qq\frac{\partial \phi}{\partial X} \\
\frac{\partial \phi}{\partial T} &\simeq 2\frac{q}{Q}\frac{\partial r}{\partial X} + \frac{\partial^2 \phi}{\partial X^2}
\end{aligned} \tag{6.15}$$

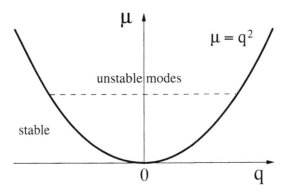

Fig. 6.1. Cellular instability: cellular patterns only exist for parameters within the parabola.

We consider modes proportional to $\exp(\eta t + iK X)$ and get from equation (6.15) the dispersion relation

$$\begin{vmatrix} \eta + 2Q^2 + K^2 & 2iQqK \\ -2(q/Q)iK & \eta + K^2 \end{vmatrix} = 0,$$

which has solutions

$$\eta(K) = -(Q^2 + K^2) \pm \sqrt{Q^4 + 4K^2 q^2}. \qquad (6.16)$$

Thus, for small K, we have two different branches:

a) the "amplitude modes," $\eta_-(K) = -2Q^2 + \mathcal{O}(K^2)$, that are damped,
b) the "phase modes," $\eta_+(K) = -K^2(1 - 2q^2/Q^2) + \mathcal{O}(K^4)$, that are nearly marginal.

The phase modes

$$\eta_+(K) = - \left(\frac{\mu - 3q^2}{\mu - q^2} \right) K^2 - 2\frac{q^4}{(\mu - q^2)^3} K^4 + \mathcal{O}(K^6) \qquad (6.17)$$

lead to an instability when

$$D_\| = \left(\frac{\mu - 3q^2}{\mu - q^2} \right) < 0$$

so when $q^2 < \mu < 3q^2$; this is the Eckhaus instability. We can thus nest this secondary instability curve in the earlier stability diagram (Fig. 6.2).

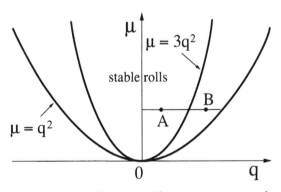

Fig. 6.2. Eckhaus instability diagram. The pattern generating primary instability occurs inside the outer parabola, but the resulting pattern is stable only within the inner parabola.

6.3.2 Nonlinear phase equation for the Eckhaus instability

We now derive the phase equation of the Eckhaus instability in order to describe the dynamics of the pattern through the evolution of its slowly varying phase. As said above, this is possible in the long wavelength limit because we have two types of perturbations with different time scales (see Fig. 6.3); adiabatic elimination of the amplitude modes leads to an evolution equation for the phase.

For a better understanding, let us consider again the linear equations (6.15). For spatially homogeneous perturbations, as already observed, the amplitude perturbation r decays exponentially whereas the phase perturbation ϕ is neutral. For perturbations slowly varying in space, r is non-zero only because it is forced by phase-gradients; thus, the amplitude follows adiabatically the phase-gradients, and the dominant balance in the evolution equation for r in (6.15) is

$$2Q^2 r \approx -2Qq\frac{\partial\phi}{\partial X}.$$

Substituting this into the phase equation of (6.15) we obtain

$$\frac{\partial\phi}{\partial T} \approx \left(1 - \frac{2q^2}{Q^2}\right)\frac{\partial^2\phi}{\partial X^2} \equiv D_{\|}\frac{\partial^2\phi}{\partial X^2}.$$

We see that the Eckhaus instability shows up as a negative diffusivity in the phase equation; the small-scale flow, generated by the primary instability, acts as a negative diffusivity for large scale perturbations, when the primary pattern wavenumber is far enough from the critical one.

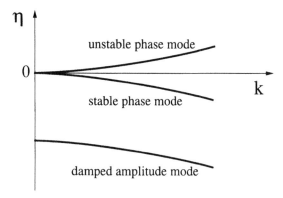

Fig. 6.3.   Amplitude and phase perturbations have different time scales in the long wavelength limit.

The higher order linear terms of the phase equation can be obtained from higher order balances in (6.15), or in a simpler way from the Fourier–Laplace transform of the small K expansion of the dispersion relation (6.17),

$$\frac{\partial \phi}{\partial T} = D_{\|} \frac{\partial^2 \phi}{\partial X^2} - \kappa \frac{\partial^4 \phi}{\partial X^4},$$

where $\kappa = 2q^4/Q^6$. In the vicinity of the Eckhaus instability onset, $D_{\|} \simeq 0$, then $\kappa \simeq 3/4\mu$.

We next derive the leading order nonlinear term of the phase equation. Let us first see from symmetry arguments what form the nonlinear term might have. Translational invariance, $x \rightarrow x + x_0$, implies that the phase equation should be invariant under $\phi \rightarrow \phi + \phi_0$. This precludes any terms explicitly depending on ϕ such as ϕ^2 or $\phi\phi_X$. Reflection invariance $(x \rightarrow -x)$ implies $\phi \rightarrow -\phi$, eliminating terms such as $\phi_X\phi_X$. The lowest order term possible is therefore $\phi_X\phi_{XX}$. The weakly nonlinear phase equation would then be (if the coefficient of this nonlinear term does not vanish)

$$\frac{\partial \phi}{\partial T} = D_{\|} \frac{\partial^2 \phi}{\partial X^2} - \kappa \frac{\partial^4 \phi}{\partial X^4} + g \frac{\partial \phi}{\partial X} \frac{\partial^2 \phi}{\partial X^2}. \tag{6.18}$$

Factoring the diffusive term, $(D_{\|} + g\phi_X)\phi_{XX}$, we see that the nonlinearity acts as an effective space-dependent diffusivity; however, we will see that it cannot compensate the negative diffusivity in the unstable regime, and does not saturate the Eckhaus instability which is thus a subcritical one whatever the sign of g. To find the coefficient g one can proceed formally with a multiple-scale expansion from the nonlinear equations for r and ϕ. There is, however, a much simpler procedure (Kuramoto, 1984): we first notice that $\phi = pX$ is a particular solution of (6.18), that simply represents a homogeneous roll-solution of wavenumber $k_c + \epsilon(q + p)$; its linear stability is thus governed by the dispersion relation

$$\eta = -D_{\|}(q + p)K^2 + \mathcal{O}(K^4).$$

We can also compute η by linearization of (6.18) near $\phi = pX$; we get

$$\eta = -\left[D_{\|}(q) + gp \right] K^2 + \mathcal{O}(K^4).$$

Identifying the two expressions to leading order in p, we obtain

$$g = \frac{\partial D_{\|}}{\partial q}.$$

We see that, at the Eckhaus instability onset $(D_{\|} = 0)$, all the coefficients of the higher order terms of the phase equation (6.18) diverge as $\mu \rightarrow 0$, showing that, as already mentioned, the phase approximation becomes invalid, because near the primary cellular instability the amplitude mode also becomes neutral and thus cannot be eliminated. The full amplitude equation (6.12) should be used to capture the correct behavior. However, one expects that for long wavelength perturbations of rolls, the form of the phase equation (6.18) remains valid along the Eckhaus instability curve, even for rather large values of $r - r_c$ out of the range of validity of the amplitude equation.

Let us now analyze the behavior of the phase equation (6.18) in the vicinity of the Eckhaus instability onset. If we look at the linear dispersion relation for a mode $\exp(\eta T + iKX)$ we find (see Fig. 6.4) that when $D_{\parallel} < 0$ there is a band of unstable modes for $0 < K < K_*$.

We will focus on the neutral mode at $K = K_*$ since it is the first to go unstable in an experiment with periodic boundary conditions (the first unstable mode is such that $K_* = 2\pi/L$, where L is the size of the periodic domain). The finite geometry thus delays the instability until $D_{\parallel} = -|D_*|$ when the mode K_* becomes neutral. The pattern is then unstable to the compression mode of wavenumber K_*; let us now study the finite amplitude behavior of this mode.

We obtain this amplitude equation as usual, by inserting a mode

$$\phi = a(T) \exp(iK_*X) + \text{c.c.}$$

The resulting equation for $a(T)$ is

$$\frac{da}{dT} = -(D_{\parallel} - D_*)K_*^2 a + 2\frac{\mu}{9}g^2K_*^2|a|^2a\,.$$

We see that there is no stabilization from the nonlinear term (since its coefficient is positive). This is shown on the bifurcation diagram (Fig. 6.5). Thus linearly stable rolls ($D_{\parallel} > D_*$) can be nonlinearly unstable to finite

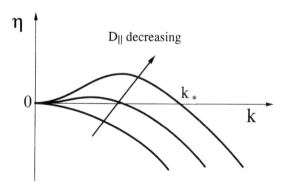

Fig. 6.4. Dispersion relation for the phase equation.

amplitude perturbations, a localized compression or dilatation of the pattern for instance. We will evaluate the critical size of the perturbation that generates the finite amplitude Eckhaus instability; this is analogous to a nucleation energy in first order phase transitions (Kramer and Zimmermann, 1985).

We look first for stationary solutions of the phase equation (6.18). Setting $\partial\phi/\partial T = 0$, we obtain by integrating (6.18)

$$\lambda = D_{\parallel}\psi - \kappa\frac{\partial^2\psi}{\partial X^2} + \frac{g}{2}\psi^2 \tag{6.19}$$

where ψ is the phase gradient,

$$\psi = \frac{\partial\phi}{\partial X},$$

i.e., the variation of the local wavenumber, and λ is a constant of integration related to the wavenumber at infinity. We can recast the problem as the motion of a particle of mass κ in a potential $U(\psi)$ such that

$$\kappa\frac{\partial^2\psi}{\partial X^2} = -\frac{dU}{d\psi}$$

$$U(\psi) = \lambda\psi - \frac{D_{\parallel}}{2}\psi^2 - \frac{g}{6}\psi^3. \tag{6.20}$$

The potential U can be simplified by eliminating the quadratic term with $\tilde\psi = \psi + D_{\parallel}/g$ to obtain

$$U(\tilde\psi) - U_0 = \left(\lambda + \frac{D_{\parallel}^2}{2g}\right)\tilde\psi - \frac{g}{6}\tilde\psi^3,$$

where U_0 is a constant. This potential is indicated in Fig. 6.6. The solutions are graphically obvious but it should be recalled that these correspond to the X dependence of steady solutions. Nothing has been said as

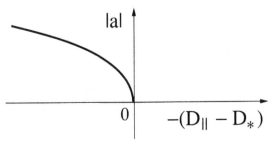

Fig. 6.5.   Bifurcation diagram of Eckhaus instability.

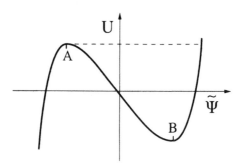

Fig. 6.6.   Potential well of stationary solutions of the phase equation.

yet about their stability. We have two stationary solutions homogeneous in space (points A and B in Fig. 6.6), and oscillatory solutions around B that represent patterns with a periodically modulated wavenumber. One solution of particular interest is the limiting solitary wave solution, where one imagines the particle traveling from the local maximum to the left and taking an infinite time to return. This motion is shown in Fig. 6.7 and we see that it corresponds to a localized compression of the rolls.

We now try to find a Lyapunov functional in order to study the stability of the above stationary solutions. To wit, we multiply equation (6.18) by $\partial\phi/\partial T$ and integrate over a wavelength in X. Integrating by parts we obtain

$$\int_0^L \left(\frac{\partial\phi}{\partial T}\right)^2 dX =$$

$$\int_0^L \left[-D_\parallel \frac{\partial\phi}{\partial X}\frac{\partial^2\phi}{\partial X\partial T} - \kappa \frac{\partial^3\phi}{\partial X^2\partial T}\frac{\partial^2\phi}{\partial X^2} - \frac{g}{2}\left(\frac{\partial\phi}{\partial X}\right)^2\frac{\partial^2\phi}{\partial X\partial T}\right] dX.$$

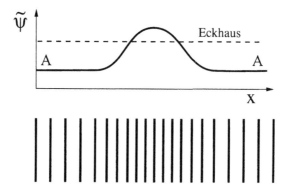

Fig. 6.7.   Solitary compression wave in the roll pattern.

Thus,

$$\frac{d\mathcal{F}}{dT} = -\int_0^L \left(\frac{\partial\phi}{\partial T}\right)^2 dX \le 0,$$

where

$$\mathcal{F} \equiv \int_0^L \left[\frac{D_\parallel}{2}\left(\frac{\partial\phi}{\partial X}\right)^2 + \frac{\kappa}{2}\left(\frac{\partial^2\phi}{\partial X^2}\right)^2 + \frac{g}{6}\left(\frac{\partial\phi}{\partial X}\right)^3\right] dX.$$

$\mathcal{F}[\phi(X,T)]$ is decreasing during the phase dynamics but is not a proper Lyapunov functional because it is not bounded from below. Moreover, we should take into account an additional constraint imposed by the conservation form of the phase equation (6.18) or of the equation for the phase gradient ψ

$$\frac{\partial\psi}{\partial T} = \frac{\partial^2}{\partial X^2}\left[D_\parallel\psi - \kappa\frac{\partial^2\psi}{\partial X^2} + \frac{g}{2}\psi^2\right]. \tag{6.21}$$

This imposes the constraint

$$\frac{d}{dT}\int \psi dX = 0 \tag{6.22}$$

on the evolution, i.e., the conservation of the mean wavenumber. Therefore, the form of the phase equation shows that a pattern cannot evolve by continuously changing its wavelength.

When $\mathcal{F}[\phi(X,T)]$ has a local minimum, we can study the dynamics by minimizing this functional where the above constraint enters with its Lagrange multiplier λ,

$$\mathcal{F}_\lambda = \int_0^L \left[\frac{D_\parallel}{2}\psi^2 + \frac{\kappa}{2}\left(\frac{\partial\psi}{\partial X}\right)^2 + \frac{g}{6}\psi^3 - \lambda\psi\right] dX.$$

Thus,

$$\mathcal{F}_\lambda = \int_0^L \left[\frac{\kappa}{2}\left(\frac{\partial\psi}{\partial X}\right)^2 - U(\psi)\right] dX, \tag{6.23}$$

and, to minimize $\mathcal{F}_\lambda$, we should maximize U. This means that the local maximum of U in Fig. 6.6 is metastable whereas the local minimum is unstable. Indeed, points A and B in Fig. 6.6 correspond to the same points on the Eckhaus stability diagram (see Fig. 6.2). Note that the oscillatory solutions about B are unstable. The solitary wave solution represents the "critical nucleus," i.e., the critical localized perturbation to the linearly stable pattern represented by A, that generates the finite amplitude Eckhaus instability. The corresponding value of $\mathcal{F}_\lambda$ minus its evaluation for the homogeneous stable pattern, represents the "energy

barrier" of a pattern changing process from the homogeneous pattern A;
this barrier vanishes at the linear Eckhaus instability onset. These points
will be further covered next by studying the amplitude equation (6.12)
directly.

6.3.3 Localized solutions of the one-dimensional Ginzburg–Landau equation

We now investigate in more detail the amplitude equation (6.12). We
first rescale the time scale so that (6.12) can be written as

$$A_T = \pm A + \frac{\partial^2 A}{\partial X^2} - |A|^2 A, \qquad (6.24)$$

where the plus or minus sign corresponds to the sign of μ. As already
mentioned in section 5, (6.24) is a variational problem, i.e.,

$$\frac{d}{dT} \mathcal{L}[A] = -2 \int_0^L \left| \frac{\partial A}{\partial T} \right|^2 dX \leq 0,$$

$$\mathcal{L}[A] = \int_0^L \left[\left| \frac{\partial A}{\partial X} \right|^2 \mp |A|^2 + \frac{1}{2} |A|^4 \right] dX. \qquad (6.25)$$

The Lyapunov functional $\mathcal{L}$ is a decreasing function and has a minimum
value for the homogeneous solution ($A_X = 0$) that maximizes $U(A) =
\pm A^2 - \frac{1}{2} A^4$. For $\mu > 0$, the stationary solutions of (6.24) are

$$A = Q \exp(iqX), \qquad \text{with} \qquad Q^2 = 1 - q^2, \qquad (6.26)$$

and the Lyapunov functional (6.25) may be rewritten as

$$\mathcal{L} = \int_0^L l(q) dX,$$

where

$$l(q) = -\frac{1}{2} (q^2 - 1)^2.$$

We check that the periodic pattern with the critical wavenumber k_c, i.e.,
$q = 0$, corresponds to the absolute minimum of $l(q)$. The inflection points
of $l(q)$ correspond to $l''(q) = -2(3q^2 - 1) = 0$, which is the Eckhaus
instability limit $q_E^2 = 1/3$. In solid state physics $l(q)$ is the free energy
density and $l''(q) < 0$ corresponds to negative compressibility. Thus,
the Eckhaus instability corresponds to a negative compressibility of the
periodic pattern (Fig. 6.8).

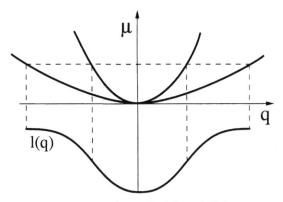

Fig. 6.8.   Plots of $\mu(q)$ and $l(q)$.

We now investigate the wavenumber changing process. Substituting $A = Re^{i\phi}$ into the amplitude equation (6.24) and equating real and imaginary parts we get

$$\frac{\partial R}{\partial T} = R - R^3 + \frac{\partial^2 R}{\partial X^2} - R\left(\frac{\partial \phi}{\partial X}\right)^2,$$

$$R\frac{\partial \phi}{\partial T} = 2\frac{\partial R}{\partial X}\frac{\partial \phi}{\partial X} + R\frac{\partial^2 \phi}{\partial X^2}.$$

(6.27)

For stationary solutions the second equation gives

$$R^2 \frac{\partial \phi}{\partial X} = h,$$

(6.28)

where h is a constant. This conservation of "angular momentum" arises from the rotational invariance of the amplitude equation. The constraint (6.28) is the reason why the pattern-wavenumber changing process cannot occur by homogeneously modifying the wavenumber at constant amplitude.

Substituting (6.28) into (6.27), we get for stationary solutions

$$\frac{\partial^2 R}{\partial X^2} = -R + R^3 + h^2/R^3,$$

which is of the form

$$\frac{\partial^2 R}{\partial X^2} = -\frac{\partial V}{\partial R},$$

$$V(R) = \frac{1}{2}R^2 - \frac{1}{4}R^4 + \frac{h^2}{2R^2},$$

(6.29)

thus corresponding again to the motion of a particle in the potential $V(R)$ (Fig. 6.9, top). Bounded solutions exist for $h^2 < 4/27$; the two extrema

$R_\pm$ of V correspond to homogeneous stationary solutions, such that

$$h^2 = R_\pm^4 (1 - R_\pm^2).$$

Their stability can be determined using the Lyapunov functional (6.25); R_+ is the stable solution corresponding to the point A in Figs. 6.2 and 6.6, and R_- is the unstable one corresponding to B.

Figure 6.9 (bottom) shows the phase space of the stationary solutions of the amplitude equation (6.24). The homoclinic orbit passing through R_+ corresponds to the solitary wave solution that describes a localized compression (or dilatation) of the pattern, already found within the phase equation approach. The constraint (6.28) shows that the amplitude R decreases in these regions. When the localized compression (or dilatation) of the pattern is too large, the amplitude vanishes and this allows the annihilation (or creation) of a pair of rolls; this wavenumber changing mechanism involves length-scales comparable to the wavelength of the primary pattern, and is out of the range of validity of the amplitude equation (6.12) or of the phase equation (6.18). However, we were able

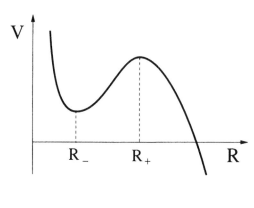

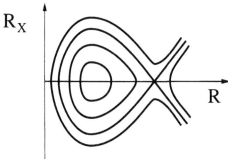

Fig. 6.9. Top: Potential well $V(R)$ for stationary solutions of the amplitude equation (6.24). Bottom: Corresponding phase space.

to obtain the qualitative behavior of the Eckhaus instability from these long-wavelength approximations.

6.4 The zigzag instability

6.4.1 Torsion mode of a one-dimensional pattern

Like the Eckhaus instability, the zigzag instability, shown in Fig. 6.10 is associated with the broken translational invariance which a stationary roll-solution exhibits. In this case however the perturbation to the rolls is a transverse one, and can be written in the long-wavelength limit,

$$u(x, y, t) = u_0 \left[x + \phi(Y, T) \right] + u_\perp(X, Y, T).\tag{6.30}$$

To investigate the dynamics of this torsion mode in the framework of the amplitude equation, we should use

$$\frac{\partial A}{\partial T} = \mu A + \left(\frac{\partial}{\partial X} - \frac{i}{2k_c} \frac{\partial^2}{\partial Y^2} \right)^2 A - |A|^2 A,\tag{6.31}$$

obtained in section 5.

Figure 6.11 shows that rotation of the rolls leads to a decrease in the wavelength of the pattern, and hence an increase in the wavenumber. The zigzag mode locally corresponds to a rotation of the pattern, so we can deduce that it will result in a local increase of the wavenumber. Referring to Fig. 6.8, we see that for $q < 0$ an increase in q corresponds to a decrease in $l(q)$, the Lyapunov functional, so in this region the pattern is unstable to zigzag perturbations. However, for $q > 0$, $l(q)$ increases as q increases, demonstrating that the pattern is stable to local rotations. The zigzag

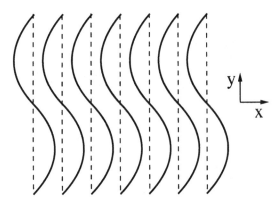

Fig. 6.10. The form of the zigzag instability: unperturbed rolls (dashed lines) and zigzag mode (solid lines).

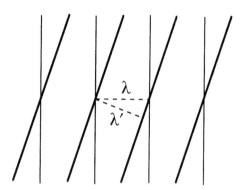

Fig. 6.11.   The effect of rotation of the roll planform.

instability thus occurs for $k < k_c$ in the framework of (6.31). Indeed, if we look for a perturbation varying slowly in y,

$$A = [Q + r(Y, T)] \exp i[qX + \phi(Y, T)],$$

we get to leading order

$$\frac{\partial \phi}{\partial T} \simeq D_\perp \frac{\partial^2 \phi}{\partial Y^2}, \qquad \text{where} \qquad D_\perp = \frac{q}{k_c}.$$

6.4.2   Nonlinear phase equation for the zigzag instability

Using symmetry considerations, translational invariance, that implies the invariance under $\phi \to \phi + \phi_0$, x-reflection symmetry, that implies the invariance under the transformation $X \to -X$, $\phi \to -\phi$, and y-reflection symmetry, that implies the invariance under the transformation $Y \to -Y$, we obtain to leading nonlinear order, the phase equation

$$\frac{\partial \phi}{\partial T} = D_\perp \frac{\partial^2 \phi}{\partial Y^2} - \kappa \frac{\partial^4 \phi}{\partial Y^4} + g \left(\frac{\partial \phi}{\partial Y}\right)^2 \frac{\partial^2 \phi}{\partial Y^2}. \qquad (6.32)$$

Note the scaling $\phi \sim \mathcal{O}(1)$, contrary to the Eckhaus instability where ϕ should be small. Rearranging (6.32) results in

$$\frac{\partial \phi}{\partial T} = \left[D_\perp + g \left(\frac{\partial \phi}{\partial Y}\right)^2\right] \frac{\partial^2 \phi}{\partial Y^2} - \kappa \frac{\partial^4 \phi}{\partial Y^4}.$$

Now, $D_\perp < 0$ for instability; if the instability is to be saturated, the effective diffusivity must be positive, and hence g must be positive. Let us compute g, using a similar method as for the Eckhaus instability: first, note that $\phi = pY$ is a particular solution of (6.32) that corresponds to a tilted roll pattern. Writing, $\phi = pY + \tilde{\phi}$, and linearizing in $\tilde{\phi}$, results in

$$\frac{\partial \tilde{\phi}}{\partial T} = (D_\perp + gp^2) \frac{\partial^2 \tilde{\phi}}{\partial Y^2} - \kappa \frac{\partial^4 \tilde{\phi}}{\partial Y^4}.$$

However, the tilted roll-solution is simply another roll pattern with a new wavenumber, q', where $q' = q + p^2/2k_c$. So we can also write

$$\frac{\partial \tilde{\phi}}{\partial T} = D_\perp(q') \frac{\partial^2 \tilde{\phi}}{\partial Y'^2} - \kappa \frac{\partial^4 \tilde{\phi}}{\partial Y'^4},$$

and matching the two previous equations gives

$$g = \frac{1}{2k_c} \frac{\partial D_\perp}{\partial q} = \frac{1}{2k_c^2}.$$

Hence $g > 0$, and the zigzag instability is supercritical in the framework of the amplitude equation (6.31).

We now use the phase equation (6.32) to study the pattern generated by the zigzag instability in the supercritical regime. If we let $\phi \propto \exp(\eta T + iKY)$ and linearize about $\phi = 0$, we obtain

$$\eta = -D_\perp K^2 - \kappa K^4.$$

This dispersion relation is similar to that of the Eckhaus instability (see Fig. 6.4); it might be thought that the mode with the maximum growth-rate, k_{max}, corresponds to the characteristic length-scale of the zigzag instability in the supercritical regime, but a nonlinear analysis will show that this is not so.

The stationary solutions of (6.32) satisfy

$$\lambda = D_\perp \psi - \kappa \frac{\partial^2 \psi}{\partial Y^2} + \frac{g}{3} \psi^3,$$

where

$$\psi = \frac{\partial \phi}{\partial Y},$$

is the phase gradient, and λ is a constant of integration related to the wavenumber at infinity. This can be written as

$$\kappa \frac{\partial^2 \psi}{\partial Y^2} = -\frac{dU}{d\psi}$$

$$U(\psi) = \lambda \psi - \frac{D_\perp}{2} \psi^2 - \frac{g}{12} \psi^4,$$

(6.33)

showing again that the determination of the stationary solutions amounts to finding the trajectories of a particle of mass κ in the potential well $U(\psi)$. We look for a Lyapunov functional in order to determine their stability; multiplying (6.32) by $\partial \phi / \partial T$ and integrating by part gives

$$\frac{d\mathcal{F}}{dT} = -\int_0^L \left(\frac{\partial \phi}{\partial T}\right)^2 dY \leq 0$$

$$\mathcal{F} = \int_0^L \left[\frac{\kappa}{2} \left(\frac{\partial \psi}{\partial Y}\right)^2 + \frac{D_\perp}{2} \psi^2 + \frac{g}{12} \psi^4\right] dY.$$

(6.34)

Thus, $\mathcal{F}[\psi(Y, T)]$ decreases; however, the governing equation for ψ has a conservative form

$$\frac{\partial \psi}{\partial T} = \frac{\partial^2}{\partial Y^2}\left[D_\perp \psi + \frac{g}{3}\psi^3 - \kappa\frac{\partial^2 \psi}{\partial Y^2}\right], \tag{6.35}$$

which implies the constraint

$$\frac{d}{dT}\int_0^L \psi dY = 0 \tag{6.36}$$

on the phase dynamics. Thus, λ appears like a Lagrange multiplier, and we have to minimize

$$\mathcal{F}_\lambda = \int_0^L \left[\frac{\kappa}{2}\left(\frac{\partial \psi}{\partial Y}\right)^2 - U(\psi)\right] dY .$$

If we start from rolls perpendicular to the x-axis, i.e., $\psi = 0$, in the unstable regime, $D_\perp < 0$, we can take $\lambda = 0$ and the potential $U(\psi)$ is symmetric (see Fig. 6.12).

The fastest growing mode is the one with wavenumber k_{max} and generates a wavy pattern as displayed in Fig. 6.10. This pattern corresponds to an oscillatory solution about $\psi = 0$ in the potential well of Fig. 6.12; thus it is unstable. The stable pattern which satisfies the constraint (6.36) corresponds to the limiting solitary wave solution that connects the two opposite values $\psi = \pm\sqrt{3|D_\perp|/g}$, i.e., two sets of rolls with different orientations, symmetric with respect to the y-axis (see Fig. 6.13). Note that this solution cannot be observed if we assume periodic boundary conditions in Y; the dynamics lead to a two-kink pattern in that case.

Thus, in the nonlinear regime, the zigzag instability generates the largest possible domains with parallel rolls, compatible with the mean

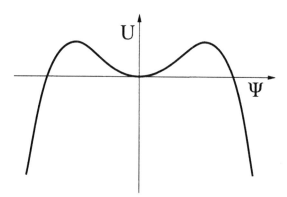

Fig. 6.12.   The potential $U(\psi)$ for $D_\perp < 0$ and $\lambda = 0$.

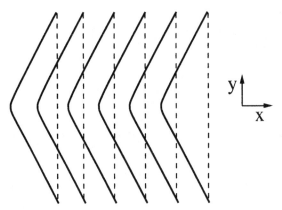

Fig. 6.13. The solitary wave solution selected by the system to preserve the mean value of ψ.

wavenumber conservation and the constraints related to boundary conditions. Although the Eckhaus and the zigzag instabilities have similar dispersion relations, their nonlinear dynamics are different: the linearly unstable perturbations cascade to short scales in the Eckhaus instability, whereas larger and larger domains are created during the nonlinear evolution of the zigzag instability.

7 Drift instabilities of cellular patterns

7.1 Introduction

Another type of secondary instability that occurs in many systems is the "drift instability." After a first bifurcation to a stationary cellular structure, $\mathbf{u}_0(x)$, further increase in the bifurcation parameter generates a secondary bifurcation to a traveling pattern of the form $\mathbf{u}_D(x \pm ct)$. The motion of the pattern in one of the preferential directions, $\pm x$, breaks the space-reflection symmetry. As we observed in section 5 (model (5.14)), a stationary bifurcation may generate a traveling pattern at onset when the system is not invariant under reflection symmetry $x \to -x$. Here, we consider systems which are invariant under reflection symmetry and give rise to a symmetric primary pattern; the reflection symmetry is spontaneously broken at finite amplitude when the static pattern undergoes the secondary drift bifurcation.

Drift instabilities of cellular patterns have been widely observed in various experimental situations. Couette flow between two horizontal coaxial cylinders with a partially filled gap (Mutabazi *et al.*, 1988), displays transitions from stationary to traveling rolls; as clearly noticed, the traveling rolls are tilted and the direction of the propagation is determined by this asymmetry. The traveling-roll state is either homogeneous in space, or there exist domains of inclined rolls with opposite tilt and thus opposite propagation direction. Similar results have been found in a film draining experiment (Rabaud *et al.*, 1990). Drift instabilities have also been observed in directional crystal growth experiments. Above the onset of the Mullins–Sekerka instability of liquid crystals, "solitary modes" propagating along the interface have been observed (Simon *et al.*, 1988). These "solitary modes" consist of domains of stretched asymmetric cells that connect two regions with symmetric cells. Similarly, domains of tilted lamellae moving transversally along the growth front have been observed during directional solidification of eutectics (Faivre *et al.*, 1989), and the relationship of the tilt direction to the one of propagation has also been emphasized. Finally, a drift instability was observed and understood as a parity-breaking bifurcation, for a standing surface wave excited parametrically in a horizontal layer of fluid contained in a thin annulus, submitted to vertical vibrations (Douady *et al.*, 1989). It was observed that, as the driving amplitude is increased, the standing wave pattern either begins to move at a constant speed in one direction, or undergoes an oscillatory instability that corresponds to a compression mode of the periodic structure (i.e., a wavenumber modulation in space and time).

On the theoretical side, it is interesting to note that a secondary bifurcation that transforms a stationary structure into a traveling one has been predicted by Malomed and Tribelsky (1984) before the experimental results quoted above. They used a Galerkin approximation for model equations of the Kuramoto–Sivashinsky type, and pointed out that the drift instability arises in that case from the coupling between the spatial phase of the basic structure with the second harmonic generation. This bifurcation has been understood in a more general way from symmetry considerations (Coullet *et al.*, 1989). Finally, a drift instability has been observed by numerical integration of the Kuramoto–Sivashinsky equation (Thual and Bellevaux, 1988).

We first propose a model that describes the drift instability of a stationary cellular pattern. We then show that a drift instability of a standing wave, like the one observed in the Faraday experiment, can be understood as a secondary bifurcation described by the evolution equations for the amplitudes of the right and left propagating waves (equations 5.39). In all cases, we will show that the basic mechanism consists of the coupling between the spatial phase ϕ of the primary pattern with the order parameter V associated with the space-reflection broken-symmetry, and we will give the general governing equations for the drift bifurcation. The notations should be considered independently in each section except in section (7.3) and (7.4).

7.2 A drift instability of stationary patterns

In their Galerkin approximation of a model equation, Malomed and Tribelsky found that the drift instability occurs when the second harmonic of the basic pattern is not linearly damped strongly enough. We thus consider a situation where two modes k and $2k$ interact resonantly,

$$
\begin{aligned}
\mathbf{u}(x,t) = &[A(X,T)\exp(ikx) + \text{c.c.}]\,\mathbf{u}_k \\
&+ [B(X,T)\exp(2ikx) + \text{c.c.}]\,\mathbf{u}_{2k} + \dots
\end{aligned}
\tag{7.1}
$$

From symmetry arguments (translational invariance in space), the evolution equations for A and B read, to third order,

$$
\begin{aligned}
\frac{\partial A}{\partial T} &= \mu A - \bar{A}B - \alpha|A|^2 A - \beta|B|^2 A \\
\frac{\partial B}{\partial T} &= \nu B + \varepsilon A^2 - \gamma|A|^2 B - \delta|B|^2 B\,.
\end{aligned}
\tag{7.2}
$$

The quadratic coupling terms describe the resonant interaction between the modes k and $2k$. Their coefficients can be taken equal to $\varepsilon = \pm1$ by appropriate scaling of the amplitudes; the coefficient of $\bar{A}B$ can be taken equal to -1, making the transformation $\mathbf{u} \to -\mathbf{u}$, if necessary. Positive values of α, β, γ and δ ensure global stability. The bifurcation diagram for problem (7.2) has been studied by several authors in the context of resonant wave interaction (see for instance, Proctor and Jones (1988) and references therein). We thus refer to these papers for the mathematical aspects and discuss equations (7.2) in the restricted context of the "drift bifurcation."

Writing

$$
A = R\exp(i\phi)\,, \qquad B = S\exp(i\theta)\,, \qquad \Sigma = 2\phi - \theta\,,
$$

we get from (7.2)

$$
\frac{\partial R}{\partial T} = (\mu - \alpha R^2 - \beta S^2)R - RS\cos\Sigma
\tag{7.3a}
$$

$$
\frac{\partial S}{\partial T} = (\nu - \gamma R^2 - \delta S^2)S + \varepsilon R^2\cos\Sigma
\tag{7.3b}
$$

$$
\Sigma_t = \left(2S - \varepsilon\frac{R^2}{S}\right)\sin\Sigma
\tag{7.3c}
$$

$$
\phi_t = S\sin\Sigma
\tag{7.3d}
$$

In the context of our study we must take $\nu < 0$ (the second harmonic is linearly damped) and increase the bifurcation parameter μ. When μ becomes positive, the null state bifurcates to an orbit of stable stationary patterns related to each other by space translation:

$$
R = R_0 \neq 0\,, \quad S = S_0 \neq 0,,\quad \Sigma = \Sigma_0 = 0\,, \quad \text{and} \quad \phi \text{ arbitrary.}
$$

A cellular pattern drifting with a constant velocity, corresponds to:

$$\frac{\partial R}{\partial T} = 0, \quad \frac{\partial S}{\partial T} = 0, \quad \frac{\partial \Sigma}{\partial T} = 0, \quad \frac{\partial \phi}{\partial T} = \text{constant} \neq 0.$$

This implies $2S - \varepsilon R^2/S = 0$, and thus $\varepsilon = 1$. So the coefficients of the quadratic terms must have opposite signs in order to observe the drift instability. Note that this means that the second harmonic does not enhance the stationary instability near onset; indeed, for $\mu \simeq 0$ and $\nu < 0$, B follows adiabatically A ($B \propto A^2$), and the quadratic terms of equation (7.2) contribute to saturate the primary instability. The stationary pattern is destabilized when $2S_0 + R_0^2/S_0$ vanishes as μ is increased. This happens if the condition $1 + \nu(2\gamma + \delta) > 0$ is satisfied, which corresponds to the condition that the second harmonic is not strongly damped ($|\nu|$ not too large). The system of equations (7.3a, b, c) then undergoes a supercritical pitchfork bifurcation. The two bifurcated stationary states are such that $R_D^2 = 2S_D^2$, $\Sigma = \pm\Sigma_D \neq 0$. Above the instability onset, ϕ increases linearly in time according to equation (7.3d). As noted earlier, this state represents traveling waves.

The bifurcation from the stationary pattern to the traveling one has the following characteristics: its order parameter, $\Sigma = 2\phi - \theta$, undergoes a pitchfork bifurcation that breaks the basic pattern reflection symmetry; the coupling with the basic pattern spatial phase ϕ induces the drift motion according to equation (7.3d), and the direction of propagation is determined by the sign of Σ. Thus the mechanism described by Malomed and Tribelsky for their model equations appears to be a general one. We expect that this k–$2k$ interaction mechanism is relevant for most of the experiments quoted above.

7.3 The drift instability of a parametrically excited standing wave

Let us now consider the drift instability of a standing surface wave, generated by parametric excitation in a horizontal layer of fluid contained in a thin annulus, submitted to vertical vibrations (Douady *et al.*, 1989). It was observed that, as the driving amplitude is increased, the standing wave pattern either begins to move at a constant speed in one direction, or undergoes an oscillatory instability that corresponds to a wavenumber modulation in space and time. We first consider the drift bifurcation and discuss the oscillatory instability next.

Close to the onset of instability, we write the surface deformation in the form

$$\xi(x, t) = A(X, T) \exp i(\omega t - kx) + B(X, T) \exp i(\omega t + kx) + \text{c.c.} + \ldots,$$
$$(7.4)$$

where A and B are the slowly varying amplitudes of the right and left waves at frequency $\omega = \omega_e/2$, where ω_e is the external driving frequency. The equations for A and B are at leading order (see section 5),

$$\frac{\partial A}{\partial T} + c\frac{\partial A}{\partial X} = (-\lambda + i\nu)A + \mu\bar{B} + \alpha\frac{\partial^2 A}{\partial X^2} + \left(\beta|A|^2 + \gamma|B|^2\right)A$$

$$\frac{\partial B}{\partial T} - c\frac{\partial B}{\partial X} = (-\lambda + i\nu)B + \mu\bar{A} + \alpha\frac{\partial^2 B}{\partial X^2} + \left(\gamma|A|^2 + \beta|B|^2\right)B \tag{7.5}$$

where λ is the dissipation ($\lambda > 0$), ν corresponds to the detuning between the surface wave frequency ω_0 and $\omega_e/2$, μ is proportional to the external forcing amplitude. The imaginary parts of β and γ describe the nonlinear frequency variation of the wave as a function of the amplitude, whereas the real parts correspond to nonlinear dissipation. α_i corresponds to dispersion.

When $\mu \geq 0$, a standing wave regime is observed. To analyze its stability we write,

$$A = \exp(S + R) \exp i(\Theta + \Phi)$$
$$B = \exp(S - R) \exp i(\Theta - \Phi),$$

and get from equations (7.5) for spatially homogeneous waves:

$$\frac{\partial S}{\partial T} = -\lambda + \cosh 2R\left[\mu\cos 2\Theta + (\beta_r + \gamma_r)\exp 2S\right] \tag{7.6a}$$

$$\frac{\partial \Theta}{\partial T} = \nu + \cosh 2R\left[-\mu\sin 2\Theta + (\beta_i + \gamma_i)\exp 2S\right] \tag{7.6b}$$

$$\frac{\partial R}{\partial T} = \sinh 2R\left[-\mu\cos 2\Theta + (\beta_r - \gamma_r)\exp 2S\right] \tag{7.6c}$$

$$\frac{\partial \Phi}{\partial T} = \sinh 2R\left[\mu\sin 2\Theta + (\beta_i - \gamma_i)\exp 2S\right]. \tag{7.6d}$$

Θ and Φ are respectively the temporal and spatial phases of the pattern. Note that the equation for Φ decouples, because of the translation invariance of the system in space. The standing wave solutions correspond to $(S_0, \Theta_0, R = 0)$. Their stability with respect to spatially homogeneous perturbations is simple to investigate from equations (7.6). We assume $\beta_r + \gamma_r < 0$ and the detuning small enough ($|\nu| < \lambda|\beta_r/\beta_i|$). Then, perturbations in S and Θ are damped. Perturbations in R and Φ obey the equations,

$$\frac{\partial \Phi}{\partial T} = 2\left[\nu + 2\beta_i\exp 2S_0\right]R + \dots \tag{7.7a}$$

$$\frac{\partial R}{\partial T} = 2\left[-\lambda + 2\beta_r\exp 2S_0\right]R + \dots \tag{7.7b}$$

When the standing wave pattern amplitude $\exp(S_0)$ is small, R is damped and the standing wave pattern is stable. As the driving amplitude is increased, S_0 increases and R becomes unstable for $\exp(2S_0) = \lambda/2\beta_r$, provided that $\beta_r > 0$. A non-zero value of R breaks the $x \to -x$ symmetry (see equation (7.4) and the expressions of A and B versus S, R, Θ, Φ). Thus R is the order parameter of the "drift bifurcation" for this standing wave problem. The coupling with the spatial phase Φ generates the drift (7.7a). The structure of the "drift bifurcation" is thus similar to that described previously for a stationary pattern. However, higher order terms in equations (7.7) show that the "drift bifurcation" is subcritical in this case. One can easily check this by noting that the drifting solution of equations (7.6), $\partial S/\partial T = \partial R/\partial T = \partial \Theta/\partial T = 0$, $\partial \Phi/\partial T \neq 0$, exists for $\exp(2S_0) < \lambda/2\beta_r$, i.e., only before the onset of the drift bifurcation. But additional terms of the form $|A|^4 A$, $|B|^4 B$, can stabilize the drifting solution, and even make the drift bifurcation supercritical.

7.4 The drift bifurcation

We observed in the above examples that the drift bifurcation consists of a secondary instability of the basic pattern that spontaneously breaks its reflection symmetry $x \to -x$. The eigenvalue λ of the corresponding eigenmode $\mathbf{u}_\pi(x)$ vanishes at the bifurcation. However, we have a persistent zero eigenvalue associated with translational symmetry; the drift instability results from the coupling between the reflection symmetry-breaking amplitude mode, and the phase mode associated with translational invariance. In the vicinity of this instability onset, we write

$$\mathbf{u}(x,t) = \mathbf{u}_0[x + \phi(X,T)] + V(X,T)\,\mathbf{u}_\pi(x) + \dots, \qquad (7.8)$$

and look for coupled equations for ϕ and V. The form of these equations is given by symmetry arguments, translational invariance in space ($\phi \to \phi + \phi_0$), and space reflection symmetry ($x \to -x$, $\phi \to -\phi$, $V \to -V$). We get, to leading orders in the gradient expansion

$$\frac{\partial \phi}{\partial T} = V, \qquad (7.9a)$$

$$\frac{\partial V}{\partial T} = \lambda V - V^3 + a\frac{\partial^2 \phi}{\partial X^2} + b\frac{\partial^2 V}{\partial X^2} + fV\frac{\partial \phi}{\partial X}$$
$$+ gV\frac{\partial V}{\partial X} + h\frac{\partial \phi}{\partial X}\frac{\partial^2 \phi}{\partial X^2} + \dots \qquad (7.9b)$$

Higher-order terms in equation (7.9a) can always be removed via a nonlinear transformation (Fauve *et al.* 1987). (The coefficient in equation (7.9a) has been scaled in V.) If the coefficients a, b are positive, the $V = 0$ solution first bifurcates when λ vanishes and becomes positive.

The *homogeneous* drifting pattern, $V_0 = \pm\sqrt{\lambda}$, $\phi_0 = V_0 T$, bifurcates supercritically; however its stability to inhomogeneous disturbances of the form $\exp(\eta T + iKX)$, is governed by the dispersion relation,

$$\eta^2 + (2\lambda - iKgV_0 + bK^2)\eta - iKfV_0 + aK^2 = o(K^2), \qquad (7.10)$$

that shows that the term $fV\partial\phi/\partial X$ destabilizes the homogeneous pattern independently of the sign of f. This may appear somewhat surprising, because at the drift instability onset, one bifurcates from the linearly stable static pattern to the linearly unstable static pattern without appearance of stable drifting patterns; this behavior traces back to the existence of subcritical localized traveling solutions of equation (7.9), that describe localized drifting regions with tilted cells, widely observed in most experiments. Thus, even when the drift bifurcation is supercritical if one only considers homogeneous patterns (as in section 7.2 for instance), the coupling term $fV\partial\phi/\partial X$ makes it generically subcritical when no restriction is imposed (Caroli *et al.*, 1992).

Another experimental observation can be understood in the framework of equations (7.9): it is the stationary pattern wavenumber selection often observed as the control parameter is increased. In most of the above quoted experiments, it is observed that a pattern wavenumber modification occurs by nucleation of a transient drifting domain that generates a phase gradient, say K, and leads to a new periodic pattern of wavenumber $k + \epsilon K$. Indeed, $V = 0$, $\phi = Kx$, is a particular solution of equations (7.9), for which the damping rate of perturbations in V is $\lambda + fK$. Consequently the drift of this new pattern is inhibited if $fK < 0$, for $|K| > \lambda/f$. The new periodic pattern thus remains stationary because of wavenumber modification. Within the framework of the model of section 7.2, this stabilization mechanism is associated with the increase of the second harmonic damping rate when the pattern wavenumber is increased.

7.5 Oscillatory phase modulation of periodic patterns

As said above, an oscillatory phase modulation of periodic patterns is observed as a secondary instability of parametrically generated surface waves (Douady *et al.*, 1989). After this instability onset, the position of the wave crests is modulated in space and time by a standing wave. This oscillatory instability is observed close to the "drift bifurcation" in the experimental parameter space. The numerical integration of equations (7.5) has shown that this oscillatory instability corresponds to a standing wave modulation of the basic pattern spatial and temporal phases, in agreement with the experimental observations.

We show that the coupling that generates the "drift bifurcation" is also a possible mechanism to describe phase modulation of periodic patterns, if the order parameter V is destabilized at a finite wavenumber. We consider equations (7.9), that govern the space dependent perturbations of the basic periodic pattern, with $\lambda \simeq -\lambda_0$, $\lambda_0 > 0$. Thus, the standing pattern is stable with respect to homogeneous perturbations. The growth rate of a perturbation of the form $\exp(\eta T + iKX)$, is governed by the dispersion relation,

$$\eta^2 + \eta\left(-\lambda + bK^2\right) + aK^2 = o(K^2). \tag{7.11}$$

Stability at short wavelength requires higher order terms (fourth order gradients). For $a > 0$ and $b < 0$ an oscillatory instability occurs first. The corresponding growth rate is displayed in Fig. 7.1; it shows that the oscillatory instability results from the interaction of the neutral mode, because of the translation invariance in space, with the slightly damped, reflection symmetry-breaking mode associated with the "drift bifurcation" before its onset value. An instability leading to a stationary modulation of the basic pattern wavelength can occur for $a < 0$ and $b > 0$.

We have thus shown that a variety of experimental observations of periodic pattern secondary instabilities, can be understood in a simple framework: the coupling of the neutral mode associated with translational invariance in space, with a reflection symmetry-breaking bifurcation. Note that a similar singularity, with two zero eigenvalues at a secondary instability onset, occurs for the oscillatory instability of convection rolls and leads to traveling waves that propagate along their axis, although the underlying physical reasons are different (Fauve *et al.*, 1987). Let us finally mention that the secondary instabilities described here fit in a general classification, proposed on the basis of symmetry arguments (Coullet and Iooss, 1989), but the present approach gives simple physical mechanisms that do generate these secondary instabilities.

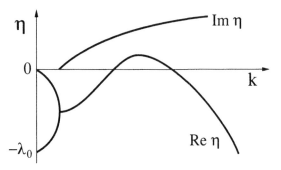

Fig. 7.1. Perturbation growth rate as a function of the wavenumber of the oscillatory instability ($a > 0$, $b < 0$).

8 Nonlinear localized structures

8.1 Different types of nonlinear localized structures

Nonlinear interactions usually transfer energy to higher harmonics, thus making wave-fronts steeper. Nonlinear effects can be balanced by dissipation; this is usually the case for shock waves (Whitham, 1974). Let us consider a simple example, the Burgers equation

$$\frac{\partial \rho}{\partial t} + \rho \frac{\partial \rho}{\partial x} = \nu \frac{\partial^2 \rho}{\partial x^2} \,. \tag{8.1}$$

We look for a traveling solution of the form $\rho = \rho(x - ct)$, that connects two constant values of ρ, ρ_1 and ρ_2 for $x \to \pm\infty$. (8.1) becomes

$$-c\rho' + \rho\rho' = \nu\rho'' \,.$$

Integrating once yields

$$-c\rho + \frac{1}{2}\rho^2 - \nu\rho' = \text{constant} \,,$$

and consequently

$$c = \frac{\rho_1 + \rho_2}{2} \,.$$

Thus, although the shape of $\rho(x - ct)$ depends on the dissipation ν, this does not affect the shock velocity that depends only on the jump in ρ.

In conservative systems nonlinear effects may be balanced by dispersion: this is the basic mechanism that gives rise to solitary waves. We have already considered soliton solutions of the nonlinear Schrödinger equation, and note that they can travel at any velocity because of Galilean invariance. This is of course not the case for all solitary waves; consider for instance the Sine–Gordon equation

$$\frac{\partial^2 u}{\partial t^2} = \frac{\partial^2 u}{\partial x^2} - \sin u \,, \tag{8.2}$$

which has kink type solitons of the form

$$u(x, t) = \pm 4 \tan^{-1} \left[\exp \frac{x - x_0 - ct}{\sqrt{1 - c^2}} \right]. \tag{8.3}$$

Unlike the solutions of the nonlinear Schrödinger equation, there is a limit speed 1 for the solutions (8.3). However, there is a more important characteristic, associated with the discrete symmetry

$$u \to u + 2n\pi \,,$$

of the Sine–Gordon equation; the soliton solution (8.3) can be considered as a localized structure, connecting in space two states $u = 0$ and $u =$

$\pm 2\pi$, related by this discrete symmetry transformation (see Fig. 8.1). This is called a topological soliton.

Similar type of localized structures, from the symmetry point of view, exist in phase transitions or in patterns generated by instabilities; they are called topological defects. As we already mentioned, the amplitudes of unstable modes are analogous to order parameters in phase transition theory; indeed, a non-zero amplitude is associated with the broken symmetry at the instability onset. Thus, in the supercritical regime, it is natural to call *defect* a region in space where the amplitude vanishes. A defect is said to be topologically stable when a slight perturbation of the amplitude in space or time does not affect its characteristic shape (for instance only slightly translates it in space). We will only illustrate this concept with very simple examples (for a review, see Mermin 1979).

Let us consider the Ginzburg–Landau equation (5.16) for a real one-space-dimensional amplitude $R(X, T)$

$$\frac{\partial R}{\partial T} = \mu R + \frac{\partial^2 R}{\partial X^2} - R^3. \tag{8.4}$$

The broken symmetry at the instability onset ($\mu = 0$) is the $R \to -R$ symmetry. Correspondingly, there exist two homogeneous solutions, $R_0 = \pm\sqrt{\mu}$, that are related by this symmetry transformation. The space dependent solution

$$R(X) = \tanh\left(\sqrt{\mu/2}\,X\right),$$

connects these two solutions (see Fig. 8.2); it is a topological defect.

This solution also exists for the complex Ginzburg–Landau equation,

$$\frac{\partial A}{\partial T} = \mu A + \Delta A - |A|^2 A. \tag{8.5}$$

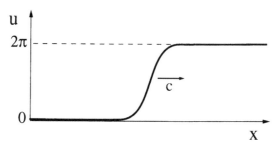

Fig. 8.1.   A kink soliton of the Sine–Gordon equation.

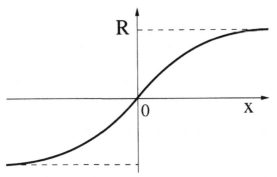

Fig. 8.2. A topological defect for the real one-space-dimensional field governed by equation (8.4).

However, it is not a topological defect since one can easily remove it with phase perturbations. On the contrary, for a two-space-dimensional field, $A(X, Y, T)$, there exist topologically stable point defects. Indeed, the complex amplitude $A(X, Y, T)$ vanishes if its real and imaginary parts vanish; at a given instant, Re $\{A(X, Y, T)\} = 0$ and Im $\{A(X, Y, T)\} = 0$ define one-dimensional curves in the $X - Y$ plane, that generically intersect at a point (see Fig. 8.3). If one slightly perturbs $A(X, Y, T)$, these curves move slightly and so does the defect, but it remains unchanged. Similarly, for a three-space-dimensional field, $A(X, Y, Z, T)$, topological defects of the complex Ginzburg–Landau equation are lines.

If equation (8.5) is modified to

$$\frac{\partial A}{\partial T} = \mu A + \Delta A + |A|^2 A - |A|^4 A, \tag{8.6}$$

the bifurcation becomes subcritical. The interesting new feature is bistability; indeed, it is clear from the bifurcation diagram displayed in Fig. 8.4

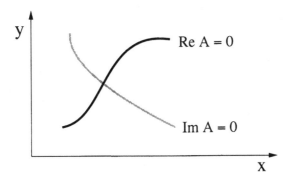

Fig. 8.3. The location of a topological defect for a two-space-dimensional Ginzburg–Landau equation.

that two stable states coexist for μ negative. Obviously, these two solutions are not related by a symmetry transformation, but we can also consider an interface that separates in space these two stable states, or a droplet that consists of a region in one state surrounded by the other one. In a system with a Lyapunov functional, i.e., a free-energy, the interface moves such that the lowest energy state increases in size. Contrary to the situation with shock waves, the interface velocity does not depend only on the energy difference between the two homogeneous solutions, but also on the shape of the interface. With non-variational systems, interesting new phenomena occur (see section 8.3).

8.2 Kink dynamics

We consider equation (8.4) in the supercritical regime ($\mu > 0$); chosing appropriate time, space and amplitude scales, we get

$$\frac{\partial R}{\partial T} = 2R(1 - R^2) + \frac{\partial^2 R}{\partial X^2} . \tag{8.7}$$

This equation admits a kink type solution or defect,

$$R = \tanh X .$$

Consider a solution consisting of several kinks (Fig. 8.5). Such a solution is unstable because the neutral translation mode, $\partial R/\partial X$, has a node; the only stable solution is the one with a single kink. What is the time evolution of such unstable structures? An interesting approach is to consider many topological defects, one far from the other, as a "gas" of such kinks and write an evolution equation for their density (Kawasaki and Ohta, 1982).

We first derive an equation describing the interaction of two kinks that are far apart. Let us seek a solution in the form

$$R(X, T) = R_1 [X_1(T)] R_2 [X_2(T)] , \tag{8.8}$$

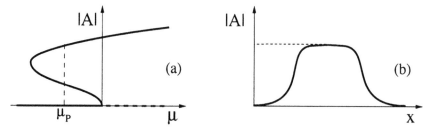

Fig. 8.4. Bistability and localized structure in the vicinity of a subcritical bifurcation

where

$$R_j\left[X_i(T)\right] = \tanh\left[X - X_i(T)\right] \tag{8.9}$$

and j is 1 or 2. Substituting (8.8) in (8.7) we obtain the equation

$$-\dot{X}_1 R'_1 R_2 - \dot{X}_2 R_1 R'_2 = 2R_1 R_2(1 - R_1^2 R_2^2) + R''_1 R_2 + 2R'_1 R'_2 + R_1 R''_2, \tag{8.10}$$

where a dot denotes time derivative and primes denote differentiation with respect to X_1 or X_2. By direct differentiation one can verify that

$$R'_i = 1 - R_i^2$$
$$R''_i = -2R_i(1 - R_i^2).$$

At $X = X_i(T)$ we have $R_i = 0$, $R'_i = 1$ and $R_j \approx \pm 1$. Substituting $X = X_i$ in (8.10) we derive the equations of motion for the kinks

$$\dot{X}_1 \approx 2\left[1 - R_2^2(X_1)\right] \approx 8\,\exp(-2|X_1 - X_2|)$$
$$\dot{X}_2 \approx -2\left[(1 - R_1^2(X_2)\right] \approx -8\,\exp(-2|X_1 - X_2|), \tag{8.11}$$

which show that each kink feels the effect of the exponentially decreasing tail of the other at large separations. We can generalize this to consider the interaction of many kinks in the nearest neighbor approximation:

$$\dot{X}_i \approx 8\,\exp(-2|X_{i+1} - X_i|) - 8\,\exp(-2|X_i - X_{i-1}|). \tag{8.12}$$

Define

$$\rho_i = \exp\left(-2|X_{i+1} - X_i|\right). \tag{8.13}$$

Then $-1/\log\rho_i$ is the "kink density". Taking the logarithm of (8.13) and differentiating with respect to time, we have

$$\dot{\rho}_i = -2(\dot{X}_{i+1} - \dot{X}_i)\rho_i, \tag{8.14}$$

and on using (8.12) and (8.13),

$$\dot{\rho}_i = -16(\rho_{i+1} - 2\rho_i + \rho_{i-1})\rho_i. \tag{8.15}$$

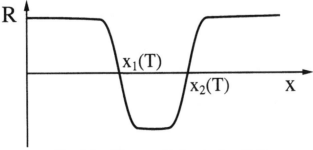

Fig. 8.5. The two-kink solution (8.8).

Finally, passing to the continuum limit, we have the equation for a "rarefied kink gas"

$$\frac{\partial \rho}{\partial T} = -16\rho \frac{\partial^2 \rho}{\partial X^2}. \tag{8.16}$$

We see that this is a nonlinear diffusion equation where the diffusion coefficient $D = -16\rho$ is negative. It therefore describes a condensation process with all the kinks tending to clump together due to mutual attraction. Let us consider a train of kinks, equispaced at a distance a apart. Then, $X_{i+1}^{(0)} - X_i^{(0)} = a$, where the superscript zero denotes the initial state. Setting $X_i = X_i^{(0)} + u_i$ and looking for normal modes, $u_n \propto \exp(\eta T - iKna)$, we have

$$\eta = 64e^{-a} \sin^2(Ka/2),$$

which shows that the mode with the largest growth rate is at $Ka = \pi$, i.e. an optical mode at twice the wavelength of the kink array.

The important point to note here, is that we can describe a system at different approximation levels corresponding to equations (8.7), (8.12) and (8.16). Depending on the problem, one can consider localized structures as "particles" and use (8.12), or use a continuous description.

8.3 Localized structures in the vicinity of a subcritical bifurcation

Consider the following simple model equation describing a subcritical bifurcation:

$$\frac{\partial A}{\partial T} = \mu A + \alpha \frac{\partial^2 A}{\partial X^2} + \beta |A|^2 A + \gamma |A|^4 A. \tag{8.17}$$

If the coefficients of this equation are real, the steady state is given by the solution which minimizes the Lyapunov functional

$$\mathcal{L} = \int_0^L \left[\alpha_r |A_x|^2 - V(|A|) \right] dX$$

$$V(|A|) = \mu |A|^2 + \frac{\beta_r}{2} |A|^4 + \frac{\gamma_r}{3} |A|^6. \tag{8.18}$$

Since we are considering a subcritical bifurcation, $\beta_r > 0$ and $\gamma_r < 0$. The three possible situations are shown in Fig. 8.6. The two states have the same energy and thus can coexist only if $\mu = \mu_P$; μ_P corresponds to the Maxwell plateau of first order phase transitions.

Now consider the situation with α and β complex so that there is no longer a Lyapunov functional for the problem. In this situation stable localized structures are possible (Thual and Fauve, 1988). The stabilization

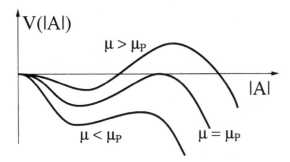

Fig. 8.6. The Lyapunov functional (8.18) for $\mu \approx \mu_P$.

can be explained by the following rough argument. Substituting

$$A = R(X,T)\exp\left[i\theta(X,T)\right]$$

in equation (8.17) and equating real and imaginary parts we have

$$
\begin{aligned}
\frac{\partial R}{\partial T} &= \alpha_r \frac{\partial^2 R}{\partial X^2} + \left[\mu - \alpha_r \left(\frac{\partial \theta}{\partial X}\right)^2\right] R + \beta_r R^3 + \gamma_r R^5 \\
&\quad - \alpha_i \left(2\frac{\partial R}{\partial X}\frac{\partial \theta}{\partial X} + R\frac{\partial^2 \theta}{\partial X^2}\right) \\
R\frac{\partial \theta}{\partial T} &= \alpha_r \left(2\frac{\partial R}{\partial X}\frac{\partial \theta}{\partial X} + R\frac{\partial^2 \theta}{\partial X^2}\right) + \alpha_i \frac{\partial^2 R}{\partial X^2} - \alpha_i R \left(\frac{\partial \theta}{\partial X}\right)^2 \\
&\quad + \beta_i R^3 + \gamma_i R^5 .
\end{aligned}
\tag{8.19}
$$

If the coefficients are real, (8.19) is solved by $\partial\theta/\partial X = 0$, and we have a variational problem for R. In the presence of a non-zero imaginary part, the amplitude and phase equations couple so that there is a non-zero $\partial\theta/\partial X$. This changes the effective value of μ in the coefficient of R in (8.19); we define

$$\mu_{\text{eff}} = \mu - \alpha_r \left(\frac{\partial \theta}{\partial X}\right)^2 .$$

The effect of the phase gradient is to decrease μ_{eff} in the outer region, thus stabilizing the zero-solution, whereas the bifurcated solution is stabilized in the core of the localized structure.

These localized structures can be obtained perturbatively in the variational and conservative limits; in the conservative limit, let us write (8.17) as

$$\frac{\partial A}{\partial T} = i\,\frac{\partial^2 A}{\partial X^2} + 2i|A|^2 A + \epsilon P(A)$$

$$P(A) = \mu A + \alpha_r\,\frac{\partial^2 A}{\partial X^2} + \beta_r|A|^2 A + \gamma_r|A|^4 A\,. \tag{8.20}$$

For $\epsilon = 0$ this is simply the nonlinear Schrödinger equation and admits the one-parameter family of solutions

$$A_s = \Delta\,\mathrm{sech}(\Delta X)\exp\left(-i\Delta^2 T\right)\,. \tag{8.21}$$

The existence of such a one-parameter family is due to the scale invariance of the nonlinear Schrödinger equation.

If ϵ is given a small but non-zero value, we look for slowly varying solitons of the form

$$A(X,T) = \Delta(T)\,\mathrm{sech}[\Delta(T)X]\exp[-i\Theta(T)]\,. \tag{8.22}$$

The temporal evolution of a soliton under the action of a perturbation $P(A)$ is a well-known problem of soliton theory and can be solved with the inverse scattering method (see for instance Lamb, 1980). The temporal evolution of $\Delta(T)$ can be found in a simpler way here: multiplying equation (8.20) by $\bar{A}$ and integrating on space leads to the evolution equation

$$\frac{d}{dT}\int |A|^2 dX = \epsilon\int\left[\mu|A|^2 + \beta_r|A|^4 + \gamma_r|A|^6 - \alpha_r|A_x|^2\right]dX\,. \tag{8.23}$$

Substituting (8.22) in (8.23), we get to leading order, an evolution equation for Δ,

$$\frac{1}{2}\frac{d\Delta}{dT} = \mu\Delta + \frac{4}{3}(2\beta_r - \alpha_r)\Delta^3 + \frac{128}{15}\gamma_r\Delta^5\,. \tag{8.24}$$

For $\alpha_r < 2\beta_r$, equation (8.24) has two non-zero solutions $\Delta_\pm$ for $\mu_s < \mu < 0$, with $\mu_s = 5(-\alpha_r + 2\beta_r)^2/96\gamma_r$. Only the larger is stable, and gives the size of the selected pulse.

The above mechanism is a rather general one: the dissipative terms of equation (8.17) stabilize one of the soliton solutions among the continuous family (8.21) and select its size by breaking the invariance associated with the corresponding conservative problem.

It is also possible to show the existence of pulse-like solutions by perturbing the variational limit in the vicinity of the Maxwell plateau (Hakim *et al.* , 1990, Malomed and Nepomnyashchy, 1990). These pulses are stable in a finite interval range in μ, which depends on the coefficients of the non-variational terms (de Mottoni and Schatzman, 1993).

The stabilization mechanism, which results from the coupling between the amplitude and the phase of the wave's complex amplitude, also works for two-dimensional fields $A(X, Y, T)$ (Thual and Fauve, 1988), solutions of (8.17) where the diffusion term is the two-dimensional Laplacian, although in the nonlinear Schrödinger equation limit, two-dimensional pulselike solutions are known to be unstable. Much remains to be understood concerning these qualitatively new effects observed beyond the variational or Hamiltonian descriptions of physical systems.

References

Arnold, V.I. (1983). *Geometrical Methods in the Theory of Ordinary Differential Equations* (Springer Verlag, New York).

Bender, C.M. and Orszag, S.A. (1978). *Advanced Mathematical Methods for Scientists and Engineers* (McGraw–Hill, New York).

Benjamin, T.B. and Feir, J.E. (1967). *J. Fluid Mech.* **27**, 417.

Busse, F.H. (1978). *Rep. Prog. Phys.* **41**, 1929.

Busse, F.H. (1981). In *Hydrodynamic Instabilities and the Transition to Turbulence*, ed. H.L. Swinney and J.P. Gollub, *Topics in Applied Physics* **45**, 97 (Springer Verlag, New York).

Caroli, B., Caroli, C., and Fauve, S. (1992). *J. Physique I* **2**, 281.

Chandrasekhar, S. (1961). *Hydrodynamic and Hydromagnetic Stability* (Clarendon Press, Oxford).

Chapman, C.J. and Proctor, M.R.E. (1980). *J. Fluid Mech.* **101**, 759.

Chiffaudel, A. and Fauve, S. (1987). *Phys. Rev.* **A35**, 4004.

Coullet, P. and Fauve, S. (1985). *Phys. Rev. Letters* **55**, 2857.

Coullet, P., Goldstein, R.E., and Gunaratne, G.H. (1989). *Phys. Rev. Lett.* **63**, 1954.

Coullet, P. and Iooss, G. (1989). *Phys. Rev. Lett.* **64**, 866.

DiPrima, R.C. and Swinney, H.L. (1981). In *Hydrodynamic Instabilities and the Transition to Turbulence*, ed. H.L. Swinney and J.P. Gollub, *Topics in Applied Physics* **45**, 97 (Springer Verlag, New York).

Douady, S., Fauve, S., and Thual, O. (1989). *Europhys. Lett.* **10**, 309.

Drazin, P.G. and Reid, W.H. (1981). *Hydrodynamic Stability* (Cambridge University Press, Cambridge).

Edwards, W.S. and Fauve, S. (1992). *C. R. Acad. Sci. Paris*, **315-II**, 417.

Edwards, W.S. and Fauve, S. (1993). *Phys. Rev. E* **47**, R 788.

Faivre, G., de Cheveigné, S., Guthmann, C., and Kurowski, P. (1989). *Europhys. Lett.* **9**, 779.

Faraday, M. (1831). *Phil. Trans. R. Soc. London* **52**, 319.

Fauve, S. (1985), *Large Scale Instabilities of Cellular Flows*, Woods Hole Oceanographic Institution Technical Report, **55**.

Fauve, S., Bolton, E. W. and Brachet, M. (1987). *Physica* **29D**, 202.

Gershuni, G.Z. and Zhukovitskii, E.M. (1976). *Convection Stability of Incompressible Fluids* (Ketter Publications).

Goldstein, R.J. and Graham, D.J. (1969). *Phys. Fluids* **12**, 1133.

Gorkov, L.P. (1957). *Soviet. Phys. JETP* **6**, 311.

Grindrod, P. (1991). *Patterns and Waves*, (Clarendon Press, Oxford).

Guckenheimer, J. and Holmes, P. (1984). *Nonlinear Oscillations, Dynamical Systems and Bifurcations of Vector Fields* (Springer Verlag, New York).

Hakim, V., Jakobsen, P. and Pomeau, Y. (1990). *Europhys. Lett.* **11**, 19.

Kawasaki, K. and Ohta, T. (1982). *Physica* **A116**, 573.

Kolodner, P., Bensimon, D. and Surko, C.M. (1988). *Phys. Rev. Lett.* **60**, 17.

Kramer, L. and Zimmermann, W. (1985), *Physica* **D16**, 221.

Kuramoto, Y. (1984). *Prog. Theor. Phys.* **71**, 1182.

Lake, B.M., Yuen, M.C., Rungaldier, M. and Ferguson, W.E. (1977). *J. Fluid Mech.* **83**, 49.

Lamb, G.L. (1980). *Elements of Soliton Theory* (Wiley, New York).

Landau, L. (1967). *On the theory of phase transitions, On the problem of turbulence*. In *Collected papers of L. Landau*, ed. D. ter Haar (Gordon and Breach, London).

Langer, J.S. (1975). in *Fluctuations, Instabilities, and Phase Transitions*, ed. T. Riste, NATO Advanced Study Institutes Series, (Plenum Press, New York).

Malkus, W.V.R. (1964). *Boussinesq Equations and Convection Energetics*, Woods Hole Oceanographic Institution Technical Report, **11**.

Malkus, W.V.R. and Veronis, G. (1958). *J. Fluid Mech.* **4**, 225.

Malomed, B.A. and Nepomnyashchy, A.A. (1990). *Phys. Rev.* **A42**, 6009.

Malomed, B.A. and Tribelsky, M.I. (1984). *Physica* **D14**, 67.

Manneville, P. (1990). *Dissipative Structures and Weak Turbulence* (Academic Press, Boston).

Melville, W.K. (1982). *J. Fluid Mech.* **115**, 165.

Mermin, N.D. (1979). *Rev. Mod. Phys.* **51**, 591.

de Mottoni, P. and Schatzman, M. (1993). preprint CNRS-URA 740, **148**.

Mutabazi, I., Hegseth, J.J., Andereck, C.D., and Wesfreid, J.E. (1988). *Phys. Rev.* **A38**, 4752.

Nayfeh A.H. (1973). *Perturbation Methods.* (Wiley–Interscience, New York).

Newell, A.C. (1974). *Lectures in Applied Mathematics* **15**, 157.

Newell, A.C. (1985). *Solitons in Mathematics and Physics* (SIAM, Philadelphia).

Newell, A.C. and Whitehead, J.A. (1969). *J. Fluid Mech.* **38**, 279.

Normand, C., Pomeau, Y., and Velarde, M.G. (1977). *Rev. Mod. Phys.* **49**, 581.

Palm, E. (1975). Nonlinear Thermal Convection, *Ann. Rev. Fluid Mech.* **7**, 39.

Pomeau, Y. and Manneville, P. (1979). *J. Phys. Lettres* **40**, 609.

Proctor, M.R.E. and Jones, C. (1988). *J. Fluid Mech.* **188**, 301.

Rabaud, M., Michalland, S., and Couder, Y. (1990). *Phys. Rev. Letters* **64**, 184.

Rabinovich, M.I. and Trubetskov D.I. (1989). *Oscillations and Waves* (Kluwer Academic Publishers, Dordrecht).

Riahi, N. (1983). *J. Fluid Mech.* **129**, 153.

Sagdeev, R.Z., Usikov D.A. and Zaslavsky G.M. (1988). *Nonlinear Physics* (Harwood Academic Publishers, London).

Schlüter, A., Lortz, D., and Busse, F. (1965). *J. Fluid Mech.* **23**, 129.

Segel, L.A. (1969). *J. Fluid Mech.* **38**, 203.

Simon, A.J., Bechhoefer J., and Libchaber, A. (1988). *Phys. Rev. Letters* **61**, 2574.

Sorokin, V.S. (1953). *Prikl. Mat. Mekh.* **17**, 39.

Spiegel, E.A. (1971). *Ann. Rev. Astron. Astrophys.* **9**, 323.

Spiegel, E.A. (1972). *Ann. Rev. Astron. Astrophys.* **10**, 261.

Stuart, J.T. and DiPrima, R.C. (1978). *Proc. Roy. Soc. Lond.* **A362**, 27.

Thual, O. and Bellevaux, C. (1988). in *Fifth Beer-Sheva Seminar on MHD Flows and Turbulence, AIAA Progress in Astronautics and Aeronautics* **112**, 332.

Thual, O. and Fauve, S. (1988). *J. Physique* **49**, 1829.

Turner, J.S. (1973). *Buoyancy Effects in Fluids* (Cambridge University Press, Cambridge).

Van der Pol, B. (1934). *Proc. I. R. E.* **22**, 1051.

Van Dyke, M. (1982). *An Album of Fluid Motion* (The Parabolic Press, Stanford).

Wesfreid, J.E. and Zaleski, S., eds. (1984). *Cellular Structures in Instabilities, Lectures Notes in Physics* **210** (Springer Verlag, New York).

Wesfreid, J.E., Brand, H. R., Manneville, P., Albinet, G., and Boccara, N., eds. (1988). *Propagation in Systems far from Equilibrium* (Springer Verlag, New York).

Whitham, G.B. (1974). *Linear and Nonlinear Waves* (Wiley, New York).

Wu, J., Keolian, R., and Rudnick, I. (1984). *Phys. Rev. Lett.* **52**, 1421.

Zippelius, A. and Siggia, E. (1982), *Phys. Fluids* **26**, 2905.

5

An introduction to the instability
of flames, shocks, and detonations

G. Joulin and P. Vidal

1 Introduction and overview

Flames, shocks and detonations are phenomena of the traveling-wave type that are frequently encountered in the course of combustion processes. They can all be characterized as rather thin, *interface*-like regions, across which the properties of the fluid into which they travel undergo rapid changes.

Before introducing them it is perhaps worth devoting a few words to combustion itself. As is widely known, combustion intervenes in many instances of daily life (cars, heating, cooking) and in numerous industrial or accidental situations (engines, furnaces, fires...); gaining insights into its intricacies may therefore be useful to better monitor appliances or for safety purposes. No less importantly, combustion science involves enough mechanisms and peculiarities to please the scientist; it may indeed be defined as that of "exothermic chemical reactions in flows with heat and mass transfer" (Williams, 1992). More precisely, the laws governing combustion, a set of equations with a nonnegligible size, contain no less than the many interplays between nonisothermal chemistry, turbulence, compressible fluid mechanics, diffusive transports, buoyancy..., each of these being a serious subject of its own, not to speak of radiative transfer and multiphase media, which we shall deliberately ignore here.

This obviously opens up an extremely rich field of investigations, as witnessed by the size of recent treatises that offer a rather broad coverage of this field (Williams, 1985; Zel'dovich *et al.*, 1985), or by the mammoth volumes which report regularly on the domain (e.g., see Combustion Institute, 1990). Within the limited format of the present contribution, a much narrower scope had to be contemplated. We chose to restrict our account to initially premixed, homogeneous reactants, and to what we consider to be amongst the most basic events, namely: flames, shocks and detonations. Furthermore, we focused on the theoretical treatments of the latter. Still, we aimed at scanning as complete a spectrum of phenomena as we could.

Even in this framework, and for the reasons sketched above, studying these traveling waves and their stability properties cannot be started from scratch. This at least necessitates recalling the basic equations for reactive fluids (Section 2); introducing weak forms of these (Section 3) will also prove useful for studying shock waves.

One of the most important attributes of a traveling wave is its speed of propagation or, better, the ratio of the latter to the speed of sound in the medium just ahead of it. This so-called Mach number, M, allows one to distinguish between subsonic ($M < 1$) and supersonic ($M > 1$) waves. This is not only a semantic problem, for $M \lessgtr 1$ leads to quite different gross variations of the fluid across a wave. As a rule of thumb, flames ($M \ll 1$) are locally chemical, diffusive and elliptic in character, whereas shock and detonations ($M > 1$ or $M \gg 1$) tend to be mechanically and energetically-controlled (Section 3), and are primarily hyperbolic in behavior.

The rather long Section 4 deals with *flames*. These are fancy creatures, as a result of a conjunction of several features: they propagate, slowly ($M \ll 1$) but autonomously, are quite sensitive and subject to various instabilities of diffusive or hydrodynamical origin; last but not least, they are very beautiful. Even though the main mechanisms responsible for propagation and appearance of cellular flame structures are presumably well identified now and their study has entered the nonlinear stage, measurements of their detailed characteristics are still very difficult because many disturbances can influence them. Flame propagation in turbulent flows still constitutes a big challenge, from any viewpoint.

Shock waves (Section 5) are rather transitional as steep waves, in the sense that they do not in general involve any significant chemistry, but are supersonic and, in most cases, compressive. They offer an opportunity to introduce some specificities brought about by acoustics in moving media, more generally by hyperbolic equations, along with new instabilities; they are also *sine qua non* building blocks of detonations.

Detonations (Section 6) are shock waves that travel in exothermically-reacting media and are followed by a very fast heat release of chemical origin which they themselves trigger. Their analysis therefore combines the difficulties pertaining to both shocks and to reacting, dynamically compressible, fluids. Needless to stress that studying them is not an easy (in fact, not yet completed) task, even in the idealized case where the equation of state of the reacting medium and the chemical rates are available. Detonation fronts have long been known to spontaneously develop three-dimensional, unsteady cellular structures which, although well characterized experimentally, are still awaiting a neat theoretical explanation.

It has to be stressed that this chapter is intended as a short *course* for graduate or post-graduate students and, more generally, as an *introductory text* for those who have not yet taken up the theory of flames, shocks and detonations and their stability properties. This led us to summarize some of the lengthiest demonstrations and skip details when necessary or emphasize other aspects. Besides, we made our own selection of themes, models and methods. To counterbalance the consequences of too personal choices, we provide the reader with enough bibliographical material to make up one's own mind.

A few last words about the notations adopted are in order. For historical reasons we followed the accepted conventions which, regretfully, are not always the same for the three kinds of traveling waves under consideration. Furthermore, if one insists upon using "readable" formulae only, it is practically impossible to employ a single set of notations to cover the whole chapter without using some fonts twice. To keep the ambiguities to a minimum, the notations are redefined wherever needed.

2 Basic equations

2.1 Conservation laws for reactive fluids

As is recalled in the introduction a few equations, which we summarize below for the sake of completeness, are needed to study the traveling waves encountered in combustion. Specifically, they belong to the broad classes: balances, equations of state, flux-force relationships and chemical laws.

2.1.1 Balances

If ρ denotes the mixture density and $\mathbf{u}$ its velocity, *mass conservation* requires:

$$\frac{\partial \rho}{\partial t} + \nabla \cdot (\rho \mathbf{u}) = 0, \qquad (2.1.1)$$

where t is the time. Let us next denote the mass of kth-species ($k = 1, \cdots, K$) in a unit volume by ρ_k, and introduce the kth mass fraction $y_k = \rho_k/\rho$. Then y_k satisfies $\sum_k y_k = 1$, by definition of ρ, and the kth reactant balance:

$$\rho \left(\frac{\partial y_k}{\partial t} + \mathbf{u} \cdot \nabla \, y_k \right) = - \, \nabla \cdot \left(\rho y_k \mathbf{U}_k^{\text{diff}} \right) + \mathcal{M}_k w_k, \qquad (2.1.2)$$

which involves the kth diffusion velocity $\mathbf{U}_k^{\text{diff}}$ (the difference between $\mathbf{u}_k$ and $\mathbf{u}$), molecular weight ($\mathcal{M}_k$) and volumetric production rate

w_k (kg/m^{3}s). From $\sum_k \rho_k \mathbf{u}_k = \rho \mathbf{u}$ and the definitions of y_k and $\mathbf{U}_k^{\text{diff}}$, and anticipating that chemical reactions meet Lavoisier's constraint $\sum_k \mathcal{M}_k w_k = 0$, one deduces the identity:

$$\sum_k y_k \mathbf{U}_k^{\text{diff}} = 0 \,. \tag{2.1.3}$$

As for $\mathbf{u}$, it has to follow *Newton's law* which, here, is written as:

$$\rho \left(\frac{\partial \mathbf{u}}{\partial t} + \mathbf{u} \cdot \nabla \mathbf{u} \right) = - \nabla \cdot (p\mathbf{I} + \mathbf{P}) + \rho \mathbf{g} \tag{2.1.4}$$

upon the assumption that the only body force ($\mathbf{g}$) is gravity, actual or fictitious; $\mathbf{P}$ is the momentum-flux tensor, $\mathbf{I}$ is the identity and, as usual, p is hydrostatic pressure. The last balance of course deals with *energy*, whose conservation principle may be expressed as:

$$\rho \left(\frac{\partial h}{\partial t} + \mathbf{u} \cdot \nabla h \right) = - \nabla \cdot \mathbf{q} + \frac{\partial p}{\partial t} + \mathbf{u} \cdot \nabla p - \mathbf{P} : \nabla \mathbf{u} \,, \tag{2.1.5}$$

where h, a function of temperature T and mass fractions, is the mixture specific enthalpy and $\mathbf{q}$ is the heat-flux vector.

2.1.2 Equations of state

Under the assumption of local thermodynamic equilibrium, one can relate p and h to the other thermodynamic variables and composition. In the context of *flames*, we shall use the perfect gas law:

$$p = \rho \mathcal{R} T / \mathcal{M} \,, \tag{2.1.6a}$$

where $\mathcal{M} = (\sum_k y_k / \mathcal{M}_k)^{-1}$ is the mean molecular weight and $\mathcal{R}$ is the gas constant. For *shocks* or *detonations*, more general laws such as

$$p = p(\rho, S, \text{composition}) \,, \tag{2.1.6b}$$

where S is the specific entropy, shall be occasionally considered. Flames will use the caloric equation of state

$$h = \sum_k \left(h_k (T_u) + \int_{T_u}^{T} C_{pk} dT \right) y_k \,, \tag{2.1.7a}$$

where $h_k (T_u)$ is the kth enthalpy of formation at some reference temperature T_u (e.g., that of the medium in which the traveling waves propagate) and C_{pk} is the kth specific heat at constant pressure; both $h_k (T_u)$ and $C_{pk}(T)$ are assumed known, e.g., from adequate data bases or molecular theories. As for the detonations, we shall occasionally employ the more general law:

$$e = e(\rho, S, \text{composition}) \,, \tag{2.1.7b}$$

which involves the mixture specific internal energy $e = h - p/\rho$.

2.1.3 Flux-force relationships

Up to now $\mathbf{P}$, $\mathbf{q}$ and $\mathbf{U}_k^{\mathrm{diff}}$ are still unrelated to the main unknowns ρ, $\mathbf{u}$, y_k, T. We shall adopt the classical momentum-flux tensor

$$\mathbf{P} = \mathbf{I}\left(\tfrac{2}{3}\mu - \mu'\right)\boldsymbol{\nabla}\cdot\mathbf{u} - \mu\left(\boldsymbol{\nabla}\,\mathbf{u} + (\boldsymbol{\nabla}\,\mathbf{u})^*\right), \qquad (2.1.8)$$

where μ (μ') is the shear (bulk) viscosity and $()^*$ denotes tensor transposition; both μ an μ' are also presumed known functions of T, composition, etc.

In the absence of radiative transfer (a pleasant assumption!) the following expression of the heat flux is accurate enough for most combustion purposes

$$\mathbf{q} = -\lambda\,\boldsymbol{\nabla}\,T + \sum_k \rho h_k y_k \mathbf{U}_k^{\mathrm{diff}} + \mathcal{R}T\sum_{i,j} X_i D_{T,j}\left(\mathbf{U}_i^{\mathrm{diff}} - \mathbf{U}_j^{\mathrm{diff}}\right)/\mathcal{M}_i D_{ij},$$
$$(2.1.9)$$

where λ is the heat conductivity, $X_i = y_i \mathcal{M}/\mathcal{M}_i$ is the ith mole fraction, $D_{T,i}$ is the coefficient of thermal diffusion of the ith species in the mixture and D_{ij} is the binary diffusion coefficient of the (i, j) pair of species. The various contributions to $\mathbf{q}$ correspond to Fourier's law, diffusion of species with unlike energy contents and Dufour effect, respectively.

Last, but not least, one has to relate the diffusion velocity to the various gradients that induce it. Unfortunately, this dependence is implicit and, when $\mathbf{g}$ is the only body force, is given by the system:

$$\sum_j \frac{X_i X_j}{D_{ij}}\left(\mathbf{U}_j^{\mathrm{diff}} - \mathbf{U}_i^{\mathrm{diff}}\right) = \boldsymbol{\nabla}\,X_i + (X_i - y_i)\frac{\boldsymbol{\nabla}\,p}{p}$$

$$+ \sum_j \frac{X_i X_j}{\rho D_{ij}}\left(\frac{D_{T,i}}{y_i} - \frac{D_{T,j}}{y_j}\right)\frac{\boldsymbol{\nabla}\,T}{T} \qquad (2.1.10)$$

which, *a priori*, is degenerate and must be supplemented by (2.1.3). The various contributions to the r.h.s. of (2.1.10) are known as Fick diffusion, barodiffusion and thermal diffusion (or Soret effect), respectively.

Once supplemented by expressions for the chemical rates w_k (to be handled later on), and for the various transport (λ, D_{ij}, $D_{T,i}$, μ, μ') or thermodynamic (C_{pi}) coefficients, equations (2.1.1) to (2.1.10) are the only ones (Williams, 1985) that need be solved in a typical problem of "exothermic chemical reactions in flows with heat and mass transfer" (Williams, 1992).

2.1.4 Reactive Euler equations

In some instances, e.g., when studying the reaction zone of detonations or the hydrodynamical fields around flames, the diffusive fluxes appearing in the balance equations are negligible. Then (2.1.1)–(2.1.5) simplify into the so-called reactive Euler equations:

$$\frac{\partial \rho}{\partial t} + \nabla \cdot (\rho \mathbf{u}) = 0 \,, \tag{2.1.11}$$

$$\rho \left(\frac{\partial \mathbf{u}}{\partial t} + \mathbf{u} \cdot \nabla \mathbf{u} \right) = -\nabla p + \rho \mathbf{g} \,, \tag{2.1.12}$$

$$\rho \left(\frac{\partial h}{\partial t} + \mathbf{u} \cdot \nabla h \right) = \frac{\partial p}{\partial t} + \mathbf{u} \cdot \nabla p \,, \tag{2.1.13}$$

$$\rho \left(\frac{\partial y_k}{\partial t} + \mathbf{u} \cdot \nabla y_k \right) = \mathcal{M}_k w_k \,, \tag{2.1.14}$$

which, parenthetically, constitute a hyperbolic system (see Section 5.2). Once these equations are solved (!!), one still has to check, *a posteriori*, that the gradients of their solutions imply negligible molecular transports.

2.2 Weak forms

Unfortunately, beside smooth solutions the Euler equations, reactive or not, admit discontinuous ones (cf. Section 5.2–5.3).

In such a situation, the question which arises is: what are the constraints imposed on the values of ρ, $\mathbf{u}$, p, $e,\ldots$ on both sides of the discontinuity surface? To answer this question, we firstly recall that the balance equations (2.1.1)–(2.1.5) are the local forms of integral conservation principles which, contrary to (2.1.11)–(2.1.14) hold even across discontinuities. Consequently, these integral laws can be used over any infinitesimal control volume containing a small element of the surface (Fig. 2.1).

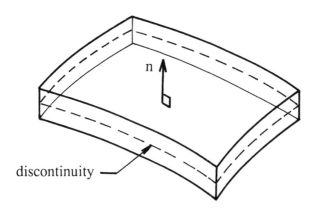

discontinuity

Fig. 2.1. Control volume for obtaining the jump conditions across discontinuities.

It is a classical, albeit lengthy exercise (Germain, 1986) to demonstrate that the discontinuous solutions to (2.1.11)–(2.1.13) meet the following jump conditions, also called the Rankine–Hugoniot relationships

$$[\rho \mathbf{n} \cdot (\mathbf{u} - \mathbf{D})] = 0 , \qquad (2.2.1)$$

$$\left[p + \rho(\mathbf{n} \cdot (\mathbf{u} - \mathbf{D}))^2\right] = 0 , \qquad (2.2.2)$$

$$[\mathbf{u} \times \mathbf{n}] = 0 , \qquad (2.2.3)$$

$$\left[h + \tfrac{1}{2}(\mathbf{u} - \mathbf{D}) \cdot (\mathbf{u} - \mathbf{D})\right] = 0 . \qquad (2.2.4)$$

The symbol $[g]$ denotes the difference between the values of g on each side of the surface, $\mathbf{n}$ is the local unit normal and $\mathbf{D}$ is the local velocity of the surface. Four remarks are in order:

i)   No condition on the mass fractions is listed, thereby allowing for a steep change in the mixture composition.

ii)   If no mass flux crosses the discontinuity, which is then termed a *contact discontinuity*, (2.2.3) has to be omitted.

iii)   The discontinuous solutions to (2.1.11)–(2.1.14) subjected the constraints (2.2.1)–(2.2.4) are called *weak*.

iv)   When used on both side of a nonzero thickness (l_T) interface, (2.2.1)–(2.2.4) are correct only to within $\mathcal{O}\left(l_T/\Lambda\right)$ corrections, where Λ is the typical length scale deduced from (2.1.11)–(2.1.14) and, provide only an estimate of the real change of the variables. In fact, (2.1.11)–(2.1.14) are themselves correct only to within the same accuracy.

3 Subsonic versus supersonic traveling waves

3.1 The (p-V) plane

It is customary to represent the relationships between the states (noted here by the dummy subscripts "+" and "−") on both sides of a discontinuity in terms of pressure p and specific volume $V = 1/\rho$. To this end, we firstly combine (2.2.1) and (2.2.2) into

$$[p]/[V] = -(\rho \mathbf{n} \cdot (\mathbf{u} - \mathbf{D}))^2 , \qquad (3.1.1)$$

which defines a straight line, called the Rayleigh line, in the $(p–V)$ plane. It connects the state ahead of the discontinuity (or front) to the final one (subscript "+"), the slope of which is related to the mass flux ($\neq 0$ for the waves considered in this chapter) crossing each front element. Because the r.h.s. of (3.1.1) is definitely negative, $[p]$ and $[V]$ must have different signs. One can thus distinguish between two kinds of discontinuous traveling waves: the compressive ($[V] < 0$) and the expansive ($[V] > 0$) ones. For a given initial ("−") state, the final one must lie on a Rayleigh line.

To specify it further, we eliminate $\mathbf{u}$ and $\mathbf{D}$ from (2.1.1)–(2.1.4) to result in:

$$h_+ - h_- = \tfrac{1}{2}(p_+ - p_-)(V_+ + V_-). \qquad (3.1.2)$$

Once caloric equations of states are provided in the forms $h = h_\pm(p, V, \text{composition})$, (3.1.2) defines a second family of curves, each of them being labeled by the mixture composition; this second set of curves is usually associated with the names of *Hugoniot*, if the composition is the same on both sides, (e.g., through shocks, Section 5) or of *Crussard*, if composition changes (e.g., in flames, Section 4, or detonations, Section 6). The initial state "$-$" being specified, the final one has to be at the intersection of a Rayleigh line and a Crussard (or Hugoniot) curve. One may note that the latter only depends upon the initial state and the properties of the mixture (composition changes included), whereas the former only depends on the mass-flux crossing the wave.

3.2 Various waves

Depending on the relative positions of the Crussard (or Hugoniot) curves and of the Rayleigh line, several configuration can be envisaged.

We firstly consider discontinuous waves in inert media, for which the composition stays fixed and, therefore, the initial state lies on the Hugoniot curve. The Rayleigh line then defines nontrival final states corresponding to either $p_+ > p_-$ (shock wave) or $p_+ < p_-$ (rarefaction wave). As is to be explained later on (Section 5.2), the second principle of thermodynamics (increase in specific entropy) in general rules out discontinuous rarefaction waves.

If composition is now allowed to be different on both sides of the wave, there is the possibility that the Rayleigh line does not intersect the Crussard curve, because the latter does not contain the initial state (Fig. 3.1). By same token multiple or double intersection can be obtained, each of them defining an admissible final state. By definition $p_+ > p_-$ ($p_+ < p_-$) will define a detonation (deflagration).

Under the additional assumptions (Courant and Friedrichs, 1948)

$$\left.\frac{\partial h}{\partial V}\right|_p > 0 \quad \text{and} \quad \left.\frac{\partial h}{\partial p}\right|_V > 0,$$

where the derivatives are evaluated at constant composition, the Crussard curve satisfies $d^2p/dV^2 > 0$ and the only possibilities are: i) no intersection, ii) tangency of Rayleigh and Crussard curves (at CJ and CJ′) and iii) two intersections at S and W (s, w) on the detonation (deflagration) branch. The condition of tangency at CJ or CJ′ define

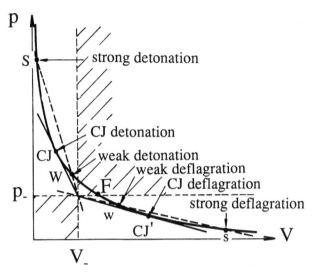

Fig. 3.1. Detonation and deflagration branches in $(p\text{–}V)$ plane

the so-called *Chapman–Jouguet* final states, and the deflagrations with $p_+ \approx p_-$ are traditionally termed *flames*.

3.3 Shocks, detonations and deflagrations

It is a simple matter to show that the specific entropy S has the extremum along the Crussard curve at the CJ or CJ' points. At these points the slope of the Crussard curve dp/dV is indeed given by (3.1.1), i.e.:

$$\frac{dp_+}{dV_+} = \frac{p_+ - p_-}{V_+ - V_-} .$$
(3.3.1)

On the other hand differentiating (3.1.2) at *fixed composition* and *initial state* yields

$$dh_+ = V_+ dp_+$$
(3.3.2)

once (3.3.1) is used. Comparing (3.3.2) with the Gibbs–Duhem identity (the differentiated form of $h = h_+(S_+, p_+)$ at fixed composition)

$$dh_+ = T_+ dS_+ + V_+ dp_+$$
(3.3.3)

implies the sought after result

$$dS_+ = 0 \qquad \text{at CJ and CJ'} .$$
(3.3.4)

Therefore $-V_+^2 dp_+/dV_+$ is simply the frozen sound speed c_+^2, where:

$$c^2 = \left. \frac{\partial p}{\partial \rho} \right|_S .$$
(3.3.5)

Comparing this result to (3.3.1), then (3.1.1), shows that the Mach number

$$M_+ = \frac{\mathbf{n} \cdot (\mathbf{u} - \mathbf{D})_+}{c_+} \tag{3.3.6}$$

is unity at the CJ and CJ$'$ points: Chapman–Jouguet deflagrations or detonations are sonic with respect to the fluid in its final state. Let us denote the r.h.s. of (3.1.1) by $-j^2$; differentiating (3.1.1) at fixed initial state yields

$$dp_+ + j^2 dV_+ = (V_- - V_+) \, d\left(j^2\right) . \tag{3.3.7}$$

From the equation (2.2.4) differentiated once, one next has

$$dh_+ + j^2 V_+ dV_+ = \tfrac{1}{2}\left(V_-^2 - V_+^2\right) d\left(j^2\right) , \tag{3.3.8}$$

which, when combined with (3.3.7) and the Gibbs–Duhem identity (3.3.3) to eliminate dp_+ leads to:

$$T_+ dS_+ = \tfrac{1}{2}\left(V_- - V_+\right)^2 d\left(j^2\right) \tag{3.3.9}$$

whereby

$$\frac{d\left(j^2\right)}{dS_+} > 0 . \tag{3.3.10}$$

On the other hand, we consider the identity

$$\frac{d\left(M_+^2\right)}{dp_+} = -\frac{d\left(j^2\right)}{dp_+}\left.\frac{\partial V_+}{\partial p_+}\right|_{S_+} - j^2 \left.\frac{\partial^2 V_+}{\partial p_+^2}\right|_{S_+} . \tag{3.3.11}$$

At the Chapman–Jouguet points $dS_+ = 0$, hence $d\left(j^2\right) = 0$ (from (3.3.9)). Therefore, at CJ or CJ$'$,

$$\frac{d\left(M_+^2\right)}{dp_+} = -j^2 \left.\frac{\partial^2 V_+}{\partial p_+^2}\right|_{S_+} .$$

Because $M_+ = 1$ at CJ points, this implies that $M_+ < 1$ when $p_+ > p_{CJ}$ if

$$\left.\frac{\partial^2 V_+}{\partial p_+^2}\right|_{S_+} > 0 .$$

Although not obligatory, the last condition is fulfilled by most materials.

We thus obtain the so-called Jouguet rules, which state that the normal flow velocity $\mathbf{n} \cdot (\mathbf{u} - \mathbf{D})$ relative to the discontinuity is

Supersonic:

• ahead of detonations (and shocks)
• behind a weak detonation (point W in Fig. 3.1)
• behind a strong deflagration (point s)

Subsonic:

• ahead of deflagrations (e.g., flames)
• behind a strong detonation (point S)
• behind a weak deflagration (points w or F)

Because they follow from (2.2.1)–(2.2.4), hence from the most general conservations equations in integral forms, the above conclusions are in fact exact for any *steady, planar* traveling wave, *whatever its thickness.*

4 Flames

4.1 Phenomenology

The following recipe for a "hands-on" experiment is accessible, even if not advisable, to everybody: leave town gas to leak out of an oven and wait for an hour (say). Nothing happens, until a match is used nearby. Provided he or she has good reflexes and a marked fondness for original observations, the amateur experimentalist will undoubtedly notice a blue, surface-like wave, propagating almost radially from the match and trying stubbornly to fill the whole kitchen with very hot gases.

This thin, blue wave is a typical premixed flame, here of town gas and air. Accumulated experience will certainly reveal that flame propagation takes place at a rather reproducible, mixture-dependent, velocity if the atmosphere is still to begin with, that the flame surface often is corrugated ultimately and displays a pebbled appearance, presumably due to some instability. Waving hands during the propagation is likely to modify the flame-shape evolution, plausibly due to a phenomenon of hydrodynamical stirring (that some hydrodynamics is at work is soon felt by the windows of the kitchen!).

Though interesting in itself and close to other situations which one has to understand better, such as combustion in most piston engines or the consequences of large spills of natural gas, the above experiment is not ideally suited for careful measurements, and is too complex as a first step to consider. To try to understand the basics of flame phenomena, the scientists soon invented a variety of burners which allow for more ample (and less risky) observations. Perhaps the simplest one (conceptually) is the plug-flow burner sketched in Fig. 4.1 (top): the reactive, gaseous premixture (methane and air, say, like in the above hazard) is injected uniformly, e.g., through a porous disc or a honeycomb, in a vertical tube. The injection rate is carefully chosen so as to maintain the flame that propagates counter-streamwise at a fixed, remote enough, location in the tube.

If, for a given mixture, there exists anything like a definite flame speed u_L relative to the incoming flow, a flat wave could be obtained. In some cases of downward propagations, this is in effect what happens (Fig. 4.1, bottom), thereby allowing u_L to be measured ($u_L \leq 13$ cm/s, typically) along with the thickness of the flame (a few 10^{-4} m for reactants at room pressure and temperature). In other circumstances, the flame spontaneously evolves to a more-or-less regularly structured, cellular interface (Fig. 4.2).

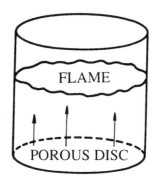

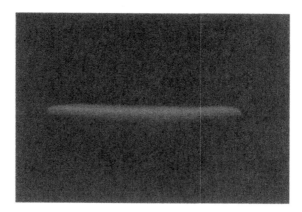

Fig. 4.1. Top: Sketch of a cylindrical plug-flow burner. Bottom: Side view of a 10 cm wide, stable steady flame. Courtesy of J. Quinard (Marseilles); see also Quinard (1984).

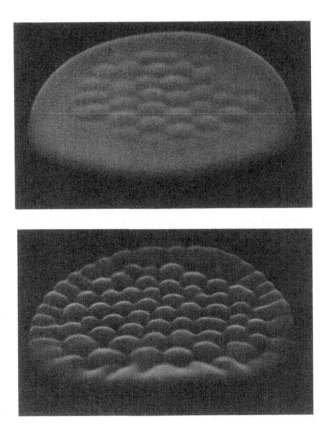

Fig. 4.2. Oblique view (from below) of cellular flames propagating downward in a plug-flow burner. Top: near marginal cells. Bottom: slightly more mature ones. Reproduced with kind permission of J. Quinard; see also Quinard (1984).

In the case of upward propagations (Fig. 4.3), bullet-shaped flames are usually observed that are sometimes covered by elongated substructures oriented along meridians, whereas experiments at microgravity almost never lead to flat flames (Dunsky, 1992), except when the flame is very close to the injection plane.

Another popular burner is Bunsen's (Fig. 4.4), i.e., an open circular tube out of which the fresh gases flow at a velocity u_∞ higher than u_L. Ideally, if injection is uniform and the flame is somehow anchored all along the tube rim, a conical flame with semi-angle $\arcsin(u_L/u_\infty)$ should result. This is nearly what is observed in many instances (Fig. 4.4, left), even though the tip is often rounded (or even open) rather than sharp. With some well chosen mixtures, however, the flame evolves to a pyramidal shape (the so-called polyhedral flames; Fig. 4.4, middle and

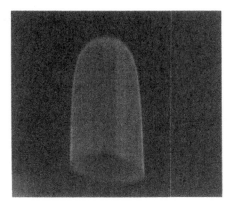

Fig. 4.3. Bullet-shaped steady flame propagating upward in a tube (10 cm in diameter). Private communication from Q. Wang & P.D. Ronney (Princeton).

right) that may even rotate about the tube axis. Clearly an instability again is at work. As the experiments demonstrate, this spontaneous symmetry-breaking phenomenon is noticeably mixture-dependent and tends preferably to show up with mixtures of heavy hydrocarbons and air that are deficient in oxygen.

Another type of experiment, which is closer to the hazard evoked previously, simply consists of filling a tube with the reactants, waiting until the medium is (presumably) at rest, then igniting it from the upper, open end, preferably over the entire cross-section. A flat flame

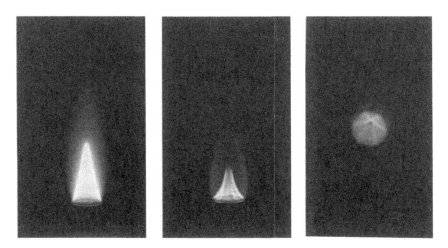

Fig. 4.4. Various Bunsen-burner flames of butane and air. Left: Regular case. Middle: Polyhedral flame. Right: Oblique view of the same polyhedral flame. Private communication from B. Deshaies (Poitiers).

propagating towards the closed end could result, but this is seldom so, especially when fast-enough flames are involved ($u_L > 20$ cm/s, say). Instead, a pebbled front often develops (Fig. 4.5).

In some cases, clearly audible sound is emitted, the flame suddenly flattens as it propagates, then gets covered with fine cells (Fig. 4.6, top) and next accelerates rapidly into a turbulent-like state while making a loud noise (Fig. 4.6, bottom).

All these phenomena are clearly connected with instability mechanisms. Along with the propagation itself, these are the subject matter of the rest of the present section; it will end by introducing a few amongst the many yet unsolved problems of combustion theory.

4.2 Minimal model and isobaric approximation

In the absence of simplification, solving the full set (2.1.1)–(2.1.10) of conservation equations is a formidable task, even for the fastest computers (Giovangigli, 1988). If anything is to be understood about flames by use of theoretical methods, this is certainly best started in the framework of a minimal model which, while doing no damage to the physics and incorporating the most salient features, does not retain every detail of the mechanisms involved and takes advantage of the realistic parameter values pertaining to the "flame" phenomenon. Afterwards, quantitative improvements can be sought for, if needed.

As is to be confirmed later on, the diffusive transports of heat and species play crucial roles in flames. Equations (2.1.10), which are linear in

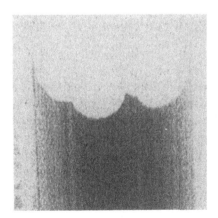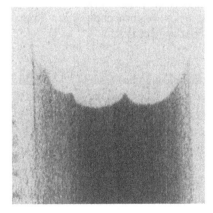

Fig. 4.5. Tomographic snapshots of a bumpy unsteady flame propagating into a premixed quiescent premixture (dark region) down a circular tube (10 cm in diameter). Courtesy of G. Searby (Marseilles).

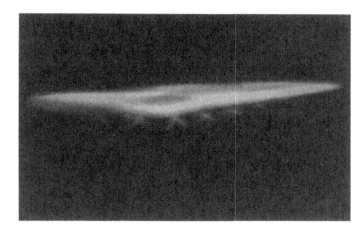

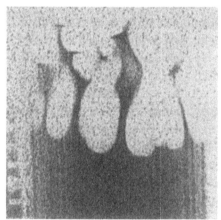

Fig. 4.6. Top: A singing, flattened flame propagating down a tube, just after the onset of acoustic-driven wrinkling. Bottom: Tomographic view of the same flame, a few milliseconds later. Reproduced with permission of G. Searby; see also Searby (1991a).

the diffusion velocities $\mathbf{U}_i^{\mathrm{diff}}$, could in principle be solved to express these in terms of the various gradients which induce them (∇y_i, ∇p, ∇T), a heavy task. It will soon transpire that barodiffusion (species migration due to pressure gradients) is quite negligible in flames, owing to the small ∇p involved. Species diffusion caused by ∇T (the so-called Soret effect) cannot be discarded so easily, due to the strong temperature gradients encountered in flames (10^6–10^7 K/m, typically). It is experimentally known, however, that the coefficients $k_{ij} \equiv D_{T,i}/D_{ij}$ often are less than 10^{-1} in absolute value. Guided by this fact, we are to totally neglect thermal diffusion of species and write $k_{ij} = 0$. This also automatically suppresses the Dufour effect (heat flux due to concentration gradients), whose actual influence on real flames is virtually nonexistent. We further simplify the matter by assuming that the initial mixtures in which the flames propagate consist of an inert gas in large excess and of a few reactants present by small amounts only; this is nearly exactly the case for fuel-lean flames using air as oxidizer, owing to the high concentration of nitrogen and to the fact that oxygen and nitrogen have very similar properties, apart from reactivity. Consequently the diffusion velocities of the active components and combustion products simplify into:

$$\mathbf{U}_i^{\text{diff}} \simeq -D_i \, \boldsymbol{\nabla} \, y_i / y_i \,, \qquad (4.2.1)$$

where D_i stands for $D_{i,\text{diluent}}$; the concentration profile for the diluent itself can be estimated *a posteriori*, because all the mass fractions in a mixture must add up to unity. Equation (4.2.1) merely is Fick's law which is exact for binary mixtures or, most importantly, when all the diffusion coefficients are equal. In general, however, using it is a rather *ad hoc* assumption, which can be relaxed once the physics is understood. Lastly, because the mixture mainly consists of an inert gas, the mixture specific heat (at constant pressure) C_p, its heat conductivity λ, and the viscosities (μ, μ') are nearly those of the diluent and may be taken to be functions of temperature only; we even take the same constant specific heat C_p for all species and a constant molecular weight $\mathcal{M}$ for the mixture, these qualitatively unimportant simplifications being introduced for analytical convenience only.

Even once reduced as above, the conservation equations still "contain" practically all the combustion phenomena, including shock waves and detonations, and turbulence. It is now time to exploit the particularities of flames. To this end, we first consider the momentum balance

$$\rho \left(\frac{\partial \mathbf{u}}{\partial t} + (\mathbf{u} \cdot \boldsymbol{\nabla}) \mathbf{u} \right) = - \, \boldsymbol{\nabla} \, p \,, \qquad (4.2.2)$$

in which the viscous and buoyant contributions to acceleration are provisionally omitted. Let l_{T} be a typical flame thickness, over which the thermal effects force velocities and pressure to also vary. We shall admit (in fact, anticipate) that:

i)   the changes in $\mathbf{u}$ about a flame are comparable to the speed u_L at which a steady planar flame would propagate towards the fresh mixture;

ii) in unsteady flames, the time scale of flame-structure evolutions is not much shorter that l_{T}/u_L (10^{-4} s, typically).

Then, (4.2.2) leads to the estimate $p - p_{\text{u}} \sim \rho u_L{}^2$ for the pressure change over the upstream one (subscript "u"). Using the perfect gas law one next obtains

$$(\rho T/\rho_u T_u) - 1 \sim u_L{}^2/(p_u/\rho_u) \,.$$

Up to a factor of order unity, the fractional change of ρT in a flame is thus proportional to the Mach-number-squared based upon u_L and the upstream speed of sound $c_u \sim (p_u/\rho_u)^{1/2}$. In practice, u_L typically ranges from 5.10^{-2} m/s to 10 m/s, thereby leading to a very small variation of ρT within the flames. Consequently, one may then safely use the simplified equation of state

$$\rho T \simeq \rho_u T_u = p_u/(\mathcal{R}/\mathcal{M}) \,, \qquad (4.2.3)$$

given that ρ and T do *not* vary by small fractions; in (4.2.3) p_u may still be a function of time. Let us now consider the energy balance (2.1.5). Using the assumption $C_{pi} = C_p$ and still ignoring the viscous effects one has:

$$\rho C_p \left(\frac{\partial T}{\partial t} + \mathbf{u} \cdot \boldsymbol{\nabla} T\right) = \boldsymbol{\nabla} \cdot (\lambda \boldsymbol{\nabla} T) + \frac{\partial p}{\partial t} + \mathbf{u} \cdot \boldsymbol{\nabla} p - \sum_{i=1}^{N} w_i h_i (T_u) \quad (4.2.4)$$

once (2.1.2) and (2.1.7a) are employed. From the preceding estimate for pressure variations and the assumption that $\mathbf{u} \sim u_L$, it is deduced that

$$\rho C_p \mathbf{u} \cdot \boldsymbol{\nabla} T \sim \rho u_L C_p (T - T_u)/l_T \,, \qquad \mathbf{u} \cdot \boldsymbol{\nabla} p \sim \rho u_L{}^3/l_T \,. \qquad (4.2.5)$$

Therefore $\mathbf{u} \cdot \boldsymbol{\nabla} p$ and the convection term $\rho C_p \mathbf{u} \cdot \boldsymbol{\nabla} T$ are in a typical ratio of $u_L{}^2/C_p (T - T_u) \sim u_L{}^2/(p_u/\rho_u)$: the term $\mathbf{u} \cdot \boldsymbol{\nabla} p$ is negligible compared to the convective flux of enthalpy at the limit of zero Mach numbers. The energy equation then acquires the simplified form:

$$\rho C_p \left(\frac{\partial T}{\partial t} + \mathbf{u} \cdot \boldsymbol{\nabla} T\right) = \boldsymbol{\nabla} \cdot (\lambda \boldsymbol{\nabla} T) + \frac{dp_u}{dt} - \sum_{i=1}^{N} w_i h_i (T_u) \,. \qquad (4.2.6)$$

When the upstream pressure p_u is time-independent, (4.2.4) is a mere thermal balance; in some other circumstances, e.g., when considering the flames which propagate, say, in the cylinders of car engines, dp_u/dt has to be retained and can lead to new phenomena: $dp_u/dt < 0$ acts as heat losses and may quench a flame. It is left to the reader's interest, as an exercise, to check that restoring the viscous terms in the momentum and energy balances does not modify the above estimates, given that the Prandtl numbers such as $\mu C_p/\lambda$ are of order unity in gases. One may also check that barodiffusion is indeed negligible, as previously claimed.

To summarize, we list below the equations describing what we shall from now onward call "isobaric flames"

$$\frac{\partial \rho}{\partial t} + \boldsymbol{\nabla} \cdot (\rho \mathbf{u}) = 0 \,, \qquad (4.2.7a)$$

$$\rho\left(\frac{\partial \mathbf{u}}{\partial t} + \mathbf{u} \cdot \nabla \mathbf{u}\right) = -\nabla \cdot (p\mathbf{I} + \mathbf{P}) + \rho\mathbf{g}, \tag{4.2.7b}$$

$$\rho C_p\left(\frac{\partial T}{\partial t} + \mathbf{u} \cdot \nabla T\right) = \nabla \cdot (\lambda \nabla T) + \frac{dp_u}{dt} - \sum_{i=1}^{N} w_i h_i (T_u), \tag{4.2.7c}$$

$$\rho\left(\frac{\partial y_i}{\partial t} + \mathbf{u} \cdot \nabla y_i\right) = \nabla \cdot (\rho D_i \nabla y_i) + w_i, \tag{4.2.7d}$$

$$\rho T = \rho_u T_u = p_u/(\mathcal{R}/\mathcal{M}). \tag{4.2.7e}$$

Of course the terminology "isobaric" only means that pressure is considered uniform *in the energy balance and the equation of state* (within and around the flame); pressure gradients, albeit small, must obviously still be accounted for in the momentum equations to make the gas move! Acoustics excepted, equations (4.2.7) can capture virtually all the flame-related phenomena, at least qualitatively; in particular they still contain turbulence!

Our final assumption deals with chemistry. Any quantitatively realistic description of flames would need to incorporate a detailed chemical network; with the most frequently encountered mixtures (hydrocarbon/air, say) this means accounting for dozens of species and some fifty to a few hundred chemical reactions. This is definitely too much for the presently available analytical tools and further simplifications are needed. A key observation is that the various reactions often proceed at quite different paces, due to microscopic phenomena or owing to the topology of the chemical network. For example, only the fastest of parallel reactional paths effectively matters, whereas the slowest step of sequential reactions is the limiting one (for further details, see Clavin (1985) and the references therein or Peters and Rogg, 1993). This has the result that only a few elementary steps effectively control the overall rates of heat release and of main-reactant consumption. The other reaction paths are either bypassed or follow adiabatically, thereby allowing one to build reduced chemical schemes.

We shall not develop this procedure and, instead, acknowledge that most aspects of premixed flame dynamical behaviors can be qualitatively captured upon retaining a single, overall irreversible reaction as the burning process

$$A + B \to \text{Products} + \text{heat}$$

where A (B) is the deficient (abundant) reactant such as methane (oxygen) in methane/air flames. Owing to an assumed markedly-unbalanced stoichiometry in the fresh medium, we are even to neglect the changes in concentration of B wherever chemical activity takes place (see

Section 4.3.7, however). Finally, we shall assign the Arrhenius structure:

$$w = \frac{\rho}{t_{\text{coll}}} \, y_A \, \exp(-E/\mathcal{R}T) \tag{4.2.8}$$

to the consumption rate of A. E will henceforth represent the activation energy and y_A the mass fraction of A; t_{coll} stands for a representative collision time (between A and B). The above form of w is chosen for simplicity, analytical tractability and because it yields reasonably accurate predictions for the fractional changes in burning speed (more generally, in chemical activity).

4.3 The basic eigenvalue problem

4.3.1 The steady planar flame

The following remark will greatly ease our analysis of flame propagation: quite often, the thickness l_T of a flame is markedly smaller than its other dimensions, such as the size of the wrinkles, the dimensions of confinement. One may then speak of a flame front. To understand the basic mechanisms of propagation, one may next envisage to locally approximate it by a flat flame which, if evolving slowly enough, may also be viewed as quasi-steady: the approximately conical flames anchored on a Bunsen burner (Fig. 4.4) or, to a far better accuracy, the disk-shaped one stabilized in a carefully controlled, uniform ascending premixture (Fig. 4.1, right) are safely amenable to such an idealization.

This naturally leads to the concept of a steady, planar premixed flame propagating in a uniform mixture, a genuine paradigm in combustion theory. In a coordinate system attached to such a flame, with the x-axis normal to the front and pointing towards the burned medium, all the profiles of temperature, mass fraction, pressure... should be functions of x only. We shall denote the mixture velocity in the x-direction, i.e., normal to the flame, by $u(x)$.

The conservation equations (4.2.7) corresponding to the minimal model of isobaric flame then simplify as follows. Continuity equation indicates $d(\rho u)/dx \equiv 0$; hence ρu is a constant throughout the whole flame, which one may relate to the yet unknown *laminar flame speed* u_L relative to the fresh gases (subscript "u") by the self-explanatory formula:

$$\rho u \equiv \rho_u u_L \,. \tag{4.3.1}$$

Energy conservation leads to:

$$\rho u C_p \frac{dT}{dx} = \frac{d}{dx} \left(\lambda \frac{dT}{dx} \right) + Q w \,, \tag{4.3.2}$$

where the heat of reaction $Q > 0$ stands for $h\left(T_u\right)_{\text{reactant}} - h\left(T_u\right)_{\text{product}}$, whereas the reactant balance gives

$$\rho u \frac{dy_A}{dx} = \frac{d}{dx}\left(\rho D_A \frac{dy_A}{dx}\right) - w. \tag{4.3.3}$$

The reaction rate w is assigned the Arrhenius form (4.2.8) which is repeated for convenience:

$$w = \frac{\rho}{t_{\text{coll}}} y_A \exp(-E/\mathcal{R}T). \tag{4.3.4}$$

As for the equation of state, it simply expresses that ρT identically equals $\rho_u T_u$, so that the ratio $u(x)/T(x)$ is also a constant. Finding the temperature profile and u_L therefore immediately gives the velocity profile $u(x)$. As a result, the x-momentum equation is not coupled with (4.3.2) and (4.3.3) and merely serves to compute $p(x)$ *a posteriori*, once $u(x)$ is found. This decoupling is particular to the (nearly-)isobaric flames considered here and to one-dimensional fronts propagating in quiescent mixtures. In the present paragraph, the x-wise momentum balance will simply be ignored. Attention is to be focused on the two "master" equations (4.3.2) and (4.3.3) which we shall attempt to solve for $T(x)$, $y_A(x)$, and the yet unknown scalar ρu, which plays the role of an *eigenvalue*. To this end, one needs boundary conditions. The following values

$$\rho(-\infty) = \rho_u, \quad T(-\infty) = T_u, \quad y_A(-\infty) = y_u \tag{4.3.5}$$

are assumed given; at the burned side one has $y_A(+\infty) = 0$ for w to vanish, and

$$T(+\infty) = T_b \equiv T_u + \frac{Q}{C_p} y_u. \tag{4.3.6}$$

The latter self-explanatory result immediately follows from (4.3.2) and (4.3.5) upon integration from $-\infty$ to $+\infty$, where all the diffusive fluxes vanish, and elimination of the integrated source term.

4.3.2 Order of magnitude estimates

An instructive preliminary step is to perform an order of magnitude analysis, as to evaluate the flame gross features. To this end we estimate l_T and ρu by requiring that the convective and chemical terms in (4.3.2) both be comparable to heat conduction. Writing $dT/dx \sim (T_b - T_u)/l_T$, $d^2T/dx^2 \sim (T_b - T_u)/l_T^2$, $\lambda \sim \lambda_u$, $w \sim w_{\max}$, one arrives at

$$(\rho u)C_p \frac{T_b - T_u}{l_T} \sim \lambda_u \frac{T_b - T_u}{l_T^2} \sim \frac{(T_b - T_u)C_p}{y_u} w_{\max}, \tag{4.3.7}$$

where $Qy_u = C_p (T_b - T_u)$ is also employed. This immediately gives:

$$l_T = \frac{\lambda_u}{(\rho u)C_p} \equiv \frac{D_{th}}{u_L} , \qquad (4.3.8)$$

where D_{th} is the upstream heat diffusivity; l_T may thus be called the convection/conduction flame thickness. Next, one has: $u_L{}^2 \sim D_{th}w_{max}/\rho_u y_u$ (Fig. 4.7).

The last formula is best interpreted once one notices that $\rho_u y_u/w_{max}$ measures the chemical time t_{ch} evaluated where chemical activity is maximum; as is to be show later, this takes place at $T \simeq T_b$ if the activation temperature $E/\mathcal{R}$ is large enough compared to T_b. Thus:

$$u_L{}^2 \sim D_{th}/t_{ch} . \qquad (4.3.9)$$

Results identical to (4.3.8) and (4.3.9) are obtained on writing that during the chemical time t_{ch} the flame travels by conduction over a distance given by both $l_T \sim u_L t_{ch}$ and $l_T{}^2 \sim D_{th}t_{ch}$. Accordingly, a mixture element has just time to burn when crossing the flame... a sensible viewpoint. One may next invoke an elementary result from kinetic theory of gases, according to which D_{th} is related to a typical collision time t_{coll} and to the sound speed (upstream, say) c_u by $D_{th} \sim c_u{}^2 t_{coll}$. This allows one to transform (4.3.9) into:

$$M^2 = (u_L/c_u)^2 \sim t_{coll}/t_{ch} . \qquad (4.3.10)$$

The flame Mach number M squared is thus simply the probability of reaction in a collision. Reactive collisions between reaction partners being very rare events, those flames which propagate by conduction and chemistry should therefore be very markedly subsonic, which is indeed true: the practical range of u_L is between 5 cm/s (lean methane/air) and

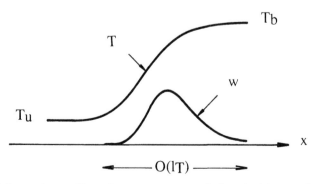

Fig. 4.7. Schematic profiles of temperature and chemical heat release across a steady flame.

10 m/s (slightly rich H_2/O_2). As an increase in temperature quickly shortens the chemical times (reactive collisions get more probable), preheating the mixture is bound to make the flame propagate faster, a result of great importance. Because the collision times and D_{th} are approximately inversely proportional to pressure, (4.3.9) suggests that u_L should roughly be pressure-independent, which is nearly the case in practice.

4.3.3 Cold boundary difficulty

Though quite appealing physically and leading to predictions that are routinely verified in experiments ($l_T \sim 1/p$, u_L almost independent of p but very sensitive to T_b), the arguments which lead to (4.3.8) and (4.3.9) contain a basic flaw: it was implicitly assumed that a solution existed, which is *not* the case! The chosen production term w is indeed nonvanishing at $T = T_u$, with the net result that the reactive premixture amply had time to fully react when traveling from $x = -\infty$ to any finite distance from the flame. This is the famous (or infamous!) "cold boundary difficulty." To show that the paradox is only apparent (Zel'dovich *et al.*, 1985; Berestycki *et al.*, 1991), it is convenient to define the cold boundary chemical time $t_u \equiv \rho_u y_u/w_u$ and to compare it to the transit time $l_T/u_L \sim t_{ch}$: $t_u/t_{ch} = w_{max}/w_u$. If w is given by the Arrhenius law (4.3.4) and if maximum activity occurs at $T \simeq T_b$, then the ratio:

$$\frac{t_u}{t_{ch}} \sim \exp\left(\frac{E}{\mathcal{R}T_u} - \frac{E}{\mathcal{R}T_b}\right)$$

is exceedingly large when the activation temperature is markedly above T_u. E being typically a few electronvolts for quantum-mechanical reasons, $E/\mathcal{R}$ is a few 10^4 K, so that such an assumption is realistic. With $T_u = 300$ K, $T_b = 2000$ K, $E/\mathcal{R} = 2.10^4$ K, one arrives at $t_u/t_{ch} \sim e^{56}$. Using the conservative estimates $u_L \sim 1$ m/s and $D_{th} \sim 10^{-4}$ m^2/s, one obtains $t_{ch} \sim 10^{-4}$ s from (4.3.9), then $t_u \sim 10^{21}$ s (one day $\simeq 10^5$ s): with such low initial temperatures and large activation energies, and if injected at a moderately large distance (less than 10^3 km, say) upstream of this thin front ($l_T \sim 10^{-4}$ m), the mixture has ample time to reach the flame practically unreacted. The conclusions could be different if the mixture was preheated, for example by an adiabatic compression or by a shock wave, to temperatures significantly higher than 300 K. At any rate, a mathematical difficulty exists, and it must be coped with, before one may attempt to solve any eigenvalue problem. To achieve this goal, we provisionally restrict our ambition to mixtures for which the dimensionless grouping

$$\text{Le} = D_{th}/D_A \tag{4.3.11}$$

called the *Lewis number*, is unity. Some gaseous reactive systems, such as dilute mixtures of methane in air, do meet the condition $Le = 1$ very accurately. Arbitrary Lewis numbers will be dealt with at a later stage. When $Le = 1$, the unsteady versions of (4.3.2) and (4.3.3) can be lumped into

$$\rho \left(\frac{\partial}{\partial t} + u \frac{\partial}{\partial x} \right) (C_p T + Q y_A) = \frac{\partial}{\partial x} \left(\frac{\lambda}{C_p} \frac{\partial}{\partial x} (C_p T + Q y_A) \right), \quad (4.3.12)$$

which, when combined with the initial condition $(C_p T + Q y_A)_{t=0} = C_p T_u(0) + Q y_u(0)$ and the requirement of vanishing diffusive fluxes at $x = \pm\infty$ (4.3.12) leads to the conclusion that

$$C_p T + Q y_A \equiv C_p T_u(0) + Q y_u(0) \quad (4.3.13)$$

provided this condition is met initially throughout space.

Therefore, even though T_u is now time-dependent, the burned gas $(y_A = 0)$ temperature stays fixed to $T_u(0) + Q y_u(0)/C_p$ (we still continue to call this T_b). We are now left with the two equations:

$$\frac{\partial \rho}{\partial t} + \frac{\partial}{\partial x}(\rho u) = 0 \quad (4.3.14)$$

$$\rho C_p \left(\frac{\partial T}{\partial t} + u \frac{\partial T}{\partial x} \right) = \frac{\partial}{\partial x} \left(\lambda \frac{\partial T}{\partial x} \right) + Q w(T, y_A) \quad (4.3.15)$$

in which it is understood that $C_p(T_b - T)/Q$ was substituted for y_A. These equations must be supplemented by the requirement that ρT be constant (pressure is) and by:

$$\rho_u C_p \frac{dT_u}{dt} = Q w \left(T_u, \frac{C_p}{Q}(T_b - T_u) \right), \quad \text{at} \quad x = -\infty \quad (4.3.16)$$

which defines the temperature evolution in the fresh mixture. We next introduce a scaled temperature increment θ defined by $\theta (T_b - T_u(t)) \equiv (T - T_u(t))$, which is to vary between zero (fresh side) and one (burned gases), along with reduced variables

$$\xi = C_p(\rho u)_{x=0} \int_0^x \frac{\rho \, dx}{\lambda \rho} \quad (4.3.17)$$

$$\tau = \frac{C_p}{\lambda_u \rho_u} \int_0^t (\rho u)_{x=0}^2 \, dt \quad (4.3.18)$$

Then, under the *provisional* assumption $\lambda \sim T$ (hence $\lambda \rho = \text{const.}$), equation (4.3.15) becomes:

$$\frac{\partial \theta}{\partial \tau} + \sigma(\tau)\xi \frac{\partial \theta}{\partial \xi} + \frac{\partial \theta}{\partial \xi} = \frac{\partial^2 \theta}{\partial \xi^2} + \Lambda W(\theta; \tau) \quad (4.3.19)$$

with $\sigma(\tau) \equiv d\left(\log(\rho u)_{x=0}\right)/d\tau$. $W(\theta;\tau)$ has the definition:

$$W(\theta;\tau) = \frac{\lambda \rho t_{\text{coll}}(T_{\text{b}})}{\lambda_{\text{b}} \rho_{\text{b}} t_{\text{coll}}(T)} \frac{E^2 (T_{\text{b}} - T_{\text{u}})^2}{R^2 T_{\text{b}}^4} (1 - \theta) \times$$

$$\left(\exp\left[\frac{E}{RT_{\text{b}}} \frac{T - T_{\text{b}}}{T}\right] - \exp\left[\frac{E}{RT_{\text{b}}} \frac{T_{\text{u}} - T_{\text{b}}}{T_{\text{u}}}\right] \frac{t_{\text{coll}}(T)}{t_{\text{coll}}(T_{\text{u}})} \right)$$
$$(4.3.20)$$

which is in fact valid, as well as that of Λ, for *any* $\lambda(T)$:

$$\Lambda \equiv \frac{\lambda_{\text{b}} \rho_{\text{b}} \, e^{-E/RT_{\text{b}}} R^2 T_{\text{b}}^4}{(\rho u)_{x=0}^2 C_p t_{\text{coll}}(T_{\text{b}}) E^2 (T_{\text{b}} - T_{\text{u}})^2} \qquad (4.3.21)$$

It is important to notice that $W(\theta;\tau)$ vanishes at $T = T_{\text{u}}$, i.e., $\theta = 0$. Consequently, if for some reason $\partial\theta/\partial\tau$ and $\sigma(\tau)$ may be safely neglected, one arrives at a well-posed eigenvalue problem (Zel'dovich *et al.*, 1985) for $\theta(\xi;\tau)$ and Λ

$$\frac{\partial\theta}{\partial\xi} \simeq \frac{\partial^2\theta}{\partial\xi^2} + \Lambda W(\theta;\tau), \qquad (4.3.22)$$

$$\theta(-\infty;\tau) = 0 = 1 - \theta(+\infty;\tau), \qquad (4.3.23)$$

in which τ only appears as a parameter and could then be omitted. This is precisely how the extreme smallness of $t_{\text{ch}}/t_{\text{u}}$ comes about to help us.

Assuming that all the time derivatives in (4.3.19) are induced by the slow drift of $T_{\text{u}}(t)$ and anticipating that one is to study (4.3.2) and (4.3.3) in the limit of large *Zel'dovich numbers*

$$\beta \equiv \frac{E}{RT_{\text{b}}} \frac{T_{\text{b}} - T_{\text{u}}(0)}{T_{\text{b}}} \gg 1 \qquad (4.3.24)$$

one is readily convinced that $\partial\theta/\partial\tau$ and $\sigma(\tau)$ are exponentially small in β, because $t_{\text{ch}}/t_{\text{u}}$ is, and therefore fully negligible against the power series in β which are to be involved when solving (4.3.19) asymptotically. We shall no longer distinguish between $(\rho u)_{x=0}$ and $\rho_{\text{u}} u_L$, or $T_{\text{u}}(0)$ and T_{u}.

4.3.4 An auxiliary model

Before proceeding to the resolution proper, it is interesting to pause for a remark. If the order-of-magnitude estimate (4.3.9) was correct, it would indicate that the eigenvalue is $\mathcal{O}\left(1/\beta^2\right)$ in the limit $1/\beta \to 0$. This is not exactly true, and Λ will soon prove to vary in a different way with β, but (4.3.9) nevertheless captures the principal (exponential) dependence of u_L on the activation energy and the corresponding strong sensitivity on reaction temperature. To guess the magnitude of Λ when β gets large, we momentarily cease to study (4.3.22) and, instead, spend some time with the auxiliary problem:

$$\frac{d\theta}{d\xi} = \frac{d^2\theta}{d\xi^2} + \Lambda\beta\theta^{1+\beta}\left(1-\theta^{\beta}\right),\tag{4.3.25}$$

$$\theta(-\infty) = 0 = 1 - \theta(+\infty).\tag{4.3.26}$$

It has the same structure as (4.3.22) and (4.3.23), apart from a different source term; that featuring in (4.3.25) is plotted in Fig. 4.8 for various values of the "peakness parameter" $1/\beta$. In much the same way as the "exact" reaction rate it is strictly positive when $0 < \theta < 1$ and vanishes at both ends; it gets more and more peaked near the final reduced temperature $\theta = 1$ as $1/\beta$ decreases to zero and is exponentially small in the distinguished limit defined by: $\beta \to \infty$, $\theta \neq 1$ fixed. Besides strikingly resembling (4.3.22), the above auxiliary model also has the advantage that its so-called ZFK solution (Zel'dovich and Frank-Kamenetskii, 1938), characterized by a conduction/convection balance at the cold side, is exactly known; it corresponds to:

$$\theta(\xi;\beta) = e^{\xi}\left(1 + e^{\beta\xi}\right)^{-1/\beta},\qquad \Lambda = 1 + 1/\beta\tag{4.3.27}$$

up to an arbitrary shift in ξ. This is readily checked upon substitution, thanks to the identity $d\theta/d\xi = \theta(1-\theta^{\beta})$ which holds in this case.

In the limit $1/\beta \to 0$, the above solution to our auxiliary model acquires three distinct asymptotic expansions, depending on the value of ξ.

1) If ξ is kept fixed and strictly negative, the limit $1/\beta \to 0$ makes $\theta(\xi;\beta)$ degenerate to:

$$\theta_{\text{upstream}} = e^{\xi} + \text{transcendentally-small terms}.\tag{4.3.28}$$

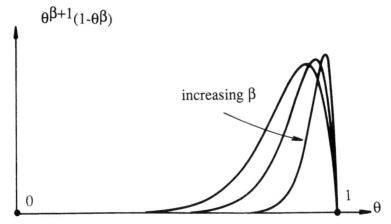

Fig. 4.8.   The rate function of the auxiliary model (4.3.25), for different β's.

This "upstream" outer profile merely is the solution to the source-free version of (4.3.25) and corresponds to an almost chemistry-free balance between conduction and diffusion; this is characteristic of the ZFK types of solution.

2) If ξ is kept fixed and strictly positive, taking the limit $1/\beta \to 0$ yields:

$$\theta_{\text{burnt}} \equiv 1 - \text{transcendentally-small terms}. \tag{4.3.29}$$

Note that (4.3.28) and (4.3.29) coincide to leading order when $\xi = 0$, but that patching them defines a profile with a corner at $\xi = 0$ (Fig. 4.9).

3) Lastly, if the stretched variable $\eta \equiv \beta\xi$ is kept fixed to investigate what happens close to the corner at $\xi = 0$, the limit $1/\beta \to 0$ leads to

$$\theta_{\text{inner}} = 1 + \sum_{m \geq 1} \frac{1}{\beta^m (m!)} \left(\eta - \log\left(1 + e^\eta\right)\right)^m. \tag{4.3.30}$$

Clearly (4.3.30) has the form of an inner expansion, which presents some distinctive features. First of all one may notice that $d\theta/d\xi = \mathcal{O}(1) \ll d^2\theta/d\xi^2 = \mathcal{O}(\beta)$ in the inner layer. Diffusion dominates convection there, which is logical since $\xi = \mathcal{O}(1)$ was the range of convection/conduction balance, by definition; this in turn implies a conduction/chemistry balance in the $\eta = \mathcal{O}(1)$ region, hence the need for having an eigenvalue Λ of order one. It is also worth recording the

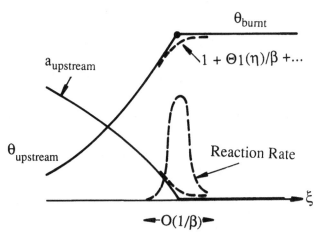

Fig. 4.9. Outer (solid curves) and inner (dashed lines) profiles of θ, a and reaction rate vs. reduced abscissa ξ, for a peaked production term $(1/\beta \to 0)$.

limiting behaviors of $\theta_{\text{inner}}(\eta)$ as $|\eta| \to \infty$, order by order:

$$\theta_{\text{inner}}(\eta \to -\infty) = 1 + \frac{\eta}{\beta} + \frac{\eta^2}{2\beta^2} + \frac{\eta^3}{6\beta^3} + \dots \qquad (4.3.31)$$

$$\theta_{\text{inner}}(\eta \to +\infty) = 1. \qquad (4.3.32)$$

Owing to the definition $\eta = \beta\xi$, one concludes that the inner expansion asymptotically matches the outer ones to all algebraic orders in β; after all, θ_{inner}, θ_{upstream} and θ_{burnt} are asymptotic expansions of the same function $\theta(\xi; \beta)$!

4.3.5 Asymptotic method

Equipped with this information it is now time to return to the more physical model (4.3.22) or, even better, to its obvious generalization corresponding to a nonunit Lewis number (Bush and Fendell, 1970):

$$\frac{d\theta}{d\xi} = \frac{d^2\theta}{d\xi^2} + \Lambda W, \qquad (4.3.33)$$

$$\frac{da}{d\xi} = \frac{1}{\text{Le}} \frac{d^2 a}{d\xi^2} - \Lambda W, \qquad (4.3.34)$$

where $a \equiv y_A/y_u$. The boundary conditions are:

$$\theta(-\infty) = 0 = 1 - \theta(+\infty), \qquad a(-\infty) = 1 = 1 - a(+\infty) \qquad (4.3.35)$$

and $W(\theta; a)$ reads

$$W(\theta; a) = \beta^2 a \exp\left(\frac{\beta(\theta - 1)}{1 + \gamma(\theta - 1)}\right) \frac{\lambda\rho}{\lambda_b\rho_b} \frac{t_{\text{coll}}(T_b)}{t_{\text{coll}}(T)}. \qquad (4.3.36)$$

with $\gamma = (T_b - T_u)/T_b < 1$. It has to be again stressed that the notion of a "steady" planar flame is an intermediate-asymptotics concept (Barenblatt, 1979) valid over times that are exponentially shorter than t_u but still very large compared to $D_{\text{th}}/u_L{}^2$; this is why all the exponentially small terms featuring in (4.3.19–20) have been omitted from (4.3.33–36). Guided by the results gained from the auxiliary model (4.3.25–32), the following working hypotheses are assumed to hold in the limit $\beta \to \infty$:

1) The flame structure still consists of three zones: an upstream preheat zone, where chemistry is frozen (because the mixture is too cold), a burnt region where the reactant is exhausted, a reaction layer of $\mathcal{O}(1/\beta)$ thickness where convection is negligible at first order and where $\theta - 1$ and a are both small.
2) Asymptotic matching holds between the partial solutions pertaining to adjacent zones.

3) Λ is $\mathcal{O}(1)$, to achieve a chemistry-diffusion balance in the reaction layer*.

Equations (4.3.33–36) are now to be solved in each of the upstream, burnt gas and reaction zones; matching procedures will provide one with the eigenvalue Λ, then u_L.

Burned gas region ($\xi > 0$). By analogy with (4.3.28), we are led to use the following profiles to all algebraic orders in β:

$$\theta_{\text{burnt}} = 1, \qquad a_{\text{burnt}} = 0. \qquad (4.3.37)$$

Upstream zone ($\xi < 0$). Locating the reaction layer about $\xi = 0$, one obtains the chemistry-free profiles:

$$\theta_{\text{upstream}} = e^{\xi}, \qquad (4.3.38)$$

$$a_{\text{upstream}} = 1 - \left(1 + \frac{1}{\beta} C_1 + \frac{1}{\beta^2} C_2 + \ldots\right) e^{\text{Le}\,\xi}, \qquad (4.3.39)$$

where the C_n's are $\mathcal{O}(1)$ integration constants. The choice of power series in $1/\beta$ may seem arbitrary in (4.3.39); to some extent it is, but will be justified *a posteriori*. To ensure leading order matching, we also assumed $\lim_{\beta \to \infty} \theta_{\text{upstream}}(0) = 1$; by exploiting the translation invariance of (4.3.33)–(4.3.36) one may next take $\theta_{\text{upstream}}(0) = 1$ to all orders in β, by adequately choosing the definition of the origin $\xi = 0$ inside the reaction layer. Once this is done, translation invariance cannot be invoked any longer to completely fix the upstream profile of $a(\xi)$; we nevertheless anticipate that $\lim_{\beta \to \infty} a_{\text{upstream}}$ equals a_{burnt} at $\xi = 0$, but the remaining integration constant $C_1/\beta + C_2/\beta^2 + \ldots$ cannot be eliminated *a priori*, when Le $\neq 1$.

Reaction layer ($\eta \equiv \beta\xi = \mathcal{O}(1)$). For a fixed η, we tentatively use:

$$\theta_{\text{inner}} = 1 + \Theta_1(\eta)/\beta + \Theta_2(\eta)/\beta^2 + \ldots,$$
$$a_{\text{inner}} = 0 + A_1(\eta)/\beta + A_2(\eta)/\beta^2 \ldots, \qquad (4.3.40)$$
$$\Lambda = \Lambda_1 + \Lambda_2/\beta + \Lambda_3/\beta^2 + \ldots,$$

by analogy with (4.3.30) and given that the equations for θ and a are coupled. Substitution of (4.3.40) in the conservation equations and collecting the terms of order β yields:

$$0 = \frac{d^2\Theta_1}{d\eta^2} + \Lambda_1 A_1 e^{\Theta_1}, \qquad (4.3.41)$$

---

* Hence $\sigma = \mathcal{O}(t_{\text{ch}}/t_u)$, $\partial\theta/\partial\tau = \mathcal{O}(\sigma^2)$ in (4.3.19) and the last term in (4.3.20) is $\mathcal{O}(\beta^2\sigma) \gg \sigma$.

$$0 = \frac{1}{\text{Le}} \frac{d^2 A_1}{d\eta^2} - \Lambda_1 A_1 \, e^{\Theta_1} , \qquad (4.3.42)$$

thereby implying dominant balances between reaction, conduction and diffusion. More generally, the conservation equations acquire the following structure at order β^{2-n} $(n > 1)$:

$$\frac{d\Theta_{n-1}}{d\eta} - \frac{d^2\Theta_n}{d\eta^2} = \frac{1}{\text{Le}} \frac{d^2 A_n}{d\eta^2} - \frac{dA_{n-1}}{d\eta}$$
$$= e^{\Theta_1} \left(\Lambda_n A_1 + A_n \Lambda_1 + A_1 \Theta_n \Lambda_1 \right) + f_n , \qquad (4.3.43)$$

where the f_n's are known polynomial expressions of $\Lambda_1, \ldots \Lambda_{n-1}$, $\Theta_1, \ldots \Theta_{n-1}$, $A_1, \ldots A_{n-1}$ that have e^{Θ_1} as a factor. As for the boundary conditions, they are provided by the requirement of matching. At $\eta \to \infty$ one must have $\Theta_n = A_n = 0$, whereas:

$$\Theta_n (\eta \to -\infty) = \frac{1}{n!} \eta^n + o(1) , \qquad \forall n \geq 1$$
$$-A_1 (\eta \to -\infty) = \text{Le} \, \eta + C_1 + o(1) , \qquad (4.3.44)$$
$$-A_2 (\eta \to -\infty) = \frac{\text{Le}^2 \, \eta^2}{2} + C_1 \, \text{Le} \, \eta + C_2 + o(1) ,$$

etc. Let us now proceed to the resolution proper. Combined with the downstream boundary conditions the leading order equations obviously yield the first integral $\Theta_1 \, \text{Le} + A_1 \equiv 0$, whose compatibility with (4.3.44) implies $C_1 = 0$. It next gives:

$$0 = \frac{d^2\Theta_1}{d\eta^2} - \Lambda_1 \, \text{Le} \, \Theta_1 \, e^{\Theta_1} . \qquad (4.3.45)$$

Multiplying by $2 \, d\Theta_1/d\eta$ and integrating once leads to

$$0 = \left(\frac{d\Theta_1}{d\eta} \right)^2 - 2\Lambda_1 \, \text{Le} \int_0^{\Theta_1} u \, e^u \, du , \qquad (4.3.46)$$

where the integration constant was deduced from $\Theta_1(+\infty) = 0$. Considering now the fresh-side limit $\eta \to -\infty$, in which $\Theta_1 \to -\infty$ and $d\Theta_1/d\eta \to +1$ (cf. (4.3.44)), imposes a *well-defined* value to Λ_1:

$$2\Lambda_1 \, \text{Le} = 1 \quad \left(= 1 \Big/ \int_0^{-\infty} e^{+v} \, v \, dv \right) \qquad (4.3.47)$$

thereby determining a leading order expression for the flame speed, to be commented on later. Once Λ_1 is known, the inner profiles are obtained upon a further integration of:

$$\frac{d\Theta_1}{d\eta} = \left(\int_0^{\Theta_1} u \, e^u \, du \right)^{1/2} \qquad (4.3.48)$$

which gives $\eta(\Theta_1)$ upon quadrature. The higher orders can be handled similarly. From (4.3.43) and the conditions $\Theta_n(\infty) = A_n(\infty) = 0$ the first integral

$$\text{Le } \Theta_n + A_n \equiv \text{Le} \int_0^{\Theta_1} (\Theta_{n-1} + A_{n-1}) \frac{d\eta}{d\Theta_1} d\Theta_1$$

is deduced. It allows one to compute C_n upon specialization at the $\eta = -\infty$ side and to eliminate A_n from (4.3.43), thereby yielding:

$$\frac{d\Theta_{n-1}}{d\eta} = \frac{d^2\Theta_n}{d\eta^2} - \left(\text{Le } \Lambda_1 e^{\Theta_1} (\Theta_1 + 1) \Theta_n + g_n + \Lambda_n \text{ Le } \Theta_1 e^{\Theta_1} \right)$$

where g_n is known. Thanks to the differentiated form of (4.3.45), this can also be written as:

$$0 = \frac{d}{d\eta} \left[\frac{d\Theta_n}{d\eta} \frac{d\Theta_1}{d\eta} - \Theta_n \frac{d^2\Theta_1}{d\eta^2} \right] - \left(\Lambda_n \text{ Le } \Theta_1 e^{\Theta_1} + h_n \right) \frac{d\Theta_1}{d\eta} \qquad (4.3.49)$$

where h_n also is known, basically in terms of Θ_1. The last equation is easily integrated once; it is left as an exercise to the interested reader to show that using the matching conditions (4.3.44) ultimately fixes Λ_n in terms of already computed quantities. The algebra soon gets boring as n increases beyond 2, but it poses no difficulty in principle. Fortunately, computing Λ_1 then Λ_2 turns out to be enough to yield a quite accurate estimate of Λ: with $T_b = 5T_u$ (a conservative figure) the fractional error in Λ is then less than 10% for $\beta > 4$ and less than 1% if $\beta > 10$ (Bush and Fendell, 1970); the fractional errors on u_L are twice as small ($\Lambda \sim 1/u_L^2$).

4.3.6 Discussion

Once Λ is found, the definition (4.3.21) allows one to obtain the laminar flame speed u_L:

$$u_L^2 = \left(\frac{D_{\text{th}} \, e^{-E/\mathcal{R} T_b}}{\beta^2 t_{\text{coll}}(T_b)} \right) (2 \text{ Le}) \left(\frac{\lambda_b \rho_b}{\lambda_u \rho_u} \right) \left(1 - \frac{2 \text{ Le } \Lambda_2}{\beta} + \dots \right). \qquad (4.3.50)$$

Comparison with (4.3.9) reveals that the approximate reasoning following equation (4.3.7) was not much in error: its main defect was only to ignore the thinness of the reaction zone, and the fact that $a = y_A/y_u$ is $\mathcal{O}(1/\beta)$ there, instead of $\mathcal{O}(1)$. In particular, (4.3.50) and (4.3.9) agree to predict $u_L^2 \sim \exp(-E/\mathcal{R} T_b)$ and suggest that shifting the temperature in the reaction zone by $\delta T \ll T_b$ could modify the burning speed from u_L to $m \, u_L$, with:

$$m^2 \simeq \exp \left(\frac{\beta \delta T}{T_b - T_u} \right). \qquad (4.3.51)$$

That $\delta T/(T_b - T_u) = \mathcal{O}(1/\beta)$ is enough to change the rate of heat release by an $\mathcal{O}(1)$ factor, when $\beta \gg 1$, is the basic reason why flames are so sensitive to many thermal influences, even if only local ones: changes in mixture composition, losses to boundaries (see Section 4.6), front or flow curvature (see Section 4.5), etc. This gives access to many experimental ways of measuring the effective activation energy.

We briefly consider another example, namely a flame which is subject to weak heat losses, e.g., radiative ones, that originate from the bulk of the mixture. This situation can be mimicked by inserting a heat-loss term $-C_p(T - T_u)\rho/t_c$ in the r.h.s. of the energy balance (4.2.6); t_c stands for a cooling time and serves to measure the heat loss intensity. If small, the resulting δT can be estimated as:

$$-\delta T \sim (T_b - T_u)\, D_{th}/t_c u_c{}^2 ,$$

where $D_{th}/u_c{}^2$ is the time of transit of a particle across the loss-affected flame and u_c is the corresponding burning speed. Still denoting u_c/u_L by m, this estimate can be rewritten as:

$$-\delta T \sim (T_b - T_u)\, D_{th}/t_c u_L{}^2 m^2 . \tag{4.3.52}$$

Combining (4.3.51) and (4.3.52) leads to an approximate equation for m:

$$m^2 \log \frac{1}{m^2} \sim \frac{\beta D_{th}}{t_c u_L{}^2}. \tag{4.3.53}$$

As is seen in Fig. 4.10, a too lean (or rich, or diluted) mixture cannot sustain a flame when bulk losses, such as water or carbon-dioxide radiation exist. At the extinction point $\delta T/(T_b - T_u)$ is only $-1/\beta$, so that extinction is not obtained by pumping the whole available chemical energy but, instead, it results from the nonlinearity of the Arrhenius law and the feed-back of m upon δT; to wit, one could suppress the downstream contribution of heat losses without altering (4.3.53). The appearance of turning points and hysteresis, along with the corresponding abrupt changes in m, are very frequent in combustion and almost always have a similar origin, i.e., the convexity of the Arrhenius law and a result-dependent temperature (Buckmaster, 1993).

Given the simplicity of the reasoning which led to (4.3.53), one may wonder why we developed the somewhat heavier machinery of asymptotic expansions, as we did to compute u_L. Besides pedagogical reasons, we did it so because this more formal approach allows one to compute the $\mathcal{O}(1)$ dimensionless constants such as those missing in (4.3.52) and (4.3.53), and it can be extended to arbitrary order in $1/\beta$ to improve accuracy if needed. Nevertheless, qualitative estimates based upon (4.3.51) are almost always unavoidable prerequisites to any quantitative analysis.

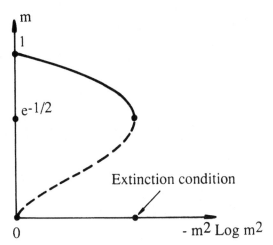

Fig. 4.10. Typical response curve (burning velocity versus heat-loss intensity) of a nonadiabatic flame, equation (4.3.53).

A few further remarks are due. Firstly, that the asymptotic method picked a single value for the flame speed u_L. Such an issue was unclear *a priori*, because the production term $W(\theta; a)$ given by (4.3.20) has a finite, positive slope $\partial W/\partial\theta$ at $\theta = 0$. At least when Le $= 1$, this suggests the possibility of a continuum of solutions (See Chapter 2 for an analogous problem). This is indeed what happens here (Nicoli *et al.*, 1990), and the flame speed singled out by the asymptotic method corresponds to a minimum admissible value. Flames with velocities larger than (4.3.50) display a convection/chemistry balance in the low temperature range ($\theta \ll 1$), whose very existence is at the origin of the continuum of flame speeds. The influence of conduction progressively diminishes as the speed increases, to become entirely negligible in the limit of very fast flames. Those solutions are considered spurious, however, for they need very special initial far-upstream temperature profiles to be established; furthermore, accounting for heat-losses which are so weak that $t_c = \mathcal{O}(t_u)$ could suppress them.

Similar features already existed in the auxiliary model (4.3.25). It is a good exercise on matched asymptotic expansions to try to construct, for a fixed value of β, the flame structure corresponding to slightly different velocities from what (4.3.27) predicts, especially if β is large. (*Hints:* work in phase-plane (θ, θ_ξ) and try power series, but remember that logarithms may crop up unexpectedly; at least when $\beta = 2$ four zones are needed.) The limit of high velocities is regular.

Concerning now the mixtures with nonunit Lewis numbers, no definitive answer about the spectrum of solutions seems to be available, even though cases of nonuniqueness are well established (H. Berestycki, private communication) and the possibility of a continuous spectrum is quite likely; we shall admit that the physically relevant solutions still correspond to (4.3.50) when an Arrhenius law with large activation energy is employed.

4.3.7 Generalizations

In Section 4.2 above, the restriction to Fickian diffusion was introduced in a rather *ad hoc* way. We now briefly outline how the model can be improved and see whether the asymptotic method can handle cross transports. Consider thermal diffusion, for example. If the mixture is sufficiently diluted with an inert carrier, the reactant balance is modified into:

$$\frac{da}{d\xi} = \frac{1}{\mathrm{Le}} \frac{d}{d\xi} \left(\frac{da}{d\xi} + s_T a \frac{d}{d\xi} \log T \right) - \Lambda W(\theta; a) ,$$

where s_T is proportional to $D_A{}^T / D_{A,\mathrm{inert}}$. Clearly the reaction zone is unmodified at the leading order, for $a = \mathcal{O}(1/\beta)$ there. In the upstream zone, the temperature profile is unchanged, whereas a_{upstream} now is governed by a linear equation which, when integrated once, reads

$$\mathrm{Le}(a - 1) = \frac{da}{d\xi} + s_T a \frac{d}{d\xi} \log T ,$$

and is readily solved since θ, hence T, are known. Clearly $da/d\xi$ is not much changed at the entrance of the reaction zone, because a_{upstream} almost vanishes there, so that the flame speed will not depend on thermal diffusion at the leading order in β and one can compute Λ analytically. Insofar as steady planar flames are concerned, incorporating $s_T \neq 0$ does not alter the picture drastically; the only qualitative change due to non Fickian diffusion is that the mass fraction of a scarce reactant can overshoot its upstream value if $\mathrm{Le} > 1$ and $s_T > 0$. Provided that all the Lewis numbers involved are constant, the method can also handle multicomponent diffusion, because only constant-coefficient equations — even if coupled— need be solved in the preheat zone. We shall no longer consider these effects here, nor thermal diffusion.

Changing the reaction order with respect to the deficient reactant, from unity ($w \sim y_A$) to any n_A ($w \sim y_A{}^{n_A}$), does not create any major difficulty because, if the eigenvalue Λ is adequately defined, the inner problem can be reduced to studying:

$$0 = \frac{d^2 \Theta_1}{d\eta^2} + \Lambda_1 \left(-\Theta_1 \, \mathrm{Le} \right)^{n_A} e^{\Theta_1}$$

instead of (4.3.45); consequently, $1 = 2\Lambda_1 \, \mathrm{Le}^{n_A} (n_A!)$ is the result generalizing (3.4.47). Similarly, explicitly accounting for the reactant B in excess, in addition to deficient reactant A, ultimately leads to the equation

$$0 = \frac{d^2\Theta_1}{d\eta^2} + \Lambda_1 \left(-\Theta_1 \, \mathrm{Le_A}\right)^{n_A} \left(\beta b(\infty) - \mathrm{Le_B} \, \Theta_1\right)^{n_B} e^{\Theta_1}$$

which can also be integrated by quadrature; the final scaled mass fraction $b(\infty)$ of the abundant reactant is easily computed *a priori* from the initial mixture composition and an overall balance across the flame. Upon integration, the last equation gives access to the variations of u_L with initial stoichiometry; except if $b(\infty) \leq \mathcal{O}(1/\beta)$, the abundant reactant does not play any qualitatively new role in steady flames.

Flames propagating in solids (e.g., in rocket boosters) are a bit different at first glance, because they do not involve any reactant diffusion ($\mathrm{Le} = \infty$), with the result that $a_{\mathrm{upstream}} \equiv 1$ to all algebraic orders in β. As further consequences $a_{\mathrm{inner}} = A_0(\eta) + A_1(\eta)/\beta + \ldots$, where $A_0(\eta) = \mathcal{O}(1)$, and $\Lambda = \Lambda_0/\beta + \ldots$; $A_0(\eta)$ is then governed by a convection/reaction balance:

$$\frac{dA_0}{d\eta} = -\Lambda_0 A_0 \, e^{\Theta_1},$$

which implies $A_0 \equiv d\Theta_1/d\eta$; the resulting inner energy balance

$$0 = \frac{d^2\Theta_1}{d\eta^2} + \Lambda_0 \frac{d\Theta_1}{d\eta} e^{\Theta_1}$$

is immediately integrated once to give $\Lambda_0 = 1$ and the problem of finding u_L is again solved analytically.

4.4 Jumps across the reaction layer

It is more important to notice what follows, especially in the context of flame instabilities. Even when the flame is curved and/or unsteady, its reaction layer may still be considered as locally nearly-planar and quasi-steady, at least provided the front is not too highly curved nor too rapidly varying; its structure can be solved once for all and the whole reaction layer can then be replaced by a reaction sheet and jump relationships for the outer profiles.

More precisely, the following constraints are assumed to hold:

i) The local radii of curvature of the reaction layer strictly exceed $\mathcal{O}\left(l_T/\beta\right)$; the concept of a thin reaction layer is then valid.

ii) The typical evolution time of the whole flame structure is longer than $\mathcal{O}\left(l_T/\beta u_L\right)$.

iii)   the normal temperature gradient on the burned side of the reaction
      layer is $\mathcal{O}\left(T_b/\beta l_T\right)$ at most.
iv)   at $\mathcal{O}(l_T)$ distances from the reaction layer, the reduced temperature
      (θ) and mass fraction profiles (a) can be expanded in asymptotic
      power series of $1/\beta \to 0^+$

$$(\theta, a) = (\theta_0, a_0) + (\theta_1, a_1)/\beta + \ldots \tag{4.4.1}$$

Then defining the stretched variable $\eta \equiv \beta n \rho_u u_L C_p / \lambda_b$ to be now
counted along the normal coordinate n to the reaction sheet, one may
seek the inner temperature and mass fraction profiles in the form:

$$(\theta, a) = (1, 0) + (\Theta_1, A_1)/\beta + \ldots \tag{4.4.2}$$

in which we anticipated that $\Theta_0 = 1$ ($\mathcal{O}(1)$ changes in Θ_0 would in general
lead to a violation of i) to iv)). Due to matching requirements, this implies
that $\theta_0\left(n = 0^{\pm}\right) = 1$, $a_0\left(n = 0^{\pm}\right) = 0$. The inner functions A_1 and Θ_1
satisfy

$$\frac{\partial^2 \Theta_1}{\partial \eta^2} = -\frac{1}{\mathrm{Le}} \frac{\partial^2 A_1}{\partial \eta^2} = -\frac{1}{2\,\mathrm{Le}} A_1 \, e^{\Theta_1} \tag{4.4.3}$$

from which one deduces

$$\Theta_1 + A_1/\mathrm{Le} = c_1 + c_2 \eta \tag{4.4.4}$$

with integration "constants" c_1 and c_2 which in general depend on
time and on the coordinates along the reaction layer. Due to iii) c_2
vanishes, since matching requires it to equal $(\partial \theta_0/\partial n)_{n=0+}$; therefore
$\Theta_1 + A_1/\mathrm{Le} \equiv \Theta_1(\eta = +\infty) = \theta_1\left(n = 0^+\right)$ and (4.4.3) is integrated
once to give:

$$\frac{\partial \theta_0}{\partial n}\left(n = 0^-\right) = -\frac{1}{\mathrm{Le}} \frac{\partial a_0}{\partial n}\left(n = 0^-\right) = \exp\left(\tfrac{1}{2}\theta_1\left(n = 0^+\right)\right) \tag{4.4.5}$$

upon use of obvious matching conditions. In general, the first equality in
(4.4.5) cannot be met if $\mathrm{Le} - 1 = \mathcal{O}(1)$, because the outer problems for
θ_0 and $a_0 - 1$ are then too unlike in the preheat zone, even if they fulfill
identical upstream boundary conditions. As (4.4.5) is only a leading order
result, one is led to impose a restriction on the *reduced Lewis number*

$$\mathrm{le} \equiv \beta\left(1 - \frac{1}{\mathrm{Le}}\right) = \mathcal{O}(1) \tag{4.4.6}$$

with the consequence that:

$$a_0 + \theta_0 - 1 \equiv 0 \tag{4.4.7}$$

if all the boundary conditions for θ and a are identical to leading order.
Next one may *define* the reaction sheet as the surface along which the

outer profiles of mass fraction intersect (hence, $a_i = 0$ at $n = 0^\pm$, for any i); θ_1 is then shown from (4.4.4) to be continuous at the flame sheet.

Eliminating the production term between the equations for Θ_2 and A_2, and integrating the result once, lead to the conclusion that the η-derivative of $\Theta_2 + A_2 + \text{le}\,\Theta_1$ is independent of η. Compatibility of this first integral with the matching conditions leads to:

$$\frac{\partial \theta_1}{\partial n}(0^-) + \frac{\partial a_1}{\partial n}(0^-) + \text{le}\,\frac{\partial \theta_0}{\partial n}(0^-) = \frac{\partial \theta_1}{\partial n}(0^+) . \qquad (4.4.8)$$

It is now convenient to introduce the scaled *excess enthalpy* $H \equiv a_1 + \theta_1$ whose value at $n = 0$ is $\theta_1(0)$. The above results can be summarized into the following *jump relations* which hold all along the flame sheet (Joulin and Clavin, 1979):

$$\theta_0 = 1 , \qquad (4.4.9)$$

$$[H] = 0 , \qquad (4.4.10)$$

$$[\mathbf{n} \cdot \nabla H] = \text{le}\ \mathbf{n} \cdot \nabla\, \theta_0|_{\text{fresh side}} , \qquad (4.4.11)$$

where $[f] \equiv f(\text{burnt side}) - f(\text{fresh side})$, ∇ refers to coordinates scaled by $l_T \lambda_b / \lambda_u$ and

$$(\mathbf{n} \cdot \nabla\, \theta_0)^2 \big|_{\text{fresh side}} = \exp(H) \qquad (4.4.12)$$

to be compared with (4.3.51).

Under assumption (4.4.6) flame problems are thus reduced to solving outer *chemistry-free* equations, for θ_0, H and the *leading order* fluid mechanics, that are endowed with (4.4.9)–(4.4.12) as conditions at the front. The only place where chemistry enters them is (4.4.12), so that switching from the Arrhenius law to another one is very easy. The aforementioned outer field-equations only contain two important parameters (besides the Prandtl numbers) namely: the *reduced Lewis number* le defined in (4.4.6) and the *density contrast* γ:

$$\gamma \equiv (\rho_\text{u} - \rho_\text{b})/\rho_\text{u} < 1 , \qquad (4.4.13)$$

which is mainly brought about by the equation of state (and by λ/λ_u, μ/μ_u):

$$\rho/\rho_\text{u} = (1 - \gamma)\,/(1 - \gamma + \gamma\theta_0) + \mathcal{O}(1/\beta) . \qquad (4.4.14)$$

Both for training purposes and to illustrate situations in which the "θ_0-H-le" formalism works well, we briefly examine below two elementary applications; for further examples, see Buckmaster and Ludford (1982).

The first one deals with the simplest nonplanar flame, namely: the so-called Zel'dovich flame ball (Zel'dovich *et al.*, 1985; Deshaies *et al.*, 1981), which is nothing but a steady, convection-free spherical flame in equilibrium with reactive cold gases. Though a bit counterintuitive at first glance, such an object exists (we do not advise the reader to try to obtain one with town gas and air in his/her kitchen), and R shall denote the radius of its reaction sheet in units of l_T. From the heat equation with convection omitted one readily deduces, assuming $\lambda = $ constant, that:

$$
\begin{aligned}
\theta_0 &= R/r\,, & H &= H(R)R/r & r &\geq R \\
\theta_0 &= 1\,, & H &= H(R) & r &\leq R
\end{aligned}
\qquad (4.4.15)
$$

where r stands for the current radius. Conditions (4.4.9)–(4.4.11) give $H(R) = -\mathrm{le}$, then (4.4.12) yields the sought after flame ball size:

$$
R = \exp\left(\mathrm{le}\,/2\right)\,, \qquad (4.4.16)
$$

and the problem is solved. This configuration is interesting for many reasons. Firstly, it explicitly shows how a nonunit Lewis number leads to changes in reaction temperature $(\sim H(R)\,(T_b - T_u)\,/\beta)$ when a flame is curved and how $\mathrm{Le}-1 = \mathcal{O}(1)$ would lead to exponential, hence in general too violent to be coped with, changes in chemical source power. Secondly, it demonstrates that the burning speeds (zero, here) of the curved and the flat flames differ, even if $\mathrm{Le} = 1$. Lastly, because such an equilibrium spherical flame is unstable to radial disturbances (Deshaies and Joulin, 1983) and, in a sense, plays the role of critical nucleus: whereas smaller ones shrink, larger flames would expand autonomously... leading to the kind of events summarized in Section 4.1. This radial instability is an important ingredient of the dynamics of a flame which is initiated by a localized source of energy such as a spark (Joulin, 1985). Once properly stabilized by heat loss mechanisms, spherical flames also intervene when modeling particular combustion modes that show up at microgravity in the form of globules (e.g., see Buckmaster and Joulin, 1991, and the references therein, or Ronney *et al.*, 1992).

Our second example precisely deals with steady flames subject to temperature-dependent volumetric heat losses in the case already evoked at the end of Section 4.4, i.e., steady planar fronts. As is clear from (4.4.12) heat losses that are capable to shift H by an $\mathcal{O}(1)$ amount along the reaction zone will affect the flame speed. One may then implant an $\mathcal{O}(1)$ heat-loss term in the excess-enthalpy balance and inquire about the consequences. A key remark is that only θ_0 will appear in the leading-order heat-loss term. Still assuming $\lambda = $ const., we shall therefore consider the system:

$$
m\frac{d\theta_0}{d\xi} = \frac{d^2\theta_0}{d\xi^2}\,, \qquad (4.4.17)
$$

$$
m\frac{dH}{d\xi} = \frac{d^2H}{d\xi^2} + \mathrm{le}\,\frac{d^2\theta_0}{d\xi^2} - h\theta_0 F(\theta_0)\,, \qquad (4.4.18)
$$

where m is the nonadiabatic to adiabatic flame-speed ratio, h is a scaled heat loss power and F is normalized to have $F(1) = 1$. One immediately finds

$$\theta_0(\xi \leq 0) = e^{m\xi}, \qquad\qquad \theta_0(\xi \geq 0) = 1, \qquad\qquad (4.4.19)$$

$$mH(0) = \frac{dH}{d\xi}(0^-) - h \int_{-\infty}^{0} \theta_0 F(\theta_0)\, d\xi + \text{le } m. \qquad (4.4.20)$$

As for the burned side, it gives:

$$\frac{dH}{d\xi}(0^+) = -h/m, \qquad\qquad (4.4.21)$$

once the exponentially growing solutions are discarded (they cannot be matched with the trailing, thicker cooling zone). Then (4.4.11) furnishes the excess-enthalpy at $\xi = 0$:

$$H(0) = -\frac{h}{m^2}\left(1 + \int_0^1 F(u)\, du\right), \qquad (4.4.22)$$

in which the upstream and downstream contributions clearly show up (the Lewis number does not). Finally (4.4.12) gives access to the relationship between m and h (Joulin and Clavin, 1976):

$$h\left(1 + \int_0^1 F(u)\, du\right) + m^2 \log m^2 = 0 \qquad (4.4.23)$$

thereby confirming that $h = \mathcal{O}(1)$ was the right working assumption to get something qualitatively new. (*Exercise:* what is the right choice of heat loss intensity to study nonadiabatic steady spherical flames ? *Hint:* in a first step study losses proportional to θ.) Equation (4.4.23) fully confirms the qualitative reasoning presented at the end of section 4.3.6 and determines the $\mathcal{O}(1)$ constants missing from (4.3.52) and (4.3.53), as claimed previously. The methods to be introduced in the next section can be employed to demonstrate that $m < e^{-1/2}$ in (4.4.23) corresponds to flames which spontaneously evolve to $m > e^{-1/2}$ or to extinction $m = 0$ (Joulin and Clavin 1979).

4.5 Diffusive instabilities

4.5.1 Linear analysis

There is ample experimental evidence (see Section 4.1) that, even if the fresh mixtures are quite still, the flame fronts only seldom want to stay steady and planar. Instead, obviously because of instabilities of some sort, they evolve to more or less complicated patterns. The standard manner to get insights into phenomena of this kind is to investigate the spontaneous evolution of small disturbances superimposed to a known background solution, here the steady planar flame studied in the preceding paragraph, and to try to determine which mechanisms make them grow or decay, when this happens, etc.

In the case of flames this good-looking program leads to a technical difficulty.   While the solution of (4.3.2)–(4.3.4) is perfectly known asymptotically, nobody has so far been able to exactly solve analytically the linearized conservation equations on the fresh side of the reaction sheet, even in the asymptotic limit $\beta \to \infty$. The high-order differential system to be solved indeed has coefficients which, there, vary with streamwise distance. In fact, what precludes analytical resolution mostly is the coupling between the heat equation and hydrodynamics, via changes in density (4.1.14) which follow those of temperature. To go further, a numerical resolution could be envisaged; this has been done (Jackson and Kapila, 1984). Complementarily, one could first try to understand partial mechanisms, a synthesis being postponed to a later stage.

The following remark will prove to be useful: even though the existence of changes in temperature, hence in density, was explicitly accounted for when establishing (4.4.9)–(4.4.12) in the limit $\beta \to \infty$, the resulting problem remains nonempty (see below) in the limit of small density contrasts, that is for $\gamma \equiv (\rho_\mathrm{u} - \rho_\mathrm{b})/\rho_\mathrm{u} = (T_\mathrm{b} - T_\mathrm{u})/T_\mathrm{b} \to 0$. In a sense, the changes in ρ mostly affect the fluid mechanics whereas those in T control the reaction rate. What survives once the additional limit $\gamma = 0$ is taken is a differential system in which the energy and reactant balances are no longer coupled to hydrodynamics. Consequently the flow field is not affected by the reaction-sheet motion and may remain one-dimensional, which we assume here. In the limit $\gamma = 0$, it is therefore enough to consider:

$$\frac{\partial \theta}{\partial \tau} + \frac{\partial \theta}{\partial \xi} = \Delta \theta \,, \tag{4.5.1}$$

$$\frac{\partial H}{\partial \tau} + \frac{\partial H}{\partial \xi} = \Delta H + \mathrm{le}\,\Delta \theta \,, \tag{4.5.2}$$

in which the subscript of θ_0 was dropped to simplify the formulae, together with the conditions

$$\theta = H = 0 \quad \text{at} \quad \xi = -\infty \,, \tag{4.5.3}$$

$$H \quad \text{bounded at} \quad \xi = +\infty \,, \tag{4.5.4}$$

and, at the reaction sheet:

$$\theta = 1 \,, \tag{4.5.5}$$

$$[H]_-^+ = 0 = [\mathbf{n}\cdot\boldsymbol{\nabla}\,H + \mathrm{le}\,\mathbf{n}\cdot\boldsymbol{\nabla}\,\theta]_-^+ \,, \tag{4.5.6}$$

$$\mathbf{n}\cdot\boldsymbol{\nabla}\,\theta|_{\text{fresh side}} = \exp(H/2) \,. \tag{4.5.7}$$

By construction, this system is hydrodynamics-free and, hopefully, what can be gained from its study is a better understanding of the way the heat and species transfers affect the dynamics of the reaction sheet, which is a free boundary: whereas (4.5.1)–(4.5.6) in principle allow one to compute θ and H whatever the reaction-sheet dynamics, the latter is in fact constrained by equation (4.5.7).

The steady, one-dimensional solutions, noted $\bar{\theta}(\xi)$ and $\bar{H}(\xi)$ from now onward, are readily found:

$$(\bar{\theta}(\xi), \bar{H}(\xi)) = \begin{cases} \left(e^{\xi}, -\text{le } \xi \, e^{\xi}\right) & \xi \leq 0 \\ (1, 0) & \xi \geq 0 \end{cases} \tag{4.5.8}$$

We next consider the situation where the reaction sheet has the equation $\xi = \varphi(\tau, \boldsymbol{\zeta})$ (Fig. 4.11) where φ has a small amplitude (δ, say) and $\boldsymbol{\zeta} = (\zeta, \eta)$ represent transverse Cartesian coordinates. That $\varphi \neq 0$ causes θ and H to depart from $\bar{\theta}$ and $\bar{H}$; specifically one writes

$$(\theta, H) = (\bar{\theta}, \bar{H}) + \varphi(\tau, \boldsymbol{\zeta}) \, (\theta', H') + O(\delta^2), \tag{4.5.9}$$

θ' and H' being $\mathcal{O}(1)$ functions of ξ, and linearize the system (4.5.1)–(4.5.7) in the limit $\delta \to 0$.

Once this is done each transverse Fourier mode of $\varphi(\tau, \boldsymbol{\zeta})$ evolves independently of the other ones, so that one may concentrate on one of them by writing $\varphi(\tau, \boldsymbol{\zeta})$ as $\delta \exp(s\tau + i\mathbf{k} \cdot \boldsymbol{\zeta})$, where the growth/decay rate s is to be found in terms of the wavenumber $k = (\mathbf{k} \cdot \mathbf{k})^{1/2}$ and the parameter le. Equations (4.5.1) and (4.5.2) are linear from the beginning and simply give:

$$(s + k^2)\theta' + \frac{d\theta'}{d\xi} = \frac{d^2\theta'}{d\xi^2}, \tag{4.5.10}$$

$$(s + k^2)H' + \frac{dH'}{d\xi} = \frac{d^2H'}{d\xi^2} + \text{le}\left(\frac{d^2\theta'}{d\xi^2} - k^2\theta'\right). \tag{4.5.11}$$

The boundary conditions (4.5.5)–(4.5.7) require more care, as they hold at $\xi = \varphi(\tau, \boldsymbol{\zeta})$, not at $\xi = 0$. They can be transferred to the latter location,

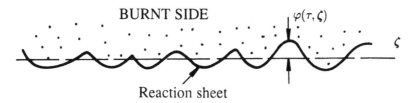

Fig. 4.11. Wrinkled flame (schematic) and coordinate system.

however. For example, the condition $\theta(\tau, \varphi(\tau, \zeta), \zeta) = 1$ transforms into:

$$\bar{\theta}\left(\varphi(\tau, \zeta)\right) + \varphi(\tau, \zeta)\theta'\left(\varphi(\tau, \zeta)\right) + \mathcal{O}(\delta^2) = 1$$

which, upon Taylor expansions of $\bar{\theta}$ and θ' gives $\bar{\theta}(0) = 1$ (already known) at $\mathcal{O}(1)$ and:

$$\left.\frac{d\bar{\theta}}{d\xi}\right|_{0^-} + \theta'|_{0^-} = 0 \qquad (4.5.12)$$

at $\mathcal{O}(\delta)$. Similarly the two equations (4.5.6) respectively give

$$\left.\frac{d\bar{H}}{d\xi}\right|_{0^-} + H'|_{0^-} = H'|_{0^+}, \qquad (4.5.13)$$

$$\left.\frac{d^2\bar{H}}{d\xi^2}\right|_{0^-} + \left.\frac{dH'}{d\xi}\right|_{0^-} + \mathrm{le}\left(\left.\frac{d^2\bar{\theta}}{d\xi^2}\right|_{0^-} + \left.\frac{d\theta'}{d\xi}\right|_{0^-}\right) = \left.\frac{dH'}{d\xi}\right|_{0^+}, \qquad (4.5.14)$$

because $\bar{H}(\xi \geq 0) \equiv 0$. Finally, (4.5.7) yields:

$$\left.\frac{d^2\bar{\theta}}{d\xi^2}\right|_{0^-} + \left.\frac{d\theta'}{d\xi}\right|_{0^-} = \tfrac{1}{2}H'(0^+). \qquad (4.5.15)$$

One has next to solve (4.5.10) (4.5.11). The temperature disturbance θ' obviously is a linear combination of $\exp(\sigma'\xi)$ and $\exp(\sigma''\xi)$ where σ' and σ'' are the roots of $\sigma^2 = \sigma + (s + k^2)$ viz:

$$2\sigma' = 1 + \left(1 + 4\left(s + k^2\right)\right)^{1/2}, \qquad \sigma'' = 1 - \sigma'.$$

We shall restrict ourselves to situations where $\Gamma \equiv (1 + 4(s + k^2))^{1/2}$ has a real part $\mathcal{R}e(\Gamma)$ which exceeds or equals one, so that $\mathcal{R}e(\sigma') \geq 0$ and $\mathcal{R}e(\sigma'') \leq 0$. Then θ' cannot contain σ'':

$$\theta' = \begin{cases} \theta'(0^-)\, e^{\sigma'\xi} & \xi \leq 0, \\ 0 & \xi \geq 0. \end{cases} \qquad (4.5.16)$$

Equation (4.5.11) can be solved similarly to give:

$$H' = \begin{cases} \left(H'(0^-) + \mathrm{le}\left(\sigma'^2 - k^2\right)\Gamma^{-1}\theta'(0^-)\,\xi\right) e^{\sigma'\xi}, \\ H'(0^+)\, e^{\sigma''\xi}, \end{cases} \qquad (4.5.17)$$

where $H'(0^-)$ and $H'(0^+)$ are readily interpreted as integration constants. Substitution of (4.5.16) and (4.5.17) into (4.5.12)–(4.5.14) determines these as functions of s, k and le. The last equation (4.5.15) has still to be fulfilled, however; it gives the sought after relation between s, k, and le (Sivashinsky, 1977a), to be studied with the constraint $\mathcal{R}e(\Gamma) \geq 1$:

$$2\Gamma^2(\Gamma - 1) + \mathrm{le}\,(\Gamma - 1 - 2s) = 0. \qquad (4.5.18)$$

Accounting for stoichiometry effects (Joulin and Mitani, 1982) would merely amount to redefining le in (4.5.18). We also note that:

$$H'(0^+) \sim \text{le}(\Gamma - 1 - 2s) . \tag{4.5.19}$$

Our last task is to study the dispersion relation (4.5.18).

Figure 4.12 summarizes the results in the (k, le) plane. In the band $-2 < \text{le} < 32/3$, the planar flame is stable with respect to the thermal and diffusional phenomena, the only ones retained here $(\gamma = 0)$. When k is small, s also is; more specifically, expanding (4.5.18) yields:

$$s = -k^2 \left(1 + \tfrac{1}{2}\text{le}\right) + \ldots \tag{4.5.20}$$

That a branch exists for which s vanishes when k does is a consequence of the translational invariance of the problem along the ξ-axis: in effect, if $\bar{\theta}(\xi)$ is a steady solution, a slightly shifted profile $\bar{\theta}(\xi + \delta) \simeq \bar{\theta}(\xi) + \delta d\bar{\theta}/d\xi + \ldots$ also is. Therefore, $s = 0$ is an eigenvalue if $k = 0$ and the corresponding eigenvector of the linearized equations is $(d\bar{\theta}/d\xi, d\bar{H}/d\xi)$.

The result (4.5.20) can be re-written in terms of the reaction-sheet shape $\varphi(\tau, \zeta)$ as:

$$\frac{\partial \varphi}{\partial \tau} = \left(1 + \tfrac{1}{2}\text{le}\right)\Delta\varphi + \ldots \tag{4.5.21}$$

or, in terms of dimensional variables $t = \tau l_T/u_L$, $\mathbf{z} = \zeta l_T$ and $\Phi(t, \mathbf{z}) = l_T\varphi(\tau, \zeta)$, as:

$$\frac{\partial \Phi}{\partial t} = D_{\text{th}} \left(1 + \tfrac{1}{2}\text{le}\right)\Delta\Phi + \ldots \tag{4.5.22}$$

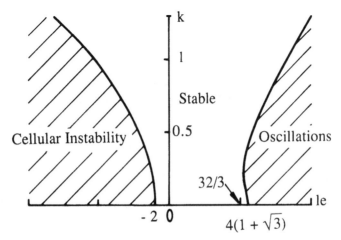

Fig. 4.12. Stability diagram of nearly-planar, constant density flames.

where, in both (4.5.21) and (4.5.22), $\Delta(.)$ represents transverse Laplacians.

Interestingly enough, (4.5.22) says that the local speed of a gently curved flame varies proportionally to front curvature, at least within the constant-density model (Barenblatt *et al.*, 1962). Furthermore, the proportionality coefficient (a diffusivity) depends on the Lewis and Zel'dovich numbers through the single grouping le $= \beta(1 - \text{Le})/\text{Le}$. This can be understood as follows:

Consider the case of le $= 0$ first, for which the reaction sheet is seen as an isothermal surface: this is why (4.5.22) is simply the heat equation in this particular situation. If le $\neq 0$, a new phenomenon comes into play. As is sketched in Fig. 4.13, wrinkling induces transverse heat fluxes which tend to heat (cool) the mixture at the crests (troughs), thereby tending to increase (decrease) the flame speed, a scalar, proportionally to the mean curvature, which is the first relevant local scalar characterizing the deformation. Simultaneously, but at a different rate if Le $\neq 1$, transverse fluxes of reactant tend to deplete (enrich) the mixture, thereby acting in quite the opposite way. The net change in reaction temperature is governed by this competition (hence the presence of Le -1 in factor of $\Delta\Phi$). Though *small* when Le $\simeq 1$, this temperature shift produces a noticeable effect on the flame speed, which is very *sensitive* to temperature if $\beta \gg 1$ (hence the term le $\Delta\Phi$ and the factor of $1/2$).

In general, flames of light fuels (Le < 1) tend to be less efficiently stabilized by curvature effects than heavy ones (Le > 1). This is patent when le < -2, in which case the "effective diffusivity" showing up in (4.5.22) is negative, which means that the flame gets unstable due to too intense transverse fluxes of reactant. This explains the leftmost region of cellular instability sketched in Fig. 4.12. When le is only slightly less than the crossover value of -2 (weak instability), (4.5.20) needs be generalized

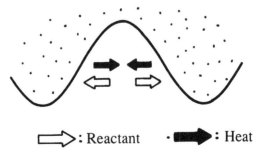

$\Longrightarrow$: Reactant $\blacktriangleright$: Heat

Fig. 4.13. Transverse diffusive fluxes generated by wrinkling ahead of the reaction sheet.

into:

$$s = -k^2 \left(1 + \tfrac{1}{2}\,\text{le}\right) - k^4 b(\text{le}) + \ldots, \qquad b(\text{le}) \geq 0. \qquad (4.5.23)$$

This confirms that only a band of unstable wavenumbers $0 \leq k^2 \leq \mathcal{O}(-1 - \text{le}/2)$ exists (Fig. 4.14), thanks to the stabilizing influence of the k^4-terms. The latter also have a simple physical origin: when le $\neq 0$, temperature is not constant along the reaction sheet, but varies proportionally to le $\Delta\varphi$ in first approximation; in turn this generates secondary transverse heat fluxes that also contribute to the temperature variation itself, proportionally to le $\Delta\Delta\varphi$. This purely dissipative effect is always stabilizing, but disappears if le $= 0$ (no change in reaction temperature).

What happens to the flame when the diffusive in cellular instability sets in will be discussed a bit later. First we need say a few words about the other instability domain sketched in Fig. 4.12. Contrary to the previous one, this instability is of the oscillatory type, for beyond the marginal value le $= 32/3 = 10.66\ldots$ a band of unstable wavenumbers appears around a finite value k_c and the imaginary part of s does not vanish there. This new instability is rather difficult to explain in physical terms, because it is not quasi-steady. Suffice it to say that its roots lie in the space dependent phase shifts involved in any fast enough diffusive processes: even though transverse conduction tends to stabilize the wrinkles and to win over diffusion (le > 0) its influence may well be felt too late where it is needed. When le reaches or exceeds $4(1 + \sqrt{3}) \simeq 10.9$, the band of unstable wavenumbers reaches $k = 0$, so that a pulsating

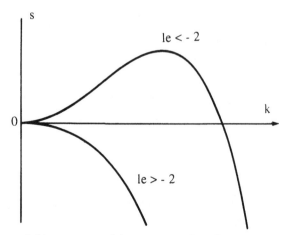

Fig. 4.14. Growth/decay rate (s) versus reduced wavenumber (k) in the neighborhood of the marginal boundary of cellular instability (le $= -2$).

mode of wrinkled propagation tends to settle. We shall not comment further on these new instabilities, because a threshold value of le ~ 10 exceeds well what can be reached in gases. One may mention, however, that the inclusion of heat losses in the flame model lowers the needed value of le to realistic figures (Joulin and Clavin, 1979).

Moreover, as (4.5.18) predicts $0 < s = \mathcal{O}(\text{le}^2)$ when le $\to \infty$, the assumptions $0 < \text{Le} -1 = \mathcal{O}(1)$ and $\beta \to \infty$ cannot be considered simultaneously for unsteady flames, because the reaction layer can no longer be viewed as quasi-steady.

4.5.2 Beyond linearity

Now that the linear stability boundaries are determined, the question which arises next is: what does the flame do if these boundaries are crossed? In the case of a slightly unstable flame of light fuels (le ≤ -2) some progress can be made analytically. Restricting ourselves to a single transverse coordinate ζ, we first note that (4.5.23) implies the following linear evolution equation for the flame shape φ:

$$\frac{\partial \varphi}{\partial \tau} = \left(1 + \tfrac{1}{2}\,\text{le}\right) \frac{\partial^2 \varphi}{\partial \zeta^2} - b(\text{le}) \frac{\partial^4 \varphi}{\partial \zeta^4} + \dots \qquad (4.5.24)$$

This can also be viewed as a way to linearly relate the ξ-wise local flame speed to the local front shape. Because the time and space scales involved in the instability are very long when le is close to -2, a more general local law is expected to govern the evolution of markedly wrinkled flames as well (Kuramoto, 1980)

$$\frac{\partial \varphi}{\partial \tau} = F\left(\frac{\partial \varphi}{\partial \zeta}, \frac{\partial^2 \varphi}{\partial \zeta^2}, \frac{\partial^3 \varphi}{\partial \zeta^3}, \dots\right).$$

The yet unknown function F cannot contain φ as argument, owing to the translational invariance $\varphi \to \varphi + \text{constant}$; similarly the odd derivatives can only contribute to it in the form of even powers, due to the $\zeta \to -\zeta$ symmetry. Therefore, for long-waved wrinkles one should have:

$$\frac{\partial \varphi}{\partial \tau} = A\frac{\partial^2 \varphi}{\partial \zeta^2} - B\frac{\partial^4 \varphi}{\partial \zeta^4} - C\left(\frac{\partial \varphi}{\partial \zeta}\right)^2 + \dots \qquad (4.5.25)$$

where A, B and C are constants. The first two are readily found ($A = (1 + \text{le}/2)$, $B = b(\text{le})$), because (4.5.25) and (4.5.24) ought to coincide in the linear limit. C also is easy to compute, by considering a *planar* flame propagating at unit normal speed and making a small angle ($\equiv \partial \varphi / \partial \zeta$) to the ξ axis: elementary geometry gives $2\partial \varphi / \partial \tau = -(\partial \varphi / \partial \zeta)^2 + \dots$ for the change in flame speed along the ξ-axis, so that $C = 1/2$. Upon rescaling,

(4.5.25) can then be brought into the canonical Kuramoto–Sivashinsky (KS) equation

$$\dot{\Phi} + \tfrac{1}{2}(\Phi')^2 + \Phi'' + \Phi'''' = 0, \tag{4.5.26}$$

the dot and the primes denoting time and space derivatives, respectively. An alternative, more systematic, way of arriving at the same leading-order result is to remark that le $= -2 - \sigma$ leads to $\mathcal{O}\left(\sigma^{1/2}\right)$ unstable wavenumbers and to $\mathcal{O}\left(\sigma^2\right)$ growth rates when $0 < \sigma \ll 1$. The requirement that all the terms in (4.5.25) be of same order of magnitude allows one to anticipate the following form: $\varphi(\tau, \zeta) = \sigma\Phi\left(\sigma^2\tau, \sigma^{1/2}\zeta\right) + o(\sigma)$. Plugging this ansatz into (4.5.1)–(4.5.7) and using the linear results to estimate the forms of $\theta - \bar{\theta}$ and of $H - \bar{H}$, one ultimately arrives (Sivashinsky, 1977a) at the 2-D version

$$\dot{\Phi} + \tfrac{1}{2}|\Phi|^2 + \Delta\Phi + \Delta\Delta\Phi = 0 \tag{4.5.27}$$

of the KS equation, at the leading order in σ.

Both (4.5.26) and (4.5.27) no longer contain any parameter and, to go further, one has to resort to numerics and computers. Numerical integrations of these equations with periodic boundary conditions at the edges of large enough domains reveal that Φ often ultimately evolves to a state of chaotic, statistically stationary motion, displaying unsteady cells of sizes comparable to the mostly amplified wavelength and which constantly merge or split without resuming the same pattern twice (Chaté, 1989); for further details, e.g., on the sequence of bifurcations encountered when enlarging the domain of integration, see Hyman *et al.* (1986).

Direct numerical simulations of two coupled reaction-diffusion equations (the unsteady forms of (4.3.33) and (4.3.34) with Le < 1 and a large but finite activation energy) do confirm that a regime of chaotic wrinkling emerges if $\beta(1 - \text{Le})$ is large enough and positive (Denet, 1988), as predicted by the KS equation.

It is worth stressing that the nonlinearity featuring in (4.5.26), which ultimately prevents the amplitude of the unstable modes from growing without limit, is of purely *geometrical* origin. It is brought about by the existence of an intrinsic burning speed ($\simeq u_L$) at which a nearly-flat flame propagates *normally* to itself and is basically the mechanism of front rotation which a Bunsen flame (Fig. 4.4) uses to stay steady even though the incoming flow velocity u_∞ exceeds u_L. If considered alone, a Huygens' type of propagation would of course generate sharp crests that "consume" flame area and therefore tend to reduce the amplitude of wrinkling. This effect has a net stabilizing influence, but nevertheless would tend to create very large curvatures at crests: it is precisely the role of higher-order curvature to take care of that. As for the second derivative appearing in (4.5.26), it provides the flame with a mechanism of spontaneous wrinkle growth. When le > -2, the latter mechanism is absent, and the leading order reduced evolution equation obviously has Burgers' form:

$$\frac{\partial\varphi}{\partial\tau} + \frac{1}{2}\left(\frac{\partial\varphi}{\partial\zeta}\right)^2 = \left(1 + \frac{le}{2}\right)\frac{\partial^2\varphi}{\partial\zeta^2} + \dots \qquad (4.5.28)$$

A few last words about (4.5.27) itself, which was obtained under the assumption that the flame is flat on transverse average. Clearly, the physics locally at work in the KS equation still operates even if the front has a more general, large scale geometry (e.g., nearly-spherical, expanding flames). Equation (4.5.27) can then be generalized to express the local normal flame velocity relative to the fresh medium (a scalar) as a function of the true scalars which characterize the local, weakly curved geometry. To leading order in the limit le $+2 \to 0$, the evolution acquires the form (Frankel and Sivashinsky, 1988):

$$\mathbf{n} \cdot (\mathbf{0} - \mathbf{D})/u_L = 1 - \left(1 + \tfrac{1}{2}\,\text{le}\right)\left(\frac{1}{r_1} + \frac{1}{r_2}\right) + 4\Delta_s\left(\frac{1}{r_1} + \frac{1}{r_2}\right), \qquad (4.5.29)$$

where $r_{1,2}$ are the scaled principal radii of curvature of the flame front at the point where $\mathbf{D} \cdot \mathbf{n}$ is computed, and $\Delta_s(.)$ is the so-called Beltrami–Laplace operator, i.e., the surface Laplacian when expressed in terms of the two-dimensional metrics of the flame shape itself. Clearly, (4.5.29) is free from any reference to a system of coordinates. Except in the case of nearly cylindrical or spherical fronts, it has so far not been studied in detail.

4.6 A conductive instability

To further illustrate how instability mechanisms can be qualitatively discussed in the framework of a hydrodynamic-free model, we could have devoted a section to those showing up when flames propagate in some premixed solids (e.g., of the kind used in rocket boosters); these are obviously devoid of any motion in the fresh medium. We decided not to do so, because such phenomena are somehow extreme cases of what happens in flames of heavy fuels; in effect, being free from species diffusion, solids formally correspond to Le $= \infty$. By the same token, however, they are not directly amenable to the preceding formalism, given that the assumptions $\beta \to \infty$ and $0 < (1 - 1/\text{Le}) = \mathcal{O}(1)$ are not usually compatible when unsteady flames are considered. The previous analysis can nevertheless be adapted when Le $= \infty$ to account for a finite (albeit large enough) β upon phenomenologically postulating an appropriate thin reaction-sheet compatible with what quasi-steady jump relationships would give (Matkowsky and Sivashinsky, 1978). The net result is a critical value for $\beta(1 - \text{Le}^{-1})_{\text{Le}=\infty}$, which obviously plays the same role for solids as le $= \beta(1 - \text{Le}^{-1})$ did for gases: if $\beta > 8$ (a pretty large number, so that a description of the reaction layer in terms of jumps is accurate) an oscillatory system of finite-wavelength wrinkles may grow, analogously to what happened with gases when le $> 32/3 \simeq 10.6$; when $\beta > 2(2 + \sqrt{5}) \simeq 8.5$, one-dimensional disturbances may also grow. We shall not comment on this kind of generalization any further.

Instead we chose to focus some attention on an instability which may occur with gaseous reactants even if Le $= 1$. It has its root in the purely conductive exchanges between a flame and a boundary, and it is therefore of a markedly different nature from what we have discussed so far. This will also be an opportunity to exhibit a situation in which the "temperature + excess enthalpy" formalism is not very convenient. Specifically we consider a flame held downstream from a cooled planar porous burner (Fig. 4.15), through which the reactant is injected uniformly at a known rate; in other words the total reactant flux-density $J \equiv \rho u y_A - \rho D_A \, dy_A/dx$ is given at the burner surface $x = 0$. From J and the total mass flow rate ρu one can define a reference mass fraction $y_u = J/\rho u$, which actually is that of the reactant deeply enough in the porous solid and in the tank where the mixture is stored; then one computes T_b by the usual formula $T_b = T_u + Q y_A/C_p$, where T_u is the burner temperature $(T(x = 0) \equiv T_u)$. We shall denote by m the ratio $(\rho u)/\rho_u u_L$, where ρ_u is computed from T_u and the nearly uniform ambient pressure, and introduce the same reduced variables θ, a, τ, ξ and ζ as in paragraph 4.5.

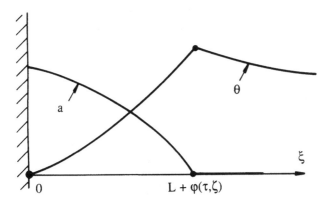

Fig. 4.15. Outer ξ-wise profiles of reduced temperature (θ) and reactant mass fraction (a) downstream of a cooled flat burner (at $\xi = 0$).

Even though the reduced equations corresponding to a unit Lewis number:

$$\frac{\partial \theta}{\partial \tau} + m \frac{\partial \theta}{\partial \xi} = \Delta \theta + \Lambda W(\theta, a) , \qquad (4.6.1)$$

$$\frac{\partial a}{\partial \tau} + m \frac{\partial a}{\partial \xi} = \Delta a - \Lambda W(\theta, a) , \qquad (4.6.2)$$

look quite alike, it is no longer convenient to introduce the excess enthalpy $\theta + a - 1$ as we did in Section 4.5, because the boundary conditions for θ and for a at the burner ($\xi = 0$) are of the Dirichlet and mixed types respectively:

$$\theta = 0 , \qquad m(a - 1) = \frac{\partial a}{\partial \xi} , \qquad \text{at} \quad \xi = 0 . \qquad (4.6.3)$$

Fortunately, a few simplifications survive. For example a thin reaction layer still exists when $\epsilon \to 0$, which we now locate at $\xi = L + \varphi(\tau, \zeta)$, i.e., about the equilibrium flame location L, and jump conditions analogous to (4.4.9)–(4.4.12) may be shown to still hold for the outer, chemistry-free profiles at $\xi = L + \varphi$, since they are consequences of *local* matchings between reactive profiles and nonreactive ones:

$$\beta(\theta - 1) \le \mathcal{O}(1) , \qquad\qquad a = 0 , \qquad (4.6.4)$$

$$[\mathbf{n} \cdot \nabla \theta + \mathbf{n} \cdot \nabla a]_-^+ = \mathcal{O}\left(1/\beta^2\right) , \qquad (4.6.5)$$

$$(\mathbf{n} \cdot \nabla \theta)^2_{\text{fresh side}} = \exp \beta(\theta - 1)\left(1 + \mathcal{O}(1/\beta)\right) . \qquad (4.6.6)$$

As the attentive reader undoubtedly noticed it, these relationships are written in a slightly different manner from (4.4.9)–(4.4.12), even though their physical content is unchanged. This is not done to confuse him or her but, rather, to emphasize the following technical difficulty. In the present problem, power series in $1/\beta$ are not enough. Instead the asymptotic expansions of the θ and a profiles at $\mathcal{O}(1)$ distances from the reaction sheet location, where (4.6.4)–(4.6.6) hold, will also include powers of $1/\log \beta$, $(\log \log \beta)/\log \beta$ and their combinations with powers of $1/\beta$ itself. Nevertheless the new gauge functions do not interfere with the power series when matching the outer profiles with the reactive ones. As a consequence the above listed relationships are still valid, provided the coefficients of $1/\beta^n$ are allowed to contain terms involving the milder functions $(\log b)^{-p}$, $(\log \log \beta)/(\log \beta)^p \ldots (p > 0)$.

Still labeling the steady profiles outer by overbars, one has:

$$(\bar{\theta}, \bar{a}) = \left(\bar{\theta}(L) \left(e^{m\xi} - 1\right) \middle/ \left(e^{mL} - 1\right), 1 - e^{m(\xi - L)}\right) \qquad (4.6.7)$$

for $0 \le \xi \le L$ and $(\bar{\theta}, \bar{a}) = (\bar{\theta}(L), 0)$ otherwise. Equation (4.6.5) leads to $\bar{\theta}(L) = 1 - e^{-mL} + \mathcal{O}\left(\beta^{-2}\right)$, which obliges one to only consider situations where $e^{-mL} \leq \mathcal{O}(1/\beta)$ for the sake of compatibility with (4.6.4). This kind of constraint is the basic reason why terms involving $\log \beta \ldots$ come about. Next, (4.6.6) gives:

$$m^2 = \exp\left(\beta\, e^{-mL}\right)(1 + \mathcal{O}(1/\beta)) \qquad (4.6.8)$$

thereby relating the steady stand-off distance L to the reduced injection rate m. m is less than one: the flame looses heat to the burner, even if only a little, hence propagates counter-streamwise more slowly than usually until the losses make the propagation velocity equilibrate the injection rate, thereby fixing L. One may next note that L goes to infinity for $m \to 1^-$ (blow-off limit) and is minimum when $\beta m L\, e^{-mL} = 2$ (hence the appearance of iterates of $\log \beta$).

The stability analysis proceeds along the same line as in Section 4.5. $\varphi(\tau, \zeta)$ is assigned a small amplitude (δ, say), (θ, a) are written as $(\bar{\theta}, \bar{a}) + \varphi(\tau, \zeta)(\theta', a') + \mathcal{O}(\delta^2)$ and attention is again focused on a single mode $\exp(s\tau + i\mathbf{k}\cdot\zeta)$ of φ, where s is to be found by solving (4.6.1)–(4.6.6). This in turn entails solving equations which have the structure:

$$\left(s + k^2\right) F + m\frac{dF}{d\xi} = \frac{d^2F}{d\xi^2} \qquad (4.6.9)$$

with boundary conditions (4.6.3) and the linearized forms of (4.6.4)–(4.6.6) imposed at $\xi = 0$ *and* $\xi = L$. One easily imagines the difficulties caused by constraints such as $e^{-mL} \leq \mathcal{O}(1/\beta)$ or $mL\, e^{-mL} = \mathcal{O}(1)$, since the solutions to (4.6.9) involve combinations of

$$\exp\left(m\xi(1 \pm \Gamma)/2\right), \quad \Gamma = \left(1 + 4\left(s + k^2\right)m^{-2}\right)^{1/2}, \quad \mathcal{R}e\,\Gamma \geq 1.$$

For example, $\exp\left(m\xi(1 + \Gamma)/2\right)$ ranges from $\mathcal{O}(1)$ at $\xi = 0$ to an $\mathcal{O}\left(\beta^{m(1+\Gamma)/2}\right)$ (hence dependent on $s + k^2$) value at $\xi = \log(\beta)$. As far as the present linear analysis is concerned, such an ordering problem does not cause much trouble, since (4.6.9) can be solved exactly whatever ξ, s, m and k. The situation is far less simple in nonlinear treatments (Joulin, 1982b) and it becomes definitely messy if one attempts to also include small hydrodynamic motions in the comparatively wide region "between" flame and burner. The boundary conditions and the linearized jumps ultimately yield the "leading"-order dispersion relation (Buckmaster, 1983):

$$\Gamma^2(\Gamma - 1) + 0 = \frac{-2\beta\Gamma^3}{1 + \Gamma}\, e^{-mL\Gamma} + \mathcal{O}(1/\beta)\,. \qquad (4.6.10)$$

Given that Le $= 1$ is assumed here, (4.6.10) obviously matches (4.5.18) when $L \to \infty$. A last word of caution about (4.6.10): for its explicitly-written r.h.s. to be significant, one must impose $Re(\Gamma) < 2$, to ensure that $\beta \exp(-Lm\Gamma)$ be larger than $\mathcal{O}(1/\beta)$.

Let us first concentrate on planar evolutions $(k = 0)$ and on small values of s, to simplify (4.6.10) into

$$2sL/m + \beta mL\, e^{-Lm}\, e^{-2sL/m} \simeq 0 \qquad (4.6.11)$$

upon the working assumption that $\Lambda \gg 1$. The grouping sL was given the benefit of the doubt $(s \ll 1$ and $L \gg 1)$. This was a sensible precaution, because (4.6.11) predicts an oscillatory conductive instability as soon as

$$\beta\, e^{-mL}\, Lm \geq \frac{\pi}{2} \qquad (4.6.12)$$

the corresponding value of sL/m being simply $i\pi/4$; the very result $\beta Lm\, e^{-mL} = \mathcal{O}(1)$ confirms that $L \gg 1$ and $s \ll 1$ at the onset of instability. To understand the origin and the name of this instability, one may first note that (4.6.12) can be rewritten as a delay-differential equation:

$$\frac{d\varphi}{d\tau}(\tau') + \frac{\beta mL\, e^{-Lm}}{2}\varphi(\tau' - 2) \simeq 0, \quad \tau' \equiv \tau m/L. \qquad (4.6.13)$$

The appearance of a delay in conduction/diffusion/reaction interactions is at first glance astonishing. To really see where it comes from, consider the "auxiliary" problem:

$$\frac{\partial F}{\partial \tau} + m\frac{\partial F}{\partial \xi} = \frac{\partial^2 F}{\partial \xi^2},$$
$$\qquad (4.6.14)$$
$$F(0,\tau) = e^{i\omega\tau}, \qquad F \text{ bounded on } -\infty < \xi < +\infty$$

When $\omega \ll m^2$ and $\xi\omega < \mathcal{O}(1)$, this is easily shown to have

$$F(\xi \leq 0,\tau) \simeq F(0,\tau + \xi/m)\, e^{m\xi},$$
$$F(\xi \geq 0,\tau) \simeq F(0,\tau - \xi/m)$$

as solutions: at a distance $|\xi|$ from the origin, the F-signals exhibit a phase shift proportional to $|\xi|m^{-1}$, those propagating counter-streamwise being exponentially damped with $|\xi|m$. Because a small flame displacement φ is equivalent to a small temperature disturbance at $\xi = L$ when the boundary condition (4.6.4) is transferred there by the linearization procedure, the origin of the delay $\tau' = 2 = 1 + 1$ is now clear. It is due to the finite phase velocity of thermo-convective waves propagating back *and* forth between the burner and the reaction layer: when the latter is shifted at time $\tau' = 0$, it cannot "feel" any retroaction of the burner as long as $\tau' < 2$. The non-delayed, quasi-steady version of (4.6.13) does not yield any instability.

At the expense of a somewhat heavier formalism (Joulin, 1982a), (4.6.13) can be extended to the nonlinear domain to produce the leading-order result:

$$\frac{d\varphi}{d\tau'} + \frac{\beta m L\, e^{-mL}}{2}\left(1 - e^{-\varphi(\tau'-2)}\right) = 0\,, \qquad (4.6.15)$$

which is a form of Hutchinson's equation (Hale, 1977). An exponential in φ appears above, basically because an $\mathcal{O}(1)$ shift in reaction-sheet location induces an $\mathcal{O}(1)$ multiplicative correction to the conductive heat loss to the burner $\left(e^{-mL} \to e^{-m(L+\varphi)}\right)$, the exp(.) function itself originating from the damping of conduction/convection signals that propagate counter-streamwise; as for the constant term in (4.6.15), it simply measures the difference ($\simeq m - 1$) between the injection rate and $\rho_u u_L$, so that (4.6.15) could be interpreted by a quasi-steady kinematic argument... if the (now-explained) delay was absent. When integrated numerically, (4.6.14) ultimately yields saw-tooth relaxation oscillations in $\varphi(\tau')$ as soon as $\beta m L \exp(-mL)$ exceeds $\pi/2$ sufficiently (Fig. 4.16).

If wrinkling with $k \ll 1$ is retained, (4.6.11) generalizes into:

$$sL/m + \frac{\beta m L}{2} e^{-mL}\, e^{-2sL/m - 2k^2 L/m}$$
$$\simeq -Lm\left(1 + \tfrac{1}{2}\,\text{le}\right) k^2\,, \qquad (4.6.16)$$

where the influence of $\text{le} \neq 0$ is restored for sake of completeness. Beside a Lewis-number-dependent curvature effect already present in (4.5.20), (4.6.16) includes a specific influence of the burner which is now multiplied by $\exp\left(-2k^2 L/m\right)$. This factor simply accounts for the dissipation

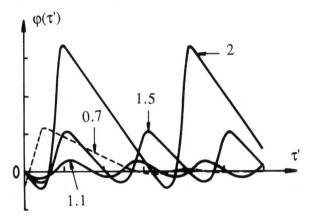

Fig. 4.16. Typical oscillations in flat-burner flame location, vs. time τ', for various values of $2\beta Lm\, e^{-mL}/\pi$, and a damped trajectory (0.7).

by transverse conduction of the traveling temperature signals which we mentioned above and can easily be understood upon adding $-k^2 F$ to the r.h.s. of (4.6.14); of course, the burner preferentially affects the longest wrinkles. Equation (4.6.16) can also be generalized to the nonlinear range (Joulin, 1981); it then acquires the form:

$$\dot\varphi + \tfrac{1}{2}|\nabla\varphi|^2 + \frac{\beta m L\, e^{-mL}}{2}\, j\left(1 - e^{-\varphi(\tau'-2)}\right) = \left(1 + \tfrac{1}{2}\mathrm{le}\right)\Delta\varphi \quad (4.6.17)$$

in terms of suitably-scaled variables. Here $j(.)$ denotes the spatial convolution with a normalized Gaussian, which accounts for the $\exp\left(-2k^2 L/m\right)$ factor featuring in (4.6.16). Clearly (4.6.17) is nothing but (4.5.28), once complemented by the influence of the burner. This nonlocal equation (in space and time) has not yet been numerically integrated, but it can be shown by the "envelope" formalism (see Chapter 4) that φ is governed by a complex Ginsburg–Landau equation when $\beta m L\, e^{-mL} \geq \pi/2$, and that it may develop phase turbulence if the Lewis number is above some positive value (Joulin, 1982b).

Incorporating some hydrodynamical contribution into the model, while still retaining the mechanism of conductive instability, is anything but a simple task: even though small, the conductive interaction between flame and burner cannot be neglected so that hydrodynamics is affected by small temperature changes and, conversely, small gas motions may have a nonnegligible cumulative effect on conduction when $L \gg 1$. Besides the mathematical challenge, completing this analysis would be worth doing because careful experiments on unstable flat-burner flames are quite feasible (El Hamdi *et al.*, 1987). Burner problems may also be viewed as first steps towards the interactions between gaseous flames and pyrolizing premixed solids of the kind used in rockets (Clavin and Lazzimi, 1992). A prerequisite step is to study the hydrodynamic instabilities of burner-free flames.

4.7 Hydrodynamic instability

4.7.1 The Landau–Darrieus model

In Sections 4.5 and 4.6 we focused our attention on a restricted, constant-density model which was absolutely devoid of any hydrodynamics; it nevertheless provided us with valuable insights into the couplings between unsteady diffusive/conductive processes and the strongly temperature-dependent reaction rate. What we envisage here is a sort of counterpoint. Specifically, we consider a nearly flat flame whose wrinkles have infinitely long wavelengths Λ compared to the conduction/convection thickness l_T, the range of thermal effects. As a consequence, each flame element may be considered as planar and, locally, it propagates at the *known*

normal speed u_L *relative to the fresh gases.* In such situations of infinite
Peclet numbers Λ/l_T the mixture temperature, hence its density, may be
considered piecewise uniform and, because the gases have $\mathcal{O}(1)$ Prandtl
numbers, one may also forget about viscosity.

Apart from u_L itself, the only ingredients which survive in this so-called
LD model (Landau, 1944; Darrieus, 1938) are the Euler equations and a
discontinuous density change from ρ_u to $\rho_b < \rho_u$: in other words we are
to study the system:

$$\nabla \cdot \mathbf{u} = 0, \tag{4.7.1}$$

$$\rho\frac{\partial \mathbf{u}}{\partial t} + \rho(\mathbf{u} \cdot \nabla)\mathbf{u} = -\nabla p, \tag{4.7.2}$$

where ρ stands for ρ_u or ρ_b, on each side of the flame. We are using
dimensional forms here, because no natural reference length exists in the
LD model. We shall denote by $x = \Phi(t, \mathbf{z})$ the equation of the flame
front shape, in such a way that $\Phi \equiv 0$ represents the steady unperturbed
solution. Far upstream ($x \to -\infty$), the velocity vector $\mathbf{u}$ reduces to
its x-component. Far downstream ($x \to +\infty$), both $\mathbf{u}$ and p must
remain bounded. Finally, across the flame front, the Rankine–Hugoniot
relationships (2.2.1)–(2.2.3), repeated for convenience,

$$[\rho\mathbf{n} \cdot (\mathbf{u} - \mathbf{D})]_-^+ = 0, \tag{4.7.3}$$

$$[\mathbf{u} \times \mathbf{n}]_-^+ = 0, \tag{4.7.4}$$

$$[p + \rho(\mathbf{n} \cdot (\mathbf{u} - \mathbf{D}))^2]_-^+ = 0 \tag{4.7.5}$$

hold, where $\mathbf{n} \cdot \mathbf{D}$ is the normal flame velocity in the chosen frame of
reference and $[f]_-^+ \equiv f(\text{burnt side}) - f(\text{fresh side})$.

In principle, (4.7.1)–(4.7.5) allow one to compute $\mathbf{u}$ and p, whatever
the flame shape dynamics. The latter is constrained, however, by the
kinematic relationship

$$\mathbf{n} \cdot (\mathbf{u} - \mathbf{D})_{\text{fresh}} = u_L, \tag{4.7.6}$$

which expresses that, locally, each flame element propagates at the known
normal speed u_L relative to the fresh gas it consumes. Equations (4.7.3)
and (4.7.6) also imply:

$$[\mathbf{u} \cdot \mathbf{n}]_-^+ = u_L(\rho_u/\rho_b - 1). \tag{4.7.7}$$

The steady planar flame ($\Phi \equiv 0$) corresponds to $\bar{\mathbf{u}} = (u_L, 0)$ and
$\bar{p} = p_u$ in the fresh medium, and to $\bar{\mathbf{u}} = (u_L\rho_u/\rho_b, 0)$ and $\bar{p} =$
$p_u - \rho_u^2 u_L^2(1/\rho_b - 1/\rho_u)$ in the burned one. Following the general normal-
mode procedure we next assign to Φ a small amplitude (d, say) compared
to the involved wavelengths. This causes $\mathbf{u}$ and p to slightly depart from

the above steady profiles $\bar{\mathbf{u}}$ and $\bar{p}$; specifically one writes

$$(\mathbf{u}, p) = (\bar{\mathbf{u}}, \bar{p}) + \Phi(t, \mathbf{z})(\mathbf{u}', p') + \ldots \tag{4.7.8}$$

and linearizes the system (4.7.1)–(4.7.6) in the limit $d/\Lambda \to 0$. Once this is done, one may again focus on one particular transverse mode of Φ by writing $\Phi = d \exp(St + i\mathbf{K} \cdot \mathbf{z})$, where the growth rate S has to be found. Using the decomposition $\mathbf{u}' = (u', \mathbf{v}')$, where u' is the x-component of $\mathbf{u}'$ and $\mathbf{v}'$ denotes the transverse ones, one obtains:

$$\frac{du'}{dx} + i\mathbf{K} \cdot \mathbf{v}' = 0, \tag{4.7.9}$$

$$\rho S u' + (\rho_u u_L)\frac{du'}{dx} = -\frac{dp'}{dx}, \tag{4.7.10}$$

$$\rho S \mathbf{v}' + (\rho_u u_L)\frac{d\mathbf{v}'}{dx} = -i\mathbf{K}p', \tag{4.7.11}$$

from (4.7.3)–(4.7.5). The continuity equation gives $\mathbf{K} \cdot \mathbf{v}' = i du'/dx$ and, after scalar multiplication by $\mathbf{K}$, (4.7.11) then yields

$$p'K^2 = -\left(\rho S + \rho_u u_L \frac{d}{dx}\right)\frac{du'}{dx},$$

where $K = (\mathbf{K} \cdot \mathbf{K})^{1/2}$. Substitution of this pressure disturbance in the x-wise momentum balance (4.7.10) furnishes a closed equation for the x-wise variations of u':

$$\left(\rho_u u_L \frac{d}{dx} + \rho S\right)\left(\frac{d^2}{dx^2} - K^2\right)u' = 0. \tag{4.7.12}$$

When the latter is solved, p' and $\mathbf{v}' \cdot \mathbf{K}$ will be easily obtained as indicated just above. Anticipating that S is to have a positive real part, one gets

$$u' = u'(0^-)\,e^{Kx}, \quad x < 0, \tag{4.7.13}$$

$$u' = \left(u'(0^+) - C\right)e^{-Kx} + C\,e^{(-S\rho_b x/\rho_u u_L)}, \quad x \geq 0, \tag{4.7.14}$$

where $u'(0^-)$, $u'(0^+)$ and C are three integration constants. As is easily checked, the contributions to u' proportional to $\exp(\pm Kx)$ correspond to potential flow disturbances, whereas that in $\exp(-S\rho_b x/\rho_u u_L)$ gives the rotational part; the latter does not exist on the fresh side, because vorticity is merely convected by the flow and vanishes in the uniform incoming mixture. One can also check that the pressure disturbance is a harmonic function on both sides of the front.

We now have to account for the hydrodynamic jump relationships across the flame. In the linear approximation, $\mathbf{n} \cdot \mathbf{D}$ is simply Φ_t and the normal vector reads $\mathbf{n} \simeq (1, i\mathbf{K}\Phi)$. As a result linearized forms of

(4.7.3)–(4.7.5) are provided:

$$[u']_{0-}^{0+} = 0 \,, \tag{4.7.15}$$

$$[p']_{0-}^{0+} = 0 \,, \tag{4.7.16}$$

$$[\mathbf{v}' \cdot \mathbf{K}]_{0-}^{0+} = -iK^2 u_L \frac{\gamma}{1-\gamma} \,. \tag{4.7.17}$$

This furnishes three relations for the three integration constants $u'(0^-)$, $u'(0^+)$ and C, since u', p' and $\mathbf{K} \cdot \mathbf{v}'$ can all be expressed in terms of them. A simple linear algebra ultimately gives

$$((2-\gamma)S + 2Ku_L)\,u'(0^-) = \frac{\gamma}{1-\gamma}K^2 u_L{}^2 \,, \tag{4.7.18}$$

where again $\gamma \equiv (\rho_u - \rho_b)/\rho_u$, for the x-wise velocity disturbance just ahead of the flame; next:

$$((1-\gamma)S - u_L K)\,C = (2-\gamma)Su'(0^+) \tag{4.7.19}$$

for the vortical component in the burned gases; note that $C \ll u'(0)$ and the flow is almost potential if, for some reason, $S \ll Ku_L$. Employing the linearized version of (4.7.6), viz.

$$S = u'(0^-) \tag{4.7.20}$$

yields the sought after dispersion relation. As expected from a dimensional analysis (see Chapter 1) S is of the form:

$$S = Ku_L\Omega(\gamma) \tag{4.7.21}$$

because this is the only way (Landau, 1944) one can express a true scalar having the dimension of (second)$^{-1}$ in terms of the four parameters u_L, $\mathbf{K}$, ρ_u, ρ_b which characterize a wrinkle in the linearized Landau–Darrieus model. Owing to (4.7.18) (4.7.20) the proportionality coefficient $\Omega(\gamma)$ is a solution to the quadratic:

$$(2-\gamma)\Omega^2 + 2\Omega = \gamma/(1-\gamma) \,, \tag{4.7.22}$$

which always has a positive root as soon as $\gamma > 0$, i.e., $\rho_u > \rho_b$. For future reference we quote the limiting behavior $\Omega(\gamma) = \gamma/2 + o(\gamma)$ as $\gamma \to 0$, so that $C \ll u'(0)$ in this limit.

Equation (4.7.20) merely states that the front moves in exactly the same way as the fresh gases do just ahead of its elements; the clue to the origin of the LD instability is thus to understand why $u - \bar{u}$ and Φ are in phase. As sketched in Fig. (4.17), the large fresh-to-burnt increase in $\mathbf{u} \cdot \mathbf{n}$ given by (4.7.7) and the constancy of tangential components accounted for by (4.7.4)) cause the streamlines to be deflected across the

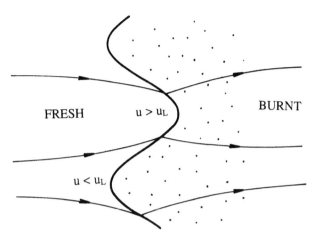

Fig. 4.17. Schematic streamlines about a thin wrinkled flame when density changes are retained.

flame. As the flow is parallel to the x-axis at both $x = \pm\infty$, the fresh mixture is accelerated at the flame crests ($\Delta\Phi < 0$) and slowed down at the troughs ($\Delta\Phi > 0$); therefore $u'\left(0^-\right)$ and Φ are in phase and, owing to the constant flame speed *relative to the gas*, an instability sets in. The above argument is quasi-steady and, as such, it may sound somewhat doubtful. It is nevertheless a quasi-steady mechanism which brings in the positive r.h.s. of (4.7.22) and implies $\Omega > 0$. This is best shown by examining a potential-flow model (Clavin and Sivashinsky, 1987), in which only (4.7.15,17,20) are retained at the front. The growth rate obtained in this way is still of the form (4.7.21), apart from a different coefficient $\Omega_{\mathrm{pot}}(\gamma)$ given by $2\Omega_{\mathrm{pot}} = \gamma/(1 - \gamma)$.

A comparison with (4.7.22) confirms that a potential, hence quasi-steady, mechanism is at the origin of the r.h.s. of (4.7.22) and of the LD instability. Contrary to a widespread belief, vorticity generation at the flame is somehow a side effect here, caused by the need of satisfying (4.7.16) as well, which in fact tends to *moderate* the instability ($\Omega_{\mathrm{pot}} \geq \Omega$). The dynamics implied by (4.7.22) is second-order in time, however, and it corresponds to the linear equation for $\Phi(t, z)$:

$$(2 - \gamma)\frac{\partial^2\Phi}{\partial t^2} + 2u_L I\left(\frac{\partial\Phi}{\partial t}, \mathbf{z}\right) + \frac{\gamma}{1 - \gamma}u_L{}^2\Delta\Phi = 0\,, \qquad (4.7.23)$$

where the linear *Landau–Darrieus operator* $I(.,\mathbf{z})$ is the multiplication by $K = (\mathbf{K} \cdot \mathbf{K})^{1/2}$ in the Fourier space ($\mathbf{K}$) conjugate to $\mathbf{z}$:

$$I\left(e^{i\mathbf{K}\cdot\mathbf{z}}, \mathbf{z}\right) = K\,e^{i\mathbf{K}\cdot\mathbf{z}}\,. \qquad (4.7.24)$$

Note that $K^2 = \mathbf{K} \cdot \mathbf{K}$ implies $I \circ I \equiv -\Delta$. An alternative form of (4.7.23) is the system of two equations:

$$\frac{\partial \Phi}{\partial t} = V \,,$$

$$(2 - \gamma)\frac{\partial V}{\partial t} + 2u_L I(V, \mathbf{z}) = -\frac{\gamma}{1 - \gamma} u_L{}^2 \Delta \Phi \,,$$

(4.7.25)

which respectively have kinematic and dynamical origins. The potential-flow model would give $\partial \Phi / \partial t = V$ and $2(1 - \gamma)V = \gamma u_L I(\Phi, \mathbf{z})$, which does not display any inertial effect. At any rate, since γ is actually definitely positive ($\rho_u > \rho_b$), both models agree to predict that any flame should be unstable, especially when K is "large." Obviously this is contrary to many experiments; furthermore, as $K = \infty$ implies $S = \infty$, the model is not well-posed dynamically. The too crude LD model needs be improved so as to include a more realistic description of flames.

To date, several broad classes of effects have been shown to modify the picture: nonconstancy of local flame speed, nonlinearity, large-scale front geometry, body forces, influence of boundaries, large scale flow geometry. This is the subject of the next sections.

4.7.2 A phenomenological curvature effect

Markstein (1951, 1964) was the first to realize that (4.7.6) constitutes the weakest point of the previous Landau–Darrieus model. In effect, the assumption of a constant local flame speed $\mathbf{n} \cdot (\mathbf{u} - \mathbf{D})_{\text{fresh}}$ is clearly untenable when the front is so highly curved that its (principal-) radii of curvature (R_1, R_2) get comparable to the conduction/convection thickness $l_T = D_{\text{th}}/u_L$: this is precisely what equation (4.5.22) of Section 4.5 indicated... in the limit $\gamma = 0$. Markstein therefore amended (4.7.6) phenomenologically into:

$$\mathbf{n} \cdot (\mathbf{u} - \mathbf{D})_{\text{fresh}} = u_L \left(1 - \mathcal{L} \left(\frac{1}{R_1} + \frac{1}{R_2}\right)\right) \,,$$

(4.7.26)

where $\mathcal{L}$, the so-called *Markstein length*, is assumed known and is taken positive if $R_i > 0$ means that the corresponding center of curvature lies in the burned gases. On dimensional grounds $\mathcal{L}$ should be proportional to l_T. This is indeed the case for $\gamma = 0$, because the *Markstein number* $\text{Ma} = \mathcal{L}/l_T$ then acquires the simple form: $\text{Ma} = 1 + \text{le}/2$ (see equation (4.5.21)).

In the context of a linearized theory, Markstein's modification amounts to replacing (4.7.15) and (4.7.16) respectively by:

$$[u']_{0^-}^{0^+} = u_L \frac{\gamma}{1 - \gamma} K^2 \mathcal{L} \,,$$

(4.7.27)

$$[p']_{0^-}^{0^+} = -2\rho_u u_L{}^2 \frac{\gamma}{1-\gamma} K^2 \mathcal{L} , \qquad (4.7.28)$$

because $1/R_1 + 1/R_2 \simeq \Delta\Phi$ for weakly curved flames, whereas (4.7.17) is left unchanged. The calculation proceeds along the same lines as in the original LD analysis, to again end up with an expression for $u'(0^-)$. Once substituted in the linearized form:

$$S = u'(0^-) - \mathcal{L}u_L K^2 \qquad (4.7.29)$$

of (4.7.20), this finally yields a new growth/decay rate S in the form: $S = u_L K \Omega(\gamma, K/K_{\text{neutral}})$, where Ω is the solution to the modified quadratic:

$$(2-\gamma)\Omega^2 + 2(1+\mathcal{L}K)\Omega = \frac{\gamma}{1-\gamma}\left(1 - \frac{K}{K_{\text{neutral}}}\right) \qquad (4.7.30)$$

in which K_{neutral} is defined by $\mathcal{L}K_{\text{neutral}} = \gamma/2$. As sketched in Fig. (4.18) the most dramatic consequence of Markstein's treatment is to bring in a cut-off for the hydrodynamic-generated instability. The stabilization mechanism is readily understood, e.g., by considering a wrinkle crest: whereas streamline focusing tends to make the amplitude of wrinkling increase, the enhanced flame speed due to curvature acts in the opposite way ($\mathcal{L} > 0$); the former contribution to $\partial\Phi/\partial t$ wins if $K/K_{\text{neutral}} \ll 1$, and the Landau–Darrieus result is then recovered, whereas the latter one ultimately dominates as K increases, because it involves a second spatial derivative.

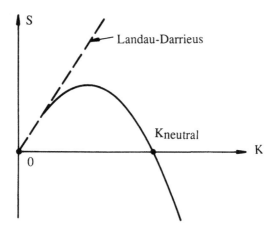

Fig. 4.18.   Dispersion relation when density changes and the Markstein curvature effect are accounted for; the dashed line corresponds to equation (4.7.21).

Before explaining how to handle the curvature effects in a more rational way, we briefly pause to remark that a potential-flow model employing (4.7.4) and (4.7.7) as jump relations would give: $2\Omega_{\mathrm{pot}}(1-\gamma) = \gamma(1 - K/K_{\mathrm{neutral}})$, with the same K_{neutral} as above; if, for some reason (steady states ?), the inertial effects get negligible, employing the potential-flow model could well be of some qualitative value.

4.7.3 Multiscale analysis

Though only phenomenological in nature, Markstein's model clearly indicates the right qualitative way to eliminate the "ultraviolet" pathology of the Landau–Darrieus formulation and the need to account for the finiteness of $l_{\mathrm{T}} = D_{\mathrm{th}}/u_L$. As said previously, this cannot in general be done by analytical means if the flame is curved and/or unsteady (the Zel'dovich flame ball, Section 4.5, or counterflow-burner configurations, Section 4.10, are a few exceptions). Fortunately, the actual flames often are weakly curved and evolve slowly, which allows for an approximate analysis. More specifically, we assume that:

i) the characteristic length scale Λ (e.g., wavelength) over which the flame shape changes noticeably satisfies $l_{\mathrm{T}}/\Lambda \ll 1$;
ii) the typical time scale of flame front evolution is not shorter than $\mathcal{O}(\Lambda/u_L)$;
iii) the flow field which the flame encounters also is characterized by the scales Λ and Λ/u_L; this is the case when the fresh flow gradients are due to the front wrinkling itself (see equation (4.7.12)).

On both sides of the reaction sheet (Section 4.4) one may then identify two distinct regions (Fig. 4.19):

• The *flame front* proper, with $\mathcal{O}(l_{\mathrm{T}})$ thickness. The locally dominant mechanisms are quasi-steady convective/diffusive transports of heat and excess-enthalpy along the normal $\mathbf{n}$ to the reaction sheet; in a frame of reference attached to the latter, unsteadiness, curvature effects and tangential transports introduce $\mathcal{O}(l_{\mathrm{T}}/\Lambda)$ corrections to one-dimensional steady profiles and are thus amenable to perturbative approaches.
• The *hydrodynamical zones*, of sizes $\mathcal{O}(\Lambda)$, which are dominantly governed by piecewise-incompressible, unsteady Euler equations, with $\mathcal{O}(l_{\mathrm{T}}/\Lambda)$ viscous corrections. Except if the front wrinkles have $o(1)$ amplitude to wavelength ratios d/Λ, the hydrodynamical zones are out of reach of the analytical tools, mainly because the burnt gas flow is rotational when the density contrast $\gamma = (\rho_{\mathrm{u}} - \rho_{\mathrm{b}})/\rho_{\mathrm{u}}$ is of order unity.

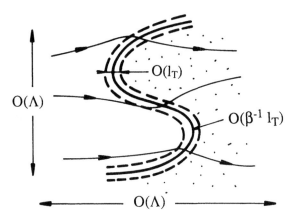

Fig. 4.19. The three length scales of a flame such that $\beta \to \infty$, $\gamma = \mathcal{O}(1)$, and $\Lambda/l_T \gg 1$.

Proceeding in a way similar to what we did when handling the reaction layer (Section 4.4), one may then treat the flame front once for all and replace it by a reactive discontinuity equipped with Hugoniot relationships that are modified by curvature, unsteadiness and tangential transports, and endowed with a *kinematic law* which tells one how the discontinuous front propagates relative to the fresh gas flow field $\mathbf{u}_{\text{fresh}}$.

Being a scalar $\mathbf{n}\cdot(\mathbf{u}-\mathbf{D})_{\text{fresh}}$ can in principle be expressed in terms of u_L and all the scalars which characterize the local flame shape, its evolution and the local structure of $\mathbf{u}_{\text{fresh}}$. As shown by Clavin and Joulin (1988) this phenomenological approach leads one to write:

$$\mathbf{n} \cdot (\mathbf{u} - \mathbf{D})_{\text{fresh}} =$$

$$u_L \left(1 - \mathcal{L}_c \left(\frac{1}{R_1} + \frac{1}{R_2} \right) + \mathcal{L}_s \frac{1}{u_L} \mathbf{n} \cdot \nabla \, \mathbf{u}_{\text{fresh}} \cdot \mathbf{n} \right) + \dots \quad (4.7.31)$$

to lowest order in l_T/Λ (compare to (4.7.26)); if the ambient pressure p_u and the fresh density ρ_u were to vary with time, one should also include $\text{div}\, \mathbf{u}_{\text{fresh}}$ and $d \log p_u/dt$ in the r.h.s. of (4.7.31). The Markstein lengths $\mathcal{L}_c$ and $\mathcal{L}_s$ are yet to be determined.

If the flame front location is *defined* as that of the reaction sheet and $\mathbf{n} \cdot (\mathbf{u} - \mathbf{D})_{\text{fresh}}$ and $\nabla \mathbf{u}_{\text{fresh}}$ represent fresh flow quantities extrapolated at the front location, one may show that:

$$\mathcal{L}_c = \mathcal{L}_s \equiv \mathcal{L}. \quad (4.7.32)$$

The quickest way to show that the "curvature" and "strain" Markstein lengths are then identical is to investigate an adiabatic, steady spherical flame held around a point-source of fresh mixture in an unbounded hot atmosphere. The methods developed in Section 4.4 may be used to show that $\mathbf{n} \cdot (\mathbf{u} - \mathbf{D})_{\text{fresh}} \equiv u_L$ whatever the flame radius. For the $\mathcal{O}(D_{\text{th}}/R)$ terms to disappear from (4.7.31), (4.7.32) must hold because $1/R_1 + 1/R_2 = -2/R = \mathbf{n} \cdot \nabla \mathbf{u}_{\text{fresh}} \cdot \mathbf{n}/u_L$ in this configuration. One can compute $\mathcal{L}$ upon specializing (4.7.31) to the flat flame shown in Fig. 4.33 which is analytically accessible when the incoming flow only weakly diverges. In accord with the direct calculations (Matalon and Matkowsky, 1982; Clavin and Joulin, 1983) one then obtains an analytical expression for the Markstein number $\text{Ma} = \mathcal{L}/l_{\text{T}}$:

$$\text{Ma} = M' + M'',$$

$$M' = \frac{1}{1 - \gamma} \int_0^1 \chi(\theta)\, d\theta\,, \qquad\qquad (4.7.33)$$

$$M'' = \frac{\text{le}}{2} \int_0^1 \chi(\theta) \log(1/\theta)\, d\theta\,,$$

where $\chi(\theta)$ stands for $\lambda\rho/\lambda_{\text{u}}\rho_{\text{u}}$ when expressed in terms of $\theta = (T - T_{\text{u}})/(T_{\text{b}} - T_{\text{u}})$; consistently with (4.5.21), $\text{Ma} \rightarrow 1 + \text{le}/2$ when $\gamma \rightarrow 0$. Provided that the definition of le is generalized to $\text{le}_{\text{A}} + (\text{le}_{\text{B}} - \text{le}_{\text{A}})/(2 + \beta b(\infty))$, where $b(\infty)$ is the final concentration of the abundant reactant B, the above formula can encompass near-stoichiometric burning as well (Garcia-Ybarra *et al.*, 1984). The last authors also incorporated dilution effects $(y_{\text{u}} = \mathcal{O}(\beta^{-1}))$, thermal diffusion $(s_{T,i} = \mathcal{O}(\beta^{-1}))$ and differences in specific heats $(C_{pi} - C_p = \mathcal{O}(\beta^{-1}C_p))$ which all contribute $\mathcal{O}(1)$, but rather small, additive corrections to Ma because they affect the perturbed excess-enthalpy profiles. With $\gamma = 5/6$ $(T_{\text{b}} = 6T_{\text{u}})$ and $\lambda \sim T^{3/4}$ theory predicts $\text{Ma} \simeq 4.5$ and $\text{Ma} \simeq 7$ for lean methane/air and propane/air flames, respectively, in good agreement with what direct measurements yield (Deshaies and Cambray, 1990; Cambray, 1992). Apart from lean flames of hydrogen (Le = 0.3 in air) Ma is now positive for most mixtures, at least when heat loss effects are negligible (Nicoli, 1985).

One has to stress that (4.7.32) holds in one-reaction models if and only if the aforementioned definitions of $\mathbf{u}_{\text{fresh}}$ and the flame location are adopted; otherwise $\mathcal{L}_{\text{s}} \neq \mathcal{L}_{\text{c}}$. Similarly, once the front location is defined, the fractional change in burning speed depends on whether it is measured relatively to the fresh or to the burned medium (the Markstein number involved in $\mathbf{n} \cdot (\mathbf{u} - \mathbf{D})_{\text{burnt}}\rho_{\text{b}}/\rho_{\text{u}}$ has no factor $1/(1 - \gamma)$ in the analog of M' and may easily change sign!). Only the consistency between definitions and results constitutes the physical law (this is akin to choosing Gibbs' surfaces when studying the thermodynamics of interfaces, Nozières, 1992).

Knowing Ma is not enough to compute the flame evolution. One also needs $\mathbf{u}_{\text{fresh}}$, hence $\mathbf{u}_{\text{burnt}}$ and boundary conditions at the now discontinuous flame front. Equations (4.7.3)–(4.7.5) are no longer valid, owing to the finiteness of l_T/Λ and the resulting accumulation effects.

The demonstration leading to corrected versions of (4.7.3)–(4.7.5) is far too lengthy to be reproduced here; for a detailed account, see Aldredge (1990). It entails solving the equations for θ_0 (the limit of $(T - T_u)/(T_b - T_u)$ as $\beta \to \infty$) and the leading order excess-enthalpy H featuring in (4.4.9)–(4.4.12), thus the full Navier–Stokes equations, at $\mathcal{O}(l_T)$ distances along the normal $\mathbf{n}$ to the reaction-sheet, under the assumptions i)-iii). Employing the jump relationships (4.4.9)–(4.4.12) and procedures of matching between the flame profiles and the outer fields of $\mathbf{u}_{fresh}$, $\mathbf{u}_{burnt}$ and pressure p ultimately indicates how the latter quantities vary across the now discontinuous flame front. Even the results (see Aldredge, 1990) are too involved to be listed here. We only give two samples, which are the leading-order-corrected versions of (4.7.15) and (4.7.17):

$$[u']_{0-}^{0+} = l_T M'' \frac{\gamma}{1 - \gamma} \left(K^2 u_L - i\mathbf{K} \cdot \mathbf{v}' (0^-) \right), \qquad (4.7.34)$$

$$[\mathbf{v}' \cdot \mathbf{K}]_{0-}^{0+} = -iK^2 \frac{\gamma}{1 - \gamma} u_L$$
$$+ \Pr l_T \frac{\lambda_b}{\lambda_u} \left[\frac{d \, \mathbf{v}' \cdot \mathbf{K}}{dx} \right]_{0-}^{0+} + l_T \{\ldots\}, \qquad (4.7.35)$$

where $\{\ldots\}$ is the value at $x = 0^-$ of

$$S \left(\frac{\mathbf{v}' \cdot \mathbf{K}}{u_L} - iK^2 \right) \int_0^1 \left(\chi(\theta) - \frac{\lambda}{\lambda_u} \right) d\theta + \Pr \frac{\lambda_b}{\lambda_u} \left(iK^2 u' + \frac{dv'}{dx} \cdot \mathbf{K} \right). \qquad (4.7.36)$$

These and the analogous equation for $[p']$ may therefore be used to predict the growth rate S of a $2\pi/K$-periodic harmonic disturbance $\left(\Phi(t, \mathbf{z}) \sim e^{St + i\mathbf{K} \cdot \mathbf{z}} \right)$ of a planar front ($\Phi \equiv 0$).

The last step to compute S is to solve the linearized, piecewise-incompressible Navier–Stokes equations on both sides of the front. Employing the boundary conditions (4.7.34)–(4.7.36), then the linearized kinematic law (4.7.31) ultimately leads (Pelcé and Clavin, 1982; Clavin and Garcia-Ybarra, 1983) to a growth rate in the form $S = u_L K\Omega$, with:

$$(2 - \gamma + \gamma K l_T M'')\Omega^2$$
$$+ 2 \left(1 + \frac{K l_T (\text{Ma} - \gamma M')}{1 - \gamma} \right) \Omega = \frac{\gamma}{1 - \gamma} \left(1 - \frac{K}{K_{\text{neutral}}} \right), \qquad (4.7.37)$$

which can be compared to (4.7.30). The neutral wavenumber is now given by:

$$(l_T K_{\text{neutral}})^{-1} = \frac{\lambda_b}{\lambda_u} + \text{Ma} \, \frac{2+\gamma}{\gamma} - 2M' + (2\text{Pr} - 1) \int_0^1 \left(\frac{\lambda_b - \lambda}{\lambda_u}\right) d\theta,$$

(4.7.38)

where $\text{Pr} = \mu C_p / \lambda$ is the assumed-constant Prandtl number. The dispersion relation corresponding to (4.7.37)–(4.7.38) is quite similar to that plotted in Fig. 4.18.

It must be recalled that, to obtain the above results, we *postulated* slow transverse and temporal variations of the flame front structure, which seems to contradict what (4.7.38) yields, i.e., an $\mathcal{O}(1)$ value for $l_T K_{\text{neutral}}$. However, with realistic values of γ (= 5/6, say), Pr (= 3/4) and le (> 0), and $\lambda \sim T^{3/4}$, one deduces $l_T K_{\text{neutral}} \leq 10^{-1}$, hence $\Lambda_{\text{neutral}} \geq 60 l_T$, so that the assumptions leading to (4.7.38) are reasonably accurate, especially if Ma is "large" such as in lean flames of heavy fuels. In fact, (4.7.38) is the most accurate analytical result obtained to date on K_{neutral}, as can be checked upon comparisons to numerical analyses of the linear stability problem (Jackson and Kapila, 1984).

4.7.4 Beyond linearity

As in Section 4.5, the question which follows a linear stability analysis concerns the fate of the unstable modes once they have reached a nonnegligible amplitude. This can be, and has been, investigated numerically by direct simulations of the starting governing equations listed at the end of Section 4.2 (Denet, 1988). On the analytical side, good insights into the problem can be gained by considering the formal limit where the density contrast $\gamma \equiv (\rho_u - \rho_b)/\rho_u$ is assumed small.

As equations (4.7.37) and (4.7.38) indicate, the linear stability analysis leads to the following simplified form of the (weak-)growth rate when γ is small and $K = \mathcal{O}(K_{\text{neutral}})$:

$$S = \frac{\gamma}{2} u_L K \left(1 - K/K_{\text{neutral}}\right) + \mathcal{O}\left(\gamma^2 u_L K\right)$$

(4.7.39)

with $2 K_{\text{neutral}} l_T = \gamma/(1 + \text{le}/2)$ here, because, owing to (4.5.21), the Markstein number simply is $(1 + \text{le}/2)$ in this limit. This simplified dispersion relation is equivalent to the linear evolution equation for the amplitude of wrinkling $\Phi(t, \mathbf{z})$:

$$\frac{\partial \Phi}{\partial t} = u_L \frac{\gamma}{2} \left(I(\Phi, \mathbf{z}) + \frac{\Delta \Phi}{K_{\text{neutral}}}\right) + \mathcal{O}\left(\gamma^2 u_L K_{\text{neutral}} \Phi\right),$$

(4.7.40)

where all explicitly written terms are $\mathcal{O}\left(\gamma u_L K_{\text{neutral}} \Phi\right)$ when $\mathbf{z} K_{\text{neutral}} = \mathcal{O}(1)$. Recall that (4.7.40) merely expresses how a Markstein-like curvature effect linearly competes with the weak local flow field disturbance induced by the whole distorted front just ahead of each of its elements. For a given front shape, hence a given $I(\Phi, \mathbf{z})$, (4.7.40) may thus be interpreted locally by a kinematic argument. It is therefore natural to compare it to the nonlinear eikonal equation

$$\frac{\partial \Phi}{\partial t} = u_L \left(1 - \left(1 + |\nabla\Phi|^2\right)^{1/2}\right) \simeq -\frac{u_L}{2}|\nabla\Phi|^2, \qquad (4.7.41)$$

which governs locally a slightly tilted element of interface propagating at constant normal speed u_L against a uniform flow with x-wise velocity u_L. We basically proceeded in the same way to guess the coefficient C in (4.5.25). Clearly, (4.7.41) possesses what is needed in (4.7.40), and conversely, so that combining the two presumably produces:

$$\frac{\partial \Phi}{\partial t} + \frac{u_L}{2}|\nabla\Phi|^2 = u_L\frac{\gamma}{2}\left(I(\Phi, \mathbf{z}) + \frac{\Delta\Phi}{K_{\text{neutral}}}\right) + \dots \qquad (4.7.42)$$

to leading order in γ, as the contributions (4.7.40) and (4.7.41) to $\partial\Phi/\partial t$ are both small. Obtained here in an heuristic way, equation (4.7.42) is in fact correct. This can be shown in a more systematic way if one realizes, as Sivashinsky (1977) first did, that all the terms explicitly written in (4.7.43) balance if Φ satisfies the scaling law:

$$\Phi = l_{\mathrm{T}}\phi\left(\gamma^2 u_L t/l_{\mathrm{T}}, \gamma\mathbf{z}/l_{\mathrm{T}}\right) + o(l_{\mathrm{T}})$$

when $\gamma \ll 1$. Using the ansatz into the Navier–Stokes equations and the jumps (4.7.34)–(4.7.36) ultimately yields a nondimensional form of (4.7.42), viz:

$$\dot{\phi} + \tfrac{1}{2}|\nabla\phi|^2 = \Delta\phi + I(\phi, \mathbf{Z}) \qquad (4.7.43)$$

in terms of obviously-scaled variables. The next order in the small-γ expansion has also been computed; it merely modifies the coefficients in (4.7.42) slightly (Clavin and Sivashinsky, 1987), but does not alter the structure of (4.7.43). It is also worth noting that the potential-flow model would also formally give (4.7.43) at the two first orders in the limit $\gamma \to 0$.

Equation (4.7.43) is called the Michelson–Sivashinsky (MS) equation, after its inventor (Sivashinsky, 1977b) and the author of a first numerical study on it in the field of combustion theory (Michelson and Sivashinsky, 1977); actually, a formally identical equation appeared previously as a model-equation for plasma instability (Ott and Sudan, 1969) and was briefly studied numerically in that context (Ott *et al.*, 1973).

Let us first concentrate on the one-dimensional version of (4.7.43)

$$\dot{\phi} + \tfrac{1}{2}\phi_Z{}^2 = \phi_{ZZ} + I(\phi, Z) \qquad (4.7.44)$$

and on its solutions which are $2\pi/\kappa_{\text{box}}$-periodic along the Z-axis. Numerical integrations, for example by pseudo-spectral methods, reveal that ϕ *often* evolves ultimately to a "steady" pattern propagating at constant speed $U > 0$ towards the fresh side ($\phi < 0$), i.e., to:

$$\phi_{\text{steady}} = -UT + F(Z), \tag{4.7.45}$$

where T is the scaled time involved in (4.7.44). The fractional increases in mean flame speed ($\sim U$) and in flame front length are related in a simple way, as is shown upon taking an average (noted $\langle . \rangle$) of (4.7.44) over the transverse coordinate: $U = \langle F_Z{}^2 \rangle / 2$. It is also worth stressing that $F(Z)$ often appears in the form of a single fold with wide parabola-like troughs, with a spatial period which takes on the maximum admissible value $2\pi/\kappa_{\text{box}}$ whatever κ_{box}, and with a single, sharp crest per period (Fig. 4.20).

In fact, these features only show up numerically when κ_{box} is not too small, but still markedly smaller than the neutral wavenumber (unity, here); otherwise ($\kappa_{\text{box}} < 1/30$, say), the main fold gets covered by apparently-chaotic smaller subwrinkles that seem to be "emitted" from the troughs, where the average pattern is the least curved, and that travel transversely to the main crest where they are "absorbed" (Gutman and Sivashinsky, 1990).

Numerical integration of (4.7.43) with periodic boundary conditions along the edges of a square domain (Michelson and Sivashinsky, 1982) lead to similar results in the long-time limit, except that the steady pattern $\Phi(\mathbf{Z})$ which emerges roughly has the shape of a Tuareg tent (Fig. 4.21), and that the corresponding increase in mean flame speed is higher than in the one-dimensional case.

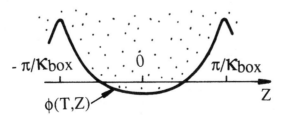

Fig. 4.20. Steady, $2\pi/\kappa_{\text{box}}$-periodic flame front pattern generated by the one dimensional Michelson–Sivashinsky equation (4.7.44).

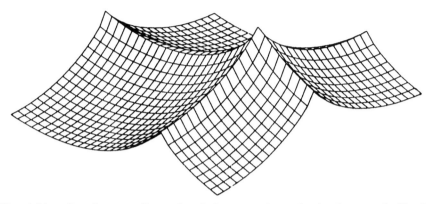

Fig. 4.21. Steady, two-dimensional flame surface obtained numerically from equation (4.7.43) with periodic boundary conditions along a square contour. Reproduced with kind permission of G.I. Sivashinsky (Tel-Aviv); see also Michelson and Sivashinsky (1982).

4.7.5 Complex flames

Surprisingly enough, nearly all the properties revealed by numerical studies of (4.7.44) can be explained analytically. The basic reason is that (4.7.44) belongs to a class of equations, originating from plasma physics (Lee and Chen, 1982), for which an infinite number of exact solutions is available. To explain how this comes about, we first focus on (4.7.44) again, provisionally ignore any periodicity in Z, and consider the following identity:

$$I\left(\log\left(Z - Z_\alpha\right), Z\right) = i \operatorname{Sign}\left(\mathcal{I}m\left(Z_\alpha\right)\right) \frac{d}{dx} \log\left(Z - Z_\alpha\right), \qquad (4.7.46)$$

which holds for any strictly complex number Z_α. Identity (4.7.46) can be demonstrated upon contour integration in the complex Z-plane, because the Fourier transform of $\log\left(Z - Z_\alpha\right)$ vanishes for $\kappa < 0$ or $\kappa > 0$ depending upon the sign of $\mathcal{I}m\left(Z_\alpha\right)$, and because $I(., Z)$ is the multiplication by $|\kappa|$ in the Fourier space (κ) conjugate to Z: for such logarithmic functions, the Landau–Darrieus operator is akin to an imaginary convective operator. By direct substitution, it is then easily found that (4.7.44) possesses nonperiodic solutions of the form:

$$\phi(T, Z) = -2 \sum_{\alpha=1}^{2N} \log\left(Z - Z_\alpha(T)\right) + \text{constant}, \qquad (4.7.47)$$

where N, an integer, is arbitrary but fixed. The Z_α's can be identified to the movable poles of ϕ_Z in the complex Z-plane and, of course, must appear in conjugate pairs for ϕ to be real when Z is. Furthermore they are subject to constraints for (4.7.44) to be satisfied, namely the $2N$ complex ODEs in which the LD "imaginary convection" explicitly shows up:

$$\dot{Z}_\alpha = 2 \sum_{\beta \neq \alpha} \frac{1}{Z_\beta - Z_\alpha} - i \operatorname{Sign}\left(\mathcal{Im}\left(Z_\alpha\right)\right), \qquad \alpha = 1, \ldots 2N. \qquad (4.7.48)$$

The simple demonstration of (4.7.48) is left as an exercise to entertain the nearly-sleepy reader (*hint*: decompose in simple fractions the nondiagonal terms coming from $\phi_Z{}^2$). An expression like (4.7.48) will be called a "pole decomposition" of $\phi(T, Z)$, and (4.7.47) will be termed the "pole equations." The last term in the r.h.s. of (4.7.48) clearly comes from the LD instability and it tends to make real singularities appear; the first one is what remains from nonlinearity after partial cancellation with the second derivative of ϕ (curvature must smooth the crests generated by Huygens propagation).

Once values are assigned to N and the initial locations $Z_\alpha(0)$, integrating the $2N$-body problem (4.7.48) automatically yields an exact unsteady solution to (4.7.44), via (4.7.47). In general the integration of (4.7.48) must be done numerically. A noticeable exception corresponds to $N = 1$, in which case one may write $Z_1 = A + iB$, $Z_2 = A - iB$ ($B > 0$ without loss of generality) to obtain:

$$\phi = -2 \log\left((Z - A)^2 + B^2\right) + \text{constant}. \qquad (4.7.49)$$

Then $\dot{A} = 0$, due to transverse translational invariance and $\dot{B} = 1/B - 1$ (hence $B(\infty) = 1$). Due to the superposition in (4.7.47) and the constancy of N, the above elementary crest plays the role of a soliton for the MS equation, since the poles are indestructible.

The detailed interaction of many of them is difficult to grasp analytically, but the following effect, first evidenced by Thual *et al.* (1985), greatly eases one's understanding. When two poles Z_α and Z_β are momentarily close to each other in the same half complex plane, (4.7.48) simplifies to $\left(\dot{Z}_\alpha - \dot{Z}_\beta\right)(Z_\beta - Z_\alpha) \simeq 4$ during the "collision," assumed to occur at $T \approx T_c$, whereby:

$$(Z_\alpha - Z_\beta)^2 \simeq -8(T - T_c) + \text{a small } complex \text{ constant}. \qquad (4.7.50)$$

Accordingly, Z_α and Z_β tend to attract (repel) each other parallel to the real (imaginary) axis (*exercise*: prove this); their complex conjugates do the same thing simultaneously. When viewed from the real axis, this represents nothing but a coalescence of two elementary crests. The same process ultimately leads to a "parade" of $2N$ poles that are aligned and have the same real part: this is the basic mechanism by which all the crests tend to consolidate into a giant one that absorbs the smaller ones. In a sense the evolution of pole-decomposed, but random-like, initial conditions towards a single fold may be viewed as a process of aggregation of "sticky" objects, the crests. It would be useful to see the extent to which the methods of statistical physics developed for other processes of aggregation, e.g., of vortices or droplets (Carnevale *et al.*, 1990; Derrida *et al.*, 1990), can be adapted to the present case. For an attempt to handle such coalescence problems, see Cambray *et al.* (1994).

Accounting for a $2\pi/\kappa_{\text{box}}$-periodicity along the real Z-axis amounts to replacing (4.7.47) by:

$$\phi(T, Z) = -h(T) - 2 \sum_{\alpha=1}^{2N} \log \sin \left(\tfrac{1}{2} \kappa_{\text{box}} \left(Z - Z_\alpha(T) \right) \right) , \qquad (4.7.51)$$

where the Z_α's satisfy new pole-equations (Thual *et al.*, 1985):

$$\dot{Z}_\alpha = \kappa_{\text{box}} \sum_{\beta \neq \alpha} \cot \left(\tfrac{1}{2} \kappa_{\text{box}} \left(Z_\beta - Z_\alpha \right) \right) - i \operatorname{Sign} \left(\mathcal{I}m \left(Z_\alpha \right) \right) , \qquad (4.7.52)$$

$$\alpha = 1, \dots 2N .$$

The demonstration of (4.7.52) proceeds similarly to that yielding (4.7.48). However, some care needs be exercised when generalizing (4.7.46) to periodic functions, because it then involves additive constants; the trigonometric identity

$$\cot \left(Z - Z_\alpha \right) \cot \left(Z - Z_\beta \right) =$$
$$- 1 + \cot \left(Z_\alpha - Z_\beta \right) \left(\cot \left(Z - Z_\alpha \right) - \cot \left(Z - Z_\beta \right) \right)$$

(which also involves an additive constant) replaces the decomposition of a rational function into simple fractions. As for $h(T)$, it is given by the rather simple formula:

$$\dot{h} = 2N \kappa_{\text{box}} \left(1 - N \kappa_{\text{box}} \right)$$

(*exercise*: prove it; *hint*: collect the various aforementioned additive constants). One must realize that $h(T)$ does not equal $\langle -\phi_T \rangle$ because genuinely unsteady fronts are continuously distorted; concerning the latter quantity, further calculations give:

$$\langle -\phi_T \rangle = 2N \kappa_{\text{box}} \left(1 - N \kappa_{\text{box}} + \frac{1}{N} \sum_{\alpha=1}^{N} \mathcal{I}m \left(\dot{Z}_\alpha \right) \right) , \qquad (4.7.53)$$

where the summation only extends over the Z_α's with positive imaginary parts. In general (4.7.52) has to be integrated numerically once N and $Z_\alpha(0)$ are given. The case $N = 1$ again is a noticeable exception which, for $Z_1 = A + iB$, $Z_2 = A - iB$ and $B > 0$, leads to $\dot{A} = 0$ and to:

$$\dot{B} = \kappa_{\text{box}} \coth \left(\kappa_{\text{box}} B \right) - 1 . \qquad (4.7.54)$$

A steady state with $B < \infty$ can be reached only if $\kappa_{box} < 1$. This was expected since $\kappa_{box} \geq 1$ would mean that no linearly-unstable mode of ϕ fits in the integration domain. Moreover, the steady flame shape ϕ pertaining to $N = 1$ reads as:

$$\phi_{steady} = -2\kappa_{box} (1 - KB) T$$
$$- 2\log (1 - \cos (\kappa_{box}(Z - A)) \operatorname{sech} (\kappa_{box}B)) \tag{4.7.55}$$

up to a shift; $\kappa_{box}B \to \infty$ corresponds to a sinusoidal wrinkle of very small amplitude and to a very small final fractional increase in flame speed. The analogy between (4.7.54) and the linear analysis of (4.7.44) is best shown by introducing the peak-to-peak amplitude Γ defined by $2\Gamma = \max(\phi) - \min(\phi) \sim \log \coth (\kappa_{box}B/2)$; (4.7.55) is then transformed into:

$$\dot{\Gamma} = 2\sinh(\Gamma/2)\kappa_{box} (1 - \kappa_{box} \cosh(\Gamma/2)) \tag{4.7.56}$$

and $\Gamma \to 0$ clearly makes (4.7.56) resume a form equivalent to the linear result. Interestingly, $\kappa_{box} \to 0$ gives $\kappa_{box}B(\infty) \to 0$; although $\Gamma(\infty)$ diverges in this limit, it does so at too low a rate ($\Gamma(\infty) \sim \log (1/\kappa_{box})$) compared to the wavelength $2\pi/\kappa_{box}$ to prevent $U = \langle \phi_Z{}^2 \rangle /2$ from vanishing again. If $\kappa_{box} > 1$, B goes to $+\infty$, Γ decays exponentially in time, and $U = 0$.

When many pairs of poles are involved in each spatial period $2\pi/\kappa_{box}$, the mechanism of pole alignment still operates, because $\cot(Z) \sim 1/Z$ when $|Z| \ll 1$ (mod $2\pi/\kappa_{box}$). Steady parades of poles with the same real part (mod $2\pi/\kappa_{box}$) are generically obtained in the long-time limit, thereby explaining why a single fold with maximum admissible wavelength ultimately survives in most spectral integrations of (4.7.44). Contrary to the nonperiodic situations, the number of poles participating to such steady alignments is constrained. This can be understood as follows:

Using $Z_\alpha = A \pm iB_\alpha$, with $B_\alpha > 0$, in (4.7.51) leads to real ODEs for the B_α; each of them involves $(2N - 1)$ hyperbolic cotangents. Considering the equation for the uppermost steady pole B_{2N} and setting $\dot{B}_{2N} = 0$ leads to a necessary condition for B_{2N} to stay at a fixed value (Thual *et al.*, 1985); the condition is $\kappa_{box}(2N - 1) - 1 \leq 0$ and exists because a $\coth(.)$ function with a positive argument always exceeds unity. N has to be an integer, however, whereas κ_{box} is any real number less than unity; equality is reached when the uppermost pole is at imaginary infinity.

We recently verified, by a *spectral* integration of (4.7.44), that the steady flame shapes coincide very nicely with those given by (4.7.51), provided that the number N_{max} of aligned pairs of poles is given by

(Joulin and Cambray, 1992):

$$N_{\max}(\kappa_{\text{box}}) = \text{Int}\left(\frac{1}{2}\left(1 + \frac{1}{\kappa_{\text{box}}}\right)\right), \qquad (4.7.57)$$

where Int(.) denotes the integer-part function; $N_{\max}$ gives the maximum flame speed allowed by $\kappa_{\text{box}}(2N - 1) \leq 1$. The corresponding function $U(\kappa_{\text{box}})$ is given by (4.7.53) and (4.7.57) (Fig. 4.22). If $2(N_{\max} + M)$ poles are initially present, one may analytically check that $2M$ of them are ejected to $\pm i\infty$ when $T \to \infty$; their terminal velocities are precisely what is needed to reconcile (4.7.53), with $N = N_{\max} + M$ and the time-derivatives included, and (4.7.53) itself with $N = N_{\max}$ but without the derivatives!

When $\kappa_{\text{box}} \to 0$, $N_{\max} \sim 1/2\kappa_{\text{box}} \to \infty$, but the aligned poles get more and more densely packed. Replacing the sums in (4.7.51) and (4.7.52) by principal-part integrals over B_β and assuming $N = N_{\max}$, Thual *et al.* (1985) ultimately obtained a limiting form for ϕ_{steady} which may be written as follows, up to shifts in ϕ or Z:

$$\kappa_{\text{box}}\phi + UT = -\frac{2}{\pi}\int_{-\infty}^{+\infty} \log\sin\left(\frac{Z\kappa_{\text{box}} - iy}{2}\right)\log\coth\frac{|y|}{4}\,dy. \quad (4.7.58)$$

Of course this defines a shape which is real when Z is. More importantly, it is a scale-invariant one: no "Markstein effect" survives on the cell-length scale when $\kappa_{\text{box}} \to 0$, except *at* the very sharp crests. The value of U corresponding to the above ϕ_{steady} simply is the small-κ_{box} limit of $2N_{\max}\kappa_{\text{box}}(1 - N_{\max}\kappa_{\text{box}})$, viz: $U = 1/2$. For future reference, we

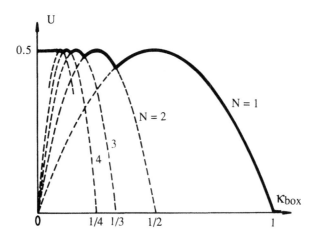

Fig. 4.22. Wrinkling-induced fractional increase in area of steady $2\pi/\kappa_{\text{box}}$-periodic flames, as is given by (4.7.53) for various wavenumbers (κ_{box}) and numbers (N) of pole pairs; the heavy line corresponds to (4.7.57).

note that (4.7.58) may be differentiated twice, then the resulting integral computed at $Z = \pi/\kappa_{\text{box}}$ to give an estimate for the curvature of ϕ_{steady} about each of its minima:

$$\sigma = +\kappa_{\text{box}}/\pi. \tag{4.7.59}$$

At least in steady, spatially-periodic, one-dimensional situations, the MS equation may be almost considered under analytical control. A last comment may be put forward, however. To obtain (4.7.58) in the limit $\kappa_{\text{box}} \to 0$, Thual *et al.* (1985) assumed $N = N_{\text{max}} \simeq 1/2\kappa_{\text{box}}$. As far as steady profiles are looked for, one could equally assume $2\kappa_{\text{box}}N = \alpha$, with $0 < \alpha < 1$ fixed, thereby possibly obtaining another limiting solution parametrized by α (Joulin, 1991) and suggesting that the "inviscid" MS equation possesses a continuum of solutions when $\kappa_{\text{box}} = 0^+$. On the other hand the MS equation (4.7.44) only admits at most a countable number of steady pole-decomposed solutions, even when $\kappa_{\text{box}} \ll 1$. This sounds rather reminiscent of what happened recently with other free-boundary problems (e.g., see Combescot *et al.*, 1986)... but remains to be proven here. As for the two-dimensional configurations, it is a simple matter to remark that (4.7.43) is separable (Joulin, 1987) in the sense that it admits solutions in the form $\phi(T, Z) + \psi(T, Y)$, where Y denotes the second transverse Cartesian coordinate and $\psi(T, Y)$ also satisfies the one-dimensional MS equation. Surprisingly enough, a superposition of independent orthogonal patterns fits very well the numerical results reproduced in Fig. 4.21, and the fractional increase in flame-speed due to wrinkling is simply twice what (4.7.44) predicts.

4.7.6 Influence of large-scale front geometry

It now remains to briefly explain why steady flames with wide, locally nearly-planar troughs may exist when κ_{box} is moderately small, even though planar flames are linearly unstable to any disturbance characterized by a reduced wavenumber κ less than marginal $(= 1)$, and why a too small κ_{box} may numerically lead to the appearance of chaotic-like subwrinkles. To this end we are to invoke an effect originally identified by Zel'dovich *et al.* (1980) in the context of combustion and first recognized by Pelcé (1989) as of tantamount importance for many growth phenomena. It can certainly be exposed in more general terms, but we found it convenient to start from the MS equation (4.7.44), whose general solution is now written as $-UT + F(Z) + f(T, Z)$, where $-U$ and $F(Z)$ can be computed through (4.7.51) (4.7.53). $f(T, Z)$ represents subwrinkles superimposed to the basic shape $F(Z)$ and of course evolves according to:

$$\dot{f} + \tfrac{1}{2}f_Z^2 + F_Z f_Z = f_{ZZ} + I(f, Z). \tag{4.7.60}$$

This differs from the usual MS equation by the presence of a convective-like term generated by the underlying large-scale shape $F(Z)$. About its parabola-like troughs, $F(Z)$ is well approximated by $F_{min} + \sigma Z^2/2$, especially if the positive curvature σ satisfies $\sigma \ll 1$. Motivated by the form that (4.7.60) then takes, one may propose the new equation (Joulin, 1989):

$$\dot{f} + \tfrac{1}{2}f_Z^2 + \sigma Z f_Z = f_{ZZ} + I(f, Z) \qquad (4.7.61)$$

as a model to get better insights into the dynamics of finite amplitude wrinkles superimposed to a nearly-parabolic steady flame. Once linearized for small enough f's, (4.7.61) admits solutions in the form $f(T, Z) = \Gamma(T, \kappa) \exp(iq(T)Z)$ and, given an initial wavenumber κ, the small amplitude $\Gamma(T, \kappa)$ is governed by:

$$\dot{q} = -\sigma q, \qquad q(0) = \kappa, \qquad \dot{\Gamma}/\Gamma = |q| - q^2. \qquad (4.7.62)$$

The amplitude growth rate is thus similar to what the linearized MS equation would predict... once $\dot{\Gamma}/\Gamma$ is expressed in terms of an instantaneous wavelength $2\pi\,e^{\sigma T}/|\kappa|$ which is *stretched* by the *nonconstant* "tangential velocity" σZ (Fig. 4.23). The instantaneous wavenumber $q(T)$ is driven exponentially rapidly from κ to the marginal mode $q = 0$ and, as a consequence, $\Gamma(T, k)$ saturates to the *finite* value

$$\Gamma(\infty, \kappa) = \Gamma(0, \kappa) \exp\left(\frac{|k|}{\sigma}\left(1 - \frac{|k|}{2}\right)\right). \qquad (4.7.63)$$

In other words, the existence of an underlying mean curvature "kills" the Landau–Darrieus instability: it is now clear why steady $2\pi/\kappa_{box}$-periodic solutions to the MS equation may exist even when κ_{box} is small. Next, one may note that $\Gamma(\infty, \kappa)/\Gamma(0, \kappa)$ is maximum if κ coincides with the nontrivial marginal mode ($\kappa = 1$) of the linearized MS equation :

$$\Gamma(\infty, \kappa) = \Gamma(\infty, 1) \exp\left(-(|\kappa| - 1)^2/2\sigma\right), \qquad (4.7.64)$$

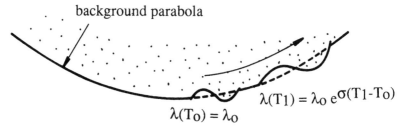

$$\lambda(T_1) = \lambda_0\, e^{\sigma(T_1 - T_0)}$$

$$\lambda(T_0) = \lambda_0$$

Fig. 4.23. The phenomenon of wavelength stretching for disturbances of a nearly-parabolic flame front.

so that the most dangerous band corresponds to $|\kappa| - 1 = \mathcal{O}\left(\sigma^{1/2}\right)$. Obviously, for the preceding linear results to hold, $\Gamma(\infty, \kappa)$ must stay small enough to yield $f_Z \ll 1$ when $T \gg 1$. Otherwise, for example if some disturbance with $|\kappa| - 1 = \mathcal{O}\left(\sigma^{1/2}\right)$ initially exists and has

$$\Gamma(0, \kappa) \geq \Gamma_c \equiv \exp(-1/2\sigma). \qquad (4.7.65)$$

$\Gamma(T, \kappa)$ reaches $\mathcal{O}(1)$ values and nonlinearity comes into play ultimately, thereby opening the possibility of qualitatively new phenomena. As can be shown from (4.7.61), again by using the pole-decomposition method (Joulin, 1989), an initial disturbance amplitude that significantly exceeds Γ_c over the $\kappa \simeq 1$ "dangerous" range leads to crests whose height ultimately diverges proportionally to T while these are convected away from the troughs of the main fold, if $\sigma < 1/4$ (*exercise*: prove this; *hint*: use $Z\,e^{-sT}$ as new variable in (4.7.61) and consider a two-pole periodic pattern).

Γ_c decreases very rapidly as σ does, and it can easily be exceeded as a result of an *external* noise, e.g., the round-off errors that inevitably pollute any numerical integration of the MS equation itself. In the situation investigated by Gutman and Sivashinsky (1990), i.e. (4.7.44) with periodic boundary conditions and $\kappa_{\text{box}} = 1/80$, Γ_c was in effect as low as $\mathcal{O}\left(10^{-55}\right)$; this tiny figure follows from combining (4.7.65) and (4.7.59) to yield: $\Gamma_c \simeq \exp\left(-\pi/2\kappa_{\text{box}}\right)$.

Quite obviously the above arguments are very robust, because they depend only quantitatively on the exact dispersion relation pertaining to nearly-flat fronts, and because any wide-enough trough may be approximated locally by a parabola. This way of "suppressing" the Landau–Darrieus instability should therefore operate in reality as well, explaining why weakly curved flames propagating along wide tubes (compared to $2\pi/K_{\text{neutral}}$) may do so steadily despite their *nearly*-flat local appearance (Fig. 4.3), even when buoyancy exerts a destabilizing influence (see Section 4.8).

As for the origin of the germ $\Gamma(0, \kappa)$ leading to secondary wrinkles which may ultimately grow, thermal noise could be enough if $\sigma \ll 1$, thereby providing one with a means to evaluate the maximum tube diameter that allows for smooth flames (Zel'dovich *et al.*, 1980). Of course a very weak residual turbulence could also do this and is probably at the origin of the "bumps" which are shown in Fig. 4.5. We finally mention an alternative way of interpreting the above results. Equation (4.7.65) may also be viewed as giving the curvature which a locally parabolic trough must exceed, in order to resist a noise of given intensity ($\sim \Gamma(0, \kappa)$) in the wavenumber range $\kappa \sim 1$: insufficiently-curved, wider folds would presumably be destroyed by secondary wrinkles, thereby indicating that the noise energy available at near-marginal wavenumbers could well contribute to controlling the size of the biggest viable folds which flames may exhibit when propagating in slightly turbulent incoming flows (Clavin, 1988); the response of unstable flames to external hydrodynamic forcing is the subject of active research (Cambray and Joulin, 1994), which tend to support this viewpoint.

4.8 Body forces

4.8.1 Buoyancy

As summarized previously, one more or less knows how to handle the high wavenumber wrinkles of an initially flat flame in a physically satisfactory, even if not yet mathematically rigorous, way. It is time to worry again about the low wavenumber range for the following reason: experiments show that flat flames freely propagating in uniform flow fields may exist (Fig. 4.16), which contradicts what (4.7.30) or (4.7.37) indicate. Since such horizontal flat flames are invariably found when the hot light gases lie above the cold, heavier ones, one is naturally led to inquire about the influence of gravity.

When normal to the mean flame front, buoyancy cannot affect the basic steady planar flame because the x-wise momentum equation is not coupled to the heat and reactant balances in this case, but it can modify the rate of growth/decay (S) of wrinkles. In the context of an extended LD model, for which $\mathbf{n} \cdot (\mathbf{u} - \mathbf{D})_{\text{fresh}}$ would still equal u_L, a dimensional argument generalizes (4.7.21) into:

$$S = u_L K \Omega \left(g, \frac{g\gamma}{K u_L{}^2} \right) , \tag{4.8.1}$$

where g is the intensity of gravity along the x-axis (g is negative for flames propagating downward). In (4.8.1) $K u_L{}^2/g\gamma$ is a Froude number and its very definition indicates that the effective influence of gravity will be comparatively larger when the considered wrinkle wavenumbers get comparable to or less than $g\gamma/u_L{}^2$. The LD analysis is readily extended to account for a nonvanishing g: suffice it to realize that the basic planar flame no longer corresponds to a piecewise *uniform* pressure profile $\bar{p}(x)$ but, instead, to $d\bar{p}/dx = \rho g$, with $\rho = \rho_u$ or $\rho_b < \rho_u$. Accordingly the linearized jump condition (4.7.16) is replaced by:

$$[p']_{0-}^{0+} = g \left(\rho_u - \rho_b \right) \qquad \left(\equiv - \left[\frac{d\bar{p}}{dx} \right]_-^+ \right) \tag{4.8.2}$$

when it is transferred at $x = 0$, instead of (4.7.16). Once this single modification is accounted for, the calculation proceeds along the same lines as in Section 4.7, because a constant g does not affect the perturbed field equations. This ultimately gives the quadratic equation in Ω (Markstein, 1964):

$$(2 - \gamma)\Omega^2 + 2\Omega = \frac{\gamma}{1 - \gamma} + \frac{g\gamma}{K u_L{}^2} , \tag{4.8.3}$$

which is obviously compatible with (4.7.22). Note that (4.8.3) is equivalent to the system:

$$\frac{\partial \Phi}{\partial t} = V ,$$

$$(2 - \gamma)\frac{\partial V}{\partial t} + 2u_L I(V, \mathbf{z}) = -\frac{\gamma}{1 - \gamma} u_L{}^2 \Delta\Phi + \gamma g I(\Phi, z) , \tag{4.8.4}$$

to be compared with (4.7.25). The resulting value(s) of S is (are) plotted in Fig. 4.24.

Two cases have to be distinguished, depending on the sign of gravity. If $g > 0$ (upward propagation), buoyancy enhances the growth rate, due to Rayleigh–Taylor's instability; the effect is comparatively more marked when $K \leq \mathcal{O}\left(g\gamma/u_L{}^2\right)$, whereas $K \gg g\gamma/u_L{}^2$ leads to a practically-pure, albeit shifted, Landau–Darrieus effect. If $g < 0$ two new features are worth noticing: Ω vanishes at $K_g = |g|(1 - \gamma)/u_L{}^2$, gets negative if K is slightly less than K_g, and complex with a negative real part if K is small enough. The occurrence of (damped-) oscillations is brought about by the two outmost terms in (4.8.3) and, as such, is strongly reminiscent of gravity waves, for which having some "inertia term" obviously is mandatory. As for the "damping," it is provided by propagation itself.

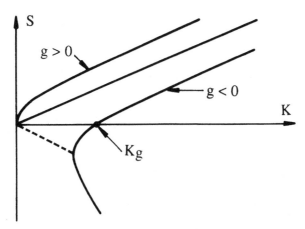

Fig. 4.24. Influence of gravity on the Landau–Darrieus growth rate S; for $g < 0$ (downward propagation) S may be complex and only the real part is plotted (dashed line).

It is a simple matter to include $g \neq 0$ in Markstein's modification of the Landau–Darrieus model; a comparison between (4.8.3) and (4.7.30) soon convinces one that the corresponding $S = u_L K \Omega \left(\gamma, K/K_{\text{neutral}}, g\gamma/K u_L^2 \right)$ is given by:

$$(2 - \gamma)\Omega^2 + 2(1 + \mathcal{L}K)\Omega = \frac{\gamma}{1 - \gamma} \left(1 - \frac{K}{K_{\text{neutral}}} \pm \frac{K_g}{K} \right), \qquad (4.8.5)$$

the "plus" sign pertaining to upward propagation (*exercise*: prove (4.8.5)).

Quite interestingly, $g < 0$ leads to a range of wrinkle wavenumbers with positive growth rates which, if any, must be sandwiched between K_g and K_{neutral} (Fig. 4.25). If $K_g/K_{\text{neutral}} \equiv |g|(1 - \gamma)/K_{\text{neutral}}u_L^2$ exceeds $1/4$ and $g < 0$, the wrinkles are headed to decay whatever their wavenumber... and a flat, horizontal, downwardly propagating flame is then allowed. In earth-based experiments this corresponds to $u_L < 13$ cm/s, approximately (Fig. 4.1, bottom).

When $|g|(1 - \gamma)/K_{\text{neutral}}u_L^2$ is gradually decreased, a band of unstable wavenumbers appears around $K_* = (K_g K_{\text{neutral}})^{1/2}$; presumably the resulting mode/mode interactions (e.g., see Manneville, 1991) are qualitatively responsible for the hexagonal-like cell patterns displayed in Fig. 4.2.

The more detailed multiscale analysis leading to (4.7.37) has been generalized (Pelcé and Clavin, 1982) to account for buoyancy. The resulting equation for Ω is again: $A\Omega^2 + B\Omega + C = 0$ where A and

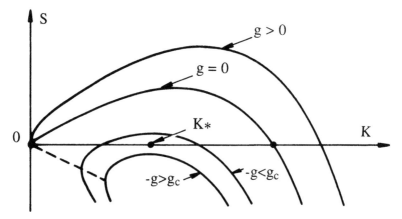

Fig. 4.25. Combined influences of Landau–Darrieus instability, curvature effects and buoyancy on the dispersion relation of a nearly-planar flame; the dashed line again stands for the real part of a complex S.

B are the same as in (4.7.37) whereas C has the form:

$$C = -\frac{\gamma}{1-\gamma}\left(1 \pm K_g l_T M'' - \frac{K}{K_{\text{neutral}}} \pm \frac{K_g}{K}\right)$$

with K_{neutral} given by the more accurate expression (4.7.38). Clearly, this qualitatively leads to the same conclusions as (4.8.5), but a marked quantitative improvement is achieved, to the extent that the improved expressions now compare rather favorably with the experiments (Quinard, 1984).

The summarized above treatment of buoyancy-affected flames has been extended to the nonlinear range, by again considering the small-γ limit. Upon the assumption that $K = \mathcal{O}(K_{\text{neutral}}) = \mathcal{O}(\gamma/\mathcal{L})$ in (4.8.5), it is readily seen that the growth rate S simplifies into

$$S = \frac{\gamma}{2} u_L \left(K - \frac{K^2}{K_{\text{neutral}}} \pm K_g\right) + \mathcal{O}\left(\gamma^3 u_L \mathcal{L}^{-1}\right) \qquad (4.8.6)$$

if $K_g = \mathcal{O}(K_{\text{neutral}})$, in the limit of $g \to 0$. This limit is not uniform over the whole range of K; to wit, the simplified expression (4.8.6) ceases to be valid when K becomes of the order of $\gamma^2 \mathcal{L}^{-1}$ or less and, as a result, predicts a nonvanishing S at $K/K_{\text{neutral}} = 0$. To restore translational invariance, (4.8.6) must thus be supplemented by: $S(K = 0) = 0$. It is not too difficult an exercise to adapt Sivashinsky's original analysis of finite-amplitude wrinkles and obtain a generalized MS equation for the nondimensionalized flame shape ϕ (Rakib and Sivashinsky, 1987):

$$\dot{\phi} + \tfrac{1}{2}|\nabla\phi|^2 = \Delta\phi + I(\phi, \mathbf{Z}) + G\left(\phi - \langle\phi\rangle\right) \qquad (4.8.7)$$

in which $G \sim g/K_{\text{neutral}} u_L^2$ and $\langle.\rangle$ denotes an average over the transverse coordinate(s); as mentioned above, the last term in (4.8.7) is required to ensure translational invariance. Presumably, equation (4.8.7) is capable of mimicking the evolution to dome-shaped flames propagating upward ($G > 0$) (Fig. 4.3). As shown by Denet (1993), it can capture the transition from flat flames to hexagonal cells if G crosses $-1/4$ from below (Michelson and Sivashinsky, 1982) and possibly their phase dynamics (... a good exercise on multiscale techniques, given that the Landau–Darrieus operator is not local).

4.8.2 Parametric instability

The most accurate dispersion relation obtained so far, i.e., (4.7.37) with
buoyancy included, can be re-written as an ordinary differential equation

$$A(K)\frac{d^2\psi}{dt^2} + u_L K B(K)\frac{d\psi}{dt} + u_L{}^2 K^2 C(K,g)\psi = 0 \qquad (4.8.8)$$

for the Fourier transform in space $\psi(\mathbf{K}, t)$ of the amplitude of wrinkling
$\Phi(\mathbf{z}, t)$. Interestingly enough (4.8.8) is still valid when the acceleration
due to gravity g is a slow-enough *function of time* for the front structure
to be regarded as *quasi-steady* at $\mathcal{O}\,(l_T)$ distances from the reaction sheet.
As a consequence of a non-Galilean transformation, this is what happens
when a flat, yet unperturbed flame is convected, as a whole, normally
to itself by a uniform (or nearly-so) flow with a time-dependent velocity,
such as that resulting from planar acoustic waves traveling back and forth
along the tube in which the flame propagates.

To mimic such circumstances one may assign the following form $g(t) =$
$g_0 + u_a\omega_a \cos(\omega_a t)$, where g_0 (earth gravity), u_a (acoustic velocity) and
ω_a (frequency) are given constants, to the effective gravity "felt" by the
flame. Equation (4.8.8) then becomes

$$A(K)\frac{d^2\psi}{dt^2} + u_L K B(K)\frac{d\psi}{dt} + u_L{}^2 K^2 \left(C\left(K, g_0\right) + C_a \cos\left(\omega_a t\right)\right)\psi = 0\,,$$
$$(4.8.9)$$

where $C_a = u_a\omega_a\gamma\left(1 - K l_T M''\right)/K u_L{}^2$. Equation (4.8.9) is linear, of the
second-order and has a time-periodic coefficient; it may thus possibly be
subject to a *parametric instability*, leading to a spontaneous growth of
$|\psi|$ if K, g_0, C_a, ω_a and the mixture properties are adequately chosen.

This kind of event is what may occur to the oscillation amplitude of
a swing, the length of which is modulated by a child, or, closer to the
present situation, to the wrinkling amplitude(s) of an open air/liquid
interface when the horizontal vessel in which the liquid is contained is
made to oscillate vertically by some auxiliary mechanical device; the
latter is known as the Faraday instability (1831).

Before seeing whether events of this sort effectively show up in flames,
let us briefly recall a few results about the canonical *Mathieu equation*:

$$\frac{d^2 y}{dz^2} + (a - 2q\cos(2z))y = 0\,, \qquad (4.8.10)$$

in which the notations y, z, a, q we employ here are purely local and
conventional, and therefore should not be assigned the meaning they
have had so far in the chapter. It has been demonstrated (Floquet, 1883)
that the general solution to (4.8.10) has the structure:

$$y(z) = \alpha_+\, e^{\mu z}\, F(z) + \alpha_-\, e^{-\mu z}\, F(-z)\,, \qquad (4.8.11)$$

where $\alpha_\pm$ are constants, $F(.)$ is π-periodic and μ is of the form $\mu = i\beta$ or $\mu = \alpha + im$, with β and α real and m integer. Depending on μ, two distinct behaviors are therefore encountered:

- If $\mu = i\beta$, with a non-integer β, (β integer is a marginal situation), $y(z)$ stays bounded for any z; this is usually referred to as a "stable" solution.
- If $\mu = \alpha + im$, with an integer m, $y(z)$ diverges exponentially when $|z| \to \infty$; this is called an "unstable" solution and may be re-written as:

$$y(z) = \alpha_+ \, e^{\alpha z} \, G(z) + \alpha_- \, e^{-\alpha z} \, G(-z) \qquad (4.8.12)$$

with a π-periodic function G if m is even and a 2π-periodic one if m is *odd*. In the latter case, the form of $y(z)$ that emerges as $z \to +\infty$ has an exponentially growing amplitude and a basic period (2π) which is *twice* the period of forcing... a rather reliable signature of a parametric instability.

The "growth" rate μ can be determined, numerically in general, as a function $\mu(a, q)$ of the constants a and q featuring in (4.8.10), for example by plugging (4.8.12) into (4.8.10), expressing F as a Fourier series in z and working out the 3-term recursion formula satisfied by the Fourier coefficients (Abramovitz and Stegun, 1970).

A typical Mathieu diagram is plotted in Fig. 4.26. As is illustrated in Chapter 4 or in Cole and Kevorkian (1981), the stable/unstable boundaries can be determined analytically by multiscale techniques if q is small; the case of large q's requires WKB types of analyses (Rochwerger,

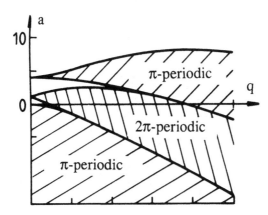

Fig. 4.26. Typical Mathieu diagram; the hatched regions correspond to unstable solutions of (4.8.10).

1991). If the damped Mathieu equation

$$\frac{d^2y}{dz^2} + 2f\frac{dy}{dz} + (a - 2q\cos 2z)y = 0 \qquad (4.8.13)$$

with $f \geq 0$, is considered next to (4.8.9), the solutions are modified into:

$$y(z) = \alpha_+\, e^{(\mu - f)z}\, F(z) + \alpha_-\, e^{-(\mu + f)z}\, F(-z) \qquad (4.8.14)$$

because (4.8.13) can be brought into the canonical Mathieu equation upon provisional use of $y(z)\exp(fz)$ as unknown: a is then replaced by $a - f^2$. Accordingly, the unstable solutions to (4.8.13) obtain when $f + \mu\,(a - f^2, a)$ has a negative real part. As sketched in Fig. 4.27 the stability channels widen as f increases and include the whole *positive* a-axis.

How can one exploit this background knowledge in the context of flames? In an obvious way, because (4.8.9) is exactly a damped Mathieu equation, up to scalings, with coefficients f, a and q which depend on K, g_0, u_a, ω_a, and the mixture properties (γ, Ma, u_L). If the latter characteristics make the pair (a, q) fall in the domains of unstable solutions to (4.8.13), wrinkles of $2\pi/K$ wavelengths will spontaneously grow in amplitude. For a given mixture, and fixed values of g_0 and ω_a, repeating the calculation of (a, q) with various K's and intensities (u_a) of forcing leads (Searby and Rochwerger, 1991) to stability diagrams of the sort depicted in Fig. 4.28.

Three remarkable properties are predicted. Firstly that the wavenumber range of gravity-affected Landau–Darrieus instability tends

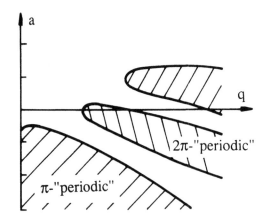

Fig. 4.27. Stability diagram of the damped Mathieu equation (4.8.13).

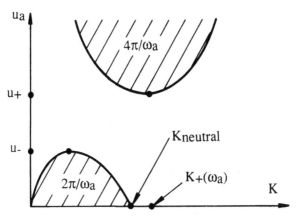

Fig. 4.28. Stability diagram of an acoustically excited flame, equation (4.8.9), for a fixed frequency $(2\pi/\omega_a)$ of forcing and a given mixture. Note the different oscillation frequencies in the instability domains, and the window of stability corresponding to $u_- < u_a < u_+$.

to shrink as u_a increases from zero to u_-; as they grow the corresponding wrinkles oscillate with the *same period as the forcing* $(2\pi/\omega_a)$, because the computed values of (a, q) happen to lie in the lowest instability domain of Fig. 4.27. On the other hand, if u_a increases above u_+, a new kind of wrinkles may grow, but those now throb *at half the frequency of forcing*. Finally, if $u_+ > u_-$ and $u_- < u_a < u_+$, the flame is predicted to *flatten* when the acoustic forcing is switched on.

All the above predictions agree qualitatively well with experiments, e.g., on flames propagating downward (or even upward!) against a uniform mean flow and acoustically excited by a loud-speaker (Searby and Rochwerger, 1991). Such careful experiments on mixtures with $u_+ > u_-$ also demonstrated good quantitative agreement with theory about the instability threshold $u_+(\omega_a)$ (Fig. 4.29) and the corresponding wavenumber $K_+(\omega_a)$ ... even for forcing frequencies ω_a as large as $10 u_L/l_T$, but this almost perfect agreement is obtained at the expense of keeping the Markstein number $Ma = \mathcal{L}/l_T$ fixed to be about 4.5 ($\simeq M'$ in (4.7.33)), rather irrespective of the limiting reactant (Searby, 1991b). This is somehow inconsistent with *a priori* computations (Cambray *et al.*, 1987) or some independent measurements (Deshaies and Cambray, 1990) of the value of $\mathcal{L}/l_T$ pertaining to the heaviest fuels used (e.g., propane). As recent researches indicate it (Joulin, 1994a), this has to do with ω_a being too large for (4.7.34–36) and (4.7.31) to be valid. When $\omega_a = \mathcal{O}(u_L/l_T)$, the Markstein lengths become unequal and frequency-dependent; furthermore, all the fuel-lean flames tend to behave as if $Le = 1$ in the limit of high frequencies.

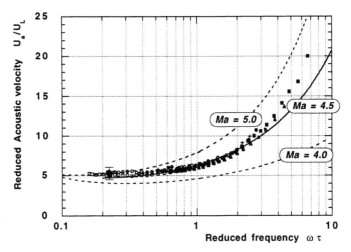

Fig. 4.29. Experimental onset (symbols) of parametric instability ($u_a = u_+$ in Fig. 4.28) for acoustically excited propane/air flames, versus acoustic frequencies; the lines correspond to theory, equation (4.8.9), for different Markstein numbers Ma. Reproduced with kind permission of the authors (Searby and Rochwerger, 1991) and of Cambridge University Press.

Two final remarks are due. Firstly, that the aforementioned acoustic waves may well be generated by the flame evolution itself, most notably by its changes in surface area (Pelcé and Rochwerger, 1991), thereby leading to an autonomous nonlinear oscillator, the subject of active current researches. Secondly, that all the above phenomena have their root in the very appearance of a second-order derivative in (4.7.23), hence in the $(2 - \gamma)\Omega^2$ term featuring in Landau and Darrieus' result (4.7.22). A more complete mechanistic interpretation of (4.7.23) itself is thus warranted... but still misses, presumably because vorticity is distributed in the whole burned gas region, when $\omega_a \ll u_L/l_T$. Clearly, potential-flow effects are not enough of an explanation.

To the best of our knowledge, nobody has to date worked out the nonlinear amplitude equation(s) associated with $u_a \geq u_+$ nor tried to investigate the influence of multifrequency forcings similar to those recently used in the context of liquids subject to Faraday's instability (Edwards and Fauve, 1992). The case of random forcings could possibly be also interesting to investigate, in connection with the physics of disordered one-dimensional systems (Luck, 1992).

4.9 Hydrodynamic influence of boundaries

In Section 4.6 we indicated how conductive interactions with the surface of a porous flat-burner select the stand-off distance of a planar flame and alter some of its stability properties. Our linear analysis was restricted to injection rates $m\rho_u u_L$ satisfying $m < 1$, for otherwise the basic, nearly flat flame would be blown-off to downstream infinity.

However, some experiments (Schimmer and Vortmeyer, 1977) suggest that a markedly wrinkled front could nevertheless stay at a finite distance off the burner when the reduced inlet velocity m exceeds unity (within some range). The phenomenon is not totally unexpected, because a wrinkled adiabatic front propagates faster than a planar one, owing to its increased surface area, but one still has to inquire about the mechanisms which select the stand-off distance and maintain the flame there. This is what we now try to elucidate, in the framework of a *hydrodynamical* model.

To this end we consider the linearized Landau–Darrieus formulation of Section 4.7 one more time, with one important modification: the velocity disturbances u' *and* $\mathbf{v}'$ are now assumed to vanish at a finite, albeit $\gg D_{th}/u_L$ and yet undetermined, distance d upstream of the front. One is entitled to also impose a condition on $\mathbf{v}'$ in the context of Euler's equations because the high-Reynolds-number flow we consider here comes *from* the boundary, thereby blowing any viscous boundary-layer and precluding a tangential slip of the outer flow (Culick, 1966). By the same token, the flow ahead of the flame is now rotational if wrinkles are present (*exercise*: prove this; *hint*: assume the contrary and exhibit a contradiction).

On dimensional grounds, the growth rate S pertaining to a harmonic, infinitesimal disturbance must have the form:

$$S = u_L K \Omega(\gamma, Kd), \qquad (4.9.1)$$

where K is again the transverse wavenumber. Detailed calculations (Joulin, 1987) yield Ω as the root of a transcendental equation.

$$(2 - \gamma + \gamma H)\Omega^2 + 2\Omega = \frac{\gamma}{1 - \gamma}(1 - H), \qquad (4.9.2)$$

$$H(\Omega, Kd) \equiv \frac{e^{-Kd}}{1 - \Omega}\left(2e^{-Kd\Omega} - (1 + \Omega)e^{-Kd}\right). \ (4.9.3)$$

The presence of exponentials in (4.9.2) and (4.9.3) comes from imposing boundary conditions (on u' and $\mathbf{v}'$) to the solutions of (4.7.12) at a finite distance from the front.

For $Kd \to \infty$, Ω resumes the form given by (4.7.22), as it should be. When Kd is finite, $\Omega(\gamma, Kd)$ is less than $\Omega(\gamma, \infty)$, especially as $Kd \to 0$, in which case:

$$S = u_L K \frac{\gamma}{2(1-\gamma)} (Kd)^2 + \dots \qquad (4.9.4)$$

Because the reactant flows normally to it at a prescribed rate, a too close burner (on the wavelength scale) inhibits the streamline divergence that would lead to a LD instability. The inequality $\Omega < \Omega(\gamma, \infty)$ is patent in the small-γ limit:

$$S = u_L K \frac{\gamma}{2} \left(1 - e^{-Kd}\right)^2 + \dots \qquad (4.9.5)$$

and it qualitatively suggests what could maintain a wrinkled flame about a fixed location when $m > 1$: a shorter (longer) stand-off distance would reduce (enhance) the disturbance growth rate S, hence the degree of wrinkling and the resulting surface area ($\partial S/\partial d > 0$). The admissible stand-off distances are such that the instability-induced increase in effective flame speed just equilibrates the excess $(m - 1)u_L$ of injection velocity over u_L. Finding d therefore entails solving a nonlinear problem of flame wrinkling. In general, this is beyond the presently available analytical tools (the flow field is vortical on both sides of the flame) except possibly in one case: the small-γ limit.

The analysis leading to (4.7.44) can be extended (Joulin, 1987) to account for a finite d. It yields an evolution equation of the Michelson–Sivashinsky type (apart from a differentiation)

$$\dot{\psi} + \psi \psi_Z = \psi_{ZZ} + J(\psi, Z; \lambda(T)) \qquad (4.9.6)$$

for the slope $\psi \equiv \phi_Z$ of the flame shape distortion about its mean location d. Note that we restored curvature effects in (4.9.6) for the sake of completeness, that $1/K_{\text{neutral}}$ is used as unit of length and that the nonlinearity is basically the same as in (4.7.44). The linear operator $J(., Z; \lambda)$ involves the reduced stand-off distance $\lambda = dK_{\text{neutral}}$, and is defined in terms of any transverse Fourier component by:

$$J\left(e^{i\kappa Z}, Z; \lambda\right) = e^{i\kappa Z} |\kappa| \left(1 - e^{-\lambda|\kappa|}\right)^2 \qquad (4.9.7)$$

consistently with (4.9.6); $J(., Z; \infty)$ is nothing but the usual Landau–Darrieus operator $I(., Z)$.

Assume that (4.9.6) is solved with periodic boundary conditions (say) over a transverse interval of $2\pi/\kappa_{\text{box}}$ length and that it admits a steady nontrival solution for any fixed value of λ belonging to some range. Processing the solution would yield the fractional increase in flame-surface area $\left\langle \phi_Z^2 \right\rangle /2$ in the form:

$$\tfrac{1}{2} \left\langle \psi^2 \right\rangle \equiv f\left(\lambda, \kappa_{\text{box}}\right) . \qquad (4.9.8)$$

The requirement that the flame sits about a defined location would then result in:

$$f(\lambda, \kappa_{\text{box}}) = \mu, \tag{4.9.9}$$

where $\mu \sim (m - 1) > 0$ measures the excess in the injection velocity over u_L and is a control parameter at one's disposal. Equation (4.9.9) therefore indicates that the *lateral* flame size can also contribute to determining the stand-off distance.

Necessarily, μ must be less than $1/2$ for a finite λ to be obtained from (4.9.9), because $U = 1/2$ is the maximum reduced increase in effective speed of burner-free flames (Fig. 4.22). Even for $\mu < 1/2$, computing λ is not a very simple affair, as is explained below. In Fig. 4.30 we plotted the marginal curve

$$\lambda = \frac{1}{|\kappa|} \log \left(\frac{1}{1 - |k|^{1/2}} \right) \tag{4.9.10}$$

associated with the linearized version of (4.9.6). Note that the curvature effects dominate the hydrodynamical one for both $|\kappa| \gg 1$, as is now usual, as well as for $|k| \ll 1$, consistently with (4.10.4).

If the flame is wide enough ($\kappa_{\text{box}} \ll 1$) any integer n such that $0 \leq n - 1 \leq \text{Int}(1/\kappa_{\text{box}})$ defines an admissible wrinkle wavelength $2\pi/n\kappa_{\text{box}}$, hence an admissible function $f(\lambda, \kappa_{\text{box}})$: in principle, the f's belonging to unequal n's do not coincide, because each wrinkle of reduced wavelength less than 2π (i.e., $L < 2\pi/K_{\text{neutral}}$) bifurcates as a different λ

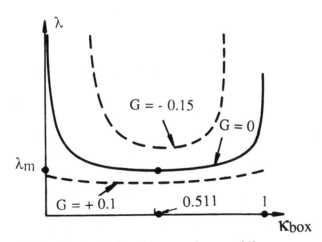

Fig. 4.30. Reduced stand-off distance, $\lambda = dK_{\text{neutral}}$, versus reduced wavenumber $\kappa_{\text{box}} = \Lambda_{\text{box}}/\Lambda_{\text{neutral}}$, in the marginal situations predicted by (4.9.10) (solid curve); the dashed lines account for buoyancy, equation (4.9.11).

(Fig. 4.30) and in general leads to a different flame speed increase when $\lambda = \infty$ (Fig. 4.22).

Accordingly, for a given mixture and a fixed lateral extent, one still has to solve a problem of *pattern selection* to effectively compute $\lambda(\mu)$ and the blow-off condition(s) of wrinkled flames; this has so far not been done, even in the small-γ limit. The candidate mechanisms for selection are many: secondary instabilities (Eckhaus...), hysteresis, influence of lateral boundary conditions... (Manneville, 1990); even very weak residual turbulence cannot be excluded *a priori*. However, one may note that the smallest λ given by (4.9.10), $\lambda_m \simeq 3$ ($d_m \simeq \Lambda_{neutral}/2$), is attained for $\kappa = 0.511$, i.e., for a wrinkle wavelength which is very close to that yielding the highest flame speed when $\lambda = \infty$ (Fig. 4.22); accordingly, cells with $\kappa \simeq 0.5$ (i.e., $\Lambda \simeq 2\Lambda_{neutral}$) are likely to give the largest f in (4.9.9) and to be selected, when μ is increased.

Next, one must acknowledge that $\Lambda_{neutral} \simeq 60 \, l_T$ for typical lean flames of hydrocarbon and air: the ratios $L = d/l_T$ corresponding to $\lambda \geq 3$ exceed 30, and it was indeed legitimate to neglect heat losses to the burner (equation (4.6.8)) to study the above configurations.

If a weak gravity is incorporated in the model, still in the limit $\gamma \ll 1$, the linearized version of (4.9.6) generalizes into:

$$\dot{\Gamma}/\Gamma = (|\kappa| + G) \left(1 - e^{-\lambda|k|}\right)^2 - \kappa^2 \qquad (4.9.11)$$

for $\psi = \Gamma(T) \exp(i\kappa Z) \ll 1$. Besides the Landau–Darrieus mechanism, the buoyancy effects (and presumably, the response to periodic acceleration) are also affected by the proximity of a burner's surface, to the extent that stable flames may be obtained for upward propagations (Fig. 4.30). The corresponding nonlinear equation is easily constructed (Joulin, 1987): suffice it to generalize the definition (4.9.7) of the operator $J(\cdot)$ so as to include the acceleration due to gravity ($\sim G$) in (4.9.6).

4.10 Large-scale flow geometry

Besides those sketched in Section 4.1, there is still another very convenient laboratory device, namely: the so-called counterflow burner, in which the flame front is sandwiched between two wide streams that impinge on one another. In its simplest version, a counterflow burner involves a stream of fresh reactants and one of burnt, hot gases (Fig. 4.31).

Provided the two incoming streams reasonably corresponds to suitably chosen self-similar solutions to the Euler equations ($u = g(x)$, $\mathbf{v} = \mathbf{z}f(x)$), there exists one solution of the entire combustion problem for which the flame front is parallel to the $(z - y)$ plane, sits about the abscissa

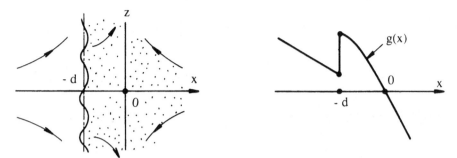

Fig. 4.31. Left: Schematic counterflow-burner wrinkled flame. Right: Unperturbed x-wise velocity.

$x = -d < 0$ and very plausibly correspond to the observations (Fig. 4.33). Such flat flames prove to be remarkably stable, a property routinely exploited in laboratories for measurement purposes and which we now try to explain, even if only qualitatively.

To simplify the matter and best display the mechanisms at work we assume that the fresh stream corresponds to:

$$\mathbf{u} = (-\sigma x + c, \sigma z, 0) \qquad 2p/\rho_{\mathrm{u}} = -\left((\sigma z)^2 + (c - \sigma x)^2\right) \qquad (4.10.1)$$

far upstream of the front. This defines a two-dimensional, potential flow in the $x - z$ plane, and $s \geq 0$ measures its rate-of-strain. Also for sake of simplicity, we again consider the Landau–Darrieus model of a discontinuous front equipped with a known Markstein length $\mathcal{L}$.

The unspecified constant c featuring in (4.10.1) is related to the stand-off distance d of the unperturbed flame, so that the basic x-wise profile $\bar{u}(x)$ reads as:

$$\bar{u}(x) = u_L - \sigma(x + d), \qquad x \leq -d. \qquad (4.10.2)$$

Finding d itself entails solving the Euler equations in the slab $-d \leq x \leq 0$. We seek the solutions to them in the forms

$$\bar{\mathbf{u}} = \left(g(x), -z\frac{dg}{dx}, 0\right), \qquad \bar{p} = -\frac{\rho_{\mathrm{u}}}{2}\sigma^2 z^2 - \frac{\rho_{\mathrm{b}}}{2}g^2(x) + p_{\mathrm{b}}, \qquad (4.10.3)$$

which automatically satisfy continuity, x-wise momentum balance and ensure the existence of a z-independent pressure jump across the flame; p_{b} is an integration constant. The x-wise velocity $g(x)$ is found from the z-wise momentum equation:

$$\left(\frac{dg}{dx}\right)^2 - g\frac{d^2g}{dx^2} = \frac{\rho_{\mathrm{u}}}{\rho_{\mathrm{b}}}\sigma^2. \qquad (4.10.4)$$

The solution to (4.10.4) which vanishes at the stagnation plane $x = 0$ has the structure:

$$g(x) = -\sigma\ell \, (\rho_{\mathrm{u}}/\rho_{\mathrm{b}})^{1/2} \sin(x/\ell), \qquad (4.10.5)$$

where the length ℓ is an integration constant. Note that (4.10.5) defines a vortical flow, since $d^2 g/dx^2 \neq 0$. The condition $g(-d) = u_L \rho_{\mathrm{u}}/\rho_{\mathrm{b}}$ used together with the continuity of tangential velocity across the front ($dg/dx = \sigma$ at $x = -d + 0$) ultimately yield:

$$\frac{\sigma d}{u_L} = \left(\frac{\rho_{\mathrm{u}}}{\rho_{\mathrm{b}}}\right)^{1/2} \left(1 - \frac{\rho_{\mathrm{b}}}{\rho_{\mathrm{u}}}\right)^{-1/2} \arccos\left((\rho_{\mathrm{b}}/\rho_{\mathrm{u}})^{1/2}\right). \qquad (4.10.6)$$

Constraining the jump in pressure across the front to be the same as in unstrained flames merely fixes the constant p_{b} in (4.10.3). The reduced stand-off distance $\sigma d/u_L$ given by (4.10.6) tends to unity when $\rho_{\mathrm{u}}/\rho_{\mathrm{b}} \to 1$, consistently with a purely kinematic argument, and goes to infinity with $\rho_{\mathrm{u}}/\rho_{\mathrm{b}}$ like $(\pi/2)(\rho_{\mathrm{u}}/\rho_{\mathrm{b}})^{1/2}$.

To compute the evolution of an infinitesimally wrinkled flame front of equation $x + d = \Phi(t, \mathbf{z}) \ll 1$, the standard procedure is to linearize the jump relationships, do the same to the Euler equations, solve them, and finally employ the kinematic constraint (4.7.26). Let us first examine the latter, which may be written as:

$$\left(u - \frac{\partial \Phi}{\partial t} - \mathbf{v} \cdot \nabla \Phi\right) = u_L \left(1 + |\nabla \Phi|^2\right)^{1/2} \left(1 - \mathcal{L}\left(\frac{1}{R_1} + \frac{1}{R_2}\right)\right) \qquad (4.10.7)$$

in the (x, z, y) coordinate system. Once linearized (4.10.7) reads as:

$$\frac{\partial \Phi}{\partial t} + \sigma\left(z\frac{\partial \Phi}{\partial z} + \Phi\right) = V(\mathbf{z}, t) + u_L \mathcal{L}\Delta\Phi. \qquad (4.10.8)$$

Finding a linearized evolution equation for $\Phi(t, \mathbf{z})$ is again reduced to expressing the x-wise velocity disturbance just ahead of the front, $V(t, \mathbf{z})$, in terms of $\Phi(t, z)$ itself.

Contrary to the classical Landau–Darrieus situation (Section 4.7) nobody has so far succeeded in solving this elliptic (?) problem exactly, due to a purely technical reason: one does not know how to solve the linearized Euler equations in the burned gases when the basic flow field is given by (4.10.3) and (4.10.5), especially if the disturbance transverse wavevector $\mathbf{K}$ lies in the $(x - z)$ plane (if $\mathbf{K}$ belongs to the $x - y$ plane, the problem is comparatively less difficult to solve, e.g., numerically; see Kim and Matalon, 1990). The fresh side and the region $x > 0$ do not pose problems, for the perturbed flow is potential there.

The only known exception is the small-γ limit ($\rho_{\mathrm{b}} \to \rho_{\mathrm{u}}$) when σ is taken in the range $\sigma = \mathcal{O}\left(u_L \gamma^2/\mathcal{L}\right)$, for which the perturbed flow field can be taken potential on both sides of the flame at the relevant leading orders in γ. $V(t, \mathbf{z})$ has then the same leading-order expression as for $\sigma = 0$, viz: $2V(t, \mathbf{z}) = \gamma u_L I(\Phi, \mathbf{z}) + \ldots$. Combined with (4.10.8) this implies

$$\frac{\partial \Phi}{\partial t} + \sigma \left(z \frac{\partial \Phi}{\partial z} + \Phi \right) = u_L \frac{\gamma}{2} I(\Phi, \mathbf{z}) + u_L \mathcal{L} \Delta \Phi + \ldots \qquad (4.10.9)$$

Let us first examine how harmonic disturbance of the form $\Phi = \Gamma(t) \exp(iKy)$ evolves. Equation (4.10.9) readily gives:

$$\Gamma_t/\Gamma = u_L \frac{\gamma}{2} |K| \left(1 - \frac{|K|}{K_{\text{neutral}}} \right) - \sigma, \qquad (4.10.10)$$

where, here, $2K_{\text{neutral}} = \gamma/\mathcal{L}$.

If $\sigma > \sigma_c = u_L K_{\text{neutral}} \gamma/8$, the wrinkle under consideration is damped, whereas $0 \leq \sigma \leq \sigma_c$ allows for a band of unstable modes (Fig. 4.32). The instability threshold is obtained upon equating two times: $1/\sigma$ (coming from flow divergence) and $8/\gamma u_L K_{\text{neutral}}$ (coming from intrinsic flame dynamics). The stabilizing mechanism is best understood once specialized to nearly-flat flames ($K \ll \sigma/\gamma u_L$) for which curvature effects do not play any significant role: if the front locally shifts upstream (downstream) of $x = -d$, it encounters a too fast (slow) flow and recedes to its equilibrium position. This is a purely kinematic process, whose origin lies in (4.10.8).

We next envisage wrinkles that harmonically undulate along the z-axis. Specifically, Φ is written as $\Gamma(t) \exp(iq(t)z)$, by analogy with the case of nearly-parabolic flames propagating in uniform far-upstream flows

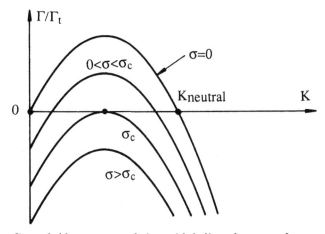

Fig. 4.32. Growth/decay rate of sinusoidal disturbances of a stretched flame, vs. wavenumber K.

(Section 4.7). Equation (4.10.9) then leads to:

$$q_t/q = -\sigma\,, \qquad \Gamma_t/\Gamma = u_L \frac{\gamma}{2}|q|\left(1 - \frac{|q|}{K_{\text{neutral}}}\right) - \sigma\,, \qquad (4.10.11)$$

to be compared to (4.10.10). Whatever $q(0)$ and $\sigma > 0$, $q(t)$ is driven exponentially rapidly by the tangential velocity σz to $q = 0$, so that ultimately $\Gamma_t/\Gamma \simeq -\sigma < 0$ and the wrinkle disappears. This rather unfortunate fate is due to a combination of wavelength stretching and of long-wave damping, both of which originate from the kinematic relationship (4.11.8).

As a practical consequence, flames held in between axisymmetric, opposed streams are linearly stable, at least in the small-γ limit: axisymmetric, potential stagnation-point flows of x-wise "strength" σ can indeed be thought of as superpositions two orthogonal, two-dimensional ones with "strength" $\sigma/2$ each, with the implication that any flame wrinkle involved experiences exponentially fast wavelength stretching. In our opinion, this may well be a good reason why counterflow burners yield so perfectly flat fronts (Fig. 4.33) despite the Landau–Darrieus mechanism (Joulin and Sivashinsky, 1992).

A few final remarks are in order. First of all, the above arguments are based on small-amplitude analyses, which could be misleading (Sivashinsky *et al.*, 1982); due to the unavoidable nonlinearities (*exercise*: guess a plausible nonlinear model equation generalizing (4.10.9); *hint*:

Fig. 4.33. An actual counterflow-burner axisymmetric flame (laser tomography of fresh mixture and intrinsic flame luminosity). Courtesy of P. Cambray and B. Deshaies (Poitiers); see also Deshaies and Cambray (1990).

read the preceding pages) a mere metastability, with a threshold that decreases very rapidly as σ does, could well be the right result when linear stability is predicted; some experiments (S. Sohrab, Northwestern Univ., unpublished) indeed show that flames held in axisymmetric stagnation-point flows can exhibit star-shaped wrinkles if σ is small enough, possibly due to nonlinearity. Lastly, this is almost the end of Section 4 and it is not surprising to encounter partially-answered problems at this stage!

4.11 Prospects

What we have summarized so far, together with the quoted references, could possibly suggest that most of the basic mechanisms at work in laminar gaseous premixed flames are now qualitatively elucidated. This is only partly true (a euphemism). Admittedly the last fifty of a million years of research on combustion produced important theoretical insights into the field, e.g., about the mechanisms of propagation or stability. However, even beside the chemical aspects, which constitute a world in themselves (see Peters and Rogg, 1993), many mysteries remain to be unraveled, some of which are presented below. It is not necessary to dig very deep to display them.

Consider the so-called V-flames when, as is done very often, these are idealized as wedge-shaped discontinuities anchored along an infinitely-thin wire that is stretched across a uniform stream of fresh reactants (Fig. 4.34). Elementary trigonometry gives the wedge semi-angle, hence a simple way of measuring u_L once u_∞ is known; so far, so good.

It is a bit more difficult to figure the burned-gas flow field, which is rotational (if $\rho_u \neq \rho_b$ and $u_\infty \neq u_L$) because the normal *and* the tangential velocity components are prescribed along the front. The corresponding stream function $\Psi(x, z)$ satisfies $\Delta\Psi = \varpi(\Psi)$ where $\varpi(.) \neq 0$ is still unknown. If self-similarity of the first kind (Barenblatt,

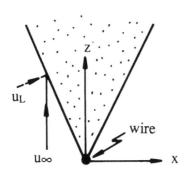

Fig. 4.34. Idealized sketch of a 2-D, V-flame anchored along an infinitely thin wire.

1979) holds in the above length-invariant situation, one ought to have $\Psi = (x^2 + z^2)^{1/2} u_L g(x/z)$ and $\varpi(\Psi) = u_L{}^2 c/\Psi$ where c is a pure number (A. Liñan, Madrid Univ., private communication). Ψ and c can be calculated but, whenever $\rho_u > \rho_b$ and $u_\infty > u_L$, the resulting pressure field logarithmically diverges at the origin and at infinity. This is incompatible with a constant pressure jump across the flame: V-flames cannot be idealized as in Fig. 4.34, the incoming flow cannot be uniform, etc. The model erroneously discarded a few length ratios (e.g., built upon l_T, gravity, wire radius, lateral dimensions...) which, however small, still play a crucial role; to date, nobody has succeeded in quantifying it theoretically. The above example is typical of the difficulties often encountered when trying to model flames as thin fronts. The contacts of flames with (adiabatic) walls, or the "regular" Bunsen flame idealized as a wedge or a cone lead to similar difficulties. Even though recent studies achieved intriguing success about polyhedral fronts in the framework of a constant-density model (Fig. 4.35), suggesting that hydrodynamics may well be an inessential mechanism, it is really disappointing not to be able to account for fluid motions even in the basic axisymmetric configuration. In general, the hydrodynamics of flames poses problems when the front exhibits sharp crests, i.e., quite often due to instabilities.

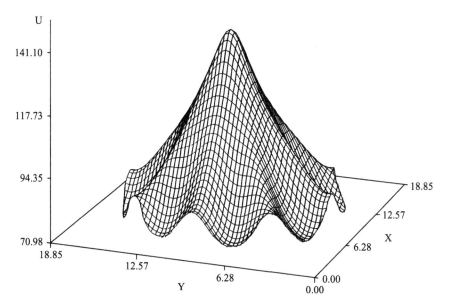

Fig. 4.35. A steady solution of the KS equation (4.5.27) when integrated with Neumann conditions on the first and third radial derivatives imposed along a circle (Compare to Fig. 4.4b). Courtesy of G.I. Sivashinsky; see also Gutman *et al.* (1993).

In fact, we do not even have any general, reliable theoretical method to compute the evolution of finite-amplitude wrinkles. The small-γ limit is particularly useful –qualitatively– in that respect, but alternative methods would be welcomed. A potentially rewarding toy is the potential-flow model (Clavin and Sivashinsky, 1987) because, upon use of appropriate Green's functions, it yields an integral equation for the evolution of markedly-wrinkled, flame(-like) closed fronts (Frankel, 1991) in the form $\mathbf{n} \cdot (\mathbf{u} - \mathbf{D}) = u_L(1 - \mathcal{L} \operatorname{div} \mathbf{n})$, with:

$$\mathbf{u} \cdot \mathbf{n} = \frac{u_L \gamma}{2(1 - \gamma)} \left(1 + \frac{1}{\nu\pi} \mathbf{n} \cdot \int_\Sigma \frac{\mathbf{r} - \mathbf{r}'}{|\mathbf{r} - \mathbf{r}'|^{\nu+1}} d^\nu \Sigma' \right) \tag{4.11.1}$$

and where $\nu = 1$ or 2 is the number of surface dimensions. Being structurally equivalent to the Michelson–Sivashinsky equation (4.7.43) for nearly planar fronts at two orders in the small-γ limit and giving similar results (Frankel and Sivashinsky, 1995) for expanding flames, (4.11.1) could provide very valuable guides when investigating the interplays between large-scale front (or flow) geometry and smaller wrinkles.

At any rate, because it does not include any "inertial" effect, equation (4.11.1) is not a panacea; for example it cannot account for gravity. As shown in Section 4.8, a Rayleigh–Taylor type of buoyancy effect promotes the growth of wrinkles whenever the fresh gases tend to fall "down" into the lighter, hot products. If, for some reason, the mean flame speed rapidly increases relatively to fresh gases that are assumed at rest, the resulting apparent gravity felt in a frame attached to the mean front location precisely has the right sign to enhance wrinkle growth. This can happen because u_L increases, as during combustion in enclosures or due to changes in mixture composition, ... or can be induced by the growth of wrinkles itself. One can therefore imagine a feedback between apparent acceleration $g(t)$ and surface area variations. A simple model dynamical system to mimic this is easily constructed on the basis of the previous sections:

$$\frac{\partial \Phi}{\partial t} + \frac{u_L}{2} \left(|\nabla\Phi|^2 - \left\langle |\nabla\Phi|^2 \right\rangle \right) = u_L \mathcal{L} \Delta\Phi + V ,$$

$$(2 - \gamma)\frac{\partial V}{\partial t} + 2u_L I(V, \mathbf{z}) = -\frac{\gamma}{1 - \gamma} u_L{}^2 \Delta\Phi + \gamma g(t) I(\Phi, \mathbf{z}) ,$$

$$g(t) = \frac{d}{dt} \left(u_L \left(1 + \tfrac{1}{2} \left\langle |\nabla\Phi|^2 \right\rangle \right) \right) .$$

$$\tag{4.11.2}$$

Here $\Phi(t, \mathbf{z})$ represents the front distortions *about the mean location*. Note that a fluctuating component, such as $u_a \omega_a \cos(\omega_a t)$, could have been added to $g(t)$ to simulate acoustic forcing and to account for the possibility of some parametric instability.

If the initial conditions (or u_a, ω_a) lead to g positive and large, instability is promoted and a runaway in both g and the surface-area is conceivable; as current researches will check, this could offer a qualitative explanation of the sudden increase in flame speed and area corresponding to Fig. 4.6 (bottom). Alternatively, milder initial conditions will presumably lead to a steady state (which can be found *via* the pole-decomposition method). Lastly, still with the same mixture (γ, $\mathcal{L}$, u_L are given) initial conditions such that $g < 0$ would lead to wrinkles which are headed to flatten. Because the middle equation (4.11.2) contains $\partial V/\partial t$, the neutral shape $\Phi \equiv 0$ may be crossed, thereby allowing for an interchange of troughs and crests. Hopefully, this is at the origin of the so-called "tulip-flame" phenomenon (Fig. 4.36), which in effect is characterized by such an interchange, and is still waiting for a neat theoretical explanation.

Finally we envisage a spherical flame expanding in an initially quiet premixture from, say, a match (see Section 4.1). As soon as its mean radius $R(t)$ noticeably exceeds $\mathcal{L}$, the flame obviously has $dR/dt = u_b \equiv u_L \rho_u/\rho_b$, a potential flow outside and burned gases at rest. A slight extension of Landau's dimensional argument (see equation (4.7.21), Section 4.7) indicates that, in the absence of any curvature effect, the amplitude $\Gamma(t)$ of a wrinkle characterized by fixed, large enough, angular harmonics of ranks n' and n'' should evolve according to:

$$\Gamma_t/\Gamma \simeq u_L\Omega(\gamma)\,(n/u_b t)\,, \qquad n = \left(n'^2 + n''^2\right)^{1/2} \qquad (4.11.3)$$

because $n/u_b t$ plays the role of an instantaneous wavenumber (Fig. 4.37).

Accordingly $\Gamma(t)$ grows algebraically, $\Gamma(t)/\Gamma(t_0) = (t/t_0)^{n(1-\gamma)\Omega(\gamma)}$, i.e., much slower than when the flame is nearly planar. If a Markstein length

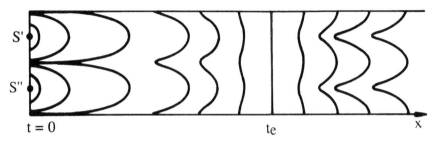

Fig. 4.36. Schematic "tulip-flame" phenomenon: two twin 2-D flames initiated by line sources (S', S'') at the closed end of a channel merge to form a wrinkled front which subsequently undergoes a sudden exchange of troughs and crests at $t = t_e$.

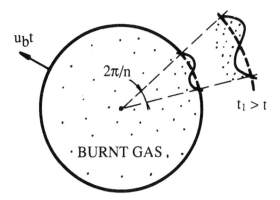

Fig. 4.37.   Expansion induced wavelength stretching of angular normal modes.

is accounted for, (4.11.3) is basically* modified into the WKB form of the growth rate pertaining to nearly flat flames, viz:

$$\Gamma_t/\Gamma = u_L\Omega(\gamma)\frac{n}{u_b t}\left(1 - \frac{n}{u_b t K_{\text{neutral}}}\right) \qquad (4.11.4)$$

whereby indicating that the relative amplitude $\Gamma(t)/u_b t$ is minimum when $t = t_n$, with $u_b t_n K_{\text{neutral}} = n^2\Omega(1 - \gamma)/(n\Omega(1 - \gamma) - 1)$. In turn t_n is minimum for

$$n = n_* \equiv 2/(1 - \gamma)\Omega(\gamma)\,, \qquad (4.11.5)$$

the corresponding time and wrinkle wavenumber being:

$$t_* = 4/\Omega(\gamma)u_L K_{\text{neutral}}\,, \qquad K_* = K_{\text{neutral}}/2\,. \qquad (4.11.6)$$

At first decreasing, the departure from sphericity, i.e., $\Gamma(t)/u_b t$, could grow when $t > t_*$ and should first appear in the form of corrugations whose size corresponds to the "most dangerous" wavenumber $K = K_{\text{neutral}}/2$ (Fig. 4.18). These predictions are basically unchanged by more sophisticated treatments (Istratov and Librovich, 1969), including those which employ the analogs of (4.7.34–36) and the Navier–Stokes equations as starting points (Bechtold and Matalon, 1987). Unfortunately, though seemingly correct qualitatively, this kind of analysis poses problems. The predicted values of n_* ($\simeq$ 10–15) are markedly off the experimental findings ($n_* \simeq$ 60–100) of Groff (1982) and the flame size $u_b t_*$ at the onset of instability is underpredicted by a factor which exceeds 10. Given the simplicity of the unperturbed configuration, we do not have a neat

---

* The actual formula is a bit more involved, but this does not change our arguments if n is large.

explanation for such a discrepancy. Admittedly, Groff's experiments were conducted in an enclosure, thereby bringing about acceleration effects and an increase in $K_{neutral}u_L$, but, here, both influences would tend to shorten t_*. Being a stabilizing mechanism, nonlinearity would act in the right way. It is certainly already at work at the visual onset of instability because the observed cells have well delineated polygon-like shapes (Fig. 4.38); moreover, a nearly-spherical shape $(\Gamma(t)/u_b t \ll 1)$ does not necessarily imply small amplitudes of wrinkling and negligible nonlinearities.

Recent analyses of the small-γ limit tend to support this viewpoint but, clearly, the question is far from being settled; even the role of external noise may be invoked (Joulin, 1994b).

Apart from a brief allusion at the end of Section 4.7, we deliberately avoided embarking on a detailed discussion of the possible influences of incoming velocity fluctuations upon the dynamics of potentially or actually unstable fronts. This was anything but a lack of interest: the problem has very close connections with propagations in (moderately) turbulent media, a subject matter of tantamount practical importance (engines, furnaces, hazards...) and is one of the most challenging

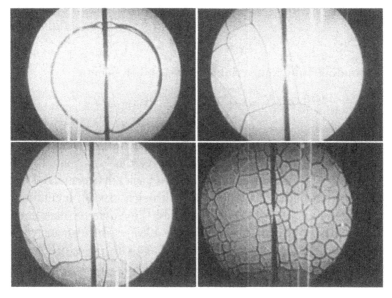

Fig. 4.38. Successive strioscopic records of a spark-initiated expanding propane/air flame in an enclosure (vertical bars = electrodes, outer circle = window edge). Top-left: 36.10^{-3} s after ignition. Top-right: 158.10^{-3} s. Bottom-left: 170.10^{-3} s. Bottom-right: 182.10^{-3} s. Courtesy of E.G. Groff (General Motors Labs., Warren); see also Groff (1982).

questions of combustion theory. Simply, very little is known on the topic, the only exception dealing with flames that are linearly stable due to a strong enough buoyancy effect (Searby and Clavin, 1986; Aldredge, 1990) and are consequently amenable to perturbative approaches.

Most studies of "noisy flames" have to date been conducted in the framework of passive propagations, i.e., upon complete neglect of the Landau–Darrieus mechanism. A simplified version of such models, corresponding to weak forcings, is the forced Burgers equation (compare with (4.5.28)):

$$\frac{\partial \Phi}{\partial t} + \frac{u_L}{2}|\nabla \Phi|^2 = u_L \mathcal{L}\Delta\Phi + u_e(\mathbf{z}, t)\,, \qquad (4.11.7)$$

where the random function $u_e(\mathbf{z}, t)$ has presumed known spectral properties and is meant to represent the turbulent, x-wise velocity fluctuations felt by the flame. In the context of other growth phenomena (Krug and Spohn, 1992), (4.12.7) is also known as the KPZ equation (Kardar *et al.*, 1986). It has recently been the subject of numerous investigations (Grossmann *et al.*, 1991, and references therein), but with forcings $u_e(\mathbf{z}, t)$ that have little to do with turbulent velocity fluctuations.

At any rate (4.11.7) neglects any hydrodynamical instability, which is likely to be a major deficiency when $u_e \leq \mathcal{O}(u_L)$: recent numerical integrations of a forced Michelson–Sivashinsky equation, i.e., (4.11.7) with a Landau–Darrieus contribution $\gamma u_L I(\Phi, \mathbf{z})/2$ added to the r.h.s., indeed suggest that the hydrodynamical instability may play a significant role in moderately-turbulent flames as well (Cambray and Joulin, 1992; Cambray *et al.*, 1994). Quite certainly, analyzing such model problems in detail, e.g., by spectral integrations, pole-sparkling (Joulin, 1988) or cellular automata (Garcia-Ybarra *et al.*, 1992), is a prerequisite before we master the couplings among hydrodynamics-related instability mechanisms, curvature effects, evolving large-scale geometry and incoming turbulence, such as those at work in Fig. 4.39.

Even when this is done, the story will not stop then. Though provisionally offering a challenge, Fig. 4.39 is probably child's play compared to what is encountered in applications..., not to mention the specificities brought about by partial unmixedness of reactants, radiation, multiphase fuels (Buckmaster, 1993), chemical quenching, transition to detonation, hollowed fronts (Dold *et al.*, 1991), etc. However, the situation is evolving very rapidly, as witnessed by the works of Joulin (1994b), Joulin *et al.* (1994), Minaev (1994), Filyand *et al.* (1994), Rahibe *et al.* (1995), and Cambray *et al.* (1996) on expanding flames (Fig. 4.38). As a starting point, all of them use an evolution equation of the Michelson–Sivashinsky type (adapted to the cylindrical geometry), for

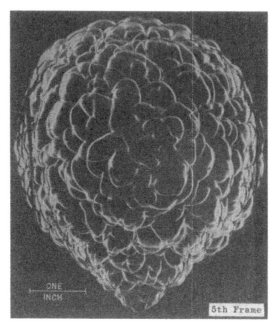

Fig. 4.39. Slightly rich propane/air flame expanding in an (initially) isotropic turbulent premixture (integral scale $\simeq 3.5 \times 10^{-3}$ m). Private communication from R.A. Strehlow; see also Palm-Leis and Strehlow (1969).

which exact solutions can again be obtained by the pole-decomposition method introduced in Section 4.7.5. One of the main current issues is to understand whether external forcing is a *sine qua non* condition for explaining the appearance of cells, and/or the ultimate acceleration of unconfined diverging flames (radius $\sim t^{3/2}$), and the progressive fractalization of the front evidenced in the experiments (Gostinsev *et al.*, 1988), even when the fresh medium seems perfectly at rest before ignition. Mastering all the mechanisms involved in Fig. 4.39 analytically is not yet done, but the corresponding flame dynamics is now close to the range of the numerics, at least in the framework of suitable evolution equations.

5 Shock waves

5.1 Phenomenology

Combustion has been employed by man for about 10^6 years but shock waves have been used for even longer, $\sim 3 \times 10^6$ years, in the process of manufacturing flint implements: adequately banging a stone against another one may produce a flake (Fig. 5.1). The *shock*, induced by the

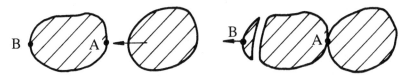

Fig. 5.1. Early shock use.

impact at A, travels to B, is reflected there into a release wave and the resulting pressure drop causes spallation.

Closer to us, *sonic* booms, rifle cracks, experiments in *supersonic* wind tunnels (Fig. 5.2), the bow wave produced by the Earth's motion through the solar wind,... are all events associated with the shock wave phenomenon: "shock waves are very general and, if not ubiquitous, at least pervasive" (Duvall and Graham, 1977). Combustion is not an exception to the rule. A flame propagating in a semi-infinite tube, from its closed end into a reactive gas initially at rest, necessarily produces, after a certain time, a precursor shock *wave* (Landau and Lifshitz, 1987) that can be intense enough to initiate chemical reactions in the gas, which ultimately lead to a *detonation*. The study of detonations is postponed to Section 6. Here, we shall deal only with shocks in inert media.

At this point, any attentive reader will have certainly noticed that we have not yet defined what a shock wave precisely is. Indeed, this is very difficult to do without sacrificing generality. We shall content ourselves

Fig. 5.2. Shock attached to a wedge (Laboratoire d'Etudes Aéro. Therm., Poitiers).

with the following practical definition: shocks are very steep waves which build up in *compressible media* when the boundary conditions (such as moving projectiles) force information (e.g., pressure changes) to be transmitted at a velocity higher than the frozen *speed of sound* (the speed at which infinitesimal pressure disturbances travel in a medium without changing its composition) in the unperturbed medium. Defining a shock more precisely, basically amounts to describing how it is formed and to analyzing the properties of specific *hyperbolic* systems of equations (Section 5.2). Suffice it to say that shock fronts shall be considered here as *hydrodynamic discontinuities* in *inviscid media*. The inner structure of a shock front will be briefly alluded to, just to point out that, for studying it, the model of a continuous medium should in general be abandoned.

Shock waves are usually observed to have a smooth front, at least piecewise (Fig. 5.2). However, a shock wave may theoretically have a corrugated front (D'yakov, 1954, 1957; Iordanskii, 1957; Kontorovich, 1958, 1959) or may split into two (or more) other waves (Bancroft *et al.*, 1956; Zel'dovich *et al.*, 1966; Gardner, 1963; Fowles, 1981; Kuznetsov, 1985, 1986) when the thermodynamic properties of the shocked medium satisfy specific conditions.

5.2 Shock formation

Continuous functions describing smooth changes in the properties of a system are commonly obtained by integrating one or several differential equations. However, in the approximation we adopt in this section, where diffusion transports of energy, mass and momentum are neglected, (cf. Section 2), such continuous functions are not necessarily valid everywhere within the domain of definition of the system because surfaces may exist through which some properties change discontinuously. Such surfaces of discontinuity are called shocks. Roughly speaking, one may say that a shock forms when too many changes in the system occur at the same locus of the domain. From a formal point of view, the latter event appears as an almost inescapable consequence of the hyperbolic nature of the nonlinear differential equations. In the following paragraphs we sketch how to make both these ends meet.

A relatively simple description of the mathematical theory of hyperbolic equations is possible when attention is restricted to the case of systems of first order partial differential equations in two independent variables, say x and t

$$\frac{\partial \mathbf{g}}{\partial t} + \mathbf{C}(x, t, \mathbf{g})\frac{\partial \mathbf{g}}{\partial \mathbf{x}} = \mathbf{h}(x, t, \mathbf{g}) \qquad (5.2.1)$$

with suitable boundary conditions. An extended presentation of this theory would take us too far afield, the interested reader is referred to such texts as Courant and Friedrichs (1948), Courant and Hilbert (1953), Whitham (1974), Brun (1974). The above system is termed hyperbolic if the matrix $\mathbf{C}$ has real, distinct eigenvalues. One can then define particular lines of the (x,t) plane, the slopes of which are the reciprocals of these eigenvalues. Such lines, called *characteristic curves*, are the corner stones for building solutions to (5.2.1) and for understanding shock formation. To exemplify the essential meaning of characteristic curves, we devote the next section to the scalar homogeneous version of (5.2.1), which involves a single unknown real function. This study reveals itself as an instructive exercise before addressing one of the problems of interest in this course, the formation of a shock in a compressible fluid flow. The latter problem is part of the classical "piston problem" that we describe in section 5.2.2.

5.2.1 Preliminary example

We begin with a study of the nonlinear scalar equation

$$\frac{\partial g}{\partial t} + c(g)\frac{\partial g}{\partial x} = 0, \quad g(x, t=0) = g_0(x), \quad t \geq 0, \tag{5.2.2}$$

to reveal the essential features of nonlinear hyperbolic problems. To this end we consider the function $g(x,t)$ at each point of the (x,t) plane and write its total differential as

$$dg = \frac{\partial g}{\partial t}dt + \frac{\partial g}{\partial x}dx. \tag{5.2.3}$$

We then define the directional derivative of g along some arbitrary curve $x(t)$ as

$$\frac{dg}{dt} = \frac{\partial g}{\partial t} + \frac{\partial g}{\partial x}\frac{dx(t)}{dt}. \tag{5.2.4}$$

A comparison of (5.2.2) and (5.2.4) shows that, if g is a solution, we must have

$$\frac{dg}{dt} = 0 \tag{5.2.5}$$

along particular curves $x_c(t)$, the slopes of which in the (x,t) plane are the reciprocals of

$$\frac{dx_c(t)}{dt} = c(g). \tag{5.2.6}$$

From (5.2.5) and (5.2.6) we infer that g, and consequently the slope $c(g)$, have constant values along the curves $x_c(t)$. Therefore, these curves are straight lines in the (x,t) plane... Along the particular straight line $x_c(t)$ that intersects the x-axis at $(x = X, t = 0)$, g and $c(g)$ are given by the initial condition (5.2.2)

$$g(x = X, t = 0) = g_0(X). \tag{5.2.7}$$

Thus the solution to (5.2.2) is obtained in parametric, Lagrangian form

$$g(x,t) = g_0(X), \qquad x = x_c(X,t) = X + tc\,(g_0(X)), \qquad (5.2.8)$$

where $x_c(X,t)$ is the equation for the set of curves $x_c(t)$. A specified value of X defines one line $x_c(t)$. The lines represented by (5.2.8) are the characteristic curves for the problem (5.2.2).

In more complicated instances, such as the ones associated with nonhomogeneous equations, the characteristics are not straight and the solutions are not constant along them. Here, we may nevertheless say that a specific value of g, say $g_0(X^*)$, is carried along the particular line $x_c(X^*,t)$. We can then interpret the quantity $c(g)$ as a "transmission velocity" for the information g: for the sake of definiteness, we associate the time with the variable t and the distance with x. In the (x,g) plane, the point of coordinates $(X^*, g_0(X^*))$ on the initial profile of g has coordinates $(X^* + tc\,(g_0(x^*))\,, g_0(X^*))$ at time $t > 0$. The profile of g at time t can be constructed by translating each point $(X, g_0(X))$ of the initial profile through a distance $tc\,(g_0(x))$ parallel to the x-axis. Thus, we find that when $dc/dg > 0$ the parts of the profile with higher values of g move faster than points with lower values. When $dc/dg < 0$, the latter conclusion is reversed. When c does not depend on g, the problem is linear and each point of the profile is translated through the same distance ct, so that the profile does not change in shape. The phenomena of transportation of the information at a finite speed along characteristic curves is typical of hyperbolic equations. This particular feature explains why intuition is so difficult to use in hyperbolic systems and why causality arguments often replace order of magnitude estimates. In our example, the nonlinear behavior of the solution is due to the particular dependence of c on g which governs the distortion in the (x,g) plane of the profile of g. Here, when $dc/dg > 0$, segments of this profile with a negative slope will eventually yield a multivalued solution for g. Such a behavior, called breaking, occurs at the time t_B when the slope of g first becomes infinite (Fig. 5.3).

To find this breaking time t_B, we first consider the last of expressions (5.2.8) as an implicit equation for $X(x,t)$ and differentiate it with respect to x at fixed t. This operation leads to

$$\frac{\partial X}{\partial x}(x,t) = \frac{1}{1 + t\frac{dc}{dg}\,(g_0(X))\,\frac{dg_0}{dx}(X)}, \qquad (5.2.9)$$

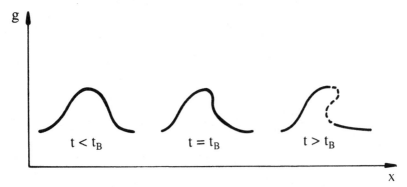

g

$t < t_B$ $t = t_B$ $t > t_B$

x

Fig. 5.3. Breaking process.

where x is given by (5.2.8). We then differentiate (5.2.8) with respect to x at fixed t so that the slope of the profile of g at any fixed time is

$$\frac{\partial g}{\partial x}(x,t) = \frac{dg_0}{dx}(X)\frac{\partial X}{\partial x}(x,t) = \frac{\frac{dg_0}{dx}(X)}{1 + t\frac{dc}{dg}(g_0(X))\frac{dg_0}{dx}(X)}, \qquad (5.2.10)$$

where, again, x is given by (5.2.8). We recall that $g_0(X)$ and $dg_0/dx(X)$ represent the values of g and $\partial g/\partial x$ calculated at point $x = X$ and time $t = 0$ so that (5.2.10) also reads

$$\frac{\partial g}{\partial x}(x,t) = \frac{\frac{\partial g}{\partial x}(X,0)}{1 + t\frac{dc}{dg}(g(X,0))\frac{\partial g}{\partial x}(X,0)}. \qquad (5.2.11)$$

Thus, by analyzing expressions (5.2.10) we find again that, when $dc/dg > 0$, the segments of the profile of g with a negative (positive) slope steepen (flatten out) when time increases. In this case, the slope becomes infinite when t and x respectively tend to the values

$$t_e(X) = \frac{-1}{\frac{dc}{dg}(g_0(X))\frac{dg_0}{dx}(X)}, \qquad x_e(X) = X + t_e(X)c(g_0(X)). \quad (5.2.12)$$

The actual breaking occurs when $t \to t_B = t_e(X_B)$ where X_B is the value of X that minimizes $t_e(X)$. In other words we have $t \to t_B = \min_X [t_e(X)]$. The corresponding abscissa, x_B, is obtained by setting $X = X_B$ in (5.2.12). In the (x,t) plane, the point (x_B, t_B) is located on the characteristic line originating from the point $(x = X_B, t = 0)$. As previously explained, a conflict among several values of the function $g(x,t)$ exists when breaking occurs. Since g is constant along the same characteristic line, we are led to the conclusion that breaking must be associated with the crossing of different characteristic lines. The condition for two adjoining characteristic lines to intersect is (Fig. 5.4)

$$x_c(X,t) = x_c(X + dX, t) \qquad \text{or} \qquad \frac{\partial x_c}{\partial X} = 0\,. \tag{5.2.13}$$

Equation (5.2.8) then leads to the condition $t = t(X)$, with:

$$t(X) = \frac{-1}{\frac{dc}{dg}\,(g_0(X))\,\frac{dg_0}{dX}(X)}\,. \tag{5.2.14}$$

Only positive values for $t(X)$ are allowed (5.2.2); this imposes:

$$\frac{dc}{dg}\,(g_0(X))\,\frac{dg_0}{dX}(X) < 0\,. \tag{5.2.15}$$

As expected in this case, (5.2.14) turns out to coincide with conditions (5.2.12) for the slope of g in the (x,g) plane to become infinite. Indeed, equations (5.2.12) define a parametric equation for a particular curve of the (x,t) plane. This curve is an envelope formed by the intersection points of those characteristics which meet condition (5.2.15). Clearly, the breaking time t_B is the minimum value of t along the envelope. Elementary differential geometry results show that the envelope has a cusp at point (x_B, t_B) when the quantity $d/dx\,[dc/dg\,(g_0(x))\,dg_0/dx(X)]$ is continuous.

We have demonstrated that the continuous solution (5.2.8) to the initial value problem described by equation (5.2.2) is not necessarily valid for all $t > 0$. When condition (5.2.15) is satisfied, that is when $c\,(g_0(x))$ is a decreasing function of x, the solution (5.2.8) does not yield a unique determination beyond the breaking time t_B. If, for physical reasons, one

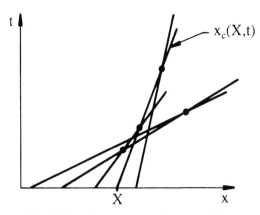

Fig. 5.4.   Intersecting characteristics.

insists upon retaining single-valued functions only, one may consider the existence of discontinuous solutions as a possible way out. Consequently, we now address the specific problem of calculating the position $x_s(t)$ of the discontinuity locus for $t \geq t_B$. We firstly have

$$x_s(t) = x_B + \int_{t_B}^{t} D(t)dt, \qquad D(t) = \frac{dx_s(t)}{dt}. \qquad (5.2.16)$$

The latter formula shows that it is sufficient to know the quantity $D(t)$, the velocity of the discontinuity locus. Before calculating D, we take for granted, as is the case, the existence of some global principle of conservation. We want to show that our differential equation (5.2.2) is a local expression of such a global principle. A conservation law asserts that the variation between two instants t_1 and t_2 of a quantity H, defined within a domain V, is equal to the sum of all fluxes ϕ that cross the boundary ∂V of V during the lapse of time (t_1, t_2)

$$H(t_2) - H(t_1) = \int_{t_1}^{t_2} \phi \, dt \qquad (5.2.17)$$

when there is no production nor dissipation for H in V, which we admit here. Denoting the density of H in V by h, the density of ϕ on ∂V by $\mathbf{f}$, and the outward unit vector normal to ∂V by $\mathbf{n}$, the latter equation becomes

$$\left[\int_V h(\mathbf{x}, t) dv \right]_{t_1}^{t_2} = - \int_{t_1}^{t_2} dt \int_{\partial V} \mathbf{n} \cdot \mathbf{f}(x, t) d\sigma. \qquad (5.2.18)$$

The differentiation of (5.2.18) with respect to t leads to the global conservation equation

$$\frac{d}{dt} \int_V h(\mathbf{x}, t) \, dV = - \int_{\partial V} \mathbf{n} \cdot \mathbf{f}(x, t) d\sigma. \qquad (5.2.19)$$

We next assume that V is a stationary domain in which h and $\mathbf{f}$ are continuous and differentiable. Then a direct application of the flux–divergence theorem and of the time derivative of a volume-integral leads to

$$\int_V \left(\frac{\partial h}{\partial t} + \nabla \cdot \mathbf{f} \right) dV = 0. \qquad (5.2.20)$$

Dividing by the volume of V and making it shrink to zero size yield

$$\frac{\partial h}{\partial t} + \nabla \cdot \mathbf{f} = 0. \qquad (5.2.21)$$

Clearly, for our one-space dimension problem, the global and local conservation equations (5.2.19) and (5.2.21) become respectively

$$\frac{d}{dt}\int_{x_2}^{x_1} h(x,t)dx = f\left(x_2,t\right) - f\left(x_1,t\right), \qquad \frac{\partial h}{\partial t} + \frac{\partial f}{\partial x} = 0. \qquad (5.2.22)$$

Upon the assumption that the two functions $h(x,t)$ and $f(x,t)$ are continuous and differentiable, one can always define two other continuous differentiable functions $g(x,t)$ and $c(x,t)$ such that

$$h(x,t) = h(g), \quad f(x,t) = f(g), \quad c(x,t) = c(g) = \frac{df}{dg}\Big/\frac{dh}{dg} \qquad (5.2.23)$$

thereby obtaining a local version of (5.2.19) with the same structure as (5.2.2):

$$\frac{\partial g}{\partial t} + c(g)\frac{\partial g}{\partial x} = 0.$$

We now are in a position to calculate the velocity D because, contrary to local equations, the first of eqs. (5.2.22) is valid even if there is a discontinuity between the boundaries x_1 and x_2. Thus, let x_2 and x_1 be stationary boundaries such that $x_2 < x_s(t) < x_1$ and denote u_L, u_R, u_2 and u_1 the values at time t of an arbitrary function $u(x,t)$ calculated on the left and right sides of the discontinuity $x_s(t)$ and on the boundaries x_2 and x_1, respectively. Then, (5.2.22) can be written as

$$f_2 - f_1 = \frac{d}{dt}\int_{x_2}^{x_s(t)} h(x,t)\,dx + \frac{d}{dt}\int_{x_s(t)}^{x_1} h(x,t)\,dx. \qquad (5.2.24)$$

Carrying out the differentiations yields

$$f_2 - f_1 = h_{\rm L}D + \int_{x_2}^{x_s(t)} \frac{\partial h}{\partial t}\,dx - h_{\rm R}D + \int_{x_s(t)}^{x_1} \frac{\partial h}{\partial t}\,dx. \qquad (5.2.25)$$

Upon the further assumption that $\partial h/\partial t|_R$ and $\partial h/\partial t|_L$ are bounded, the above integrals tend to zero in the limit $x_2 \to x_s(t)$. The velocity of the discontinuity is thus given by

$$D = \frac{f_2 - f_1}{h_2 - h_1}, \qquad (5.2.26)$$

where the subscripts 2 and 1 now denote quantities evaluated on the left and right sides of the discontinuity respectively. An approximation for D can be useful in situations where the jumps in f and h are small.

Expanding c and D in Taylor series in $h_2 - h_1$ and using (5.2.23) leads to

$$c = \frac{df}{dh}$$

$$c_2 = c_1 + \left.\frac{d^2 f}{dh^2}\right|_1 (h_2 - h_1) + o\,(h_2 - h_1) \tag{5.2.27}$$

$$D = \left.\frac{df}{dh}\right|_1 + \frac{1}{2}\left.\frac{d^2 f}{dh^2}\right|_1 (h_2 - h_1) + o\,(h_2 - h_1)$$

Recognizing that $o\,(h_2 - h_1) = o\,(g_2 - g_1)$, we obtain

$$D = \frac{1}{2}\,(c_1 + c_2) + o\,(g_2 - g_1) \ . \tag{5.2.28}$$

We conclude this paragraph by emphasizing two very important points. The first one is to acknowledge that a given *conservative form* (5.2.22) defines unambiguously a differential equation (5.2.2), but that the converse statement is not true: the global conservation principle (5.2.17) defines explicitly two functions h and f and leads to a unique function c (5.2.23) and a unique discontinuity velocity D (5.2.26). In other words, there is only one differential equation (5.2.2) corresponding to the conservation equation (5.2.22). On the other hand, one cannot construct a unique conservation equation (5.2.22) from a mere knowledge of a differential equation (5.2.2), because there is an infinite number of functions f and h that define a given function c. For example, consider the case where

$$f(g) = \frac{1}{n+2}g^{n+2} , \qquad h(g) = \frac{1}{n+1}g^{n+1} . \tag{5.2.29}$$

We have $c(g) = g$ whatever n, we also have

$$D = \frac{n+1}{n+2}\frac{g_2^{n+2} - g_1^{n+2}}{g_2^{n+1} - g_1^{n+1}} , \tag{5.2.30}$$

which clearly shows that different values of n produce different values of D. Thus, a preliminary definition of some global principle of conservation is a necessary condition before solving properly a hyperbolic equation. The second point is to acknowledge that the latter rule is in no way a sufficient condition because several piecewise continuous solutions to the same initial value problem that can be constructed from the conservative form (5.2.22) are not unique. For example, consider the problem where $f(g) = g$ and $h(g) = \frac{1}{2}g^2$ (Brun, 1974). We have from (5.2.22,23,26)

$$\frac{\partial g}{\partial t} + \frac{\partial}{\partial x}\left(\tfrac{1}{2}g^2\right) = 0 , \qquad a(g) = g , \tag{5.2.31}$$

$$D = \frac{1}{2}(g_1 + g_2) .$$ (5.2.32)

Now, let us select the particular initial condition

$$g(x,0) = g_0(x) = 0 .$$ (5.2.33)

A theoretically viable solution to this inviscid Burgers problem is a piecewise continuous function (Fig. 5.5) defined by

$$
\begin{aligned}
&\text{domain 1:} \quad t \geq 0, \quad x \in \,]-\infty, x_{2S}(t)[\,\cup\,]x_{1S}(t), +\infty[\\
&\qquad\qquad g(x,t) = 0, \\
&\text{domain 2:} \quad t > t_0, \quad x \in \,]x_0, x_{1S}(t)[\\
&\qquad\qquad g(x,t) = 2g^*, \\
&\text{domain 3:} \quad t > t_0, \quad x \in \,]x_{2S}(t), x_0[\\
&\qquad\qquad g(x,t) = -2g^* .
\end{aligned}
$$ (5.2.34)

x_0 and t_0 represent the coordinates of the intersection point of three theoretically admissible discontinuity loci defined by the functions $x_{1S}(t)$, $x_{2S}(t)$ and $x_{3S}(t)$ which, for any $t \geq t_0$, are such that

$$
\begin{array}{llll}
x \geq x_0 & x_{1S}(t) = x_0 + g^*(t - t_0) & D_1 = g^*, \\
x \leq x_0 & x_{2S}(t) = x_0 - g^*(t - t_0) & D_2 = -g^*, & (5.2.35) \\
x = x_0 & x_{3S}(t) = 0 & D_3 = 0 .
\end{array}
$$

g^* is an arbitrary positive constant and therefore we have an infinity of piecewise continuous solutions to the initial value problem described by equations (5.2.31) and (5.2.33). Thus, additional conditions are needed to single out a specific discontinuity. To get some insight into this issue,

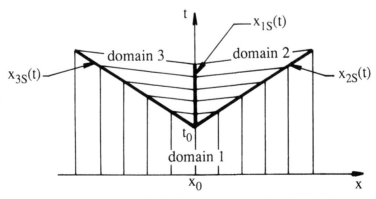

Fig. 5.5. Theoretical solutions to equation (5.2.31).

we write the parametric equations for the characteristics associated with
the example (5.2.31)–(5.2.33):

$$\begin{array}{lll}
\text{domain 1:} & x_{1c}(X,t) = X\,, & \\
\text{domain 2:} & x_{2c}(X,t) = X + 2g^*(t - t_0)\,, & X = x_0\,, \qquad\qquad (5.2.36) \\
\text{domain 3:} & x_{3c}(X,t) = X - 2g^*(t - t_0)\,, & X = x_0\,.
\end{array}$$

We next note that as t increases, only the discontinuities $x_{1S}(t)$ and
$x_{2S}(t)$ receive information from the initial data line $t = 0$ through the
characteristics $x_{1c}(t)$. The third discontinuity line $x_{3S}(t)$ receives it from
the two other $x_{1S}(t)$ and $x_{2S}(t)$ through characteristics $x_{2c}(t)$ and $x_{3c}(t)$
respectively. Therefore the values of $g(x,t)$ calculated at $t = t^*$, say
in the neighborhood of $x_{3S}(t)$, only depend on values of g calculated at
$t \geq t^*$. Consequently, if t represents a physical quantity such as time,
we are facing the unacceptable situation where the causality principle
is violated because the configuration of the system depends on its own
future. The discontinuity line $x_{3S}(t)$ is not admissible for it cannot be
constructed from any physically realistic process. We deduce that an
admissible discontinuity line should satisfy here the geometrical condition

$$\left(\frac{dx_c}{dt}\right)_1 > \frac{dx_s}{dt} > \left(\frac{dx_c}{dt}\right)_2. \qquad\qquad (5.2.37)$$

Such an admissible discontinuity can be called a shock. Criterion (5.2.37)
can be stated as follows (Lax, 1973):
> "The characteristics starting on either side of the discontinuity
> curve when continued in the direction of increasing t intersect
> the line of discontinuity."

This also amounts to saying that the shock velocity is greater than the
information-transmission velocity in the domain ahead of the shock, and
lower than the information transmission-velocity in the domain behind
the shock. The above criterion is not restricted to the example dealt with
in this paragraph but applies equally in more general situations. In the
case of a shock propagating in compressible media, *Lax criterion* (5.2.37)
is strictly equivalent to the *subsonic–supersonic* condition for the shock
velocity so that it is compatible with the general principle of increase
of entropy from which the subsonic-supersonic condition for shocks of
arbitrary strength is inferred (Fowles, 1975).

5.2.2 The piston problem

We consider a column of compressible fluid, for example contained in
a pipe, of semi-infinite length along the x-axis. We assume the pipe
closed at the left end by a piston that smoothly accelerates, to the right,
into the fluid. We further assume that the fluid is at rest and uniform

before the piston starts moving. This motion transmits to the fluid small
compressional waves that originate from the piston face and advance
through the medium already set in forward motion by the preceding
waves. In most materials, the velocity of such small compressional waves
increases as the density increases; consequently, the waves emitted at
the piston face tend to overtake and reinforce their predecessors. The
spatial profiles of particle velocity, density and pressure between the
piston and the foremost point of the perturbation front steepen with
time so that, neglecting such dissipative processes as viscous transfer
and heat conduction, they eventually admit a segment with an infinite
slope and a shock front forms (Section 5.2.1). The *second principle of
thermodynamics* implies that the shock velocity is supersonic with respect
to the fluid ahead of its front and subsonic with respect to the fluid
behind it (Fowles, 1975). Accordingly, if the piston accelerates or keeps
a constant velocity so does the shock front. If the piston decelerates
however, the small waves sent forward are *rarefaction waves*. Here again,
in classical materials and just as for compressional waves, rarefaction
waves overtake and weaken the shock front. If the deceleration of the
piston persists, the shock keeps being eroded until its velocity becomes
sonic: the shock has disappeared and transformed back into a small
acoustic wave.

Let ρ, p, u and e denote the specific mass, the pressure, the particle
velocity relative to a laboratory-fixed frame and the specific internal
energy, respectively. We will use the sub/superscript "0" to denote the
initial state. Since we consider a nonreactive flow the energy is a function
of only two independent thermodynamic variables so that the equation
of state of the fluid can be equivalently written as

$$e = e(p, \rho) \qquad \text{or} \qquad e = e(p, S) \qquad \text{or} \qquad e = e(\rho, S) \,. \qquad (5.2.38)$$

It is convenient to introduce the sound speed c, defined by

$$c^2 = \left.\frac{\partial p}{\partial \rho}\right|_S = \frac{\frac{p}{\rho^2} - \left.\frac{\partial e}{\partial \rho}\right|_p}{\left.\frac{\partial e}{\partial p}\right|_\rho} \,. \qquad (5.2.39)$$

Upon neglecting diffusive transports (inviscid fluid flow), the conservation
equations then reads

$$\frac{\partial \mathbf{h}(\mathbf{g})}{\partial t} + \frac{\partial \mathbf{f}(\mathbf{g})}{\partial x} = 0 \,,$$

where

$$\mathbf{g} = (\rho, u, p) \,, \qquad (5.2.40)$$

$$\mathbf{h} = \left(\rho, \rho u, \rho \left(\tfrac{1}{2}u^2 + e(p, \rho)\right)\right), \tag{5.2.41}$$

$$\mathbf{f} = \left(\rho u, p + \rho u^2, \rho u \left(\tfrac{1}{2}u^2 + e(p, \rho) + p/\rho\right)\right). \tag{5.2.42}$$

The solution to this system must be subjected to the boundary conditions at the piston face and the initial conditions. Denoting the piston trajectory in the (x, t) plane by $x_\mathrm{p}(t)$ we have

$$u\left(x = x_\mathrm{p}(t), t = \tau\right) = \frac{dx_\mathrm{p}}{d\tau}. \tag{5.2.43}$$

Since the fluid is initially at rest we also have

$$u_0 = u(x, t = 0) = 0. \tag{5.2.44}$$

The system of conservation laws (5.2.40–42), which is a vectorial extension of the scalar equation (5.2.22), implies that the entropy of any fluid particle is a constant (local conservation of entropy). Indeed (5.2.40) and (5.2.41) can be used to transform (5.2.42) into

$$\frac{d^0 e}{dt} = -p \frac{d^0 V}{dt}, \tag{5.2.45}$$

where V and $d^0 g_i/dt$ denote the specific volume $(1/\rho)$ and the directional derivative of g_i along the path $x_0(t)$ of a fluid particle in the (x, t) plane, respectively:

$$\frac{d^0 g_i}{dt} = \frac{\partial g_i}{\partial t} + u \frac{\partial g_i}{\partial x}, \qquad \frac{dx_0}{dt} = u. \tag{5.2.46}$$

(5.2.45) can also be directly derived by applying the first principle of thermodynamics to an inviscid fluid particle in adiabatic evolution. Since, in addition, the fluid is nonreactive, the total differential of (5.2.38) (Gibbs' identity) reduces to

$$de = T\, dS - p\, dV. \tag{5.2.47}$$

Restricting the variations in (5.2.47) to a particle path and comparing the result with (5.2.45) finally yields

$$\frac{d^0 S}{dt} = 0, \tag{5.2.48}$$

which shows that a fluid particle carries a constant specific entropy. One observes that (5.2.48), (5.2.38) and (5.2.39) imply that e, p and c only depend on ρ along a *particle path*, outside of discontinuities. We now differentiate (5.2.38) and use (5.2.48) to transform (5.2.45) into

$$\frac{d^0 p}{dt} - c^2 \frac{d^0 \rho}{dt} = 0 \quad \text{or} \quad \frac{\partial p}{\partial t} + u \frac{\partial p}{\partial x} + \rho c^2 \frac{\partial u}{\partial x} = 0. \tag{5.2.49}$$

System (5.2.40–42) can thus be rewritten as

$$\frac{\partial \mathbf{g}}{\partial t} + \mathbf{C}(\mathbf{g})\frac{\partial \mathbf{g}}{\partial x} = 0, \qquad \mathbf{C}(\mathbf{g}) = \begin{pmatrix} u & \rho & 0 \\ 0 & u & 1/\rho \\ 0 & \rho c^2 & u \end{pmatrix}. \qquad (5.2.50)$$

The system (5.2.50) is the vectorial generalization of (5.2.2). It is hyperbolic because the matrix $\mathbf{C}$ has real and distinct *eigenvalues* u, $u+c$, $u-c$. Denoting by $d^+ g_i/dt$ and $d^- g_i/dt$ the directional derivatives of g_i along the curves $x_+(t)$, and $x_-(t)$ of the (x,t) plane defined by

$$\frac{dx_+}{dt} = u + c \qquad \text{and} \qquad \frac{dx_-}{dt} = u - c, \qquad (5.2.51)$$

the system (5.2.50) is found equivalent to

$$\frac{d^0 S}{dt} = 0, \qquad (5.2.52)$$

$$\frac{d^+ u}{dt} + \frac{1}{\rho c}\frac{d^+ p}{dt} = 0, \qquad (5.2.53)$$

$$\frac{d^- u}{dt} - \frac{1}{\rho c}\frac{d^- p}{dt} = 0, \qquad (5.2.54)$$

where the entropy S can be read as a specified function of p and ρ (cf. (5.2.38)). Equations (5.2.52–54) and curves $x_0(t)$, $x_+(t)$ and $x_-(t)$ are respectively the characteristic equations and the characteristic curves of our piston problem. $x_+(t)$ and $x_-(t)$ represent the locus in the (x,t) plane of the forward-facing and backward-facing acoustic waves and are commonly referred to as the $(\mathbf{C}^+)$ and $(\mathbf{C}^-)$ characteristics. The classical model for the shock formation that we present here makes use of the assumption that the entropy is uniform throughout the domain between the piston and the foremost point of the perturbation front. Actually it is enough to assume $\partial S/\partial x = 0$ at $t = 0$, so that $\partial S/\partial x = 0$ for $x_p(t) < x < c_0(t - \tau)$. This implies that entropy is constant along any curve of the (x,t) plane:

$$\frac{d^i S}{dt} = 0, \qquad i = 0, +, -, \ldots, \qquad (5.2.55)$$

which, together with (5.2.38) shows that p, e and c only depend on ρ along the characteristics. Therefore (5.2.52) is identically satisfied and equations (5.2.53) and (5.2.54) can be integrated into the form

$$u + I(\rho) = -2r_+(X) \qquad \text{along} \qquad \frac{dx_+}{dt} = u + c(\rho), \quad (5.2.56)$$

$$u - I(\rho) = 2r_-(Y) \qquad \text{along} \qquad \frac{dx_-}{dt} = u - c(\rho), \quad (5.2.57)$$

where

$$I(\rho) = \int \frac{dp}{\rho c} = \int c^2(\rho) \frac{d\rho}{\rho}. \qquad (5.2.58)$$

The quantities r_+ and r_- are called the *Riemann invariants*. They are constant along their respective characteristics but vary from one to another. The parameters X and Y are labels attached to those lines. We now recall the initial condition that the medium is at rest and uniform before being reached by the perturbation front so that all the C^- characteristics originate from a constant state $(u(x,t) = u_0 = 0,$ $\rho(x,t) = \rho_0)$ and carry the same value $-2r_0 = I(\rho 0)$. Such a flow is called a simple wave. We then infer from (5.2.56) that u and ρ, hence p, e and c, are constant along the same C^+ characteristic. We deduce

$$u(X) = r_0 - r_+(X), \qquad I(\rho(X)) = -(r_0 + r_+(X)). \qquad (5.2.59)$$

These equations imply that the C^+ characteristics are straight lines in the (x,t) plane... Since all the C^+ lines originate from the piston face, the parameter X can be conveniently chosen as the piston position at time τ, $X = x_p(\tau)$, so that the parametric equation for the C^+ lines is

$$x_+(\tau, t) = x_p(\tau) + (t - \tau)(u(x_p(\tau)) + c(\rho(x_p(\tau)))). \qquad (5.2.60)$$

We finally write the solution to system (5.2.40–42) that satisfies the boundary condition at the piston face and the initial condition as

$$u(x, t) = \frac{dx_p}{d\tau}(\tau) \qquad (5.2.61)$$

along lines $x = x_+(\tau, t)$. When the expression for the equation of state is specified, we can use (5.2.38) and (5.2.39) and the isentropy assumption to express c in terms of ρ. Then we can employ (5.2.59) to obtain the sound speed as an explicit function of $X = x_p(\tau)$ along any C^+ characteristic. Following the procedure used in paragraph 5.2.1, we now look for the conditions for a shock to be formed and consequently write the conditions for the C^+ characteristics to intersect. This condition is here

$$\left.\frac{\partial x_+}{\partial \tau}\right|_t = 0. \qquad (5.2.62)$$

Expression (5.2.60), once subjected to the latter condition, yields the following parametric equation for the possible locus in the (x, t) plane of the envelope formed by the intersections of C^+ characteristics:

$$t_e(\tau) = \tau + \frac{c(\dot{x}_p(\tau))}{\Gamma(\tau)\ddot{x}_p(\tau)}, \qquad x_e(\tau) = x_+(\tau, t = t_e(\tau)), \qquad (5.2.63)$$

where $\dot{x}_p$ and $\ddot{x}_p$ denote the first and second derivatives of $x_p(\tau)$, respectively. Physical values for $t_e(\tau)$ exist when $t_e > \tau$, i.e., $\Gamma\ddot{x}_p > 0$.

The quantity Γ, defined by $d(u+c)/du$, is often referred to as "the fundamental derivative of gasdynamics" (Thomson, 1971). Using (5.2.54) and (5.2.39) we obtain

$$\Gamma = \frac{d(u+c)}{du} = \frac{V^3}{c^2}\frac{\partial^2 p}{\partial V^2}\bigg|_S = \frac{\rho}{c}\frac{\partial c}{\partial \rho}\bigg|_S + 1. \tag{5.2.64}$$

Therefore we finally deduce that, when the acceleration $\ddot{x}_p$ of the piston is positive, an envelope can exist if and only if

$$\Gamma > 0 \qquad \text{i.e.,} \qquad \frac{\partial^2 p}{\partial V^2}\bigg|_S > 0. \tag{5.2.65}$$

For the overwhelming majority of compressible materials, Γ is positive, and compression shocks form in such fluids pushed by a piston. However there are non classical materials such as n-decane $C_{10}H_{22}$ (Lambrakis and Thomson, 1972) or some fluorocarbons (Cramer, 1989), for which a $(p\text{–}V)$ domain exist where Γ is negative. In this case, rarefaction shocks may form in fluids pulled by a piston. The importance of condition (5.2.65), known for a long time (Duhem, 1909), is not restricted to shock problems. More details can be found in Thomson (1971) and Thomson and Lambrakis (1973). Condition (5.2.65) implies that the forward-facing acoustic perturbations generated at the piston face catch up with their predecessors and that the sound speed increases with density. To put things more clearly, the condition for compressive shock formation is not that the sound speed c be an increasing function of the density ρ (or the pressure p) but that the forward-facing acoustic wave speed $u+c$ be an increasing function of the particle velocity u (or ρ or p). Indeed, definitions (5.2.65) show that c increases (decreases) with ρ when Γ is greater (smaller) than 1 but that $u+c$ increases (decreases) with u (or ρ or p) when Γ is greater (smaller) than 0. Also, note that differentiating (5.2.61) at fixed t with respect to x leads to

$$\frac{\partial u}{\partial x} = \frac{d\dot{x}_p}{d\tau}\bigg/\frac{\partial x_+}{\partial \tau} \qquad \text{or} \qquad \frac{\partial u}{\partial x} = \frac{\ddot{x}_p}{(t-\tau)\Gamma\ddot{x}_p - c}, \tag{5.2.66}$$

which shows that the gradient of u tends to infinity when (5.2.63) is met, i.e., when t tends to $t_e(\tau)$. Thus, when condition (5.2.65) is satisfied, we can conclude that

- if the initial acceleration of the piston vanishes (i.e., $\ddot{x}_p(0) = 0$), the C^+ characteristic $x_+(0,t)$ belongs to the envelope at an infinite time (i.e., $t_e(0) \to \infty$);
- if the piston acceleration is strictly positive for all positive times (i.e., $\ddot{x}_p(\tau) > 0$, for any $\tau > 0$) then the time $t_e(\tau)$ at which the characteristic

$x_+(\tau, t)$ belongs to the envelope increases indefinitely as τ tends to infinity;

• if $\ddot{x}_p$ vanishes at a time $\tau = \tau_1 > 0$, $t_e(\tau)$ tends to infinity as $\tau \to \tau_1$.

In any case, the function $t_e(\tau)$ has a minimum t_B for some value of τ: the shock forms at the point $(x_+(t_B, t_B = t_e(t_B)), t_B)$. Using the same technique as in paragraph 5.2.1, one can show that the shock velocity D is equal to

$$D = \frac{[f]}{[h]} = \frac{[\rho u]}{[\rho]} = \frac{[p + \rho u^2]}{[\rho u]} = \frac{\left[\rho \left(\frac{1}{2}u^2 + e + p/\rho\right)\right]}{\left[\frac{1}{2}u^2 + e\right]}, \qquad (5.2.67)$$

where $[g]$ denotes the jump in g through the shock. Relations (5.2.67) also apply for multidimensional curved shocks in which case case u and D must be replaced by $\mathbf{u} \cdot \mathbf{n}$ and $\mathbf{D} \cdot \mathbf{n}$ respectively, the tangential components of $\mathbf{u}$ being conserved. Clearly, the relations (5.2.67) are the conservation laws through a discontinuity introduced in Section 2.

5.2.3 Comments

Throughout Section 5.2.2, the flow between the piston face and the perturbation front has been considered isentropic; by contrast, the transformation through a compressive shock front is not. We restrict our demonstration to the case where the shock has a finite, but still small, strength. We first recall the Hugoniot relation

$$e - e_i = \tfrac{1}{2}(p + p_i)(V_i - V) \qquad \text{or} \qquad \frac{-\delta e}{\delta v} = \frac{1}{2}(p + p_i), \qquad (5.2.68)$$

where δg denotes the small jump $g - g_i$ and the subscript i refers to the state of the medium in front of the shock. The states (i) and (0) may be different because the shock is not necessarily formed at the head of the perturbation front. We consider the pressure p as a function of S and V (cf. equation (5.2.38)). Since the variations δg are small, we can expand p in a Taylor series in δV and δS around p_i. We obtain

$$-\frac{\delta e}{\delta V} = \frac{1}{2}(p + p_i) = p_i + \frac{1}{2}\left.\frac{\partial p}{\partial V}\right|_S (\delta V) + \frac{1}{6}\left.\frac{\partial^2 p}{\partial V^2}\right|_S (\delta V)^2 \dots$$

$$+ \frac{1}{2}\left.\frac{\partial p}{\partial S}\right|_V (\delta S). \qquad (5.2.69)$$

We similarly expand the specific energy $e(S, V)$ in a Taylor series in δV and δS. Using Gibbs' identity (5.2.47), and anticipating that δS is to be very small, we find

$$-\frac{\delta e}{\delta V} = p_i + \frac{1}{2}\left.\frac{\partial p}{\partial V}\right|_S (\delta V) + \frac{1}{6}\left.\frac{\partial^2 p}{\partial V^2}\right|_S (\delta V)^2 \dots - T_i \frac{\delta S}{\delta V}. \qquad (5.2.70)$$

Specializing (5.2.69) and (5.2.70) and neglecting high powers of δV, we have

$$\delta S = S - S_i = -\frac{1}{12 T_i} \frac{\partial^2 p}{\partial V^2}\bigg|_{S,i} (\delta V)^3 = \frac{c_i^2}{4 T_i} \Gamma_i (\delta p)^3, \tag{5.2.71}$$

which shows that a weak compressive shock wave ($\delta V < 0$, $\delta p > 0$) produces a third order increase of *entropy* ($\delta S > 0$) when Γ_i, i.e., $\partial^2 p/\partial V^2|_{S,i}$ is positive. Conversely, (5.2.71) also shows that if Γ_i is positive (negative), then a small-strength shock is compressive (expansive) because any physically admissible transformation is such that the associated entropy change is positive.

We finally look for an approximation for the velocity of a weak shock wave, for example the velocity of the shock just after its formation by an accelerating piston. The procedure is similar to that used in Section 5.2.1. We transform the general expressions (5.2.67) for the shock velocity and obtain

$$\frac{D - u}{V} = \frac{D - u_i}{V_i}, \qquad p + \left(\frac{D - u}{V}\right)^2 = p_i + \left(\frac{D - u_i}{V_i}\right)^2, \tag{5.2.72}$$

which lead to

$$-\frac{p - p_i}{V - V_i} = \left(\frac{D - u}{V}\right)^2 = \left(\frac{D - u_i}{V i}\right)^2. \tag{5.2.73}$$

Since we consider small changes, the left hand side of (5.2.73) can be expanded as

$$-\frac{p - p_i}{V - V_i} \equiv -\frac{\delta p}{\delta V} = -\frac{\partial p}{\partial V}\bigg|_{S,i} - \frac{1}{2} \frac{\partial^2 p}{\partial V^2}\bigg|_{S,i} (\delta V) + \cdots$$

or

$$-\frac{\partial p}{\partial V} = -\frac{\partial p}{\partial V}\bigg|_S + \frac{1}{2} \frac{\partial^2 p}{\partial V^2}\bigg|_S (\delta V) + \cdots, \tag{5.2.74}$$

where the term $(\delta S/\delta V) = \mathcal{O}((\delta V)^2)$ was neglected (cf. (5.2.71)). We then recall the definition (5.2.39) and use (5.2.73) to obtain

$$(D - u_i)^2 = c_i^2 - \frac{1}{2} V_i^2 \frac{\partial^2 p}{\partial V^2}\bigg|_{S,i} (\delta V) + \cdots \tag{5.2.75a}$$

and

$$(D - u)^2 = c^2 + \frac{1}{2} V^2 \frac{\partial^2 p}{\partial V^2}\bigg|_S (\delta V) + \cdots \tag{5.2.75b}$$

Adding the two equations yields the approximation

$$D = \tfrac{1}{2}((u+c) + (u_i + c_i)) + o(\delta V).\tag{5.2.76}$$

Relation (5.2.76) says that the velocity of a weak shock wave is the mean of the forward facing acoustic wave velocities on each side of the shock surface. We also infer from (5.2.75) that the velocity of a weak compressive shock wave satisfies the inequalities

$$u + c > D > u_i + c_i \qquad \text{when} \qquad \left.\frac{\partial^2 p}{\partial V^2}\right|_S > 0.\tag{5.2.77}$$

We emphasize that the first of inequalities (5.2.77) also holds for shocks of arbitrary strength (Fowles, 1975). The demonstration does not require the second inequality (5.2.77) (i.e., (5.2.65)) to be satisfied and only uses the principle of increase in entropy during the shock transformation and the classical thermodynamic stability conditions $\partial p/\partial V|_S < \partial p/\partial V|_T < 0$.

So far, the shock waves have been viewed as surfaces of discontinuity embedded in perfect fluids. This is only an approximation and it is now time to wonder how the existence of nonzero heat conductivity and viscosities change the picture. For the sake of brevity we actually only address the following question : what is the characteristic thickness (ℓ) of a steady, planar shock front propagating in a gas at slightly supersonic velocity D?

Anticipating that the gas evolution inside the shock front is nearly isentropic (Landau and Lifshitz, 1987) one approximates dp/dx by $c^2 d\rho/dx$. Once the continuity equation $\rho u = \text{const.}$ is made use of, the momentum balance normal to the shock acquires the form:

$$\rho u\left(1 - \frac{c^2}{u^2}\right)\frac{du}{dx} = \mu\frac{d^2 u}{dx^2}\tag{5.2.78}$$

in a shock-fixed frame of reference. Inside the shock front $1 - c^2/u^2$ is $\mathcal{O}(1 - c_0/D)$ so that ℓ is estimated from (5.2.78) as $\mu/\rho u(D/c_0 - 1)$. We next recall that, in gases $\mu/\rho = \mathcal{O}(c_0^2 t_{\text{coll}})$, where t_{coll} is the representative collision time between molecules. Therefore, since $u = \mathcal{O}(c)$, we find that

$$\ell = \mathcal{O}\left(\frac{c_0 t_{\text{coll}}}{D/c_0 - 1}\right).\tag{5.2.79}$$

In other words, only the inner structure of very weak shocks ($D \simeq c_0$) may be described by the Navier–Stokes equations. Otherwise ($D/c_0 - 1 = \mathcal{O}(1)$) their thickness gets comparable to the mean free path ($\sim c_0 t_{\text{coll}}$) and even describing the gas as a continuous medium is not appropriate any longer.

5.3 Majda and Rosales' model problem

Before handling the problem of two-dimensional stability of a plane shock wave (cf. Section 5.4), we found it convenient to devote a few lines to a model system imagined by Majda and Rosales (1983) which will serve to introduce some of the oddities of multidimensional acoustics. To this end we consider the scalar $u(t, x, z)$ which obeys

$$\frac{\partial^2 u}{\partial t^2} = \nabla^2 u \qquad (5.3.1)$$

in terms of dimensionless variables $t(> 0)$, $x(> 0)$ and z (any real). This wave equation is endowed with the boundary condition:

$$\frac{\partial u}{\partial t} + \frac{1}{H}\frac{\partial u}{\partial x} = 0 \qquad \text{at} \quad x = 0 \qquad (5.3.2)$$

in which H is a control parameter. Our present task is to get insights into the behavior of u for $t > 0$. To this end we remark that $u(t, x, z)$ has a compact support in the (x, z) plane if $u(0, x, z)$ meets this condition (the u-signals travel at finite speed). It is then convenient to define the norm $N(t)$ of u by:

$$2N(t) = \int_0^\infty dx \int_{-\infty}^\infty dz \left(\left(\frac{\partial u}{\partial t}\right)^2 + \nabla u \cdot \nabla u \right), \qquad (5.3.3)$$

When N is finite, its time derivative satisfies

$$\frac{dN}{dt} = \iint dx\, dz\, \nabla \cdot \mathbf{W}, \qquad \mathbf{W} = \left(\frac{\partial u}{\partial t}\frac{\partial u}{\partial x}, \frac{\partial u}{\partial t}\frac{\partial u}{\partial z} \right). \qquad (5.3.4)$$

Using the flux-divergence theorem, then (5.3.4), one has:

$$\frac{dN}{dt} = \frac{1}{H} \int_{-\infty}^\infty \left(\frac{\partial u}{\partial t}\right)^2 dz. \qquad (5.3.5)$$

Accordingly,

- for $H < 0$, u decreases in amplitude, and the system (5.3.1–2) is stable against all localized disturbances;
- for $H > 0$, some energy is injected in the $x > 0$ domain through the boundary $x = 0$.

To go further, we now perform a normal mode analysis and seek elementary solutions to (5.3.1–2) in the form :

$$u = f(x) \exp(st + ikz), \qquad k \text{ real}. \qquad (5.3.6)$$

Upon the *assumption of boundedness* at $x = \infty$, one finds f so that

$$u = \exp\left(st + ikz - x\sqrt{s^2 + k^2} \right) \qquad (5.3.7)$$

from the wave equation. The boundary condition (5.3.2) then imposes

$$s = \frac{1}{H}\sqrt{s^2 + k^2}, \tag{5.3.8}$$

which is the sought-after dispersion relation to be studied with the constraint that $(s^2 + k^2)^{1/2}$ has a positive real part. One then deduces that

$$s^2 = k^2/\left(H^2 - 1\right), \tag{5.3.9}$$

whereby $s = |k|\sqrt{H^2 - 1}$ is an adequate root when $H > 1$. In other words, when $H > 1$ the system is unstable, consistently with what (5.3.5) predicts. The last question to be answered of course is "what does happen when $0 < H < 1$?" Certainly the boundary injects some energy in the system ($H > 0$) but the resulting u does not decay when $x \to \infty$. This led Majda and Rosales to look for elementary solutions in the form:

$$u = f(x)exp(i(kz - \omega t)) \qquad \text{with} \qquad \omega, k \text{ real}. \tag{5.3.10}$$

From the wave equation one deduces that:

$$f(x) = \exp\left(\pm ix\sqrt{\omega^2 - k^2}\right) \tag{5.3.11}$$

and the boundary condition (5.3.2) yields:

$$-i\omega H \pm i\sqrt{\omega^2 - k^2} = 0. \tag{5.3.12}$$

For ω to be real, $\omega^2 > k^2$ is needed. Solving (5.3.12) for ω gives $\omega = \pm|k|\sqrt{1 - H^2}$, which is indeed an adequate answer when $0 < |H| < 1$. When z is fixed, u acquires the form:

$$u = g(z)\exp\left(\pm i\sqrt{\omega^2 - k^2}\, x \pm i\omega t\right)$$

and the group velocity $\partial(\pm\omega)/\partial\sqrt{\omega^2 - k^2} = H$ along the x-axis is positive when $H > 0$, so that energy is radiated from the boundary. Therefore the system (5.3.1–2) admits the so-called *radiating boundary modes* in the range $0 < H < 1$ which, in a sense, correspond to emission of sound by the boundary. The situation is similar to what a supersonic stream flowing past a sinusoidally corrugated wall would produce. On the other hand, for a fixed x, e.g., $x = 0$, $u(t, x, z)$ has the form

$$u = f(0)\exp\left(i\left(kz \pm |k|t\sqrt{1 - H^2}\right)\right), \tag{5.3.13}$$

which corresponds to transverse traveling waves that are neither damped nor amplified; the corresponding transverse sound speed $c_{||} = \sqrt{1 - H^2}$. Such traveling waves have already been encountered when studying the diffusive instabilities of flames (Section 4.5) but only when the control parameter (le in that case) took on a very special value (le $= 32/3$);

here they are allowed *over a whole range* of H which is adjacent to the unstable range ($H > 1$) predicted by the normal mode analysis. Similar phenomena will be encountered when studying shock waves (Section 5.4). Note that $\phi(t, z) \equiv u(t, 0, z)$ satisfies the transverse wave equation:

$$\frac{\partial^2 \phi}{\partial t^2} = c_\|^2 \frac{\partial^2 \phi}{\partial z^2}. \qquad (5.3.14)$$

5.4 D'yakov–Kontorovich's instabilities

The Hugoniot curve has not always as regular a shape as that sketched in Fig. 5.6 (left): discontinuities, bumps, inflection points or slope changes of sign are sometimes found, for example in some porous materials or when the intrinsic properties of the material upon consideration allow for elastic failure or phase transformation (Fig. 5.6, right).

Theoretically, shock states defined by points that belong to some segment of such a nonregular Hugoniot curve exhibit interesting dynamical features: a (plane) shock wave can split into several other stable waves (*break-up*) or its front surface can undulate indefinitely (emission of sound). The conditions for occurrence of these phenomena can be obtained by applying to a shock discontinuity the so-called theory of decay and branching of discontinuities (DBD theory, Kuznetsov, 1985; Landau and Lifshitz, 1987). A straightforward normal mode approach, similar to what was employed for flames (cf. Section 4.5–4.10), can also be used, which defines conditions under which fortuitous small wrinkles on the shock wave front would grow exponentially with time (linear instability limits) or persist without being either damped or amplified (*neutral radiating modes*). Indeed these conditions are the same as the ones derived with the DBD theory. The normal mode approach

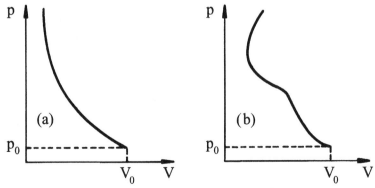

Fig. 5.6.   Regular (left) and nonregular (right) Hugoniot curves.

however does not allow one to identify which physical events should ultimately be connected to a violation of the linear stability limits. It nevertheless provides a systematic way for defining all possible types of linear perturbations present in compressible inviscid nonreactive media and is also a prelude to linear treatments of detonation instability problems.

5.4.1 The normal mode analysis for step-shock instability

This linear analysis deals with the problem of the corrugation (transverse) instability of a plane shock front associated with a one-dimensional, plane, constant velocity, nonreactive shock wave. Such a wave can only be induced by a constant velocity supporting piston. Consequently, the reference flow between this piston and the unperturbed shock front is constant and uniform (cf. Section 5.2.3). This also means that the particle velocity equals the piston velocity. When the latter is specified one can derive the shock front velocity from the jump relations (5.2.67) and the material's equation of state. The wave may be assumed to travel leftward in a laboratory-fixed frame, where the unshocked medium is at rest, so that the particle velocity is positive and constant in a frame attached with the unperturbed shock. The infinitesimal three-dimensional perturbations under consideration over this one-dimensional, time-independent reference flow are to be supposed harmonic with respect to the two coordinates (z) transverse to the direction of propagation. Therefore, it is always possible to find a rotation in the z-plane that transforms the three-dimensional unsteady equations into two-dimensional ones (the so-called *Squire transformation*). We let ρ, u, v, p and t denote the specific mass, the horizontal and vertical particle velocity components relative to a nonperturbed, shock-fixed frame, the pressure, the time respectively; x is the distance between the perturbed shock front and a given fluid particle (Fig. 5.7).

The reference flow (overbar) is then defined by the three dependent variables $\bar{g} = (\bar{\rho}, \bar{u}, \bar{v} \equiv 0, \bar{p})$ and is described by the three following conservation laws

$$
\begin{aligned}
\text{mass} \qquad & \bar{\rho}\bar{u} = M_1, \\
\text{momentum} \qquad & \bar{p} + \bar{\rho}\bar{u}^2 = M_2, \\
\text{energy} \qquad & \bar{h} + \tfrac{1}{2}\bar{u}^2 = M_3.
\end{aligned} \qquad (5.4.1)
$$

(The latter system is also valid for any one-dimensional, plane, steady, adiabatic flow of a reactive inviscid fluid, cf. Section 6.5). The enthalpy h can be considered as a function of p and ρ when the medium is nonreactive (incomplete equation of state). The M_i, $i = 1, 3$ are constant quantities

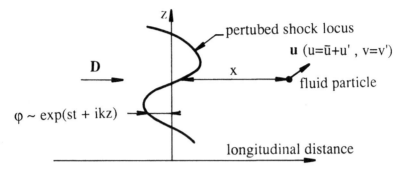

Fig. 5.7. Notations for step-shock instabilities.

that are specified here by the boundary conditions (the jump relations) at the shock front ($M_1 = \rho_0 \bar{D}$, $M_2 = p_0 + \rho_0 \bar{D}^2$, $M_3 = h_0 + \frac{1}{2}\bar{D}^2$). The subscript "0" denotes the medium ahead of the shock front and $\bar{D}$ is the constant velocity of the relative flow before it enters the unperturbed shock front. The normal mode analysis can then be formulated as a classical eigenvalue problem and proceeds in three steps.

The first step solves the linearized equations that describe the perturbed relative flow, between the shock front and the piston, so as to identify the different modes. The two-dimensional, unsteady, nonreactive, perturbed flow is defined by the four dependent variables $\mathbf{g} = (\rho, u, v, p)$ and is described by the four *Euler* conservation equations (cf. Section 2.2). Linearizing these equations around a constant state in space and time according to the rule:

$$\mathbf{g} = \bar{\mathbf{g}} + \phi \mathbf{g}' \, ; \qquad (\mathbf{g}', \bar{\mathbf{g}}) = \mathcal{O}(1) \, , \quad \phi = \mathcal{O}(\epsilon) \, , \quad \epsilon \ll 1, \qquad (5.4.2)$$

where $\phi(t, z)$ denotes the perturbed shock position, leads to a set of differential equations with constant coefficients for $\mathbf{g}'$. This set consequently admits elementary solutions of the form

$$\mathbf{g}'(x) = \mathbf{g}'' \exp(qx) \, , \quad \phi = \epsilon \exp(st + ikz) \, , \quad \mathbf{g}'' = \mathcal{O}(1) \text{ constant} \quad (5.4.3)$$

which, after insertion, yields a linear homogeneous system of four equations for the four perturbations $\mathbf{g}'$. Setting the determinant of the latter system to zero to discard the trivial solution identifies three roots which, in turn, define four independent particular solutions $\mathbf{g}'_j$, $j = 1, 4$. The general perturbation is some linear combination, yet to be determined, of these four particular solutions:

$$\mathbf{g}'(x) = \sum_{j=1}^{4} C_j \mathbf{g}''_j \exp(q_j x) \, . \qquad (5.4.4)$$

The second step introduces the shock boundary conditions in the problem. Upon linearization, the jump relations provide the value at the perturbed shock front (i.e., at $x = \phi$) of the general perturbation, g'_H (Fig. 5.7), which can be conveniently written as

$$(\phi g')_H = \phi g'_H, \quad g'_H = -sB_H h, \tag{5.4.5}$$

where B_H is a scalar function of the reference shock state and the h_j, $j = 1, 4$ are functions of s, k, and the reference shock state. Identifying (5.4.5) with (5.4.4), we obtain a linear non-homogeneous system of four equations in the four unknowns C_j, $j = 1, 4$, where the columns of the matrix are the g''_j, $j = 1, 4$, and the inhomogeneous term is g'_H. The (linear) instability then appears when perturbations exist which grow exponentially with time, that is when $\mathcal{R}e(s) > 0$. Solutions of the form (5.4.3), however, are physically valid only if they remain spatially bounded, in particular at downstream infinity. In other words, such solutions must satisfy a *boundedness condition*. This can be shown to be implied by the causality principle which states here that the only source of instability is the shock wave itself. In other words, the physically admissible solution must fulfill the double condition

$$\mathcal{R}e(q) < 0 \quad \text{when} \quad \mathcal{R}e(s) > 0. \tag{5.4.6}$$

One of the four particular solutions g'_j is thus discarded, for it would describe a perturbation that would grow exponentially both with t and x. The constant, say C_4, associated with this particular solution must therefore be set to zero. We then are left with a linear homogeneous system of four equations for the four unknowns C_j, $j = 1, 3$ and s (or sB_H), where the columns of the matrix are the g''_j, $j = 1, 3$ and $B_H h$ (or h). As before, the determinant of this system must be set to zero in order to avoid a trivial solution. This operation leads to the dispersion relation $s(k)$ of the problem. The growth/decay rate $s(k)$ thus appears as the eigenvalue of the problem expressing the compatibility of the solutions of the perturbed-flow equations, subjected to the causality principle, with the boundary conditions at the perturbed shock front.

The third step is the analysis of the dispersion relation and leads to the linear instability and emission conditions in terms of the material properties and the shock velocity.

Let us now detail these steps. After linearization according to (5.4.2) around a one-dimensional, plane, constant state, the two-dimensional, unsteady nonreactive Euler equations yield

$$\frac{d\rho'}{dt} + \bar{\rho}\frac{\partial u'}{\partial x} + \bar{\rho}\frac{\partial v'}{\partial z} = 0,$$

$$\bar{\rho}\frac{du'}{dt} + \frac{\partial p'}{\partial x} = 0,$$

$$\bar{\rho}\frac{dv'}{dt} + \frac{\partial p'}{\partial z} = 0,$$ (5.4.7)

$$\frac{d}{dt}\left(p' - \bar{c}^2\rho'\right) = 0.$$

d/dt and $\bar{c}$ denote the total time particle derivative and the sound speed $(\partial p/\partial \rho)^{1/2}|_S$ respectively (S, the specific entropy).

The equation of state for our nonrelaxing medium may be written, for example, as $f(p, \rho, S) = 0$ so that we also have

$$p' - \bar{c}^2\rho' - \overline{\frac{\partial p}{\partial V}}\bigg|_\rho S' = 0 \qquad (5.4.8)$$

upon linearization. Since the coefficients of the derivatives in the system (5.4.7) are constant and uniform, one may look for elementary solutions in the form

$$\phi = \epsilon\exp(st + ikz),   \qquad (5.4.9)$$
$$\mathbf{g}'(x) = \mathbf{g}''\exp(qx),   \qquad (5.4.10)$$

where s, q and k denote the *complex frequency*, the complex longitudinal and transverse wavenumbers respectively. The transverse *wavenumber* k is a pure real number, for otherwise perturbations would grow exponentially as $z \to \pm\infty$. Such an event is not physically admissible here, because the reference flow is assumed laterally unbounded. After insertion of (5.4.10) in (5.4.7), one obtains the following homogeneous system

$$\begin{bmatrix} s + \bar{u}q & \bar{\rho}q & i\bar{\rho}k & 0 \\ 0 & \bar{\rho}(s + \bar{u}q) & 0 & q \\ 0 & 0 & \bar{\rho}(s + \bar{u}q) & ik \\ -\bar{c}^2(s + \bar{u}q) & 0 & 0 & s + \bar{u}q \end{bmatrix}\begin{bmatrix} \rho'' \\ u'' \\ v'' \\ p'' \end{bmatrix} = 0. \quad (5.4.11)$$

For the solution to this system not to be trivial its determinant must vanish, which leads to

$$(s + \bar{u}q)^2\left((s + \bar{u}q)^2 - \bar{c}^2\left(q^2 - k^2\right)\right) = 0. \qquad (5.4.12)$$

One may then distinguish among three roots q_i, $i = 1, 3$. Mode 1 is associated with the double root $q_1 = q_2 = -s/\bar{u}$ and describes a *contact discontinuity* also called here an *entropy-vortex wave*, carried along in the fluid with the relative particle velocity $\bar{u}$. Indeed, inspection of (5.4.11) shows that

$$p'' = 0 \quad \Rightarrow \quad \rho'' = \left.\overline{\frac{\partial\rho}{\partial S}}\right|_p S'',$$

$$\text{div}\,(\mathbf{u}') \sim q_1 u'' + ikv'' = 0 \quad \Rightarrow \quad \text{curl}\,(\mathbf{u}') \sim (q_1, k) \times \mathbf{u}'' \neq 0.$$

$|S'|$ is necessarily different from zero because entropy cannot be uniform behind a corrugated shock (Hayes, 1957). The latter solution is one-fold degenerate and can therefore be considered as a linear combination of the two following independent particular solutions, which form a basis of the corresponding eigenspace:

$$\begin{aligned}
\mathbf{g}_1'' &= (\rho_1'', u_1'', v_1'', p_1'') = \rho_1''(1, 0, 0, 0)\,,\\
\mathbf{g}_2'' &= (\rho_2'', u_2'', v_2'', p_2'') = u_2''\,(0, 1, -is/\bar{u}k, 0)\,.
\end{aligned} \tag{5.4.13}$$

Note that $\mathbf{g}_2''$ is identically zero when *transverse perturbations* are neglected $(k = 0 \Rightarrow u_2'' = 0)$ and that q_1 and q_2 satisfy the boundedness (or causality) principle (5.4.6). Modes 3 and 4 are associated with the two roots q_3 and q_4 of

$$(s + \bar{u}q)^2 = \bar{c}^2 \left(q^2 - k^2\right) \neq 0\,. \tag{5.4.14}$$

They define acoustic perturbations (sound waves) because system (5.4.10) and equation (5.4.8) show successively that

$$p'' - \bar{c}^2 \rho'' = 0 \quad \text{and} \quad S'' = 0\,.$$

The two corresponding independent particular solutions $\mathbf{g}_3''$ and $\mathbf{g}_4''$ can be written as

$$\begin{aligned}
\mathbf{g}_j'' &= (\rho'', u'', v'', p'')_j \\
&= u_j'' \left(-\frac{\bar{\rho}\,(s + \bar{u}q_j)}{q_j \bar{c}^2}, 1, \frac{ik}{q_j}, -\frac{\bar{\rho}\,(s + \bar{u}q_j)}{q_j}\right), \quad j = 3, 4\,.
\end{aligned} \tag{5.4.15}$$

Equations (5.4.13) and (5.4.15) define the four particular solutions of (5.4.11), the general solution of which is thus given by

$$\mathbf{g}'(x) = \sum_{j=1}^{4} C_j \mathbf{g}_j'' \exp\,(q_j x)\,. \tag{5.4.16}$$

The C_j, $j = 1, 4$ are arbitrary constants that can absorb ρ_1'', and $u_{2,3,4}''$. Introducing now the perturbed shock position

$$\phi = \epsilon \exp\,(st + ikz)\,, \quad \mathcal{I}m(k) = 0\,, \tag{5.4.17}$$

linearizing the jump relations (cf. Section 2.2), we obtain the shock value $\mathbf{g}_H'$ (subscript H) of the perturbation amplitudes, to be identified with expression (5.4.16) evaluated at $x = \phi$

$$\mathbf{g}'_H = -sB_H\mathbf{h}$$

$$\mathbf{h} = \left(\frac{2\rho_0\bar{D}H}{\bar{u}^2}, 1 - H, \bar{D}(1 + H)\frac{ik}{s}, -2\rho_0\bar{D}\right), \quad B_H = \frac{1 - \bar{u}_H/\bar{D}}{1 + H}.$$

$$(5.4.18)$$

The coefficient $-H$ is defined as the ratio of the slopes in the (p, V) plane $(V = \rho^{-1}$ the specific volume) of the *Rayleigh line*, $-(\rho_0 D)^2$, and the Hugoniot curve $(dp/dV)_H$ associated with the nonperturbed shock-state

$$H = (\rho_0\bar{D})^2 / \overline{(dp/dV)_H}.$$

$$(5.4.19)$$

A direct examination of the expressions for roots q_3 and q_4 reveals that either q_3 or q_4 violates causality. One thus sets C_3 or C_4 to zero in (5.4.16), say C_4, and leave $q = q_3$ or q_4 as it is in the solution to derive the dispersion relation. The analysis of this relation is then carried out conjointly with the causality principle (5.4.6) through some adequate change of variables. By identifying (5.4.16), where C_4 is set to zero, with the boundary condition (5.4.18), we obtain a linear homogeneous system for the C_j, $j = 1,3$ and $s\left[1 - (\bar{u}_H/\bar{D})\right]/\left[1 + H\right]$. The determinant of this system must vanish because the latter quantities are different from zero. This defines the dispersion relation in the form:

$$\begin{vmatrix} 1 & 0 & -\bar{\rho}\left(s + \bar{u}q\right)/q\bar{c}^2 & 2\rho_0\bar{D}H/\bar{u}^2 \\ 0 & 1 & 1 & 1 - H \\ 0 & -is/\bar{u}k & ik/q & \bar{D}(1 + H)ik/s \\ 0 & 0 & -\bar{\rho}\left(s + \bar{u}q\right)/q & -2\rho_0\bar{D} \end{vmatrix} = 0, \quad (5.4.20)$$

or explicitly

$$2\left(s/\bar{u}\right)\left(\left(s/\bar{u}\right)^2 - k^2\right) - \left(\left(s/\bar{u}\right) + q\right)\left(\left(s/\bar{u}\right)^2 - \left(\bar{D}/\bar{u}\right)k^2\right)\left(1 + H\right) = 0$$

$$(5.4.21)$$

in which the subscript "H" is omitted and where q is the root of (5.4.14) that satisfies (5.4.6). We recall that $\mathcal{I}m(k) = 0$. By using successively the transformations

$$q = (\omega/\bar{c})\cos\theta, \quad ik = (\omega/\bar{c})\sin\theta, \quad r\exp(i\psi) = \cotan(\theta/2),$$

$$Z = m\left(\frac{1 + \cos\theta}{1 - \cos\theta}\right)^\nu \quad \text{or} \quad \nu\cos\theta = \frac{Z - m}{Z + m}, \quad (5.4.22)$$

where $\nu = \pm 1$, $m = (1 - \bar{M})/(1 + \bar{M})$, and $\bar{M} = \bar{u}/\bar{c} < 1$ (Mach number of the relative unperturbed shocked flow), we can turn the dispersion relation into equivalent quadratic equations for $\nu\cos\theta$ and Z:

$$A(\nu\cos\theta)^2 - 2B(\nu\cos\theta) + C = 0, \quad (5.4.23a)$$

and

$$(A - 2B + C) \left(\frac{Z}{m}\right)^2 - 2(A - C) \left(\frac{Z}{m}\right) + (A + 2B + C) = 0, \quad (5.4.23b)$$

where

$$A = \frac{4\bar{M}^2}{1 + H},$$

$$B = \bar{M} \left(3 + \frac{\bar{M}^2}{1 + H}\right), \qquad (5.4.24)$$

$$C = 2 \left(1 + \frac{\bar{M}^2}{1 + H}\right) - \left(1 + \frac{\bar{M}^2 \bar{D}}{\bar{u}}\right).$$

The causality principle (5.4.6) can be rewritten as

$$\mathcal{R}e(k) \sin \psi > 0, \qquad (5.4.25)$$

so that the corresponding condition for linear instability is now

$$(\nu - \bar{M}) r^2 < \nu + \bar{M} \qquad \text{or} \qquad m\,r^{2\nu} < 1. \qquad (5.4.26)$$

Recalling that

$$r^2 = \left| r^2 \exp(2i\psi) \right| = \left(\cotan \frac{\theta}{2}\right)^2 = \frac{1 + \cos \theta}{1 - \cos \theta} \qquad (5.4.27)$$

we finally write the causal linear instability condition (5.4.26) as

$$|Z| < 1. \qquad (5.4.28)$$

Subjecting now the solutions of (5.4.23) to the latter condition reveals that the shock is linearly stable when

$$-1 < H < 1 + 2\bar{M}. \qquad (5.4.29)$$

These conditions correspond by their very nature to the general case where Z (i.e., θ, s and q) is a complex solution of (5.4.23) for which the amplitudes of small wrinkles on the shock front surface decay with time. These equations however, also have pure imaginary solutions that consequently describe perturbations, the amplitudes of which neither grow nor decrease with time. The ripples keep indefinitely emitting waves without being damped or amplified. These waves are sound waves since the wavenumber q in the dispersion relation is the solution of the acoustic mode equation (5.4.14). The variable θ in (5.4.22) then represents the angle between the sound wave vector (q, ik) and the normal to the unperturbed shock. Acoustic emission occurs if sound waves can leave the shock surface. In the unperturbed shock-fixed frame, this is expressed by

$$\bar{u} \pm \bar{c} \cos \theta > 0 \qquad \text{or} \qquad -\bar{M} < \nu \cos \theta < 1. \qquad (5.4.30)$$

Subjected to the latter constraints, the solution of (5.4.23) leads to the following conditions for sound emission to occur

$$\frac{1 - \bar{M}^2 \left(1 + V_0/\bar{V}\right)}{1 - \bar{M}^2 \left(1 - V_0/\bar{V}\right)} \equiv r_i < H < 1 + 2\bar{M} . \qquad (5.4.31)$$

Note that $-1 < r_i < 1$ (Fig. 5.8).

When (5.4.31) is met, the "growth" rate s is purely imaginary, viz.:

$$s = i\nu|k|c_\| , \qquad (5.4.32)$$

where $c_\|$ is given by:

$$c_\| = \frac{\bar{c}}{\sin \theta} \left(1 + \bar{M}\nu \cos \theta\right) . \qquad (5.4.33)$$

The very form of (5.4.32) could have been guessed beforehand: as the basic unperturbed shock does not bring in any reference length, repeating Landau's dimensional argument (Section 4.7.1) necessarily leads to a "growth" rate (with dimension (second)$^{-1}$) of the form (5.4.32). The apparent indeterminacy about the choice "$\nu = \pm 1$" merely reflects the invariance of the basic solution under the symmetry $z \to -z$ (it does not affect the first of (5.4.23), whose root is the whole grouping $\nu \cos \theta$). Furthermore, when (5.4.32) holds, the instantaneous shape of the corrugated shock front is given by :

$$x = \epsilon \exp \left(ik \left(z \pm c_\| t\right)\right) . \qquad (5.4.34)$$

When many transverse harmonics are involved, (5.4.34) must be generalized into:

$$x = \phi \left(z \pm c_\| t\right) , \qquad (5.4.35)$$

so that the front shape $\phi(x, t)$ satisfies the wave equation (5.3.14). The corresponding perturbation of the flow field then has the structure

$$\mathbf{g}' = \sum_{m=1}^{3} \mathbf{C}_m \phi_z \left(z \pm c_\| t \pm \lambda_m x\right) , \qquad (5.4.36)$$

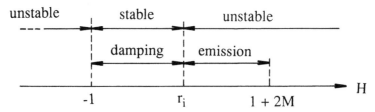

Fig. 5.8. Shock stability limits.

where the $\mathbf{C}_m$ are constant vectors and $\lambda_m = iq_m/k$. As (5.4.13) indicates, q is indeed imaginary for the outgoing acoustic mode ($q = q_3$) when s is; this also holds for the entropy and the shear modes, because $q_1 = q_2 = -s/u$ in such cases. At fixed t, each $\mathbf{C}_m$ contributes to $\mathbf{g}$ a disturbance which is constant along the lines $z = \pm\lambda_m x$; specifically λ_1 and λ_2 correspond to shear/entropy modes, whereas λ_3 is associated with an outgoing acoustic mode. In particular, if ϕ makes an angle to the z-direction, the aforementioned modes correspond to weak discontinuities in the flow field. This combination of a ramped shock, shear/entropy discontinuity and a weak outgoing shock is known as a Mach stem (Fig. 5.9).

According to the above standpoint, the possibility of weak Mach stems in the range (5.4.31) was suggested from the normal mode analysis and the superposition principle (equations. (5.4.35–36)). An alternative approach was developed by Kuznetsov (1986): postulating the existence of such steady (in an adequate frame) four-wave configurations, this author directly worked out the compatibility between the various regions of the flow from the Rankine–Hugoniot relationship (Section 2). Under the assumption of "shallow angles" of deflection for the leading shock, conditions for the weak Mach stems to be viable were deduced: they coincide with (5.4.31). This is not fortuitous but merely reflects the fact that the transversally-traveling waves evidenced by the normal mode analysis are not dispersive (5.4.32) in models which do not involve a reference length.

5.4.2 Comments

The stability limits derived in the preceding paragraph admit a nice geometrical interpretation (see for example: Cowperthwaite, 1968; Fowles, 1976). We first recognize that the slopes in the (p, V) plane ($V = \rho^{-1}$, the specific volume) of the isentrope curve (S_V), the Rayleigh

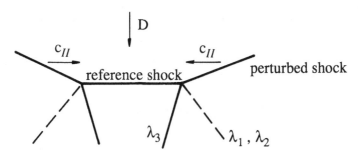

Fig. 5.9. Pair of Mach stems headed to collide.

line (R_V) and the Hugoniot curve (H_V) are given respectively by (overbars omitted)

$$(S_V): \quad \left.\frac{\partial p}{\partial V}\right|_S = -(\rho c)^2 < 0, \tag{5.4.37}$$

$$(R_V): \quad \left.\frac{dp}{dV}\right|_R = -(\rho_0 D)^2 = -(\rho u)^2 < 0, \tag{5.4.38}$$

$$(H_V): \quad \left.\frac{dp}{dV}\right|_H = \left.\frac{\partial p}{\partial V}\right|_S \frac{2 - M^2 g\left(\dfrac{V_0}{V} - 1\right)}{2 - g\left(\dfrac{V_0}{V} - 1\right)}, \tag{5.4.39}$$

so that the control parameter H (cf. equation 5.4.19) is expressed by

$$H = -M^2 \frac{2 - g\left(\dfrac{V_0}{V} - 1\right)}{2 - M^2 g\left(\dfrac{V_0}{V} - 1\right)}. \tag{5.4.40}$$

The quantity g, called the *Grüneisen coefficient*, is defined by

$$g = \frac{V}{\partial e/\partial p|_V} = \frac{V}{T}\left.\frac{\partial p}{\partial S}\right|_V, \tag{5.4.41}$$

T being the temperature and e the specific internal energy. It is also convenient to acknowledge that the slopes in the (p, u_L) plane (u_L is the particle velocity with respect to a laboratory-fixed frame) of the isentrope curves ($S_u^\pm$), the Rayleigh line (R_u) and the Hugoniot curve (H_u) are given respectively by

$$(S_u^\pm): \quad \pm \left.\frac{\partial p}{\partial u_L}\right|_S = -\rho c, \tag{5.4.42}$$

$$(R_u): \quad \left.\frac{dp}{du_L}\right|_R = \rho_0 D = \rho(D - u_L) = \rho u > 0, \tag{5.4.43}$$

$$(H_u): \quad \left.\frac{dp}{du_L}\right|_H = 2\rho_0 D/(1 - H). \tag{5.4.44}$$

We recall that a shock state is represented by the intersection of the Hugoniot curve and the Rayleigh line. Equations (5.4.38–39) and (5.4.43–44) are straightforwardly derived from the differentiation of the jump relation and of an arbitrary $e(p, V)$ equation of state. Equation (5.4.37) is the definition of the sound speed in nonrelaxing media. Equation (5.4.42) is derived from the characteristic forms of the Euler equations for one-dimensional unsteady nonreactive flow of an inviscid fluid (cf. Section 5.2.2).

$$\frac{d^\pm p}{dt} \pm \rho c \frac{d^\pm u_L}{dt} = 0, \tag{5.4.45}$$

where the total time derivatives d^+/dt and d^-/dt are derivatives along lines of the (x,t) plane, the slopes of which are $(u_L + c)^{-1}$ and $(u_L - c)^{-1}$ respectively. We recall that these lines are the forward and backward characteristic lines and that the quantity $u_L + c$ ($u_L - c$) is the velocity with respect to a laboratory-fixed frame of an acoustic perturbation that travels in the same direction as a fluid particle (in the opposite direction). We then infer from equations (5.4.37–38) that

$$M = \frac{u}{c} = 1, \quad \text{i.e.,} \quad D = u_L + c, \tag{5.4.46}$$

and

$$\left.\frac{dp}{dV}\right|_H < 0, \quad \text{i.e.,} \quad H < 0, \tag{5.4.47}$$

whenever any two of the three curves (S_V), (R_V) or (H_V) are tangent. In other words, the necessary condition for the isentrope curves, the Rayleigh line or the Hugoniot curve to be tangent in the (p, V) plane is that the flow be sonic and that the slope of the Hugoniot be negative. We also infer from equations (5.4.42–44) that $M = 1$ when (S_u) and (R_u) are tangent and that $H = 1 \pm 2M$ when (S_u) and (H_u) are tangent. In other words, the necessary condition for the isentrope curves and the Rayleigh line to be tangent in the (p, u_L) plane is that the flow be sonic. The tangency constraint for the isentrope curves and the Hugoniot curve here leads to a double solution. The minus sign must be chosen whenever $M = 1$ in order to satisfy (5.4.46–47). Consequently we have $H = -1$ when the isentrope curve and the Hugoniot curve are both tangent to the Rayleigh line in the (p, u_L) plane and $H = 1 + 2M$ when the isentrope curve and the Hugoniot curve are tangent curves that intersect with a non zero angle the Rayleigh line in the (p, u_L) plane. These values of H are thus the same as the limits of linear stability (5.4.29). It is now important to note that the violation of the lower linear stability condition corresponds to that of one of the basic subsonic–supersonic conditions for shock, the second law of thermodynamics, namely $D < u_L + c$ or $M < 1$. This result can be directly obtained by using expression (5.4.40) to check that $H > -1$ is equivalent to $M < 1$.

As a conclusion, let us now examine the case of a shock propagating in a medium governed by a polytropic equation of state in a situation where the initial pressure p_0 is negligible. We have

$$e - e_0 = \frac{pV}{\gamma - 1} \quad \text{so that} \quad g = \gamma - 1. \tag{5.4.48}$$

The equations for the Hugoniot curve and its slope are respectively

$$\frac{V}{V_0} = \frac{\gamma - 1}{\gamma + 1}, \quad \left.\frac{dp}{dV}\right|_H \to -\infty \quad \text{or} \quad H \to 0^-. \tag{5.4.49}$$

We also have

$$\frac{pV_0}{D^2} = \frac{2}{\gamma + 1}, \quad \frac{u}{D} = \frac{V}{V_0} = \frac{\gamma - 1}{\gamma + 1}, \quad M^2 = \frac{\gamma - 1}{2\gamma}, \tag{5.4.50}$$

and we find that the lower boundary r_i for neutrally radiating modes (5.4.31) is equal to H. We thus observe that it is not possible to conclude whether strong shocks in polytropic media are stable or not. Fowles (1981) however, has shown that neutrally radiating modes do not occur in perfect gas (a particular case of polytropic media) by considering the effect of the initial pressure on the behavior of H and r_i: The condition for acoustic emission (5.4.31) is first expressed as

$$H = \frac{-2M^2 + M^2 g(V_0/V - 1)}{2 - M^2 g(V_0/V - 1)} > r_i = \frac{1 - M^2(1 + V_0/V)}{1 - M^2(1 - V_0/V)}. \tag{5.4.51}$$

Using the equation for the balance of momentum through the shock and the definition of the sound speed the latter condition is turned into

$$-V^2 \left.\frac{\partial p}{\partial V}\right|_S = c^2 < V(p - p_0)(1 + g). \tag{5.4.52}$$

For the ideal gas equation of state we have

$$p(V, S) = A(S)V^{-\gamma}, \tag{5.4.53}$$

so that (5.4.47) becomes

$$\left.\frac{\partial p}{\partial V}\right|_S = -\frac{\gamma p}{V} > -(1 + g)(p - p_0)/V \quad \text{or} \quad \gamma p_0 < 0. \tag{5.4.54}$$

Thus neutrally radiating modes cannot exist in ideal gas since neither γ nor p_0 can be negative. Fowles (1981) then speculates that real gas effects might lead to instabilities. This point will be evoked again in Section 6.3.

5.5 Prospects

It is conceptually important to recall (cf. Section 5.2) that a shock front does not propagate itself, in the sense a flame does: this nonautonomous object has to rely on some external agency (e.g., a "piston") to survive, the removal of which often leads to a decay into acoustic waves. A shock front is the result of the flow behind it and, to a large extent, is always forced, so that no information on its shape or velocity can be obtained without knowing its history or, equivalently, the entire rear flow. Consequently, when considering shock fronts as hydrodynamic discontinuities, one must admit that *instabilities* necessarily derive from particular supports (rear or lateral boundary conditions that would vary with time or space) or from an intrinsic instability of the flow behind the wave front. Examples of the latter are the D'yakov's linear instabilities

(Section 5.4) that are associated with the thermodynamic properties of the fluid. The corrugations recently observed on the shock front of a Taylor–Sedov blast wave created by an explosion-point source (Grun *et al.*, 1991) are likely to belong to this wide class of instabilities.

Another class of shock instability problems is defined when the assumption of a hydrodynamic discontinuity is dropped. Considering the shock front as a dissipative layer, one obviously faces various possible structural instabilities that would act on length scales small compared to the shock front thickness (e.g., molecular transport phenomena, magnetogas- and/or plasma-dynamics).

These studies are very intricate in nature (see for instance Woods, 1971) and, unfortunately, cannot be presented in this overview chapter.

6 Detonations

6.1 Phenomenology

Detonations are traveling waves involving combinations of shocks followed by *chemical reactions* that they themselves trigger. They have been identified as such only comparatively recently (Berthelot and Vieille, 1881, 1882). They are amongst the most violent mechanical–chemical waves, which certainly does not facilitate their study (Fig. 6.1). To wit, the power released by 1 m^2 of a *detonation* front propagating in a solid explosive is in the range of 10^{14} watts, a non negligible figure, considering that the whole earth surface receives energy from the sun at a rate of about 4×10^{16} watts (Fickett and Davis, 1979).

Getting pictures of detonations requires good reflexes from the photographer: these waves are pretty thin (a few 10^{-4} m at most) and their speed relative to the unreacted material typically ranges from 10^3 m/s in gases to 10^4 m/s in the fastest solid explosives. Nevertheless experimentalists have managed to meet and survive the challenge (Fig. 6.1): in gases, typical burned-to-unburned pressure ratios are around 30, whereas pressure levels exceeding 10^5 bars are commonplace with initially condensed explosives. Needless to stress that studying detonations by means of burners is rather impractical in laboratories.

The *theory* of detonations is far from complete but, fortunately, different levels of description may be envisaged. As far as propagation speed (noted D from now on) is concerned, Chapman's (1899) and Jouguet's (1905, 1906, 1917) (CJ) modeling of a detonation as a mere discontinuity, endowed with a simple condition on the final state, remains by far the best one, for it predicts D within a few percent in most cases

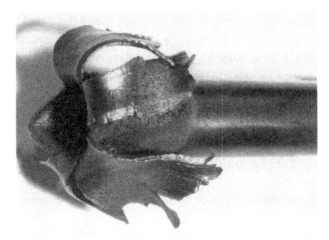

Fig. 6.1.   Consequence of a detonation propagating in a condensed explosive confined in a steel tube.   The thickness of the tube wall is about 2 mm (Laboratoire d'Energétique et Détonique, Poitiers, 1990).

(Section 6.2). Given the inherent complexity of the actual phenomenon, and also the underlying uncertainties on the equation of state of the detonation products, reaching such levels of accuracy is a remarkable achievement. This result can be put in perspective with Landau's and Darrieus' modeling of flames which gives no access to the burning speed.

However, a fresh-to-burnt conversion by chemical means cannot be infinitely fast (Vieille, 1900): One of the purposes of the Zel'dovich (1940), Von Neumann (1942) and Doering (1943) (ZND) model of one-dimensional steady planar detonation waves is precisely to provide the *CJ model* with a reference time and to describe how necessarily finite reaction rates ensure the chemical conversion behind an inert leading shock wave front (Section 6.5). Besides its physically plausible mechanism, the *ZND model* is also the basis from which current linear and nonlinear stability analysis are developed. Indeed, experiments reveals that most detonation waves in homogeneous explosives are anything but quasi-planar or quasi-steady; instead they exhibit unsteady *cellular structures* that involve the propagation of *transverse shock waves*, as sketched in Fig. 6.2a.

The widely used method for visualizing the crest (e.g., A and B) trajectories is to coat the lateral boundaries (the walls of the experimental set up in general) with a soft material, such as a fine soot, which is eroded by the violent fluid mechanical interactions taking place at each crest (Fig. 6.2b,c,d). The same thing can be done for radially expanding

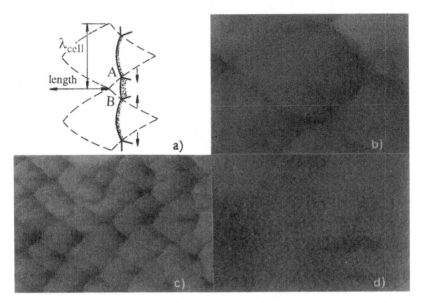

Fig. 6.2. a) Schematic sketch of a structured detonation (A,B crests); b) to d) soot prints of crest trajectories for increasing initial pressure ($2H_2 + O_2 + Ar$ mixtures); reproduced with kind permission of Desbordes (Poitiers, 1990).

hemispherical detonations (Fig. 6.3) in which case the crest imprints follow approximately *logarithmic spirals* (Section 6.4).

Another way of giving access to the structure is to stretch a very thin plastic film normal to the mean direction of propagation, then to

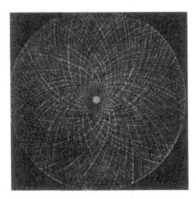

Fig. 6.3. Soot prints of crest trajectories for cylindrically expanding detonation; reproduced with kind permission of the Combustion Institute (see also Lee *et al.*, 1965).

record (photographically) its deformation as the detonation impinges on it: polygon-like imprints are obtained (Fig. 6.4).

Except in very special circumstances the patterns obtained in these ways are rather irregular.   However, systematic experiments amply demonstrate that the average *cell size* λ_{cell} (Fig. 6.2a) varies as a typical *chemical length* $\lambda_{cell} \sim 1/p$, $\lambda_{cell} \sim \exp(E/RT)$ where p and T are the shock pressure and temperature (Fig. 6.5) corresponding to an ideal plane wave; the cells' *aspect ratio A* (length/width) varies much more slightly if not at all ($A \simeq 0.6$) (Desbordes, 1990).

This experimental evidence strongly suggests that the observed patterns are related to some chemical-affected instability mechanisms. Most of the theoretical studies on detonation stability are now conducted within the ZND model framework, taking the Euler equations for compressible reacting inviscid fluid flows as a basis. These apparently simple equations, however, lead to a complicated problem: the investigation of linear stability amounts to constructing, point by point, the dispersion relation upon integration of a set of ordinary differential equations with non-constant coefficients (cf. Section 6.6).

The objective of the following paragraphs is to provide the reader with only the basic aspects of detonation theory; the reference models on the one hand, an overview of the current methods and main results associated with the detonation instability problem on the other.

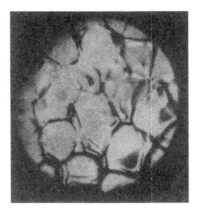

Fig. 6.4.   Head-on visualization of polyhedral detonation cells; reproduced with kind permission of Presles (Poitiers, 1987).

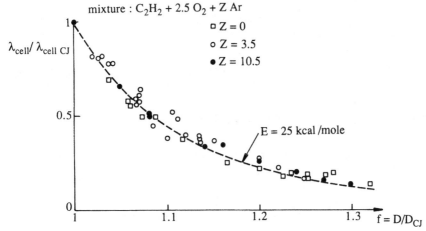

Fig. 6.5. Dependence of the reduced mean cell size λ_{cell} of overdriven detonation (Section 6.5) on the overdrive degree f. (Symbols: measurements, dashed line: best fit assuming an Arrhenius law for the induction time); reproduced with kind permission of Desbordes (Poitiers, 1990).

6.2 Chapman–Jouguet model and sonicity condition

This model was independently proposed by Chapman (1899) and Jouguet (1905) in order to account for the existence of a very fast propagation regime of combustion, with an astonishingly constant velocity, the so-called detonation wave. This particular regime was identified in the course of experiments on the propagation of flames in tubes filled with combustible gas performed in the 1880s by Mallard and Le Chatelier (1881) and Berthelot and Vieille (1881, 1882). Chapman and Jouguet (CJ) postulated, in somewhat different but equivalent terms, that the detonation wave is a steep plane front sweeping over the unburnt (fresh) gas and changing it instantaneously into a state of chemical equilibrium and that the wave front velocity D_{CJ} is equal to the local forward facing acoustic perturbations:

$$D_{CJ} = u_B^{CJ} + c_{eB}^{CJ}, \qquad (6.2.1)$$

$$c_e^2 = \left.\frac{\partial p}{\partial \rho}\right|_{S,A=0}. \qquad (6.2.2)$$

The latter condition is Jouguet's celebrated *sonicity condition* which provides an enlightening and simple explanation for the wave front autonomy, i.e., for its independence upon the phenomena that occur behind the front. c_e, u and D denote the equilibrium sound speed and the particle and wave front velocities in a laboratory-fixed frame, respectively. A denotes the *chemical affinity* vector, which vanishes at *chemical equilibrium* (Prigogine and Defay, 1954). The subscripts or superscripts "B" and "CJ" denote quantities evaluated on the burnt side of the detonation front and in the particular state corresponding to the Chapman–Jouguet (CJ) postulates, respectively. In the next section we will briefly present the CJ model and, more particularly, we will show that the CJ postulates are compatible with a constant wave front velocity. Further comments are presented in Section 6.2.2.

6.2.1   Basic aspects of fully reactive, plane discontinuity fronts

The transition across a fully reactive, plane discontinuity front is described by the same set of overall conservation equations (jump relations) as that describing the transition across a nonreactive shock front (Section 2.2). For simplicity we assume that the medium ahead of the wave front, denoted by the index 0, is uniform and constant. We have

$$\rho_0 D = \rho_B \left(D - u_B \right) , \qquad (6.2.3a)$$

$$p_B - p_0 = \rho_0 D u_B , \qquad (6.2.3b)$$

$$e_0 + \frac{p_0}{\rho_0} + \frac{1}{2} D^2 = e_B + \frac{p_B}{\rho_B} + \frac{1}{2} \left(D - u_B \right)^2 , \qquad (6.2.3c)$$

where p, ρ and e denote the pressure, the density and the specific internal energy, respectively. The difference between detonation and nonreactive shocks lies in the fact that the chemical nature of the gas now changes across the wave front. Therefore the caloric equation of state for the burnt gas (the dependence of the specific internal energy e on local state) is different from that prevailing in the unburnt one: one must take into account the energy released by the chemical processes. Assuming exothermic reactions we have

$$e(p, V) = E(p, V) - Q , \qquad (6.2.4)$$

where $V = \rho^{-1}$ is the specific volume; $Q > 0$ the heat of complete chemical equilibrium reaction and $E(p, V)$ is the internal energy function for the burnt gas. Expressions (6.2.3–4) define a system of four equations for the four unknown variables $g_B = (\rho_B, u_B, p_B, e_B)$ and one extra parameter D, the initial state p_0, V_0, $e_0(p_0, V_0)$ being given. We thus have

$$g_B = g_B(D) \qquad (6.2.5)$$

For a specified value of D, the solution to the system (6.2.3–4) is conveniently represented in the $(p - V)$ plane as the intersection of the Rayleigh line (R) and of the *Crussard curve* (C) (Fig. 6.6). We recall that the equations for these two curves are obtained by eliminating the particle velocity u from (6.2.3a) and (6.2.3b), and u and D from (6.2.3c) respectively. The Crussard curve (C) for plane fully reactive discontinuity fronts is the analog of the Hugoniot curve (H) for plane nonreactive shock fronts. (C) represents the locus of chemical equilibrium states that can be reached through adiabatic reactive shock processes. The essential difference between (H) and (C) is that the latter does not intersect (R) at (p_0, V_0): (C) lies above (H) because of the presence of the heat release term $Q > 0$ in (6.2.4). A direct consequence is that the solution to system (6.2.3–4) is not unique but double. For values of D greater than a value D_{CJ} for which (R) is tangent to (C) at point A (Fig. 6.6) there exists two intersection points S and W. To demonstrate that the CJ postulates of chemical equilibria and sonic wave front define a unique constant wave front velocity represented by point A, we assume that the flow behind the wave front is one-dimensional and unsteady, that molecular transport phenomena can be neglected, and that the fluid evolution is adiabatic.

The equations expressing the balance of mass, momentum, and energy are

$$\frac{d^0 \rho}{dt} + \rho \frac{\partial u}{\partial x} = 0 \,, \tag{6.2.6}$$

$$\rho \frac{d^0 u}{dt} + \frac{\partial p}{\partial x} = 0 \,, \tag{6.2.7}$$

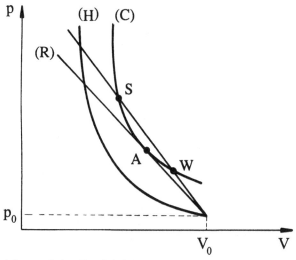

Fig. 6.6. Positions of the Rayleigh line, Hugoniot and Crussard curves in the $(p\text{--}V)$ plane.

$$\frac{d^0 e}{dt} + p\frac{d^0 V}{dt} = 0 . \tag{6.2.8}$$

x, t and $d^0 g/dt$ denote the abscissa, the time and the total derivative of g along particle paths $x_0(t)$

$$\frac{d^0 g}{dt} = \frac{\partial g}{\partial t} + u\frac{\partial g}{\partial x}, \quad \frac{dx_0}{dt} = u . \tag{6.2.9}$$

We then let A_k, λ_k, and e denote the chemical affinity and progress variable of one of the $K \geq k$ chemical reactions involved in the combustion process, and the specific internal energy of the off-(chemical) equilibrium medium (2.1.7b), respectively (Prigogine and Defay, 1954). We choose to express e and $\mathbf{A}$ as functions of the $2 + K$ independent thermodynamic variables p, V, and $\boldsymbol{\lambda}$. The conditions for a fluid particle to be in a state of chemical equilibrium are:

$$\mathbf{A}(p, V, \boldsymbol{\lambda}) = 0 \tag{6.2.10}$$

The associated equilibrium composition $\boldsymbol{\lambda}_e$ and equation of state e_e are

$$\begin{aligned}
\boldsymbol{\lambda} &= \boldsymbol{\lambda}_e(p, V), \\
e &= e_e(p, V) = e(p, v, \boldsymbol{\lambda} = \boldsymbol{\lambda}_e(p, V)) .
\end{aligned} \tag{6.2.11}$$

Thus, the differential of the equilibrium equation of state e_e is

$$de = \frac{\partial e_e}{\partial p}dp + \frac{\partial e_e}{\partial V}dV ,$$

$$\frac{\partial e_e}{\partial p} = \frac{\partial e}{\partial p} + \sum_{k=1}^{K} \frac{\partial e}{\partial \lambda_k}\frac{\partial \lambda_{ek}}{\partial p}, \tag{6.2.12}$$

$$\frac{\partial e_e}{\partial V} = \frac{\partial e}{\partial V} + \sum_{k=1}^{K} \frac{\partial e}{\partial \lambda_k}\frac{\partial \lambda_{ek}}{\partial V} .$$

Restricting the variations in (6.2.12) to a particle path and eliminating the variation of e owing to (6.2.8) yield the following expression for the energy balance $(\rho = V^{-1})$

$$\frac{d^0 p}{dt} - c_e^2 \frac{d^0 \rho}{dt} = 0 , \tag{6.2.13}$$

$$c_e^2 = \frac{p + \partial e_e/\partial V}{\partial e_e/\partial p}V^2 . \tag{6.2.14}$$

Identifying the differential (6.2.12) of the e_e equilibrium equation of state and the Gibbs identity $de = TdS - pdV - \mathbf{A}d\boldsymbol{\lambda}$ subjected to the equilibrium conditions (6.2.10) shows that the definitions (6.2.2) and (6.2.14) of the equilibrium sound speed c_e are identical.

We now transform the conservation equations, (6.2.6,7,13), by changing the total time derivative along lines which at time t are parallel to the front path $x_s(t)$ in the (x, t) plane. Using the definitions

$$\frac{d^0 g}{dt} = \frac{\partial g}{\partial t} + \frac{dx_0}{dt} \frac{\partial g}{\partial x}, \qquad \frac{dx_0}{dt} = u, \tag{6.2.15}$$

$$\frac{d^s g}{dt} = \frac{\partial g}{\partial t} + \frac{dx_s}{dt} \frac{\partial g}{\partial x}, \qquad \frac{dx_s}{dt} = D, \tag{6.2.16}$$

we build the transformation rule

$$\frac{d^s g}{dt} = \frac{d^0 g}{dt} + (D - u) \frac{\partial g}{\partial x}. \tag{6.2.17}$$

By forming the combination $c_e^2 \times$ (6.2.6) $+ (D - u) \times$ (6.2.7) $+$ (6.2.13) we then directly find

$$\frac{d^s p}{dt} + \rho(D - u) \frac{d^s u}{dt} = -\rho c_e^2 \left(1 - \left(\frac{D - u}{c_e} \right)^2 \right) \frac{\partial u}{\partial x}. \tag{6.2.18}$$

Specializing the latter relation to points located on the front path $x_s(t)$ we can write, owing to (6.2.3), that

$$\left. \frac{d^s g}{dt} \right|_{\mathrm{B}} = \frac{d\mathbf{g_B}}{dD} \frac{d^s D}{dt}. \tag{6.2.19}$$

We thus obtain the wave front acceleration

$$\frac{d^s D}{dt} = F(D) \left(1 - \frac{D - u_{\mathrm{B}}(D)}{c_{e\mathrm{B}}(D)} \right)^2 \left. \frac{\partial u}{\partial x} \right|_{\mathrm{B}}, \tag{6.2.20}$$

$$F(D) > 0. \tag{6.2.21}$$

It is clear that the conditions for the front to propagate at constant velocity are

$$D = u_{\mathrm{B}} + c_{e\mathrm{B}}, \tag{6.2.22}$$

or

$$\left. \frac{\partial u}{\partial x} \right|_B = 0. \tag{6.2.23}$$

So far no specific assumption has been made on the nature of the rear boundary. We may unrestrictively imagine the presence of a piston whose velocity u_{p} is either negative or positive. Thus, the wave front velocity will be constant if

$$u_{\mathrm{p}} \leq u_{\mathrm{B}}^{\mathrm{CJ}}(D),$$

in which case we have

$$D = u_{\mathrm{B}}^{\mathrm{CJ}}(D) + c_{e\mathrm{B}}^{\mathrm{CJ}}(D) = D_{\mathrm{CJ}} \qquad \text{and} \qquad \frac{\partial u}{\partial x} \geq 0, \tag{6.2.24}$$

or if

$$u_{\mathrm{p}} \geq u_{\mathrm{B}}^{\mathrm{CJ}}(D),$$

in which case we have

$$D = [u_{\mathrm{p}}(D)]^{-1} \geq D_{\mathrm{CJ}} \qquad \text{and} \qquad \frac{\partial u}{\partial x} = 0. \qquad (6.2.25)$$

In other words, if the piston velocity u_{p} is lower than a minimum value $u_{\mathrm{B}}^{\mathrm{CJ}}$, then the wave front velocity is that of the local forward facing acoustic perturbations and the flow profile between the piston and the front is time dependent. In this case u_{p} need not be a constant and the wave can be considered autonomous (e.g., self-sustained). If the piston velocity u_{p} exceeds the value $u_{\mathrm{B}}^{\mathrm{CJ}}$ however, then the situation is similar to that described in Section 5.2: the front velocity is constant if u_{p} is a constant and its value D is determined owing to the jump relations for a specified value of u_{p}. The flow profile between the piston and the front is flat. The wave is then termed *overdriven* (Fig. 6.7b).

For values of D greater than D_{CJ} only the state corresponding to point S (Fig. 6.6) must be retained. The other point, W, is excluded on physical grounds: for example, one can show that it describes a non-physical wave, the front speed of which, D, would increase (decrease) upon decreasing (increasing) the piston velocity u_{p} (cf. (6.5.19)).

Let us now demonstrate that there exists only one value D_{CJ} for the velocity of the autonomous (CJ) detonation wave. In the (p, V) plane, the

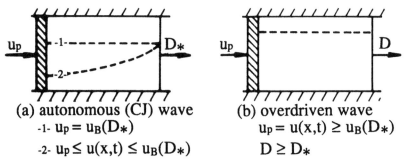

(a) autonomous (CJ) wave
-1- $u_p = u_B(D_*)$
-2- $u_p \leq u(x,t) \leq u_B(D_*)$

(b) overdriven wave
$u_p = u(x,t) \geq u_B(D_*)$
$D \geq D_*$

Fig. 6.7. Autonomous and overdriven fully reactive discontinuities. (Dashed lines are particle velocities profiles.)

slopes of the Rayleigh line (R) and Crussard curve (C) are respectively

$$\left.\frac{dp}{dV}\right|_R = -\left(\frac{D}{V_0}\right)^2,$$

$$\left.\frac{dp}{dV}\right|_C = -\left.\frac{dp}{dV}\right|_R + \frac{2T}{V_0 - V}\left.\frac{dS}{dV}\right|_C. \tag{6.2.26}$$

From the $p = p(S,V)$ equation of state for the media in a chemical equilibrium state we also deduce that

$$\left.\frac{dp}{dV}\right|_C = \left.\frac{\partial p}{\partial V}\right|_S + \left.\frac{\partial p}{\partial S}\right|_V \left.\frac{dS}{dV}\right|_C. \tag{6.2.27}$$

The definition of the sound speed (6.2.2), the mass conservation equation (6.2.3a) and the sonicity condition (6.2.1) respectively lead to

$$\left(\frac{c_e}{V}\right)^2 = -\left.\frac{\partial p}{\partial V}\right|_S, \quad \left(\frac{D}{V_0}\right)^2 = \left(\frac{D-u}{V}\right)^2, \quad c_e^2 = (D-u)^2. \tag{6.2.28}$$

Bringing together equations (6.2.26), (6.2.27) and (6.2.28) indicates that the sonicity condition can only be satisfied at points where the Rayleigh line (R) and the Crussard curve (C) are tangent. Under the additional assumption that the first derivative of (C) is monotonous, there exists only one point (A, cf. Fig. 6.6) where the tangency condition is satisfied so that D_{CJ} is unique. We emphasize that both (R) and (C) are tangent to an isentrope at point A. Indeed the last of conditions (6.2.28) and the constraint $dS = 0$ are equivalent; it is interesting to note that this result is similar to the one derived in Section 5.4.2 for nonreactive shock front.

We now conclude this paragraph with an example of CJ state and wave speed calculations in the simple case of a medium governed by a *polytropic* equation of state, again upon neglect of the initial pressure. Expression (6.2.4) specializes to

$$e(p,V) - e_0 = \frac{pV}{\gamma - 1} - Q, \tag{6.2.29}$$

and we obtain

$$\frac{p_B V_0}{D^2} = \frac{u_B}{D} = 1 - \frac{V_B}{V_0} = \left(2\frac{e_B - e_0}{D^2}\right)^{1/2} = \frac{1 + L}{\gamma + 1}, \quad M_B^2 = \frac{\gamma - L}{\gamma(1 + L)}, \tag{6.2.30}$$

where M_B and L are defined by

$$M_B = \frac{D - u_B}{c_e}, \quad L^2 = 1 - \left(\frac{D_{CJ}}{D}\right)^2, \quad D_{CJ}^2 = 2\left(\gamma^2 - 1\right)Q. \tag{6.2.31}$$

On can then infer the CJ state by equating M_B to one or, equivalently, D to D_{CJ} or L to 0 in (6.2.30):

$$\frac{p_B^{CJ} V_0}{D_{CJ}^2} = \frac{1}{\gamma + 1}, \qquad \frac{D_{CJ} - u_B^{CJ}}{D_{CJ}} = \frac{V_B^{CJ}}{V_0} = \frac{\gamma}{\gamma + 1} \qquad (6.2.32)$$

Indeed D_{CJ} (6.2.31) represents the velocity of the autonomous detonation wave. The structure of the last expression in (6.2.31) could have been directly anticipated on dimensional grounds: when the initial pressure p_0 is neglected, the heat of reaction Q is the only term left in the equations that can be used to construct a characteristic velocity. In other words, we have $p_0 \ll \rho_0 D^2 \Rightarrow D_{CJ} \approx Q^{1/2}$.

6.2.2 Comments

Despite its simplicity and the crudeness that characterizes it, the CJ model remains to date the best tool for calculating the characteristics of detonation of explosive substances. However, these calculations require a precise knowledge of the equation of state of the combustion products, which is not necessarily the case for condensed explosives. For gaseous explosives at low initial pressure, the use of the perfect gas equation of state has led to calculated values of detonation velocity in excellent agreement with experimental measurements ($< 1\%$, typically). Conversely, the CJ theory may be used to probe yet unknown equations of state (Jones, 1949; Manson, 1958). It is important to note that this model of a detonation wave does not consider the kinetic features of the actual combustion process since, by assumption, the length of the reaction zone is neglected. Therefore this model of wave does not essentially differ from a nonreactive shock wave, apart from the existence of a positive heat of reaction term Q, which brings in the possibility of a self-sustained wave with a characteristic velocity $Q^{1/2}$.

6.3 Analogs of D'yakov–Kontorovich's instabilities

Our objective here is to investigate the linear and weakly nonlinear instability properties of a detonation when the latter is described in the simplest way, namely as a plane, fully reactive discontinuity (cf. Section 6.2). This will prove to be an instructive exercise before carrying out such an analysis in the framework of more sophisticated models, those of Brun (Section 6.4) and of ZND (Section 6.5 and 6.6), specifically.

6.3.1 Linear analogs

Clearly, in the case where the wave is overdriven (i.e., $D/D_{CJ} > 1$) D'yakov–Kontorovich's criterion (cf. Section 5.4) can be used in the linear domain: the reference flow behind the wave front is still in a uniform constant-state and the burnt medium, being nonreactive, is governed by an equation of state with only two independent variables (see (6.2.29–30) for example). This criterion states that exponentially growing modes cannot exist when

$$-1 < H < 1 + 2M , \qquad (6.3.1)$$

and that neutrally stable linear modes (acoustic emission) exist when

$$r_i < H < 1 + 2M . \qquad (6.3.2)$$

M is the Mach number of the flow relative to a wave front-fixed frame and $-H$, the control parameter, is now the ratio of the slope of the Rayleigh line to that of the Crussard curve (Section 6.2). The expressions for H and r_i are:

$$H = \frac{-2M^2 + gM^2\left((V_0/V) - 1\right)}{2 - gM^2\left((V_0/V) - 1\right)}, \qquad r_i = \frac{1 - M^2\left((V_0/V) + 1\right)}{1 + M^2\left((V_0/V) - 1\right)},$$
$$(6.3.3)$$

where g and V_0/V are the Grüneisen coefficient and the ratio of the initial specific volume (unburnt state) to that of the detonation products (index B omitted), respectively. For simplicity's sake we evaluate H and r_i in the case of a polytropic medium and we neglect the initial pressure. The equation of state is given by the expression (6.2.21) and, using the results derived in Section 6.2 and 5.4, we obtain

$$r_i = H = \frac{L - 1}{L + 1}, \quad 1 + 2M = 1 + 2\left(\frac{(\gamma - L)}{\gamma(1 + L)}\right)^{1/2}, \qquad (6.3.4)$$

where:

$$L = \left(1 - \left(\frac{D_{CJ}}{D}\right)^2\right)^{1/2}, \quad 0 < L \leq 1, \quad D_{CJ}^2 = 2\left(\gamma^2 - 1\right)Q. \qquad (6.3.5)$$

We thus observe that H is negative and contained in the interval $(-1, 0)$ so that no exponentially growing mode exists. However, as already noted for nonreactive, strong shocks in polytropic media (Section 5.4), we find that H and r_i have the same value so that neutrally stable linear modes cannot be evidenced without using a more general equation of state. A more throughout computation of H and r_i reveals that $-1 < H < r_i$ when a nonvanishing $p_0/\rho_0 D^2$ is retained; therefore, no radiating boundary wave is obtained in any case. Nevertheless, such traveling disturbances look tantalizingly similar to what is observed in experiments (Fig. 6.2b–d). It is therefore natural to inquire about their fate when they have finite, even if small, amplitudes.

6.3.2 Beyond linearity

The above-stated task was undertaken by Majda and Rosales (1983) in the case of waves traveling along the front in one direction only, say rightwardly, along the z-axis. We shall denote the grouping $z - c_{\parallel}t$ by ζ. It is out of the question to reproduce Majda and Rosales' account in detail; within the imparted size of our essay, we shall content ourselves with an outline of the method and a discussion of the results.

By analogy with the case of nonreactive discontinuities (Section 5.4.1) it is clear that the instantaneous shape $\phi(z,t)$ of the corrugated front is in first approximation given by $x = \phi(\zeta)$, see (5.4.35), for disturbances of $\mathcal{O}(\epsilon)$ amplitude-to-wavelength ratios ($\epsilon \ll 1$). Still for $\epsilon \to 0$, the flow field disturbances should have the form (5.4.36), approximately. However, the various modes contributing to **g** are constant along the lines $z - c_{\parallel}t - \lambda_m x = $ const. in first approximation only: the latter lines are not exactly straight nor parallel when ϵ is finite. As a consequence, the corresponding modes of **g** are headed to interfere, or interact, in a far-field defined by $X = x/\epsilon = \mathcal{O}(1)$. Furthermore, like in any weakly-nonlinear problem of acoustics (Whitham, 1974) the transverse propagation speed of disturbances is $c_{\parallel}$ to only within $\mathcal{O}\left(\epsilon c_{\parallel}\right)$ corrections; the "slow" time $T = t/\epsilon$ is again needed to describe the flow field and front-shape evolution.

To summarize, the five variables z, x, X, t, T are all involved so that describing the front shape evolution necessitates using a multi-scale formalism (Cole and Kevorkian, 1981). With the additional proviso that the only *incoming waves* (presumably of order $\phi_\zeta \phi_{\zeta\zeta}$) are those generated by the interaction of the primary, *outgoing waves* due to $\phi_\zeta \neq 0$ (see 5.4.36), Majda and Rosales ultimately found (after a few rescalings and some pleasant algebra!) that $\phi(T, \zeta)$ is described by a nonlocal, nonlinear evolution equation of the form

$$\frac{\partial \phi}{\partial T} + a_1 \left(\frac{\partial \phi}{\partial \zeta}\right)^2 + a_2 \int_0^\infty \frac{\partial \phi}{\partial \zeta}(\zeta + \beta\xi, T)\frac{\partial^2 \phi}{\partial \zeta^2}(\zeta + \xi, T)\, d\xi = 0 \quad (6.3.6)$$

where a_1, a_2, and $\beta > 1$ are positive constants which only depend on the unperturbed configuration.

Besides a quadratic nonlinearity $(\partial \phi/\partial \zeta)^2$ which is reminiscent of what pure geometry gave for flames (4.5.26, 4.5.28, 4.7.43), this equation also includes a nonlocal term (the integral) which accounts for the secondary incoming waves. Very likely this term would be absent if the flow behind the discontinuity was sonic. If one had $\beta = 1$ (6.3.6) would be a mere inviscid Burgers equation for the slope $q = \partial \phi/\partial \zeta$ (e.g., (5.2.31)).

Equation (6.3.6) admits constant solutions $\partial \phi/\partial \zeta = $ const. which correspond to slightly tilted planar detonations; as expected the nonlocal term vanishes when $\phi_{\zeta\zeta}$ does. Quite remarkably it also admits solutions for which $\partial\phi/\partial\zeta$ is piecewise uniform:

$$\frac{\partial\phi}{\partial\zeta} = \begin{cases} q_+ & \text{when} \quad \zeta - CT > 0, \\ q_- & \text{when} \quad \zeta - CT < 0, \end{cases} \tag{6.3.7}$$

where C is a constant (the $\mathcal{O}(\epsilon)$ difference between $c_{\parallel}$ and the actual propagation speed). Upon substitution of (6.3.7) into (6.3.6) one finds

$$C = (a_1 - a_2)\, q_+ + a_1 q_- \,. \tag{6.3.8}$$

Because the pattern corresponding to (6.3.7) represents a triple point (with a slope mismatch $q_+ - q_-$), (6.3.8) can be observed to be nothing but Kuznetsov's result (Section 5.4) once adapted to fully reactive discontinuities. Majda and Rosales (1984) integrated (6.3.6) numerically in the case of compact-support initial conditions and found that, in most cases, $\phi(T, \zeta)$ tends to develop angles as times elapses, the prototype of which being locally given in (6.3.7); simultaneously, a weak shock and a slip line appear in the flow behind the leading discontinuity (see end of Section 5.4.1). It can therefore be said that Majda and Rosales model reproduces the essential features of the formation of triple points.

However one must keep in mind that this model only considers one-sided transverse propagations; to the best of our knowledge an analysis incorporating both directions has yet to be done. Furthermore, the above model does not account for any intrinsic reference length.

6.3.3 Comments

The latter developments are among the simplest for understanding nonlinear wave interactions in a nonreactive compressible medium. However, they are unlikely to realistically describe the actual mechanisms of detonation instabilities because cellular structures are commonly observed in combustible gas, the detonation products of which are accurately governed by a perfect-gas equation of state. Indeed, detonation instabilities must be attributed another origin than the peculiarities of the equation of state of the combustion products. With a theory based upon a fluid model where molecular transports are neglected (Section 6.5.2), there is no other effect to consider than chemical processes. These would also have the merit of bringing in a reference length. How such an effect can be used for studying detonation instabilities is the object of Sections 6.4, 6.5 and 6.6.

6.4 Brun's model for autonomous diverging waves

As evidenced in experiments, detonation fronts are not plane and do not have a unique constant velocity D_{CJ} (Section 6.2). For example, the detonation wave that propagates along the symmetry axis of a tube containing a liquid explosive or along that of a cylindrically-shaped stick of solid explosive has a curved front and a velocity which depends on the tube or stick diameter (Fig. 6.8). Furthermore, there exists a critical value of the diameter below which no propagation of the detonation is possible (Campbell and Engelke, 1976; Fay, 1959). This effect is often referred to as the diameter effect.

In gaseous and many liquid explosives, a three-dimensional unsteady cellular structure is superimposed on the wave front. All these features are experimentally compatible with a connection to the kinetics of the chemical heat release. Describing the diameter effect and, more generally speaking, investigating the relationship between chemical kinetics and curved detonation properties in a simple and physical manner, is what *Brun's model* (1989) is applied to. In his model, the detonation wave is described as a curved, *partially reactive*, discontinuity endowed with a sonicity condition. Thus, in a way akin to the CJ model, the wave is *autonomous* because the wave front velocity is equal to that of the local forward facing acoustic perturbations:

$$\mathbf{D} = D\,\mathbf{n} = \mathbf{u}\cdot\mathbf{n} + c_b\,, \qquad (6.4.1)$$

$$c^2 = \left.\frac{\partial p}{\partial \rho}\right|_{\lambda,S}\,, \qquad (6.4.2)$$

where c, $\mathbf{u}$ and $\mathbf{D}$ denote the frozen sound speed, the particle velocity and the wave front velocity vectors, both defined in a frame fixed to the explosive ahead of the front. The subscript "b" denotes quantities evaluated on the reactive (burning) side of the discontinuity front and $\mathbf{n}$ is the unit vector normal to the wave front. In addition to the above

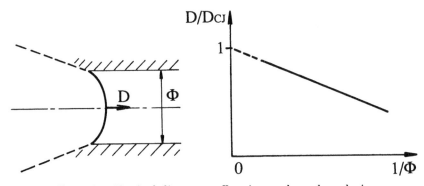

Fig. 6.8.   Typical diameter effect in condensed explosive.

"frozen" sonicity condition, the model is based upon the four following assumptions:

1) A single-step kinetic process describes the chemical decomposition. The quantity λ in (6.4.2) is a scalar, called the *progress variable*, that represents the extent of the reaction or, equivalently here, the mass fraction of burnt explosive.
2) The various gradients at the reactive side of the wave front are finite.
3) Molecular transport phenomena are neglected in the reactive flow.
4) The density and the pressure of the inert explosive are uniform and constant.

Notice that assumption 2 was implicitly assumed to hold in the CJ model (Section 6.2.1).

6.4.1 Brun's equation

The first step of the derivation is to notice that the value, at the partially reactive sonic wave front, of any of the dependent variables u (modulus of the particle velocity), p (pressure), ρ (density), λ (progress variable) or of any function of these variables like e (specific internal energy) or c (frozen sound speed) is again only a function of the local velocity D. Denoting such a value by g_b, we have

$$g_b = g_b(D) \,. \tag{6.4.3}$$

Indeed, the three conservation equations (Section 2.2) for mass, normal momentum and energy through the discontinuity front, the equation of state for the reactive medium and the sonicity condition define a system of five equations in the five unknowns p_b, ρ_b, u_b, λ_b and e_b. In the frame of the unburnt explosive, u_b is normal to the wave front because the tangential component to the front of the momentum equation is identically satisfied so that $\mathbf{u_b} \cdot \mathbf{n}$ is equal to u_b. Thus, the above mentioned conservation equations and the sonicity condition (6.4.1) are

$$\rho_0 D = \rho_b \left(D - u_b \right) ,$$
$$p - p_0 = \rho_0 D u_b , \tag{6.4.4}$$
$$e_0 + \frac{p_0}{\rho_0} + \frac{1}{2} D^2 = e_b + \frac{p_b}{\rho_b} + \frac{1}{2} \left(D - u_b \right)^2 ,$$
$$D = u_b + c_b , \tag{6.4.5}$$

where the subscript "0" denotes the unburnt initial medium. Introducing the equation of state $e = e(p, \rho, \lambda)$, we have $c = c(p, \rho, \lambda)$ because of the identity

$$c^2 = \left.\frac{\partial p}{\partial \rho}\right|_{\lambda S} = \frac{\frac{p}{\rho^2} - \left.\frac{\partial e}{\partial \rho}\right|_{\lambda p}}{\left.\frac{\partial e}{\partial p}\right|_{\lambda \rho}} . \tag{6.4.6}$$

Clearly, equations (6.4.4) to (6.4.6) prove (6.4.3). A direct consequence is that the time derivative of g_b is related to the velocity and the acceleration of the front:

$$\frac{dg_b(D)}{dt} = \frac{dg_b(D)}{dD}\frac{dD}{dt} . \tag{6.4.7}$$

The second step is to relate the time derivatives of the g's to the properties of the reactive flow. We firstly introduce the notation $d^V g/dt$ for the time derivative of the quantity $g(x,t)$ associated with an object moving with velocity V:

$$\frac{d^V g}{dt} = \frac{\partial g}{\partial t} + \mathbf{V}\cdot\boldsymbol{\nabla} g . \tag{6.4.8}$$

We then infer the identities

$$\frac{d^{V_1} g}{dt} = \frac{d^{V_2} g}{dt} + (\mathbf{V}_1 - \mathbf{V}_2)\cdot\boldsymbol{\nabla} g , \tag{6.4.9}$$

$$\frac{d^{V_1} g}{dt} = \frac{d^{V_2} g}{dt} \quad \text{if } \mathbf{V}_1 = \mathbf{V}_2 \quad \text{and} \quad \boldsymbol{\nabla} g \text{ bounded.} \tag{6.4.10}$$

Therefore, at the wave front, the sonicity condition (6.4.5), assumption 2 and equation (6.4.10) lead to

$$\left.\frac{d^+ g}{dt}\right|_b = \left.\frac{d^D g}{dt}\right|_b = \frac{dg_b}{dt} = \frac{dg_b}{dD}\frac{d^D D}{dt} , \tag{6.4.11}$$

where $d^+ g/dt$ denotes the total time derivative of g along the acoustic velocity field $u+c$. We next write the Euler equations for a reactive fluid as

$$\frac{d^u \rho}{dt} + \rho\,\boldsymbol{\nabla}\cdot\mathbf{u} = 0 , \tag{6.4.12}$$

$$\rho\frac{d^u \mathbf{u}}{dt} + \boldsymbol{\nabla} p = 0 , \tag{6.4.13}$$

$$\frac{d^u p}{dt} - c^2\frac{d^u \rho}{dt} = \rho c^2 \sigma w . \tag{6.4.14}$$

Like the sound speed c (6.4.6), the thermicity coefficient σ is expressed as a function of the variables p, ρ and λ when an $e(p,\rho,\lambda)$ equation of state is specified. The *chemical decomposition rate* w is also a function of these variables:

$$\rho c^2 \sigma = \left. \frac{\partial p}{\partial \lambda} \right|_{e\rho} , \tag{6.4.15}$$

$$\frac{d^u \lambda}{dt} = w(p, \rho, \lambda) > 0 . \tag{6.4.16}$$

Evaluating the derivatives in (6.4.9) at the wave front, we have

$$(\boldsymbol{\nabla} \cdot \mathbf{u})_b = u_b \, \boldsymbol{\nabla} \cdot \mathbf{n} + \mathbf{n} \cdot (\boldsymbol{\nabla} \, \mathbf{u})_b \tag{6.4.17}$$

because $\mathbf{u}_b = u_b \mathbf{n}$. Then, performing the combination $c_b^2 \times (6.4.12)$ $+ (c_b \mathbf{n}) \times (6.4.13) + (6.4.14)$, we obtain

$$\left. \frac{d^+ p}{dt} \right|_b + \rho_b c_b \left. \frac{d^+ u}{dt} \right|_b = \rho_b c_b^2 \, (\sigma_b w_b - u_b \, \boldsymbol{\nabla} \cdot \mathbf{n})_b . \tag{6.4.18}$$

The last step is to transform the above equation into an *evolution equation* for the wave front shape. We firstly recall that the scalar $\frac{1}{2} \, \boldsymbol{\nabla} \cdot \mathbf{n}$ represents the local mean *curvature* $\frac{1}{2} \, (1/R_1 + 1/R_2)$ of the front, so that, using (6.4.7) and (6.4.11), we turn (6.4.18) into

$$\frac{d^D D}{dt} - 2 c_\parallel^2(D) \left(K(D) - \frac{1}{2} \left(\frac{1}{R_1} + \frac{1}{R_2} \right) \right) = 0 , \tag{6.4.19}$$

where

$$c_\parallel^2(D) = \frac{\rho_b c_b^2 u_b}{dp_b/dD + \rho_0 D \, du_b/dD} , \qquad K(D) = \frac{1}{2} \frac{\sigma_b w_b}{u_b} . \tag{6.4.20}$$

Introducing now the equation of the *wave front surface*

$$x = F(y, z, t) \tag{6.4.21}$$

and the identities

$$q_y = \frac{\partial F}{\partial y} , \quad q_z = \frac{\partial F}{\partial z} , \quad N^2 = 1 + q_y^2 + q_z^2 , \tag{6.4.22}$$

$$-\left(\frac{1}{R_1} + \frac{1}{R_2} \right) = \frac{\partial}{\partial y} \left(\frac{q_y}{N} \right) + \frac{\partial}{\partial z} \left(\frac{q_z}{N} \right) , \quad D = \frac{1}{N} \frac{\partial F}{\partial t} , \tag{6.4.23}$$

we finally write Brun's equation (6.4.19) as

$$\frac{\partial D}{\partial t} - \frac{D}{N} \left(q_y \frac{\partial D}{\partial y} + q_z \frac{\partial D}{\partial z} \right) - c_\parallel^2(D) \left(\frac{\partial}{\partial y} \left(\frac{q_y}{N} \right) + \frac{\partial}{\partial z} \left(\frac{q_z}{N} \right) \right)$$
$$- 2 c_\parallel^2(D) K(D) = 0 . \tag{6.4.24}$$

Equation (6.4.24) is a quasilinear, second order evolution equation for the front shape F. It is a local one in contrast to (6.3.6) and is hyperbolic when $c_{\|}^2(D) > 0$, which we henceforth assume. From equation (6.4.19) one infers the existence of a characteristic state, from now on denoted by the symbol "$*$", for which the wave front is plane and has a constant velocity D_* such that

$$K\,(D_*) = 0\,. \tag{6.4.25}$$

Strictly speaking, D_* is not the same velocity as D_{CJ}, that of the CJ model (Section 6.2), because the sonicity condition in the latter model was based upon the equilibrium sound speed (cf. (6.2.2)) while here this condition uses the frozen sound speed (6.4.2). However, it is important to note how Brun's model is in the direct line of the CJ model: curvature and acceleration effects are accounted for by letting the chemical heat release not be fully accomplished beyond the curved sonic locus. A curvature–velocity relationship $2K(D) = 1/R_1 + 1/R_2$ is thus defined at any unaccelerated point of the curved front and, in the plane wave limit, for which the discontinuity front is fully reactive, a characteristic constant velocity D_* emerges (6.4.25).

6.4.2 Linear stability analysis

As in Sections 5.4 and 6.3, we wish here to consider whether small wrinkles on an initially plane discontinuity front will exponentially grow with time. When studying the linear stability of the CJ detonation wave (Section 6.3) we met the situation where neutrally radiating modes should be excluded on a plane, fully reactive discontinuity front propagating in a perfect gas. This disappointing result suggested that an additional effect such as a chemical process had to be inserted in the analysis. Thus, Brun's evolution equation for partially reactive curved discontinuity fronts provides an opportunity for trying to analyze the effect of a (vanishingly-small) chemical heat release on the linear stability of a plane detonation wave front. We restrict the analysis to two-dimensional perturbations and denote the direction normal to the plane-wave direction of propagation by z. We firstly linearize equation (6.4.24) around the reference state "$*$" (6.4.25) by setting

$$F = D^*t + \phi(t, z)\,, \qquad |\phi_z| = \mathcal{O}(\epsilon)\,, \qquad 0 < \epsilon \ll 1\,, \tag{6.4.26}$$

to obtain the following telegraph-like equation

$$\frac{\partial^2 \phi}{\partial t^2} + 2\left(-K'_* C_*^2\right)\frac{\partial \phi}{\partial t} = C_*^2 \frac{\partial^2 \phi}{\partial z^2}\,, \tag{6.4.27}$$

$$K'_* = \frac{dK}{dD}\,(D = D_*)\,; \qquad C_* = c_{\|}\,(D = D_*)\,,$$

which, in the context of Brun's model, is the analog of the Landau–Darrieus result (4.7.23) or of (5.3.14). We thus observe, since C_*^2 is positive (cf. Section 6.4.1), that exponentially growing modes can exist when $K_*' > 0$, that neutrally radiating modes appear at the limit $K_*' = 0$ and that stability is ensured when $K_*' < 0$. Let us now specialize the analysis to a polytropic equation of state, neglecting the initial pressure. We have

$$e(p, \rho, \lambda) - e_0 = \frac{p}{(\gamma - 1)\rho} - \lambda Q, \qquad (6.4.28)$$

where Q is the specific heat of reaction. We find

$$\frac{c_b}{\gamma D} = \frac{\rho_0}{\gamma \rho} = \frac{p_b}{\rho_0 D^2} = \frac{u_b}{D} = \frac{1}{\gamma + 1}, \qquad (6.4.29a)$$

$$\lambda_b = \frac{\gamma + 1}{2\gamma^2 \sigma_b} = \left(\frac{D}{D_*}\right)^2, \qquad (6.4.29b)$$

$$\left(\frac{c_\|(D)}{D}\right)^2 = \frac{\gamma}{3(\gamma + 1)} > 0, \qquad (6.4.30)$$

where $D_* = \left(2\left(\gamma^2 - 1\right)Q\right)^{1/2}$, the solution to (6.4.25), coincides with the CJ velocity value if a choice of a global irreversible chemical kinetic process is adopted. As is clear from (6.4.27), the entire discussion on detonation stability in Brun's model is tied up with the sign of K_*', which is obviously that of $dw\left(p(\lambda), \rho(\lambda), \lambda\right)/d\lambda$ at the frozen sonic point of the plane detonation. Regardless of the exact expression of w (reversible or not) one must have $dw/d\lambda|_* < 0$ for λ to relax towards its equilibrium value λ_*. For $\lambda \approx \lambda_*$, one indeed has

$$\frac{d^u \lambda}{dt} = w \approx (\lambda - \lambda_*) \left.\frac{dw}{d\lambda}\right|_*. \qquad (6.4.31)$$

Therefore $K_*' < 0$ and stability of a planar detonation is always obtained in Brun's model.

6.4.3 Comments

Brun's partially reactive, sonic, curved discontinuity calls for three remarks. The first of them is that the quantity $c_\|$ that appears in the evolution equation (6.4.24) can be interpreted as the velocity of propagation of acoustic perturbations along the wave front (Fig. 6.9; from Brun, 1989).

The envelope of the successive loci of the perturbation forms a cone (OFF'), the apex O of which makes an angle ϕ with the direction of the

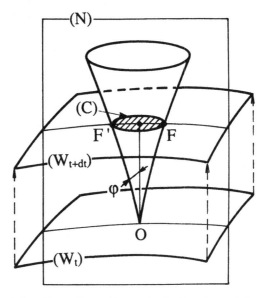

Fig. 6.9.   Spreading of a perturbation issued from O.

wave propagation such that

$$\tan\phi = \frac{c_\parallel}{D}. \tag{6.4.32}$$

Translating the cone (OFF') at points F and F' defines a detonation cell-like pattern, the aspect ratio of which (width/length) is given by (6.4.32). When the reacting explosive is described by the expressions (6.4.29) and (6.4.30), the aspect ratio is a constant equal to

$$A = \left(\frac{\gamma}{3(\gamma+1)}\right)^{1/2}, \tag{6.4.33}$$

so that, for a non-constant-velocity wave front, the segment OF is a curve that intersects all the radii issued from the point O with the same angle ϕ. In the case of an expanding spherical or cylindrical wave, such a curve is by definition a logarithmic spiral, which is a fairly good approximation of experimental crest-imprint paths (see Fig. 6.3). When γ varies from 1 to 3, A ranges from 0.4 to 0.5 which is also nicely close to the experimental observations, even though a bit underestimated ($A_{exp} \sim 0.6$, Section 6.1). This suggests that the observed detonation instabilities involve some "transverse acoustics" in addition to chemistry.

The second remark concerns the basic meaning of modeling the wave as a curved, sonic discontinuity. The underlying idea is that the mean radius of curvature R of the wave front is locally much greater than a characteristic length L representative of the distance of the curved shock front, responsible for the initiation of the chemical processes, to the sonic locus, that is to say, the rear boundary of the domain of dependence of the shock. This situation is analogous to what was done for flames in paragraph 4.7.3. Denoting the ratio L/R by ϵ, one observes that Brun's treatment consists in setting ϵ to zero which permits the use of the jump relations to obtain the flow solution at the sonic locus. It is interesting to observe that Brun's equation reduces to Whitham's equation (1974; the so-called A-M relation) when chemistry is absent from the problem. However the nature of the approximations in the approaches of Brun and of Whitham is different: the flow at a nonreactive shock cannot be strictly sonic (Section 5.2). When ϵ is considered as a small, nonzero, parameter the integration of the reactive flow equations between the shock front and the sonic locus is unavoidable. The latter approach was pioneered by Wood and Kirkwood (1954) and Bdzil (1981) and generalized to the three-dimensional unsteady case by Stewart and Bdzil (1988) with systematic use of the matched asymptotic expansion method. This approach results in a mean shock curvature–shock celerity compatibility relationship, parabolic by nature.

The last remark follows from the second one. By its very nature, Brun's model cannot consider the actual shock initiation and progress of the chemical reaction (cf. Section 6.4.1). These restrictions are not important for the usual engineering purpose of diverging wave front tracking but prevent the model from accounting for the experimental and numerical findings that evidence a flow structuration located in the close vicinity of the detonation shock front. This is why we now proceed to the description of the one-dimensional plane ZND detonation model, its basic features in the next section, a summary of its linear and nonlinear two-dimensional instabilities in Section 6.6.

6.5 Zel'dovich–Von Neumann–Doering model

This model was independently developed by Zel'dovich (1940), Von Neuman (1942) and Doering (1943), (ZND), during WWII and is based upon an idea put forward by Vieille (1900). The detonation is a combustion zone initiated by a nonreactive compressive shock front. Through the shock, the pressure p and the density ρ are raised suddenly and then decrease during the combustion process while the entropy S keeps increasing because of the chemical decomposition. Let us assume that, in a laboratory-fixed frame, the wave is propagating to the left with velocity D, along the x-axis (Fig. 6.10). The ZND structure of the one-dimensional plane detonation can then be represented as the assembly of two subdomains. The first one (I) is the *reaction zone* and extends to the right of the flat shock front. The second one (II) is the nonreactive zone, located between the end of the reaction zone (ERZ) and the right boundary which, without loss in generality, can be conveniently pictured as a moving piston.

In the case of a constant-velocity wave, to which the rest of the paragraph is devoted, the flow in the reaction zone is assumed to be *steady-state*. This means that any of the dependent variables, say g, is only a function of the particular combination of the independent variables x_L (the position of the fluid particle in a laboratory-fixed frame) and t (the time) representing the distance between the initiating shock and the fluid particle under consideration:

$$g\left(x_L, t\right) = g(x), \qquad x = Dt - x_L. \tag{6.5.1}$$

The matching of the steady-state reaction zone with domain (II) is a difficult problem that cannot be studied without saying more about the kinetics of the chemical decomposition process. A careful presentation of this subject is found in Fickett and Davis (1979). In the following lines, we consider the simplest possibility, that of a global irreversible chemical process. The situation is then analogous to that met for fully reactive discontinuities (Section 6.2). The flow beyond the reaction zone can be either time-dependent or constant-state, depending on whether the piston velocity u_p is lower or greater than a minimum value u_p^0. When u_p is lower than u_p^0, the ERZ surface moves with the local forward-facing acoustic perturbation velocity and is equal to the shock-front velocity. The latter has a constant particular value, D_0, even if u_p is not constant. This wave is autonomous because its velocity does not depend upon that of the piston. When u_p is constant and greater than u_p^0, the ERZ surface moves with a constant velocity, again equal to the shock-front velocity D but

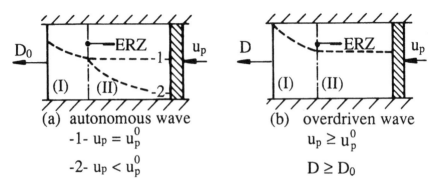

(a) autonomous wave

-1- $u_p = u_p^0$

-2- $u_p < u_p^0$

(b) overdriven wave

$u_p \geq u_p^0$

$D \geq D_0$

Fig. 6.10. One-dimensional plane ZND detonation (dashed lines are particle velocity profiles).

lower than the local forward-facing acoustic perturbation velocity. The particle velocity on this locus is equal to that of the piston and D is greater than D_0. The wave is termed overdriven because the shock-velocity depends upon that of the piston: the domain between the shock front and the piston is entirely subsonic; hence all the perturbations originating from the piston catch up with the shock front. In the autonomous-wave flow, the perturbations cannot go further than the ERZ sonic surface. In both cases, the value D of the shock velocity is determined by the value of the particle velocity at the end of the reaction zone. As in the previous paragraphs we assume that the initial properties of the explosive medium are constant and uniform.

6.5.1 The steady-state, one-dimensional, plane reaction zone

In this paragraph, we summarize the mathematics of the steady-state reaction zone. We firstly demonstrate that the steady-state assumption is compatible with a constant-velocity plane shock and then show how to calculate the latter velocity. To this end, we again consider an inviscid adiabatic reactive fluid and consequently write the one-dimensional plane conservation equations (Section 2.2) as

$$\frac{d^0 \rho}{dt} + \rho \frac{\partial u}{\partial x_L} = 0 , \qquad (6.5.2)$$

$$\rho \frac{d^0 u}{dt} + \frac{\partial p}{\partial x_L} = 0 , \qquad (6.5.3)$$

$$\frac{d^0 e}{dt} + p \frac{d^0 V}{dt} = 0 , \qquad (6.5.4)$$

where e is the specific internal energy, $V = \rho^{-1}$ the specific volume and u the particle velocity defined in a laboratory-fixed frame. Considering a chemical kinetic scheme with n independent reactions, the extents of which are described by the K progress variables λ_k, we introduce an $e(p, V, \boldsymbol{\lambda})$ equation of state that we differentiate along a particle path $x_L^0(t)$ $(dx_L^0/dt = u)$ to obtain an equivalence for the adiabaticity equation (6.5.4)

$$\frac{d^0 p}{dt} - c^2 \frac{d^0 \rho}{dt} = \rho c^2 \boldsymbol{\sigma} \cdot \mathbf{w} , \qquad (6.5.5)$$

where c, $\boldsymbol{\sigma}$ and $\mathbf{w}$ are state-dependent functions denoting the frozen sound speed, the thermicity vector and the chemical decomposition rate vectors:

$$c^2(p, \rho, \boldsymbol{\lambda}) = \left.\frac{\partial p}{\partial \rho}\right|_{\lambda S} = \frac{\frac{p}{\rho^2} - \left.\frac{\partial e}{\partial \rho}\right|_{\lambda p}}{\left.\frac{\partial e}{\partial p}\right|_{\lambda \rho}},$$

$$\rho c^2 \sigma_k(p, \rho, \boldsymbol{\lambda}) = \left.\frac{\partial p}{\partial \lambda_k}\right|_{e, \rho, \lambda_{j \neq k}},$$

$$\frac{d^0 \lambda_k}{dt} = w_k(p, V, \boldsymbol{\lambda}).$$

(6.5.6)

At the wave front, $\boldsymbol{\lambda}$ is equal to zero since the shock is responsible for the chemical decomposition process. Let us perform the combination $c^2 \times$ (6.5.2) $+ (D - u) \times$ (6.5.3) $+$ (6.5.4), where D is the shock velocity. We find

$$\frac{d^S p}{dt} + \rho(D - u)\frac{d^S u}{dt} = \rho c^2 \left(\boldsymbol{\sigma} \cdot \mathbf{w} - (1 - M^2)\frac{\partial u}{\partial x_L}\right), \qquad (6.5.7)$$

where $M = (D - u)/c$ and $d^S g/dt$ denote the Mach number of the flow relative to a shock-fixed frame and the time derivative of any dependent variable g along a path $x_S(t)$ parallel to that of the shock $(dx_S/dt = D)$ at fixed t. We now recall that the value of g at the shock front (subscript "H") can be expressed as function of the shock velocity D. Thus evaluating (6.5.7) at the shock, we obtain the wave front acceleration

$$\frac{d^S D}{dt} = F(D)\left(\boldsymbol{\sigma}_H \cdot \mathbf{w}_H - \left(1 - M_H^2\right)\left.\frac{\partial u}{\partial x_L}\right|_H\right), \qquad (6.5.8)$$

$$F(D) = \frac{\rho_H c_H^2}{dp_H/dD + \rho_0 D du_H/dD}, \qquad g_H = g_H(D), \qquad \lambda_H = 0.$$

Hence, the condition for the shock to propagate at constant velocity is that the shock-value of the particle velocity gradient be

$$\left.\frac{\partial u}{\partial x_L}\right|_H = \frac{\boldsymbol{\sigma}_H \cdot \mathbf{w}_H}{1 - M_H^2}. \qquad (6.5.9)$$

We next show that the steady-state assumption necessarily leads to gradients satisfying (6.5.9). From (6.5.1), we deduce

$$\frac{\partial g}{\partial x_L} = -\frac{dg}{dx}, \qquad \frac{\partial g}{\partial t} = D\frac{dg}{dx},$$

$$\frac{d^0 g}{dt} = \frac{\partial g}{\partial t} + u\frac{\partial g}{\partial x_L} = q\frac{dg}{dx}, \qquad q = D - u, \qquad (6.5.10)$$

where q denotes the particle velocity with respect to the shock-fixed frame (relative particle velocity). Inserting the above identities in the conservation equations (6.5.2), (6.5.3) and (6.5.4) yields a system of three equations in the three unknown derivatives dp/dx, dq/dx and $d\rho/dx$

$$q\frac{d\rho}{dx} + \rho\frac{dq}{dx} = 0 , \tag{6.5.11}$$

$$\rho q\frac{dq}{dx} + \frac{dp}{dx} = 0 , \tag{6.5.12}$$

$$\frac{dp}{dx} - c^2\frac{d\rho}{dx} = \frac{\rho c^2 \boldsymbol{\sigma}\cdot\mathbf{w}}{D - u} , \tag{6.5.13}$$

which in particular leads to

$$-\frac{dq}{dx} = \frac{\partial u}{\partial x_{\mathrm{L}}} = \frac{\boldsymbol{\sigma}\cdot\mathbf{w}}{1 - M^2} . \tag{6.5.14}$$

Evaluated at the shock, the latter expression obviously reduces to (6.5.9). We now turn to the problem of calculating the shock velocity and firstly notice that the system (6.5.4)-(6.5.11)-(6.5.12) can be integrated in the form

$$\rho q = C_1 , \quad p + \rho q^2 = C_2 , \quad h + \tfrac{1}{2}q^2 = C_3 , \tag{6.5.15}$$

where h is the specific enthalpy ($h = e + pV$) and the C_j's are constants determined by the shock boundary conditions. We have $C_1 = \rho_0 D$, $C_2 = p_0 + \rho_0 D^2$, $C_3 = h_0 + \tfrac{1}{2}D^2$. Together with an $h(p, V, \boldsymbol{\lambda})$ equation of state, the latter equations (6.5.15) define a system of four equations in the three unknowns p, ρ, q and the K unknowns λ_k, with the velocity D and the initial-state data p_0 and ρ_0 as parameters. Therefore p, ρ and q can be expressed as functions of (λ_k), D, p_0 and ρ_0. The integrated form (6.5.15) is similar to that of the jump relations, thus defining a one-parameter (D) family of waves (k); hence the solution can be represented in the (p, V) plane as a point of intersection of the Rayleigh line (R) and of a so-called partially reactive Hugoniot curve (H_i) which corresponds to the fraction $\boldsymbol{\lambda}\cdot\mathbf{Q}$ of the total available chemical energy. In the case of a single irreversible reaction, to which we now restrict our presentation, we obtain, as in Section 6.2, a double solution on the fully-reactive Hugoniot (H_1) ($\lambda = 1$, Crussard curve) with a minimum characteristic solution when (H_1) and (R) are tangent (Fig. 6.11).

As an example, let us consider the usual case of a polytropic medium with initial pressure neglected. The equation of state is

$$h - h_0 = \frac{\gamma p}{(\gamma - 1)\rho} - \lambda Q , \tag{6.5.16}$$

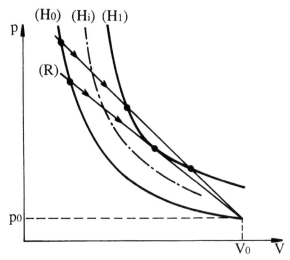

Fig. 6.11. The $(p\text{-}V)$ plane for the ZND detonation model and a single irreversible chemical process. Arrows indicate the path of the fluid particles in the steady-state reaction zone.

so that the equation of the partially reactive Hugoniot curve is

$$\frac{pV_0}{D_0^2} = \frac{\lambda}{(\gamma + 1)(V/V_0) - (\gamma - 1)} \qquad (6.5.17)$$

and the solution to system (6.5.15–16) is

$$\frac{u}{D} = \frac{p}{\rho_0 D^2} = \frac{1 + \ell}{\gamma + 1}, \quad \frac{\rho_0}{\rho} = \frac{V_0}{V} = \frac{q}{D} = \frac{\gamma - \ell}{\gamma + 1}, \quad M^2 = \frac{\gamma - \ell}{\gamma(1 + \ell)}, \qquad (6.5.18)$$

where

$$\ell = (1 - \lambda/f)^{1/2}, \qquad f = \left(\frac{D}{D_0}\right)^2, \qquad D_0^2 = 2\left(\gamma^2 - 1\right)Q,$$

f being the *degree of overdrive*. According to the mechanism previously described, the particle velocity u at the end of the reaction zone associated with an overdriven wave is that u_p of the piston. We thus find that the shock velocity is given by

$$D = \frac{\gamma + 1}{2} u_p \left(1 + \left(\frac{1}{\gamma + 1}\frac{D_0}{u_p}\right)^2\right). \qquad (6.5.19)$$

Considering the positive values of D we observe that D/D_0 is minimal and equals one when u_p/D_0 is equal to $1/(\gamma + 1)$. This value of u_p is precisely the critical value u_p^0, below or for which the wave is autonomous, because one checks that M is equal to one when u_p is equal to u_p^0. The same considerations apply to the CJ model. Under the choice of the polytropic equation of state (6.5.16) and of a single irreversible process, the particular value D_0 is the sought-after velocity of the autonomous ZND plane detonation wave, and has the same expression as those associated, under the same conditions, to the CJ and Brun's models. To compute the profile of the dependent variables in the steady-state reaction zone, an expression for the chemical decomposition rate is required because these variables are functions of λ in this domain. The solution associated with a given degree of overdrive f can thus be implicitly written as

$$t(\lambda) = \int_0^\lambda \frac{d\lambda'}{w\left(p\left(\lambda'\right), \rho\left(\lambda'\right), \lambda'\right)}, \qquad (6.5.20a)$$

or

$$x(\lambda) = \int_0^\lambda \frac{q\left(\lambda'\right) d\lambda'}{w\left(p\left(\lambda'\right), \rho\left(\lambda'\right), \lambda'\right)}. \qquad (6.5.20b)$$

The total reaction time t_* and the reaction zone length x_* are $t(1)$ and $x(1)$ respectively. Considering the following expression for a single irreversible kinetic

$$w(p, \rho, \lambda) = k(p, \rho, \lambda)(1 - \lambda)^\nu \qquad (6.5.21)$$

with k and ν positive function and positive constant respectively, we observe that t_* and x_* are infinite when $\nu \geq 1$ and finite when $\nu < 1$. In the case where k is a constant and $D = D_0$, we have the expressions

$$t(\lambda) = \frac{1 - (1 - \lambda)^{1-\nu}}{k(1 - \nu)}, \qquad (6.5.22)$$

$$\frac{(\gamma + 1)x(t)}{D_0} = \gamma t - \frac{1 - (1 - kt(1 - \nu))^a}{k\left(\frac{3}{2} - \nu\right)}, \qquad (6.5.23)$$

$$a = \frac{\frac{3}{2} - \nu}{1 - \nu}, \qquad x_* = \frac{D_0}{k}f(\gamma, \nu), \qquad (6.5.24a)$$

$$f(\gamma, \nu) = \frac{\gamma(3 - 2\nu) - 2(1 - \nu)}{(\gamma + 1)(1 - \nu)(3 - 2\nu)}. \qquad (6.5.24b)$$

Equations (6.5.22) are valid even when $D > D_0$.

6.5.2 Comments

We conclude this summary of the main features of the plane detonation
ZND model with a few words on the autonomous wave, for which the
ERZ surface moves with the local forward-facing acoustic perturbations.
More specifically, we address the problem of calculating the gradients
on this locus under the assumption of a global irreversible chemical
decomposition process. The general expression for the decomposition
rate of such a process is given by (6.5.21). The "sonic" surface is, *a
priori*, a surface of discontinuity for the gradients because it separates
a time-dependent domain (II) from a steady-state one (I) (Fig. 6.10).
On the "steady-state" side of the ERZ surface (subscript "ST"), the
gradients are undetermined: considering the particle velocity gradient
expression (6.5.14), we observe that its numerator and denominator
vanish at the same time ($w = 0$, $M = 1$). Gradient discontinuities
and undeterminacy are not restricted to the particular case of the
one-dimensional, plane, steady-state detonation but should really be
considered within the framework of the theory of hyperbolic equations.
Indeed, the latter indicates that the only possible loci for discontinuities
in the state-variable derivatives are characteristic surfaces (or envelopes
of), provided that the state variables remain constant through them. On
such surfaces, as a rule, the derivatives are undetermined and a non-
local (e.g., space-time) information must be brought into the analysis to
remove the undeterminacy. Here the non-local information is represented
by the steady-state assumption which allows one to express the gradients
as functions of the state variables at any point of the reaction zone (see
(6.5.14) and (6.5.18) for example). In the particular case where the
properties of the medium are described by the equation of state (6.5.16),
we obtain, on the steady-state side of the ERZ surface (subscript ST),
the following limit:

$$\left.\frac{\partial u}{\partial x_L}\right|_{\mathrm{ST}} = \frac{\nu k_{\mathrm{ST}}}{\gamma}\left(1 - \lambda_{\mathrm{ST}}\right)^{\nu-1/2}, \qquad \lambda_{\mathrm{ST}} = 1. \qquad (6.5.25)$$

We thus observe that the gradient is infinite when $0 < \nu < 1/2$, finite
when $\nu = 1/2$ and vanishes when $\nu > 1/2$. Brun (1989) has shown that
this result also holds for an arbitrary equation of state. Because p, ρ,
c, u, T... can all be expressed in terms of λ (equations (6.5.1–15)), the
typical thickness l of a steady ZND detonation is given by:

$$l = \mathcal{O}\left(c\,t_{\mathrm{ch}}\right)$$

since the fluid velocity relative to the leading shock wave is comparable
to the local speed of sound c. It is instructive to compare l to the
convection/conduction length $l_T = \mathcal{O}\left(D_{\mathrm{th}}/c\right)$ built upon the local heat

diffusivity D_{th}. Using $D_{th} \sim c^2 t_{coll}$, where t_{coll} is a local collision time, one deduces that:

$$l/l_T = \mathcal{O}\left(t_{ch}/t_{coll}\right) .$$

Reactive collisions being scarce, t_{ch} is much longer than t_{coll} and $l \gg l_T$. As a consequence, if the Prandtl and Lewis numbers are of order unity, one may neglect the diffusive transports of heat, mass and momentum in comparison to their convective counterparts. In other words, the reactive Euler equations are enough to study the flow field behind the leading shock of a steady and planar ZND detonation. This is also true for unsteady detonation waves when, in a shock-fixed frame of reference, the structure of the flow field does not vary significantly on times that are shorter than t_{ch}.

6.6 Chemistry-related instabilities

As Sections 6.3 and 6.4 demonstrated, one cannot explain the appearance and development of cellular detonation structures without taking into account far-from-equilibrium chemistry. In other words, one has to investigate how disturbances, which may be assumed small to begin with, evolve when superimposed on the basic one-dimensional, steady ZND solutions (Section 6.5); the next logical step is to investigate how nonlinearity would "saturate" the instabilities. Unfortunately, even the linear step is hampered by great technical difficulties because of a familiar difficulty (Section 4.5.1): once again, the linearized equations to be solved have nonconstant coefficients (Section 6.6.1). Erpenbeck (1970, and the (many) references therein) was the first to consider the problem within a sound mathematical framework and to obtain a few (albeit important) results in the case of *Arrhenius kinetics*. We choose not to explain his contribution, a rather delicate blend of subtle complex-plane methods and of numerical analysis which is beyond the scope of the present introductory course; nevertheless one has to do Erpenbeck justice by acknowledging the importance of his work which almost pushed the studies of detonation stability to the forefront of what the available analytical tools permitted at his time (the computers were also somewhat slower than nowadays). Instead we present the comparatively simpler normal-mode approach (Section 6.6.1-2) which is employed in more recent treatments of the problem and gave the most complete result to date (Section 6.6.3).

6.6.1 The linearized system

As in Section 5.4 and Section 6.5, we assume that the wave travels leftward and horizontally in a laboratory-fixed frame. We shall restrict our presentation to the case of a reactive medium undergoing a global irreversible chemical decomposition process; $\mathbf{g} = (\rho, u, v, p, \lambda) = (g_n, \ n = 1, 5)$ will denote the vector of dependent variables (ρ density, u and v the horizontal and vertical component of the particle velocity, p pressure, λ progress variable). For the same reasons as those given in Section 5.4, only a two-dimensional problem need be considered (hence a single v). We shall use overbars and primes to refer to the unperturbed reference state and to any perturbation, respectively. We recall that the reference state is described by a one-dimensional planar, steady-state, reactive and adiabatic flow which sustains a constant velocity shock (ZND model, Section 6.5). We firstly write the equations for the two-dimensional unsteady reactive and adiabatic flow in an unperturbed shock-fixed frame. The resulting system has the same form as in a laboratory-fixed frame because the equations under consideration are invariant under Galilean transformations. Let x, z and t denote the horizontal and vertical coordinates and the time, respectively. Thus, x represents the (positive) distance between the unperturbed shock and the fluid particle upon consideration. We have

$$
\left(\mathbf{I} \frac{\partial}{\partial t} + \mathbf{A}_1(\mathbf{g}) \frac{\partial}{\partial x} + \mathbf{A}_2(\mathbf{g}) \frac{\partial}{\partial z} \right) \cdot \mathbf{g} = \mathbf{d}(\mathbf{g}) , \tag{6.6.1}
$$

with $\mathbf{d} = (0, 0, 0, \rho c^2 \sigma w, w)$ and:

$$
\mathbf{A}_1 = \begin{bmatrix} u & \rho & 0 & 0 & 0 \\ 0 & u & 0 & \rho^{-1} & 0 \\ 0 & 0 & u & 0 & 0 \\ 0 & \rho c^2 & 0 & u & 0 \\ 0 & 0 & 0 & 0 & u \end{bmatrix} \quad \text{and} \quad \mathbf{A}_2 = \begin{bmatrix} v & 0 & \rho & 0 & 0 \\ 0 & v & 0 & 0 & 0 \\ 0 & 0 & v & \rho^{-1} & 0 \\ 0 & 0 & \rho c^2 & v & 0 \\ 0 & 0 & 0 & 0 & v \end{bmatrix}
$$

We next linearize the above system around the one-dimensional planar reference state $\bar{\mathbf{g}} = (\bar{\rho}, \bar{u}, 0, \bar{p}, \bar{\lambda})$ according to the rule

$$
\mathbf{g} = \bar{\mathbf{g}} + \phi \mathbf{g}' , \quad (\bar{\mathbf{g}}, \mathbf{g}') = \mathcal{O}(1) , \quad \phi = \mathcal{O}(\epsilon) , \quad \epsilon \ll 1 , \tag{6.6.2}
$$

where $\phi(t, z)$ denotes the perturbed shock position. We obtain

$$
\mathbf{A}(\bar{\mathbf{g}}) \cdot \mathbf{g}' = 0 ,
$$
$$
\mathbf{A} = \mathbf{I} \frac{\partial \phi}{\partial t} + \phi \mathbf{A}_1'(\bar{\mathbf{g}}) \frac{\partial}{\partial x} + \mathbf{A}_2'(\bar{\mathbf{g}}) \frac{\partial \phi}{\partial z} + \mathbf{B}'(\bar{\mathbf{g}}) \phi , \tag{6.6.3}
$$

where the $\mathbf{A}'_i$ are inferred from the $\mathbf{A}_i$ by replacing g_n by $\bar{g}_n$, and the matrix $\mathbf{B}'$ reads

$$\mathbf{B}' = \begin{bmatrix} \dfrac{\partial \bar{u}}{\partial x} & \dfrac{\partial \bar{p}}{\partial x} & 0 & 0 & 0 \\[2mm] \dfrac{\partial \bar{p}}{\partial x}\Big/\bar{\rho}^2 & \dfrac{\partial \bar{u}}{\partial x} & 0 & 0 & 0 \\[2mm] 0 & 0 & 0 & 0 & 0 \\[2mm] \bar{\pi}_\rho & \dfrac{\partial \bar{p}}{\partial x} & 0 & \bar{\pi}_p & \bar{\pi}_\lambda \\[2mm] -\bar{w}_\rho & \dfrac{\partial \bar{\lambda}}{\partial x} & 0 & -\bar{w}_p & -\bar{w}_\lambda \end{bmatrix}$$

where

$$\bar{\pi}_\alpha = \frac{\partial \bar{u}}{\partial x}\frac{\partial \bar{\rho}\bar{c}^2}{\partial \bar{\alpha}} - \left(\bar{\rho}\bar{c}^2 \bar{\sigma}\bar{w}\right)_\alpha, \qquad \bar{w}_\alpha = \frac{\partial \bar{w}}{\partial \bar{\alpha}}, \qquad \bar{\alpha} = \bar{\rho}, \bar{p}, \bar{\lambda}.$$

We now recall (cf. Section 6.5) that the $\bar{g}_n$'s and the $\partial \bar{g}_n/\partial x$'s are functions of $\bar{\lambda}$ and that $\bar{\lambda}$ can be expressed in terms of x only, once the chemical decomposition rate $w(p, \rho, \lambda)$ is specified. Accordingly, equations (6.6.2) admit elementary solutions (the so-called normal modes) of the form

$$\mathbf{g}' = \mathbf{g}'(x), \quad \phi = \epsilon \exp(st + ikz), \quad \mathcal{I}m(k) = 0. \tag{6.6.4}$$

After insertion of (6.6.4) in (6.6.3) we obtain the following differential system

$$\frac{d\mathbf{g}'}{dx} = \mathbf{M} \cdot \mathbf{g}', \tag{6.6.5}$$

where

$$\mathbf{M} = \mathbf{M}(s, k, x) = (\mathbf{A}'_1)^{-1} \cdot (s\mathbf{I} + ik\mathbf{A}'_2 + \mathbf{B}').$$

System (6.6.5) has to be integrated from the unperturbed shock position ($x = 0^+$) to downstream infinity, where $\bar{\lambda} = 1$. The initial (shock) value $\mathbf{g}'_\mathrm{H}$ of $\mathbf{g}'$ is obtained by linearizing the jump relations (cf. Section 2.2) at the nonreactive perturbed shock. We ultimately find

$$\mathbf{g}'_\mathrm{H} = -\frac{\partial \bar{\mathbf{g}}}{\partial x}\bigg|_{0^+}$$

$$- s\frac{1 - (\bar{u}_\mathrm{H}/\bar{D})}{1 + H}\left(\frac{2\rho_0 \bar{D}H}{\bar{u}_\mathrm{H}^2}, 1 - H, i\frac{\bar{D}(1 + H)k}{s}, -2\rho_0\bar{D}, 0\right). \tag{6.6.6}$$

The subscripts "0" and "H" denote the initial and the unperturbed-shock states, respectively, and $-H$ is the ratio of the Hugoniot curve slope to that of the Rayleigh line (cf. Section 5.5). $\bar{D}$ is the unperturbed wave velocity. For a given initial state, we have $\mathbf{g}'_\mathrm{H} = \mathbf{g}'_\mathrm{H}(s, k)$ once the equation of state is specified. The presence of the gradient term $\partial \bar{g}/\partial x|_{0+}$ in the boundary condition (6.6.6) is due to the fact that the shock conditions must be shifted from the perturbed shock position $(x = \phi)$, the only locus where strictly speaking they apply, to the unperturbed shock position $(x = 0^+)$ which represents the origin of our relative reference frame. This is analogous to what happened for flames (Section 4.5). In a perturbed shock-fixed frame the gradient term would not appear in (6.6.6) but the system (6.6.5) would not be homogeneous because the transformation of the conservation laws from a laboratory-fixed frame to an accelerating one is not Galilean.

6.6.2   The hot-boundary condition

Now, how to get the dispersion relation $s(k)$ from an integration (in general performed numerically) of (6.6.5–6)? To achieve this goal, it is convenient to anticipate the existence of a relation

$$G\left(s, k, g'_{b1}, \ldots, g'_{b5}\right) = 0 \qquad (6.6.7)$$

involving the components $g'_{bn}, n = 1 \ldots 5$, of g' at the end of the reaction zone $(x \to +\infty$, subscript "b"). Equations (6.6.5–7) depend implicitly on the parameters of the equation of state and of the chemical rate, along with those defining the initial state (p_0, ρ_0) of the medium; they also depend on the unperturbed detonation velocity $\bar{D} \geq D_0$ (Section 6.5). The construction of $s(k)$ can thus be viewed as a 3-step shooting problem, with (6.6.7) as target:

1) Pick a (new) set of control parameters
2) Make a (new) choice for s and k and integrate (6.6.5), with (6.6.6) as initial condition at $x = 0^+$, towards $x = +\infty$.
3) Check whether the condition (6.6.7) is met at a "large enough" x.
    3.1) if "No" return to step 2;
    3.2) if "Yes" stop or go back to step 1.

The crux of the method is thus to find an explicit form for (6.6.7). To this end we note that the matrix $\mathbf{M}$ showing up in (6.6.5) tends to a constant $\mathbf{M}_b$ as $x \to \infty$ (cf. Section 6.5.2). Then, the equation for $\mathbf{g}'$ admits solutions in the form

$$\mathbf{g}'_b = \mathbf{g}''_b \exp(qx) \qquad (6.6.8)$$

as $x \to +\infty$, where the constants q and $\mathbf{g}''_b$ must be found. Inserting (6.6.8) into (6.6.5) yields the homogeneous system $(\mathbf{M}_b - q\mathbf{I}) \cdot \mathbf{g}''_b = 0$;

hence the condition $\det(\mathbf{M_b} - q\mathbf{I}) = 0$ or, explicitly

$$\begin{vmatrix} \bar{s} & \bar{\rho}_b\bar{q} & i\bar{\rho}_b k & 0 & 0 \\ 0 & \bar{\rho}_b\bar{s} & 0 & q & 0 \\ 0 & 0 & \bar{\rho}_b\bar{s} & ik & 0 \\ -\bar{c}_b^2\bar{s} & 0 & 0 & \bar{s} & -\bar{\rho}_b\bar{c}_b^2\bar{\sigma}_b\bar{w}_\lambda \\ 0 & 0 & 0 & 0 & \bar{s} - \bar{w}_\lambda \end{vmatrix} = 0, \qquad (6.6.9)$$

where $\bar{w}_\lambda \equiv (\partial\bar{w}/\partial\bar{\lambda})$, and $\bar{s} = s + q\bar{u}_b$ where q appears as an eigenvalue. Equation (6.6.9) can be expanded as

$$\bar{s}^2\left(\bar{s}^2 - \bar{c}_b^2\left(q^2 - k^2\right)\right)(\bar{s} - \bar{w}_\lambda) = 0. \qquad (6.6.10)$$

Condition (6.6.10) is clearly the reactive analog of (5.4.12). Solving (6.6.10) leads one to distinguish among four roots and five independent asymptotic modes $\mathbf{g}_{bj}''$, $j = 1, \ldots, 5$. The one-fold degenerate root $q_1 = q_2 = -s/\bar{u}_b$ defines both entropy (q_1) and shear (q_2) disturbances and corresponds to the particular eigenmodes

$$\mathbf{g}_{b1}'' = \rho_1''(1, 0, 0, 0, 0), \qquad (6.6.11)$$
$$\mathbf{g}_{b2}'' = u_2''(0, 1, -is/k\bar{u}_b, 0, 0). \qquad (6.6.12)$$

The next two roots q_3, q_4 define acoustic modes evolving at frozen composition and are solutions to:

$$(s + q\bar{u}_b)^2 - \bar{c}_b^2\left(q^2 - k^2\right) = 0, \qquad (6.6.13)$$

the associated eigenmodes being $(j = 3, 4)$:

$$\mathbf{g}_{bj}'' = u_j''\left(-\bar{\rho}_b\left(s + q_j\bar{u}_b\right)/\bar{c}_b^2 q_j, 1, ik/q_j, -\bar{\rho}_b\left(s + q_j\bar{u}_b\right)/q_j, 0\right). \qquad (6.6.14)$$

The last root $q_5 = (\bar{w}_\lambda - s)/\bar{u}_b$ defines the "chemical" mode, with eigenvector:

$$\mathbf{g}_{b5}'' = u_5''\left(-\bar{\rho}_b\alpha_1/\bar{u}_b, 1, i\alpha_3, -\bar{\rho}_b\bar{u}_b\alpha_4, \alpha_5/\bar{\sigma}_b\bar{u}_b\right) \qquad (6.6.15)$$

where the α_i's are the dimensionless coefficients

$$\alpha_4 = (1 - s/\bar{w}_\lambda)^{-1}, \qquad \alpha_1 = \alpha_4\left((1 - s/\bar{w}_\lambda)^2 - (k\bar{u}_b/\bar{w}_\lambda)^2\right),$$
$$\alpha_3 = \alpha_4\left(k\bar{u}_b/\bar{w}_\lambda\right), \qquad \alpha_5 = \alpha_1 - \alpha_4\bar{M}_b^2,$$

and $\bar{M}_b = \bar{u}_b/\bar{c}_b$.

The general asymptotic solution $\mathbf{g}_b'$ is thus given by

$$\mathbf{g}_b' = \sum_{j=1}^{5} C_j\mathbf{g}_{bj}''\exp(q_j x), \qquad (6.6.16)$$

where the C_j's are arbitrary constants that can absorb ρ_1'' and u_j'', $j = 2\ldots5$. For (6.6.16) to be admissible, it must yield a solution $\mathbf{g}_{bj}'$

which is bounded at $x = \infty$. To select the adequate $\mathbf{g}'_{bj}$, we restrict our attention to unstable situations. One must then impose:

$$\mathrm{Re}(q) < 0 \qquad \text{when} \qquad \mathrm{Re}(s) > 0 \,. \qquad (6.6.17)$$

Clearly only the modes 3 or 4 can violate (6.6.17) and, consequently, C_3 or C_4 must be set to zero in (6.6.16). For sake of definiteness, we assume that $C_4 = 0$ (once s is obtained, one will have to check that q_4 is the root which does not meet (6.6.17)). Then (6.6.16) can be rewritten as

$$C_1 \mathbf{g}'_{b1} + C_2 \mathbf{g}'_{b2} + C_3 \mathbf{g}'_{b3} + C_5 \mathbf{g}'_{b5} - \mathbf{g}'_b = 0 \,, \qquad (6.6.18)$$

or, equivalently, into a homogeneous system $\mathbf{F} \cdot \mathbf{X} = \mathbf{0}$ for the trivially non-zero vector $\mathbf{X} = (C_1 \exp(q_1 x), C_2 \exp(q_2 x), C_3 \exp(q_3 x), C_5 \exp(q_5 x), -1)$, where the columns of the matrix $\mathbf{F}$ are the $\mathbf{g}''_{bj}$, $j = 1, 2, 3, 5$ and $\mathbf{g}'_b$. Setting its determinant to zero provides one with the sought-after condition (6.6.7) in the form:

$$\det(\mathbf{F}) = \begin{vmatrix} 1 & 0 & -M_b^2 (1 + s/q\bar{u}_b) & -\alpha_1 & -\rho'_b/\bar{\rho}_b \\ 0 & 1 & 1 & 1 & -u'_b/\bar{u}_b \\ 0 & -is/k\bar{u}_b & ik/q & i\alpha_3 & -v'_b/\bar{u}_b \\ 0 & 0 & -(1 + s/q\bar{u}_b) & -\alpha_4 & -p'_b/\bar{\rho}_b\bar{u}_b^2 \\ 0 & 0 & 0 & \alpha_5 & -\bar{\sigma}_b\lambda'_b \end{vmatrix} = 0$$

or, explicitly :

$$m_1 \left[\left(1 + \frac{s}{q\bar{u}_b} \right) \left(\frac{u'_b}{\bar{u}_b} - i\frac{k\bar{u}_b}{s}\frac{v'_b}{\bar{u}_b} \right) + \left(1 + \frac{k^2\bar{u}_b}{qs} \right) \frac{p'_b}{\bar{\rho}_b\bar{u}_b^2} \right]$$

$$+ \bar{\sigma}_b\lambda'_b \left[\left(1 + \frac{k^2\bar{u}_b}{qs} \right) - m_2 \left(1 + \frac{s}{q\bar{u}_b} \right) \right] = 0 \qquad (6.6.19)$$

with

$$m_1 = \left(1 - \frac{s}{\bar{w}_\lambda} \right)^2 - \left(\frac{k\bar{u}_b}{\bar{w}_\lambda} \right)^2 - \bar{M}_b^2 \,, \qquad m_2 = 1 - \frac{s}{\bar{w}_\lambda} + \frac{k^2\bar{u}_b^2}{s\bar{w}_\lambda} \,.$$

Recall that q in (6.6.19) stands for the root q_3 (or q_4) which fulfills (6.6.17). It is instructive to note that, by allowing $\bar{w}_\lambda$ to go to zero, (6.6.19) resumes the relation (5.4.21) of the D'yakov–Kontorovich problem (once $\bar{u}_b, u'_b \ldots$ are identified with $\bar{u}_H, u'_H \ldots$).

6.6.3 Partial linear results

So far the preceding numerical program has been applied to the particular case of a polytropic perfect gas ($p/\rho \sim T$, $\gamma \equiv C_p/C_V = \mathrm{const.}$, Q the heat of reaction per unit mass of the mixture) which undergoes a first-order, irreversible chemical reaction of the Arrhenius type, i.e.,

$$w = \tau^{-1}(1 - \lambda) \exp(-E/RT) \,, \qquad (6.6.20)$$

where E is the activation energy. Even in such a simple framework, it is a four-dimensional space of control parameters which must be explored: beside γ, E/RT_0 and Q/RT_0, all the results are parametrized by the degree of overdrive $f = D/D_{CJ} \geq 1$ pertaining to the basic profiles, in addition to the current wavenumber $|k|$.

Lee and Stewart (1990) restricted their attention to planar disturbances ($k = 0$) and clearly demonstrated that, even in such instances, ZND detonations equipped with Arrhenius kinetics may be unstable. For a fixed γ, the trends are as follows. Increasing E/RT_0 and/or reducing f may ultimately lead to oscillatory instability (i.e., a pulsating detonation velocity). This happens first as a Hopf bifurcation (a pair of complete-conjugate growth rates crosses the real s-axis) and, at the onset, $|\mathcal{I}m(s)|$ is of course proportional to the reciprocal chemical-time. When E/RT_H is increased further, the number of unstable branches of $s(k)$ is found to grow, possibly limitlessly; in general, the Chapman–Jouguet, ZND detonations ($f \to 1^+$) tend to be the least stable. The critical values of $1/f$ and of E/RT_H are found to increase with decreasing $\gamma - 1$ and/or Q/RT_0.

The more recent computations of Bourlioux and Majda (1992) revealed that allowing for wrinkles ($k \neq 0$) leads to instabilities in even wider conditions than in the planar case (this also held for flames, Section 4.5). At the onset, the growth rates s were invariably found to be imaginary and the corresponding wavenumber k_c finite (and proportional to a reciprocal chemical length, l_{ch}): *traveling waves* propagating in both transverse directions are thus predicted in such instances. In most cases the instability is confined to a finite band of wavenumbers around k_c (Fig. 6.12); deep enough in the instability domains, this band may include the origin $k = 0$. As shown analytically by Erpenbeck (1966) through a delicate WKB analysis (Stokes lines,...), the range of instability can even extend to arbitrarily short wavelengths, under the necessary condition that $c^2(x) - u^2(x)$ monotonically increases with streamwise coordinate x or exhibits a single maximum. As for the influence of γ, E/RT_0, Q/RT_0 and f, they are qualitatively the same as in the planar configurations.

6.7 Recent results and prospects

The attentive reader has now certainly realized that the problems of detonation stability are still awaiting a neat theoretical explanation, as we announced in the Introduction. Therefore, in the absence of any certitude, we shall henceforth content ourselves with mentioning a few attempts and/or suggestions.

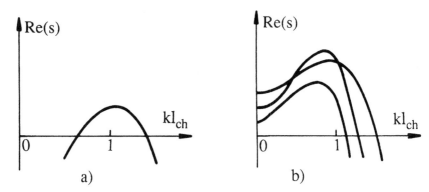

Fig. 6.12. Two dispersion relations (sketched from Bourlioux and Majda (1992)); a) $\gamma = 1.2$, $E/RT_0 = 10$, $Q/RT_0 = 50$, $f = 1.2$; b) same as a) except $E/RT_0 = 50$.

As mentioned earlier, the linear stage already poses difficulties of technical origin, because the linearized equations (6.6.5) have variable coefficients even when the *a priori* simple Arrhenius kinetics is retained. It may thus be worth paying more attention to simpler models involving (piecewise-)constant coefficients, in a way similar to what was done for flames (Section 4.5). The so-called *square wave model* (Zaidel, 1961) belongs to this class: the leading shock is followed by a thermally-neutral reaction whose development is controlled by an Arrhenius rate; once the corresponding progress-variable (a sort of a timer) reaches unity (end of induction), the entire chemical heat is released instantaneously. Unfortunately, Erpenbeck (1963b) showed that the square-wave model leads to growth rates with arbitrarily high real and imaginary parts, presumably because the heat-release region is too thin. A variant of this model (Abouseif and Toong, 1982), where the latter region is thickened, leads to a finite number of unstable roots when "realistic" control parameters are employed. Exploiting the limit of large activation energies $(E/RT_H \rightarrow \infty)$, was also attempted (Buckmaster and Nevis, 1988). In this limit, the basic profiles tend to be square-shaped with a heat-release region of $\mathcal{O}(RT_H/E)$ relative thickness and, not surprisingly, unstable modes with increasing growth rates were found more and more numerous/violent as E/RT_H approaches infinity. Just as was done for flames $(E(\text{Le}-1)/RT_b$ kept of order unity), it might prove necessary to account for the fact that $\gamma - 1$ is often as small as RT_H/E, or even less; the consequences of exploiting distinguished limits (e.g., $E(\gamma - 1)/RT_H$ kept of order unity) have not yet been contemplated. Similarly, the limit of small heat release could be of interest. In the last resort, one may try to set up the brute-force Galerkin-method machinery which is known

(Manneville, 1990) to sometimes yield very accurate results... when physically realistic basis functions are available beforehand; however, this implies that information has already been obtained!

Besides, it is legitimate to ask the following question: to what extent is it necessary to determine analytically (or numerically) the linear dispersion relation in order to tackle the nonlinear domain? After all, the structure of the KS or MS equations (4.5.26, 4.7.44) of flame theory could have been written down without an exact knowledge of the burning speed or of the exact dispersion relation. Actually, such situations are often encountered in the physics of pattern formation (e.g., convective cells) where the sole qualitative shape of $s(k)$ close to the onset of instability, together with the symmetries of the problem, are enough to guess the structure of relevant weakly nonlinear equations (e.g., Cross and Newell, 1984). In our case, more precisely when $s(k)$ looks like in Fig. 6.12a, something can be done by the "envelope" formalism (see Chapter 4), witness the work of Roytburd (1991): restricting attention to waves that propagate along a single transverse direction, the latter author got a complex Ginzburg–Landau (GL) equation to describe the modulation of basically sinusoidal traveling waves. It would at least be desirable to encompass waves traveling in both transverse directions, a situation for which the envelope formalism leads to two coupled complex GL equations (Manneville, 1990). Whether such analyses need be pursued to longer time scales is yet unclear but, undoubtedly, the modulated sine waves so obtained do not resemble what processing the experiments suggests (Fig. 6.2a).

An alternative, perhaps complementary, way could be to acknowledge that the dynamics of detonation structures has distinctive features of "surface acoustics" (Fig. 6.2b–d, 6.3). Accordingly, a good starting point could be a wave equation written in as intrinsic a form as possible: Brun's equation (6.4.19) provides one with very interesting building blocks (in particular appealing *geometrical nonlinearities*) but its middle term needs be amended so as to account for a finite range of unstable wavenumbers. A courageous attempt in this direction was recently published by Borissov and Sharypov (1991); the detonation patterns which result look tantalizingly similar to the experimental ones.

Finally, a pessimistic but still conceivable possibility could be that the phenomenon of detonation structures belongs to the "no theory" category (Langer, 1991) because it might not be reducible to less than the compressible reactive Euler equations. In effect, one has to acknowledge that, to date, only carefully conducted direct numerical simulations (Bourlioux and Majda, 1992) succeeded in realistically mimicking the experiments (cell shapes, triple-point trajectories...). This approach will likely take the lead, insofar as one is able to interpret the (then numerical) experiments.

The latter conclusion is however likely to be overpessimistic, as witnessed by the recent contributions by He and Clavin (1995) for unsteady, overdriven, plane detonations, and by Stewart *et al.* (1995) for cellular detonations.

He and Clavin (1995) use such distinguished limits as nearly equal specific heats (i.e., a small $\gamma - 1$, as suggested above) and large heats of reaction. Importantly, they also introduce a generalized Arrhenius rate which distinguishes between a large activation energy for the induction zone and a finite one for the heat-release zone. The resulting integral equation for the detonation speed $D(t)$ describes the instability of unsteady, overdrive, plane detonations rather well. The associated mechanism originates essentially from the time delay between $D(t)$ and the profile variations in the heat-release zone.

Stewart *et al.* solve the structure of a quasi-CJ detonation iteratively, under the assumption of long waves of front wrinkling and of slowly varying reaction zone. Also, curvature and acceleration effects are assumed to act on the same scale. The result is an intrinsic, local evolution law relating the curvature, the normal velocity, and the normal acceleration of the detonation shock front; it is different from Brun's (6.4.19) but of the same type (i.e., hyperbolic). Once endowed with appropriate boundary conditions, this law shows good prospects for reproducing triple points such as those in Fig. 6.2.

Acknowledgements

Claude Godrèche kindly invited one of us (GJ) to give a course (Notions d'hydrodynamique compressible, voire réactive) at the 1991 Beg-Rohu School on Condensed Matter Physics; he next agreed that our written contribution be slightly different in content from this mostly classical subject matter. It is a pleasure to thank him here. Paul Manneville spent uncountable hours to improve the scientific content of our text and put it in its final format. His constructive criticisms and perseverance are greatly appreciated. We are also indebted to Mrs A. Barreau and M. Dupuy, who typed and retyped our incessantly evolving hieroglyphs, and to the numerous colleagues who kindly lent us first-hand material for the illustrations.

References

Abouseif, G.E., Toong, T.Y. (1982). *Comb. and Flame* **45**, 67.

Abramovitz, M., Stegun, I.A. (1970). *Handbook of Mathematical Functions*, 9th Edition (Dover Publications, New York).

Aldredge, R.C. (1990). "Theory of premixed flame propagation in large-scale turbulence," Ph. D. dissertation, Princeton University.

Bancroft, D., Peterson, E., Minshall, S. (1956). *J. Appl. Phys.* **27**, 291.

Barenblatt, G.I. (1979). *Similarity, Self-Similarity and Intermediate Asymptotics* (Consultant Bureau, New-York).

Barenblatt, G.I., Zel'dovich, Ya. B., Istratov, A.G. (1962). *Prikl. C. Mekh. Fiz.* **2**, 21.

Bdzil, J.B. (1981). *J. Fluid. Mech.* **108**, 195.

Bechtold, J.K., Matalon, M. (1987). *Comb. and Flame* **67**, 77.

Berestycki, H., Larrouturou, B., Roquejoffre, J.M. (1991). In Fife *et al.* (1991).

Berthelot, M., Vieille, P. (1881). *C.R. Hebd. Acad. Sci., Paris* **93**, 18.

Berthelot, M., Vieille, P. (1882). *C.R. Hebd. Acad. Sci., Paris* **94**, 149.

Borissov, An. A., Sharypov, O.V. (1991). In *Dynamic Structure of Detonation, in Gaseous and Dispersed Media* (Kluwer, Dordrecht).

Bourlioux, A., Majda, A.J. (1992). *Comb. and Flame* **90**, 211.

Brun, L. (1989). "Une théorie de la détonation dans les explosifs condensés fondée sur l'hypothèse de Jouguet," *French Atomic Energy Commission Report* CEA-DAM, CEV, DPM, DO89023.

Brun, L. (1974). "Equations hyperboliques et applications à la dynamique des fluides compressibles," *Cours de l'Ecole Polytechnique*, Paris (France).

Brun, L., Chaissé, F. (1989). "A theoretical analysis of the sonic point properties in a plane detonation wave," in *Ninth Symp. (Int.) on Detonation*, p. 757, OCNR 1132891-7.

Buckmaster, J.D. (1983). *SIAM J. Appl. Math.* **43**, 1335.

Buckmaster, J.D. (1993). *Annual Rev. Fluid. Mech.* **25**, 21.

Buckmaster, J.D., Joulin, G. (1991). *J. Fluid. Mech.* **227**, 407.

Buckmaster, J.D., Ludford, G.S.S. (1982). *Theory of Laminar Flames* (Cambridge University Press, Cambridge).

Buckmaster, J.D., Nevis, J. (1988). *Phys. Fluids* **31**, 3571.

Bush, W.B., Fendell, F.E. (1970). *Combust. Sci. Tech.* **1**, 420.

Cambray, P. (1992). "Markstein number for methane-air mixtures," p. 185 in the book of abstracts of *24th Symp. (Int.) on Combustion* (The Combustion Institute).

Cambray, P., Joulin, G. (1992). "On moderately forced unstable premixed flames," *24th Symp. (Int.) on Combustion*, p. 61 (The Combustion Institute).

Cambray, P., Joulin, G. (1994). *Combust. Sci. and Tech.* **95**, 405.

Cambray, P., Deshaies, B., Joulin, G. (1987). *AGARD reprint* **422**, 36.1.

Cambray, P., Joulain, K., Joulin, G. (1994). *Combust. Sci. and Techn.* **103**, 265.

Cambray, P., Joulain, K., Joulin, G. (1996). *Combust. Sci. and Techn.* **112**, 273.

Campbell, A.W., Engelke, R. (1976). "The diameter effect in high-density, heterogeneous explosives," in *Sixth Symp.(Int.) on Detonation* ONR-DN (ACR-221), p. 642 (Washington D.C.).

Carnevale, G.F., Pomeau, Y., Young, W.R. (1990). *Phys. Rev. Lett.* **64**, 2913.

Chapman, D.L. (1899). *Phil. Mag.* **47**, 90.

Chaté, H. (1989). "Transition vers la turbulence via l'intermittence spatio-temporelle," Doctoral Thesis, Université P. et M. Curie, Paris (France).

Clavin, P. (1985). *Prog. in En. and Comb. Sci.* **11**, 1.

Clavin, P. (1988). In *Disorder and Mixing* E. Guyon, J.P. Nadal, Y. Pomeau, eds., p. 293, NATO ASI Series (Kluwer, Dordrecht).

Clavin, P., Garcia-Ybarra, P.L. (1983). *J. Mécanique. Théor. et Appl.* (France) **2** 245.

Clavin, P., Joulin, G. (1983). *J. Phys. Lett. (France)* **1**, L1.

Clavin, P., Joulin, G. (1988). In *Turbulent Reactive Flows*, Lecture notes in engineering **40**, p. 213 (Springer-Verlag, New York).

Clavin, P., Lazzimi, D. (1992). *Combust. Sci. Tech.* **83**, 1.

Clavin, P., Sivashinsky, G.I. (1987). *J. Physique (France)* **48**, 193.

Cole, J.D., Kevorkian, J. (1981). *Perturbation Methods in Applied Mathematics*, Appl. Math. Sci. series **34** (Springer-Verlag, New York).

Combescot, R., Dombre, T., Hakim, V. (1986). *Phys. Rev. Lett.* **56**, 2036.

Combustion Institute (1990). Editor of 23rd Symposium International on Combustion (Pittsburgh, PA).

Courant, R., Friedrichs, K.O. (1948). *Supersonic Flow and Shock Waves* (Wiley-Interscience, New York).

Courant, R., Hilbert, D. (1953). *Methods of Mathematical Physics* (Wiley-Interscience, New York).

Cowperthwaite, M. (1968). *J. Franklin Inst.* **285**, 275.

Cramer, M.S. (1989). *Phys. Fluids A* **1**, 1894.

Cross, M.C., Newell, A.C. (1984). *Physica D* **10**, 299.

Culick, R.L. (1966). *AIAA J.* **4**, 1462.

D'yakov, S.P. (1954). *Zh. Eksp. Teor. Fiz.* **27**, 288.

D'yakov, S.P. (1957). *Zh. Eksp. Teor. Fiz.* **33**, 948; English translation (1958), *Sov. Phys. JETP*, **9**, 729.

Darrieus, G. (1938). "Propagation d'un front de flamme: essai de théorie des vitesses anormales de déflagration par développement spontané de la turbulence," unpublished work presented at *La Technique Moderne*, (Paris).

Denet, B. (1988). "Simulation numérique d'instabilités de fronts de flamme," Doctoral Thesis, Université de Provence, Aix–Marseilles (France).

Denet, B. (1993). *Europhys. Lett.* **21**, 299.

Derrida, B., Godrèche, C., Yekutieli, I. (1990). *Europhys. Lett.* **12**, 385.

Desbordes, D. (1990). "Aspects stationnaires et transitoires de la détonation dans les gaz: Relation avec la structure "cellulaire" du front," Doctoral Thesis, Université de Poitiers (France).

Deshaies, B., Cambray, P. (1990). *Combust. Sci. Tech.* **82**, 361.

Deshaies, B., Joulin, G. (1983). *Combust. Sci. Tech.* **31**, 75.

Deshaies, B., Joulin, G., Clavin, P. (1981). *J. de Mécanique* **20**, 691.

Doering, W. (1943). *Ann. Phys.* **43**, 521.

Dold, J.W., Hartley, L.J., Green, D. (1991). In Fife *et al.* (1991)

Duhem, P. (1909). *Z. Physik Chem. (Leipzig)* **69**, 169.

Dunsky, C.M. (1992). "Microgravity observations of premixed laminar flame dynamics," *24th Symp. (Int.) on Combustion* (The Combustion Institute).

Duvall, G.E., Graham, R.A. (1977). *Rev. Mod. Phys.* **49**, 253.

Edwards, W.S., Fauve, S. (1992). *C.R. Acad. Sci.*, Paris (France) **315**, série 2, 417.

El Hamdi, M., Gorman, M., Mapp, J.W., Blackshear, J.I. (1987). *Combust. Sci. Tech.*, **55**, 33.

Erpenbeck, J.J. (1963). *Phys. Fluids* **6**, 1368.

Erpenbeck, J.J. (1963b). "Structure and stability of the square-wave detonation," *Ninth Symp. (Int.) on Comb.*, p. 442 (The Combustion Institute).

Erpenbeck, J.J. (1966). *Phys. Fluids* **9**, 1293.

Erpenbeck, J.J. (1970). *Phys. Fluids* **13**, 2007.

Faraday, M. (1831). *Phil. Trans. Roy. Soc. London* **121**, 319.

Fay, J.A. (1959). *Phys. Fluids* **2**, 283.

Fickett, W., Davis, W.C. (1979). *Detonation* (University of California Press).

Fife, Liñan, Williams, eds. (1991). *Dynamical Issues in Combustion Theory*, IMA vol. in Math. and Appl. **35** (Springer-Verlag, New-York).

Filyand, L., Sivashinsky, G.I., Frankel, M.L. (1994). *Physica D* **72**, 110.

Floquet, G. (1883). *Annales. Ecole Norm. Sup.* **13**, 47.

Fowles, G.R. (1975). *Phys. Fluids* **18**, 776.

Fowles, G.R. (1976). *Phys. Fluids* **19**, 227.

Fowles, G.R. (1981). *Phys. Fluids* **24**, 220.

Frankel, M. (1991). In Fife *et al.* (1991).

Frankel, M., Sivashinsky, G.I. (1988). *Physica* **D30**, 207.

Frankel, M., Sivashinsky, G.I. (1995). "On the fingering instability in nonadiabatic low Lewis number flames," *Phys. Rev. E* (submitted).

Garcia-Ybarra, P.L., Lopez-Martin, A., Antoranz, J.C., Castillo, J.L. (1994). *Transport Theor. and Stat. Phys.* **23**, 173.

Garcia-Ybarra, P.L., Nicoli, C., Clavin, P. (1984). *Combust. Sci. Tech.* **40**, 41.

Gardner, C.S. (1963). *Phys. Fluids* **6**, 1366.

Germain, P. (1986). *Mécanique*, vol. I & II (École Polytechnique and Ellipses – Édition Marketing, Paris).

Giovangigli, V. (1988). "Structure et extinction de flammes laminaires prémélangées," Doctoral Thesis, Université P. et M. Curie, Paris (France).

Godrèche C. ed. (1991). *Solids Far from Equilibrium* (Cambridge University Press, Cambridge).

Gostinsev, Yu.A., Istratov, A.G., Shulenin, Yu. V. (1988). *Comb. Expl. Shock Waves* **24**, 563.

Groff, E.G. (1982). *Comb. and Flame* **48**, 51.

Grossmann, B., Guo, H., Grant, M. (1991). *Phys. Rev.* **43**, 1727.

Grun, J., Stamper, J., Manka, C., Resnick, J., Burris, R., Crawford, J., Ripin, B.H. (1991). *Phys. Rev. Lett.* **66**, 2738.

Gutman, S., Sivashinsky, G.I. (1990). *Physica* **D43**, 129.

Gutman, S., Axelbaum, R.L., Sivashinsky, G.I. (1994). *Combust. Sci. Tech.* **98**, 57.

Hale, J. (1977). "Theory of Functional Differential Equations" *Appl. Math. Sci.* **3** (Springer-Verlag, New York).

Hayes, W.D. (1957). *J. Fluid. Mech.* **2**, 595.

He, L., Clavin, P. (1995). *C.R. Acad. Sci. (Paris)* **320**, 365.

Hyman, M., Nicolaenko, B., Zaleski, S. (1986). In *Spatio-temporal Coherence and Chaos in Physical System*, edited by Bishop, Grüner, Nicolaenko (North Holland, Amsterdam).

Iordanskii, S.V. (1957). *Prikl. Mat. Mekh.* **21**, 465.

Istratov, A.G., Librovich, V.B. (1969). *Acta Astron.* **14**, 453.

Jackson, T.L., Kapila, A.K. (1984). *Combust. Sci. Tech.* **41**, 191.

Jones, H. (1949). "On the properties of gases at high pressure which can be deduced from explosion experiments," *3rd Symp. on Comb.*, p. 590 (The Williams and Wilkins Company).

Jouguet, E. (1905). *J. Math.* (France), 347.

Jouguet, E. (1906). *J. Math.* (France), 6.

Jouguet, E. (1917). *Mécanique des explosifs* (O. Douin et fils, Paris).

Joulin, G. (1981). *Combust. Sci. Tech.* **27**, 83.

Joulin, G. (1982a). *Comb. and Flame* **46**, 271.

Joulin, G. (1982b). in *Problèmes Non Linéaires Appliqués*, p. 147 (edited by INRIA, Paris).

Joulin, G. (1985). *Combust. Sci. Tech.* **43**, 99.

Joulin, G. (1987). *Combust. Sci. Tech.* **53**, 315.

Joulin, G. (1988). *Combust. Sci. Tech.* **60**, 1.

Joulin, G. (1989). *J. Physique (France)* **50**, 1069.

Joulin, G. (1991). *Zh. Exp. Teor. Fiz.* (USSR) **100**, 428.

Joulin, G. (1994a). *Comb. Sci. and Techn.* **96**, 219.

Joulin, G. (1994b). *Phys.Rev E* **50**, 2030.

Joulin, G., Cambray, P. (1992). *Combust. Sci. Tech.* **81**, 243.

Joulin, G., Cambray, P., Joulain, K. (1994). "On the mechanisms of wrinkling for expanding flames," Proceedings of the Zel'dovich Memorial, Vol. I, p. 212; A.G. Merzhanov and S.M. Frolov, eds. (Russian section of the Combustion Institute, Moscow).

Joulin, G., Clavin, P. (1976). *Acta Astron.* **3**, 223.

Joulin, G., Clavin, P. (1979). *Combust. Sci. Tech.* **35**, 139.

Joulin, G., Mitani, T. (1982). *Comb. and Flame* **40**, 235.

Joulin, G., Sivashinsky, G.I. (1992). "On the hydrodynamic stability and response of premixed flames in stagnation-point flames," *24th Symp. (Int.) on Comb.*, p. 37 (The Combustion Institute).

Kardar, M., Parisi, G., Zhang, Y.C. (1986). *Phys. Rev. Lett.* **56**, 889.

Kim, Y., Matalon, M. (1990). *Combust. Sci. Tech.* **69**, 85.

Kontorovich, V.M. (1958). *Zh. Eksp. Teor. Fiz.* **33**, 1525; English translation: *Sov.Phys. JETP* **6**, 1179.

Kontorovich, V.M. (1959). *Akust. Zh.* **5**, 314; English translation (1959): *Sov.Phys. Acoustics* **5**, 320.

Krug, J., Spohn, H. (1992). In Godrèche (1991), p. 479.

Kuramoto, Y. (1980). In *Dynamics of Synergetic Systems*, H. Haken, ed., p. 134 (Springer-Verlag, New York).

Kuznetsov, N.M. (1985). *Zh. Eksp. Teor. Fiz.* **88**, 470; English translation: *Sov. Phys. JETP* **61**, 275.

Kuznetsov, N.M. (1986). *Zh. Eksp. Teor. Fiz.* **90**, 744; English translation: *Sov. Phys. JETP* **63**, 433.

Lambrakis, K.C., Thomson, P.A. (1972). *Phys. Fluids* **15**, 933.

Landau, L.D. (1944). *Acta Physicochem.* (USSR) **19**, 77.

Landau, L.D., Lifshitz, E., (1987). *Fluid Mechanics* (Pergamon Press, New York).

Langer, J.S. (1991). In Godrèche (1991), p. 297.

Lax, P.D. (1973). "Hyperbolic systems of conservation laws and the mathematical theory of shock waves," in *CBMS Regional Conference series in Applied Mathematics Proceedings* (SIAM, Philadelphia).

Lee, H.I., Stewart, D.S. (1990). *J. Fluid. Mech.* **216**, 103.

Lee, J.H., Lee, B.H.K., Shanfield, I. (1965). *10th Symposium (Int.) on Combustion*, p. 805 (The Combustion Institute).

Lee, Y.C., Chen, H.H. (1982). *Phys. Scripta* **T2**, 41.

Luck, J.M. (1992). *Systèmes Désordonnés Unidimensionnels*, Aléa Saclay (Cambridge University Press, Cambridge).

Majda, A., Rosales, R. (1983). *SIAM J. Appl. Math.* **43**, 1310.

Majda, A., Rosales, R. (1984). *Studies in Appl. Math.* **71**, 117.

Mallard, E., Le Chatelier, H.L. (1881). *C.R. Hebd. Acad. Sci.* (Paris) **93**, 145.

Manneville, P. (1990). *Dissipative Structures and Weak Turbulence* (Academic Press, Boston).

Manson, N. (1958). *C.R. Acad. Sci. (Paris)* **246**, 2860.

Markstein, G.H. (1951). *J. Aeron. Sci.* **18**, 199.

Markstein, G.H. (1964). *Nonsteady Flame Propagation* (Pergamon Press, Oxford).

Matalon, M., Matkowsky, B.J. (1982). *J. Fluid Mech.* **124**, 239.

Matkowsky, B.J., Sivashinsky, G.I. (1978). *SIAM J. Appl. Math.* **35**, 465.

Michelson, D.M., Sivashinsky, G.I. (1977). *Acta Astron.* **4**, 1207.

Michelson, D.M., Sivashinsky, G.I. (1982). *Comb. and Flame* **48**, 211.

Minaev, S.S. (1994). "Theoretical analysis of nonlinear effects in hydrodynamic instability of premixed flames," Proceedings of the Zel'dovich Memorial, Vol.II, p. 302, S.M. Frolov, ed. (Russian section of the Combustion Institute, Moscow).

Nicoli, C. (1985). "Dynamique des flammes prémélangées en présence de mécanismes contrôlant les limites d'inflammabilité," Doctoral Thesis, Marseilles (France).

Nicoli, C., Clavin, P. and Liñan, A. (1990). In *Spatial Inhomogeneities and Transient Behavior in Chemical Kinetics*, Gray *et al.*, eds., p. 317 (Manchester University Press, Manchester).

Nozières, P. (1991). In Godrèche (1991), p. 1.

Ott, E., Sudan, R.N. (1969). *Phys. Fluids* **12**, 2388.

Ott, E., Manheimer, W.M., Book, D.L., Boris, J.P. (1973). *Phys. Fluids* **16**, 855.

Palm-Leis, A., Strehlow, R.A. (1969). *Comb. and Flames* **13**, 111.

Pelcé, P., ed. (1989). *Dynamics of Curved Fronts* (Academic Press, Boston).

Pelcé, P., Clavin, P. (1982). *J. Fluid. Mech.* **124**, 219.

Pelcé, P., Rochwerger, D. (1991). *J. Fluid. Mech.* **239**, 293.

Peters, N., Rogg, B. (1993), *Reduced Kinetic Mechanisms for Combustion Applications,* (Springer-Verlag, New York).

Presles, H.-N., Guerraud, C., Desbordes, D. (1987). *C.R. Acad. Sci. Paris* **304** Série II, n° 13, 695.

Prigogine, I., Defay, R. (1954). *Chemical Thermodynamics* (Longmans Green, London)

Quinard, J. (1984). "Limites de stabilité et structure cellulaire dans les flammes de prémélanges," Doctoral Thesis, Université de Marseilles (France).

Rahibe, M., Aubry, N., Sivashinsky, G.I., Lima, R. (1995). "The formation of wrinkles in outwards propagating flames," *Phys. Rev. E* (in press).

Rakib, Z., Sivashinsky, G.I. (1987). *Combust. Sci. Tech.* **54**, 69.

Rochwerger, D. (1991). "Contrôle paramètrique de fronts de croissance," Doctoral Thesis, Université de Provence, Aix–Marseilles (France).

Ronney, P.D., Whaling, K.N., Abhud-Madrid, A., Gatto, J.L., Pisowicz, V.L. (1992). "Stationary premixed flames in spherical and cylindrical geometries," *AIAA Journal* (submitted).

Roytburd, V. (1991). In *Fluid Dynamical Aspects of Combustion Theory*, edited by M. Onofri and A. Tesei, Pitman Research Notes in Mathematics Series **223**, p. 184 (Longman, Harlow).

Schimmer, H., Vortmeyer, D. (1977). *Comb. and Flame* **41**, 17.

Searby, G. (1991a). *Combust. Sci. Tech.* **81**, 221.

Searby, G. (1991b). In *Nonlinear Phenomena Related to Growth and Forms*, eds. M. Ben Amar, P. Pelcé, P. Tabeling, NATO ASI Series, (Plenum Press, New York).

Searby, G., Clavin, P. (1986). *Combust. Sci. Tech.* **46**, 167.

Searby, G., Rochwerger, D. (1991). *J. Fluid. Mech.* **231**, 529.

Sivashinsky, G.I. (1977a). *Acta Astron.* **4**, 1177.

Sivashinsky, G.I. (1977b). *Comb. Sci. Tech.* **15**, 137.

Sivashinsky, G.I. (1983). *Annual Rev. Fluid. Mech.* **15**, 179.

Sivashinsky, G.I., Law, C.K., Joulin, G. (1982). *Combust. Sci. Tech.* **28**, 155.

Stewart, D.S., Aslam, T.D., Yao, J. (1995). "The evolution of detonation cells," at *The 15th Int. Coll. on the Dynamics of Explosions and Reactive Systems*, Boulder, Colorado.

Stewart, D.S., Bdzil, J.B. (1988). *Comb. and Flame* **72**, 311.

Thomson, P.A., Lambrakis, K.C. (1973). *J. Fluid. Mech.* **60**, 187.

Thual, O., Frisch, U., Hénon, M. (1985). *J. Physique* (France) **46**, 1485.

Vieille, P. (1900). *C.R. Hebd. Acad. Sci.* (Paris) **131**, 413.

Von Neumann, J. (1942). O.S.R.D. Report #549; also in *John Von Neumann Collected Works* **6**, edited by A.J. Taub (Pergamon Press, Oxford, 1963).

Whitham, G.B. (1974). *Linear and Nonlinear Waves* (Wiley-Interscience, New York).

Williams, F.A. (1985). *Combustion Theory* (Benjamin/Cummings, Menlo-Park).

Williams, F.A. (1992). Plenary Lecture of *24th Symp. (Int.) on Comb.* (The Combustion Institute).

Wood, W.W., Kirkwood, J.G. (1954). *J. Chem. Phys.* **22**, 1920.

Woods, L.C. (1971). *J. Plasma Physics* **6**,615.

Zaidel, R.M. (1961). *Proc. Acad. Sci. USSR, Phys. Chem. Sec.* **136**, 167.

Zel'dovich, Ya.B., Frank-Kamenetskii, D.A. (1938). *Acta Physicochem.* (USSR) **9**, 341.

Zel'dovich, Ya.B. (1940). *Zh. Eksp. Teor. Fiz.* **10**, 542 (NACA TM, 1261, 1960).

Zel'dovich, Ya.B., Barenblatt, G.I., Librovich, V.B., Makhviladze, G.M. (1985). *Mathematical Theory of Combustion and Explosions*, English translation (Plenum Press, New York).

Zel'dovich, Ya.B., Istratov, B., Kidin, A.G., Librovich, V.B. (1980). *Combust. Sci. Tech.* **24**, 1.

Zel'dovich, Ya.B., Raizer, Yu.P. (1966, Vol.I; 1967, Vol.II). *Physics of Shock Waves and High-Temperature Hydrodynamic Phenomena* (Academic Press, New York).

Index